Flora Indica

*Being a Systematic Account of the Plants
of British India, Together with Observations
on the Structure and Affinities of their
Natural Order and Genera*

JOSEPH DALTON HOOKER
THOMAS THOMSON

CAMBRIDGE
UNIVERSITY PRESS

CAMBRIDGE UNIVERSITY PRESS

Cambridge, New York, Melbourne, Madrid, Cape Town,
Singapore, São Paolo, Delhi, Tokyo, Mexico City

Published in the United States of America by Cambridge University Press, New York

www.cambridge.org
Information on this title: www.cambridge.org/9781108037495

© in this compilation Cambridge University Press 2011

This edition first published 1855
This digitally printed version 2011

ISBN 978-1-108-03749-5 Paperback

CAMBRIDGE LIBRARY COLLECTION

Books of enduring scholarly value

Life Sciences

Until the nineteenth century, the various subjects now known as the life sciences were regarded either as arcane studies which had little impact on ordinary daily life, or as a genteel hobby for the leisured classes. The increasing academic rigour and systematisation brought to the study of botany, zoology and other disciplines, and their adoption in university curricula, are reflected in the books reissued in this series.

Flora Indica

Sir Joseph Hooker (1817–1911) was one of the greatest British botanists and explorers of the nineteenth century. He succeeded his father, Sir William Jackson Hooker, as Director of the Royal Botanic Gardens, Kew, and was a close friend and supporter of Charles Darwin. His journey to the Himalayas and India, during which he collected some 7,000 species, was undertaken between 1847 and 1851 to increase the Kew collections; his account of the expedition (also reissued in this series) was dedicated to Darwin. In 1855 he published *Flora Indica* with his fellow-traveller Thomas Thomson, who became Superintendent of the East India Company's Botanic Garden at Calcutta. Lack of support from the Company meant that only the first volume of a projected series was published. However, the introductory essay on the geographical relations of India's flora is considered to be one of Hooker's most important statements on biogeographical issues.

Cambridge University Press has long been a pioneer in the reissuing of out-of-print titles from its own backlist, producing digital reprints of books that are still sought after by scholars and students but could not be reprinted economically using traditional technology. The Cambridge Library Collection extends this activity to a wider range of books which are still of importance to researchers and professionals, either for the source material they contain, or as landmarks in the history of their academic discipline.

Drawing from the world-renowned collections in the Cambridge University Library, and guided by the advice of experts in each subject area, Cambridge University Press is using state-of-the-art scanning machines in its own Printing House to capture the content of each book selected for inclusion. The files are processed to give a consistently clear, crisp image, and the books finished to the high quality standard for which the Press is recognised around the world. The latest print-on-demand technology ensures that the books will remain available indefinitely, and that orders for single or multiple copies can quickly be supplied.

The Cambridge Library Collection will bring back to life books of enduring scholarly value (including out-of-copyright works originally issued by other publishers) across a wide range of disciplines in the humanities and social sciences and in science and technology.

FLORA INDICA:

BEING

A SYSTEMATIC ACCOUNT

OF THE

PLANTS OF BRITISH INDIA,

TOGETHER WITH

OBSERVATIONS ON THE STRUCTURE AND AFFINITIES OF
THEIR NATURAL ORDERS AND GENERA.

BY

J. D. HOOKER, M.D., R.N., F.R.S., ETC.,

AND

THOMAS THOMSON, M.D., F.L.S.,

SURGEON H.E.I.C.S.

VOLUME I.

RANUNCULACEÆ TO FUMARIACEÆ,

WITH

An Introductory Essay.

LONDON :
PRINTED FOR THE AUTHORS.
PUBLISHED BY W. PAMPLIN, 45, FRITH STREET, SOHO.
1855.

TO

SIR W. J. HOOKER, K.H., F.R.S.,

LL.D., D.C.L. OXON., ETC. ETC. ETC.,

TO WHOM THE AUTHORS ARE INDEBTED FOR THEIR EARLIEST

INSTRUCTION IN THE SCIENCE OF BOTANY,

FOR THAT ASSISTANCE AND ENCOURAGEMENT WHICH ALONE HAS

ENABLED THEM TO PURSUE IT IN AFTER LIFE,

AND IN WHOSE UNRIVALLED HERBARIUM AND LIBRARY

THE ' FLORA INDICA ' HAS BEEN COMMENCED,

This Work is Dedicated

BY HIS AFFECTIONATE SON, AND PUPIL,

JOSEPH D. HOOKER

AND

THOMAS THOMSON.

ROYAL GARDENS, KEW,
February, 1855.

PREFACE.

———◆———

The object, scope, and design of this Work, together with the motives that induced us to commence it, are all detailed in the Introductory Essay.

It will be seen that we anticipated considerable difficulty in our proposed attempt to establish the genera and species of the 'Flora Indica' on a sound and philosophical basis, and to unravel their synonymy. The result has proved that we underrated the difficulties of the task, for the number of plants described is very much smaller than we hoped to have accomplished, and in many of the genera the species are not satisfactorily limited. This has arisen from many causes, to two of the most important of which, as suggestive of improvements that may be introduced into botanical science, we shall briefly allude.

In the first place, a critical study of the vast number of well-selected specimens that we possess of most of the plants, has enlarged those already extended views of the variability of species which we have professed in our Introductory Essay. In every case, the more specimens we examined, and especially if taken from different individuals, the greater the difficulty in framing diagnoses. This has shaken our confidence in the sufficiency of the descriptions we have drawn up from few specimens; and it proves that the characters of exotic plants, in systematic works, being unavoidably those of *individuals*, and not of *species*, have been far too much relied on as affording means of identification.

The other great obstacle has been the immense number of works, and especially of periodicals, we have had to consult: 120 authors' names are attached to the 430 species described, and the completion of the 'Flora Indica' will require a reference to upwards of 1000 volumes. We would now therefore call the attention of our fellow-botanists to the fact, that the time is rapidly approaching, when the difficulty of obtaining access to the necessary periodicals must render the effectual study of botany impossible; and that the practice of naturalists sending their several papers to different periodicals, and above all to local ones, or to such as embrace many branches of science, is one of the greatest obstacles to the study of natural history in the present day. We have found it impossible to obtain access to several journals of local or of ephemeral interest, and it would be well if isolated naturalists paused before they sought to establish such, or to send their contributions where they must be inevitably overlooked.

After a careful review of the state of botanical literature, in this country at any rate, we have no hesitation in saying that the Transactions of well-established Associations for the furtherance of natural science, diffuse most effectually the labours of naturalists. This is because these societies are supported by persons whose interest it is to disseminate their publications at the smallest possible delay and cost; and, what is of great importance, all papers communicated are subjected to a system of supervision before publication, which ensures their being worthy of it.

We need not say, that while urging the propriety of centralization within reasonable limits, we are far from wishing to see the natural and physical sciences entirely separated. In

a large scientific community there is always a Society established for the furtherance of such researches as have a very
wide-spread interest, not confined to the branch they especially illustrate, nor even to the class of sciences under which
they rank : but researches of such importance are necessarily
rare, and the Transactions in which they are embodied are
universally accessible.

It is the intention of the Authors to continue the ' Flora
Indica,' one of them in the Hookerian Herbarium, the other
at the Calcutta Botanic Gardens. The propriety, however,
of pursuing the attempt to complete the history, etc., of each
Indian genus and species, is, in the present state of science,
very doubtful. Considering how little is accurately known
of the outlines of Indian Botany, and how extensive our materials are, it may be better to ensure accuracy in the most
important identifications only, and to omit quoting such works
as are not worth referring to. In this we shall be guided by
the opinions of those botanists who may honour us by con-
sulting our labours critically.

CONTENTS OF INTRODUCTORY ESSAY.

———◆———

D. *Enumeration and Description of the Provinces of India, as referred to in the ' Flora Indica.'*

Primary Divisions :—1. Hindostan ; 2. Himalaya ; 3. Eastern India ; 4. Afghanistan.

I. HINDOSTAN.

PROVINCES OF HINDOSTAN.

II. THE HIMALAYA.

Map I. to face p. 82 of Introductory Essay.
Map II. to be placed at the end of the Introductory Essay.

1. *Systematice* plantas suas disponit verus Botanicus ;
   Nec absque ordine easdem enumerat.
2. *Frutificationis* principium in theoretica dispositione agnoscit ;
   Nec dispositionem secundum Herbam immutat.
3. *Genera* naturalia assumit ;
   Nec erronea ob speciei notam aberrantem conficit.
4. *Species* distinctas tradit ;
   Nec e Varietatibus falsas fingit.
5. *Varietates* ad species reducit ;
   Nec eas pari passu cum speciebus obambulare finit.
6. *Synonyma* præstantissima indagat et seligit ;
   Nec acquiescit in quacunque obvia nomenclatura.
7. *Differentias* characteristicas inquirit ;
   Nec inania nomina specifica præponit veris.
8. *Plantas* vagas ad Genera amandare studet ;
   Nec rariores obvias fugitivis oculis adspicit.
9. *Descriptiones* complectentes differentias essentiales compendiose sistit ;
   Nec naturalissimam structuram oratorio sermone ebuccinat.
10. *Minimas* partes attente scrutatur ;
    Nec ea quæ maxime illustrant, flocci facit.
11. *Observationibus* ubique plantas illustrat ;
    Nec in vago nomine acquiescit.
12. *Oculis* propriis quæ singularia sunt observat ;
    Nec sua solum, ex Auctoribus, compilat.

LINNÆUS, *Philosophia Botanica.*

INTRODUCTORY ESSAY.

—◆—

IN the following pages it is our intention not only to explain the objects of the Flora Indica, and our reasons for undertaking it, but also to dwell upon a considerable number of topics having a direct bearing upon the study of Systematic Botany, and upon the correct appreciation of which must depend the progress which the student may make in this department of science. As however the principal aim of our labours is to further the study of Botany in India, we shall confine ourselves as much as possible to those points which it is more particularly essential for the Indian botanist to understand well, and we shall illustrate them by a reference to the plants of that country. The chief subjects treated of in this Essay will therefore be :—

1. The object, scope, and design of the Flora Indica, and our motives for undertaking it.

2. General considerations connected with the study of systematic and descriptive botany.

3. The influence of variation, the origin of species, specific centres, hybridization, and geographical distribution, on the views taken by ourselves of species, and of the right manner in which they should be treated, and in which their affinities should be developed. We consider these theoretical points to be inseparable from a philosophical study of plants, and we believe it to be essential that systematic authors should

b

explain the principles by which they are guided in the execution of similar works to this.

4. An historical summary of the labours of our predecessors in Indian botany, whether as authors or collectors, and some account of the materials at our disposal.

5. A sketch of the meteorology and climate of India, the excessive complexity of whose seasons offers the most formidable obstacle to the student's appreciation of the prominent features of its vegetation.

6. An attempt to divide the area embraced in the Flora Indica into physico-geographical or geographico-botanical districts. This is intended to serve the double purpose of giving a slight sketch of the physical characters and vegetation of these provinces, and of adopting such a carefully-selected system of nomenclature, as shall be available for assigning intelligible localities to the species in the body of the Flora, and such as may be easily committed to memory, or found with little trouble on any map. We have long deplored the defective geographical nomenclature adopted in almost every work treating of the Natural History of India, and the fact that "E. Ind." or "Ind. Or." is considered in most cases sufciently definite information as to the native place of any production found between Ceylon and Tibet, or Cabul and Singapur; and we hope that the present attempt to remedy so important a defect will be received with indulgence.

I. *Object, scope, and design of the Flora Indica.*

Our object, in the work here commenced, is to present a systematic account of the vegetable productions of British India, arranged according to natural principles, and based upon a careful examination of all the materials within our reach. Besides the descriptions of the Orders, Genera, and Species, all matters of importance connected with anatomical, structural, morphological, and physiological points, will, wherever it is practicable, be treated of, and in other cases pointed out as subjects worthy of future attention. Geographical distribution, and the effect of climate, soil, and exposure, have been made the objects of our special study, and will in all cases be particularly noted. With regard to economic botany, it is obviously impossible to do more than briefly enumerate, under their respective species, the various products which have been used in the arts: for detailed accounts of their value, we must refer our readers to the many excellent works on those subjects, which have been published by Indian botanists.

Our work is intended to facilitate the progress of economists, by supplying their great desideratum, a critical description of the plants which yield the products they seek. We have had a considerable experience both in medical and economic botany, and we announce boldly our conviction, that, so far as India is concerned, these departments are at a standstill, for want of an accurate scientific guide to the flora of that country. Hundreds of valuable products are quite unknown to science, while of most of the others the plants are known only to the professed botanist. The mass must indeed always remain so: just as the refinements of the laboratory and the calculations of the mathematician must ever be mysteries to the majority of manufacturers and navigators, whose operations are based on the sciences in question. It is a mistake to suppose that it can be otherwise; or that those who are engaged in forwarding a science so extensive and abstruse as philosophical botany, can command the time to become so familiar with the details of the commercial value of vegetable products, as to be safe referees on these subjects. On the other hand, it is equally a mistake to suppose that those who devote themselves to the collection of economic products, can possess the experience and botanical knowledge necessary to render their identifications of tropical plants trustworthy in the eyes of men of science*. It is therefore as a strictly

* For proof of this we have only to refer to the pages of any book on medical or economic botany; and to the fact, first indicated in these pages, that the celebrated Bikh Poison, about which so much has been written, is produced

scientific work that we offer this commencement of the Flora Indica to the public; but though the advancement of abstract science is indeed its primary object, yet as we yield to none in our estimate of the value of economic botany, we confidently trust that, as pioneers in this department also, our labours will be found of material service.

On this account we need scarcely offer an apology for our partial use of Latin, which is necessary, as well for economy of space, as because we are labouring for the benefit of Continental botanists as well as English ones, and because we write under a sense of the obligation the former have rendered us, by having published in Latin (instead of French or German, or still less familiar languages) the many valuable memoirs on economic and scientific Indian botany, which we owe to their exertions. When the flora of India is established on a scientific foundation, it will be desirable that a compendious English version of such a work as ours should be provided for the use of those who do not pursue science for its own sake, but yet are desirous of availing themselves of its results: at present such an undertaking would be premature.

Had it been possible to take up the economic plants of British India by themselves, and to present a history of them to the English reader, we should at once have devoted ourselves to the task, with the certainty of obtaining an amount of encouragement which a so-called paying work is sure to command, but which one of a more scientific nature is not thought worthy of receiving. We should however only be deceiving the public, were we to propose a scheme which, in the present deplorably backward state of scientific Indian botany on the one hand, and the confusion of Indian economic botany on the other, is literally impracticable. Dr. Royle's great work, published twenty years ago, is the only one on Indian plants that attempts to combine practical with scientific botany; but five volumes of its size would not bring the

in the Himalaya by the common *Aconitum Napellus* of Europe and North America, as well as by other species of the genus.

subject there treated of up to the present state of our knowledge: the difficulties have increased fourfold, from scientific
botany not having advanced *pari passu* with the economic
branch; and so long as the plants themselves remain undescribed, it is obviously impossible to recognize what are useful,
or so to define them that they shall be known by characters
that contrast with those of the useless. Our principal aim
however being purely botanical, the most insignificant and
useless weed is as much the object of our attention as the
Teak, Sal, and Tea: in the vegetable kingdom, and in the
great scheme of nature, all have equal claims on our notice,
and no one can predicate of any, its uselessness in an economic point of view.

Every one who has studied Indian plants, whether for economic purposes or for those of abstract science, must have
felt the want of a general work which should include the
labours of all Indian botanists, to be a very serious inconvenience. Our own experience in India has convinced us of
this; for we found it impossible to determine the names of
many of the most ordinary, and, in an economic point of view,
often most valuable forms; and every day's additional experience in the preparation of this volume has served to show
more and more clearly, that whilst such a work is wanting satisfactory progress is impossible. At present the student has
to search in general systematic works, for the descriptions
of species; and as all of these are imperfect, a multitude of
scattered papers must be consulted for the additions which
have from time to time been made. These too have unfortunately so often been published without reference to preceding works of a similar nature, that the same plant has been
described as new by many successive botanists, ignorant or
neglectful of the labours of their predecessors.

A general flora of India must comprise a careful study of
all previously published materials, so as to blend them into an
harmonious whole, and to establish Indian botany on a secure
basis of observation and accurate description. Such a task is,

however, the labour of a lifetime, and although we have undertaken its commencement, we cannot hope to bring it to a
conclusion; our progress in it must depend entirely upon circumstances at present beyond our control; but we have no
doubt that when we are compelled to abandon the undertaking, the necessity for the completion of such a work will induce some one to follow in our steps, and to lend a helping
hand to the compilation of a further portion of so indispensable an aid to botanical research.

We should however be wrong, were we to convey the impression that this arduous undertaking has wholly originated
with ourselves: on the contrary, the conviction has for some
years been general among botanists, that the collections accumulated in this country were so ample, that the time had
fully come for the preparation and publication of a Flora Indica; and when it was known that we had returned from
India with large and important materials, we were invited
by all the most illustrious names in the science to combine a
revision of the labours of our predecessors with the publication of our own discoveries.   Many of our friends considered
that for such an undertaking we possessed greater advantages
and facilities than had ever before been available to any botanist.   Our collections were most extensive, having been
formed over a very wide extent of country, with a knowledge
of the great variability of species, of the chief forms of which
we were desirous of making our specimens illustrative; they
were moreover accompanied by an extensive series of drawings and dissections from the life, and by voluminous notes, indicative of distribution, habit, structure, etc.   It was known
that we intended to distribute our plants, which ought not to
be done without a careful examination, for the purpose of determining their names.   During this examination much of the
most laborious part of the preparation of a flora must necessarily be undergone; and we were urged to put our results
on record for the benefit of science.   Nor must we omit, in
the enumeration of the advantages we enjoyed, a free access

to the rich herbarium and library of Sir William Hooker, and its vicinity to a metropolis containing other collections (especially the Wallichian Herbarium) indispensable to an Indian botanist.

Under a combination of so many favourable circumstances, we felt it our duty to undertake the task proposed to us. Not, however, having at our command the necessary funds, the subject was brought before the British Association at the meeting of 1851, and being most favourably received by its members, the Directors of the East India Company were strongly memorialized on behalf of an undertaking in which it was expected that they would feel the deepest interest. In reply to this recommendation, the Court declined promoting the object, but expressed a willingness to take its merits into consideration on its completion. The President of the British Association, in communicating to us this answer, at the same time intimated to us the hopes of his colleagues that we should at least commence the work. This we did, but, we must confess, with a feeling of discouragement, for the unfavourable answer of the Court materially retarded our progress, our private resources not being sufficient to provide such assistance as would have relieved us from the mechanical labours of arranging, distributing, and writing tickets, which have in consequence hitherto occupied more than three-fourths of our time. The difficulty of the task has also far exceeded our anticipations, as we were not prepared for so large a proportion of Indian plants proving identical with those of other parts of the world. This has obliged us, in every large genus, to have recourse to a critical study of the European, Siberian, Chinese, and Japanese floras, which has elucidated results totally unexpected by ourselves and fellow-botanists, and at the same time of extraordinary interest and importance to the science of Botanical Geography.

As we are anxious to render each portion of the work as complete in itself as possible, and are desirous of enlisting in the cause such of our fellow-botanists as may be willing to

work up those Natural Orders with which they are most familiar, the Flora Indica, when completed, will probably consist of a series of monographs.　In the commencement now
offered to the public, we have arranged the principal Natural
Orders in the mode of sequence usually adopted in systematic
works, altering the places of a few of the smaller ones, whose
botanical affinities we conceive to have been misunderstood.

We consider it important that the Flora Indica should embrace as wide an area as possible, as we are firmly convinced
that no species can be properly defined, until it has been examined in all the variations induced by those differences in
climate, locality, and soil, which an extensive area alone affords.　As also the flora of an area cannot be worked out
without a knowledge of the botany of the countries surrounding it (with which it has many plants in common), it follows
that the greater the area embraced, the more fully will it illustrate the habits, forms, and variations, of the species comprised within it.　For this reason we have extended the limits
of our Flora from Persia to the Chinese dominions.

II. *General considerations connected with the study of
Systematic Botany.*

It may seem almost chimerical to look forward to a time
when all the species of the vegetable world shall have been
classified upon philosophical principles, and accurately defined ; and it must be confessed that the present state of descriptive botany does not hold out much prospect of the reali-
zation of so very desirable an object.　This, we think, is in a
great measure due, not to any want of students willing and
anxious to take up the subject, but rather to a gradually increasing misapprehension of the true aim and paramount importance of systematic botany, and of the proper mode of
pursuing the study of the laws that govern the affinities of
plants.　We are therefore desirous, at the outset of a work
which is devoted to these subjects, of explaining our views on

them; and as we trust that our work will fall into the hands of many beginners who are anxious to devote themselves usefully to the furtherance of botanical science, but who have not an opportunity of acquiring in any other way its fundamental principles, we shall make no excuse for dwelling at some length on the subject. We are also anxious to refute the too common opinion (which has been productive of much injury to the progress of botany) that the study of system presents no difficulties, and that descriptive botany may be undertaken by any one who has acquired a tolerable familiarity with the use of terms.

There can be no doubt that any observant person may readily acquire such a knowledge of external characters, as will in a short time enable him to refer a considerable number of plants to their natural orders; though even for this first step more knowledge of principles is required, than to make an equal advance in the animal kingdom: but to go beyond this, —to develop the principles of classification, to refer new and obscure forms to their proper places in the system, to define natural groups and even species on philosophical grounds, and to express their relations by characters of real value and with a proper degree of precision, demand a knowledge of morphology, anatomy, and often of physiology, which must be completely at command, so as to be brought to bear, when necessary, upon each individual organ of every species in the group under consideration. To follow the laws that regulate the growth of all parts of the plant, especially the structure of stems, the functions of leaves, the development and arrest of floral organs, and the form, position, and minute anatomy of the pollen and ovule, and to trace the whole progress of the ovule and its integuments to their perfect state in the seed, ought all to be familiar processes to the systematic botanist who proceeds upon safe principles; but no progress can be made by him who confines his attention chiefly to the modifications of these organs in individual plants or natural orders.

To many all this may appear self-evident, and we should

c

fear to be censured for stating truisms, did not the annals of
natural science present too many instances of the reckless-
ness with which genera, orders, and even so-called natural
systems, have been instituted by tyros without the smallest
practical acquaintance with structure and affinities. We do
not refer merely to the vagaries of a Rafinesque, a Bowditch,
or a Blanco, though a botanist so eminent as Endlicher has
thought it necessary to encumber his pages with characters of
genera which must remain for ever enigmatical, unless some
happy chance should make us acquainted with the specimens
of the authors; we have in view more well-meaning persons,
who have the progress of science at heart, but who, by defective definitions and erroneous classification, crowd our books
with imperfectly defined genera and with groups and subdivisions of no practical value.   A knowledge of the relative
importance of characters can only be acquired by long study;
and without a due appreciation of their value, no natural group
can be defined.   Hence many of the new genera which are
daily added to our lists rest upon trivial characters, and have
no equality with those already in existence.   A proneness to
imitation leads to a gradual increase in their numbers, without a corresponding increase of sectional groups.   Indeed,
even when the sectional groups are well defined, and the genera in themselves natural, a too great increase in the number
of genera is detrimental, by keeping out of view those higher
divisions which are of greater importance. The modern system
of elevating every minor group, however trifling the characters
by which it is distinguished, to the rank of a genus, evinces,
we think, a want of appreciation of the true value of classification.   The genus is the group which, in consequence of our system of nomenclature, is kept most prominently before the mind,
and which has therefore most importance attached to it*.

* We may make our meaning more clear by a few examples. The genus
Ficus is surely more natural than the subgenera *Pogonotrophe, Covellia, Uro-
stigma,* into which it has been subdivided. So with the genera *Anemone, Hedyotis, Erica, Andromeda,* and others which have been split into many by
modern botanists. Mr. Brown has, in all his works, laboured to keep this

The rashness of some botanists is productive of still more detrimental effects to the science in the case of species; for though a beginner may pause before venturing to institute a genus, it rarely enters into his head to hesitate before proposing a new species. Hence the difficulty of determining synonymy is now the greatest obstacle to the progress of systematic botany; and this incubus unfortunately increases from day to day, threatening at no very distant period so to encumber the science, that a violent effort will be necessary on the part of those who have its interests at heart, to relieve it of a load which materially retards its advancement. The number of species described is now so very great, and the descriptions are scattered through such a multitude of books, that even after long research it is difficult to avoid overlooking much that is already known; and when botanists with limited libraries and herbaria institute new species, it is almost certain that the latter will be found to have been already characterized. To such an extent is this carried, that we could indicate several works, in which one half and even more of the species are proposed in ignorance of the labours of other botanists. Indian Botany unfortunately, far from forming an honourable exception in this particular, presents a perfect chaos of new names for well-known plants, and inaccurate or incomplete descriptions of new ones.

It must be remembered too that the Linnean canon, by which twelve words were allowed for a specific character, is now becoming quite inadequate to the requirements of the science; and that the brief descriptions, which are now so generally substituted for definitions, unless prepared with the greatest skill, as well as care, and after an inspection of very numerous specimens, seldom express accurately the essential characters of a plant. It is indeed becoming more and more evident, that in the great majority of instances no definition is sufficient to enable inexperienced botanists to determine

important principle in view, and to impress it upon others; he has, however, failed to check the prevalent tendency to the multiplication of genera.

with accuracy the species of a plant, even when the whole genus is well known; much more is this the case in genera, many of whose species are yet undiscovered; and most of all, in those where the forms, though sufficiently well known, are liable to much variation. In the last case their determination becomes a special study; and when attempted without access to authentic specimens, leads to inextricable confusion, and its evil effects are not confined to specific botany, but extend to all departments.

The pages of our Indian Flora will supply numerous illustrations of these remarks, and we would direct the attention of those commencing the study to the lesson to be derived from these instructive errors; for where the first botanists of the day have failed, beginners cannot be expected to succeed. It cannot be too strongly impressed upon all students of botany, that it is only after much preliminary study, and with the aids of a complete library, and an herbarium containing authentic specimens of a very large proportion of known species, that descriptive botany can be effectively carried out; and it would be well for science if this were fully understood and acted upon.

The prevailing tendency on the part of students of all branches of natural history, to exaggerate the number of species, and to separate accidental forms by trifling characters, is, we think, clearly traceable to the want of early training in accurate observation, and of proper instruction in the objects and aim of natural science. Students are not taught to systematize on broad grounds and sound principles, though this is one of the most difficult processes, requiring great judgment and caution; or, what is worse, they are led by the example if not by the precepts of their teachers, to regard generic and specific distinctions as things of little importance, to be fixed by arbitrary characters, or according to accidental circumstances. As a consequence, the study of systematic botany is gradually taking a lower and lower place in our schools; and, being abandoned by many of those who are

best qualified to do it justice, it falls into the hands of a class
of naturalists, whose ideas seldom rise above species, and who,
by what has well been called *hair-splitting*, tend to bring the
study of these into disrepute.

It will generally be found that botanists who confine their
attention to the vegetation of a circumscribed area, take a
much more contracted view of the limits of species, than
those who extend their investigations over the whole surface
of the globe. This is partly, no doubt, owing to the force of
bad example; and partly to the fact that the student who
takes up the study of the flora of his native country, finds
that the species are all tolerably well known, and that no
novelty is to be discovered. There is therefore a natural tendency to make use of trifling differences, from the scope which
they afford for minute observation and critical disquisition;
whilst the more close comparison of the few species which
come under his investigation, leads the local botanist to attach
undue importance to differences which the experienced observer knows may be safely attributed to local circumstances.
To this tendency there can be no limit, when the philosophy
of system is not understood; the distinctions which appeared
trifling to botanists a quarter of a century ago, are at the present day so magnified by this class of observers, that they
constantly discover novelties in regions which have been thoroughly well explored; considering as such, forms with which
our predecessors were well acquainted, and which they rightly
regarded as varieties*.

Another result of the depreciated state of systematic botany is, that intelligent students, being repelled by the pueri-
lities which they everywhere encounter, and which impede
their progress, turn their attention to physiology before they
have acquired even the rudiments of classification, or an elementary practical acquaintance with the characters of the na-

---

* Many of the species which have been revived in modern times, were indicated by Haller, Ray, Tournefort, and other ancient botanists, but were reduced
to the rank of varieties, when the science was reformed by Linnæus.

tural orders of plants.   Unfortunately, in botany, as in every
other branch of natural science, no progress can be made in
the study of the vital phenomena except the observer have a
previous accurate acquaintance with the various modifications
under which the individual organs of plants appear in the different natural orders, and such an appreciation of the comparative value, structural and morphological, of these modifications, as can only be obtained by a careful study of the affinities of their genera and species.   Ignorance of these general
laws leads to misinterpretation of the phenomena investigated
by the physiologist, and to that confusion of ideas which is so
conspicuous in the writings of some of the astute physiological observers of the day.

The modern system of botanical instruction attempts far
too much in a very limited space of time, and sends the student forth so insufficiently grounded in any branch of the
science, that he is unprepared for the difficulties which he
encounters, let his desire to progress be ever so great.   The
history of botanical discovery, and the philosophy of its advance, form instructive chapters for the student in any department of natural science.   In Professor Whewell's ' History of the Inductive Sciences,' the subject is ably sketched
for the information of the general reader; and it is there
shown that the most important contributions to the progress
of the science have been purely physiological questions, investigated with consummate judgment by our most eminent
systematists.   We owe to Linnæus the establishment of the
doctrine of the sexuality of plants; and we find by the writings of the same great naturalist, that besides foreseeing many
physiological discoveries, he preceded Goethe in the discovery
of morphology, a doctrine which, more than any other, has
tended to advance scientific botany.   A third great discovery,
that of the nature of the ovule, and the relation of the pollen-
tube to the ovary, received its principal illustration at the
hands of Brown, our chief systematist, and of Brongniart, also
a practised botanist.

It should not be forgotten, that the relative importance of physiology is very different in the animal and vegetable kingdoms. In the former, structure and function operate so directly upon one another, that the great groups are, to a certain extent, defined by well-marked external characters, which are at once recognizable by the student, and are familiar, or at least intelligible, to those even who have paid no attention to natural history. In the vegetable kingdom this is by no means the case: the processes of assimilation and secretion present but little of that complication which renders the study of animal physiology so important; they are, on the contrary, uniform almost throughout its whole extent, and moreover so simple in their *modus operandi*, that this very simplicity prevents their being rightly understood. In consequence, even the two great classes of Monocotyledons and Dicotyledons are not distinguishable without considerable practice and study; and were we dependent upon actual inspection of the organs whence the essential characters of these two groups are drawn, for the means of recognizing them, Systematic Botany would be an impracticable study.

Herein lies one great obstacle which meets the beginner on the very threshold of his botanical studies : he sees the great divisions of the animal kingdom to be recognizable by mere inspection, and that familiar characters are also natural, and available for purposes of classification : the very names of the groups convey definite information, and to a great extent give exact ideas. Birds, fishes, reptiles, etc. are all as natural as they are popular divisions; but what have we in the vegetable kingdom to guide the student through the two hundred and fifty natural orders of flowering-plants? As with a new language, he must begin from the very beginning, and also avail himself of artificial means to procure as much superficial knowledge of structure and affinity as shall enable him to see that there is a way through the maze. Hence the obvious necessity of an artificial system of some sort to the beginner, who has, at the same time, to master a terminology, which, if not so complex as that of zoology, is more difficult at the outset, from the want of standards of comparison between the organs of plants and those he is familiar with in himself as a member of the sister kingdom. Applying these remarks to practice, the botanical student finds that he has much to un-learn at the very outset; in many cases he has misapplied the terms root, stem, leaf, etc., and contracted most erroneous ideas of their structure and functions; while he is startled to find that the popular divisions of plants into trees, shrubs, and herbs,—leafy and leafless, water and land, erect, climbing, or creeping,—are valueless even as guides to the elements of the science.

It is not however to be supposed, because pure physiology is of secondary importance to the right understanding of the affinities of plants, that botany is therefore a less noble or philosophical study than zoology; since we find anatomy, de-velopment, and morphology, occupying a very far higher rank in proportion. Being deprived, as he is in most cases, of all technical aids to the determination even of the commoner exotic natural families, the systematist is compelled to com-mence with the knife and microscope, and can never relinquish these implements. Systematic Botany is indeed based upon development; and no one can peruse, however carelessly, the most terse diagnosis of a natural order or genus of plants, without being struck with the variety and extent of know-ledge embodied as *essential* to its definition and recognition. Not only are the situation and form, division or multiplica-tion, relative arrest or growth, of the individual organs ex-actly defined, in strictly scientific and scrupulously accurate language, but the development of each is recorded from an early stage: the vernation and stipulation of the leaves; the æstivation of the young calyx and corolla, and their duration relatively to other organs; the development and cohesion of the stamens; the position and insertion of the anther; its pollen; the cohesion or separation of the carpels, and the stages of their development from the bud to the mature fruit, and from the ovule to the ripe seed, are all essential points; all, however minute, must in many cases be actually inspected before the position of a doubtful genus can be ascertained in the Natural System; and this is not the exception, but the rule.

The necessity for acquiring so extensive and detailed a knowledge indicates a power of variation in those organs from which the natural characters are drawn, that defeats any attempt to render one, or a few of them only, available for the purposes of classification; and hence it is that the study of morphology, or the homologies of the organs, becomes indispensable to the systematist: by this he reduces all anomalies to a common type, tests the value of characters, and develops new affinities. The number, form, and relative positions of organs may supply technical characters, by which observers of experience recognize those natural orders under which a great number of plants arrange themselves; but a knowledge of structure and anatomy alone enable the botanist to progress beyond this, and to define rigidly: whilst the study of development affords him safe principles upon which to systematize and detect affinities, and morphology supplies the means of testing the value of the results, and reveals the harmony that reigns throughout the whole vegetable world.

Physiology, again, is a branch of botany very much apart from these: its aim is the noblest of all, being the elucidation of the laws that regulate the vital functions of plants. The botanical student of the present day, however, is too often taught to think that getting up the obscure and disputed speculative details of physiology, is the most useful elementary information he can obtain during the short period that is given him to devote to botany*; and that, if to this he adds the scru-

* As we are writing in the hope of being useful to our medical brethren amongst others, we may be excused from remarking here, that it is not to the credit of our medical curriculum, that, travel where we will, we find the medical man deploring his inability to apply the knowledge of botany obtained at his college, to any useful purpose. The little he has learned about the names and functions of organs he might easily have acquired at school, and thus have been prepared to devote the whole period of his botanical studies to the practical ap-

tiny of a few of the points under a microscope, he has made real progress as an observer. This, we maintain, is no more botany, than performing chemical experiments is chemistry, or star-gazing, astronomy. A sound elementary knowledge of vegetable physiology is essential to the naturalist, and should indeed be a branch of general education, as it requires nothing but fair powers of observation and an ordinary memory to acquire it. For the student to confine his attention to this knowledge of the vegetable world, and to try and improve upon it by crude experiments of his own, undertaken in ignorance of the branches of pure botany we have enumerated, is a very rational amusement, but nothing more.

A review of the progress of the science in England during the last fifty years, proves indisputably, that more botanists were made by the thorough grounding in classification to which all students were formerly subjected, than by the present method of commencing instruction with anatomy and physiology, organic chemistry, the use of compound microscopes, and similar abstruse subjects, which are mysteries to the majority of students. The latter are indeed, in too many cases, perfectly ignorant of the elements of natural science, and require some practical acquaintance with plants and their organs, before they can appreciate the relations of the different branches of botany to one another, or discriminate between what it is essential to understand first, and what is better acquired afterwards. Were the elements of science taught at schools, this would not be so: we should then have the student presenting himself at the botanical lectures fully prepared for the more difficult branches of the science, and for making that progress in them for which the professor's aid is indispensable. A sound practical knowledge of system we hold to be an essential preliminary to the study of the physiology of

plication of the Natural System, as illustrated by medicinal plants and their properties. The botanical class would not then be considered, as it now universally is, as time thrown away, and an interference with the legitimate studies of the medical student,—an opinion also shared by many of the professors.

plants,—a study which requires also a practical acquaintance with organic chemistry, consummate skill in handling the dissecting knife, and command over the microscope, a good eye, a steady hand, untiring perseverance, and above all, a discriminating judgment to check both eye, hand, and instrument. A combination of these rare qualities makes the accomplished vegetable physiologist, and their indispensability gives physiology its pre-eminence in practice.

III. *Subjects of Variation, Origin of Species, Specific Centres, Hybridization, and Geographical Distribution.*

It has been with no desire of obtruding our views upon our readers that we have ventured to discuss these obscure subjects with relation to Indian plants, but from a conviction, that in the present unsatisfactory state of systematic botany it is the duty of each systematist to explain the principles upon which he proceeds ; and we do it not so much with the intention of arguing the subject, as of pointing out to students the many fundamental questions it involves, and the means of elucidating them.

To every one who looks at all beneath the surface of descriptive botany, it cannot but be evident that the word *species* must have a totally different signification in the opinion of different naturalists ; but what that signification is, seldom appears except inferentially. After having devoted much labour in attempting to unravel the so-called species of some descriptive botanist, we have sometimes been told that the author considers all species as arbitrary creations, that he has limited the forms he has called species by arbitrary characters, and that he considers it of no moment how many or how few he makes. So long as this opinion is founded on conviction, we can urge no reasonable objection against its adoption ; but it is absolutely necessary that the principle should be avowed, and that those who think the contrary should not have to waste time in seeking for nature's laws in the works of naturalists who seek to bind nature by arbitrary laws. So again with regard to specific centres; except we are agreed with an author as to whether the same species has been created in one or more localities, and at one or more times, we shall be at cross purposes when discussing points and principles relating to identity of species and geographical distribution.

Great differences of opinion have from the earliest days of science always existed on the nature of species. The prevalent opinion has undoubtedly at all times been, that a species is a distinct creation, distinguishable from all others by certain permanent characters. Many eminent philosophers, however, have taken a contrary view; of these the best known have been Lamarck, and more recently the anonymous author of the ' Vestiges of Creation.' Into the arguments on either side it is not now our intention to enter; indeed we could not do so without occupying more space and time than are at our disposal. A most masterly view of the present state of the question will be found in Sir C. Lyell's ' Principles of Geology,' where the arguments of Lamarck and others are stated with great fairness, and answered by the author, whose opinion is decided in favour of species being definite creations. In this we are disposed to, agree, having seen no argument which is sufficient to alter the *à priori* conclusion to which facts appear to point, that it is more probable that species should have been created with a certain degree of variability, than that mutability should be a part of the scheme of nature. This however is pre-eminently a question for systematists. Long and patient observation in the field, and much practice in sifting and examining the comparative value of characters, can alone give the experience which will warrant the expression of a decided opinion on a question of so much difficulty.

It cannot be doubted that the general acceptance which the doctrine of the mutability of species has met with amongst superficial naturalists, has originated in a reaction from early impressions of the absolute fixity of characters. The student

who is taught that species are definite creations, constant
and unchangeable, without being cautioned as to their power
of variation within certain limits, finds, when he begins to observe for himself, that he has constant difficulty in determining their limits, and that abler judges than himself are equally
at fault. The more books he consults, the greater are the
discrepancies he meets with; if he has recourse to gardens,
he there finds species still more sportive; and if he travels,
he meets with a change of form under every climate; till at
last, perplexed and mortified, he gives up the study of specific
botany, and becomes a convert to the belief that species are
the arbitrary creations of systematists. And such must be
the result in the great majority of instances, while each observer has to acquire for himself that familiarity with the
amount of variation to which organized beings are subject,
which alone will render him a sound systematist. For so long
as our early education does not teach us this important principle, so long shall we find beginners refusing to accept the
conclusions arrived at by abler botanists.

Even if we admit the hypothesis that the existence of species
as definite creations is inconsistent with facts, it does not necessarily follow that the study of systematic botany is fruitless;
for such a supposition involves the operation of laws which
govern the variations of plants, and in accordance with which
they remain fixed for a longer or shorter period; and such
laws it becomes the duty of the systematist to develop. The
advocates for their agency principally base their belief upon
hybridity, and variability induced by climatic influences; but
we shall attempt to show, that all the legitimate conclusions
which can be drawn from a study of these phenomena are opposed to the theory of universal mutability.

A. *On the effects of Hybridization.*

Recent experiments have led to the following results :—

1. It is a much more difficult operation to produce hybrids,
even under every advantage, than is usually supposed. The

number of species capable of being impregnated even by skilful management, is very few; and in nature the stigma exerts a specific action, which not only favours and quickens the operation of the pollen of its own species, but which resists and retards the action of that of another; so that the artist has not only to forestall the natural operation, but to experience opposition to his conducting the artificial one.

2. Even when the impregnation is once effected, very few seeds are produced, still fewer of these ripen, and fewest of all become healthy plants, capable of maintaining an independent existence; this is a very important point, for under the most favourable influences the average number of seeds that are shed by a healthy plant in a state of nature come to nothing, chiefly owing to the pre-occupation of the soil and the wants of the animal creation.

3. The offspring of a hybrid has never yet been known to possess a character foreign to those of its parents; but it blends those of each, whence hybridization must be regarded as the means of obliterating, not creating, species.

4. The offspring of hybrids are almost invariably absolutely barren, nor do we know an authenticated case of the second generation maturing its seeds.

5. In the animal kingdom hybrids are still rarer in an artificial state, are all but unknown in a natural one, and are almost invariably barren.

On the other hand, it is often argued that hybrids are common in gardens, and that their occurrence in a state of nature cannot be denied; and that if the permanence of one such hybrid be admitted, the whole fabric of species is shaken to its foundation. Such summary conclusions are however opposed to philosophical caution : the whole subject is one that cannot be cleared up by a consideration of exceptional cases; it must be argued upon broad principles, and unfortunately no argument has ever been adduced that has not been taken in evidence on both sides of the question. This is especially the case with hybridization, which, in so far as it can produce a form distinct from either parent, does, in one sense, create what may temporarily pass for a species; and in so far as the hybrid combines the characters of both parents, it temporarily obliterates the distinctive characters of each. All, then, that we could legitimately conclude from these facts is, that were hybrids of universal occurrence, they would have obliterated all traces of species, but that, exceptional in art, and not proven if not almost impossible in nature, they cannot be assumed to have produced any appreciable result.

There are, however, other points connected with the subject of hybridity, which are of practical importance to the systematist; and in the first place, the fact of its being generally assumed by continental botanists that hybrids do occur in nature, must not be overlooked. Thus we have so-called hybrid gentians in the Jura, and hybrid thistles in Germany; whence the possibility of similar productions occurring in India is to be borne in mind. It is, however, a singular fact, that these hybrids are vouched for only in genera most notoriously apt to vary, and mainly by hair-splitting botanists. In the course of our extended wanderings, it has been our habit to acquaint ourselves with the plants as we gathered them, and so to observe their differential characters in the field, that we were never at a loss for the means of understanding one another when alluding to any particular species; yet we never met with a plant that suggested to us even a suspicion of hybridization. Dr. Wallich, whose tropical experience is probably greater than that of any other botanist whatever, and whose mind and eyes were always open to seize characters and discriminate species, makes the same remark. Griffith, a man of singular powers of observation, and whose experience was very great, never alludes to the subject; nor is the existence of hybrids in nature ever noticed in the pages of Roxburgh, Jack, Wight, or Gardner (of Ceylon)*. It is very true that

* M. Jordan has not unfrequently, it would appear, found that seeds collected on particular species have produced a different form, and he has not hesitated to infer that the ovules of the plant had been impregnated by a different

all this proves nothing; but when we add the tacit acquiescence of Robert Brown, and of all other botanists who have lived amid a tropical vegetation, and devoted themselves to its study, it will not be considered surprising that we should suspect such evidence as has hitherto been adduced by local observers only, and in very limited areas.

The subject of hybridization is however well worthy of the attention of the tropical botanist; and both in his garden and in the field, he should keep his attention always alive to the importance of observing every phenomenon that may bear upon its agency, and should institute operations that will throw light upon the subject.

B. *On Variation of Species.*

Although the researches of naturalists have not hitherto led to the detection of those laws in obedience to which many species of plants vary much in one climate and less in others, or remain constant throughout many climatic conditions, they indicate the operation of certain general laws, whose effects are as follows :—

1. Contiguous areas, with different climates, are peopled by different species of plants, and not by the same under different forms. 2. Similar climates in distant areas are not peopled by the same or even similar species, but generally by different natural orders of plants. 3. Both contiguous and remote areas contain a certain admixture of species common to two or all of them, which retain their individuality under every change of climate.

These are generally admitted facts; there are however exceptions, upon which are based the arguments for attributing to climatic effects the creation of many species from one variable type. Careful observation reveals many such exceptions; and the tendency which plants display to revert to one typical

species. The contrary inference, that species are subject to a certain amount of variation, does not seem to have occurred to him.

form, is often the only guide we have to their origin. To us it appears that but one legitimate conclusion may be drawn from the facts; and that, taking the broadest view of the case, while it is difficult, on the one hand, to reconcile the acknowledged tendency of varieties and hybrids to revert to their original state, with the fact that the floras of remote areas, possessing similar climates, are permanently and prominently different in their main elements; on the other, it is equally remarkable that the majority of the plants found wild or cultivated in all climates, are not specifically changed by any; and this, whether they are of species that have been thus widely spread for ages, or such as have been introduced by man in later times.

In the Botanical Gardens at Calcutta many thousands of plants from all parts of the world have been cultivated with more or less success, and some have become denizens of the soil; but in no instance has such a change of character been produced as could justify the suspicion that specific marks might be obliterated by even such violent contrasts of climate as Calcutta and Australia, or Calcutta and the Cape of Good Hope, afford. On the contrary, the seedlings seem infallibly to resemble their parents for generation after generation, altered perhaps in size, and more frequently in habit, and accommodating themselves to the seasons of India, but remaining true to their botanical characters.

With regard to the specific effects of climate on plants, they are extremely difficult of appreciation, the observer seldom having the opportunity of becoming familiar with the same species under very different climatic influences, at one and the same time. This is, however, an essential point, for nothing is so fallacious as recollections of the habit and general appearance even of very familiar plants. We have ourselves repeatedly gathered some of the commonest English weeds in foreign countries without recognizing them, though they differed in no respect, even of habit, from those we had been familiar with from childhood,—so deceptive are the ef-

e

fects of local circumstances and temporary associations, which give a foreign colouring to everything surrounding them.

The following remarks on the relation between climate and the development of species in India, though crude, may prove suggestive to those enabled to pursue this subject. Although India presents greater contrasts of climate than any other area of equal size in the world, we do not find that those genera and species, which prevail over all its parts, are so variable in any respect as are the plants of some countries which enjoy a more uniform climate; as an example, we may say that the species forming the flora of New Zealand are, as a whole (proportionately to the extent of the flora), far more variable than those of the mountains or plains of India. Could this fact be expanded, and, being confirmed in a wider survey, be proved to be of general application, it would be one of the most important data to start from in the investigation of those laws that regulate the development of varieties; but we are not prepared to say that a comparison of the species which inhabit the excessive climates of different parts of India with those that inhabit the uniform climates, supports this view: for instance, the central or temperate regions of the Himalaya, where perennial humidity and coolness prevail, are not peopled by very variable genera and species, whilst the alpine regions that are characterized by an excessive climate are so, and the annuals of the hot plains are peculiarly sportive in stature, habit, hairiness, foliage, and number and form of their smaller organs.

Another point, intimately connected with the question of the power of climate in producing change in species, is the relation that exists between the climate of an area, and the number of species that inhabit it; and this affords a fertile and most interesting field of inquiry in India, where so many climates may be met with in a comparatively limited area. A few facts have appeared to us worthy of notice, though as yet far from well established: as that the equable climate met with on the cool parts of the Khasia mountains and temperate regions of the Himalaya, and on the hot humid coasts of Bengal and the Malay peninsula and islands, produce an abundance of well-marked species of plants, whilst the dry, hot, lower hills of Central India, with contrasted seasons, produce comparatively few, and none presenting any great difficulties to the systematist; as also that the plains of the Gangetic valley and of the peninsula, which have marked seasons, are comparatively poor in species, whilst those of the Cape, Australia, and South America, also having decided summer heat and winter cold, abound in species. Such discrepancies prove how subtle an element climate is, and how extremely cautious the naturalist should be in generalizing upon its effects. They especially warn us not to consider the influence of climate as paramount in determining the distribution of species or prevalence of forms. We learn from them also that the *primâ facie* evidence in favour of definite creations is not to be lightly put aside; and they suggest the propriety of instituting observations in proportional botany, as that branch of the science may be called, which develops the relations between the number of orders, genera, and species, contained in an area, and its climate and other physical characters.

And now that we are on the subject of variation, it appears advisable to impress upon the Indian botanist the value of studying its phenomena in the field. We pledge our experience that he will find it the most profitable department of systematic botany he can pursue; and that the result of his investigations will be that he will take a wide and extended view of the variations of species, consistently with their still possessing certain definable limits. We shall offer a few remarks on this point under two heads :—variation of parts of the same individual, and variation between different individuals of the same species.

1. *Variation in organs of the same individual plant.* From the luxuriance of the vegetation with which the Indian botanist is so often surrounded, and the rapidity of its development, he has advantages for pursuing this inquiry that observers in colder climates do not possess. In general terms,

the most important groups of phenomena requiring elucidation and careful description are,—1. The changes that accompany the growth of individual organs from the seedling state
to the decaying plant. 2. Variations in the same organs, as
displayed in different parts of the same individual. 3. Variations in the development and distribution of the sexual organs in plants with unisexual flowers, and in bisexual plants.

It is to our neglect, and often to our ignorance, of the
changes in form that so many organs undergo during the different stages of the life of the individual, or of the different
form under which they appear in different parts of the same
individual, that we owe so many of the spurious species
which crowd the pages of our systematic works; and it is
to the want of that early training to habits of observation in
the field, which we have so strenuously advocated, that is to
be attributed the rarity of that power of discrimination between essential and non-essential characters, which alone can
make an observer a sound systematist. We therefore earnestly recommend to the Indian botanist the detailed study
of individuals and their organs*, with the view of determining their limits of variation. In relative size especially, the
observer will find immense variation; for, unlike the animal
creation, proportional dimensions are of small moment in the
vegetable kingdom. This fact, so familiar to the botanist of
experience, is always a puzzle to the zoologist, who fancies he
perceives a vagueness and want of exactness in all botanical
writings (except in those of the too numerous class that make
a parade of measuring to lines organs that vary by inches),
that contrasts unfavourably with descriptive zoology. Sym-
metry again is only a relative term amongst plants, for even
such leaves as grow in pairs are never alike, and often differ
much in form, texture, and colour; whilst the various sepals,
petals, etc., of an individual flower, never so exactly correspond, as the relative members of an animal do; and there are

* In Wight and Arnott's 'Prodromus,' p. xxi., this point is especially dwelt
upon, and a warning given to beginners, which has been too little attended to.

still greater differences between these organs, when taken from different flowers. And however carefully we investigate the anatomy of a plant, we never fail to find similar deviations from ideal regularity prevailing; for even the number of ovules (when more than two) varies in the different cells of one ovarium, as do the number of ovaria in flowers that bear several*. As regards variations in the floral organs, these are apparently more likely to occur, the less the individual parts deviate from the normal type (the leaf), of which they are modifications; as if the more complete adaptation to a special function rendered them less liable to casual variation. We find, for instance, that the carpels of Ranunculaceous plants vary much in shape, while those of *Umbelliferæ* and *Compositæ* are almost constant; and that the sepals of *Rosa* and *Pæonia* present remarkable variations of form, while those of *Dianthus* and *Kalanchoe*, which are united into a tube, retain their form, with scarcely any modification, in each species†.

2. *Variation between different individuals of the same species.* This is a more fertile source of spurious species than that last treated of, and, in our opinion, the neglect of its effects has mainly contributed to such a multiplication of species in the vegetable kingdom, as botanists unfamiliar with large herbaria and exotic plants are slow to believe; and to the exaggerated estimates of the supposed known extent of the vegetable creation that gain common credence. We feel safe in saying

* It is hardly necessary to allude to the desirability of studying the various forms induced by artificial causes: the browsing of cattle on shrubs, for instance, which is almost invariably followed by an abnormal state of foliage on the subsequently developed shoots, has been a prolific source of bad species; while there is scarcely an operation of man that does not tend to produce change in the vegetation surrounding him.

† The shape of floral leaves and bracts is, in general, much less constant than that of the perianth. It is important to bear this in mind in many families of plants. We could especially notice, as an instance, *Coniferæ*, in which the scales of the cone are very generally relied on as affording specific characters. If botanists who have an opportunity would examine and record the degree of variation which occurs in the shape of the scales of the cones of the individual trees, in the Indian species of Pine, especially *Abies Webbiana*, and its variety *A. Pindrow*, a great benefit would be conferred upon science.

that the number of known plants is swelled one-third beyond its due extent, by the introduction of bad species founded on habit, and on accidental varieties produced by soil, exposure, etc. This subject admits of classification under two heads, to neither of which can we be expected to devote much space in this Essay.

1. There are accidental variations due to no apparent causes or to very fluctuating ones, as colour of flowers and leaves, odour, hairiness (to a great degree), development of parts, strength of medicinal or other properties, hardness and various properties of wood, and many others. 2. More permanent deviations that accompany change of locality, and affect more or less all the individuals inhabiting a certain area: these may often be traced to physical causes, and give rise to races and stocks, which are more or less permanent under cultivation and changed conditions, such as habit, hardiness, and duration of life and of foliage (evergreen or deciduous), predilection for certain soils and exposures, and other characters which are more or less obviously induced by operations that have extended through a series of generations.

Gregarious plants, in all states, whether wild or cultivated, and field-crops in particular, offer excellent opportunities of studying these phenomena. Nor are these remarks applicable to herbaceous or shrubby plants only: even in this country the variations of the recently introduced Deodar are already attracting attention to the question of its specific diversity from the Cedar of Lebanon and that of North Africa*.

---

* As regards the specific differences between the common Cedar and Deodar, we think the question still open to discussion. We have no fixed opinion on the subject, and in the present incomplete state of our knowledge we recommend caution. The prominent difference strongly urged is founded on error; *i.e.* that the scales of Cedar-cones are persistent and those of the Deodar deciduous; the fact being that the Cedars at Kew and elsewhere scatter their cone-scales whenever a warm summer ripens their wood. As to the differences of timber, that of the Cedar is so very variable as to throw suspicion on the value of this character; and other trees, as we have elsewhere said, present immense difference. The odour and quality of Cedar-wood varies according to the circumstances under which the trees have been grown. Length and colour of leaf, and habit,

The varieties that may be selected from a plantation of seedling Spruce, Larch, or Yew plants are innumerable; but so led away are observers by dominant ideas as to the form and habit that plants should assume, that similar differences in other species are seldom put down to a similar power of varying, as *à priori* they should be, but are taken as evidence of specific difference. To this proneness to attach undue importance to variation, we owe the separation of *Pinus Pindrow* from *Webbiana*, *P. Khutrow* or *P. Morinda* from *P. Smithiana;* nor is this all, for species have been made of the commonest English plants which grow in the Himalaya, because they present differences of habit when compared with English individuals, but which plants, if compared with continental specimens of the same species, are found to be identical with them : to such an extent has this been carried, that of the several hundred European plants found in India, there is hardly a species that has not had one (and many, more) new names given to it.

The differences in the properties of plants and in the colour and durability, etc. of woods, demand a short notice, because the idea is too prevalent that these are very unvarying diagnostic properties of species. That some woods are always good, and some as constantly worthless, is incontestable; but this applies chiefly to those of very remarkable hardness or density or weight, or other very unusually marked quality; and even of these, the Teak, Sissoo, Sal, etc., each vary much in quality, whilst the wood of other kinds is singularly variable, as of the Indian Pines, Oaks, Laurels, Ebonies, etc. With regard to the Pines, this is very much to be attributed to the soil and climate, and consequent rapidity of growth

are so sportive in the Deodar, that we have seen many specimens of it that are as unlike what we call the typical Deodar, as they are unlike the Cedar; and others that approach the latter very closely. There are very slight differences in the shape of the cone-scales of the Deodar, Cedar, and Algerine Cedar, which have never been indicated, and may be of value: but we doubt their proving so, from the fact of the Algerine Cedar, in this respect, approaching the Himalayan, and thus uniting all three.

and development of resinous qualities. Thus the wood of the English-grown Lebanon Cedars differs greatly in colour, hardness, and odour; and the Swiss Larch and Scotch Pine, when planted in England, yield very inferior timber compared to what they do in their native forests. The wood of the English Oak grown at the Cape of Good Hope is worthless, as is that of the American Locust-tree, and indeed of most American timber-trees, in England. The varieties of Oak* wood in our own climate are no less notoriously different; and the endless discussions that have arisen as to the relative properties of timber-trees, and the specific differences between the plants that produce them, may to a great extent all be traced to the same cause.

With regard to the development of medicinal properties they vary extremely in the same species. Of this the most conspicuous Indian examples are presented by the Opium Poppy, Mudar (*Calotropis*), and the *Cannabis sativa*, the common Hemp of England, which yields Bhang and Chirris in varying quantities, and of different quality, very much in proportion to the humidity of the soil and climate it grows in. The *Digitalis* grown in the Himalaya is said to have proved almost inert, and so with other plants which have been cultivated for medical and economic purposes, as the Tea and many English fruits and vegetables.

We have reserved habit as the last point to which we shall allude in connection with this subject, though we believe it to be of all others the most deceptive, as indicating specific difference. Habit is a thing which every one thinks he appreciates, but which no two persons similarly appreciate; each individual's conception of it depending on his own knowledge and experience, usually on first impressions, and often on preconceived ideas which become dominant. Like all other vague terms, it is used with as much confidence by a gardener to

* We do not here allude to the difference between *Quercus pedunculata* and *sessiliflora*, but to that between the wood of the same species or variety, as grown in different climates.

discriminate varieties, as by the botanist to distinguish species. The student should be on his guard to avoid being led
astray by dominant ideas on this subject, and fancying that
the aspect of a species to which he is most accustomed is the
typical one of its race. Let him examine well, in their native
forests, the Pines (those most variable of plants). Let him
compare *Pinus longifolia* from a deep dell in the humid atmosphere of Kumaon, Nipal, or Sikkim, with the same tree
growing on a sandstone rock in the arid climate of the Panjab. Let him contrast the Larch of Switzerland or the Tyrol,
with that cultivated in our English plantations, or the common
Scotch fir of the sandy plains of North Germany, with the
same tree on the higher Alps; or attempt to give limits to the
variations of the Yew-tree everywhere, whether wild or cultivated. Our Junipers, Willows, Birches, and Roses, will afford
in abundance similar instances of great mutability of form,
with no modification of essential characters; and the gardener
makes of one and the same species, or even variety, a standard
or espalier, a tree or shrub, an erect or decumbent plant.
Most of these instances, and many others, must be familiar to botanists; yet we believe we shall meet with few sup-
porters in the opinion we have formed, and to which direct
observation has led us, that habit alone, when unaccompanied
by characters, in the organs of reproduction especially, is of
no specific weight whatever.

As we write, a hundred instances of protean habit in Indian plants crowd upon our memory. The common Yew, which
is indigenous throughout the whole length of the Himalaya
and in the Khasia mountains, wherever it grows in the deep
forests is a tall tree, with naked trunk, rivalling in dimensions the giant pines and oaks with which it is surrounded;
on the skirts of the same forests it is a lax, almost prostrate
bush, while on open slopes it becomes a stout, dense, tabular-
branched tree. The Rose, Spiræa, and Berberry of the Western Himalaya are truly protean in character, being abundant
in all situations,—whether forming underwood in forest, or

f

growing on open slopes. The common Junipers defy all attempts at circumscription by habit, and so do the Cotoneasters. The Himalayan Box (*Sarcococca*), like that of Europe, is now an undershrub and now a tree. The *Hippophae* and *Myricaria* of Western Tibet, which are first met with as trees, as they ascend to colder regions dwindle down to little shrubs, stunted and almost prostrate; while *Ephedra*, an erect shrub, two feet high, on the Indus, at 7000 feet, in the more humid climate of Kunawur sends out long, lax, whip-like branches, and at 15,000 feet is scarce an inch long. Let any one recal to mind the gigantic Sal, with tapering trunk, in the Terai forest, and the gnarled tree it becomes on dry slopes; or contrast the noble Sissoo near a village in Upper India with the slender, pale, and apparently sickly (yet really robust and healthy) inhabitant of the gravelly banks of streams at the base of the Himalaya; or the wild Jujube, an undershrub, not a foot high, with the same plant cultivated as a spreading tree. Many figs have straight, erect, unsupported trunks, in open dry places, yet in humid forests the same species send down thousands of roots from their branches, like the Banyan. Most of the Indian annuals are, in like manner, multiform; being tall, slender, and delicate, in moist grassy places, during the rains, and prostrate and wiry in open spots, and at a drier season: this is especially the case with the little *Cassiæ* of the Mimosoid group, with various *Indigoferæ* and *Alysicarpi*, and even with *Æschynomene*.

The universal recognition of the importance of habit, as a character upon which to found specific distinction, is the more surprising, when we consider how many well-marked varieties are distinguished mainly by habit, and, though very permanent when the plants are increased by cuttings or grafts, soon disappear when they are raised from seed. The weeping birch and ash are good instances of this, as well as the Lombardy poplar—a diœcious tree, of which one sex only is known, and that in cultivation, and which appears to be nothing more than a tapering state of *Populus nigra*, accidentally produced, and perpetuated by cuttings. Similar examples are afforded by all our domestic fruit-trees, among which, by a practised eye, many different sorts can be recognized at once.

In conclusion, the majority of our readers will smile when we add that the general impression of persons of intelligence, that they know our common English trees at first sight, is to a great degree illusory : we have all an ideal Oak, Elm, Poplar, etc., and we call the specimens that do not come up to that ideal abnormal, and representations of such we say are not characteristic; but let any one keep a watch upon himself in the fields, parks, or forests of countries not his own, yet tenanted by trees specifically the same as those of his own, and we venture to assert that he will find his preconceived ideas fall to the ground in very many cases. We do not mean to say that he will not recognize a park oak, churchyard yew, or weeping willow; but we do assert that he will not recognize by habit the same oak at the Cape of Good Hope, where it is now abundant, or the same yew in a thick forest; and we may add that no Himalayan traveller within our experience has, on his return to England, ever recognized the Deodar at Kew Gardens *by habit* to be the plant of those mountains, and that, on the contrary, we have frequently had the Cedar of Lebanon pointed out as that tree.

It is very much to be wished that the local botanist should commence his studies upon a diametrically opposite principle to that upon which he now proceeds, and that he should endeavour, by selecting good suites of specimens, produced under all variations of circumstances, to determine *how few*, not *how many* species are comprised in the flora of his district. The permanent differences will, he may depend upon it, soon force themselves upon his attention, whilst those which are nonessential will consecutively be eliminated. There is no better way of proving the validity of characters than by attempting to invalidate them. The unavoidable tendency of the human mind, when occupied with the pursuit of minute differences, is to seize on them with avidity, and to relinquish them with regret; hence the irresistible desire to rest contented with a character, however bad, so long as it is obtained with difficulty, and in the observer's opinion is tolerably constant. It is strange that local naturalists cannot see that the discovery of a form uniting two others they had previously thought distinct, is much more important than that of a totally new species, inasmuch as the correction of an error is a greater boon to science than is a step in advance.

C. *Geographical Distribution.*

This, which is in very many respects the most interesting branch of botany, has made very little real progress of late years, owing to the confused state of Systematic Botany; for we do not consider rudely cataloguing the ill-defined species of limited areas, or loosely defining geographical regions by the supposed prevalence of certain natural orders or forms of vegetation, as calculated to advance directly the philosophy of distribution, however useful such regions are to the beginner, or such catalogues to the systematist.

If we take India as the area for examination, we are met at the outset by difficulties that plainly indicate the backward state of Indian Botany. Beginning with the first requirement of the student of geographical distribution, we are literally perfectly ignorant of the numerical value of a single important Indian natural order of plants: turning to their numerical proportions, there are no sufficient data for saying which of the five largest orders in the vegetable kingdom is the most abundant in India, viz. *Leguminosæ, Compositæ, Gramineæ, Orchideæ,* or *Rubiaceæ,* nor in what climates each most prevails; still less do we know how the important tribes of these natural orders are distributed, or what physical features of temperature, elevation, and moisture they indicate, or to what other floras their relative predominance allies that of India. There is no work that pointedly indicates the natural orders peculiar to India, and still less the genera and species. With regard to the European genera, which in some parts literally form the mass of the flora, we find them but vaguely indicated in our best authorities; and the European and British species have, as we have said already, been almost invariably described as new, without examination or comparison, and many of them more than once or twice. Yet all these elements must be approximately settled before we can attempt a solution of those great questions involved in Botanical Geography, which place it as a philosophical study in the foremost ranks of science: we allude to the laws which govern the development, progression, and distribution of forms and species; the connection of these laws, not only with one another, but with physical features; and their modifications by geological change. We must know at what rate European and African plants disappear in advancing eastwards in India, and Malayan ones in following an opposite direction; how the Chinese, Japanese, and North American genera and species mingle with western forms along the Himalaya and Khasia; and the exact amount of Arctic and Siberian plants, which are spread all over the loftier Himalayas, and descend the valleys of the Indian watershed. And lastly, there are extraordinary anomalies to unravel, or to secure on a basis of accurate observation; such as the absence of Oaks in the peninsula of Hindostan and Ceylon, though they abound on the opposite shores of the Bay of Bengal continuously from the Himalaya to Java; the want of any Pine whatever in the peninsula of Hindostan, and of *Cycadeæ* in Ceylon; and many other points of the highest interest, that have never yet attracted the attention of naturalists, and want illustration previous to explanation.

We cannot pursue these interesting subjects here, nor dare we, in our present ignorance of botanical facts, allude to the connection which we think shadowed out between the geological events that have resulted in the present configuration of the Indian continent and peninsulas, and the lines along which certain groups and species of plants have consequently been distributed.

We have already remarked that the effect of confounding
variations with specific differences has been to swell the supposed number of known plants by one-third; and we think
that, if mistaken ideas of distribution be added, we shall find
that, of the number of species enumerated in catalogues, the
proportion that are spurious amounts to at least one-half.
Thus, there are not a few botanists who have contributed a
very considerable number of such, founded solely on the fact
of their supposed isolation, and which were not even compared
with their described congeners previous to being thrust as new
into the annals of botany.   The Indian Flora swarms with
these.   In the natural order *Ranunculaceæ* alone, comprising
115 species, we have been obliged to reduce 28 supposed species*, founded exclusively on Indian specimens, to well-known
European plants, besides a multitude of others, natives of
Siberia, Persia, Western Asia, and some eastern Asiatic ones.
Of the 27 European *Ranunculaceæ* enumerated, only 4 had
previously been identified, and of 17 others all had one or
more new names, there being 28 new names in all.   When
we add, that such plants as the common English Marsh-Ma-
rigold, Monks-hood, Columbine, Pæony, Actæa, Crowfoot,
Berberry, White Waterlily, and Red Poppy, have all had
names lavished on them in virtue of their Indian birthplace,
our readers may judge for themselves of the progress that
the geographical distribution of Indian or European plants is
likely to make for some years to come†.   Of the undue im-

---

* This is a very moderate estimate, for we fully believe that future authors
will reduce many other species which we keep distinct, to English forms, especially among the *Ranunculi* and *Delphinia;* we have, however, considered it
necessary to prove absolute identity between the European and Indian individuals, before uniting them, which of course obliges us to keep separate many
plants which we fully believe to be only Indian forms of well-known western
ones.

† The converse of this is equally instructive and illustrative of the point we
wish to impress.   The Silver Cedar of our parks, so long as its habitat was unknown, was universally considered to be a variety of the Lebanon Cedar: now
that it is known to come from Algeria, and not Lebanon, it is considered a different species in standard works.

portance attached to locality, we believe that botanists have no conception. Witness the fact, that several common European garden-plants introduced into the grounds of the British Resident at Katmandu (Nipal), and thence re-imported to England, have been at once put forth in this country as new Himalayan discoveries, and specific characters invented for them. But instances of this multiplication of names are almost incredibly numerous : the common English Yew has two Himalayan names ; the *Pteris aquilina* (English Bracken), seven ; the eighteen known Indian species of *Clematis* are in Steudel's ' Nomenclator' ranked under forty names ; and we may conclude by announcing our conviction, that more than one-half of the recorded species of Indian plants are spurious, and that in many natural orders the undescribed species hardly equal in number those which require to be cancelled.

The fact that almost every Himalayan plant has a vertical range of nearly 4000 feet, and many of 8000, is in itself a suggestive one. Several hundred species are dispersed from the Levant to the Indus, and many more from the Ganges to the Chinese Sea. Such instances of distribution in tropical plants are called strange and exceptional by unreflecting botanists, who forget how many species are common to all longitudes between England and Kamtchatka, or to all the mountains of Europe; or to the Rocky Mountains of America, and those of Scotland and Norway; or to all latitudes between England and North Africa.

The subject of geographical distribution leads to questions of practical importance, upon which we have a few remarks to offer, as eminently bearing upon all questions relating to the treatment of a systematic flora : these are,—1. Its dependence on the doctrine of specific centres. 2. The power of migration as capable of effecting the present distribution. 3. The general effects of migration in producing a much wider dispersion and ubiquitous diffusion of species than is generally admitted by botanists who have not investigated tropical floras, and especially continental ones.

1. As regards specific centres, we proceed in our investigations on the assumption that all the individuals of a unisexual plant proceeded from one originally created parent, and all of a bisexual from a single pair. To discuss this subject would be out of place here: for a *résumé* of the principal facts opposed to it, as well as of those which support it, we must refer our readers to Sir Charles Lyell's ' Principles of Geology,' and to the Introductory Essay to the Flora of New Zealand. It is sufficient for our present purpose to declare, that after many years' unprejudiced careful consideration of the subject in all its bearings, during which period we have been fettered by no professed opinion to support, and have had no inculcated theory to eradicate, we have been independently led to this conclusion, as being most consonant with our very considerable experience in the field and herbarium.

2. In attributing the present dispersion to natural causes, we by no means limit them to existing ones. We have every reason to believe that many living species of plants have survived the destruction of large continents, just as many animals have ; that in short they have outlived recent geological changes, of whatever magnitude, that they have witnessed gradual but complete revolutions in the relative positions of land and sea, and consequently in the climate of the several parts of the globe. Such an antiquity is proved for shells especially, and to a greater or less degree for all tribes of the animal kingdom ; the amount of evidence depending solely on the adaptation of their dead parts to preservation in a recognizable condition. Fossil plants are specifically never thus to be identified, and our argument is hence one founded on analogy only, but supported by many facts* in distribution, not less than by the effects of such operations as we now see in progress.

* Sir Charles Lyell was the first to appreciate this most important element in geographical distribution (Principles of Geology, chap. xxxiii.) ; and Professor Edward Forbes first brought it to bear upon an existing Fauna and Flora, in his admirable Essay on the ' Distribution of the Plants and Animals of the British Islands' (in the 1st vol. of Mem. Geolog. Survey of U. K.). We

Applying this view to the Indian Flora, we may illustrate it by assuming, as an example, that the majority of the many plants common to the Himalaya and Java migrated over continuous intervening land, which has been broken up by geological causes, chiefly by subsidence; just as the partial subsidence of Java itself would effect a further dismemberment of an area now continuously peopled with plants, and which would result in a cluster of islets, having a vegetation in common. Extending this idea of submergence and emergence of land, one island may at different epochs have been continuous with different continents, from all of which it may have received immigrants. We are very far from denying the active agency of the winds and of animals in aiding distribution, and, to a limited extent, of oceanic currents also; but all the phenomena of geographical distribution, when carefully studied, are so uniform in their nature, and so harmonious, as to demand some far higher and more comprehensive agent than the desultory and intermittent motions of the elements or of animals, to produce the present grouping of plants.

There is a very curious theoretical point bearing upon the distribution of species, first enunciated, we believe, by a most accomplished observer, Dean Herbert, and which, we think, has never been sufficiently appreciated or followed out; it is, that species in general do not grow where they like best, but where they can best find room. Plants, in a state of nature, are always warring with one another, contending for the monopoly of the soil,—the stronger ejecting the weaker,— the more vigorous overgrowing and killing the more delicate. Every modification of climate, every disturbance of the soil, every interference with the existing vegetation of an area, favours some species at the expense of others. The life of a plant is as much one of strife as that of an animal, with this

cannot too strongly recommend this able and original essay to the study of our readers, as the most important contribution to the philosophy of distribution that has ever appeared. We consider the principles embodied to be sound, of universal application, and as necessary to be understood by the student of nature as are the laws of climate and the distribution of heat and cold.

difference, that the contention is not intermittent, but continuous, though unheeded by the common observer. In the common course of events, therefore, the ground occupied by a widely-distributed plant is held on a very different tenure in different places; some individuals are obliged to grow in the shade, others in the sun; and they hence flower earlier in certain places: we say of such plants that they have a power of accommodating themselves to their altered conditions, or better, that they have the power of resisting the effects of the change. Now, this power we believe to be very much underrated, specific characters being too often founded on the differences in habit induced during a plant's migration over great areas, or brought about by the change of soil and climate and surrounding vegetation, to which individuals and their successors are subjected in different parts of one and the same area.

The simple fact that, of all the functions of vegetable life, reproduction is the most uncertain in its effects and results, seems to bear upon this particular point. Some plants are never known to seed; of many, not one ovule out of a thousand ripens into a seed; not one seed out of a thousand germinates, nor one plant reproduces out of a thousand that have germinated. We are too apt to consider such facts, when applied to species or individuals, as indicating that they are not in a natural condition, whereas they appear to be the consequences of a law of nature, and ought to teach us that plants, in a state of nature, are subjected to the operation of external agents, which not only alter their habit but influence their vital functions.

In these somewhat desultory remarks on the various subjects of which we proposed treating, we have endeavoured to illustrate our great argument, the imperative necessity of checking the addition of species on insufficient grounds, and the importance of treating scientifically those that are already known. We consider it to be desirable, that for all practical purposes species be regarded as definite creations, the offspring each of but one parent or pair; we believe that they are endowed with great powers of migration, and that they have been aided in their dispersion primarily by those changes of climate, land, and sea, which accompany, or are effected by what are called geological changes, and secondarily by the elements and the animal creation. Under these convictions, we feel it imperative, on philosophical grounds as well as on those of expediency, to use every effort to reduce the vast bulk of forms we have to deal with in the Indian Flora to as few species as we can, consistently with a careful study of the structural and morphological characters of each. We shall, as a rule, banish from our minds the idea that a species is probably new because hitherto unknown to ourselves or to the Flora of India; we shall, upon principle, keep two or more doubtful species as one, carefully and prominently indicating their differences, and, when expedient, ranking them as varieties; in preference to keeping doubtful species separate till they shall be proved the same; having ample proof that in so doing we shall avoid the greater evil. We shall not think it desirable to adopt the opinions of others in preference to our own* on points where we have had the best materials to judge from. With regard to nomenclature, we shall not alter names established by Linnæus, and usually retained by subsequent botanical authors, upon the ground of their having received prior names before botany was systematized. We shall incline to adopt old established familiar names, though of doubtful applicability, in preference to giving new, even when legitimate to do so. We shall endeavour to retain the first published specific name† of a plant, even when the genus requires to be changed, and shall always give preference to priority of pub-

---

* This may to some non-botanical readers sound dogmatical, if not presumptuous; but the fact is, that a system is deeply rooted and widely spread, of keeping up known bad species in so-called deference to authorities; in nine cases out of ten, this is done to save the trouble of a re-examination, and in too many, simply to swell catalogues. The same authorities are held very cheap, when they unite what hair-splitters wish to keep separate. Witness the state of the British Flora with regard to Willows, Brambles, and Roses.

† With every wish to bind ourselves by the canons (most of which are ex-

lication, except where there are obvious reasons for the contrary, which we shall explicitly state.

Lastly, we find it necessary to say a few words regarding the employment of the native appellations of plants as specific names. These are in general very uncouth, and disagreeable to those who are unfamiliar with Indian languages; moreover, they are quite unpronounceable without special education in the mode of spelling. The only advantage which they are supposed to possess, is the identification of useful species by their means. This we believe to be an entire delusion, except in a very few exceptional cases, where the native names are so extensively known that they ought to be learned as a part of a language, and not sought for in the catalogues of scientific botany. In general they are mere local appellations, confined to a single dialect of one of the many languages of quite different roots spoken over the area the plant inhabits. Added to this, they are, in by far the greater number of cases, founded on error; and it becomes necessary for the systematist to explain, that the name which, by the laws of priority, is irretrievably placed upon the records of the science, has been misapplied, and ought to be borne by another, and frequently very different plant, or by none at all. We have therefore retained native names with great unwillingness, and have not hesitated to change them wherever it has appeared practicable without violation of established rules.

In conclusion, we may state that in all these points we have only followed the example set by Wight and Arnott in their 'Prodromus Floræ Peninsulæ Orientalis,' a work which is, as regards Indian Botany, unique; and indeed there are few systematic works in our own or any other language, that equal it for accuracy, truly philosophical views of the limits of genera, species, and varieties, and scrupulous attention to the details of nomenclature, synonymy, etc.

cellent) laid down by the British Association for nomenclature in Natural History, we have, in common with every botanist who has tried to do so, been obliged to set them aside in many instances.

IV. *Summary of the labours of Indian Botanists, and of the materials at our disposal for prosecuting the Flora Indica.*

A. *Publications of importance to Indian Botanists.*

The masterly sketch of the progress of botanical science in continental India, which is contained in the introduction to Wight and Arnott's Prodromus, a work which is in the hands of every botanist, renders it unnecessary for us to enter into such full details as would otherwise be requisite, regarding the older Indian botanists and their collections. A brief notice of some works, to which we shall frequently have occasion to refer in the course of our labours, is however desirable.

The earliest scientific work on the Flora of India is the 'Hortus Malabaricus' of Van Rheede (Governor of Malabar), which was published in Holland about the end of the seventeenth century, in twelve volumes, with figures of nearly seven hundred plants. It is a very remarkable book, from the general excellence of the plates, which are faithful representations of the plants. Malabar was for many years so little explored, that till very recently a great many of the plants figured were not familiarly known: within the last twenty years, however, its flora has been investigated by so many botanists, as to be considered nearly exhausted; and as the novelties will consist chiefly of obscure plants, we may conclude that when the collections now in Europe (particularly Wight's) are described, Rheede's plants will be all identifiable.

Rumphius' 'Herbarium Amboinense' is of much less value as a work of reference than that of Rheede, because the plates are in general much inferior. They are often greatly reduced in size, and frequently bear too little resemblance to the plants which they are meant to represent, to render it useful to quote them. The flora of Amboyna is not so well known as that of Malabar, but Blume has done much towards identifying the plants figured by Rumphius, and by so doing has done good service to the antiquarian branch of botany.

The collections of Paul Hermann, a medical man in Ceylon, have been rendered classical from having constituted the materials for the 'Thesaurus Zeylanicus' of the elder Burmann, published in Holland, and afterwards of the 'Flora Zeylanica' of Linnæus. These collections form part of the very valuable herbarium at the British Museum, and are of great service in the determination of many of the doubtful species of Linnæus.

The 'Flora Cochinchinensis' of Loureiro, though it relates to a country beyond our limits, contains so many forms identical with those of Ava and Malaya, that we shall have frequent occasion to refer to it. Father Loureiro, a native of Portugal, resided for thirty-six years in the kingdom of Cochin-China, whither he proceeded as a missionary, but finding that Europeans were not permitted to reside there without good cause, entered the service of the King, as chief mathematician and naturalist*. Though he had no acquaintance with the science of botany, the difficulty of procuring European medicines induced him to direct his attention to native drugs; and with a zeal of which we have unfortunately too few instances, he prosecuted his botanical studies, and so successfully, notwithstanding his want of early education, as to produce a work of standard value. The 'Flora Cochinchinensis' was published at Lisbon, in two volumes quarto, in 1790; and a second edition, edited by Willdenow, with a few notes, appeared in octavo, at Berlin, in 1793. As was to be expected, in a work devoted to the botany of a previously unexplored tropical region, the 'Flora Cochinchinensis' contained a great amount of novelty; but the absence of plates, and a defective terminology, caused by a want of familiarity with the labours of other botanists, render the descriptions

* He styles himself, in his own narrative, "rebus mathematicis et physicis præfectum."

often obscure, so that a number of the genera described by Loureiro have not yet been identified, while others, not being recognized, have been described as new, and re-named by subsequent botanists.

We must refer to the Introduction of Wight and Arnott for full details regarding the illustrious series of botanists*, commencing with König and ending with Wallich, who investigated with so much success the botany of continental India. The volumes of the 'Asiatic Researches,' and of most of the systematic works of the end of the last and beginning of the present century, afford ample proof of the value of their labours; but none of them brought their materials together in the form of a flora, except Roxburgh, whose 'Flora Indica' however remained in manuscript for some years after his death, in 1815. Two editions of it have been published since that period; one, which is incomplete, was edited by Drs. Carey and Wallich; it extends to the end of *Pentandria Monogynia*, but contains many additional plants not contained in Roxburgh's manuscript, and requires therefore occasionally to be quoted; the other, which is an exact reprint of the manuscript as left by its author, is in three volumes, and was published in 1832.

Besides editing this portion of the 'Flora Indica' of Dr. Roxburgh, Dr. Wallich commenced, in India, an illustrated work on Nipal plants, which was the first specimen of lithography ever produced in that country; and after his return to England, he published a series of 296 plates of plants in the 'Plantæ Asiaticæ Rariores,' a work which, with the equally valuable Coromandel plants of Dr. Roxburgh, in three folio volumes, with three hundred coloured plates, forms the principal contribution of the Indian Government to the illustration of botanical science.

The eastern or Malayan Peninsula of India was unknown botanically till it was visited by Jack, whose descriptions of

---

* Jones, Fleming, Hunter, Anderson, Berry, John, Roxburgh, Heyne, Klein, Buchanan Hamilton, Russell, Noton, Shuter, Govan, Finlayson.

Malayan plants were published in the 'Malayan Miscellanies,' and have been reproduced by Sir William Hooker in the 'Companion to the Botanical Magazine,' and by Dr. M'Clelland in the Calcutta Journal of Natural History.

Dr. William Jack was appointed to the Bengal Medical Service in 1813, and was in the earlier part of his career employed in the ordinary duties of his profession. During the Nipal War of 1814–15 he was attached to the army under General Ochterlony, and had an opportunity of seeing the outer valleys of Nipal, a country which at that time was a *terra incognita* to science. In 1818, while at Calcutta, on a visit to Dr. Wallich, he met with Sir Stamford Raffles, the Governor of the British settlements in Sumatra, who at once appreciated his great merits, and offered him an appointment on his staff, promising him every facility for the exploration of the natural history of that island. This promise was most fully kept; and under the enlightened patronage of one of the most liberal Governors whom the Indian service has ever produced, Jack devoted himself with zeal and success to researches in all branches of natural history. Unfortunately his career was a very short one, as he sank under the effects of fatigue and exposure on the 15th September, 1822, on board the ship on which he had embarked on the previous day to proceed to the Cape of Good Hope. It is evident, from his published papers, unfortunately far too few, that Dr. Jack's botanical talents were of the first order, and that he had thoroughly familiarized himself with the structure of all the remarkable forms of vegetation which presented themselves to him in the peculiarly rich and varied Malayan flora.

Wight and Arnott's 'Prodromus Floræ Peninsulæ Indiæ Orientalis' appeared in 1834. We have already characterized this work as the most able and valuable contribution to Indian botany which has ever appeared, and as one which has few rivals in the whole domain of botanical literature, whether we consider the accuracy of the diagnoses, the careful limitation of the species, or the many improvements in the definition and limitation of genera and the higher groups of plants. One volume only has been published, the work having been interrupted by Dr. Wight's return to India in 1834. It contains the whole of *Thalamifloræ*, and of *Calycifloræ* down to the commencement of *Compositæ*, including descriptions of nearly 1400 species. A smaller work, entitled 'Contributions to the Botany of India,' contains the peninsular *Compositæ*, elaborated by De Candolle; the *Asclepiadeæ*, by Wight and Arnott, with the addition of the extra-peninsular species collected by Wallich and Royle, by Dr. Wight alone; and the *Cyperaceæ* of Wallich, Wight, and Royle, by Nees von Esenbeck, with valuable annotations by Arnott. Dr. Wight has also published in 'Hooker's Botanical Miscellany' some excellent descriptions and plates of Indian plants, and Dr. Arnott has communicated various detached memoirs to the botanical periodicals of the day.

On his return to Madras Dr. Wight conceived the idea of carrying out, on a very extensive scale, an illustrated work on the plants of India, and in 1838 the 'Illustrations of Indian Botany' were commenced, and soon after were followed by the 'Icones Plantarum Indiæ Orientalis.' The former work, which is furnished with coloured plates, contains a series of memoirs on the Natural Orders, full of important information with regard to species, and valuable notes on their affinities: it terminated with the end of the second volume and the 182nd plate, in 1850. In the Icones, the letterpress usually contains only the descriptions of the species, though in the later volumes occasional general details are given, especially in those natural orders which are not included in the Illustrations. The plates of the Icones are uncoloured, and amount to 2101, a surprising number, when we bear in mind that they were commenced only fifteen years ago, and take into consideration the excellence of the execution of the later ones. In the 'Spicilegium Neilgherrense,' a third illustrated work, there are coloured copies of a portion of the plates of the Icones, with much valuable matter relative to the Nilghiri

h

Flora. This is not the place to dwell on the extraordinary exertions in the cause of science of the author of these great works. They are themselves the best proof of his wonderful energy, and show what can be accomplished by perseverance under apparently insurmountable obstacles. At the period of the publication of the earlier numbers the art of lithography was in a very rude state in India, and the plates are consequently very imperfect; but in the later volumes the improvement is great, and the outline drawings are admirably reproduced. The volumes form the most important contributions, not only to botany, but to natural science, which have ever been published in India, and they have been of the greatest service to us throughout our labours.

Besides these great works, Dr. Wight has published many minor papers in the various periodicals of the day, particularly in the ‘Madras Journal of Science,’ and in M‘Clelland’s ‘Calcutta Journal of Natural History.’

Mr. Bentham’s eminent services to Indian botany demand especial notice here; and while recording our sense of the value of his labours and our admiration of his writings, we would most strongly recommend to the student of Indian botany the careful study of his works, as those of the most industrious, able, useful, and philosophical systematic botanist of the age, who, for correct appreciation of the value and limits of genera especially, is not surpassed by any systematist. His connection with Indian botany commenced by his taking a large share of the labour of distributing the Wallichian collection in 1829, in conjunction with Dr. Wallich, and he again volunteered his services to assist that eminent botanist in the second distribution, that of 1849; he has also been actively engaged in the arrangement and naming of the extensive collections sent by Major Jenkins to Sir William Hooker, by Mr. Griffith to Dr. Lemann and Sir William Hooker, as well as by Dr. Stocks and Mr. Edgeworth to his own herbarium. Of his published works, the monographs of *Scrophularineæ* and *Labiatæ* are of standard excellence, and have been incorporated into De Candolle's Systema. These, and his Florula of the Island of Hongkong, in 'Hooker's Journal of Botany,' connect his name most intimately with the progress of Indian botany; it is however impossible here to indicate the long list of memoirs he has published, and which more or less bear upon the subjects discussed in this Essay.

Since the date of publication of Wight and Arnott's Prodromus, the great work of De Candolle, the 'Prodromus Systematis Regni Vegetabilium,' has advanced from the fourth to the thirteenth volume; and as the rich materials for the Indian Flora, especially those collected by Wallich, were communicated to its author, the Prodromus contains a very complete *résumé* of our knowledge of Indian botany up to the period of publication of each natural order. This materially facilitates the study of the Corolliflorous Orders, the most important of which have been worked up by Mr. Bentham. With regard to the Thalamiflorous and Calyciflorous Orders previous to Compositæ, these, with the exception of the Peninsular ones, have for the most part to be worked out *ab initio* for the Flora Indica; the earlier volumes of the Prodromus being to a great extent compilations, and particularly defective in all that regards the vegetation of Asia.

Next in point of botanical importance comes Dr. Royle's 'Illustrations of the Botany of the Himalayan Mountains,' in two volumes quarto, with 100 plates. This is the only book except Dr. Wallich's 'Tentamen Floræ Nepalensis,' devoted to the rich flora of these mountains; and it further contains the first and only attempt to demonstrate the prominent features of the geographical distribution of Northern Indian plants in reference to the elevations and climates they inhabit, and to the botany of surrounding countries. A vast amount of valuable miscellaneous botanical matter is here brought together, with characters of a considerable number of species. These, however, are rather to be regarded as indications of the supposed novelties in the author's herbarium, than as descriptions available for botanical purposes. This should be

carefully borne in mind by those using the systematic portion
of the work, the great merit of which resides not only in the
information it contains on the subjects mentioned above, but
also in the laborious accumulation of valuable and curious
matter relative to the medicinal, economical, and other vegetable products of India, and to their history and literature.

The volume of Messrs. Cambessèdes and Decaisne, on some
of the plants of Jacquemont's voyage, is (with the exception
of Mr. Griffith's papers, to be mentioned in connection with
his distributed herbarium,) the only remaining one of any
importance relating to Indian plants generally, that has been
published since the Prodromus of Wight and Arnott. This,
a quarto work, with 180 beautifully executed plates of Indian
plants collected by M. Jacquemont, was published at Paris in
1844. The authors, not having access either to the Wallichian or Roylean herbarium, have published as new, many
plants well known in this country, but the descriptions and
plates are of great value and botanical merit.

The catalogue of Bombay plants by Mr. Graham, published
in 1830, has unfortunately been of little use to us, the absence of descriptions rendering it impossible to identify in a
satisfactory manner the species referred to. In a thoroughly
explored country, the plants of which are accurately determined, such catalogues are of great value; but where the
flora is only partially known, and imperfectly described, they
are not to be depended on. In the present instance, internal
evidence occasionally enables us to recognize with certainty
the plant named; but more frequently it shows that the identification is erroneous, without affording that clue which a description would have given, for the rectification of the error.
This is the more to be regretted, as Mr. Graham was, we
believe, a botanist of great promise, quite able to have determined with accuracy the plants of the regions he explored.
The work contains a few descriptions, chiefly from the pen of
Mr. Nimmo, upon whom the superintendence of the work de-
volved, on the sudden death of its author during its printing.

Moon's catalogue of the plants of Ceylon is also a bare list of names. Many of these are evidently erroneously applied, so that it is impossible to make use of them. Fortunately, however, this is of little consequence, as we have no lack of specimens from Ceylon. Moon's collections were excellent; but he does not appear to have sent any specimens to Europe.

Dr. Voigt's 'Hortus Suburbanus Calcuttensis,' published at Calcutta in 1845, is, for the same reason, not available as a work of reference, nor can we refrain from expressing our regret that talents of so high an order should have been devoted to a work of so little practical use.

Dr. Lindley's invaluable 'Genera and Species of Orchideous Plants' contains descriptions of all the Indian Orchideæ collected by Wallich and his predecessors; and in the published parts of the 'Folia Orchidacea' (now in course of publication) we have a complete account of many of the genera, drawn up after a most laborious and critical examination of all the materials accessible up to the latest day. Our own collections are being thus published, and we consider ourselves highly fortunate in their falling into such able hands*. Dr. Lindley has further rendered essential service to Indian botany by numerous descriptions and figures of Indian plants that have appeared in various illustrated periodicals. He laboured indefatigably in the distribution of the great Wallichian Herbarium; his elementary books on botany, and his great work, the 'Vegetable Kingdom,' are indispensable both to botanical students and to proficients; whilst, by the scientific direction he has given to the study and practice of horticulture, as an author and as secretary of the Horticultural Society of London, he has been the means of rendering English botanists familiar with the plants of India in a living state, to an extent that would have been thought visionary a few years ago.

* The analysis of plants of this Order, in a dried state, is a work of the utmost difficulty; and we would urge upon botanists in India the necessity of drawing and describing the fresh specimens, and of preserving the flowers (as of all plants whose parts are injured by the operation of pressing and drying) in spirits or acid.

While the botany of continental India has advanced thus rapidly, equal progress has been made in the Dutch possessions by the indefatigable exertions of a succession of distinguished botanists. One of the earliest in the field, though the extent of his labours is unfortunately but little known, was Dr. Horsfield, whose researches in Java and the neighbouring islands began in 1802, and were continued till 1819. During that time he collected upwards of two thousand species, the most curious and interesting of which have been published by Messrs. Brown and Bennett, in the ' Plantæ Javanicæ rariores,' one of the most profound and accurate botanical works of the day, and one most important for the Indian botanist to study with attention.

Professor Blume, whose extraordinary labours have long since placed him at the head of Malayan botanists, was originally a student of medicine and zoology, and directed his attention to botany in the prosecution of his pharmaceutical studies. The remarkable novelty and curious forms of vegetation with which he was surrounded in Java, effectually diverted his attention from his original pursuits; and he undertook a botanical tour in that island in 1823, 1824, provided with an unusually large staff of collectors and artists; and in 1825 he commenced the ' Bijdragen tot de Flora van Nederlandsch Indie,' an octavo work, containing descriptions of an immense number of new genera and species of Javanese and other insular plants. Though very incomplete in its scope, and written in great ignorance of the labours of others, and of the necessity of detailed descriptions, this is in many respects a remarkable book, evincing a capacity for scientific botany, such as has been displayed by few at so early an age and under so great disadvantages.

On his return to Holland, Professor Blume commenced his magnificent publications on the plants of Java and others of the Malayan Islands, all of which are indispensable to the Indian botanist; very many species, and nearly all the genera of these islands, being also common to the Malayan peninsula and Eastern Bengal. The 'Flora Javæ' was commenced in 1828, and the 'Rumphia' in 1835, each of which consists of several folio volumes, illustrated with a profusion of admirable coloured plates, in many cases accompanied by anatomical details of rare excellence; these are amongst the most splendid and learned botanical works of the age, and have placed their author high in the rank of botanists. In them many of the defective parts of the Bijdragen are worked up and illustrated, and in the 'Museum Botanicum Lugduno-Batavum,' an octavo periodical, with outline plates, containing admirable analyses, commenced in 1852, we have careful descriptions of more of these, and of still other genera and species of Java, Borneo, Molucca, and Japan plants.

The Museum at Leyden is a rich store of botanical materials, which have been accumulating for many years from all the Dutch possessions in the east and west; and it is exceedingly to be regretted, for the sake of science, and the honour of the Dutch Government, which has patronized botany to an extent unsurpassed by any other country, that the enormous piles of duplicates which they possess should be withheld from the scientific institutions of Europe and America.

The beautiful folio volume of M. Korthals, 'Kruidkunde,' or Botany of the Dutch East Indian possessions, is another monument of the munificence of the Dutch Government. It contains seventy coloured plates, illustrating, amongst other natural orders, that of Nepenthaceæ.

The botanical Professors De Vriese, of Leyden, and Miquel of Amsterdam, have laboured long and successfully in Indian botany, and we owe to their industry and energy many important memoirs; and to their liberality most valuable herbaria, procured in some instances at their own cost. M. Miquel's monographs of the difficult orders Piperaceæ and Fici are standard works of essential service to us as Indian botanists, though we do not concur in the author's limitations of genera. M. Miquel has also named the Canara and Nilghiri collections distributed by Hohenacker; but any approach to

accuracy in the determination of the known species and discrimination of those which are new, was obviously impossible
without a considerable general knowledge of Indian botany,
and a comparison with English herbaria, of which Dr. Miquel
had not the opportunity of availing himself.

M. De Vriese's labours include various memoirs on Malayan
Island plants; and his recent monograph of *Marattiaceæ* is
a work of great labour, but his views of the limits of species
are wholly at variance with our experience.

Hasskarl, the author of the 'Hortus Bogoriensis' a catalogue (with occasional notes and descriptions of new species)
of the plants cultivated in the Government Botanical Garden
of Buitenzorg, near Batavia (published in Batavia in 1844),
is also author of an octavo volume of descriptions, entitled
'Plantæ Javanicæ rariores' (Berlin, 1848).

The 'Reliquiæ Hænkianæ,' of Presl, is a folio volume with
plates, devoted to the materials collected by Hænke, who was
employed in the Spanish service, and collected in America and
Manilla; the Indian plants described are few, and the descriptions and identifications far from satisfactory.

The 'Flora de Filipinas' of Father Blanco, published at
Manilla in 1837, is a botanical curiosity, written in Spanish.
The descriptions are intelligible, but, from the author's want
of acquaintance with scientific works, so many well known
plants are treated as new, that we consider it undesirable to
devote time to their identification.

Turning to the west of India, we find ourselves treading
upon the limits of other floras, that have been more or less
perfectly elucidated, in works which we have constantly quoted
in the Flora Indica: of these, the most important are the
writings of Ledebour, especially the 'Flora Rossica,' 'Flora
Altaica,' and 'Icones Floræ Rossicæ.' The 'Flora Rossica'
contains descriptions of the plants of the whole Russian dominions, which may be said to be very satisfactorily explored,
botanically, especially considering their enormous area. The
majority of our Afghan and Tibetan plants, being also natives

respectively of the Caspian steppes and North Persia on the one hand, and of Siberia on the other, have been described by Russian botanists, and especially by Ledebour, Bunge, Turczaninow, C. A. Meyer, and Fischer, besides being rendered classical by the labours of Gmelin and Pallas.

Boissier's 'Diagnoses Plantarum Orientalium,' published in the 'Annales des Sciences Naturelles,' contain descriptions of many new Persian and Levantine plants, mainly from the collections of Kotschy and Aucher-Eloy, which are also common to Western Tibet, Afghanistan, Sind, and Beluchistan. We have largely availed ourselves of the excellent descriptions in these diagnoses, though differing from their truly learned author in his estimate of the influence of climate and the effects of variation. M. Boissier's knowledge of the South European and Mediterranean flora is, we believe, unrivalled, and derived from personal experience acquired during several years spent in exploring indefatigably the Spanish, Grecian, and Oriental floras, of which we have numerous representatives in India, and we therefore record our dissent from the views of so great a botanist, on the limits of species especially, with the most sincere respect, and with considerable diffidence.

It would be out of place here to enumerate the European and Mediterranean Floras of which we have made daily use; there are few of them that we have not been obliged to consult, especially with reference to the critical discrimination of plants belonging to such genera as *Ranunculus, Delphinium, Aconitum*, etc., etc. So many of these floras are mere compilations, or made up of local varieties ranked as species, or studies of the plants of particular areas, treated of without reference to their value as members of the vegetable kingdom, that we find ourselves, when studying any of the large European genera, plunged into a maze of difficulties, to extricate ourselves from which it has been necessary to work out each species *ab initio*, and from a study of all its forms. Koch's 'Flora Germanica' for descriptions, and Reichenbach's 'Icones' for illustrations, are both accurate and useful; and in Vivi-

i

ani's ' Flora of Dalmatia' we have an excellent systematic and descriptive work, displaying enlarged views of the limits of genera and species.

It remains to allude to the labours of writers on American botany, to whom we have been indebted in an unusually great degree, considering the remoteness of that country from India. Of these, the ' Flora Boreali-Americana' of Sir William Hooker, and the unfinished ' Flora of North America,' by Torrey and Gray, are books of standard excellence : the plants described in both these great works having been critically compared with European specimens, their authors have been enabled to throw great light upon their distribution, limits, and variations, of which, however, European botanists have been slow to take advantage. Gray's ' Flora of the Northern United States' is another excellent systematic work; and the ' Illustrations of the Genera of North American Plants,' by the same admirable botanist, is one of the most able and philosophical works in the whole range of botanical literature, and one to which we have been largely indebted.

B. *Enumeration of Herbaria.*

We now proceed to enumerate the materials which we have at our disposal in the preparation of the Flora Indica. It is not possible at present to estimate with accuracy the number of species contained in each individual herbarium, as a critical examination of every one would be necessary for that purpose. We have, however, endeavoured to approximate to a correct estimate.

1. The great Wallichian Herbarium, the history of which is well known to all botanists, having been given in detail in the lithographed list of its contents, which was distributed with it, also in the ' Plantæ Asiaticæ Rariores,' and in the introduction to Wight and Arnott's Prodromus. The first set of this truly valuable collection was presented by the East India Company to the Linnean Society of London, in whose

apartments it is preserved. As all the duplicates were made up into sets, ticketed, and distributed at home and abroad, this herbarium has taken the place of a standard work of reference, and it is impossible to over-estimate its value, or the importance of the constant access which we have enjoyed to its contents. The numbers attached to each plant have been so cited by all monographists, that a reference to these, in the great majority of instances, suffices for the identification of the species; and we have therefore constantly quoted the catalogue numbers, carefully examining every specimen before doing so, in order to avoid as much as possible the risk of error. The distribution appears on the whole to have been made with much care, though the limited time allotted to its execution prevented that critical comparison without which species of difficult genera cannot be discriminated. Hence we occasionally find two or more species under the same number and letter, and far more frequently the same species under two or more numbers. It is not easy to say how many species are contained in the Wallichian collection; but the 9000 numbers may, we think, be diminished by at least one-fourth, as Dr. Wallich, being obliged to distribute without describing, very judiciously avoided uniting apparently distinct forms. For the present therefore we estimate this great collection at between 6500 and 7000 species. The named specimens of this Herbarium having been, as we have said, extensively distributed, it has been customary with botanists to retain the names given by Dr. Wallich. We have been careful to do the same ourselves for all otherwise unpublished genera and species; but where published names, accompanied with descriptions, have come in contact with them, we have considered it to be our duty to follow the generally recognized rule of priority, and to retain the published one; except, of course, in cases where the authors of these names had habitually availed themselves of the Wallichian collections, and where we feel justified in assuming that they would wish to have adopted the Wallichian name had they recognized the plant.

2. In the herbarium of the British Museum there are several small collections, which are of great importance to the Indian botanist, especially one containing many of Loureiro's plants, which are not readily recognizable, at all events as to species, by the descriptions in the ' Flora Cochinchinensis.' There are also a considerable number of specimens forwarded to Sir Joseph Banks by Roxburgh, Hamilton, and Russell, which are occasionally of use in determining the species described by Roxburgh. It contains also a fair but not a full set of the Wallichian herbarium. The British Museum also contains König's collections and manuscripts, Kämpfer's Japan and other plants, and Hermann's herbarium.

3. Dr. Wight's earlier collections, which were distributed in 1832-3, have been enumerated in detail in the ' Prodromus Floræ Peninsulæ,' and have been in part described in that work. Dr. Wight went back to India in 1834, and has, as we have already said, devoted prodigious zeal and energy to the advancement of Indian botany ; he returned to England in 1853, with enormous collections, chiefly from the mountainous parts of Southern India. To these we have been allowed the freest access; and though the mass of duplicates is as yet only partially unpacked, an admirably selected set of specimens has enabled us to determine with accuracy all his species.

4. The collections of Mr. Griffith were made in various parts of India. Their contents may be known by a reference to his posthumous notes and journals, published in Calcutta under the auspices of the Indian Government; in general terms they include collections from Malacca, Tenasserim, the Khasia Mountains, and the whole Assam Valley, Mishmi and Naga hills and the upper Irawadi, Calcutta, Bhotan, Simla, Sind, and Afghanistan. It is unfortunate that these fine herbaria should have been distributed promiscuously, without any determinate plan, and without any reference to his published notes and journals, which robs the collections of half their value, and the journals of more than half theirs. This is the more to be regretted, as Mr. Griffith's collections

were not always made with a view to extensive distribution, and he frequently could not pay the necessary attention to the preservation of specimens in a fit state for future examination, devoting his time mainly to making notes, which are of extreme value, and to a certain extent obviated the necessity of many specimens. Of these collections we believe one and the only complete set is in Calcutta, and was retained for Mr. Griffith's private use, as containing the manuscript numbers referred to in the journals; the specimens were small and poor. It is of the utmost importance that this should be transmitted to England and deposited in some safe quarter for public access. The total number of species collected by Griffith is probably not under 9000, which is by far the largest number ever obtained by individual exertions. Amongst the distribution of his miscellaneous collections were three conspicuous ones :—

a. Malacca, Tenasserim, and Afghanistan plants, distributed numbered by himself. The best sets of these went to the late Dr. Lemann, and the majority will form part of the Cambridge University Herbarium; the Afghan ones were transferred, previous to Dr. Lemann's decease, to Mr. Bentham, and are incorporated with that botanist's extensive and admirably-named herbarium. The second sets were communicated by Mr. Griffith to Sir William Hooker's herbarium. Others were sent to Dr. Gardner of Ceylon, and Dr. Wight of Madras. Of these, Gardner's were sold at his death, when Sir William Hooker purchased the Malacca specimens.

b. A distribution, through the late Dr. Lemann, of Khasia and Assam collections; of these, some were formed by Mr. Griffith, at his own expense, and others, we believe, formed part of the Assam Tea Deputation collections, and were due to the joint labours of Dr. Wallich and himself.

c. More lately there has been a distribution of Khasia, Bhotan, Mishmi, Assam, and Calcutta garden specimens, and of miscellaneous Palms, under the direction of the East India Company.

d. An immense collection of Ferns sent to Sir William Hooker by Mr. Griffith.

We believe that some of this lamented botanist's collections still remain in the vaults of the India House, but their contents are unknown to us; perhaps they contain the Irawadi collections, and those of Tenasserim and Martaban, which are a great desideratum to science.

Now that we are on the subject of Mr. Griffith's botanical labours, we feel it incumbent upon us to record our sincere regret at not being able to quote regularly the posthumously-published drawings and observations of that indefatigable naturalist. It is well known that these manuscripts were not left in a fit state for publication, and that to have edited them properly, required a very able and careful botanist, well versed in the Indian flora especially. It is a most unfortunate circumstance for the fame of Griffith, and the credit of all parties concerned, that what has been published is not available for the purposes of science. Even in the folio volume on the Palms of British East India, the materials for which were left in a tolerably perfect state, the errors of all kinds are so numerous and involved, that it cannot be consulted without the greatest caution; and, as we have said above, the specimens distributed, whether by Mr. Griffith or the East India Company, not bearing the numbers of his printed catalogue, we have, in an overwhelming number of instances, no means of identifying his plants with his notes of their locality, habit, etc., except in the rare instances where the brief descriptions contained in his 'Itinerary Notes' enable us to do so. Our own opinion of Mr. Griffith's exertions and botanical attainments is, that he has never been surpassed in India; and we wish all the more to give publicity to this opinion, because the circumstances alluded to prevent that repeated acknowledgment of the value of his writings, which would have appeared everywhere in our work, had his own been so edited as to render this possible. We cannot conclude this notice of his labours, without a regret that he was not spared, both to edit his own manuscripts, and to publish what he so often mentions to be the great ultimatum of his labours, an accurate and philosophical Flora Indiæ. For such a task he had no rival, and he justly appreciated, in common with all botanists, the paramount importance of such a work, (already far too long delayed, considering the present state of the science,) not only as being absolutely necessary to ensure further sound progress, but as the only means of checking that hasty publication of Indian plants from imperfect materials, which has now thrown the Indian Flora into so great confusion.

5. The Parisian Herbarium at the Jardin de Plantes possesses the valuable collections of the indefatigable Jacquemont, whose premature death deprived botany of an ardent and enlightened votary, whose labours would have done much to advance the science. M. Jacquemont's collections were made partly in the Gangetic plain, but mainly in the northwest Himalaya, a great part of which was first explored by him. He entered the mountains at Massúri, and explored Garhwal and Sirmur, and ascended the Satlej into Kanawer and the Tibetan province of Piti. Returning thence to the plains, he visited Lahore and the Salt-range of the western Punjab, and travelled by Jelam and Bhimbar to Kashmir. In this (at that time) unexplored province of the Himalaya he spent a whole summer, and accumulated rich collections. Leaving the mountains, he travelled through Delhi, Ajmir, and Nimach, across Malwah to Bombay, whence he went to Púnah, on the eastern slope of the range of the Ghats, and there succumbed under repeated attacks of liver-complaint, brought on by hardship and reckless exposure in the pursuit of his favourite science.

The journals of Jacquemont, which were published by the French Government, bear ample testimony to his great botanical attainments. He was evidently deeply impressed with the importance of careful observations in geographical botany, and noted with the utmost care the localities of his plants. Had he lived to work out the result of his own labours, Hi-

malayan botany would have long ago been established on a
foundation of judiciously collected facts; but unfortunately
his journals, though sufficient to show the ample means at
his disposal, were not thrown into a shape in which they are
available to science, nor would it have been possible to give
them such a form without the local knowledge which was lost
with their collector.   Other botanists have since traversed the
scenes of M. Jacquemont's labours, and, more fortunate than
he, have been enabled to reap the well-earned reward of their
exertions; but let it not be forgotten that a foreigner was the
first in the field, and but for his lamented decease, would have
stood in the very foremost rank of Indian botanists.   We are
proud to say that the Directors of the Jardin de Plantes
(through M. Decaisne's good offices) have been so liberal as
to place at our disposal a nearly complete set of these truly
valuable collections, which are accurately ticketed, so that the
exact localities are in almost every case easily determined.
Our acquaintance with many of the districts where Jacquemont travelled, will enable us to make the best use of this
valuable gift, and to give to his discoveries their well-merited
precedence.

6. Dr. Royle's extensive collections of Northern Indian and
Himalayan plants, formed the groundwork of his work already
noticed.   A detailed account of the districts investigated by
Dr. Royle, and by his collectors, will be found in the introduction to that work.   These were chiefly the Jumno-Gan-
getic Doab, the upper part of the Gangetic plain, and the
mountains of Garhwal, Sirmur, Kanawer, and Kashmir.   By
continental authors, Dr. Royle's Himalayan plants are occasionally quoted as from Nipal, a mistake which leads to erroneous conclusions, and which therefore requires to be guarded
against.   The original set of Dr. Royle's collections remains
in his own possession, and he has liberally placed it at our
disposal for examination and comparison with our own.   As
the specimens are named in accordance with his work, we
have been able in every case to identify them.   Dr. Royle

presented to the Linnean Society a similar named set, as complete as possible, together with all his duplicates, for the purpose of distribution: his intentions in this matter have, however, unfortunately not yet been carried into effect.

7. Besides the herbaria of Wallich and Royle, the Linnean Society possesses several very valuable collections of Indian plants, which have been of great service to us. These are— 1. An authentic collection of Roxburgh's plants, for the most part named. The names are chiefly Roxburgh's earlier ones, but they are in all cases identifiable with those of his Flora Indica, by means of the coloured drawings at the India House, of which copies made by Sir William Hooker, as related in detail in Wight and Arnott's Prodromus, are at our disposal. With these means of determining Roxburgh's plants, we trust that few, if any, of those contained in the orders which we have investigated will remain in obscurity. Several species not hitherto recognized either by Wallich, or by Wight and Arnott, will be found in the first part of our Flora, and the number may be expected to be increased. 2. A large collection of plants of the Bombay Presidency, chiefly from the neighbourhood of Punah, presented by Colonel Sykes to the Society. These amount to nearly a thousand species, and the specimens, though often indifferent and much injured by insects, are, in general, capable of determination. 3. The Smithian Herbarium contains a good many specimens from Hamilton and others, and is valuable as a means of determining the species described by Sir J. Smith in Rees' Cyclopædia and in the 'Exotic Botany,' where he has occasionally indicated new Indian plants. It is almost superfluous to add, that the Linnean Herbarium is the gem of the Society's possessions.

8. The collection distributed by Captain Strachey and Mr. Winterbottom consists chiefly of the plants of Kumaon and Garhwal, and of those of the adjacent parts of Tibet. Captain Richard Strachey was appointed by the Indian Government to make a scientific survey of the province of Kumaon, and

k

was occupied on the task about two years, during which time, in addition to the important investigations in physical science which occupied his attention, he thoroughly explored the flora of the province, carefully noting the range of each species. He was joined by Mr. Winterbottom in 1848, and they travelled together in Tibet. Their joint collections, amounting to 2000 species, were distributed, in 1852-3, to the Hookerian Herbarium, the British Museum, the Linnean Society, and some foreign museums; and the scientific results are now in course of publication. The beautiful preservation of the specimens, and the fullness and accuracy with which they are ticketed, render this herbarium the most valuable for its size that has ever been distributed from India; and we beg here to record our sense of the great benefit that has been rendered to botanical science by the disinterested labours of these indefatigable and accomplished collectors.

9. The herbarium of Dr. Arnott at Glasgow is particularly rich in Indian plants, and especially valuable as containing the materials from which the 'Prodromus Floræ Peninsulæ' was elaborated. Its distance has prevented our having it in our power to consult it regularly, but Dr. Arnott has been good enough to afford us his assistance in making comparisons in every case of difficulty. This has been to us a most material benefit, as we have not hesitated to apply to him in all doubtful points.

10. The extensive herbarium of Mr. Bentham, our greatest descriptive botanist, has in like manner been readily accessible to us by the kindness of its owner*. In addition to its value as an authentically-named collection,—in which respect it is, we believe, in proportion to its size, quite unrivalled,—this herbarium contains a number of important contributions from Indian botanists. We have consulted it for the orders included in the present part, and hope to continue to do so in

* Whilst these pages have been passing through the press, Mr. Bentham's Herbarium has become the property of the Royal Gardens at Kew, through the disinterested liberality of its owner.

all cases in future. Mr. Bentham has also been good enough to entrust to us his complete set of Mr. Edgeworth's plants, which are authentically named by that gentleman, and correspond with his paper on North Indian plants in the twentieth volume of the Transactions of the Linnean Society of London. We have thus had it in our power to quote the synonyms of that memoir with confidence. The benefits which we have derived from Mr. Bentham's profound knowledge and ready help, and the obligations we are under to him, are such as it is impossible adequately to express.

11. We have in like manner to thank Dr. Lindley for his generous assistance in every way, and for unlimited access to his valuable collection, which has enabled us to identify many of the species described in the ' Botanical Register,' the ' Journal of the Horticultural Society,' and other works of this excellent botanist. Dr. Lindley's herbarium contains a fine set of Penang plants, communicated by Mr. Prince, and by Mr. Phillips; and numerous specimens from Ceylon collected by Mr. Macrae.

12. The Indian collection of Colonel Munro, 39th Regiment, has also, by the liberality of its owner, been placed at our disposal. Colonel Munro's earlier collections were made in the Madras Presidency, but after his removal to Bengal he explored the vicinity of Agra, and made an extensive tour in the Himalaya from Kumaon to Simla and Kanawer.

We cannot conclude this comprehensive catalogue without an allusion to the labours of Dr. Falconer, one of the most estimable, able, and accomplished of Indian botanists; to whose liberality and good offices we were in many ways indebted as travellers in India, and are still, as workers at home. Dr. Falconer was one of the first botanists who visited Kashmir and Little Tibet, where he formed magnificent collections, as he also did in Kumaon and the Punjab, illustrating his specimens with voluminous notes and details of their structure and affinities. His collections are, we believe, still in the India House, where they have been for many years. They consti-

tute the only herbarium of importance to which we have failed
to procure access, and we are hence unable to do our friend
that justice in the body of this work, to which, as the discoverer of many of the plants described, he is pre-eminently
entitled.

13. The only other extensive collection in Great Britain is
the Hookerian Herbarium, in which our work is carried on.
This is beyond all doubt both the richest and best-named herbarium in the world, and it possesses the rare advantage of
containing an extensive series of specimens of each species
from many countries and collectors, so preserved and arranged
that all may be brought at one time under inspection.   For
these reasons (and from the extreme liberality of its owner)
the Hookerian Herbarium has been studied by most monographists at home and abroad, and possesses in consequence
an enormous proportion of authentically-named specimens,
by Arnott, Asa Gray, Bentham, Boott, Choisy, Decaisne, De
Vriese, Grisebach, Herbert, Lehmann, Liebmann, Lindley,
Meisner, Miers, Miquel, Moquin-Tandon, Meyer, Munro,
Nees von Esenbeck, etc. etc., and illustrates the published
works of these and many other botanists, to an extent that
no other herbarium does.   It is also enriched with many valuable manuscript notes, dissections, sketches, and remarks
by its possessor, and by M. Planchon, who was for some
years its curator.   It would be out of place here to give a
history of the rise and progress of the Hookerian Herbarium,
or of the sources from which it is mainly derived; though
this would form a most interesting contribution to the literature of the science, and would include a history of the progress
of systematic and descriptive botany during the last half-cen-
tury.   It is especially rich in Indian plants; and an enumeration of these, which is necessary, as they constitute a large
part of our materials, will give the reader an idea of the nature of the abundant sources from which its riches are derived.   The Indian portion of the Hookerian Herbarium comprises the undermentioned collections.

1. A good set of the Wallichian Herbarium, and some collections communicated by Dr. Wallich from Nipal, previous to his first visit to England.

2. Dr. Wight's Peninsular collections, distributed in 1832-33.

3. General and Mrs. Walker's very extensive Ceylon collections, and a smaller herbarium from Simla.

4. Dr. Gardner's Ceylon and Nilghiri plants, both numerous and good.

5. Major Champion's Ceylon plants, presented by him in 1852, along with his whole Herbarium.

6. Large collections of Ceylon plants from Mr. Thwaites. These are in course of publication by that botanist, who succeeded Mr. Gardner as superintendent of the Botanic Gardens of Peradenia, and who is now actively and ably investigating the flora of the island.

7. Mr. Griffith's Malacca, Tenasserim, Khasia, Assam, Mishmi, Bhotan, and Afghan plants.

8. Hohenacker's Nilghiri, Kurg, and Canara plants, collected by the Rev. Mr. Schmid and others, and named by Professor Miquel.

9. Admiral Sir Frederic Adams' Nilghiri plants (a small collection).

10. Sir William Norris's Penang and Malacca plants: an excellent collection.

11. Mr. Prince's Penang plants.

12. Mr. Lobb's Malacca, Tenasserim, Khasia, and Malabar collections. Mr. Lobb collected in the service of Mr. Veitch, the eminent nurseryman of Exeter; his Khasia and Malacca collections are very numerous.

13. Mr. Cuming's Malacca plants.

14. The Rev. Mr. and Mrs. Mack communicated beautiful collections from Assam and the Khasia mountains.

15. Colonel Jenkins' and Mr. Masters' Assam plants. These formed immense collections, made in various parts of the Assam valley, chiefly in the neighbourhood of Gowhatty.

16. Mr. Simon's Assam and Khasia collections consist of numerous and well-preserved specimens.

17. Mr. Law's very valuable and extensive collections from Bombay, Tanna, Dharwar, and Belgaum contain probably about 1500 species.

18. Mr. Dalzell's extensive collections from the southern Concan and Canara, many of which have been published by him in a valuable series of papers printed in the 'London Journal of Botany.'

19. Mr. Gibson's rich herbarium, chiefly collected in the Concan and Dekhan.

20. A few Bombay plants, from Mr. Nimmo.

21. Dr. Stocks's extremely valuable collections from Sind and Beluchistan, amounting to about 1500 species.

22. Captain R. Strachey and Mr. Winterbottom's magnificent herbarium, already described.

23. The Countess of Dalhousie's extensive Simla collection, formed when the late Earl of Dalhousie was Commander-in-Chief. Also, a small Penang collection by the same lady.

24. Major Madden's Simla and Kumaon plants : numerous and excellent specimens.

25. Jacquemont's superb collections already alluded to.

26. Major Vicary's small but very valuable herbarium, containing many scarce plants from Gorakpur, the Punjab, Peshawer, Sind, etc.

27. Mr. Edgeworth's collections made since his return to India in 1847; these contain his Bandelkand plants, and a very complete Multan herbarium; also some of his Himalayan plants published in the Linnean Society's Transactions.

28. Captain Simpson's Simla and Khasia plants, presented by the late Mr. Fielding.

29. Mr. Winterbottom's valuable and beautifully preserved herbarium from Kashmir, Balti, Hasora, and Gilgit : it contains excellent specimens and much novelty.

30. A small miscellaneous collection from Colonel Munro.

31. Dr. Fleming's interesting collection from the Salt-

Range of the Punjab, and another from the Marri hills between the Jelam and the Indus.

32. Mr. Lance's Kashmir and Tibet collections, communicated through Mr. Edgeworth.

33. Dr. Jameson's collections from Massuri and the Saharanpur Gardens.

As, however, we are so largely indebted to the floras of countries bordering upon India for the elucidation of our Flora, it is necessary to add that the Hookerian Herbarium is as rich in proportion in the plants of surrounding countries as it is in Indian. Of these, the most important are the following :—

A. From the Malayan Archipelago and China.

1. Cuming's magnificent Philippine Island collections, containing about 3000 species.

2. Lobb's Java, Borneo, and Philippine plants, which are very numerous and in excellent preservation.

3. Extensive Javanese collections, communicated by Professors De Vriese and Miquel.

4. Zollinger's Javanese plants.

5. Spanoghe's plants from Java and Timor (not numerous).

6. Professor Blume has communicated authentically-named specimens of a very few Javanese and Molucca natural orders : these are extremely valuable, especially the *Anonaceæ* and *Cupuliferæ*.

7. Mr. Motley's extensive Borneo collections.

8. Mr. Lowe's small collection from the same island.

9. Dr. Seemann's Malayan and Chinese collections.

10. Major Champion's Hongkong herbarium, which has been described by Mr. Bentham in the 'Florula Hongkongensis' in Hooker's Kew Journal.

11. Mr. Millett's Macao plants.

12. The Rev. Mr. Vachell's Chinese collections.

13. Captain Beechey's plants from China, collected by Messrs. Lay and Collie, and described in the 'Botany of Beechey's Voyage.'

14. Mr. Fortune's Chinese collections.

B. From countries to the west and north of India.

1. Very complete collections made by Russian botanists in Siberia, the Altai, North China, Dahuria, and indeed in the whole of the Russian possessions in Asia, chiefly from Lede-bour, Prescott, Bunge, Turczaninow, Fischer, Meyer, etc. etc.

2. Karelin and Kirilow's Soongarian and Alatau plants.

3. Szovitz's North Persian and Caspian plants.

4. Aucher-Eloy's complete collections from various parts of Persia, Asia Minor, Arabia, and the Levant.

5. Colonel Chesney's Euphrates plants.

6. Mr. Loftus's small collection from Assyria.

7. Kotschy's very extensive and beautiful North and South Persian collections, chiefly named by M. Boissier, and hence of very great value.

8. Asia Minor and Kurdistan plants from various collectors.

To these very ample materials already existing in this country have to be added our own collections, which we estimate at about 8000 species (including Cryptogamic plants), and an immense number of duplicates. Many of the species were gathered in numerous localities, so that we have it in our power to compare specimens from a great diversity of climates and soils. They may be divided into five groups :—

1. Dr. Thomson's collections made in the plains of North-west India, between 1842 and 1847, chiefly in Rohilkand, Lodiana, and the Punjab, which amount to about 1000 species.

2. Dr. Thomson's Himalayan collections, partly collected in Kumaon and Garhwal during short visits to these provinces in 1844 and 1845, but mainly consisting of the herbarium collected during a Government mission in the north-west Himalaya and Tibet, in 1847, 1848, 1849, in the course of which he visited, in 1847, Simla, Kanawer, Piti ; and in 1848 Kashmir and the Panjab Himalaya, Ladak, and the Karakoram Pass. The summer of 1849 he spent at Simla and Ladak. These amount to rather more than 2500 species.

3. Dr. Hooker's collections, made during a botanical mission to India in the years 1848, 1849, 1850, under the auspices of the Commissioners of Woods and Forests. Starting from Calcutta, Dr. Hooker proceeded first to Behar, ascended the Soane valley and crossed the Kymor range to Mirzapur, descended the Ganges, and proceeded to Sikkim. The collections made in Behar and the Gangetic valley amount to about 1000 species. Dr. Hooker spent the summer of 1848 and the greater part of 1849 in the Sikkim and the East Nipal Himalaya, during which he botanized the whole country from the plains to the Tibetan frontier, and accumulated an herbarium of 3500 species. In December, 1849, he was joined by Dr. Thomson at Dorjiling, and they proceeded together, in May, 1850, to the Khasia hills, where the summer was spent: the joint collection amounting to about 3000 species. In November of that year they visited Silhet and Cachar, descended the Megna to the Bay of Bengal, and proceeded to Chittagong, returning by the Sunderbunds to Calcutta, where they embarked for England; this journey yielded about 1000 species.

4. A large herbarium of Peninsular plants formed by Dr. Thomson's brother, the late Gideon Thomson, of Madras, mainly by means of collectors. It amounts to nearly 2000 species, gathered partly in the plain of the Carnatic (chiefly in the neighbourhood of Madras), and partly in the Nilghiri and Cúrg mountains, and in the Courtalam hills.

5. Several collections which were liberally presented to us in India. These, though not extensive, were often extremely valuable, being illustrative of little known regions. From Dr. Jameson we received Saharunpur and Massuri plants; from Dr. Fleming a collection from the Salt-range of the Panjab; from Dr. Grant, a small herbarium of Kanawer plants; from Lieutenant Parish, a set of specimens from the hills of Mandi and Kulu (in the Panjab Himalaya); and from Mr. Simons several hundred Assam species.

As all our own materials were selected with a view to future

l

publication, no pains were spared to render them as perfect an illustration as possible of the flora of their several districts. For this purpose aberrant forms and varieties were carefully collected, and a great many specimens were dried of each species. Great attention was paid to the ticketing of the specimens, so as to certify the locality and elevation from which they were obtained. In Sikkim and the Khasia hills 500 large specimens of wood were cut; and Palms, *Pandani*, Bamboos, tree-ferns, etc., were preserved entire; whilst the flowers and fruits of more than 1000 species were preserved in spirits. Many notes and dissections were also made on the spot; and we have the further assistance of a series of coloured drawings and dissections (of upwards of 1000 species) taken by Dr. Hooker from the live plants, and of a valuable portfolio of upwards of 500 drawings of Sikkim plants, executed at Dorjiling by native artists, under the superintendence and at the expense of our enlightened and lamented friend, the late J. F. Cathcart, Esq. of the Bengal Civil Service, very much in furtherance of our botanical labours. This has been presented to the Kew Museum by the liberality of his surviving sister.

V. *Sketch of the Meteorology of India.*

Climate is an extremely important element in the geographical distribution of plants; and though it is not necessary to dwell at any great length upon the general principles of Meteorology, an outline of these, as they are brought into operation in India, is requisite for the correct understanding of the transitions of vegetation in different parts of that country. The phenomena of climate in a particular area, are well known to depend not only on its latitude, but also on the configuration of its surface and on its position relative to the ocean, upon the direction of the mountain-chains and their elevation above the level of the sea, and upon the course of the winds. Temperature and humidity, the two grand elements which give the character to the climate, react naturally upon one another, so that it is not easy to determine which is the cause and which the effect.

For all practical purposes we may regard the sun as the sole source of the temperature of the surface of the globe. If the surface of our planet were uniform, the sun's heating power would be directly proportional to his altitude, and the mean temperature would diminish equably in receding from the equator. A variety of circumstances disturb this regular gradation of temperature. These are—1. The more rapid heating and cooling of land than sea, which arises in a great measure from the heat being gradually diffused throughout the ocean (by means of oceanic currents), the hot water from the tropics being thus carried into temperate regions, while the cold water of the Arctic seas occupies its place. Proximity to the ocean, therefore, promotes uniformity of temperature.—2. The elevation of the land above the level of the sea. The sun's heating power is rather augmented at great elevations; but a diminution of temperature at high levels is caused by the rarefaction of the air, and is a consequence of the law according to which, the *specific heat* of the atmosphere increasing inversely with its density, its *sensible heat* becomes absorbed as it expands. As this law is universal, it follows, that when a current of air ascends or descends, its temperature is changed to an amount exactly proportional to the change of level; and it is only when such a current is hotter than the normal temperature of the place whence it ascends, that it is a warm wind at a higher level.—3. The presence or absence of clouds. These intercept the solar rays during the day, and tend to keep the ground cool. During the night, on the contrary, clouds intercept the radiation of the heat accumulated in the earth during the day, and tend to keep the ground warm. A cloudy climate is hence an equable one, having comparatively cool days and warm nights, cool summers and mild winters.

When the sky is clear, the air in contact with the earth becomes warmed by radiation from its heated surface; and being expanded and made lighter, it immediately ascends, its place being supplied by air from colder regions. Thus, since no two places have the same temperature, and since the temperature constantly changes, even in the same place, the atmosphere is kept in constant motion.

As the amount of aqueous vapour which is capable of remaining suspended in the atmosphere is directly proportional to the temperature, ascending currents of air finally become so cooled that condensation or precipitation takes place; and the nearer to saturation the air is before it begins to ascend, the sooner it will reach a sufficiently low temperature for condensation. We can therefore understand why mountain-chains (which impede the direct course of the currents, and force them to ascend) cause precipitation of the moisture of an atmosphere which has already traversed, without any condensation, a great extent of level country.

The direction of the wind is primarily dependent upon the sun's position, and is a very complex phenomenon, in consequence of the perfect fluidity of the air. On the open sea, at a sufficient distance from land to escape its influences, the trade-winds, owing to the intertropical heat, blow with great regularity towards the equator, or rather towards a point immediately under the sun's position, varying therefore with the season of the year. Their direction is not due north and south, but more or less towards the west. This is in consequence of their retaining the momentum proper to the latitude whence they start, in their advance towards the equator, where the motion of a point on the earth's surface (due to its revolution round its own axis) is a maximum. They therefore lag behind, as it were, and appear to blow from the north-east in the northern hemisphere, and from the south-east in the southern hemisphere. The presence of land interferes with the regularity of the trade-winds; and where it occurs in large masses, it becomes so much more heated than the ocean, that it attracts the aerial currents towards itself, and hence completely changes the direction of the wind.

The whole of Continental India lies north of the equator, and considerably more than half of its area north of the Tropic of Cancer, whose position very nearly corresponds with the base of the peninsula of Hindostan. Proceeding northwards from the tropic, there is no sea nearer than the Arctic Ocean; but as we advance towards the equator the width of the land gradually diminishes both in the Madras and Malayan peninsulas. It may be observed also, that due south of India, the ocean extends without interruption beyond the Antarctic Circle, while to the eastward, not only on the equator but in the southern hemisphere, there is much land. The Eastern Archipelago, from consisting of large islands, separated by belts of sea, possesses a humid and equable climate; but the great continent of Australia, being a vast expanse of low land, becomes enormously heated when the sun is in the southern hemisphere, and presents extremes of climate. To the westward the coast-line of Beluchistan continues somewhat north of the tropic till it enters the Persian Gulf; but the great continent of Arabia advances far within the tropic; while, a little further west, Africa extends, uninterrupted by sea, far into the south temperate zone. From this relative position of land and sea, it is evident that the whole of the rain which falls in India must be derived from the southward or eastward, and that those parts only can be subject to heavy rains, towards which the sea-wind blows.

The maps of the monthly isothermals*, recently published by Dove, enable us to trace with considerable accuracy the periodical changes of temperature throughout India and the neighbouring countries. An inspection of these maps shows us that in January the isothermal lines in the northern hemisphere are nearly parallel to the equator, but that, in the southern, Africa and Australia are preternaturally hot. Till the vernal equinox, the equator of heat (or that line from which the temperature diminishes both towards the north and towards the south) lies south of the terrestrial equator; but

* See Maps of Isothermals appended to this Essay.

after the beginning of April, it advances rapidly into the northern hemisphere, and two defined regions of excessive heat (86° Fahr.) occur, one in Africa, and a smaller one in the peninsula of India. In May and June the equator of heat lies in India considerably north of the tropic, and the two regions of excessive heat, becoming united, extend uninterruptedly from North Africa, across Arabia and Persia, over all India west of the Bay of Bengal. In July, a still hotter area occurs in Nubia and Arabia, and Northern India is very little inferior in temperature, whilst Southern India becomes cooled; the heat throughout India being modified by the accession of the rains. In this month the isotherms in all parts of Asia are much curved, the convexity being towards the north; and the amount of curve increases towards the northern part of the continent.

In August the equator of heat passes through Northern India, which is still occupied by the rapidly contracting region of excessive heat. In September and October the equator of heat advances rapidly towards the south, and in November it has entirely left India, and corresponds almost exactly with the terrestrial equator, while the region of excessive heat lies in the Indian Archipelago over Borneo and New Guinea.

We see therefore that from the vernal to the autumnal equinox a great part of India is preternaturally hot, but that from October to February (inclusive) it is comparatively cool, and at the same time the continents of Africa and Australia become preternaturally hot. During the summer months therefore, or the hot season as it is commonly called in India, the wind blows from the south towards the north, while in the winter or cold season it blows from north to south. At both seasons these directions are often modified by local causes, besides being uniformly affected by the earth's rotation, and by the heating and cooling of the continent.

The monsoons or periodical winds are known in the Indian Ocean, and indeed generally throughout India, by the name of the south-west and north-east monsoon, these being their

directions at sea. At the commencement of the vernal equinox, the south-west monsoon is very local in its character,
the heat being greatest over a small region in southern India.
At the same time Arabia and the countries east of Persia are
much heated, and cause a southerly wind to blow from the
ocean west of India, towards Persia and Afghanistan, while
an east wind blows up the valley of the Ganges. After April
the northern parts of India become much hotter, and the direction of the southerly monsoon is remarkably influenced,
as has been well pointed out by Dove, by the great heat of
Tibet, Siberia, and Tartary, which, in consequence of their
cloudless climate, acquire an almost tropical temperature during the summer months, and attract the currents northwards.

I. *The south-west or summer monsoon.* This, in almost all
parts of India, is a sea wind, and is therefore loaded with vapour. On the west coast of the Madras Peninsula it comes
in contact with the range of mountains called the Western
Ghats, upon which it deposits a great part of its moisture;
in its further course it meets with no greater elevation in
southern India, the eastern parts of which are comparatively
dry. On the coasts of Orissa and Bengal the direction of this
wind is more to the north, from the heating of the continent
to the north and north-west, and much moisture is deposited
on the mountains of these provinces. In northern India the
rainy season commences later than in the Peninsula, because
it is not till June that the sun acts sufficiently energetically
on the Tibetan mountains and the plains of temperate Asia
to attract in that direction the full force of the monsoon.
This wind, after passing over the plains of Bengal, comes in
contact with the Khasia mountains, upon which, and upon
the whole chain of the Himalaya, it discharges itself in
heavy rains diminishing in amount as we advance westward,
with the increasing distance from the sea. At Calcutta the
wind, during the whole of the monsoon, from April onwards,
blows from the east of south, but after the beginning of August, when the great rain-fall in eastern Bengal has con-

siderably lowered the temperature of that province, (the arid plains of the Panjab, however, remaining excessively heated,) it becomes S.S.E., and in September still more easterly.

In the eastern (Malayan) peninsula it is probable that the direction of this monsoon is nearly from south to north; but more detailed information is required to enable us to understand the precise course of the aerial current in all parts of that Peninsula. At the commencement of the monsoon the wide and open valley of the Irawadi seems to act as a local source of attraction, to which the wind blows from both oceans. At a later season, the elevated temperature of the plain of the Ganges and the Tibetan valley of the Brahmaputra overpowers that influence, and the main atmospheric current flows over the mountains south of Assam and ascends the valleys of both these rivers in a north-westerly direction.

II. *The north-east or winter monsoon.* As a consequence of what we have stated, after the autumnal equinox, the great mass of the Himalaya becomes intensely cold, and the whole of the continent comparatively cool, while the southern hemisphere gets powerfully heated. The north-east monsoon, which results from this distribution of temperature, is the effect of a distant attraction, and therefore blows with great regularity. It is everywhere a land wind, except in the Malayan Peninsula and on the coast of the Carnatic. In Malaya it blows over a great extent of sea, and is therefore very rainy; but in the Carnatic the width of sea is not great, so that the rain-fall, though well marked, is less, and terminates long before the end of the monsoon, probably from the wind acquiring a more directly southerly direction, after the sun has reached the southern tropic.

The current which flows towards the southern hemisphere as the north-east monsoon, is replaced by an upper one which flows northward. It is from this northerly current, which arrives moisture-laden from the southern ocean, that are derived the winter snows of the Himalaya and of the mountains of Afghanistan, and the winter rains of the lower hills and of the plains at the foot of the mountains. These last are irregular in amount and period, and dependent perhaps on local disturbances of the great current, the causes of which are still obscure and require careful investigation. During the south-west monsoon, a similar return current from Siberia and Tartary probably flows almost uniformly from the northward at a very great elevation, and joins the ascending current from the plains of India.

When the causes and direction of the periodical winds are clearly indicated, there is no difficulty in understanding why it is that in some parts of India the climate is always moist, both monsoons being rainy, while in others ohe monsoon only is rainy, and in others again there is no rain at any period of the year. The only permanently rainy province is the Malayan peninsula, and the only absolutely arid ones are Sind and the neighbouring deserts of the Panjab. Throughout the greater part of India one monsoon is rainy, and that generally the south-west one, blowing from May or June till the end of September.

The amount of rain varies prodigiously in different parts of India, from almost none to six hundred inches, but the details must be reserved for notice under the several districts. It is very essential to bear in mind that the rain-fall affords no direct criterion of the humidity of any climate, for the atmosphere may be saturated with moisture without any precipitation taking place. The influence upon vegetation of the vapour suspended in the air, and thus brought in contact with every surface of the foliage, is most important, and can only be ascertained by means of daily observations with the hygrometer. This instrument is indeed, generally speaking, of far more importance to the botanist than the thermometer; the distribution of tropical plants especially, in so far as it is influenced by climate, being so by its moisture*.

* To make our meaning clearer, we may say that any part of the tropics is hot enough for the growth of a tropical plant, but that whole natural orders,

m

The normal mean temperature of the equator is stated by
Dove to be a very little below 80°, but this is somewhat
exceeded in many parts of continental India. The normal
mean temperature scarcely diminishes at all between 0° and
10° N. lat. Between 10° and 20° it diminishes 2½°; between
20° and 30°, 7°; and between 30° and 40°, 13·3°. In 20° N.
lat. therefore the diminution may be estimated at about half
a degree of temperature, and in 30° N. lat. at 1° of temperature, for a degree of latitude. In India, however, the mean
temperature does not diminish so rapidly, owing to the increase of the mass of land to the northward, which, as has
been shown, becomes excessively heated in summer. The
normal difference of temperature between summer and winter
is least at the equator, and increases with the latitude; and
this effect is enhanced in India by the increase in the mass
of land, which makes the summers hotter and the winters
colder than the average.

The phenomena of vegetation are less dependent upon the
mean temperature of the year than upon that of the season
of growth: thus, within the tropics, vegetation is active at
all periods of the year, but in the cooler temperate zone, and
at considerable elevations on the mountains of the tropics,
only during the summer season. It is therefore important in
the investigation of climate with regard to its application to
botany, to know the mean temperature of each of the four
seasons, and, if possible, that of each month.

The only other important element by which climate is affected, is elevation above the level of the sea. The diminution of temperature as we ascend (on the surface of the

genera, and individual species are extremely sensitive to the amount of moisture in the air, and its fluctuations. Some plants are confined to perennial humidity, others to perennial drought, whilst still others are dependent on accessions of heat or drought at certain fixed periods, for life and health or the
means of propagation. Comparatively few observations on temperature, and
those in certain months only, give us a sufficient approximation to the requirements of a plant in that particular, but the hygrometrical observations
should be continued throughout the year.

January

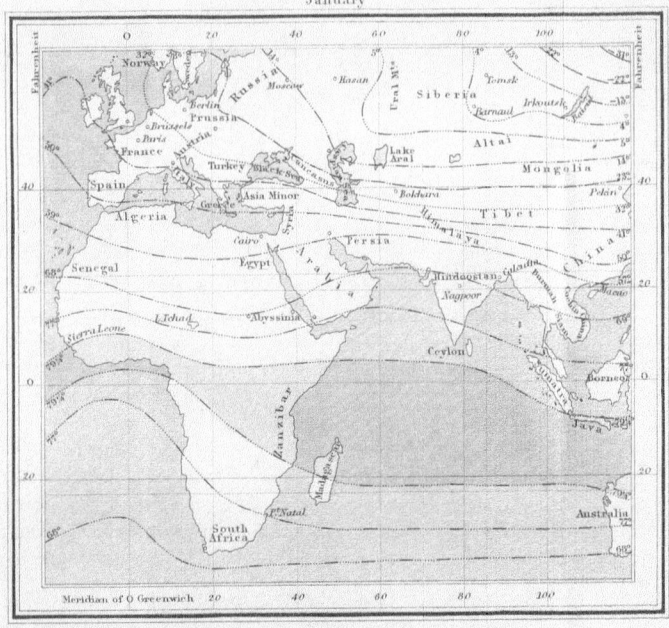

July

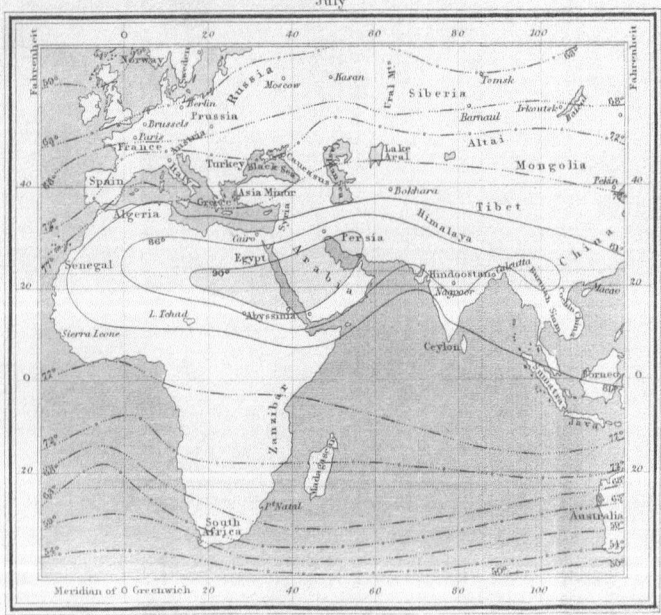

Extracted from Maps I & II of Dove on the Distrib
for Dr Hooker & Thomson.

April

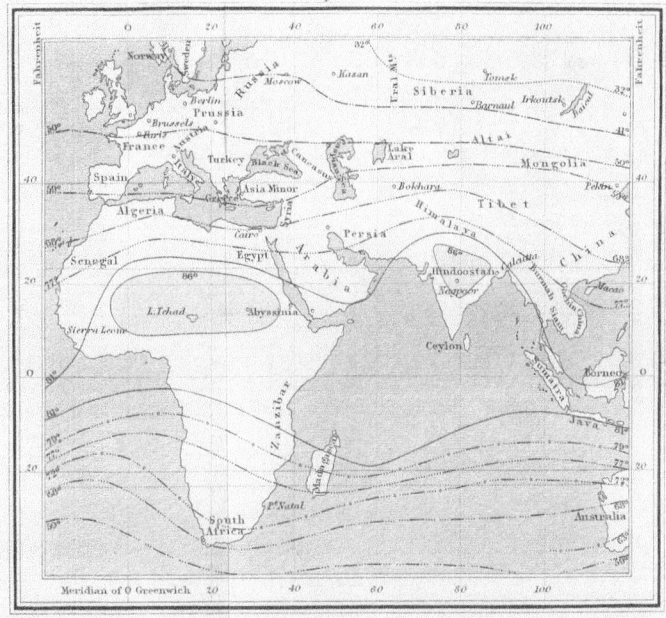

October

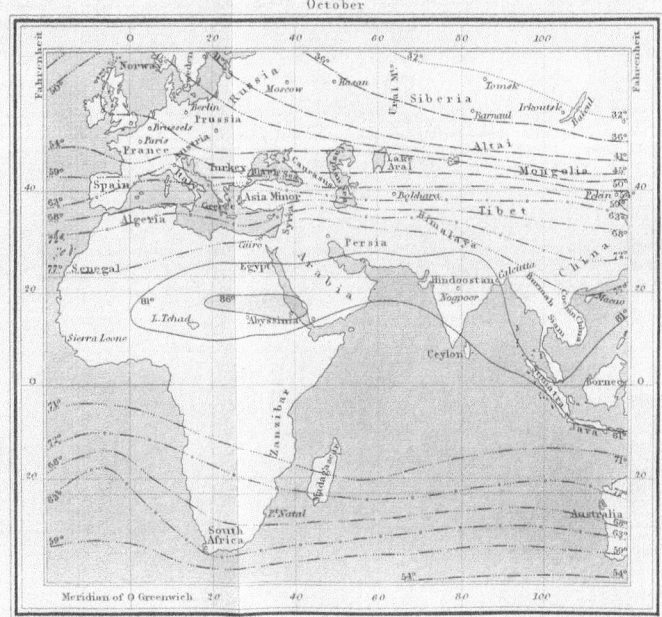

the Distribution of heat on the surface of the Globe
& Thomson "Flora Indica"

Engraved by J.& C. Walker.

earth) is usually estimated at one degree for three hundred feet. In India, it is only in the most perennially humid and densely wooded mountains, that the diminution of temperature is so rapid as this, for in the drier districts it is very much less. Thus, while in Sikkim 1° for 300 feet is the proportion for elevations below 7000 feet, on the Nilghiri Hills it is about 1° for 340 feet, in Khasia 1° for 380 feet; and the elevations of Nagpur and Ambala produce no perceptible diminution in their mean temperature, which is as great as that which would normally be assigned to them were they at the level of the sea.

When the latitude, the amount of land, the humidity, and the elevation are known, we have every element which influences climate; and as the limits between which each of these elements varies is in India considerable, it is evident that the diversity in the climate of its parts must be very great. We reserve the details of these to the following chapter, and shall confine ourselves here to pointing out the two broad divisions of climates, which it is important to bear in mind, namely, those which are excessive, and those which are equable.

An equable climate prevails in the vicinity of the equator, and in all perennially humid districts; while an excessive climate, in which the summer is very hot and the winter cold, is characteristic of the north-western regions, of the interior of the continent, and of provinces characterized by extreme drought. The northern districts of India are more excessive in climate than the southern, because they are broader expanses of land; and the western side of the great (Madras) peninsula is more equable than the eastern, because it is much more humid.

VI. *Sketch of the Physical Features and Vegetation of the Provinces of India.*

A. *Limits of the ' Flora Indica.'*

Although the main object of this Flora is the illustration

of the Botany of the British Possessions in India, we cannot restrict ourselves to these limits without omitting many important additions made by English naturalists to our knowledge of the Indian Flora; and we have hence, in assigning geographical limits to our labours, been guided as well by circumstances of botanical importance, as by natural and political boundaries. We shall therefore include,—to the north, the whole Himalaya, and as much of Tibet as is known,—to the west, Afghanistan and Beluchistan,—to the east, all the countries to the west of the chain which divides Ava from Siam, and the whole of the Malayan peninsula,—and to the south, the island of Ceylon. It is obviously impossible, even were it necessary, to define these boundaries more rigidly. By including them, we gain a point of the greatest importance botanically, in illustrating the Indian Flora, namely, a very fair representation of the Floras of Egypt, Persia, and Europe, to the west,—of Siberia to the north,—of China to the east,—and of the Malayan Archipelago to the south-east; of the union of the species, genera, or orders of which floras, that of India is mainly composed.

Lest, however, we should be thought too arbitrary in pushing our boundaries so far, we may appropriately introduce here a few remarks on the subject, which will explain our motives more fully. Till very recently, no part of the Himalaya belonged to the British Government, the province of Kumaon (between the Ganges and Kali) alone excepted; but later events have added the whole mountain region between the Ravi and Satlej, and placed the remainder of the Northwest Himalaya, including Kashmir, so much under British influence, that an account of its Flora is as essential to botanists in India and Europe, as is that of any of the British possessions. The Tibetan provinces of Ladak and Balti, which continue, as formerly, appanages of Kashmir, have recently been very completely explored botanically by several travellers, whose labours cannot be overlooked, because their herbaria contain many plants which will hereafter be found within the British boundaries, besides many others which, from being in a different state, or belonging to different varieties of others found elsewhere, are essential for the elucidation of our Flora. For the same reasons we include the Chinese Tibetan district of Guge, immediately north of Kumaon, which has been examined by Captain R. Strachey and Mr. Winterbottom, and whose Flora is identical with that of the British Tibetan valleys of Piti, and of Niti (in Kumaon).

Nipal and Bhotan again are wholly independent states; but to exclude them would be to omit all notice of the splendid labours of Wallich on the one hand (which reflect so much lustre on the liberality of a former Government of India), and of Griffith on the other, who alone has explored Bhotan. Sikkim occupies an intermediate position between Nipal and Bhotan; a considerable part of it belongs to the British, the rest is maintained by our influence and authority; and the whole presents a flora which is not only the best investigated of any district east of Kumaon, but unites the Floras of Nipal, Bhotan, East Tibet, and the Khasia mountains; being hence, in a geographico-botanical point of view, one of the most important provinces in India, if not in all Asia.

Returning to the extreme west, the political boundary of British India lies at no great distance beyond the Indus, but does not include the mountainous regions of Afghanistan, the whole of which was investigated about fifteen years ago by Griffith, who accompanied the army of the Indus on its march from Sind to Candahar and Cabul, and penetrated as far as Bamian and Saighan, forming very large collections. These, besides containing an immense number of Persian and European plants, which find their eastern limits within the British territory, are rich in Himalayan forms which advance no further west, and, what is of still greater importance, they contain many species common both to Europe and the Himalaya, but which, from presenting differences induced by local causes in these two distant countries, might not be imagined to have had a common origin, did not the Afghanistan specimens blend their characters, or show the transition between them.

The botany of our eastern frontier is less known than that of any other part of India, and, indeed, it is to it alone that we look for any considerable amount of novelty; for though the upper Assam valley and Mishmi hills have been investigated by Griffith, and Lower and Middle Ava by Wallich, their united materials are not extensive; whilst the upper valley of the Irawadi, Manipur, and the other districts east and south of Cachar, are wholly unknown. Griffith, indeed, botanized in the Húkúm valley, but his collections from that country have not hitherto been made available to botanists. The whole of the Malayan Peninsula is also included in our Flora; for though the British settlements of Penang, Malacca, and Singapur, comprise but a small proportion of the peninsula, they may be supposed to represent well the Flora of so narrow a tract of land, whose climate and physical features are almost uniform throughout.

It will thus be seen that the limits of the Flora Indica extend from the 36th parallel of north latitude to the equator, and from about the 62nd to the 105th degree of east longitude; the area of land embraced being little less than two millions of square miles. This is by far the greatest tropical or subtropical area that has ever been made the subject of one Flora; and at the same time it is the most varied, including every climate, from the burning heat and absolute drought of the deserts of Sind, to the humid jungles of the Malayan peninsula, and to the everlasting snows of the Himalaya. Europe, which (to the regret of every botanist) has never been made the subject of one Flora, considerably exceeds India in superficial area, containing three and a half millions of square miles; and it presents several geographical points which afford familiar standards of comparison for distances in India. Thus, the distance in latitude from Ceylon to Tibet is just that from Gibraltar to the Orkneys, or from the Gulf of Finland to the Morea. The greatest breadth of our limits in longitude is from

Cabul to the Irawadi, which is approximately near that from the Bay of Biscay to the Caspian Sea. The extreme breadth of India along a diagonal line is from Cabul to Malacca, and that is also about the extreme diagonal breadth of Europe from Spain to the northern termination of the Ural mountains at the Arctic Sea. We wish to press these comparisons especially upon the attention of local botanists, and of those more familiar with species of plants than with geography, for the following reason,—that on several occasions, having identified a plant of the lower Himalaya with one that inhabits an elevation of 8000 feet in Ceylon, we have been met with expressions of surprise and incredulity, by naturalists who do not for a moment hesitate to unite many species of Scotland with those of a sufficient altitude on the Sierra Nevada in South Spain; who habitually quote the Alps and Pyrenees as containing many species in common with Iceland and Norway, and even Arctic America; and who, whilst acknowledging that many of the elements of the Floras of the Pyrenees, Alps, Carpathians, Ural, Norway, Iceland, and Arctic America are identical, are prepared to deny a similar extension of species over the mountains of Ceylon, the Madras peninsula, Khasia, Himalaya, and Java.

If, on the one hand, we experience opposition to our identifications of species inhabiting localities in India sundered by considerable areas of land and sea, so, on the other, we find equal or greater difficulty in persuading a large class of our fellow-botanists of the specific identity of Indian plants with those of other better known but more distant countries; and we have hence felt anxious on this account also, so to extend the limits of our Flora, that we might meet such botanists on their own ground as it were, and trace these species continuously from those parts of the world with which they are familiar to those we know best. It is, however, impossible altogether to overcome a proneness of the human mind to regard everything from an unknown country, or that is seen surrounded with foreign associations, as itself unknown, and to banish prejudice from the domain of Systematic Botany as effectually as it has been from some allied sciences, which have fortunately been most successfully cultivated by many men of large experience and extensive attainments in collateral branches of knowledge.

B. *Necessity of dividing India into provinces; and principles according to which it is proposed to be done.*

In order to define with accuracy, and at the same time in an intelligible manner, the geographical range of the individual species comprised within our Flora, it is necessary to divide India into botanical provinces. This we have found a very much more difficult task than might have been supposed, partly from the constantly shifting political and other boundaries of our dominions and its subdivisions, and partly from the necessity of selecting as far as possible such provinces as are defined by physical features rather than by arbitrary lines. We have devoted much time to a careful study of all available information regarding the geography of British India, having had recourse in every case to original documents, in preference to the numerous maps on the physical geography of India published in this country and on the Continent, which have been compiled from these sources, and which, however conspicuous for research, are unexceptionally extremely defective, owing to their authors not having that necessary general acquaintance with the country, which alone could enable them to classify the thousands of facts they have laboriously collected, and which are represented with distorted effect in such maps.

We enter upon our task with a lively sense of our inability to meet the requirements of Botany on the one hand, and of Geography on the other; but it was imperatively necessary that we should, before any part of our Flora went to press, decide upon the geographical divisions to be adopted and the nomenclature to be employed. Though our conjoint

personal experience is very much greater than that of any other
naturalists, there are still large areas of the region under
consideration, of which we have no personal knowledge whatever : we do not therefore presume to consider our scheme as
established beyond the necessity of future modification; on
the contrary, we submit it with great diffidence to the criticism of Indian geographers, and earnestly court inquiry into
its details.

The physical features of the several provinces will be treated
in considerable detail.   This seems called for by the general
want of accurate information on Indian geography, displayed
in many valuable works on various branches of Indian
science; and this not only on the Continent, but quite as
conspicuously in England.   It perhaps arises from the fact
that no physicist or naturalist has hitherto proposed such a
classified or systematic arrangement of habitats or localities, as may be readily acquired by the professed naturalist;
though it should not be forgotten that it is primarily due to
the defective state of our education, which leaves otherwise
accomplished men so ignorant of the general features of the
geography of India, that when the demands of their profession or of science oblige them to study its details, they find
insuperable obstacles to their acquisition.   At the commencement of this essay it has been observed, that "Ind. Or." is
too often the sole indication of the native place of many ines-
timably valuable vegetable products, even in works of standard authority; and when more detailed localities are given,
they are generally copied at random from the tickets of collectors, or the catalogues of local botanists, and are in most
cases mis-spelt and equally unintelligible to the resident in
Europe and in India.   Many botanists indeed seem tacitly
to admit that there is a recognized license to overlook both
generalities and specialities in treating of Indian plants, and
with the honourable exception of Dr. Royle we do not know
of one who has written extensively, and not availed himself of this license.   Dr. Royle's great aim seems to have

n

been to break down this system, both by precept and example, and we consequently find his work unique as regards
the value of the notices it contains on the geographical distribution of the plants of North-west India; and it is with
regret that we see the information he has lavishly given too
frequently so distorted in subsequent systematic works, that
we have to refer to the original to arrive at the truth. This
is certainly from no want of accuracy in Dr. Royle's work,
or inappreciation of details, but in some measure to a due
prominence not being given to a classified arrangement of the
provinces of so extensive and varied a country, and the adoption of such a nomenclature as could be referred to, independently of the other information with which the geographical
matter is at present embodied in his writings.

In the scheme we are about to propose, we shall keep the
natural divisions (botanical provinces) as large as is consistent with our objects; and in selecting names for them, shall
endeavour to choose such as are already familiar to persons
conversant with the outlines of Indian geography, studiously
avoiding the introduction of any that have not a broader
claim to be known and used than mere botanical conveni-
ence. Under the description of each province we shall endeavour to communicate as much definite trustworthy information as we can embody, regarding its elevation, the nature
of its surface, its climate, etc.; this we have chiefly gleaned
from various periodicals and travels, Government reports, and
other sources of information, which have come under our
notice. In order, however, to avoid much repetition in our
descriptions of these provinces, it is necessary to preface our
account of them with some general remarks on the geographical distribution of Indian plants.

C. *General Remarks on the Vegetation of India.*

Before proceeding to describe the physical features, etc., of
the provinces, we shall give a very short and comprehensive

sketch of the vegetation of India, and of the relation which the Botany of its different great divisions bears to that of neighbouring or distant countries. These remarks, from the incompleteness of the data at our disposal, must necessarily be vague, and may be viewed rather as indications of results likely to be obtained than as absolutely ascertained facts.

We have already said that all the main elements of the Indian Flora exist in surrounding countries, and to this is to be attributed one of the most remarkable botanical features of so extensive an area, namely, the very limited number of peculiar families that are largely represented in it. Thus, *Aurantiaceæ, Dipteraceæ, Balsamineæ, Ebenaceæ, Jasmineæ,* and *Cyrtandraceæ* are the only Orders which are largely developed in India, and sparingly elsewhere; and of these, few contain one hundred Indian species. In this respect the Indian Flora contrasts remarkably with that of Australia, South Africa, or South America, or even with Europe, North Asia, and North America. On the other hand, India contains representatives of almost every natural family on the globe, a very few small South American, Australian, and South African Orders being the chief exceptions; and it contains a more general and complete illustration of the genera of other parts of the world than any other country whatsoever, of equal or even of considerably larger extent. It is hence not surprising that some of the large cosmopolitan families are perhaps less universally preponderant in India than in most other continents, *Compositæ* especially being deficient, as are *Gramineæ* and *Cyperaceæ* in some regions, *Leguminosæ, Labiatæ,* and Ferns in others, whilst *Euphorbiaceæ* and *Scrophulariaceæ* are universally present, and *Orchideæ* appear to form a larger proportion of the Flora of India than of any equally extensive country.

We assume the total number of Indian species included in the limits of our Flora, to be from 12–15,000, but whether this estimate is to be regarded as large or small, comparatively with other parts of the globe, we are not prepared to

say; compared with the exaggerated estimates of the Floras
of other tropical countries, which are so frequently put forth,
this number (which is certainly not too small) must appear
insignificant; nor would it be fair of us to expect credence
for it, did we not add that it is the result of the collation of
many irrefragable data, after making a large allowance for du-
bious, undescribed, and even undiscovered species. It is right
also to add, that our conviction that the estimates of other
Floras (and indeed of the Flora of the whole globe) are excessively exaggerated, is founded upon extensive personal experience, and the careful consideration of a large body of well
established facts; and we are emboldened in enforcing it, by
the sanction of Mr. Brown, with whom we have repeatedly
discussed this curious and extremely important subject.

With regard to the general diffusion of species throughout
India, we believe that there is no part of the whole area included in our Flora where a radius of ten miles produces
many more than 2000 species of flowering plants, and that
this is very rare, confined to mountainous districts, and possibly to the Khásia. It is further probable that a continuous
area, with a radius of fifty miles, containing 4000 species, is
nowhere to be found in India; if anywhere, its centre is probably in the Assam valley, in which case it would include
the Khasia, Jheels of Bengal, and the loftiest regions of the
Himalaya.

With regard to local assemblages of species in very narrow
areas, these are never very numerous, except in the pastures
of the temperate and subalpine districts, where thirty to forty,
in different stages of luxuriance, may be found within a radius
of six feet. Nearly as many may be gathered in the neighbourhood of, and upon, one moss-covered rock or tree-stump
on the damp, exposed hill-tops of the Khasia. It is almost
impossible, however, to appreciate the nicely balanced local
circumstances that determine the number of species which
will all find room, and keep it, in a limited space: much depends on the prevalence of species that combine to check the

full growth of individuals on the one hand, and that exclude gregarious species on the other. In the more humid jungles of the luxuriantly clothed parts of India, a very few species are to be found in close contiguity, but many in a moderately large area. In the drier and hilly districts of Central India we have found it difficult, especially in winter, to collect 150 species in a walk of several miles, and this where there was no apparent want of trees, shrubs, or herbs. On the other hand, during the rains we have, in the Panjab, collected eighty species, chiefly of tropical annuals, in an area of a hundred yards square; these, however, were brought together by local circumstances, and the total Flora of the country for ten miles around the same spot probably comprised less than 800 species. At 4–5000 feet elevation in the Khasia we have collected upwards of fifty species of *Gramineæ* alone, in an eight miles' walk, and twenty to thirty *Orchideæ;* but these are quite exceptional cases.

There is almost a total absence of absolutely local plants in India, at least so far as our experience serves us; but in saying this, we are only giving the result of general impressions, and of comparing the contents of our collections with those of other travellers, and with the statements of trustworthy botanists in Australia and South America.

Before dismissing this branch of our subject, we may mention that the general physiognomy of the greater part of the Indian Flora probably approximates more to that of Tropical Africa than to any other part of the globe, accompanying in both cases immense alluvial plains, bounded by deserts at certain points, and traversed by mountain-chains of moderate elevation. The more loosely timbered drier regions probably assimilate very much to the districts of Senegal, Upper Egypt, and Abyssinia; the west shores of the Madras peninsula, and the whole Malayan peninsula to the tropical African coasts; and the deserts of Sind to those of North Africa.

Besides the absence of great forests, there is in India no representative of the Catingas of Brazil, the Pampas of South

America, the Savannahs of North America, nor of those dry
plains studded with hundreds of species of flowering shrubs
and bulbous herbs, which are so characteristic of the Cape of
Good Hope and of Australia. The plains of India are indeed
everywhere extremely poor in species, and such as abound
in individuals are usually of a weedy character. The hilly
parts of moderate elevation again are far from presenting that
gorgeous display of flowers and foliage that the Brazilian
forests do. The gaudy *Cacti, Amaryllideæ, Liliaceæ,* and *Melastomaceæ,* amongst other Orders of that country, have no representatives in India similar in beauty, variety, and abundance. In fact, there are few countries in which the vegetation
of the more accessible parts presents so little beauty, or such
short seasons of bloom.

Maritime plants, again, are rare in India; nor is there a
well-marked and generally diffused littoral Flora; such, we
mean, as is composed of plants that are not absolutely sea-
side, but which never wander many miles from the ocean.

a. *On the Distribution of Indian Plants as influenced by
Climate.*

From the position of India, we have seen that its climate
(and hence its vegetation) is more generally tropical, than the
latitude under which so much of it is included would alone
indicate. The mountains, however, when above 4–5000 feet,
everywhere present more or less of a temperate vegetation,
which becomes wholly temperate at greater elevations, and
which passes into an alpine Flora over a large extent of still
loftier mountain country.

Within the limits of the strictly tropical region there is
the greatest possible difference between the vegetation of the
humid and that of the arid climates, shown not only by a
difference of species, but of genera and whole natural families, and accompanied by a corresponding dissimilarity in the
aspect of the country. Thus, the impenetrable green jungles of the equable and rainy Malayan peninsula, of Eastern

Bengal, the west coast of the Madras peninsula, and of Ceylon, contrast strongly with the drier parts of the intertropical zone, and still more so with the loosely-timbered districts of Central India, and of the base of the western Himalaya. The absolutely sterile deserts are confined to the extensive plains, which are all cut off from the rains by being placed to leeward of mountain-ranges, or by other causes. There are hence in India no vast plains clothed with gigantic timber-trees, such as cover immense areas of the American tropics; and even the valleys of the great Indian rivers, the Ganges, Nerbada, etc., are nowhere heavily timbered, but are generally absolutely destitute of forest, and extremely populous and highly cultivated*.

The tropical forests of India may be divided into those which inhabit perennially humid districts, and those which are confined to regions presenting contrasted seasons, of summer rain and winter drought.

The perennially humid forests are uniformly characterized by the prevalence of Ferns, and, at elevations below 5000–7000 feet, by the immense number of epiphytal *Orchideæ, Orontiaceæ,* and *Scitamineæ :* they contain a far greater amount of species than the drier forests, and are further characterized by *Zingiberaceæ, Xyrideæ,* Palms, *Pandaneæ, Dracæna, Piper, Chloranthus, Urticaceæ* (especially *Artocarpeæ* and *Fici*), *Araliaceæ, Apocyneæ,* shrubby *Rubiaceæ, Aurantiaceæ, Garciniaceæ, Anonaceæ,* Nutmegs, and *Dipterocarpeæ.*

The drier tropical forests of the regions with contrasted seasons, are much modified in luxuriance and extension by the winter cold in those extratropical latitudes over which they spread. In the chapter upon the meteorology of India, it is shown that though the summer heat scarcely decreases

---

* It is a much discussed question in India, whether the Gangetic plain was ever covered with forest: the best authorities consider that it never was so ; but there are others who hold the contrary opinion, and aver that the destruction of the timber has produced a great change in the climate. The absence of vegetable remains in the alluvium appears unfavourable to the latter opinion.

with the increasing latitude till the 30th degree north, the
cold of winter rapidly increases (see the map of Isothermals). Hence many tropical species, genera, and even families,
which are sensitive to cold, are comparatively local when found
beyond the tropic, as most Palms, *Cycas, Dipterocarpeæ* (except *Vatica*), *Aurantiaceæ, Connaraceæ, Meliaceæ, Myrtaceæ,
Rubiaceæ, Ebenaceæ,* and many more. Others are indifferent
to the cold of winter, provided they experience a great summer heat; these advance far beyond the tropic, and lend a
more or less tropical aspect to the Flora even of the base of
the north-western Himalaya, in 33° north. Such are many
Leguminosæ (as *Bauhinia, Acacia, Erythrina, Butea, Dalbergia, Millettia*), *Bombax, Vatica, Nauclea, Combretaceæ,
Verbenaceæ, Lagerstrœmia, Grislea, Jasmineæ,* and *Bignonia
Indica.*

Passing from the forest vegetation to that of annual plants,
we find that an immense proportion of these are uniformly
distributed throughout India, and, vegetating only during
the hot rainy season, are neither exposed to drought nor cold.
Of these some of the most conspicuous are, besides *Gramineæ* and *Cyperaceæ*, a vast number of small *Leguminosæ* and
Scrophularineæ, Sida, Corchorus, Nama, Blumea and other
Compositæ, some *Labiatæ* (as *Leucas, Anisomeles,* etc.), *Amaranthaceæ, Acanthaceæ, Convolvulaceæ, Ludwigia, Jussieua,*
etc.

Dr. Royle has well shown that this distribution of tropical
annuals and of perennial-rooted plants with annual stems is
not confined to the plains, but ascends the loftier mountain
valleys as far as the well-marked rainy season extends, and
that such plants only disappear where the accession of heat
and humidity is not sufficient in amount or regular enough
in period to stimulate their vegetative organs. Some of the
most remarkable of these extratropical examples of tropical
genera are species of *Begonia, Osbeckia, Argostemma, Plectranthus,* various *Cyrtandraceæ, Scitamineæ, Araceæ, Commelynaceæ,* and a few epiphytal *Orchideæ.*

A vegetation of a different nature from any of the above prevails in the extratropical regions of India during the cold months only ; and, though contrasting in character with that of tropical annuals, is dependent upon analogous modifications of climate for its presence. This consists of annual plants of the north temperate zone that do not appear within the tropics (except at a considerable elevation), and which owe their southward extension into India to the winter's cold, just as the summer annuals owe their northward extension to the heat. These flower when the tropical plants are torpid : they are very numerous, comprising many European and cosmo-politan genera, and even species. Besides the winter crops of the Gangetic plain, consisting of Wheat, Barley, and more rarely Oats, with various kinds of pulse, there are, of wild plants, *Ranunculus sceleratus* and *muricatus*, *Capsella Bursa-pastoris*, *Silene conica*, *Alsine media*, *Arenaria serpyllifolia*, *Euphorbia Helioscopia*, *Medicago lupulina* and *denticulata*, *Lathyrus Aphaca*, *Gnaphalia*, *Xanthium*, *Veronica agrestis* and *Anagallis*, *Heliotropium Europæum*, various *Polygona*, *Juncus bufonius*, *Butomus umbellatus*, *Alisma Plantago*, and very many *Cyperaceæ*, Grasses, and such aquatics as *Myriophyllum*, *Potamogeton natans* and *crispus*, *Vallisneria*, *Zannichellia*, *Ranunculus aquatilis*, *Lemna*, and many others.

The transition from the tropical to the temperate Flora is more rapid in ascending above the level of the plains, than in advancing northward at the same level; the change of vegetation in a few thousand feet of ascent being much greater than in as many degrees of latitude as would compensate for the decrease of temperature experienced in that ascent. In the perennially humid provinces of India the climate of the base of the mountains is even more equable than that of the adjacent plains, from the atmosphere being more loaded with moisture. Hence in these regions a warm temperate Flora (neither strictly temperate nor markedly tropical) commences at elevations of 2–3000 feet, and prevails over the purely tropical, which appears in scattered trees,

o

shrubs, etc., amongst it. This vegetation presents many peculiar features, and its total absence from the plains is not to be
accounted for by any simple law of climate. Amongst other
Orders we may mention especially *Magnoliaceæ, Ternstræmiaceæ,* subtropical *Rosaceæ* (as, *Prunus, Photinia,* etc.), *Kadsura, Sphærostema, Rhododendron, Vaccinium, Ilex, Styrax,
Symplocos, Olea, Sapotaceæ, Lauraceæ, Podocarpus, Pinus longifolia ;* with many mountain forms of truly tropical families,
as Palms, *Pandanus, Musa, Clusiaceæ,* Vines, *Vernonia,* and
hosts of others. These are instances of more or less strictly
mountain plants prevailing uniformly over many degrees of
latitude and longitude without ascending or descending much,
but which are so rarely seen on the plains, as to entitle them
collectively to a separate notice when treating of the phases of
Indian vegetation.

Advancing westward, especially in the Himalaya, we experience a drier climate, which exaggerates the effect of elevation on the vegetation, and produces besides many curious
anomalies, as a reduced mean temperature divided into two
seasons, one of heat and one of cold, which are more contrasted at these elevations than on the plains. It is obviously impossible to enter here into the details of the apparent anomalies thus caused in the distribution of plants ;
each individual species demanding a study of its natural habits
to explain its aptitude for an extended distribution in elevation, or geographical position, or its absolute restriction to a
very narrow area, or to a few spots characterized by a combination of favourable circumstances. Examples may be seen
in the *Ephedra* of the Panjab and north-western Himalaya,
which ranges from the plains to 16,000 feet ; in the genus
Marlea, which ascends from 3000 to 8000 feet in Sikkim,
and in the western Panjab, at scarcely 4000 feet, accompanies
Celtis and a species of Ash ; in a subtropical *Myrsine,* which
extends even into Afghanistan ; in *Juniperus excelsa,* found as
low as 5000 feet in Afghanistan, and which ascends to 15,000
in Tibet.

Of the tropical and subtropical plants that accompany this high summer temperature and withstand the cold of considerable elevations, are many of those mentioned towards the commencement of this section as natives of dry tropical forests with contrasted seasons, at the level of the sea or on plains raised but little above it. *Populus Euphratica*, a *Cynanchum*, *Chloris barbata*, and *Cyperus aristatus*, all of which ascend to 11,000 feet in Ladak, are other remarkable instances, as is *Pegamum Harmala*, which attains 9000 feet.

In the Himalaya the truly temperate vegetation supersedes the subtropical above 4000–6000 feet; and the elevation at which this change takes place corresponds roughly with that at which the winter is marked by an annual fall of snow. This phenomenon varies extremely with the latitude, longitude, humidity, and many local circumstances. In Ceylon and the Madras Peninsula, whose mountains attain 9000 feet, and where considerable tracts are elevated above 6–8000 feet, snow has never been known to fall. On the Khasia mountains, which attain 7000 feet, and where a great extent of surface is above 5000, snow seems to be unknown. In Sikkim snow annually falls at about 6000 feet elevation, in Nipal at 5000 feet, in Kumaon and Garhwal at 4000, and in the extreme West Himalaya lower still.

It is hence only on the Himalaya and Mishmi mountains that a purely temperate flora prevails, to the exclusion of all tropical forms; though in Ceylon, the Nilghiri mountains, and Khasia, the temperate forms are very numerous, and so prevalent on the highest summits as to render it very desirable that these heights should be subjected to a very close botanical examination. Local circumstances, again, seem to bring the temperate forms lower upon the Khasia and Nilghiri mountains than upon the Himalaya, which are further north; and of these causes the fact that the exposed flat or undulated surfaces of the Khasia are swept by violent winds, is one of the most powerful. The contrast in this respect between the Khasia and the Sikkim-Himalaya is very remarkable, many hundred species of temperate types common to both, being habitually found 1–3000 feet lower on the Khasia than in Sikkim. For the same reason many tropical types, and even species, ascend higher in Sikkim than they do in the Khasia; the warm forest-clad and sheltered Himalayan valleys at 5–7000 feet elevation, offering a very different climate to the broad grassy tops of the Khasia. Such apparent exceptions to the laws of distribution are frequent in India, rendering it very difficult for the beginner to comprehend even the general features of this branch of science, and for us to reduce them to such a system as shall be readily acquired.

It is unnecessary here to enumerate the prevalent forms of the temperate flora of India, including as they do every natural family, and almost every extensive or widely-spread genus of north Europe, Siberia, and colder temperate America, and this whether of shrubs, trees, or herbs. The exceptions become, however, the more important from their comparative paucity; of these we may mention the total absence of *Erica, Arbutus, Azalea, Fagus, Cochlearia, Cistaceæ, Tilia, Lupinus, Rhinanthus, Empetrum,* various *Umbelliferæ,* whilst we find but few species of *Hieracium, Trifolium, Centaurea, Veronica,* and *Dianthus.*

Of genera many of which have hitherto been usually considered as most characteristic of other parts of the world, but for whose maximum development we must look to the Himalaya, are *Rhododendron, Monotropa, Pedicularis, Corydalis, Nepeta, Carex, Spiræa, Primula, Cerasus, Lonicera, Viburnum,* and *Saussurea.*

Lastly, the Alpine or Arctic Flora demands a few words here, though it forms comparatively so small a feature in the vegetation of all India, that its full discussion must be reserved to our remarks on the Alpine region of the Himalaya. This, which hardly reaches its extreme upper limit at 18,500 feet above the sea, commences (as we restrict it) above the limit of trees throughout a great part of the Himalaya; it partakes in its characteristic genera of the temperate Flora, and, though fully representing the Flora of the Polar regions, contains so many types that are foreign to them (as *Gentiana, Ephedra, Valerianeæ, Corydalis*), and some which are even rare in Siberia, that it must rather be considered as a continuation of the Alpine Flora of Europe than a representation of that of the Arctic zone. It displays one remarkable feature throughout its whole extent, a comparative paucity of Cryptogamic plants ; and it is especially poor in those luxuriant mosses of tall growth and succulent habit, which form vivid and broad green tufts, loaded with rich brown capsules, and which abound both in the Alps and Polar regions. This is no doubt indirectly due to the elevation of the region, and directly to the sudden accessions of great heat and drought, which are the effects of a highly rarefied atmosphere, and which, though strongly enough marked to check the development of Mosses and Hepaticæ, are not of sufficient duration to affect phænogamic vegetation in the same degree.

b. *On the Distribution of Indian Plants as influenced by Geographical Position.*

Hitherto we have solely considered the spread of plants in India as influenced by climate, but geographical position is accompanied by such remarkable phenomena in vegetation, as to indicate other influences, which demand some notice here. The Floras of the frontier provinces of India, as we have repeatedly remarked, are identical with those of the countries that surround them, and are continuous with them, and that this should be so stands to reason ; but we sometimes see a decided affinity between the Floras of areas separated by oceans, deserts, or mountain-chains, between which it is unwarrantable to assume that a migration of the species common to both, has taken place since the interposition of the barriers in question, and which further present many natural characters in common, which neither migration (if conceded to any amount) nor climate will account for. We have already

alluded to this subject in the third chapter of this Essay
(p. 40), as one intimately connected with geological change,
and as involving questions of the antiquity of species and of
continents, which, as regards the Flora of India, we have no
materials for discussing. It would be very easy to assume a
few premises, and to suppose elevations and depressions of the
islands, oceans, plains, and mountains of India, that would
afford each area marked by a peculiar vegetation the means
of having derived its species, or its botanical features, from
another now isolated or distant region; and to extirpate
species from areas where it would, for the theory's sake, be
convenient to do so. It would also be easy to suppose climatic and other changes that would derange the whole existing order of vegetation, and to adapt the little we know of
the Geology of India to support such movements; but we consider that all such speculations are unsafe and inexpedient in
our present incomplete knowledge of any one branch of Indian science; they should be based primarily on geological
data, and mainly on palæontological evidence that has been
thoroughly sifted, should be well supported by zoological facts,
and only extended to botany after the species of plants inhabiting the whole area shall have been approximately determined. It must not be supposed that, in declining to enter
upon this subject, we are actuated by a spirit hostile to
speculative reasoning; on the contrary, were we fully acquainted with the species and distribution of Indian plants,
we would willingly throw out such suggestions as we think
an analysis of them would legitimately warrant our advancing, and wait the result of zoological and palæontological
evidence, with the hope, on the one hand, of establishing the
truth of our deductions, and, on the other, in the belief, that
if proved in the wrong, we should at any rate have erred
within reasonable limits. But at this time in particular,
when the labour of comparing and determining plants, and
accumulating exact data, is shunned by the majority of botanists; when loose theories on geographical distribution, and on

the development of species, are replacing research ; and when the data usually employed for deducing the laws of the distribution of plants consist of a compilation of raw materials from the works of travellers and local observers more or less skilled in botany, it becomes incumbent upon us, who hold that progress in this branch of botany depends on an exact knowledge of species, genera, families, and their affinities, to refrain from crude speculations as to the origin of the Indian Flora.

The following geographical alliances or affinities (if we may use the terms) of the Indian Flora, with more or less remote countries, we consider well established ; they are capable of much illustration, even in the present state of our knowledge, but it is obviously impossible to dilate upon them here.

1. *The Australian type.*—The Flora of Australia is well known to contain far more endemic species and families than any other country does, and of these a few representatives extend into India. Besides *Pittosporum* and *Scævola*, which, though more characteristic of the Australian than of other Floras, are found all over India and Africa; there are two species of *Stylidium*, which are the only extra-Australian ones known : one of these extends up the Malay peninsula to Silhet, and is also said to be found at Midnapore on the west side of the Gangetic delta; and the other is confined to the Malay peninsula. Several species of Australian genera of *Myrtaceæ* (*Leptospermum*, *Bæckia*, and *Metrosideros*) inhabit the same peninsula, besides the very remarkable genus *Tristania*, which advances to Moulmein in 17° N. lat. *Casuarina*, which is cultivated throughout India, is wild on the east coast of the Bay of Bengal as far north as Ramri ; and of *Helicia* (a Proteaceous genus) several species abound in the Malay peninsula, and one extends to Silhet, and along the base of the Himalaya to Central Nipal. *Lagenophora*, a small Australian genus of *Compositæ* (also found in New Zealand and Fuegia), has a representative in the Khasia and Ceylon. We thus see that Australian types are almost confined to a

meridian east of the Ganges; and the only important exceptions known to us are another species of *Helicia* in Ceylon,
Lagenophora in the same island, and the curious genera *Acrotrema* and *Schumacheria* of *Dilleniaceæ*, which are more nearly
allied to Australian forms of that Order than to any others,
and of which *Schumacheria* is confined to Ceylon, *Acrotrema*
being also found in the Malayan peninsula and in Malabar.

2. *The Malayan Archipelago type.*—This forms the bulk
of the Flora of the perennially humid regions of India; as of
the whole Malayan peninsula, the upper Assam valley, the
Khasia mountains, the forests of the base of the Himalaya
from the Bramaputra to Nipal, of the Malabar coast, and
of Ceylon. It is of course impossible to specify the genera
or even families of so predominant an element; to do so
would be to enumerate a very large proportion of the Indian
genera, and to except only the north temperate and the comparatively few African types. The extent, however, to which
this element predominates is not yet appreciated, nor do we
ourselves know its total amount; for constantly, during our
examination of the temperate as well as tropical plants of
the Nilghiri, Khasia, Ceylon, and the Himalaya, we find them
identical in species with Javanese mountain plants. That
botanists have neglected comparing these Indian plants with
Javanese Floras is not surprising, when it is considered how
remote Java is from any part of continental India, and that
geographical isolation is by many considered equivalent to
specific difference. We are, however, convinced, after a very
careful examination, that there are several plants, as *Gaultheria nummularia*, which extend into the North-west Himalaya, and are also found in the Javanese mountains, which
are nearly 3000 miles distant: some of these have already
been found in intermediate localities, as the *Gaultheria*,
which occurs along the whole Himalayan range, and in the
Khasia, and which will probably be found in the mountains
of the Malay peninsula and of Sumatra; and there are many
other Java plants which are more uniformly spread over the

hilly districts of India and Ceylon. Amongst the more conspicuous trees common to Java and India are *Sedgwickia cerasifolia*, Griff., a native of Assam, which is undoubtedly the *Liquidambar Altingia* of Blume; *Marlea*, which spreads into China on the one hand, and throughout the Himalaya to the mountains south of Kashmir on the other. The curious *Cardiopteris lobata* of Java is also a native of Assam, and several oaks and chesnuts, *Antidesmæ*, a willow, and *Myrica*, have already proved to be common to the Khasia and Java.

3. *The China and Japan type.*—In the Indian flora we meet with many temperate genera and species, which are also common to North America west of the Rocky Mountains, and which are foreign to Europe, to America east of that range, and to Western Siberia; besides many tropical species that are also Malayan and West Polynesian. The Chinese type is abundant in the temperate regions of the Himalaya, extending westward to Garhwal and Kumaon, but is most fully developed in Sikkim, Bhotan, and the Khasia. Amongst the most striking examples of its temperate forms in the Himalaya, are species of *Aucuba, Helwingia, Stachyurus, Enkianthus, Abelia, Skimmia, Bucklandia, Adamia, Benthamia, Corylopsis*, genera that have been considered as almost exclusively Japanese and Chinese, and of most of which there are but solitary species known in that country.

Other temperate plants common to India and China are *Microptelea parvifolia* (a species of elm) ; *Hamamelis Chinensis*, found by us in the Khasia; *Nymphæa pygmæa*, and *Vaccinium bracteatum*, both of which occur in the Khasia; and *Quercus serrata*, which is a native of Nepal, Sikkim, and the Khasia. Besides these cases of absolute identity of species, many Chinese genera may be noticed. *Illicium* inhabits the Khasia, *Thea* Assam; and *Magnolia*, Sikkim and Khasia. *Schizandreæ* are peculiarly characteristic of the Chinese Flora, but also extend into Java; *Lardizabaleæ*, which attain their maximum of development in the Himalaya, are Japanese and Chinese, a few only having hitherto been de-

p

tected in temperate South America.   Other instances are
Camellia, Deutzia, Hydrangea, Viburnum, several *Corneæ,* and
Houttuynia.

The recent able investigation of the Hongkong Flora by
Major Champion and Mr. Bentham has materially increased
our knowledge of the intimate relationship between the
Floras of China and the eastern parts of India; amongst
many instances, we may select the remarkable genus of Ferns,
Bowringia*, found in Hongkong and in the Khasia mountains; *Wikstrœmia,* a genus of *Daphneæ; Bucklandia, Enkianthus, Henslovia, Scepa, Antidesma, Benthamia, Goughia,
Myrica,* and very many others; in fact, there is scarcely a
genus in the whole Hongkong Flora that is not also Indian.
Euryale ferox, which is wild in the Gangetic delta, and is
found as far westward as Kashmir, is abundant in China; and
Nepenthes phyllamphora, a native of the Khasia mountains,
is also found at Macao, and eastward to the Louisiade Archipelago.

4. *The Siberian type.*—This is characteristic of the colder
temperate parts of Asia, and is very fully represented in the
upper temperate and alpine regions of the Himalaya, descending in the north-western and drier parts of the chain to
very low levels.   It approaches, in many respects, to the
South European vegetation, but is characterized by the predominance of *Fumariaceæ, Potentillæ, Leguminosæ,* especially
Hedysarum and *Astragaleæ,* of *Umbelliferæ, Lonicera, Artemisia, Pedicularis,* and *Boragineæ;* and by the rarity or total
absence of certain groups or genera which are especially
abundant in Europe, such as *Cistaceæ, Rosa, Rubus, Trifolium,
Erica,* Ferns, and other cryptogams.   As the Alps of Central
Asia rise gradually from the elevated tracts of Southern Siberia, and possess a very similar climate, the increasing elevation
compensating for the diminution of latitude, a very Siberian

---

* *Bowringia* of Hooker, 'Kew Journal of Botany,' vol. v. p. 237.   A name
superseded by the *Bowringia* of Bentham, in Hooker's 'Kew Journal of Botany,' vol. iv. p. 75.

Flora predominates throughout the drier regions of the Himalaya*. Siberian forms are, however, by no means confined to the drier parts of the chain, but may be observed even in the most humid regions of the Himalaya, and occasionally even on the mountains of tropical India. Thus *Artemisia* and *Astragalus,* which are perhaps the most characteristic genera of the Siberian type of vegetation, are not only abundant throughout Tibet and the interior Himalaya, but are represented by a few species in the plains of the Panjab, on the outer slopes of the western Himalaya, and even on the Khasia mountains. *Spiræa Kamtchatica, chamædrifolia,* and *sorbifolia,* and *Paris polyphylla,* are also Siberian forms which extend into the rainy Himalaya; and *Corydalis Sibirica* and *Nymphæa pumila* are remarkable instances of specific identity between Khasia and Siberian plants†.

5. *The European type.*—The extent to which European plants abound in India has never hitherto been even approximately appreciated. Dr. Royle was the first to indicate this affinity between the vegetation of the eastern and western continents of the old world; and throughout his writings we find constant evidence of his never having lost sight of this being a marked feature. Had the collections, upon which he founded his conclusions, been critically compared and worked out, the keystone to the whole system of distribution in Western Asia could not have escaped him, which does not rest so much upon a number of representative species, as

* As a few instances, besides the many *Ranunculaceæ* and *Fumariaceæ* enumerated in the pages of the present volume, we may mention *Tauscheria desertorum, Biebersteinia odora, Potentilla Salessovii, multifida,* and *bifurca, Chamærhodos sabulosa, Pyrus baccata, Astragalus contortuplicatus, densiflorus,* and *subulatus, Phaca frigida, Oxytropis diffusa, Cicer Soongaricum, Sedum quadrifidum, Artemisia Dracunculus, scoparia, Tournefortiana, fasciculata,* and *salsoloides, Saussurea latifolia* and *pygmæa, Mulgedium Tataricum, Osmothamnus fragrans* (*Rhododendron anthopogon,* Don), *Salix augustifolia,* Poputus *balsamifera, Carex microglochin, stenophylla, physodes, supina,* and, *tristis.*

† It is curious to remark that there are in Siberia a certain number of forms indicative of tropical Indian types, as, for instance, *Menispermum* and *Anandria.*

upon the fact that not only are a large proportion of annual and herbaceous species of each common to Western India and Europe, but of shrubs and trees also.

Although the progress we have hitherto been able to make in critically examining our own Indian collections is very limited, we have already established the identity of so many Himalayan plants with European ones, as to oblige us to look to a common origin for the species found in both these regions, and to seek for causes no longer in operation to account for their distribution over so extended an area. The mountain mass of Asia, as is well known, sinks to the westward of Afghanistan, rising again only in isolated peaks; and hence the Himalaya is rather ideally than really connected with the mountains south of the Caspian, and so with the Caucasian Alps on one hand, and those of Asia Minor on the other; nevertheless we find a multitude of mountain plants, and indeed many of the most conspicuous ones of Europe, ranging from the coasts of the Levant and the Black Sea to the Himalaya. Of these, again, some are confined within these limits, as *Corylus Colurna* (*C. lacera*, Wall.); others spread no further east than the North-western Himalaya, but continue westward to the south of Spain, as *Quercus Ilex, Ulmus campestris, Celtis australis* and *orientalis*; and others, again, advance eastward, spreading over the whole Himalaya, as the Walnut, Ivy, Juniper, and Yew, some of which extend into the Khasia; and two, Juniper and Yew, spread yet further across China, Mexico, and throughout North America. These European forms are almost confined to the temperate regions of India, and with them we also find abundantly the herbs and shrubs of Northern Europe, inhabiting a loftier level in the Himalaya, where they blend with the Siberian types. We cannot conceive anything more valuable or suggestive to the student of geographical distribution than an accurate list of these European plants, which may be grouped under three heads:—1. Such as are common to most parts of Europe, Northern Asia, and North America, and the Himalaya, such as the Yew,

Juniper, *Aquilegia vulgaris, Caltha palustris,* etc. 2. Those which are confined to Europe and India. These, again, belong partly to the Mediterranean Flora, as, for instance, *Celtis, Quercus Ilex, Olea Europæa, Myrtus communis,* etc.; and partly to that of Europe north of the Alps, including the greater number of herbs and small shrubs. Meanwhile we shall here confine ourselves to subjoining a list of 222 British plants which extend into India. Many of these require a more critical comparison; but we are convinced that the errors which may be detected in our enumeration are too few to invalidate the important general law. The list, indeed, is very far from complete, as we have omitted all plants regarding which we are not tolerably certain.

Thalictrum *alpinum.*
    „   *minus.*
Ranunculus *aquatilis.*
    „   *Lingua.*
    „   *sceleratus.*
    „   *arvensis.*
Caltha *palustris.*
Aquilegia *vulgaris.*
Actæa *spicata.*
Berberis *vulgaris.*
Nymphæa *alba.*
Papaver *dubium.*
    „   *hybridum.*
Fumaria *Vaillantii.*
Nasturtium *amphibium.*
    „   *officinale.*
Barbarea *vulgaris.*
Turritis *glabra.*
Cardamine *hirsuta.*
Sisymbrium *Sophia.*
    „   *thalianum.*
Alliaria *officinalis.*
Draba *incana.*
    „   *verna.*

Thlaspi *arvense.*
Hutchinsia *petræa.*
Lepidium *latifolium.*
    „   *ruderale.*
Capsella *Bursa-Pastoris.*
Silene *inflata.*
    „   *conica.*
Sagina *procumbens.*
Arenaria *serpyllifolia.*
Holosteum *umbellatum.*
Stellaria *media.*
Cerastium *vulgatum.*
Hypericum *perforatum.*
Geranium *lucidum.*
    „   *Robertianum.*
Erodium *cicutarium.*
Oxalis *Acetosella.*
    „   *corniculata.*
Ononis *arvensis.*
Medicago *lupulina.*
    „   *denticulata.*
Melilotus *officinalis.*
    „   *vulgaris.*
Trifolium *pratense.*

Trifolium *repens.*
„   *fragiferum.*
Lotus *corniculatus.*
Ervum *tetraspermum.*
„   *hirsutum.*
Vicia *sativa.*
Lathyrus *Aphaca.*
Prunus *Padus.*
„   *Avium.*
Agrimonia *Eupatoria.*
Alchemilla *vulgaris.*
Sibbaldia *procumbens.*
Potentilla *rupestris.*
„   *anserina.*
„   *verna.*
„   *reptans.*
Fragaria *vesca.*
Rubus *fruticosus.*
„   *saxatilis.*
Geum *urbanum.*
Rosa *spinosissima.*
„   *rubiginosa.*
Cratægus *Oxyacantha.*
Cotoneaster *vulgaris.*
Pyrus *Aria.*
Lythrum *Salicaria.*
Epilobium *palustre.*
„   *parviflorum.*
„   *tetragonum.*
„   *montanum.*
„   *roseum.*
„   *alpinum.*
Circæa *lutetiana.*
Myriophyllum *verticillatum.*
Hippuris *vulgaris.*
Sedum *Telephium.*
„   *Rhodiola.*
Ribes *Grossularia.*
„   *nigrum.*

Saxifraga *granulata.*
„   *cernua.*
Sium *angustifolium.*
Daucus *Carota.*
Torilis *Anthriscus.*
Scandix *Pecten.*
Hedera *Helix.*
Galium *tricorne.*
„   *Aparine.*
„   *boreale.*
Valerianella *dentata.*
Tussilago *Farfara.*
Bidens *tripartita.*
„   *cernua.*
Achillea *Millefolium.*
Artemisia *vulgaris.*
„   *maritima.*
„   *Absinthium.*
Senecio *Jacobæa.*
Lappa *major.*
Centaurea *Calcitrapa.*
Silybum *Marianum.*
Lapsana *communis.*
Cichorium *Intybus.*
Picris *hieracioides.*
Sonchus *oleraceus.*
„   *arvensis.*
Campanula *latifolia.*
Pyrola *rotundifolia.*
Erythræa *Centaurium.*
Villarsia *nymphæoides.*
Polemonium *cæruleum.*
Convolvulus *arvensis.*
Asperugo *procumbens.*
Lycopsis *arvensis.*
Lithospermum *arvense.*
Myosotis *arvensis.*
Solanum *nigrum.*
„   *Dulcamara.*

Hyoscyamus *niger.*
Orobanche *cærulea.*
Lathræa *squamaria.*
Verbascum *Thapsus.*
Antirhinum *Orontium.*
Linaria *Elatine.*
Euphrasia *officinalis.*
Veronica *Anagallis.*
  „ *Beccabunga.*
  „ *officinalis.*
  „ *verna.*
  „ *triphyllos.*
  „ *agrestis.*
Origanum *vulgare.*
Thymus *Serpyllum.*
Clinopodium *vulgare.*
Scutellaria *galericulata.*
Prunella *vulgaris.*
Nepeta *Cataria.*
Lamium *amplexicaule.*
Stachys *arvensis.*
Marrubium *vulgare.*
Verbena *officinalis.*
Utricularia *minor.*
Glaux *maritima.*
Samolus *Valerandi.*
Salsola *Kali.*
Atriplex *patula.*
Chenopodium *album.*
  „ *viride.*
Rumex *palustris.*
  „ *obtusifolius.*
  „ *Acetosa.*
Oxyria *reniformis.*
Polygonum *Bistorta.*
  „ *viviparum.*
  „ *Hydropiper.*
  „ *aviculare.*
Hippophae *rhamnoides.*

Buxus *sempervirens.*
Euphorbia *helioscopia.*
  „ *Peplus.*
  „ *exigua.*
Callitriche *aquatica.*
Parietaria *officinalis.*
Ulmus *campestris.*
Salix *purpurea.*
  „ *alba.*
Orchis *latifolia.*
Convallaria *verticillata.*
Lloydia *serotina.*
Gagea *lutea.*
Juncus *glaucus.*
  „ *lamprocarpus.*
  „ *bufonius.*
Alisma *Plantago.*
Sagittaria *sagittifolia.*
Butomus *umbellatus.*
Triglochin *maritimum.*
  „ *palustre.*
Sparganium *ramosum.*
Acorus *Calamus.*
Lemna *minor.*
Potamogeton *natans.*
  „ *perfoliatus.*
  „ *crispus.*
  „ *gramineus.*
Zannichellia *palustris.*
Eleocharis *palustris.*
  „ *acicularis.*
Scirpus *maritimus.*
Blysmus *rufus.*
Carex *incurva.*
  „ *divisa.*
  „ *remota*
  „ *atrata.*
  „ *rigida.*
  „ *ustulata.*

Carex *flava*.
„ *Pseudo-cyperus*.
„ *ampullacea*.
„ *paludosa*.
Alopecurus *pratensis*.
Polypogon *Monspeliensis*.
Agrostis *vulgaris*.
Kœhleria *cristata*.
Poa *annua*.

Poa *alpina*.
„ *nemoralis*.
„ *pratensis*.
Dactylis *glomerata*.
Festuca *ovina*.
Brachypodium *sylvaticum*.
Bromus *tectorum*.
Lolium *temulentum*.
Hordeum *pratense*.

One very remarkable result has already struck us with regard to the Himalayan distribution of European plants, namely, their rapid disappearance to the east of Kumaon. Few species, comparatively, extend into Nipal, and still fewer occur in Sikkim. Thus *Myrtus communis*,—to mention only a few instances,—is not found further east than Afghanistan; *Nymphæa alba, Marrubium vulgare, Nepeta Cataria, Potentilla reptans*, and *Trifolium fragiferum*, have not been observed beyond Kashmir; *Cratægus Oxyacantha* stops in Kishtwar; *Rubus fruticosus* in the outer hills near Jamu; and *Aquilegia vulgaris* in Kumaon. There is thus a blending of European forms with the proper Himalayan Flora in the western parts of the chain, just as, to the eastward, we find Chinese and Malayan forms intermixed with it. How far this curious fact is due to climatic or physical causes, our present data do not enable us to decide. It cannot however, we think, be disconnected from the gradually diminishing rain-fall of the more western Himalaya. We ought also not to forget that in the longitude of Kumaon there exists a great watershed, which stretches north-east as far as the sea of Japan; for, however little this point of physical structure may now affect the vegetation of the outer regions of the Himalaya, its influence during the elevation of the land must have been very considerable.

6. *The Egyptian type.*—Egypt, Southern Arabia, and the warmer parts of Persia, possess a remarkable similarity of climate to Beluchistan, Sind, and the Panjab, and at the same

time a nearly complete identity of vegetation. Many North African or Arabian forms, such as *Peganum Harmala, Fagonia Cretica, Balanites Ægyptiaca, Acacia Arabica, Alhagi, Grangea, Calotropis, Salvadora Persica*, extend throughout all the drier parts of India. Others have a less extensive range, being only found in Northern and Western India : of these, *Malcolmia Africana, Farsetia*, several species of *Cleome, Balsamodendron, Astragalus hamatus* and others, *Cucumis Colocynthis, Berthelotia, Anticharis Arabica*, spinous *Acanthaceæ, Cometes, Forskalea, Populus Euphratica, Ephedra, Salix Ægyptiaca, Crypsis*, etc. etc., may be mentioned as instances. In India, as in Africa, this peculiar vegetation passes by insensible gradations into the European Flora on the one hand, and into the tropical on the other.

7. *The Tropical African type.*—Though tropical Asia and Africa are separated by a vast expanse of ocean, there is a striking similarity in their vegetation. This is shown not only by the identity of the annual vegetation which springs up during the rainy season*, but by a great similarity in the families and genera of the trees and shrubs : *Capparis, Grewia, Sterculiaceæ, Tiliaceæ*, columnar *Euphorbiæ*, and many other *Euphorbiaceæ, Antidesma, Lepidostachys, Olacineæ, Acacia*, and *Rubiaceæ*, may be mentioned as examples.

Too little is known of the African Flora to enable any definite conclusions to be drawn as to the numerical value of this type in India, but it is evidently an important one†.

A curious affinity may also be traced between the mountain vegetation of western tropical Africa and that of the Peninsular chain, where the absence or comparative rarity of many of the principal features of the Malayan Flora has already

---

* *Polanisia, Gynandropsis, Urena, Sida, Melochia, Riedleya, Corchorus, Triumfetta, Æschynomene, Smithia, Indigofera, Dolichos, Ammannia, Cucurbitaceæ, Blumea, Vernonia cinerea, Exacum, Scrophulariaceæ, Leucas, Ocymum, Hedychium, Amomum, Gloriosa, Commelynaceæ*, Grasses, and *Cyperaceæ*.

† The *Melianthus Himalayanus*, described by Planchon, is a garden plant, introduced from the Cape of Good Hope into the Himalaya, and is not distinct from the common Cape species.

been remarked. With our present knowledge, this affinity is chiefly indicated by the occurrence of Indian natural orders or genera, such as *Stephania, Grewia, Hippocratea, Impatiens, Brucea, Zizyphus, Anogeissus, Blumea, Jasminum, Torenia;* and by the prevalence of those tribes of the larger or cosmopolitan families which are especially Indian. This is the case with *Malvaceæ, Euphorbiaceæ, Terebinthaceæ, Leguminosæ, Rubiaceæ, Asclepiadeæ, Acanthaceæ, Amaranthaceæ,* Figs, and *Orchideæ.* Few cases of specific identity are known to us, but we confidently believe that many will be found to exist. The occurrence of *Delphinium dasycaulon* of Abyssinia in the mountains of the Dekhan is one instance ; and we have little doubt, notwithstanding that M. Ach. Richard attempts to distinguish it, that *Pterolobium lacerans* is identical with the Indian species. The Indian plants, *Sponia velutina* and *Antidesma paniculata,* are also African ; and the *Celtis eriocarpa* of Decaisne appears identical with *C. vesiculosa,* Hochst., from Abyssinia. Lastly, the absence of Oaks and Pines in both countries is a very strong point of resemblance.

There are further examples of American genera, and even species, being found in India, but so few and scattered, comparatively, as to render it unadvisable to complicate our arrangement by the introduction of an American type. As conspicuous examples, it will be sufficient to indicate *Adenocaulon* and *Oxybaphus,* of which genera the Indian species were first described by Edgeworth; *Podophyllum,* the section *Stylopodium* of *Meconopsis,* and *Liquidambar. Gnetum* also is a South American genus, which has not hitherto been found in Africa; and *Lardizabala* is interesting as a Chilian genus of a small order, the rest of which is entirely East Asiatic. *Monotropa uniflora* and *Brasenia* are common to North America and India; and the curious little *Mitreola paniculata,* Wall., is remarkable as being a native of India and Brazil, and, so far as is known, of no intermediate country*.

---

* The West African and East tropical American coasts afford curious examples of a similar relationship in the identity of species of *Schmidelia,* and in the

We cannot dismiss this branch of the subject without alluding to a few anomalies in the distribution of Indian plants. Of these, the most remarkable are the prevalence of Oaks and Chesnuts throughout the Himalaya, Khasia, and Malayan Peninsula, descending to the level of the sea in East Bengal, Malaya, Sumatra, Java, and Borneo, contrasted with their total absence throughout the Peninsula of Hindostan and Ceylon. Secondly, the prevalence of *Coniferæ* (along with these Oaks), not only inhabiting high levels, but descending considerably below 4000 feet : of these, *Pinus*, *Podocarpus*, *Taxus*, and *Dacrydium*, are all found in the Malay Peninsula and Khasia, but not one in the Hindostan Peninsula or Ceylon, though these present far more extensive and loftier mountain-ranges. Thirdly, we would call attention to the absence of *Cycadeæ* in Ceylon, and to the comparative rarity of Palms and epiphytic *Vacciniaceæ* in that island and in the Peninsula of Hindostan.

D. *Enumeration and description of the Provinces of India, as they will be referred to in the ' Flora Indica*.'*

The primary divisions of Continental India are four :—
1. *Hindostan,* in the widest sense of that term, including the

representation of several curious peculiar genera. The Atlantic Islands and North America show an equally striking instance, in a representative species of the otherwise American genus *Clethra,* inhabiting Madeira ; North America and Western Europe present others in *Eriocaulon septangulare, Trichomanes brevisetum,* etc. China and Japan present similar analogies with the west coast of North America. The most curious instance of all is, however, the occurrence in New Zealand of Chilian species of *Edwardsia* and *Haloragis,* and of representatives of *Fuchsia, Calceolaria,* and other genera, which are found nowhere else throughout the Old World.

* The sources from which the published facts contained in the following pages are derived are too numerous and too well known to make it desirable to quote them. For many details regarding those districts which we have not ourselves seen, we have to thank Dr. Wallich, Dr. Wight, Dr. Gibson, Dr. Stocks, and Captain R. Strachey. The last-named gentleman has also very kindly allowed us to make use of tables of mean temperature and rain-fall, collected with great labour for his work on the Physical Geography of the Himalaya, now in the press.

whole Western (Madras) Peninsula, and the Gangetic plain
to the base of the Himalaya. 2. The *Himalaya,* a mountain chain which rises abruptly from the Gangetic plain, and
is connected with a still loftier mountain mass (of Tibet) to
the north, and beyond India. 3. *Eastern India* (India ultra
Gangem), including the kingdom of Ava and the Eastern or
Malayan Peninsula. 4. *Afghanistan.*

The direction of the great mountain barrier of India on the
north is not parallel to the Equator, the western extremity
being the most northern. Its height is immense, being nowhere below 15,000 feet, usually exceeding 17,000–18,000,
and rising in isolated peaks, or groups of peaks, to from
20,000–28,000. The Afghan mountains form a meridional
chain from the western extremity of the above, descending
parallel to the Indus, with a gradually decreasing elevation,
from above 15,000 feet, to the level of the sea, at the Arabian
Gulf. The Ava and Malayan mountains form a chain parallel
to these, which is given off from the snow-clad mountains of
East Tibet, and, though rapidly diminishing in elevation, is
continued uninterruptedly almost to the Equator.

In Europe, *Hindostan* is generally understood to comprise
the whole continent of India, from the base of the Himalaya
to Cape Comorin; but in India the term is frequently restricted to the provinces north of the Nerbada, whilst all
those to the southward of that river are called the Dekhan, or
southern provinces. In this work, however, we shall give to
the term Hindostan its most extended sense, and restrict that
of Dekhan to the elevated country north of Mysore.

A complicated system of mountain-chains gives to Hindostan its peculiar configuration; these, which may be traced by
following on a map the courses of the rivers of which they
form the watersheds, are three in number, and bear no obvious relation to one another. They are,—1. The Peninsular chain (also called Ghats and Western Ghats) extending
from Cape Comorin to the Tapti river. 2. The Vindhia
chain, which crosses the centre of Hindostan from the Gulf

of Cambay to the Ganges. 3. The Arawali mountains, extending from Hansi and Delhi to Gujerat.

1. The *Peninsular chain* is the most important of these; it forms a continuous watershed, throughout its length of upwards of nine hundred miles, scarcely deviating from a straight line, which is parallel and close to the west coast of the Peninsula, and perpendicular to the direction of the monsoons. This chain divides the Peninsula unequally into two portions, marked by different climates,—a narrow western one, including the provinces of Malabar and the Concan; and a broad eastern one, traversed consequently by all the great rivers, and including the Carnatic, Mysore, and the Dekhan. Khandesh lies to the north of the chain, and includes that portion which sinks into the Tapti valley, together with the southern (opposite) slope of the Satpura branch of the Vindhia to the north of that river.

2. The *Vindhia chain*, from the little that is known of its structure, appears to consist of two parallel ranges, connected towards their centres, where the table-land of Umarkantak is said to attain an elevation of 4500 feet; elsewhere they are separated by the great rivers Son and Nerbada, which rise together and flow in opposite directions. The more southern of these ranges is probably always the higher of the two, but it appears seldom to exceed 3000 feet. The Vindhia mountains separate the Ganges and its tributaries from those rivers (the Mahanuddy, etc.) which flow south-east to the Bay of Bengal, as also from the Tapti and Nerbada, which flow west to the Arabian Sea. To the south of the range are the provinces of Khandesh, Berar, and Orissa; and to the east and north is the Gangetic valley, extending to the base of the Himalaya, and forming one great botanical province.

3. The *Arawali chain* is the least elevated of the three: it divides the tributaries of the Indus from those of the Ganges, and may hence be regarded as a continuation of the Cis-Satlej chain of the Himalaya, which terminates, to all appearance, in the plains near Nahan in Sirmur. In like manner, the Penin-

sula of Katiwar may be considered as the southern termination of the Arawali, though separated from it by an alluvial plain, being the continuation of the watershed, and dividing the streams flowing to the Gulf of Kach (or the delta of the Indus) from those that flow into the Gulf of Cambay.

We shall now proceed to give a rapid sketch of the physical features of the provinces of Hindostan, commencing with the southernmost. These are—

1. Ceylon.	7. Khandesh.	13. Gujerat.
2. Malabar.	8. Berar.	14. Sind.
3. Concan.	9. Orissa.	15. Rajwara.
4. Carnatic.	10. Bahar.	16. Panjab.
5. Mysore.	11. Bandelkhand.	17. Upper Gangetic plain.
6. Dekhan.	12. Malwah.	18. Bengal.

1. CEYLON.

This island extends from 6° almost to 10° N. lat., and is about 200 miles long, and 150 in greatest width. It is encircled by a belt of level land, which forms extensive plains at the northern extremity; and is traversed by a meridional chain of mountains. These mountains form a narrow range towards the north, seldom exceeding 1000 feet in elevation, and sink into the plain eighty miles from that extremity; to the southward they spread out, attain nearly 9000 feet of elevation, and extend eastward from Adam's Peak to Maha Ellia (or Horton plains) and Newera Ellia. The main ridge retains, perhaps, 6000–7000 feet of mean elevation for thirty miles, and expands into elevated plains of considerable extent, from which the loftier peaks rise. To the south and east, this transverse ridge dips abruptly into a low but hilly forest-clad country, but to the north it gives off a number of meridional ranges of considerable height; these separate tributaries of the Mahawali river which flow in elevated mountain valleys.

The great extent and elevation of the high land in Southern Ceylon powerfully influences the climate of the whole island. During the south-west (or summer) monsoon the north and east parts receive but little rain, which is all deposited on the intervening heights; the belt of low land in the south is, on the contrary, abundantly moist at the same season. During the north-east (or winter) monsoon, the rain-fall on the mountains, though considerable, is less than during summer, this wind being cooler and having less capacity for moisture; but showers occur at this season throughout the northern parts of the island. During winter, heavy rain falls along the southern coast.

The difference in climate presented by the various parts of Ceylon is hence very great. In the mountainous districts, where every wind is a moisture-laden sea-wind, it is temperate, equable, and humid throughout the year. The southern parts experience the moist tropical heats of an almost equatorial climate, and this at a season when the north coasts are scorched with dry heat. The mean temperature of Trincomali hence rises to $81\frac{1}{2}°$; and its climate is so dry, that when Mr. Gardner visited it, he found there had been no rain for nine months,—both anomalous conditions, when the proximity of the ocean is considered. Kandy, again, in the centre of the island, which is only 1800 feet above the sea, and is situated in a mountain valley, has a mean temperature of about 73°, and that of Newera Ellia, elevated 7000 feet, is probably about 60°.

The coast of Ceylon is generally fringed with a belt of Cocoa-nuts, which vegetate luxuriantly in the sandy soil of the sea-shore. In the estuaries, mangroves (*Rhizophora*) inhabit the muddy swamps, accompanied with *Heritiera, Sonneratia, Lumnitzera, Avicennia,* and *Scævola,* but none of the *Phœnix paludosa* and *Nipa fruticans,* so characteristic of the Sunderbunds.

In the drier flat parts of the island, extensive sandy plains covered with short grass alternate with undulating downs, either bare or clothed with dense thickets of thorny shrubs. The plants of these parts are generally those of the Carnatic, the climate being the same.

A dense forest clothes all the humid southern and western parts of the island, composed of plants eminently character-istic of Malabar. The vegetation of the upper and lofty dis-tricts is more mixed with temperate forms, and is extremely luxuriant, containing many, and indeed composed almost ex-clusively, of the species of the great Peninsular chain. Be-sides the mountain-slopes being covered with dense forests, there are open and undulating lofty table-lands which appear, like those of the Nilghiri and Khasia, to be clothed with large clumps of shrubs, swards of grass, and a rich herbaceous ve-getation, the large trees being confined to the ravines. In these places, *Ternstrœmiaceæ, Rhododendron arboreum, Vac-cinia, Gaultheria, Symploci, Michelia, Goughia,* and *Gomphan-dra,* seem as frequent as they are on analogous elevations of the continental ranges.

Though the Flora of Ceylon (which probably does not con-tain 3000 phænogamic plants) is on the whole identical with that of the peninsula, it presents a considerable number of endemic species, and a few genera, especially tropical ones, which are not found in the peninsula. *Dilleniaceæ*, *Anonaceæ, Garciniaceæ, Balsamineæ,* are all abundant in Ceylon. Its most remarkable deficiencies are *Scitamineæ,* Oaks, Willow, *Nipa, Gnetum, Pinus, Podocarpus, Cycas.* It presents also but few Palms : amongst these the most conspicuous are Cocoa-nut (cultivated only), *Corypha umbraculifera, Borassus flabel-liformis, Phœnix farinifera, Caryota urens,* an *Arenga, Areca,* and several *Calami.* This is a remarkably small number, when the Flora is contrasted with the Malayan*.

The Cingalese Flora has been investigated by a succession of industrious botanists, but no attempt at an enumeration of

---

* The adaptation of the soil and climate of the lowest and hottest parts of Ceylon to the ripening of grapes, is a most remarkable fact connected with the cultivation of the vine. Mr. Edgar Layard (whose zoological researches in Ceylon are so well known and appreciated) informs us that at Jaffna, at the northern extreme, the grape is grown successfully. The cold weather or north-east monsoon sets in there early in November, and the " sweet water" fruits in May and in October, and the "black cluster" in September; after fruiting,

its plants has been made since the publication of Moon's inefficient catalogue. Owing to the extent and impenetrability of the forests, some novelties must still remain; and many of the species, being large timber-trees and dioecious plants, varying abundantly, require skilful analysis and observation in the country. We have already mentioned Burmann's and Linnæus's labours. Moon was the first English collector, and curator of the Government Botanical Gardens at Peradenia, near Kandy. His collections (according to Gardner, Lond. Journ. Bot. iv. 397) were extensive and good, and formed the foundation of the Peradenia Herbarium, which is now rapidly acquiring a European fame, through the successive exertions of Gardner and Thwaites, Moon's successors in charge of the garden; and of Major Champion, who resided several years in the island. Moon's plants were never distributed; but other and most extensive collections have been, of which the following are the most important:—1. Macrae's, a collector in the service of the Horticultural Society of London.—2. Colonel and Mrs. Walker's: these were both extensive and excellent, and were illustrated by many drawings and manuscripts.—3. Major Champion's, alluded to at p. 69.—4. Mr. Gardner's; abundant and good: these were in part distributed, in part sold after his decease, while a part remain in the Peradenia Herbarium. Gardner has published several papers on Ceylon plants in the ' Journal of Botany,' and in the ' Calcutta Journal of Natural History;' sometimes in conjunction with Major Champion.

Mr. Thwaites, the present able superintendent of the Peradenia Botanic Gardens, has for several years continued energetically the investigation of the flora of the island which was commenced by Mr. Gardner; bringing his great botanical acquirements, skill in analysis, and powers of observing and

an artificial winter is produced by exposing the roots, and bullocks' blood is used as manure. According to the same authority, the grape also bears well at Tangalle, at the southern extremity of Ceylon, a locality which must have a very different climate from Jaffna.

collecting when travelling, to bear upon the rich materials collected by his predecessors and himself. His exertions have already given him a prominent position amongst Indian botanists; and from his continued labours we hope to see the Cingalese Flora fully illustrated in an economical and botanical point of view.

2. MALABAR.

We shall employ this term in its widest signification, and as usually applied by older geographers, to designate the whole of the narrow belt of country (rarely above fifty miles broad) west of the great Peninsular chain, from Goa to Cape Comorin: it thus includes the British district of Malabar, besides Canara and Kúrg to the north of it, and the kingdoms of Cochin and Travancor to the south. The eastern political boundaries of these districts correspond nearly, but not uniformly, with the crest of the mountains; and though some parts of the latter are included politically in the provinces of Mysore and the Carnatic, we shall consider them all as one province botanically.

Malabar is in general hilly and mountainous; a narrow strip of low land borders the sea, frequently intersected by long sinuous salt-water creeks, and covered with Cocoa-nuts; the hills which are thrown off as spurs from the main axis often reach the sea and dip suddenly into it: they enclose well cultivated valleys, and, though generally low to the west, they rapidly rise to the east, where they join the chain.

The climate of Malabar is characterized by extreme humidity, and an abundant rain-fall during the south-west monsoon, when the temperature seldom rises above 75° (the mean of the year being 81°). In many parts the rains commence as early as the middle of March, but rarely become heavy till May, continuing thenceforward incessant till October, and depositing more than one hundred inches on the coast. In the extreme south the rain-fall is less considerable; at Quilon 77 inches, and at Trivandram 65 inches, probably from the narrowing of the land and the lower elevation of the mountains. The humidity, however, continues excessive. At Cape Comorin the amount of rain is only 30 inches. To the northward, in Canara, the climate is drier, especially in winter, and the hills are less elevated. During the north-east monsoon, from January to April, which includes the hottest season of the year throughout the province, irregular winds and showers prevail everywhere, except opposite Coimbator, where, from the great depression in the mountains, dry winds are at that season not unfrequent.

From the humid character of the Malabar climate, its luxuriant vegetation might be inferred. Hamilton tells us that it resembles Bengal in verdure, but has loftier trees and more Palms : the shores are skirted with Cocoa-nuts, and the villages surrounded with groves of Betel-nut Palms and Talipots. *Vateria Indica*, a noble Dipterocarpous tree, is abundantly planted in many parts ; Cassia, Pepper, and Cardamoms flourish wild in the jungles, and form staple products for export. The fact that the Pepper is cultivated without the screens used in other parts of India, to preserve a humid atmosphere about it, is the best proof of the dampness and equability of the climate. The low valleys are richly clothed with rice-fields, and the hill-sides with millets and other dry crops, whilst the gorges and slopes of the loftier mountains are covered with a dense and luxuriant forest.

The mass of the Flora is Malayan, and identical with that of Ceylon, and many of the species are further common to the Khasia and the base of the Himalaya. Teak is found abundantly in the forests, but the Sandal-wood occurs only on the east and dry flanks of the chain. Oaks and *Coniferæ* are wholly unknown in Malabar, but the common Bengal Willow (*Salix tetrasperma*) grows on the hills. *Gnetum* and *Cycas* both occur, the former abundantly.

The mountain-chain which forms the eastern boundary of Malabar, separating it from Mysore and the Carnatic, has, except on the eastern slopes of the most lofty parts, a very humid climate, and is therefore most appropriately noticed here. It attains its greatest elevation to the southward, and is broken up, by considerable depressions, into two or more separate masses, of which the southernmost may be called the Travancor range, whilst to the northward it is continued as the Nilghiri, Kúrg, and Nagar mountains.

TRAVANCOR.—The mountains of Travancor form an isolated mass at the extreme south of Malabar, which they separate from the districts of Tinnevelly and Madura, in the Southern Carnatic.   They are completely cut off from the mountains on the north (Nilghiri) by a remarkable depression, in 11° N. lat., which is fifteen miles wide, and is occupied by the western portion of the district of Coimbator. The Travancor group of mountains thus presents a striking analogy to the island of Ceylon in position and outline.   The main chain runs southward for 150 miles to Cape Comorin, with occasional deep depressions, and terminates in a bold precipitous mass, 3–4000 feet high, within three miles of the Cape itself. The Travancor mountains are loftiest at the extreme north of the district, where they stretch east and west for sixty to seventy miles, separating the districts of Dindigal and Madura, and rising into peaks of 8–9000 feet, which overhang the plain of Coimbator; and they retain an elevation of 5–6000 feet throughout their extent to the southward. They are generally very precipitous, and undulating or rounded grassy ridges seem to be of common occurrence at 6–7000 feet.   Of the deep depressions that intersect the Travancor range, and by which communications are kept up between the districts which it divides, that of Courtalam, in 9° N. lat., is a well-known botanical station, which, though on the eastern or Carnatic side, from its peculiar form and situation, is under the influence of the south-west monsoon, and enjoys, together with the rest of the province, a deliciously cool and equable climate.   Notwithstanding the perennial humidity, the rainfall at Courtalam is only 40 inches; on the hills around, however, it is doubtless much greater.   The Pulney or Palnai mountains west of Dindigal, the Animalaya south of Coimbator, the Shevaghiri mountains south-west of Madura, and the ranges near Courtalam, are all well-known as the scenes of Dr. Wight's indefatigable labours, which have extended to Cape Comorin itself in this direction.

There are few botanical features of Travancor not common to both Ceylon and Malabar in general. Nutmegs, coffee, and cinnamon flourish at Courtalam. The remarkable Palm, *Bentinckia*, so common on its mountains, is however not known in Ceylon. The other Palms are *Caryota urens*, an *Areca*, *Phœnix farinifera*, and one or two species of *Calamus*.

NILGHIRI AND KÚRG MOUNTAINS.—To the north of the Coimbator valley, this part of the peninsular chain rises abruptly to 8000 feet elevation as the Nilghiri range, and is continued northward as the mountains of Kúrg at nearly the same elevation. Below 6000 feet they are steep and densely wooded; above that they form undulating grassy table-lands, with scattered bushes and copsewood, from which low sloping hills arise, of which Dodabetta, the loftiest of the range, attains 8429 feet.

To the west and south, the Nilghiri mountains are precipitous; to the east, long transverse ranges covered with dense forest are given off, enclosing the lofty valleys of Mysore.

The rain-fall, which is excessive to the westward, is much diminished before reaching the axis of the chain: at Dodabetta it is 100 inches; and at Utacamand only 64 inches. The seasons are uniform throughout the year, the cold never being extreme, though frosts do occur in clear winter nights. The following abstract (which we borrow from Gardner) will afford a few data as to the temperatures of certain positions and elevations:—

	Alt.	Mean temp.
Dinhetty	6166 feet	64·0
Kotaghery	6407 ,,	63·4
Utacamand	7197 ,,	61·0
Dodabetta	8429	56·0

The monsoon is so checked by the great elevation and breadth of this range, that its east flank partakes much of the climate of Mysore, many plants of that country ascending almost to the crest of the chain, which is therefore, as Gardner informs us, wholly unsuited to the growth of Coffee.

The ravines and shady slopes near the undulating summits of the Nilghiri hills are occupied by thickets of small trees and bushes, like those of Ceylon, but probably composed of a greater number of species, all of which are equally characteristic of similar situations in the Khasia, as *Ternstrœmiaceæ*, *Michelia*, *Symplocos*, *Photinia*, *Ilex*, *Eugenia*, *Vaccinium*, *Gaultheria*, *Myrsineæ*, *Rhododendron arboreum*, *Pittosporum*, *Laurineæ*, with *Rubus*, *Cotoneaster*, *Desmodium*, *Jasminum*, *Euonymus*, *Indigofera*, *Daphne*, *Euphorbiaceæ*, *Antidesmeæ*, Willow, *Melastomaceæ*, and a vast number of others. Of forms that do not extend to Ceylon, are Willow, *Gnetum*, *Viburnum*, *Lonicera*, *Rosa*. Balsams attain their maximum in the Nilghiri and Travancor mountains; and amongst European forms are *Alchemilla*, *Potentilla*, *Gentianeæ*, and *Labiatæ*. *Agrimonia*, however, which is found both in the temperate parts of India and in Ceylon, is absent from the Nilghiri.

NAGAR.—Of this district, which lies to the north of Kúrg, comparatively little is known; politically it belongs to Mysore, but its climate and vegetation appear to be identical with that of Malabar. For the most part it consists of rounded or table-topped hills, 4–5000 feet in mean elevation, often cultivated to that height, and rising in some places to upwards of 6000 feet, the portion called Bababuden Hills being said to be 5700 feet. As with all other parts of the chain, the climate of the western parts is excessively humid: the rains at the town of Nagar (or Bednor), elevated 4000 feet on a spur to the westward of the chain, are said to last for nine months, during six of which they are so heavy that the inhabitants cannot leave their houses. The eastern parts again are more level, and drier, and resemble other districts of Mysore.

North of Nagar, and near the sources of the Warda River (in 14° N. lat.), there is a marked break in the chain, which there seems hardly to rise above the level plain of Dharwar to the eastward. Here the watershed recedes further than usual from the west coast, and two considerable rivers flow in deep ravines from the immediate vicinity of Dharwar to the Western Ocean, separated by lateral spurs which run southwest from the axis of the chain.

Dr. Buchanan Hamilton was the first after Rheede to explore the botany of Malabar. Having been deputed to that province by the Government of Madras, charged with a multiplicity of duties, he does not seem to have collected largely, nor has he published any general work on the subject. Many important botanical observations of his are, however, detailed in various publications, and especially in his Commentaries on the 'Hortus Malabaricus,' which have in part only appeared in the Linnæan Transactions. To this task he brought an extensive knowledge of tropical botany and Oriental literature.

Dr. Wight's researches, in many parts of the province, are justly celebrated throughout Europe; he has personally explored the Travancor mountains as far south as Cape Comorin, the Courtalam and Pulney hills, the neighbourhood of Quilon, and especially the Nilghiri chain, which is easily accessible from Coimbator, where he so long resided as superintendent of the Government Cotton Plantations. Dr. Gardner, when on a visit to Dr. Wight, also collected in the Nilghiri chain, as did Sir Frederic Adam, and Mr. Schmid, a missionary, a few of whose plants have been published by Zeuker.

The northern district, or Canara, has been diligently explored by Mr. Dalzell, who resided for many years at Vingorla, in the Southern Concan, and made extensive journeys. A large collection of Canara and Kúrg plants was also made by Mr. Metz*, a missionary, and distributed in Germany by Hohenacker, and named by Miquel; these are partly from the

---

* The name of Mr. Metz should be substituted for that of Mr. Schmid at p. 69 of this Essay.

neighbourhood of Mangalore, and partly from the vicinity of Mercara in Kúrg.

The mountains of Kúrg were first explored by Captain Munro and Captain Gough, who seem to have sent many plants to Dr. Wight. Copious Herbaria were also made in various parts of the chain by our own collectors. The district of Nagar seems to have been visited by Hamilton only, on his return from Canara to Mysore : his notices of it are very scanty. Dr. Wight has further published a few plants of the Bababuden hills.

A careful comparison of much of the materials comprised in these different collections, from all parts of the chain, assures us that Malabar is comparatively well explored botanically, and that there are not many more phænogamic plants to reward the labours of future investigators.

3. CONCAN.

This district extends from Goa to Daman, or very nearly to the Tapti river. Like Malabar, which it greatly resembles in general aspect, it is comprised between the western ocean and the Ghats, and consists of a narrow belt near the sea with salt-water inlets, and a succession of mountain spurs. In the northern parts of the Bombay Presidency, the chain separating the Concan from the Dekhan is called the Northern Ghats, or Siadri mountains, a term which may conveniently be extended to their whole length, and which we shall thus apply when it is necessary to particularize them. Throughout the Concan they form a continuous chain of hills, interrupted, however, by deep depressions. Throughout their length, they seem seldom rugged, but to rise often into sharp or flat-topped peaks. To the east they slope gently into the plains of the Dekhan. The summits rise to the height of 4000–5000 feet, but the mean elevation is very much less. The station of Mahabaleshwar is 4700 feet. In the latitude of Daman 20½° N.), the chain begins to sink abruptly into the Tapti valley, and changes its course, or sends off a spur of considerable elevation in an easterly direction, as the Chandor hills.

This range of the Ghats is sufficiently lofty and abrupt to produce a heavy rain-fall during the south-west monsoon; between May and September this is in some parts immense, and only rivalled by that of Malabar and the Khasia hills in East Bengal. At Mahabaleshwar, it amounts to 248 inches annually. In the Southern Concan, especially in the Sawant Wari district, the rains are as heavy as in Canara. At Bombay, the rains last from June till the end of September, and the fall is only 80 inches, which is considerably less than at any point further south on the coast. At Tannah, however, the average fall is more than 100 inches. During the north-east monsoon, which blows from November till March, the climate is dry compared with that of Malabar, the change commencing rather suddenly where the mountains are lowest and most distant from the coast. At Bombay there are regular sea-breezes in the afternoon, so that the atmosphere never becomes extremely arid.

The change of climate, marked by diminished mean temperature, a lower winter temperature, and greater dryness, which accompanies the increased distance from the Equator, has a decided influence on the vegetation. The whole Concan is hence more open than Malabar, heavy forests are rarer, many tropical Malayan forms disappear, and the most moisture-loving types of vegetation linger only in the damp recesses of the mountains. A rich cultivation replaces the forest in the valleys especially, and the dense jungles are confined more or less to the lower slopes of the main chain. In the more open parts there is a remarkable mixture of African types; instead of the luxuriant *Acanthaceæ* of Southern India, there occur spiny-leaved species, similar to Abyssinian and Arabian ones. Curious *Umbelliferæ*, allied to no others in India, accompany these, as well as a great variety of forms typical of the north tropical African vegetation. The arid flora of the Dekhan, of Marwar and Sind, however, hardly enters the Concan.

The Flora of the Bombay Presidency has only lately been

s

diligently investigated, little having been known of it up to
the date of publication of Wight and Arnott's Prodromus.
The plants of Concan were first catalogued by Mr. Graham,
assisted by Mr. Nimmo; these botanists seem to have been
diligent workers, and were correspondents of Dr. Wight, to
whom they communicated valuable discoveries.

Dr. Gibson, the energetic Conservator of Bombay Forests,
has had, owing to the nature of his duties, ample opportunities of investigating the Flora of Bombay, and we are indebted to him for a considerable Herbarium. We have also
had the opportunity of examining the excellent collections of
Dr. Stocks, who officiated for Dr. Gibson during that gentleman's visit to Europe, and to whom we have been greatly
indebted for information and assistance.

It is, however, by Mr. Law and Mr. Dalzell, that the Concan Flora has been most ably and energetically investigated.
Mr. Law resided for many years at Tannah (near Bombay),
and explored the Northern Concan, whilst Mr. Dalzell chiefly
employed himself in the Southern Concan and adjacent province of Canara.

4. CARNATIC.

In the extreme south of the Peninsula, the Carnatic extends from the eastern sea to the borders of Malabar; but
further north, where the Peninsula is wider, it comprises only
the sea-coast, the province of Mysore being interposed between
it and the great peninsular chain. The northern part of the
Carnatic is a nearly level tract, of no great width, extending
from the mouth of the Godavery to the delta of the Cavery.
It is not a perfect level, as a few low ridges project at intervals
from the Ghats; and some isolated hills of trifling elevation
occur, scattered over the surface, evidently the remnant of
former continuous ranges, which have been apparently removed by aqueous action. None of these exceed a few hundred feet in height, and they exercise no material influence
on the climate or vegetation. Much of the country is sandy,

and scarcely arable, and the inhabitants are in general so dependent on the periodical rains for their crops, that any deficiency in the rain-fall is followed by a bad harvest.

Throughout the northern Carnatic, the rain-fall during the south-west monsoon is trifling in amount; and as the sun's action is not mitigated by a cloudy sky during the hottest period of the year, as is the case in Mysore, the temperature from March till November is extremely high. In the middle of October or the beginning of November the north-east monsoon sets in, and with it a more or less abundant rain-fall. In the end of December the rains cease, from the gradual change in the direction of the wind, which makes it less directly a sea-breeze than in the earlier part of the winter season. The mean temperature of Madras is 82°, and the rain-fall does not exceed 45 inches.

In the southern Carnatic, the district of Salem, between the rivers Penar and Cavery, which is considerably more elevated than the rest of the province, may be considered a prolongation of the most elevated part of the central platform of Mysore. The table-land of Mysore dips abruptly into the plain of Salem, which has an elevation of about 1100 feet above the sea, and contains several detached masses of hills scattered over it, all rising to very considerable elevations. Of these, the most lofty are the Shíwari hills, which rise a few miles north-east of the town of Salem, in a range of densely wooded flat-topped hills. The mean height of the table-land on their summits is about 4600 feet, but the highest peak rises to 5260 feet. The Salem district, from its position opposite the Coimbator gap, and from the influence of the considerable masses of high land just mentioned, is rather more rainy than the northern Carnatic. The south-west monsoon sets in early in June, and short but heavy and frequent showers continue till September. Towards the end of October, the north-east monsoon brings a return of showery weather, with a cloudy sky. This continues till the middle of December, when the rains cease in consequence of the gradual change of the direction of the wind from north-east to due north.

The district of Coimbator has, like that of Salem, so many peculiar features, as to call for a special notice. It lies opposite the great gap in the Peninsular chain already so often referred to, and is conterminous with Malabar. Between the southern slopes of the Nilghiri mountains, and the northern face of those of Travancor, there is interposed a space of about thirty miles in width traversed by low hills. Across that depression, the south-west monsoon has almost a free passage to the eastward; but the great elevation of the mountains on both sides, and the absence of any considerable hills in the district, cause the monsoon wind to pass over without depositing much of its moisture, and, though the climate is humid, the rain-fall is very trifling. During the north-east monsoon again, the high hills of eastern Mysore and those of the Salem district intercept a considerable portion of the moisture which would otherwise reach this district. Coimbator is thus remarkable for the very small annual amount of rain, which is not more than twenty-one inches.

The district of Tanjor, which comprises the delta of the river Cavery, appears to present no remarkable features beyond those common to all tropical deltas. Its climate is more humid and cool than the remainder of the Carnatic, chiefly owing to the swampy soil.

The extreme southern portion of the Carnatic, including the districts of Madura and Tinnevelly, is separated from the remainder by a lofty transverse range of mountains, which runs from west to east, passing to the south of Dindigal. These mountains, which at their eastern extremity, where they are called Pulney (Palnai) mountains, are 6000–8000 feet in height, gradually diminish in elevation to the eastward. About five miles south of Dindigal the Serroo Mullay (Serú Malaya) hills, rise to 3500 feet, and the range sinks, about twenty miles to the eastward of Dindigal, into the plain of the Carnatic. This range of hills insulates in a very remarkable manner the districts to the south of it, which are sheltered from the south-west monsoon by the high mountains of Travancor on the west, and from the north-east monsoon by this range to the north, and by the island of Ceylon to the east. We have, therefore, in the southernmost part of India, in a latitude between 8° and 10° N., a hot, arid climate, resembling that of Egypt, like which it produces the best quality of senna and cotton, and many wild plants characteristic of the Egyptian Flora, which avoid humidity, and are not known elsewhere in the Peninsula. Of this, two remarkable instances are *Cocculus Leæba,* and *Capparis aphylla.*

As a whole, the vegetation of the Carnatic is neither rich nor varied. The climate being very arid except during the northeast monsoon, the humid flora is entirely absent. There is no forest, except on the flanks of the higher mountains, which bound the province on the west, or rise from its plains ; and there the vegetation resembles that of the drier parts of Ceylon or of the Mysore hills. The shrubby flora of the open plains consists chiefly of *Capparideæ, Rhamnaceæ, Acaciæ,* and spinous *Rubiaceæ, Alangium, Azima, Carissa* and *Calotropis gigantea, Ehretia buxifolia, Gmelina, Salvadora, Antidesma, Pisonia,* and such like shrubby plants. The only Palms are a *Calamus* and *Phœnix,* besides the commonly cultivated *Cocos, Borassus* (which characterizes dry countries), and *Areca.* Along with these, grow many shrubs which are spread over the whole of the drier parts of India, as far as the Himalaya. Many of the annual plants have an equally wide range, especially those of the rains, which are scarcely different from those of the Gangetic valley. As there is no winter, there are no northern types found in any part of the Carnatic.

The vegetation of the hilly parts of the Carnatic has yielded no peculiarities. Most of the hills are of too trifling elevation to exhibit any marked difference of mean temperature; and even the Salem range, from the isolated position of its masses, appears to present fewer peculiar features than more continuous mountain masses of even less elevation. The flanks are covered with dense bamboo jungle, and the summit is bare and grassy, except in ravines and along the streams. A detailed account of the flora of their summits is, however, a desideratum.

The vegetation of the plain of the Carnatic has been investigated by so many persons, that it is now thoroughly well known. The earliest peninsular botanists were the Danish missionaries, who originally settled at Tranquebar ; and most of the collectors who have visited the peninsula have traversed the Carnatic *en route* to the interior. It is therefore unnecessary to enumerate the names of all those who have botanized there.

5. MYSORE.

The province of Mysore is bounded on the north by the Dekhan, on the west by the mountain axis of the peninsula, and on the east and south by the low country of the Carnatic. It is usually described as a table-land enclosed between the western and eastern Ghats ; a form of expression which has doubtless originated in the fact that a considerable rise is made in entering the province from either side.

The Western Ghats, as we have already fully explained, form a chain extending in a direction parallel to the western ocean ; and Mysore, which occupies the eastern and more gentle slope of these mountains, contains the upper part of the basins of the Cavery, Penar, and Tungrabudra rivers, all of which discharge their waters into the Bay of Bengal.

Through the centre of this elevated tract, nearly in the parallel of Mangalor and Madras, is situated the watershed which separates the first of these rivers from the two latter. This is not an elevated ridge, but a rounded and often scarcely perceptible swelling, usually undulating very gently, but rising at intervals into rugged masses often more than a thousand feet above its mean elevation. The highest summits in Mysore (except in the district of Nagar) are situated on this line, and are north-east and north of Bangalor, where several peaks rise to 4000 feet, and one to 4500 feet. To the north of this range the elevation is less considerable, but the appearance of the country is the same. The rivers which flow towards the Kistna are separated by spurs of a high table-land, rarely rising into hills, so that the country appears nearly flat, except to the eastward, where it dips suddenly into the plain of the Carnatic. The elevation of Bellary is 1600 feet; Karnúl is about 1000 feet; and Cadapah, in the gorge of the Penar, where it issues from among the mountains, is only 500 feet above the level of the sea.

Another spur from the great peninsular chain forms the southern boundary of the province, separating the district of Coimbator and the basin of the Bhowani river from the upper basin of the Cavery. This range, which attains generally an elevation of nearly 4000 feet, extends in an easterly direction from the eastern slopes of the Nilghiri.

Between these two watersheds, the table-land of Mysore forms a gently undulating plain, sloping downwards, from 4000 feet at the base of the mountains, to 3000 at Bangalor, and 2400 at Seringapatam on the banks of the Cavery.

The highlands of Mysore sink everywhere abruptly into the plain of the Carnatic, except where the great rivers debouche; and the extremities of the broad flat-topped ranges which form the table-land, when viewed from a little distance, present the appearance of a continuous range of hills parallel to the coast-line, commonly known as the Eastern Ghats.

The districts of Bellary, Karnúl, and Cadapah, which occupy the northern slope of the central range of Mysore, and the higher parts of the basin of the Tungabudra and the Penar, are usually excluded from Mysore, being known as the Ceded Districts, because they were transferred from the kingdom of Mysore to the Nizam after the war in 1800, and afterwards made over to the British Government in lieu of a money-payment. As they present no physical or botanical features which would make it desirable to consider them as a separate province, we shall include them under the general name of Mysore, of which the Kistna will therefore form the northern boundary.

The climate of Mysore is much drier than that of Malabar, because the greater part of the south-west monsoon is intercepted by the lofty ranges of the Nilghiri and of Kúrg. The summer heat is however very moderate, partly on account of the elevation of the table-land, and partly because the proximity of the high central chain, which is very much cooled, produces a great amount of cloudy weather throughout the summer months. In winter the north-east monsoon is little felt in the interior, the greater part of the discharge from it being on the coast and on the line of Ghats at the border of the table-land. The winter temperature is therefore not much less than that of summer, so that the climate is very equable. The mean temperature of Bangalor is 74°, and the rain-fall 35 inches; at Bellary the rain amounts to only 22 inches. To the northward, the north-east monsoon is very little felt in the western districts, but at Cadapah there is generally heavy rain in November, and the remainder of the winter is dry. This place is so low, and so far from the mountain axis and the west coast, that the south-west monsoon is scarcely felt, even by the formation of clouds, though strong westerly winds prevail at that season. Cadapah is hence one of the hottest and most unhealthy parts of the Madras Presidency.

The vegetation of Mysore, like that of the Carnatic, is rather scanty. The level surface of the table-land is frequently very barren, and the hills are often bare or covered with low scrubby jungle. In the western part of the province, the eastern slopes of the central chain are clothed with dense forest, and the humidity is there very considerable, and the vegetation in consequence more varied, but approaching closely to that of Malabar.

The steep slopes of the eastern Ghats, which are powerfully affected by the north-east monsoon, are also in general densely wooded. Characteristic trees and shrubs are *Isora, Cedrelaceæ* and *Meliaceæ, Erythroxylon, Dipterocarpus, Myrtaceæ, Acacia Lebbek, Cassia Fistula, Pterocarpus, Butea frondosa,*

Lagerstrœmia parviflora, Terminalia, Conocarpus, Nauclea cordifolia, Diospyros, Teak, *Santalum album, Alnus integrifolia, Trophis aspera, Bambusa,* etc. etc.

The absence of winter, and the great heat of the dry season from December to June, give a predominance to arid types, especially to those which have been already indicated as intolerant of cold. Few palms are indigenous, except in the dense western forest. *Phœnix sylvestris,* however, occurs, and *Areca Catechu, Cocos,* and *Borassus* are cultivated extensively. During the more humid summer season a number of Balsams spring up ; a genus unknown at that season in the hotter and drier Carnatic.

Our earliest knowledge of the plants of Mysore is due to the indefatigable Buchanan Hamilton, in whose travels many details regarding the aspect of its vegetation will be found. It has since been partially investigated by many botanists, in particular by Heyne and by Wight, but a detailed list of its plants is still a desideratum.

6. DEKHAN.

The Dekhan embraces the whole of the country between the Kistna and the Godavery, except a very narrow belt along the Bay of Bengal, which is included in the Carnatic. To the west it is separated from the ocean by a narrow strip of land, the Concan, the crest of the mountain axis forming the (physical) boundary between the two provinces. To the north, a low range separates it from Khandesh, and further east the Godavery forms an artificial boundary between it and Berar.

The mountain-chain which forms the axis of the peninsula is considerably lower in its northern half than further south. North of Nagar, it appears to dip rather abruptly, so that between Goa and Belgaum it is very much depressed, and presents scarcely any perceptible elevation above the surface of the table-land, which is there 2500 feet. Further north, the elevation of the table-land gradually diminishes, notwithstanding the increasing width of the continent. At

t

Púnah it is 1800 feet, and the peaks of the chain attain an elevation of 4–5000 feet, but they are separated by great depressions. The table-land of the Dekhan forms the watershed between the basins of the Kistna and the Godavery, and has an average elevation of from 1800 to 2000 feet, with an undulating surface, but no mountains rising much above the mean level. Hyderabad is 1672 feet, the Cantonment of Secanderabad, close by, 1837 feet, and a hill near, 2017 feet above the level of the sea. The valley of the Godavery is of course considerably lower. The surface of the table-land, which is generally open, with little forest, but much low jungle, is at one season richly cultivated, but during the remainder of the year extremely arid and burnt up.

The abrupt escarpment of the western Ghats condenses so much of the moisture of the south-west monsoon, that the summer rains in the Dekhan are very moderate in amount, and the north-east monsoon is so much a land-wind, that but little rain falls during the cold season. The rain-fall at Hyderabad and Jaulnah averages about 43 inches; at Satara (2300 feet) it is 36 inches. The mean temperature of Púnah is 77°, and the rain-fall 24 inches. This is the average rain-fall throughout the north-western part of the district, close to the crest of the Ghats, but the amount is greater to the eastward.

Along the crest of the Ghats, the hilly tract known as the Máwal possesses a very different climate and aspect from the remainder of the Dekhan, bearing more resemblance to the Concan. This is due to its greater humidity; the depressions of the watershed, here as elsewhere, allowing the moisture-laden wind to pass to the eastern part of the chain for a considerable distance. In this district the surface is perennially green, and the vegetation much more luxuriant than further east. In the western parts of the district of Belgaum this tract is especially marked, as the depression of the mountains is there greater than elsewhere. At Belgaum the rain-fall is 50 inches, and at Dharwar 39 inches. These numbers, however, afford only a very faint indication of the degree of humidity.

The vegetation of the plain of the Dekhan is not very different from that of Mysore. The flora is not extensive, the great drought of the hot season being unfavourable to vegetation. The earliest collection of its plants was made by Colonel Sykes, and is now in the possession of the Linnean Society. In Graham's Catalogue there is an enumeration of all the plants known to him, and its flora has recently been explored by Dr. Gibson and Dr. Stocks. The green hilly tract bordering upon the Concan, being more elevated, as well as more humid than the remainder of the district, presents a peculiar vegetation. Some of its plants are apparently confined within very narrow limits, and are not known elsewhere in India.

7. KHANDESH.

This province occupies the lower part of the valley of the Tapti river, and is enclosed on the north by the Satpura range, a branch of the Vindhia, which has an elevation never exceeding 2500 feet, and is often much lower. To the south, the Ajanta range, separating Khandesh from the basin of the Godavery and the district of Aurangabad in the Dekhan, is even less elevated, rarely attaining so great an elevation as 1800 feet. To the east this province is separated by no very definite boundary from the Ellichpur district of Berar.

The valley of Khandesh is, in general, a level plain, rising gently towards the mountains on both sides. Occasional flat-topped hills are scattered over the surface, and the slopes of the Ajanta and Satpura ranges are covered with dense jungle.

The rainy season, in Khandesh, is the south-west monsoon, commencing in June. The rains are heavy and long-continued, but we have not been able to ascertain their exact amount, nor have we any definite knowledge of the flora of the province.

8. BERAR.

The province of Berar includes the districts of Ellichpur

and Nagpur, the former occupying the upper part of the basin of the Tapti, and that of its tributary the Púrna, the latter situated on a tributary of the Godavery, and therefore separated by no well-defined boundary from the north-eastern part of the Dekhan.

To the north, Berar is separated from the valley of the Nerbada by the continuation of the Satpura range, gradually increasing in height to the eastward, and attaining an elevation of 3000 feet, south of Hosungabad. The Rev. Mr. Clarke states that Chouragadh, the highest peak of the Mahadeva hills, north of Nagpur, rises to 4200 feet. The Ajanta range, on the contrary, is very inconspicuous to the eastward, as the plain on both sides slopes up to its crest; but the Gawilgarh hills, which separate the Púrna and Tapti rivers, rise in peaks to a height of 3000 feet. The eastern boundary of Berar corresponds pretty closely with the watershed of the Mahanadi river, the elevation of which is unknown. Berar is, in general, level, but the plains are separated by low ranges of naked table-topped hills, most numerous in the northern portion. Nagpur is 900 feet above the level of the sea, and Ellichpur may be conjectured to be very little more.

The rains in Berar are of short duration, but more considerable in amount than in the Western Dekhan. At Nagpur, the fall is 40 or 50 inches between June and October. The remainder of the year is dry and intensely hot, the mean temperature of Nagpur being $81\frac{1}{2}°$. The vegetation is probably identical with that of the Dekhan, but the province is botanically unknown.

9. Orissa.

Under this name we include the whole basin of the Mahanadi river. On the north, this province is bounded by the crest of the Vindhia, on the north-east by a spur descending thence towards the sea near Balasor, on the south-east by the sea, on the west by the watershed separating the Mahanadi from the tributaries of the Godavery, and on the south-west by that river from Chandah to the sea.

The physical structure of Orissa is very imperfectly known. It is in general hilly, and the ranges have probably pretty uniformly a maximum elevation of 3000 feet. They are often table-topped ridges, separated by flat broad valleys, but perhaps most frequently spreading out into elevated platforms. The table-land of Sirgujah and Chota Nagpur, which forms the northern part of the province, is an expansion of the southern branch of the Vindhia, here forming the watershed between the Mahanadi and the Sôn. It is said to have a mean height of 3000 feet, and to be covered with dense forest. The ridge which separates it from Berar presents probably, in like manner, an extensive platform of moderate elevation.

Throughout Orissa, the hills approach within a distance of the sea which varies from twenty to fifty miles, and at Vizigapatam and Ganjam they advance close to the shore. These hills (like the Ghats of Mysore further south) terminate very abruptly, and are separated from the sea by an alluvial belt, which skirts their base and advances between the different spurs, so as to form richly-cultivated valleys among the hills. The Ghats generally rise abruptly to an elevation of 1500 or 2000 feet. Their flanks are covered with dense forest, as well as the flat tops of the outer and more humid portions of the spur, but in the interior these spread out into bare table-topped ridges.

The Mahanadi being the principal river of Orissa, its valley is the lowest part of the province. It is navigable for large boats as far as Boad, a hundred miles above Kattak. It is then hemmed in for some distance by mountain-spurs, but higher up its valley expands into the level plain of Sambalpur.

The table-land of Umerkantak, in which the rivers Nerbada and Sôn take their rise, as well as one branch of the Mahanadi, is an elevated tract of dense jungle, traversed only by narrow paths, and quite removed from the great line of traffic across the continent. It is said to attain an elevation of 4500 feet; but the observations upon which this statement rests are of doubtful accuracy. Umerkantak was visited many years ago by Dr. Spilsbury, and it may be gathered from the narrative of his visit that the reports which ascribe to it an elevation of 7–8000 feet are greatly exaggerated.

The climate of Orissa is peculiar. Influenced by the hot plains of Northern Hindostan, the summer monsoon blows from the south or south-east, as in Bengal, instead of from the south-west, which is its direction in the Carnatic. It is therefore a sea-wind, and brings with it much humidity, which is deposited on the outermost hills. The coast and outer ranges are therefore extremely humid, but the valleys of the interior are much more dry. During the winter monsoon, the great heat of the dry plains of Nagpur and the Dekhan causes a sea-breeze to blow, during the day at least, all along the coast of Orissa. The hills are therefore, at this season also, damp and humid, though the rain-fall is not great in amount. In April and May there are occasional land-winds, before the heating of the great Gangetic plain changes the direction of the south-west monsoon. We possess no register of the rain-fall on the mountains of Orissa, where it would probably be found very large in amount. Along the coast the fall is much less considerable, being 50 inches at Kattak, and gradually diminishing to the southward. At Masulipatam it is only 34 inches.

The forests which cover the slopes of the outer ranges are very dense, and, though not equal in luxuriance or variety to those of Malabar and Malaya, they are richer in forms than those of Mysore, many Malabar plants not found in the Carnatic or on the Eastern Ghats recurring in these more northern jungles. Thus the wild Pepper is found there abundantly, with numerous *Zingiberaceæ* and Orchids, *Arenga saccharifera*, and perhaps *Caryota*, but apparently no other palm. Species of *Dillenia, Leea, Mimusops, Bassia, Roxburghia*, etc., also occur. The forests which cover the mountains of the interior are much drier, and are separated by open valleys, more or less under cultivation.

The botany of the coast of Orissa, and that of the forests of the Ghats, has been investigated by Roxburgh, who, during the earlier part of his Indian career, resided at Samalcotah in the northern Circars, by which name the district is usually referred to in the 'Flora Indica' of that distinguished botanist. Dr. Russel's collections were also chiefly from the same district. The vegetation of the interior of the province is quite unknown, except from a few notices in Major Kittoe's journey to the Sambalpur valley.

10. BAHAR.

The boundaries of the ancient province of Bahar have no doubt varied at different epochs, and in modern times the name is understood in a great variety of senses, being restricted at one time to a small judicial district, and at other times extended so as to include the whole of the western part of the lower provinces of the Bengal Presidency. Its employment in an arbitrary manner can therefore be productive of no inconvenience, so long as it is accurately defined. We shall therefore, in our present work, understand under the name of Bahar the whole of the northern slope of the eastern portion of the Vindhia mountains, from the borders of Bandelkhand (or rather Rewah) and Malwah to the Gangetic plain. In this way it is separated from Orissa by the watershed of the chain, and includes the districts of Palamow and Ramgarh, as well as the lower half of the valley of the Sôn.

The eastern portion of the Vindhia chain, as we have seen, is a spreading table-land, and the spurs which it sends down to the northward are similar in nature to those which run south, and separate the different valleys of Orissa. There is a great want of authentic information regarding the elevation and even the physical features of these wild and little-known countries. The elevated table-land of Chota Nagpur is said to have an average height of 3000 feet; and further west, towards the borders of Sirgujah, the surface is perhaps a little higher. The plain of Hazaribagh has a mean height of about 1800 feet; and twenty or thirty miles further east, that out of which the mountain Parasnath (an isolated peak) rises suddenly to an elevation of 4500 feet, is 1200 feet high. Parasnath is the highest known elevation in the province, though perhaps in the unknown districts to the westward the hills may rise as high or higher.

The flat-topped spurs of the Vindhia sink abruptly into the valley of the Sôn, which is bounded on the west by a line of cliffs rising 1000 feet or more above the bed of the river. Further east, the elevation is less considerable, and the table-land is broken up into a rugged hilly country, the last spurs of which approach close to the Ganges at Monghir, Bhogilpur, and Rajmahal.

The climate of Bahar resembles that of the interior of Orissa. During the south-west monsoon, from June to October, there is a moderate rain-fall, the amount of which has nowhere been determined with accuracy. Throughout the remainder of the year the province is very arid, and subject to hot winds, which blow over it from the dry plains of Rajwara and the upper Gangetic valley. At the same time, perhaps from the gentle slope, and consequent imperfect drainage in a densely wooded country, the forests (like those of Orissa) are extremely unhealthy, even in the dry season, so that Europeans cannot penetrate into their recesses, except at the height of the cold season, without great risk to life.

In all parts of the mountain districts of Bahar the open valleys are more or less cultivated, but, with rare exceptions, the soil is poor and the population scanty, and the crops very indifferent. The surface of the platforms between the valleys, when level, is often rocky and bare, but, when undulating, is covered with bush jungle, in which bamboo is very abundant. The steep slopes of the hills are covered with dense forest. The flora is very similar to that of the hills which form the eastern Ghats between the Carnatic and Mysore, or to that of the drier slopes of the central Himalaya.

Cedrela Toona, Vatica robusta, Buchanania, Semecarpus Anacardium, Cassia Fistula, Butea frondosa and *parviflora,* erect and scandent *Bauhiniæ, Acaciæ,* especially *A. Catechu, Conocarpus, Terminalia,* and *Nauclea cordifolia* are characteristic forms. All of these extend likewise to the Himalaya, but a few species have their northern limit in the mountains of Bahar and Bandelkhand, such as *Cochlospermum Gossypium, Chickrassia tabularis, Swietenia febrifuga, Boswellia thurifera, Hardwickia binata,* and *Bassia latifolia,* which are all more or less abundantly distributed throughout the province. No palm is indigenous but *Phœnix acaulis;* for the common *Calamus* of Bengal, which extends north to the base of the hills at Monghir, is not found in the interior.

The flora of the mountain Parasnath, an isolated peak which searcely attains a temperate elevation, presents few peculiar features. The upper part is however more humid than the base, and plants indicative of a moist climate, such as parasitical *Orchideæ,* Ferns, *Arum,* and others, make their appearance in small numbers. The temperate forms, *Berberis, Clematis, Thalictrum,* etc., are all Himalayan species, but most of them are widely diffused plants, extending also to the peninsula. *Vernonia divergens,* common near the summit, occurs also in Bandelkhand, and is equally abundant throughout the drier hills of the peninsula.

The Sôn valley in climate and vegetation is identical with the drier parts of the upper Gangetic valley, or the plains of Rajwara; and the low Kaimur (Kymore) range, to the north, exhibits a continuation of the features of the elevated platforms of Bandelkhand.

A part of Bahar was explored by Dr. Buchanan Hamilton, who made considerable collections in the Monghir and Rajmahal hills, and elsewhere among the mountains. Dr. Hooker also visited parts of it, but not at a favourable season; and a list of its plants has been published by Dr. M'Clelland in his geological report. It is probable that the greatest variety of form is to be met with in the more eastern hills, which, from their proximity to the Bay of Bengal, are more humid, and that to the westward the flora approaches more and more to that of the drier parts of the peninsula.   *u*

11. BANDELKHAND.

The district of Bandelkhand, including the small state of Rewah, which has the same physical features, occupies the northern slope of the Vindhia range, from the borders of Bahar on the east to Gwalior on the west. The watershed of that range is included within the province of Malwah, but long, flat-topped spurs descend towards the Jumna, separating the broad valleys of numerous rivers which flow northward. A little east of Gwalior these spurs extend almost to the Jumna, but further east they recede from the river, and, when viewed from the northward, appear to form an amphitheatre of precipices, so as to give the plain of Bandelkhand the appearance of a vast bay of the sea surrounded by sandstone cliffs, which again advance almost to the river not far from Mirzapur. The greatest width of the plain is about thirty miles, and near the hills many scattered insulated rocks occur, behind which the surface rises in a succession of steps, separated by level platforms, to the height of 2000 feet, whence it slopes gradually up to the watershed of the Nerbada, the average elevation of which is perhaps 2500 feet.

The plain of Bandelkhand near the Jumna is fertile and well cultivated, but the interior is generally barren, except in the valleys. Many lakes, which are all partly artificial, diversify the surface, and the hills are covered with low jungle. Its seasons are those usual in northern India. The rains commence in June and terminate in September, but, from the central position of the province, they are less heavy than in Malwah. The dry season is intensely hot, and there is a well marked cold season.

For our knowledge of the vegetation of Bandelkhand, we are mainly indebted to Mr. Edgeworth, who has published* a catalogue of the plants of the district of Banda. He enumerates 605 species of phænogamous plants; few of these differ from those common in the Dekhan and Gangetic plain, and the hill species are mostly common in the subtropical Hima-

* In the Journal of the Asiatic Society of Calcutta.

laya. The forests on the slopes of the higher hills are less luxuriant than in Bahar, and consist of fewer species; but *Mimusops Indica, Bassia latifolia, Cochlospermum Gossypium, Ailanthus excelsa,* and the Teak, have here their northern limit, as well as *Oxalis sensitiva, Sutera glandulosa,* and *Trichodesma Zeylanicum,* among herbaceous plants. The limited extent of the flora shows the dryness of the climate, which is also indicated by the occurrence of a few shrubby species typical of the dry flora: these are, *Capparis aphylla* (*Sodada* of Forskål), *Niebuhria oblongifolia, Althæa Ludwigii, Balanites Ægyptiaca, Alhagi Maurorum, Salvia pumila,* and *Tecoma undulata.* Several of these however occur equally in the Dekhan, so that the Sindhian and Arabian types are very few. No palms are indigenous, and Mr. Edgeworth's list includes very few ferns, and only one epiphytical orchid.

12. MALWAH.

Under this name we propose to include the whole of Central India, from Mandlah and Saugor to the borders of Gujerat. It thus comprises the whole of the basin of the Nerbada east of Gujerat, as well as the higher parts of the Vindhia hills to the north of that river, and is bounded on the south by Khandesh and Berar, on the north by Rajwara and Bandelkhand, on the west by Gujerat, and on the east by Bahar.

The Nerbada rises on the table-land of Umerkantak, the elevation of which is variously estimated at 3500–4500, or even more, feet. In the upper part of its course the river flows among low ranges of hills on the surface of the platform. Below Jabalpur its valley forms a deep excavation in the general level of the table-land of Central India, and is bounded on both sides by rugged hills, which often hem in the river pretty closely. The Satpura range on the south has a mean elevation of about 1800 feet, and the Vindhia, on the north, is only a very little more elevated; at Jabalpur the elevation of the bed of the river is 1450 feet, and at Mandlésir it is 700.

To the north of the lower Nerbada is situated the basin of

the river Mhai (Mhye), which discharges its waters into the
Gulf of Cambay, draining the whole of the western part of
Malwah. This river is not separated by any very marked
watershed from the basin of the Chambal, the sources of both
rivers being in low hills, scarcely rising above the level of the
table-land.

The Vindhia hills descend very abruptly on the south into
the valley of the Nerbada, but slope very gently to the northward. The table-land of Malwah to the north is on the
whole level, without any high ranges of mountains, but its
surface is diversified with small conical or table-topped hills,
and occasional low ridges. The general level of the crest of
ghats, or passages by which the roads ascend from the valley
of the Nerbada, is about 2000 feet, and it is but rarely that
the ridge rises to a greater elevation. Jamghat, south of
Mhow, is, according to Malcolm, 2328 feet, and Shaizgarh,
Royle tells us, is 2628. The gentle nature of the slope towards the north may be learned by a comparison of the elevations of Saugor (2050 feet), Mhow (2019 feet), Indore (1998
feet), Ujain (1698 feet), and Mahidpur (1600 feet), as given
by Malcolm. Nimach (Neemuch) still further north, but to
the west of the Chambal river, and close to the watershed separating it from the Mhai, is only 1476 feet above the level
of the sea, or not more than 800 feet above Gwalior and Agra,
the lowest part of the platform of the Ganges in the direction
in which the Chambal flows. Bhopawer, in the Mhai basin,
but close to the crest of the Vindhia range, is 1836 feet.

The table-land of Malwah is in general highly cultivated,
the soil being rich and productive, the climate mild and moist
during the hot season, and the surface well watered by numerous rivers and copious streamlets, all of which have their
sources in the crest of the Vindhia hills. The rains, which set
in early in June, with the south-west monsoon from the Bombay sea, and continue till September, are copious, especially in
the southern and western parts of the province, the average
rain-fall in the valley of the Nerbada being rather less than

50 inches. The cold season is delightful, and the hot season much more temperate than in the Dekhan, from the more northerly position and the greater humidity, as well as from the elevation of the table-land. Hot winds seldom blow, as the south-westerly wind sets in long before the commencement of the rainy season.

The valley of the Nerbada, being much below the average elevation of the table-land, is hotter and more humid than the latter. In many places it is well cultivated, but a great part is hilly, the spurs of the bounding ranges approaching close to the river, which is so much interrupted by rapids as to be scarcely navigable. The low hills are usually covered with bush-jungle, and the slopes of the more elevated ranges are clothed with much dense forest.

The flora of Malwah is scarcely known. The forests of the valley of the Nerbada may be expected to present a considerable amount of variety, but the climate and physical features do not differ sufficiently from those of Khandesh on the one hand and of Bahar on the other, to lead us to expect much novelty. Griffith has described a few remarkable new forms in a paper in the Journal of the Asiatic Society.

13. Gujerat.

The province of Gujerat separates readily into three divisions, which are very distinct in physical features. These are—1. The peninsula of Katiwar; 2. The alluvial plain along the Gulf of Cambay, from the Tapti to the Gulf of Kach; 3. The lower slopes of the Vindhia, where they dip into the plains.

Katiwar is a mountainous district traversed by two parallel ranges of hills, running east and west, which seem to be connected by a north and south axis corresponding in direction, as has been already observed, with the Arawali range. These hills, which rise into peaks about 2000 or 2500 feet in height, make the southern part of the peninsula much more humid than the northern, which participates in the climate of Sindh.

The alluvial plain through which the great western rivers debouche into the Gulfs of Kach and Cambay is perfectly flat, and in many places fertile and richly cultivated. Its seasons are very similar to those of the Concan, but a good deal less rain falls. At Baroch the average fall is about 33 inches, at Baroda it is 31 inches, at Ahmedabad only 16, and probably considerably less to the north and west of that place, where the plain is continuous with the desert of Marwar. There are occasional hot winds from the north-east and east, and the cold and hot seasons are similar to those of lower Sindh.

The hilly district of Bariah, at the western extremity of the Vindhia, participates in the general features of the lower part of the valley of the Nerbada. The hills are densely covered with forest, and very unhealthy for a considerable part of the year, especially after the close of the rainy season. The rain-fall is probably much greater than in the plain of Gujerat.

The district of Kach (or Cutch), which is separated from Katiwar by the Gulf of Kach, a narrow arm of the sea, from Sindh by the most eastern branch of the Indus, and from Marwar by the *Run* (a very singular saline and more or less marshy plain, in which the river Lúni loses itself), has a very similar climate to the peninsula of Gujerat, being like that traversed by a range of hills running from west to east. It may therefore (for our purposes) with more propriety be considered a part of Gujerat, than to belong to Sindh, to which physically as well as politically it is more nearly related. The northern districts of both Kach and Katiwar, being screened from the rain-bringing winds by the hills, are extremely arid.

Our knowledge of the vegetation of Gujerat is entirely derived from Dr. Gibson's excellent paper in the 'Bombay Medical Transactions.' On the open plain there is a very rapid transition, in advancing northward, from the Concan vegetation to that of Marwar and Sindh. Between the Tapti and Nerbada this is already well marked, and north of the latter river the Sindh vegetation of stunted *Acaciæ* and *Capparis aphylla* predominates. The forest which skirts the base of the mountains is the same which prevails all over India in those hilly districts in which there is a moderate rain-fall between June and September, and dry weather for the remainder of the year. The moisture-loving types of Malabar and the Concan do not occur, and the common trees are *Butea frondosa, Acacia Catechu, Cassia Fistula, Careya arborea,* and all those trees which are common in the tropical parts of the middle Himalaya. The same vegetation extends northward along the west face of the Arawali range, and probably on the Katiwar hills. In the valley of the Nerbada, which is more humid, a more varied flora will probably be met with.

14. SINDH.

The province of Sindh extends from the sea on the south to the borders of the Panjab on the north. Westward it is bounded by the mountains of Beluchistan, and on the east it is continuous with the desert of Marwar. Sindh is an alluvial plain watered by the various-branches of the Indus. For the most part it is perfectly level, but a few low hills (spurs from the Beluch mountains) here and there, as at Rori, Hyderabad, and Karachi, advance close to the Indus.

The climate of Sindh is perfectly arid, little or no rain falling at any period of the year. Now and then, however, exceptional seasons occur, when heavy showers fall at intervals, especially at the commencement of the south-west monsoon, at which time there is a considerable rain-fall in the mountains of Beluchistan and Afghanistan. The average rain-fall of Sindh is not more than four or five inches, but occasionally upwards of twenty inches of rain have been registered. Even with this amount of rain, however, the climate is so dry that the air does not remain humid for any length of time, the storms being transitory in duration. The heat is therefore very great, and the mean temperature probably as high as anywhere in India.

Though extremely fertile where irrigation is practicable, Sindh is, in consequence of the great dryness of the air, naturally sterile. There is no forest of large trees; and though extensive tracts near the river are covered with dense jungle, chiefly of *Acacia Arabica* and *Prosopis spicigera,* the greater part of the surface is barren of vegetation, and the driest parts are an absolute desert. In the lower part of the delta, within reach of the tides, a low jungle of mangroves occupies the swampy islets.

The vegetation of Sindh was first made known to science by Griffith, who traversed the upper part of the province on his way to Afghanistan, and has recorded in his private journals and literary notes the most characteristic plants which he observed. It has also been explored by Major Vicary, who has published in the Asiatic Society's Journal a list of its plants. For our very complete knowledge of its flora we are, however, mainly indebted to the late Dr. Stocks*, whose labours in this interesting province throw much light on Indian botany. Dr. Stocks' collections amount to little more than four hundred species, so that the flora is a very poor one. No doubt, as he has himself stated, a careful exploration of the hilly districts would considerably increase this number; but we feel confident that the novelties would be almost if not entirely western forms, and would therefore increase the proportion, already great, which these bear to forms characteristic of Eastern India vegetation.

More than nine-tenths of the Sindh vegetation, on a rough estimate, consists of plants which are indigenous in Africa. At least one-half of these are common Nubian or Egyptian plants, but which, from being indifferent to moisture, are diffused over all parts of India. As examples we may mention *Gynandropsis pentaphylla, Abutilon Indicum, Tribulus terrestris, Tephrosia purpurea, Glinus lotoides, Grangea Maderaspatana,*

---

* Since the printing of the earlier part of this Introduction, Indian botany has sustained an irreparable loss by the death of Dr. Stocks, from whose labours much was expected, and to whom we had ourselves looked for valuable assistance in the preparation of these notes on the vegetation of Western India. Fortunately for science a very complete series of his collections exists in the Hookerian and Benthamian Herbaria, accompanied by a catalogue very carefully drawn up, and many important notes, of which we have made use above.

Trichodesma Indicum, Lippia nodiflora, Solanum Jacquini, Ærua lanata, Achyranthes aspera. A smaller number, but still considerable, are tropical African, which are also widely diffused over India. Among these are many *Convolvulaceæ*, as *Batatas pentaphylla, Pharbitis Nil, Ipomœa muricata* and *reptans*, and many of the commonest Indian weeds, such as *Peristrophe bicaliculata* and several species of *Corchorus* and *Triumfetta*. A considerable proportion (perhaps one-sixth of the whole) consists of common Egyptian plants, which are too intolerant of moisture to withstand the climate of the more humid parts of India, but which extend along the Arabian and Persian coasts to Sindh, and thence to the Panjab and the drier parts of the Gangetic plain, and some even to the Dekhan and Mysore. Such are *Peganum Harmala, Cocculus Leæba, Capparis aphylla, Fagonia Arabica, Alhagi Maurorum, Acacia Arabica, Prosopis spicigera, Zizyphus Lotus,* and *Calotropis procera*, all of which extend to the drier parts of the peninsula; and *Malcolmia Africana, Corchorus depressus, Cucumis Colocynthis, Berthelotia lanceolata, Heliotropium undulatum, Salvia Ægyptiaca, Lycium Europæum, Cometes Surattensis*, several *Chenopodiaceæ*, and *Crypsis schœnoides*, which are confined to northern India. With these there occur also a few central European plants, though far fewer than in the northern Panjab, as for example *Ranunculus sceleratus, Convolvulus arvensis, Heliotropium Europæum, Rumex obtusifolius, Asphodelus fistulosus*, and *Potamogeton pectinatus* and *natans*.

Sindh also contains a considerable number of species which have not been met with elsewhere in India, but which are Arabian or Nubian plants. Such are *Zygophyllum album* and *simplex, Balsamodendron, Neurada procumbens, Aizoon Canariense, Seddera latifolia, Trichodesma Africanum, Acanthodium hirtum*, and several *Barleriæ*. A few Persian and Mesopotamian plants not yet known further west, such as *Populus Euphratica* and *Gaillonia*, occur also in the list. *Puneeria coagulans*, Stocks, is confined to Sindh, and the neighbouring province of Beluchistan. Eastern species which find their western limit in Sindh are almost entirely wanting. The following are all that are contained in Dr. Stocks' catalogue, excluding plants manifestly cultivated (such as *Tamarindus*), *Rhus Mysorensis, Zizyphus Jujuba, Hedyotis aspera, Coldenia procumbens, Salvia plebeia* (a New Holland plant), *Clerodendron phlomoides, Aristolochia bracteata,* and *Zeuxine sulcata.* There are, however, a considerable number of species which have not been met with in Egypt or Arabia, but which belong to genera characteristic of those countries, and are very closely related to Egyptian species. Instances of this kind are *Crotalaria Burhia, Dicoma lanuginosa, Leptadenia Jacquemontiana, Oxystelma esculentum, Linaria ramosissima, Streptium asperum, Solanum gracilipes, Chamærops Ritchiana.* If we add to this enumeration the coast flora of *Sonneratia, Rhizophora, Ceriops, Scævola, Ægiceras, Ipomœa Pes-capræ,* and *Avicennia,* a good general idea is given of the nature of the flora of Sindh.

15. RAJWARA.

The districts or states which are included under the general name of Rajwara lie to the north of Malwah, and to the south of the river Jumna. The whole of Marwar, including Jodhpur, Bikanir, and Jesalmir, lies in the basin of the Indus to the west of the Arawali range. The remainder of the province, consisting of the states of Mewar, Jaipur, Kotah, and Gwalior, is situated in the basin of the river Chambal, the great southern branch of the Jumna.

The Arawali mountains, as we have seen, form a continuous range, running from north-east to south-west, which traverses the whole of the province. It dips on its western side very abruptly into the plains of Marwar, which are perfectly level, and are continuous with the great sandy desert stretching west to the Indus. To the eastward, these hills give off numerous spurs, which form low ridges, separating the different branches of the Chambal. The crest of the Arawali range appears never to rise much above 3000 feet, and the head valleys are 1000 feet lower. Thus Udepur and Ajmir, both close to the crest of the range, have an elevation of about 2000 feet, and are surrounded by hills, the highest of which are about 1000 feet higher. Abu, on a spur to the east of the watershed, is said to attain 4500 feet.

Another range of hills, connected with the Arawali to the south of Udepur, passes by Nimach, and runs parallel with and west of the Chambal, as far as its junction with the Banas. The elevation of Nimach is 1476 feet, and as the surrounding hills are very low, they are perhaps not much higher than 2000 feet. The level of the country gradually sinks towards the north-east. The elevation of Agra above the sea is 670 feet, and the junction of the Jumna and Chambal is a few feet lower.

Rajwara is on the whole a barren province, a great part of it being hilly and unimprovable, but the valleys are occasionally rich and very fertile. The climate is drier than that of Malwah, and becomes very arid in the northern parts. On the western slopes of the Arawali hills there is a considerable rain-fall during the south-west monsoon, but the whole country to the eastward is sheltered by that range from the effects of the monsoon, so that the average rain-fall at Agra is only 19 or 20 inches. The plain of Marwar is even more arid, and the desert which stretches towards the Indus is as dry and sterile as the worst parts of Sindh. The mean temperature of Rajwara is higher than might have been anticipated from its elevation and latitude. At Ajmir and Nasirabad it is 76°.

The vegetation of Rajwara is not known in detail, but it probably differs little from that of the Dekhan and upper Gangetic valley. The forest-clad slopes of the Arawali range are so dry for nine months of the year, that only those trees which are tolerant of great dryness can grow there. They may therefore be expected to present a vegetation similar to that of the hills of Gujerat, or the western and drier Himalaya, where the climate is similar. The summit of Abu, like that of Parasnath, produces some epiphytical *Orchideæ* and other humid types, but their number is no doubt inconsiderable. The flora of the desert of Jesalmir resembles that of the southern Panjab.

16. PANJAB.

The Panjab extends from the northern border of Sindh and Marwar, or rather Jesalmir, to the base of the Himalaya, and from the mountains of Afghanistan, which skirt the right bank of the Indus, to the borders of the Gangetic plain. Strictly speaking, the river Satlej, or Gharra, is the south-eastern boundary of the Panjab, but politically the Cis-Satlej states have been attached to it, and for our purposes it is convenient to draw the boundary along the line which separates the waters tributary to the Ganges from those which flow towards the Indus. This line lies to the eastward of the river Gagar, whose channel may be traced by Bhatnir to the Satlej, a little above Bahawalpur, though its waters are generally absorbed by the desert long before they reach that river. It therefore includes Bahawalpur and Bhatiana, as well as the Cis-Satlej states.

The Panjab, as is well known, derives its name from the five great tributaries of the Indus by which it is traversed. These are the Jelam, the Chenab, the Ravi, the Beas, and the Satlej, all of which, uniting to form the Panjnad, join the Indus near the southern extremity of the province. The surface is on the whole level, but the north-western angle is more or less diversified with hills. West of the Indus there is only a narrow strip of level country, and here and there the hills approach close to the river. No definite physical boundary can therefore be laid down along this frontier, and the political boundary must be adopted. Practically this is of no importance, as the vegetation of the lower hills of Afghanistan is the same as that of the western Panjab.

Between the Indus and the Jelam an elevated platform of considerable elevation (at Rawil Pindi 2000 feet) abuts upon the Himalaya, and south of that town rises into a low range of hills usually known as the salt range, the southern escarpment of which crosses the *Doab** from Pind Dadan Khan in a westerly direction. The summits of this range do not rise higher than 3000 feet. East of the Jelam a very low range of hills, only a few hundred feet in height, runs parallel to that river for some distance from the Himalaya. Elsewhere the country is level, and slopes very gently down from the base of the Himalaya towards the sea. Attok, on the Indus, is elevated 1000 feet, and Lahore about 800 feet above the level of the sea. The junction of the Panjnad with the Indus is elevated about 200 feet.

The climate of the Panjab is very dry. Along the base of the Himalaya the periodical rains are well marked, occurring at the same season as elsewhere in northern India, but their quantity diminishes rapidly in advancing westward, and to the west of the Jelam they disappear. The amount of rain-fall also diminishes in receding from the mountains. At Firozpur and Lahore it is in ordinary seasons very small; and at greater distances from the Himalaya the rains may be said to cease entirely. Throughout the province, however, heavy rain usually falls at midwinter, but does not continue for any length of time.

The mean temperature of the Panjab does not differ materially from that of Agra and Delhi, but is rather lower. The absence of rain in the western and southern parts of the province raises the summer temperature very high, but the coolness of the winter months compensates for this, and reduces the mean temperature of the whole year.

The surface of the Panjab, like that of Sindh, is very fertile where water is procurable for irrigation, but elsewhere it is quite barren. Along the base of the Himalaya, from Ambala as far as the Jelam, there is a very rich belt of fertile country. At a little distance from the mountains, however, the centre of each Doab is dry and barren, and the cultivation is confined to a narrow belt along the great rivers. The soil

* Any tract of country included between two rivers which join is called in India a Doab.

is usually a hard clay, and water is only procurable at great
depths. East of the Satlej a sandy desert extends from Sirsa
as far as Marwar and the Run of Kach. The streams which
descend from the Himalaya and the western face of the Arawali hills are all dissipated before they can mingle their
waters with the Satlej, and below Bahawalpur the desert advances close to the river.

The vegetation of the Panjab varies with the climate. In
the southern part of the province, where little or no rain falls,
the flora is almost identical with that of Sindh ; but as the latitude increases and the mean temperature, and especially the
winter temperature, diminishes, we find a gradual increase of
plants characteristic of the Mediterranean flora, which is fully
represented on the mountains of Afghanistan. These are,
however, chiefly winter-flowering annuals, such as *Goldbachia
lævigata, Frankenia pulverulenta, Silene conica, Arenaria serpyllifolia, Euphorbia Helioscopia, Carthamus oxyacantha, Veronica agrestis, Poa annua*, and their number is not considerable. All the shrubby plants which give the character to the
vegetation are the same as those of Sindh. The extensive
tracts of low and scattered tree-jungle which occupy the dry
clay soil at a little distance from the river, even further to the
north and east than Lahore and Firozpur, consist chiefly of *Capparis aphylla, Acacia Arabica* and *leucophlœa, Prosopis spicigera, Zizyphus Lotus*, and *Salvadora oleoides (S. Indica*, Royle).
Cocculus Leæba, a Senegal, Egyptian, and Sindh species, climbs
over the trees. *Populus Euphratica* forms thickets along the
Satlej, as far east as Bahawalpur, along with *Tamarix Gallica*,
which, however, is generally diffused over India. *Berthelotia
lanceolata*, a low shrubby plant, which is widely diffused over
the drier parts of Asia and Africa, covers large tracts, either
quite alone or interspersed with other plants.

Nearer to the Himalaya, as the climate becomes moister,
the vegetation changes, the plants of the desert giving place
to those of the Gangetic plain. At Ludiana and Jalandhar
the shrubby vegetation is quite changed. *Butea frondosa*

has become common, accompanied by all the characteristic forms, which will be enumerated in the next section, and the dry country shrubs have quite disappeared. With the annual herbaceous vegetation the change is less marked, these districts presenting a mixed flora, the cold and hot seasons producing plants of a dry climate, while during the rains more humid types are numerous.

West of the Jelam, wherever the surface is hilly, as is usually the case, it supports a very different vegetation. *Acacia modesta*, and some other species, with a spinous *Celastrus*, form the greater part of the jungle. *Olea undulata, Rhazya stricta, Dodonæa, Reptonia (Edgeworthia* of Falconer), and other plants of the lower hills of Afghanistan, occur occasionally, and many mountain plants of the Persian flora, which descend from the hills, are here met with. Several species of *Delphinium*, described in the present part of our work, and numerous *Caryophylleæ, Geraniaceæ, Cichoraceæ, Cynaraceæ, Labiatæ, Boragineæ*, and other genera of the Oriental flora, might be enumerated as instances; but the flora of this district is still very imperfectly known, no extensive collection of its plants having reached this country. Those which we have seen were collected by Jacquemont, who explored the Salt range; by Dr. Fleming, who has more recently visited the same district, and has communicated to us a complete series of the plants which he collected; and by Major Vicary, chiefly from the neighbourhood of Peshawer.

Griffith's private journals, Jacquemont's ' Voyage aux Indes Orientales,' and Royle's ' Illustrations,' contain many interesting notes regarding the Panjab flora. Mr. Edgeworth has fully investigated the neighbourhood of Multan, and has communicated many specimens to the Hookerian Herbarium. These and our own materials give us a very complete knowledge of its vegetation.

17. Upper Gangetic Plain.

Between the Himalaya on the north and the spurs of the

Vindhia on the south, the Ganges and its tributaries flow through a broad plain, uninterrupted by any inequality of surface. The Jumna above and the Ganges below the junction of the two rivers, flow near the southern margin of the plain, occasionally washing the rocky extremities of the hills, which advance from the southward, and always at no great distance from them, so that the greater part of the plain lies to the north, between these rivers and the Himalaya. As far as the commencement of the delta of the Ganges, its surface is characterized by great uniformity of physical character; it may therefore conveniently be regarded as one botanical province, including the districts of Delhi and Agra on the left bank of the Jumna, which adjoin the Rajput states, the Doab between the Jumna and Ganges, and Rohilkhand, Oude, and Benares, with the district of Tirhut, on the left bank of the Ganges.

Though the Gangetic plain is not separated from the Panjab by any perceptible ridge, the line of separation between the two, which lies very little to the left of the Jumna, between Karnál or Jagadri, and Thanesir, is the most elevated part of the plain which lies at the base of the Himalaya. Ambala, on a branch of the Gogra, and Saharanpur, on the left bank of the Jumna, are each about 1000 feet above the level of the sea, and the high lands on the right bank of the Jumna are probably not more than fifty feet higher. Thence the plain slopes very gradually to the sea, with an average fall of about a foot a mile. Agra is 670 feet, Cawnpore 500 feet, Allahabad 305 feet, and Benares 265 feet above the level of the sea.

The mean temperature of the upper Gangetic plain varies from 78° at its lower extremity, to $72\frac{1}{2}°$ at Saharanpur, the diminution being mainly caused by the increased cold of the winter months, as the heat of summer is in all parts very great. The rains set in everywhere soon after the sun has attained its most northern limit. The rain-fall is greatest near the Himalaya, and diminishes gradually as we recede from the mountains. Along the base of the Himalaya it is greatest to the eastward, and becomes much less in the extreme west. Close to the mountains the amount of fall is not known, but at Benares it is 54 inches, at Gorakpur it is 50 inches, at Moradabad 41 inches, and at Saharunpur only 30 inches. Further from the hills the fall at Meerut is 30 inches, at Alighar 24 inches, at Fattighar 22 inches, at Panipat 25½ inches, at Delhi 21½ inches, at Agra 19½ inches, at Cawnpore 23 inches, at Allahabad 33 inches, and at Mirzapur 35 inches. These numbers present many irregularities, and are probably not to be relied on, but they suffice to show the diminution of rain as the distance from the Himalaya increases. Nor is the reduced rain-fall an accurate indication of the change of climate, as the atmosphere near the mountains is shown by the dew-point observations to be much more moist at all seasons than at a distance.

The flora of the Gangetic plain varies with the degree of humidity. The surface (except along the base of the mountains) is nowhere clothed with forest, but uncultivated tracts are usually covered with a loose bush-jungle, in which *Butea frondosa, Flacourtia sepiaria, Capparis sepiaria, Zizyphus Jujuba* and *Œnoplia, Adhatoda Vasica,* and *Carissa edulis* are among the commonest shrubs, till the climate becomes too dry for them, when they are gradually replaced by the vegetation of the Panjab region, which usually advances as far as the Jumna, and now and then penetrates a little way into the Doab; indeed several of the species which are most characteristic of the arid flora, as, for instance, *Tecoma undulata* and *Berthelotia lanceolata,* were first collected by General Hardwicke in the neighbourhood of Cawnpore. *Alhagi* is also found in the same district, and *Peganum Harmala* is recorded as a native of Monghir.

If we exclude this dry country flora, which just skirts the southern part of the plain, the vegetation of the Gangetic plain presents few peculiar features; indeed a catalogue of the plants of Rohilkhand contains very few species which are not

y

common all over India, even to the extreme south of the peninsula, in those provinces which have a similar climate. A
very few winter-flowering plants (such as *Ranunculus sceleratus*) are the only exceptions, and these are mostly wanderers
from the temperate region of the Himalaya. We have already
had occasion to direct attention to the remarkable uniformity
of the vegetation over large areas of India, and as our information becomes more precise, the sameness becomes more
striking.

A considerable portion of the flora of the peninsula does not
extend to the upper Gangetic plain, because of the increased
cold of winter, and even within the district several plants
which are common in the south-eastern portion do not extend
to the north-west. *Trichodesma Zeylanicum* is common about
Patna, but not found in Rohilkhand. *Cassytha*, which is common in Bahar, is found at Agra, but not on the north of the
Ganges. The Palmyra (*Borassus*) is cultivated as far up the
Ganges as Alighar and Shahjehanpur, but is not known at
Meerut or Moradabad. The only wild palm in the province
is *Phœnix sylvestris*.

Near the base of the Himalaya there is always a belt of
forest of considerable width; but as it is identical in vegetation with the tropical belt of the mountains, to which indeed
it owes its existence, it will be more convenient to notice it in
describing the Himalaya.

The vegetation of the upper Gangetic plain, which was first
explored by Hardwicke, Govan, and Wallich, was afterwards
illustrated in detail by Dr. Royle, whose long residence at
Saharunpur gave him ample opportunity of investigating it.
In his 'Illustrations,' the influence of the climate upon the
vegetation, and the curious transition from the humid to the
dry country flora, are first pointed out. Our own collections
are chiefly from Rohilkhand.

18. BENGAL.

The lower part of the Gangetic plain, which constitutes the

province of Bengal, differs so strikingly in climate and vegetation from the upper, that it must necessarily be regarded as a separate province. Along the sea-coast Bengal includes the whole of the delta of the Ganges, extending from Balasor to the mouth of the Fenny. It is bounded on the west by the hilly districts of Orissa and Bahar, and on the east by the Assam valley, and the Khasia, Tippera, and Chittagong hills. To the north it extends to the base of the Himalaya, but to the north-west the boundary between Bengal and the upper Gangetic plain must be an arbitrary one, the transition of climate and vegetation being gradual; it may, however, conveniently be drawn at the river Cosi. Further west the plains are screened by the Bahar hills from the direct influence of the moist air from the Bay of Bengal, and are therefore drier.

The surface of Bengal is perfectly flat, and so little elevated above the level of the river that a great part of it is under water during the rainy season. Close to the base of the Himalaya the surface is a little more elevated, but elsewhere it is everywhere intersected by watercourses, which are formed by the branching of the two great rivers, the Ganges and Brahmaputra, and of their tributaries.

The climate of Bengal is much more equable than that of the upper Gangetic plain. The rains are heavier and of longer duration; the heat of summer never rises to so excessive a temperature as in the north-west provinces of Hindostan, and the winter is much less cold. North of the Ganges, hot winds blowing from the westward towards the funnel-shaped valley of Assam occasionally traverse the plain, but they are rarely of sufficiently long continuance to affect the vegetation. South of the Ganges the delta is sheltered by the hills of Bahar, so that no hot winds blow, and the atmosphere always remains more or less humid. This humidity is no doubt primarily due to the proximity of the sea, though we learn from the dryness of Sindh, on the opposite side of the peninsula, that that alone is not sufficient to induce it; the main cause would appear to be the proximity of the enormously elevated snow-clad masses of the Himalaya, and the suddenness with which they rise out of the plain.

During the rainy season, when the wind blows from the south, and arrives saturated with moisture at the base of the mountains, a sudden condensation at once takes place; and the distance from the sea is so small, that the effect of the cooling is nearly uniform over the whole area, and does not diminish rapidly as we recede from the mountains, as in the upper provinces. During the remainder of the year, when land winds prevail, the humidity of the atmosphere must be mainly due, as has already been observed (at p. 80), to an upper return current, which is stopped by the high wall of the Himalaya, and, being cooled, sinks towards the earth, and is carried back towards the sea along with the normal current, which descends along the course of the Ganges and Brahmaputra. In support of this explanation, it may be noticed that a belt of equable climate, gradually narrowing as we advance westward, skirts the base of the Himalaya, the summers of the Terai and Himalayan valleys being less hot, and the winters moister and less cold than those of the open plain.

The rain-fall in Bengal varies from sixty to one hundred inches. It is least in the north-western part of the province, and greatest on the eastern sea-coast, near the mouth of the Megna. The mean temperature of Calcutta is 78°, which may be considered as that of the whole area.

The province of Bengal is celebrated for its fertility, and is for the most part under cultivation. The surface is perennially green, and the villages are usually buried among lofty trees; Bamboos, Figs, Mangoes, and various Palms occupying a conspicuous place. The Palms are chiefly Cocoa and Betelnut, *Phœnix*, *Borassus*, and, near the sea, *Corypha*. The two first may be considered the most characteristic cultivated plants, as they are intolerant of cold and do not extend into the drier provinces. Two species of Rattan (*Calamus Roxburghii* and *fascicularis*) are common throughout Bengal, and a third (*C. Mastersianus*), which is common in Silhet and

Assam, is found occasionally in the eastern districts. The indigenous flora is much more extensive than that of the upper Gangetic plain, comprising all the species which grow there except those belonging to the Egyptian or arid flora, besides many others which are not found to the north-west. Ferns are numerous, and a few epiphytical *Orchideæ* are found upon the trees, *Vanda Roxburghii* being the most common. One of the most remarkable forms is a species of rose (*R. involucrata*), which is common in the grassy jungles of the northern parts of Bengal. Many peninsular species which are prevented by the cold of winter from extending northward to the upper Gangetic plain are abundant in Bengal. The common shrubs are species of *Zizyphus, Adhatoda, Calotropis, Carissa, Melastoma, Alangium, Stravadium, Tetranthera, Antidesma,* and *Guatteria suberosa. Pedalium Murex, Tiaridium Indicum, Trichodesma Zeylanicum, Coldenia procumbens, Thespis divaricata,* and *Tiliacora acuminata* may be mentioned as instances of peninsular forms which are equally common in Bengal, but are not found in the upper Gangetic plain. One of the most curious natives of Bengal is *Ethulia divaricata,* a tropical African plant, which is found nowhere else in India. The flora of Bengal does not exhibit much affinity with that of the Malayan Peninsula, containing no *Cycas,* Oaks, nor Nutmegs, though these all grow in Chittagong very little to the eastward, and in the Khasia hills on the north-east frontier.

Within the influence of the tides the delta of the Ganges is covered with a dense jungle of trees peculiar to salt-marshes, called the *Sunderbunds.* This is most largely developed in the western parts of the delta, where the rise and fall of the tides are not considerable, and where there is but little influx of fresh water. To the eastward, near the mouth of the Megna, the bay is almost fresh, and its shores are muddy without vegetation. The rise and fall of the tides are here so considerable, that there is not the same facility for the growth of shrub and trees along the margin of the ocean, that there is on the banks of the creeks which traverse the Sunderbunds in the western part of the delta. There mangroves, *Sonneratia*, *Ægiceras*, and *Heritiera*, mingled with gigantic grasses and *Typha*, abound. *Nipa fruticans* fringes the watercourses, and vast tracts are covered with *Phœnix paludosa*, an elegant little palm six or eight feet in height.

The vegetation of Bengal has been well explored. The foundation of its flora was laid by Roxburgh, who was appointed in the year 1793 to the superintendence of the Calcutta Botanic Gardens, which, by his labours and those of his distinguished successors Hamilton and Wallich, became very rich in tropical plants. A complete enumeration of the plants of Bengal is found in Roxburgh's 'Flora Indica.' Griffith's 'Itinerary Notes' and Voigt's 'Hortus Suburbanus' also contain notices of many indigenous species.

II. *The Himalaya.*

To the north of the great plain of Hindostan is situated a mountain-tract of great extent, strictly defined on its plainward face, and increasing in elevation as we advance towards the interior. As a whole, this tract is extremely rugged, lofty mountain-chains being separated by deep valleys. Amid the numerous and intricate ramifications of these chains there is considerable difficulty in acquiring a definite idea of the composition of the mass. Superficial observation gives the impression that numerous ranges rise one behind another, the more distant of which are loftier than those in front; but a nearer approach shows the fallacy of this impression, and proves that the arrangement is much less simple.

A prodigiously elevated but scarcely known chain traverses Asia from east to west in about 36° N. lat. South of this chain flow two rivers, the Indus and the Brahmaputra, which, rising nearly together, run in directly opposite directions; one nearly west, the other nearly east. Throughout the greater part of their course they preserve these directions, but at last both turn abruptly south, to discharge their waters into the Indian Ocean. The chain between these rivers and the plains of India is the Himalaya, which is connected with the still loftier chain of the Kouenlun behind at the common source of these two rivers by mountains of comparatively moderate elevation, which are perhaps portions of a chain running from south-west to north-east, and forming the watershed of Asia as far as the Sea of Japan. Nothing can be more simple than this definition, which is that given by Mr. Hodgson, and we think it is the only one which will suffice. The Himalaya thus includes the whole extent of country between the Indus at Attok and the great bend of the Brahmaputra, but nothing to the west of the Indus or to the east of the Brahmaputra. The axis of the main chain of the Himalaya lies in general far back, much nearer to the two great rivers which run behind it than to the plains of India; hence the secondary chains on the south face are much more important than those on the north.

The Himalaya may be regarded as consisting of two portions, one on each side of the point of origin of the meridional ridge, by which it is connected with the Kouenlun behind. Of these the Western Himalaya is rather shorter than the Eastern, and it is better known throughout a great part of its course from its lying within British territory, while the Eastern Himalaya is for the most part Tibetan. The elevation of the chain is probably everywhere very great, no known pass across the watershed being of lower elevation than 16,500 feet, except close to the extremities of the chain. The most remarkable depressions in the inner Himalaya are the Rotang Pass between Kulu and Lahul, which is 13,000 feet, and the Zoji Pass between Kashmir and Dras, which is only 11,300 feet.

From the central axis of the chain of the Himalaya a succession of secondary ranges take their origin, which descend on the one hand towards the plains of India, and on the other towards the northern rivers. These secondary chains, on the

Indian side, separate the great rivers which flow towards the plains of India, and which, successively uniting in their courses through the plains, ultimately discharge their waters into the Indus and Brahmaputra, from which they are at first separated by the whole width of the chain of the Himalaya. The great rivers from west to east are in succession—the Jelam, the Chenab, the Ravi, the Beas, the Satlej, the Jumna, the Ganges, the Gogra, the Gandak, the Cosi, the Tista, the Monas, and the Subansiri; all of these are separated by chains at first of great elevation, but which terminate at last abruptly in the plains of India. Some of these chains are now well explored, but others, especially those in Nipal and Bhotan, are still very imperfectly known. They vary a good deal in direction, some being almost perpendicular to the main axis, while others form with it a very acute angle. They all ramify very much, giving off chains of the third order, separating the tributaries of the great rivers.

The length of the chain of the Himalaya, from the Indus to the Brahmaputra, may be estimated at about 1400 miles, while its width varies from 200 to 100 miles. Most of the lofty peaks with which we are acquainted are situated on the secondary chains, but the mean height of the main axis is probably greater. The elevation of the secondary chains diminishes, on the whole, as they approach their termination in the plains, though with a certain degree of irregularity. In length these vary considerably, according to their direction, but we must refer to the map for details of their structure and arrangement. It will be seen that their ramifications are innumerable; their flanks are in general steep, and separate deep valleys. Open plains are rare, but occur occasionally at all elevations, and there are a few inconsiderable lakes. The mean slope of the Himalaya from the plains to the axis is not more than 1 in 25, and the mean slope of the ridges of the scondary chains, which are usually very oblique, and always sinuous, must be considerably less. It is important to keep in view these numbers, which serve to correct the erroneous estimates usually formed of the steepness of these mountains. The chain does not run due east and west, its western extremity being in 35° north latitude, while the latitude of the east end is only 28° north.

Though the Gangetic and Panjab plains, from which the Himalaya rises abruptly, are for the most part devoid of trees, or covered only with scattered jungle, there is usually a belt of forest ten or twenty miles in width, along the base of the mountains, composed of the same trees which form the mass of the tropical vegetation of the lower hills.

The extension of the forest over the plain is no doubt the effect of the equable and humid climate which prevails along the base of the mountains, but the nature of the drainage is also not without its influence. The forest grows usually on slightly inclined gravelly slopes, and is succeeded on the side furthest from the mountains by a swampy tract, without trees, and covered with long grasses, called the Terai. Beyond the Terai the surface generally rises again slightly, so that the swampy tract may be regarded as a series of flat-floored valleys, skirting the base of the mountains; or rather, in a strictly scientific point of view, it consists simply of the outermost valleys themselves, and the bases of the mountains forming scarcely perceptible undulations between them.

Immediately within the mountains the first series of lateral valleys are often broad and bounded by low hills, or on one side (the southern) by low hills, and on the other (the northern) by considerably higher ones. These are known by the name of *Dhúns* (Doons); and when very open, flat-floored, and with gradually sloping beds, their true relation to the surrounding mountain-chains is not at once apparent. Sometimes they appear to be indefinitely extended east and west, in a direction parallel to the Himalayan chain; and, running from one great river to another, they appear to belong to a different order of valleys from what occur further within the mountains. This arises in some cases from the slope of their beds being so extremely gradual, that the watershed

z

between the valley that ascends from one river, to the corresponding valley that descends to the other river, can only be detected by the observation of the drainage; whence the two valleys appear to form one. Such is the case with the Dehra Dhún, which appears to form one continuous transverse valley between the Jumna and the Ganges, but which really consists of two valleys; one descending from the village of Dehra (which occupies the *col*) westerly to the Jumna, and the other descending from the same spot easterly to the Ganges.   Other Dhúns, again, are simply very broad, open valleys, differing in no physical features from those that occur in other parts of the mountains. In the Panjab-Himalaya, where the tertiary sandstones acquire a great development, two or three such valleys occur in succession before the higher mountains begin. These valleys, or Dhúns, are not, as is very generally supposed, continuous along the whole extent of the Himalaya, and interposed between the tertiary and secondary mountains. They are merely the outer series of lateral valleys, and are always of limited extent.

In the enormous chain of the Himalaya, which rises nearly from the level of the sea to perpetual snow, we have of course every variation of temperature between tropical or subtropical heat and extreme cold. The diminution of temperature is 1° for 300 feet of elevation in the more humid, and for 400 feet in the drier part of the chain. The elevation of the snow-line, at equal distances from the plains, is nearly uniform throughout the whole extent of the chain, the increase of latitude of the more westerly part being compensated for by the greater distance from the sea, and consequent diminished snow-fall. This level on the outer ranges has been determined to be about 16,000 feet, but it becomes higher on the inner ranges, and in the Tibetan Himalaya is not under 19 or 20,000 feet.

The climate of the Himalaya varies much in different parts. During the winter season the weather is generally unsettled ; for while the north-east monsoon is blowing over the lower parts of India, an upper current of south-westerly wind carries its moisture to the higher mountains, where it is condensed in the form of snow. Snow falls in the eastern parts, in severe seasons, as low as 5000 feet, and in the north-west occasionally as low as 2000 feet. The ordinary limit, however, is several thousand feet higher. After the vernal equinox, by which time the south-west monsoon has fairly set in, the sky is usually serene and the weather beautiful. To the eastward this rule is subject to frequent exceptions, the same causes which make the climate of Bengal humid at all seasons operating more markedly on the Himalaya to the northward of that province. As summer advances, the wind becomes more humid, and occasional heavy thunderstorms in the afternoons mark the approach of the rains, which set in about midsummer; considerably earlier, however, in the eastern than in the north-western Himalaya. During the rainy season, which continues almost till the autumnal equinox, when the decreasing declination of the sun changes the direction of the wind, the atmosphere is very humid, usually almost to saturation. There are, however, occasional interruptions in the rains, during which the weather is superb. The rain-fall is greatest to the eastward, and diminishes gradually in advancing westward.

As the source of the deluge of rain which falls on the Himalaya is very distant, a great part of the moisture is necessarily deposited on the first range with which the humid wind comes in contact, of sufficient elevation to cool the air to the point of saturation. The rain-fall is therefore greatest on ranges elevated from 6 to 10,000 feet, especially where these advance in considerable masses near to the plains, while isolated peaks, and ranges of lesser elevation, as well as the valleys of the great rivers, are evidently drier. As a consequence of this, all the valleys of the interior which are separated from the plains by continuous chains, attaining an elevation of 10–12,000 feet, are to a great extent sheltered by these from the rains, which fall only as occasional showers ;

while those still further back, and bounded on the plainward face by mountains rising everywhere to the level of perpetual snow, are absolutely without rain during the monsoon. In Sikkim and Bhotan, where the wide valleys are perpendicular to the axis of the chain, and correspond to the direction of the winds, the rains are heavy till we penetrate far into the interior, but great irregularities everywhere occur even in adjacent valleys; thus the transverse chain of the upper Tista makes the climate of the higher parts of the Lachen valley much drier than that of the Lachung river, though the two are only a few miles apart.

We meet, therefore, in the Himalaya, with all the modifications of climate which have already been enumerated as occurring in India, and the aspect of the mountains varies with the climate. In the permanently humid parts the mountains are covered everywhere with an uniform sombre forest, masking all inequalities of surface, and giving a dull and monotonous aspect to the scenery. This forest rises to the upper limit of trees, at 12–13,000 feet, and is succeeded by grassy pastures, which ascend to the snow-line. Forests are also plentiful where the dry season is well marked and the rains abundant; but they are there confined to the shady and moister exposures, while the sunny slopes and all the lower hills are grassy and rocky. The permanently arid mountains of the extreme west are barren and rocky, and devoid of trees at all elevations.

In the temperate valleys of the inner Himalaya, where the rain-fall is moderate in amount and the ground is permanently covered with snow during winter, and where the hot summer's sun powerfully stimulates vegetation, the mountain slopes present a delightful intermixture of beautiful forest and of luxuriant vegetation; while above the limit of trees the compact turf is enamelled with myriads of lovely flowers, nourished by the melting snows and the genial warmth of summer. To this, however, as we penetrate further into the interior, a barren, treeless climate rapidly succeeds, in which the principal vegetation occurs at the commencement of spring, when the melting snow supplies abundant moisture to small annual plants, which run their course with great rapidity, and are speedily shrivelled up by a scorching sun.

As respects climate, we have therefore two different systems of division of the Himalaya:—1, into the tropical, temperate, and alpine zones; and 2, into the exterior or rainy, the interior or intermediate, and the Tibetan or arid Himalaya.

The term tropical is not strictly applicable to any part of the chain, which is nowhere within the tropics, but we find it convenient to adopt it, and, the vegetation being strictly tropical, it can, we think, lead to no inconvenience; while the only word which could be substituted, namely *subtropical*, is required to express the transition from the vegetation of the base to that of the temperate zone. There are of course no strict lines of demarcation between the three zones first enumerated; but they are sufficient to express the three prominent changes in the vegetation which correspond to those observable in passing from the equator towards the poles, and on the whole are sufficiently distinct to be readily recognizable.

In the extreme west the tropical belt rises to about 4000 feet, and as we advance eastward its elevation gradually increases. In Kumaon it is 5000 feet, and in Nipal rather higher. In the permanently humid country to the eastward it rises still higher, tropical vegetation being found as high as 7000 feet; but the equable nature of the climate masks the effect, and carries many temperate plants much lower than that level. The alpine zone may be said to commence at the upper limit of trees, which varies from 12,000 feet in the extreme west to nearly 13,000 feet in the eastern Himalaya. A number of trees and shrubs which are peculiar to the higher part of the temperate zone, we shall generally characterize as subalpine.

The division of the Himalaya into exterior, interior, and Tibetan, corresponds in the temperate zone to very marked differences of vegetation. In the great valleys the tropical flora stretches far into the interior, and is much the same there as in the outer portion of the mountains. In the exterior Himalaya there is a well marked rainy season. The width of the belt of the exterior or humid Himalaya is much greater to the eastward than in the extreme west, the rain-fall and humidity being much less to the westward. We therefore find the plants of the interior zone advancing much nearer to the plains in the western Himalaya than they do in the eastern, where a humid or rainy climate vegetation penetrates far into the interior. In the outer zone of the eastern Himalaya, indeed, a vegetation characteristic of an equable climate prevails throughout the year, while to the westward those families which delight in humidity only make their appearance with the commencement of the rainy season, before which time no *Zingiberaceæ*, terrestrial orchids, especially *Malaxideæ, Cyrtandraceæ, Acanthaceæ,* or balsams, are to be met with.

Considered with respect to its longitudinal extent, the Himalaya, when regarded solely from a physical point of view, consists of only two divisions, a western and an eastern, corresponding respectively to the Indus and Brahmaputra. For botanical purposes, however, the chain requires to be divided into western, central, and eastern Himalaya. The kingdom of Nipal, in the middle, constitutes the whole of the central Himalaya. To the eastward lie Sikkim, Bhotan, and Abor, to the westward Kumaon and the Panjab Himalaya.

We have thus three principal series of divisions of the Himalaya, according to length, breadth, and height. Accordingly we say—

    1 (*longitudinally*). The eastern, central, and western Himalaya.

    2 (*latitudinally*). The exterior, interior, and Tibetan Himalaya.

    3 (*altitudinally*). The tropical, temperate, and alpine Himalaya.

A combination of these three modes of division will be our usual mode of defining the localities of the plants. In the great majority of cases these terms are abundantly sufficient for our purposes, the range of each species being very considerable. There are, however, many instances in which it is desirable to enter into further detail, and in such cases we shall either make use of the river valleys (a very convenient mode of indicating the regions), or of the political subdivisions usually recognized. To these we shall refer in the following remarks on the great geographical divisions, which correspond to the longitudinal divisions given above, with the addition of a fourth, namely, Tibet, which includes not only the Tibetan slope of the Himalaya,—that is to say, the ramifications which extend from its axis towards the Tibetan Brahmaputra and Indus,—but also the mountainous country to the north of these rivers, as far as the axis of the chain of the Kouenlun.

Eastern Himalaya.

In this are included the states of Sikkim and Bhotan, and the districts lying to the eastward of the latter as far as the great bend of the Brahmaputra, which we shall call collectively by the name of Abor.

1. Abor.

To the eastward of the Subansiri river there is probably only one range of any considerable elevation, and the mountains by which the Himalaya terminates in that direction perhaps nowhere attain a greater height than eight or ten thousand feet, while the valley of the Dihong or Brahmaputra is probably broad and open. These mountains are inhabited by wild and suspicious tribes, who have hitherto refused all access to the interior of their country. The climate and vegetation are probably identical with those of the Mishmi mountains, to the eastward of the Brahmaputra, which will be noticed in a future page.

2. Bhotan.

Bhotan is at present one of the least accessible parts of the
Himalaya, and is only known to us by the narratives of Turner and of Pemberton; for Mr. Bogle, who passed through
it in 1774, has left no record of his journey. Captain Turner
traversèd the most westerly part of the province, from the
plains of Bengal to the towns of Tashisudon and Panaka, and,
after a short residence in Tibet, returned by the same route to
India; he has not, in his 'Travels,' given any details of the
vegetation.

Major Pemberton, who was accompanied by Mr. Griffith,
entered Bhotan a little to the west of the meridian of Gowahatti, in Assam, and crossed a range of mountains into the
valley of the Monas river, whence he travelled in a westerly
direction across high mountains to the valley of the Pa-chu.
This river, which rises to the eastward of Chumalari, in Tibet,
has an almost due south course to the plains; but the Monas
as well as the Subansiri have a south-west course in Bhotan:
higher up they probably run south-east, and bend round to
south-west in a curve somewhat parallel to that of the Yaru
or Dihong, which afterwards becomes the Brahmaputra.

In western Bhotan the mountain-ranges are lofty and rugged, and the river-courses very deep and generally narrow.
At Panaka the Pa-chu is only 3700 feet above the sea, though
eighty miles distant from the plains; and the Monas, where
Pemberton and Griffith crossed it, is only 1400 feet, while the
range south of it attains an elevation of 9500 feet. In their
journey from the Monas to Panaka, these travellers crossed
ridges 12,400 feet in height. On their return to India they
followed Captain Turner's route.

The mountain mass which descends from the axis of the
Himalaya to separate the Monas from the Subansiri attains an
elevation of at least 24,000 feet as far south as latitude 28°.
Three peaks upon this are visible from the Khasia mountains,
and spurs descending from it were ascended to an elevation

of nearly 12,000 feet by Mr. Booth in 1849, in a district north of Bishnath, in Upper Assam, which is inhabited by a race called Duphlas. He collected some Ferns, and especially seeds of Rhododendrons, of which an account has been published by Nuttall in ' Hooker's Journal of Botany.'

Mr. Griffith's attention was of course mainly devoted to the botany of the district, and in his ' Itinerary Notes' and journals we have a mass of important information regarding the general features of the vegetation, together with a great deal of detail which will become valuable as soon as the species are determined.

The climate of Bhotan seems to be very equable, and the humidity of the winter months appears to increase to the eastward. We do not, however, possess any records of temperature or humidity, and our inferences regarding the climate are drawn from the vegetation only. The steepness with which the mountains rise, and the influence of the elevated mass of the Khasia to the south, make the lower mountains which skirt the plains of Assam, between the Godada and the Monas, drier than those nearer Sikkim, which are exposed to the full force of the monsoon, or than those further east.

The deep narrow valleys of the great rivers carry a tropical vegetation very far into the interior of Bhotan, among lofty mountains capped with almost perpetual snow. These attract to themselves so much of the moisture of the atmosphere, that the bottoms of the valleys are everywhere comparatively dry and bare of forest, which only begins at about 6000 feet of elevation, except in ravines. The outer ranges, too (except near Sikkim), even above this level are only partially wooded, the trees being arranged in clumps, among which are interspersed open grassy glades, which are compared by Griffith to those of Khasia; Oaks and Rhododendrons being extremely abundant.

On the northern face of the range which separates the Monas valley from Assam, Pines make their appearance, the first species being *Pinus longifolia* in the drier valleys below 6000

2 *a*

feet.   On the more humid ranges *Abies Brunoniana* appears
at 8000 feet, and above it *Picea Webbiana*.   *Pinus excelsa*
also occurs abundantly, as well as the Yew, and *Cupressus
funebris* is cultivated as low as 2000 feet, and a very little way
from the Assam plain.   Further in the interior *Abies Smithiana* occurs, and *Larix Griffithii* to the westward, *Pinus longifolia* being still found in the hot dry valleys.

In general features the flora of Bhotan resembles that of
Sikkim, which is much better known.   It differs principally
by containing several Khasia and eastern forms which do not
extend further west, such as *Liquidambar, Corylopsis,* and an
oak with leaves like *Robur* (*Quercus Griffithii*, H.f. et T.).
These are chiefly plants of the subtropical and lower temperate
zone; while those of the upper temperate and subalpine zone
appear, so far as we have had an opportunity of comparing
them, to be almost identical with those of Sikkim.   It must,
however, be recollected that the collections of Griffith are all
from the western parts of Bhotan, and that the eastern parts
are not at all known.

3. SIKKIM.

The province of Sikkim, though of very limited extent, is
now the best known part of the central or eastern Himalaya,
and presents many features of much interest.   It consists entirely of the basin of the river Tista, which, with its tributaries, drain the whole country.   The course of this river is for
the most part meridional, that is, perpendicular to the plains;
and the same may be said of its great tributary the Rangit
river, which joins it from the west, flowing for a short distance parallel to the plains, through a deep ravine not 1000
feet above the sea, to the north of a transverse range elevated 7–8000 feet.

The position of Sikkim, opposite to the opening of the
Gangetic valley, between the mountains of Bahar on the one
hand, and those of Khasia on the other, exposes it to the full
force of the monsoon; its rains are therefore heavy and almost

uninterrupted, and are accompanied by dense fogs and a saturated atmosphere. This weather indeed prevails throughout the year, as there are frequent winter rains, which are generally accompanied by cold fogs, and alternate with frost and snow. March and April are the driest months, and in fine seasons are often bright and clear, but the rains commence in May, to continue with little intermission till October. The bounding mountain-chains are very lofty, and snow-clad throughout a great part of their extent, but the central range which separates the Rangit from the Tista is depressed till very far in the interior. The river-valleys are also considerably depressed, but less markedly so than those of western Bhotan. The rainy winds have thus free access to the heart of the province, and sweep almost without interruption up to the base of Kanchinjanga (28,178 feet), the loftiest mountain and most enormous mass of snow in the world. The snow-level is here about 16,000 feet. Between the two principal sources of the Tista, however, the Lachen and the Lachung, a lofty snowy range is projected; and as this chain has a south-west direction, and is moreover sheltered to a considerable extent by the boundary chain between Sikkim and the Tibetan valley of Chumbi, we have in these valleys a rapid diminution of the rain-fall and an equally rapid transition to the Tibetan climate, while the level of perpetual snow rises to above 18,000 feet.

From the level of the sea to an elevation of 12,000 feet Sikkim is covered with a dense forest, only interrupted where village clearances have bared the slopes for the purposes of cultivation; and there the encroachment of the forest is with difficulty prevented by frequent fires and the incessant labour of the villagers. The forest consists everywhere of tall umbrageous trees; with little underwood on the drier slopes, but often dense grass jungle; more commonly however it is accompanied by a luxuriant undergrowth of shrubs, which renders it almost impenetrable. In the tropical zone large Figs abound, with *Terminalia, Vatica, Myrtaceæ,* Laurels, *Eu-*

phorbiaceæ, Meliaceæ, Bauhinia, Bombax, Morus, Artocarpus, and other *Urticaceæ,* and many *Leguminosæ;* and the undergrowth consists of *Acanthaceæ,* Bamboos, several *Calami,* two dwarf *Arecæ, Wallichia,* and *Caryota urens.* Plantains and tree-ferns, as well as *Pandanus,* are common; and, as in all moist tropical countries, ferns, orchids, *Scitamineæ,* and *Pothos* are extremely abundant. Few oaks are found at the base of the mountains, and the only conifers are a species of *Podocarpus* and *Pinus longifolia,* which frequents the drier slopes of hot valleys as low as 1000 feet above the level of the sea, and entirely avoids the temperate zone. The other tropical Gymnosperms are *Cycas pectinata* and *Gnetum scandens,* genera which find their north-western limits in Sikkim.

The rarity of oaks at the base of the mountains must be ascribed to the great dryness and winter's cold of that part of the chain, for we miss also other eastern types which abound in the equable and moist climate of the Malayan archipelago and peninsula, such as *Liquidambar* and nutmegs; whilst *Dipterocarpeæ,* and especially *Anonaceæ,* are exceedingly few in number. *Liquidambar* is common in the Assam jungles, and indicates their greater humidity. The same inference may be drawn with regard to the tropical belt of the Khasia, from the occurrence there of two nutmegs and numerous *Anonaceæ.*

Oaks, of which (including chesnuts) there are upwards of eleven species in Sikkim, become abundant at about 4000 feet, and at 5000 feet the temperate zone commences, the vegetation varying with the degree of humidity. On the outermost ranges, and on northern exposures, there is a dripping forest of cherry, laurels, oaks and chesnuts, *Magnolia, Andromeda, Styrax, Pyrus,* maple and birch, with an undergrowth of *Araliaceæ, Hollböllia, Limonia, Daphne, Ardisia, Myrsine, Symplocos, Rubi,* and a prodigious variety of ferns.

Plectocomia and *Musa* ascend to 7000 feet. On drier exposures bamboo and tall grasses form the underwood. Rhododendrons appear below 6000 feet, at which elevation snow falls occasionally. From 6–12,000 feet there is no apparent diminution of the humidity, the air being near saturation during a great part of the year; but the decrease of temperature effects a marked change in the vegetation. Between 6000 and 8000 feet epiphytical orchids are extremely abundant, and they do not entirely disappear till a height of 10,000 feet has been attained. Rhododendrons become abundant at 8000 feet, and from 10,000 to 14,000 feet they form in many places the mass of the shrubby vegetation. *Vaccinia*, of which there are ten species, almost all epiphytical, do not ascend so high, and are most abundant at elevations of from 5000 to 8000 feet.

The flora of the temperate zone presents a remarkable resemblance to that of Japan, in the mountains of which island we have a very similar climate, both being cold and damp. *Helwingia, Aucuba, Stachyurus,* and *Enkianthus* may be cited as conspicuous instances of this similarity, which is the more interesting because Japan is the nearest cold damp climate to Sikkim with whose vegetation we are acquainted. At 10,000 feet (on the summit of Tonglo) yew makes its appearance, but no other conifer except those of the tropical belt is found nearer the plains than the mountain Phalút, north of Tonglo, on which *Picea Webbiana* is found, at levels above 10,000 feet. *Abies Brunoniana* is first met with at 9000 feet in the Rangit valley, at Mon Lepcha, and *A. Smithiana* and *Brunoniana,* and the larch, are found everywhere in the valleys of the Lachen and Lachung rivers, above 8000 feet. The Pines are thus specifically the same as those of Bhotan, except *Pinus excelsa,* which occurs nowhere in Sikkim.

A subtropical vegetation penetrates far into the interior of the country along the banks of the great rivers; rattans, tree-ferns, plantains, screw-pines, and other tropical plants occurring in the Ratong valley, almost at the foot of Kanchinjanga, and 5000 feet above the level of the sea. With the pines, however, in the temperate zone, a very different kind of vegetation presents itself. Here those great European families which are almost entirely wanting in the outer temperate zone become common, and the flora approximates in character to that of Europe, though not to the same extent as that of the western Himalaya does. Shrubby *Leguminosæ*, such as *Indigofera* and *Desmodium*, *Ranunculaceæ* (*Thalictrum*, *Anemone*, *Delphinium*, *Aconitum*, etc.), *Umbelliferæ*, *Caryophyllæ*, *Labiatæ*, and *Gramineæ*, increase in numbers as we advance into the interior. The air becomes drier, and from the increased action of the sun the temperature does not diminish in proportion to the elevation, the summers being warmer, though the winters are colder. The forests at the same time become more open, and are spread less uniformly over the surface, the drier slopes being bare of trees, and covered with a luxuriant herbaceous vegetation. It is only in the upper part of the valley of the Tista, however, above the junction of the Lachen with the Lachung, that this change becomes marked; and from the rapidly increasing elevation, not only of the surrounding mountains, but of the floors of the valleys, it proceeds with great rapidity, and the temperate soon gives place to an alpine flora.

The subalpine zone in Sikkim scarcely begins below 13,000 feet, at which elevation a dense rhododendron scrub occupies the slopes of the mountains, filling up the valleys so as to render them impenetrable. Here the summer is short, the ground not being free of snow till the middle of June. It is, however, comparatively dry, and the alpine flora very much resembles that of the western Himalaya and (in generic types at least) the alps of Europe and western Asia; while as we advance towards the Tibetan region we have a great increase of dryness, so that a Siberian flora is rapidly developed, which at last entirely supersedes that of the subalpine zone, and ascends above 18,000 feet.

A small herbarium of Dorjiling plants was, we believe, formed by collectors sent by Griffith while in charge of the Calcutta Botanic Garden, but our knowledge of the vegetation of Sikkim is entirely derived from our own collections, which we believe to be very complete. These consist of about

2770 species of flowering plants and 150 ferns, of which the majority inhabit the temperate zone; fewer are tropical, and still fewer alpine. The prevailing natural orders are:—

Ranunculaceæ	. . . .	55
Papaveraceæ	. . . .	25
Fumariaceæ	. . . . .	16
Magnoliaceæ	. . . .	7
Malvaceæ ⎫		
Bombaceæ ⎬	. . . .	30
Tiliaceæ ⎪		
Byttneriaceæ ⎭		
Ternstrœmiaceæ	. . .	11
Aurantiaceæ	. . . .	12
Caryophylleæ	. . . .	30
Cruciferæ	. . . . .	30
Vitaceæ	. . . . .	20
Balsamineæ	. . . .	18
Acerineæ	. . . . .	6
Leguminosæ	. . . .	100
Rosaceæ	. . . . .	80
Umbelliferæ	. . . .	50
Araliaceæ	. . . .	26
Melastomaceæ	. . . .	10
Cucurbitaceæ	. . . .	20
Rubiaceæ	. . . .	80
Crassulaceæ	. . . .	16
Compositæ	. . . .	170
Ericeæ ⎫		
Vaccinieæ ⎭	. . .	60

Gentianeæ	. . . . .	38
Asclepiadeæ ⎫		
Apocyneæ ⎭	. . .	45
Scrophularineæ	. . . .	70
Labiatæ	. . . . .	90
Cyrtandreæ	. . . . .	27
Myrsineæ	. . . . .	12
Primulaceæ	. . . . .	36
Boragineæ	. . . . .	18
Acanthaceæ	. . . . .	35
Polygoneæ	. . . .	45
Euphorbiaceæ	. . . .	35
Urticeæ	. . . . .	110
Amentaceæ	. . . . .	15
Coniferæ	. . . . .	10
Laurineæ	. . . . .	30
Aroideæ	. . . . .	16
Orchideæ	. . . . .	150
Scitamineæ	. . . . .	24
Palmeæ	. . . . .	10
Smilaceæ ⎫		
Liliaceæ ⎭	. . . .	40
Junceæ	. . . . .	25
Gramineæ	. . . .	180
Cyperaceæ	. . . .	106

Central Himalaya, or Nipal.

The kingdom of Nipal extends for 500 miles along the Himalaya, from the western extremity of Sikkim to the eastern border of Kumaon, from which it is separated by the river Kali. The jealous policy of the Nipalese government has prevented our acquiring an intimate knowledge of this country, the only part to which Europeans have been allowed access (with one exception) being the capital, Kathmandu, elevated 4000 feet above the sea, and distant about thirty miles from the plains of India. Here a British Resident has resided since 1817, and several botanists have been enabled to explore its vegetation. To these the Government of Nipal, though invariably refusing permission to penetrate far into the interior, has always afforded every facility for prosecuting their researches by permitting the despatch of collectors.

Dr. Buchanan Hamilton visited Nipal in 1802, remaining for more than a year, during which time he explored the valley of Kathmandu and surrounding mountains. His plants were described by David Don in the ' Prodromus Floræ Nepalensis,' a work which should have been alluded to in conjunction with Wallich's 'Tentamen' at page 51. In 1820 Dr. Wallich arrived at Kathmandu. During his residence in the valley he laboured indefatigably in the investigation of the rich and scarcely known flora by which he was surrounded; collectors were despatched in every direction, and a great Herbarium was formed, which is well known to science. The flora of the subtropical and lower temperate zone was probably almost wholly exhausted; but the alpine zone was much less completely explored, as the task had to be confided to Bengali collectors, who dread cold, and by whom many small alpine plants would naturally be overlooked. The collectors were sent to the valley of the Gandak and the neighbourhood of the great mountain Gosainthan.

In 1845, Dr. Hoffmeister, a German traveller and botanist, visited Kathmandu, but we have not had an opportunity of learning whether or not he made any collection there. A small collection, which now forms a part of the Hookerian Herbarium, was made there by the late Mr. Winterbottom. Between the Gandak and the Kali the country has not been traversed by any European, nor had any part of eastern Nipal been visited till 1848, when Dr. Hooker, by permission of the Nipalese Government, entered it from Sikkim, visited the Tambar river, the most easterly tributary of the Aran, ascending its valley from an elevation of 1000 feet, as far as its sources in the Walanchún and Kanglachem passes (16–17,000 feet). This journey was made during winter, and therefore gave less important results botanically than would have been obtained at a more favourable season.

It is unnecessary to dwell at length on the general character of the surface of Nipal, as to do so would only be to recapitulate what has already been said regarding the Himalaya in general. Little is known of the details of the higher parts of the chain, or of the position of the axis of the Himalaya, which probably lies in general very far back. The political frontier of Tibet is usually far to the south of the axis, the upper part of the course of most of the rivers of the Indian slope of the chain belonging almost invariably to Tibet. Two giant masses project from the axis towards the Indian plain, the culminant peaks of which form a conspicuous feature from Kathmandu, and even from the Gangetic plain, so that their elevation has been approximately determined; that of Dhawalagiri being 27,600 feet, and that of Gosainthan 24,700 feet. By these masses the whole of Nipal is divided into three great river-basins,—that of the Karnali or Gogra to the westward, that of the Gandak in the centre, and that of the Kosi or Aran to the eastward*. These divisions are no doubt highly natural. For our purposes a subdivision is little necessary, from our very slight acquaintance with the flora of any part of Nipal except that in which Dr. Wallich collected, and it will suffice to distinguish eastern, central, and western Nipal, whenever it appears requisite to assign particular localities to our plants.

* See an excellent paper by Mr. Hodgson in the Journal of the Asiatic Society of Bengal, in which the importance of the river-basins as geographical divisions is forcibly pointed out. Mr. Hodgson has however misunderstood Captain Herbert's views, which are certainly the same as his own in that respect. Captain Herbert's proposition, that the line of the great peaks intersects the river-basins (and is therefore not the true axis of the Himalaya), was the first enunciation of a very important fact in physical geography, the true significance of which is not yet duly appreciated.

There are probably many mountains equally elevated with those just enumerated, but bearing a less important relation to the river systems. A very lofty peak between the Kosi and its tributary the Aran has been conjectured to be almost as lofty as Kanchinjanga, but on very imperfect data. The uniform appearance of snowy masses throughout the whole extent of Nipal, leaves no doubt, however, as to the great elevation of the axis of the chain and the mountains of the interior.

With regard to the outer mountains we have no detailed information, except of those in the immediate neighbourhood of Kathmandu, where Sheopore, on the watershed between the Gandak and the Kosi, is upwards of 10,000 feet. On the whole, if we may judge from the distribution of the rivers, the outer mountains of Nipal are probably less elevated than those of other parts of the Himalaya, the width of the river basins being comparatively great, so that the boundary ridges ramify repeatedly, and run for a considerable length without much increase of altitude. In eastern Nipal the outer and central ranges are very much lower than those of Sikkim, and the open valleys and low mountains of central Nipal indicate that the same is the case there.

The climate of Nipal has been discussed with that of the Himalaya generally. There is probably a somewhat abrupt transition from the humid winter of Sikkim to the drought which prevails at that season in the western Himalaya, as the proximity, not only to the sea, but also to the great mass of snow-clad mountains which in Sikkim advances to within sixty miles of the plains, is no doubt the cause of the superabundance of moisture in that province. We may therefore expect to find all the eastern or humid types of the subtropical Sikkim flora wanting in the forest between Kathmandu and the Gangetic plain. Accordingly, among palms, *Areca .gracilis* and *disticha*, *Licuala* and *Caryota* have disappeared, and one or two *Calami*, *Chamærops*, *Phœnix acaulis*, and *Wallichia* alone occur. With diminished humidity we find increased sun-power, to which the open nature of many of the valleys contributes in no small degree.

The principal plants of the tropical zone of Nipal belong to a less humid type than those of Sikkim, and are abundant all over the subtropical mountains of India, where a dry and wet season alternate. The commonest trees are *Moringa, Putranjiva, Bombax, Vatica robusta, Buchanania, Spondias, Butea frondosa* and *parviflora, Erythrina, Acacia Lebbek* and *stipularis, Bauhinia purpurea* and *Vahlii, Ventilago, Conocarpus, Terminalia, Nauclea cordifolia,* and *Ulmus integrifolia.*

In the plain of Kathmandu, which is elevated 4000 feet, the ground is in a great measure under cultivation, and the hills are bare of trees. The vegetation and climate are therefore subtropical, and from the position of the Kathmandu plain, close to the ridge of the spur which separates the basins of the Gandak and Kosi, its mean level is probably greater than that of many of the valleys of both rivers, and of the ridges which separate their tributaries.

In the temperate flora of central Nipal, for the same reason, the Japanese and Malayan types are much fewer; *Enkianthus, Stachyurus, Vaccinia, Aucuba, Helwingia,* several *Rubi,* and *Rhododendron Dalhousiæ* and *Edgeworthii* being all absent, while European and west Himalayan forms which are wanting in Sikkim make their appearance. In the extreme east of Nipal, in the valley of the Tambar river, Rhododendrons are scarcely less abundant than in Sikkim; but those of the temperate zone are certainly entirely wanting in that part of central Nipal from which Dr. Wallich obtained his collections, with the exception of *R. arboreum,* which is found throughout the whole Himalaya, *R. barbatum,* which extends to Kumaon, and *R. campanulatum,* which is a subalpine species. The more alpine species cannot be so positively affirmed to be absent, but it is highly probable that the number of species is not great, none having been obtained by Dr. Wallich's collectors, but such as are universally distributed throughout the Himalaya. The pines are the same as those of Sikkim, except that *Pinus excelsa* is common, and the larch is not found west of the Kosi.

In the present state of our knowledge, it is not safe to institute a comparison between the alpine flora of Nipal and that of Sikkim. Wallich's collections show us that the species are on the whole the same. There is evidently a very gradual change as we advance westward, partly owing, it may be presumed, to increase of latitude and of summer drought, and partly to more obscure causes which regulate the distribution of plants. The elucidation of these will, we trust, be one of the most important results of this work when completed, but with our present imperfect knowledge of species the subject cannot be approached. The occurrence of Siberian types in small numbers among Wallich's alpine plants shows that the climate to the North becomes at last arid, exactly as elsewhere in the Himalaya.

Though unable to indicate with any approach to precision the number of Nipalese genera and species that are common to the Eastern and Western Himalaya respectively, we have collected a few instances of Himalayan species that we believe find their limits in Nipal. Of these the majority of the Western Himalayan forms that advance no further east are of European and Oriental genera or even species, as :—

Caltha *palustris*.   Rosa *moschata*.
Delphinium *vestitum*.   Ulmus *campestris*.
Cratægus *Pyracantha*.

Others are more peculiarly Himalayan :—

Chamærops *Martiana*.   Potentilla *atro-sanguinea*.
Quercus *lanata*.   „   *Nipalensis*.
Stranvæsia *glaucescens*.   Spiræa *Kamtschatica*.
Rosa *Lyellii*.

Of these the *Stranvæsia*, though not found further eastward in the Himalaya, occurs in the Khasia, and perhaps the *Chamærops* may be the same as the Khasian species. The *Spiræa Kamtschatica* is a native of Eastern Siberia.

The number of Eastern Himalayan and Khasian forms that

advance no further to the westward will, we do not doubt, prove very much larger, as the following list of species already identified proves :—

Aconitum *palmatum.*
Manglietia *insignis.*
Magnolia *sphenocarpa.*
Michelia *excelsa.*
„   *lanuginosa.*
Sphærostemma *elongatum.*
Stephania *hernandifolia.*
Berberis *Wallichiana.*
„   *angulosa.*
Meconopsis *simplicifolia.*
„   *Nipalensis.*
„   *Wallichii.*
Corydalis *juncea.*
Pyrus *Indica.*
„   *foliolosa.*
Cotoneaster *rotundifolia.*
Eriobotrya *elliptica.*
Photinia *dubia.*
„   *integrifolia.*
Rubus *rugosus.*
„   *calycinus.*
Cerasus *rufa.*
„   *acuminata.*
Neillia *thyrsiflora.*

Sanguisorba *decandra.*
Panax *Pseudo-ginseng.*
Hedera *polyacantha.*
Toricellia *tiliæfolia.*
Wightia *gigantea.*
Schœpfia *fragrans.*
Gaultheria *fragrantissima.*
Pieris *formosa.*
Edgeworthia *Gardneri.*
Eriosolæna *Wallichii.*
Cinnamomum ? *caudatum.*
Benzoin *Neesianum.*
Phœbe *paniculata.*
Tetranthera *sericea.*
„   *elongata.*
„   *oblonga.*
Sphærocarya *edulis.*
Helicia *robusta.*
Corylus *ferox.*
Quercus *serrata.*
„   *Arcaula.*
„   *lamellosa.*
Podocarpus *macrophylla.*
Larix *Griffithii.*

A considerable number of tropical forms also creep along the base of the Himalaya as far west as the valley of Nipal, which have not been collected in Kumaon or west of it, as :—

Dillenia *speciosa.*
„   *aurea.*
Saccopetalum *tomentosum.*

Parabæna *sagittata.*
Cocculus *mollis.*
Castanea *Indica.*

and a species of Calamus.

Western Himalaya.

The mean elevation of the western Himalaya is not mate-

rially less than that of the eastern, for the passes over the
principal chains are quite as lofty, though none of the peaks
attain the extreme altitude of Kanchinjanga or Dhawalagiri.
The highest mountain west of Nipal is Nanda Devi in Kumaon, 25,750 feet, but there are many peaks above 20,000
in all parts of the range.   The last great peak is Dayamar,
north-west of Kashmir, the height of which is 20,000 feet,
beyond which the chain dips rapidly to the Indus.

The main chain of the western Himalaya, commencing near
the great peak of Kailas, north of the lake Mansarowara, runs
to the south of and parallel to the Indus, which it separates
first from the Satlej, then from the Chenab, and latterly from
the Jelam.   To the eastward this chain is entirely Tibetan,
but north-west of Piti it separates Lahul and Kishtwar from
the Tibetan districts of Parang and Zanskar; still further
west it separates Kashmir from Dras, and finally terminates
at the great bend of the river Indus.

The primary ramifications of the main chain are three in
number.   One (the Cis-Satlej Himalaya) is given off close to
the great lakes, and separates the Satlej basin from that of
the Ganges and its tributaries, terminating in the plains of
Hindostan near Nahan. A second (the Cis-Chenab Himalaya)
branches off from the main chain near the lake Chumoreri in
Tibet, and separates the basin of the Chenab from those of
the Beas and Ravi, terminating in the plain of the Panjab a
little east of Jamu.   The third principal branch of the chain
separates the Chenab from the Jelam.

Our knowledge of the Western Himalaya is so much more
definite than that which we possess regarding Nipal and the
eastern provinces, that it is necessary to adopt a more minute
subdivision. The following districts will be frequently referred
to, and described in detail at a future page :—

1. Kumaon.

2. Garhwal.

3. Simla; including Sirmur and Basehir and a number of
petty states, extending from the Jumna to the Satlej.

4. Kunawar; the upper part of the Satlej basin to the Tibetan districts of Piti and Guge.

5. Kulu; including Mandi and other petty states in the basin of the Beas.

6. Chamba; the basin of the Ravi.

7. Lahul; the highest and subtibetan course of the Chenab.

8. Kishtwar; the middle part of the Chenab basin.

9. Jamu; the lower part of the Chenab basin, including Banahal.

10. Rajaori; the states between Kashmir and the plains.

11. Kashmir.

12. Hazara or Marri.

In consequence of the increased distance from the sea, and partly also from the great obliquity of many of the great mountain ranges, the rain-fall in the Western Himalaya is much less considerable than it is in the Central and Eastern. The rain-fall also diminishes, *cæteris paribus*, regularly and gradually from east to west, but the amount varies so much with local circumstances that, unless used with proper caution, absolute numbers are apt to mislead. Thus, while the average rain-fall at Naini Tal, elevated 6500 feet on the last spurs of the Gagar overhanging the plains of Rohilkhand, is 88 inches, at Almora, elevated 5500 feet, but fifteen miles further from the plains, only 34 inches fall. The fall at Naini Tal may however be compared with that of Dorjiling (125 inches), for in both these localities there is no considerable amount of higher land interposed between them and the plains of India. The rain-fall at Masuri and at Simla is materially less.

The vegetation of the Western Himalaya alters with the climate, presenting a very gradual transition from the flora of Nipal to that of the arid Afghan hills. This is the case equally in the tropical, temperate, and alpine zones of vegetation, and in the interior as well as in the exterior Himalaya.

In the tropical zone of Kumaon a dense forest skirts the base of the mountains, corresponding in all its features with that which we have indicated as prevalent in similar localities in Nipal. The forest is most luxuriant where the higher mountains overhang the plains, and becomes stunted or disappears entirely where a great river debouches on the plain. In Garhwal, west of the Ganges, the forest which skirts the Siwalik hills is less extensive, but many parts of the Dehra Dhún are densely wooded. A species of *Calamus* which grows in its jungles marks the western limit of that genus along the Himalaya. West of the Jumna the vegetation changes rather suddenly. A similar change has already been indicated at the same place in the plain's vegetation (page 161), but the forest belt close to the mountains, being always more humid than the plain at a distance from them, their vegetation is never the same. The gigantic *Bombax*, and the lofty trees of *Nauclea, Lagerstrœmia, Conocarpus, Terminalia, Sterculia,* and others, and the scandent species of *Butea, Bauhinia, Millettia, Ventilago,* etc., have however disappeared, and spinous bushes or stunted trees of *Zizyphus Jujuba, Butea frondosa, Cassia Fistula, Acacia Arabica* and *Catechu,* form the greater part of the jungle, mixed with *Diospyros cordifolia, Adhatoda Vasica,* and *Isora corylifolia.* In the extreme west, *Acacia modesta* becomes very abundant, and beyond the Jelam the flora is identical with that of the lower Afghan hills.

The tropical vegetation advances far within the mountains, ascending the valleys of the great rivers, and corresponding in character with the forest belt without, but often rather drier. In eastern Kumaon the humid valley of the Sarju is filled with dense forest. The curious palm *Wallichia oblongifolia* has there its western limit, and a pepper, a *Pothos,* an arborescent *Aralia,* and a few other plants indicative of humidity, still linger in its recesses. The valley of the Ganges is much drier and contains little forest, and the tropical portions of the Jumna and the Satlej are quite bare. In the Satlej valley, Afghan forms, unknown further east, begin to make their appearance,—*Paliurus* and *Olea cuspidata* being the most conspicuous. To these are added, in the Chenab valley, *Acacia modesta, Zizyphus Lotus*, and a spiny *Celastrus*, which west of the Jelam form the great mass of the tropical vegetation. Of tropical fruits, the orange and plantain are cultivated in all the hot valleys of the Panjab Himalaya; and the mango extends to the Indus, and perhaps beyond it. The pomegranate, both wild and cultivated, is abundant in the subtropical jungles, even as far west as Lower Kishtwar.

In the temperate zone of the outer Western Himalaya, the commonest trees of the drier exposures are *Rhododendron arboreum, Andromeda ovalifolia, Quercus incana* and *dilatata*; and the prevailing shrubs are species of *Berberis, Rosa, Spiræa, Rubus*. All of these occur throughout the whole of the chain from Kumaon to the Indus, but to the westward they seem restricted within gradually narrower limits, and in the extreme west are found only in moist and shady woods, which in Kumaon and Garhwal they carefully avoid. To the eastward they are accompanied by many other trees which gradually disappear: thus *Quercus lanata* and *Betula cylindrostachya* are not found west of the Ganges, and *Carpinus viminea* has not been observed west of the Satlej.

In the valleys of the temperate zone and on the lower slopes of the hills the forest is usually very different: *Celtis, Alnus, Populus ciliata, Prunus Padus, Æsculus*, and two species of *Acer* are common trees as far west as the Jelam, or perhaps the Indus. Most of them indeed seem to occur in the humid forests of the Hindu Kúsh, north of Jelalabad. *Benthamia floribunda* and a *Hydrangea* extend from the Eastern Himalaya as far as the Satlej, but have not been found further west, and many species of *Lauraceæ* advance to the Indus.

The influence of climate is much more perceptible on the herbaceous vegetation of the temperate region, and especially on the annual plants which spring up during the rainy season, than on the trees and larger shrubs, which may be presumed to have greater powers of resistance. Hence the *Scitamineæ*, epiphytical and terrestrial *Orchideæ, Araceæ, Cyrtandraceæ*,

2 c

Melastomaceæ, and *Begoniæ*, which form so conspicuous a part of the vegetation of the humid eastern Himalaya, occur in very small numbers in Kumaon, rapidly diminish to the westward, and scarcely extend beyond the Satlej. *Streptolirion* and *Adenocaulon*, two of Mr. Edgeworth's most remarkable discoveries in the Simla Himalaya, which there find their western limit, are in like manner Sikkim forms. *Balanophora* also extends west as far as the Satlej, while *Colquhounia* and *Heterophragma* have not been found west of Kumaon.

The cultivation of fruit-trees affords a remarkable exemplification of the difference between the climate of the Eastern and Western Himalaya. In Sikkim no European fruit of any kind, save the strawberry, comes to perfection; even the peach, the only commonly cultivated tree, does not ripen its fruit, and the apricot, the most abundant Western Himalayan fruit, is unknown. In central Nipal, apples, figs, peaches, quinces, and apricots, all ripen, but hardly arrive at perfection. Towards the interior of Kumaon apricots and all the above fruits become abundant, with the pear and cherry; and from Kumaon westward, vineyards and large orchards form a conspicuous feature in the scenery of all interior temperate valleys.

Of the cerealia, Wheat and Barley are the staple crops (as throughout Northern India); the various millets and rice are however cultivated in hot valleys at all elevations below 5–6000 feet, with occasionally maize and sugar-cane. Buckwheat is grown at 5–8000 feet, and the various *Amaranthaceæ* of the Eastern Himalaya extend also to the Western. The cultivation of Tea on the slopes of the outer ranges of Kumaon and Kulu appears to be increasing with great rapidity, and promises to be eminently successful.

The coniferous trees which are common to the Eastern and the Western Himalaya are—1. *Pinus longifolia*, which is found on drier exposures from 7000 as low as 2000 feet, and extends to the mountains of Hindu Kúsh. 2. *P. excelsa*, which occurs in all parts of the Himalaya (except Sikkim),

as well as in Balti (in Western Tibet) and in Afghanistan.
3. *Abies Smithiana,* which also inhabits all parts of the Himalaya, extending into Afghanistan. 4. *A. Brunoniana,* which is not found further west than the upper part of the valley of the Kali, in Eastern Kumaon. 5. *Picea Webbiana,* the most alpine of all the species which ranges from Bhotan to Kashmir : it covers the mountains, between 8000 and 12,000 feet, with a sombre forest, appearing equally at home in the humid climate of Sikkim and on the arid mountains of Upper· Kunawar. 6. *Juniperus recurva.* 7. *J. Wallichiana.* 8. *J. excelsa.* 9. *Taxus baccata.* The two first of the junipers, and the yew, are found in all parts of the Himalaya.

Two species only are confined to the Eastern Himalaya, namely, *Larix Griffithii* and *Podocarpus macrophylla ;* but *Pinus Sinensis,* so common in Khasia, will perhaps prove to be a native of Eastern Bhotan. The Western Himalaya has four species which are not found in Nipal or the Eastern Himalaya. These are—1. *Pinus Gerardiana,* a native of Afghanistan, of Hasora, north of Kashmir, and of the drier valleys of the Himalaya as far as the Satlej. 2. *Cedrus Deodara,* which is scarcely indigenous in Eastern Kumaon, and ranges from Garhwal to Afghanistan. The deodar is closely allied to, if not identical with the cedar of Lebanon, which extends from Syria and the Taurus to the Atlas mountains. 3. *Cupressus torulosa,* which is probably the wild state of the common cypress; it is a rare plant in the Himalaya, but is found at Niti, near Simla, and at Naini Tal, and may perhaps occur in Western Nipal. 4. *Juniperus communis,* found in all the drier parts of the chain from Afghanistan and Kashmir to Kumaon.

There is no abrupt transition from the flora of the outer temperate Himalaya to that of the interior. The amount of rain-fall diminishes very gradually as we ascend the great valleys, and the diminution of humidity is accompanied by the appearance of new types of vegetation. This transition is most observable in the Satlej and Chenab valleys, which lie so obliquely to the axis of the chain that they have a long course through a moderately dry climate. The valleys of the other rivers (except the Jelam) are much more perpendicular to the axis, and the humid vegetation passes almost immediately into an alpine and Tibetan flora, without the intervention of a dry temperate flora.

It must not be supposed that the vegetation of the interior temperate Himalaya is altogether, or even in a great measure, different from that of the outer ranges. A very large proportion of the species is the same throughout both regions, consisting of western forms, to which even heavy rain at one season is not injurious so long as a great portion of the year is dry, but whose progress to the east is stopped as soon as the humidity becomes permanent. The rains' vegetation of the outer mountains is, however, entirely absent from the interior, and its place is taken by such Tibetan forms as are not entirely intolerant of moisture. The presence of *Pinus Gerardiana, Ephedra, Quercus Ilex, Ribes Grossularia,* and *Dianthus,* may be considered as indicating that the rains are very trifling in amount in average seasons. *Pinus longifolia* disappears, with *Rhododendron arboreum* and its associated plants; but all the other pines continue to the upper limit of trees, or to the borders of Tibet. The cultivation of the vine is only carried on in this inner region, the rainy season of the outer mountains preventing the ripening of grapes.

West of the Ravi the rain-fall has so much lessened even on the outer hills, that it is only on the first range which rises into the temperate zone, that the normal West Himalayan vegetation (*Quercus incana,* etc.) occurs; while the valleys immediately north of it, when sheltered by hills rising continuously to 9000 or 10,000 feet, present many of the features characteristic of the interior Himalaya. The presence or absence of *Quercus incana, Rhododendron arboreum,* and *Andromeda ovalifolia,* on the one hand, and of *Pinus Gerardiana* and *Ephedra* on the other, may be regarded as a fair criterion of the two extreme climates; but there are many valleys in the extreme west from which both classes of plants are absent, or in which these exterior Himalayan trees are found along with forms common in Kunawar and Kishtwar. *Fothergilla involucrata* (first observed by Falconer, in Kashmir) is a curious instance of a tree plentiful in all parts of the temperate zone, from Kashmir to the Ravi, but not found further east.

The alpine flora of the Western Himalaya presents the same gradual transition from humid and eastern types to the characteristic forms of Western Asia, which we have observed in the tropical and temperate zones. The mountains of Eastern Kumaon are rich in beautiful Nipal forms, such as *Cyananthus, Meconopsis, Codonopsis,* various gentians, saxifrages, and many others; but their number rapidly diminishes as we advance westward, and the vegetation of the higher Alps of Kashmir is almost identical even in species with that of the mountains of Afghanistan, Persia, and Siberia.

For our earliest knowledge of the vegetation of the Western Himalaya we are indebted to Dr. Govan, who seems to have explored some parts of Sirmur and Garhwal, and to General Hardwicke, who travelled in Garhwal and communicated plants to Roxburgh and Wallich. The Wallichian Herbarium contains specimens from both these travellers, and also from the Gerards, who collected in the Simla hills and in Kunawar. Dr. Wallich's travels extended only to Hardwar and Dehra Dhún, but he also distributed extensive collections made in the interior of Kumaon by Blinkworth and others.

The list of botanists who have investigated the flora of the Western Himalaya, includes the names of Royle, Jacquemont, Falconer, Griffith, Munro, Edgeworth, Madden, Strachey, Winterbottom, and Fleming; but we have already (pp. 60–70) entered into such details regarding their labours, as to render it unnecessary to dwell upon them here. Mr. Edgeworth collected in Kumaon, Garhwal, Simla, and Kunawar, and he has recently communicated to the Hookerian Herbarium a valuable set of plants from Chamba and Kulu, and an interesting collection made by Captain Hay in the little known district of Lahul.

The botanical provinces of the Western Himalaya may be divided into two principal groups, characterized both by their climate and geographical position. Of these, the first group consists of seven provinces, all bounded on the south by the plains of India, and through which the Himalayan rivers that water them flow in a direction at right angles to the course of the mountains. The second group of provinces consists of five beyond the Satlej, most of which lie to the northward of the first group, and follow a line parallel to them. These are the upper valleys of some of the same rivers as flow through the first group of provinces, and owe their existence as distinct regions in physical geography to the fact elsewhere indicated (page 168), that the courses of the upper parts of the larger rivers of the Western Himalaya are parallel to the axis of the chain.

The great elevations of the secondary chains (or spurs of the main chain) that divide the upper group of provinces from the lower, forming the southern boundary of the upper, prevents the access of humid winds to them, which, together with the greater elevation of their valleys, makes their climate very different.

It is to be borne in mind that the necessity of thus dividing the North-western Himalaya beyond the Satlej into two parallel lines of provinces does not indicate any great difference between this part of the Himalaya and that to the eastward; for, as we have repeatedly remarked, the heads of all the larger Himalayan rivers are in an arid climate. The upper valleys of most of these rivers are too small to constitute provinces, but it cannot be doubted that when the physical features of such large rivers as the Subansiri, Aran, etc., come to be explored, their upper valleys will be found to constitute provinces with a climate and vegetation intermediate in character between those of the Himalaya and Tibet.

The two groups of provinces of the Western Himalaya we propose are :—

First Group.	Second Group.
1. Kumaon.	8. Kunawar (north of Simla).
2. Garhwal.	9. Lahul (north of Kulu).
3. Simla.	10. Kishtwar (north of Chamba
4. Kulu.	and Jamu).
5. Chamba.	11. Kashmir (north of Rajaori).
6. Jamu.	12. Marri (between the Jelam
7. Rajaori.	and Indus).

The observations we have to offer upon the vegetation of these are very fragmentary, as the majority of the natural orders have still to be worked out; we shall however endeavour, after describing the physical features of each, to give as many examples as we can of the peculiarities of their floras, as will show the importance of the study and the means of prosecuting it. Their complete elucidation must be left for local botanists.

1. KUMAON.

Kumaon, as at present limited, is bounded on the east by the Kali, separating it from Nipal; on the west by the Alaknanda branch of the Ganges, and its western feeder, the Mandakni; on the north by the axis of the Cis-Satlej Himalaya, and on the south by the upper Gangetic plain. The elevation of the Terai at its base varies from 600 to 1000 feet; the mountains of the outer ranges rise to 7000 in many places, and in the interior attain 10,000, while still further north many rise above 20,000, and a few above 24,000 feet. The loftiest, as elsewhere in the Himalaya, are never on the axis of the chain, which is still further north, and whose great mean elevation may be judged of from that of the passes over it. Of these, proceeding from the eastward, the Lankpya Pass is 18,000 feet, the Lakhur 18,400, the Balch 17,700, the Niti 16,800, and the Mana 18,760. Almora, the capital of the province, is elevated 5500 feet, the lake of Naini-tal 6500, of Bhim-tal 4000. Binsar, a mountain of the interior region and a well known botanical station, is elevated, we believe, about 7500.

For further particulars we must refer to Captain R. Strachey's account of the provinces of Kumaon and Garhwal in the Journal of the Geographical Society of London (May, 1851).

The vegetation of Kumaon appears to afford rather a rapid transition from the humidity of Nipal to the drier provinces further west. Its flora, according to Strachey's and Winterbottom's excellent collections, includes fully two thousand flowering plants,—a much larger number than are to be found in an equal area anywhere to the westward, though considerably fewer than to the eastward.

Amongst the natural orders we have examined in detail, the following species find their eastern limit in Kumaon, so far as is at present known :—

Thalictrum *pauciflorum*.	Corydalis *Govaniana*.
Oxygraphis *polypetala*.	„ *flabellata*.
Ranunculus *hirtellus*.	Pyrus *baccata*.
„ *arvensis*.	Rosa *pimpinellæfolia*.
Trollius *acaulis*.	Rubus *saxatilis*.
Aquilegia *vulgaris*.	Potentilla *alpestris*.
Delphinium *denudatum*.	Geum *urbanum*.
„ *incanum*.	Spiræa *sorbifolia*.
„ *ranunculifolium*.	Daphne *oleoides*.
„ *Kashmirianum*.	Celtis *eriocarpa*.
Aconitum *Lycoctonum*.	Corylus *Colurna*.
„ *heterophyllum*.	Quercus *lanata*.
Pæonia *officinalis*.	Cedrus *Deodara*.
Papaver *dubium*.	Cupressus *torulosa*.
Meconopsis *aculeata*.	Juniperus *communis*.

Of Eastern Himalayan plants which have not hitherto been traced to the westward of Kumaon there are :—

Clematis *grewiæflora*.	Thalictrum *elegans*.
„ *acuminata*.	„ *glyphocarpum*.

Thalictrum *Punduanum*.
Oxygraphis *glacialis*.
Ranunculus *flaccidus*.
Trollius *pumilus*.
Magnolia *Champaca*.
Michelia *Kisopa*.
Miliusa *velutina*.
Sabia *parviflora*.
Corydalis *chærophylla*.
Rubus *reticulatus*.
„ *paniculatus*.
„ *peduncularis*.
Potentilla *polyphylla*.
„ *monanthos*.
Cerasus *Nipalensis*.
Hedera *serrata*.
„ *æsculifolia*.
„ *terebinthacea*.
„ *parasitica*.

Aralia *Leschenaultii*.
Panax *fragrans*.
Olax *nana*.
Camphora *glandulifera*.
Phœbe *pallida*.
„ *lanceolata*.
Litsæa *lanuginosa*.
Dodecadenia *grandiflora*.
Daphnidium *pulcherrimum*.
„ *bifarium*.
Goughia *Himalensis*.
Henslovia *heterantha*, Bl.
Salix *Lindleyana*.
Elæagnus *conferta*.
Carpinus *viminalis*.
Castanea *tribuloides*.
Abies *Brunoniana*.
Wallichia *oblongifolia*.
Chamærops *Martiana*.

2. GARHWAL.

This province, which is bounded on the west by the Tons, presents a continuation of the physical features of Kumaon, though it is on the whole a less elevated country, and consists chiefly of the basins of the Bhagiratti and Jumna rivers. Its comparatively short northern frontier is formed by the continuation of the Cis-Satlej chain, and, judging from the elevation of the principal passes (15,000 to 16,000 feet), its mean elevation is not much less than Kumaon. The level of the plains at the foot of the hills is 1000 feet, both at Hardwar and Saharunpore, and of the Dehra Dhún, within the first range of hills, 2300 at the village of Dehra. The station of Masuri is 7000 feet; Kedarnath, a well-known botanical station in the interior, is 11,800; the valley of the Bhagiratti at Tirhi, 2300; and Khalsa, at the junction of the Tons and Jumna, is only 1700. There are few plants common to Ku-

2 *d*

umaon and Garhwal that are not also found in Simla; those
that have hitherto occurred to us are—

Delphinium *cæruleum.*
Clematis *Nipalensis.*
Aconitum *ferox.*
Berberis *umbellata.*
Gaultheria *repens* (*nummularia,*
    Don).
Monotropa *uniflora.*
Pieris *villosa.*
Celtis *Roxburghii.*
Antidesma *diandrum.*

Stranvæsia *glaucescens.*
Rosa *sericea.*
Rubus *biflorus.*
    „   *alpestris.*
    „   *nutans.*
Potentilla *microphylla.*
Hedera *tomentosa.*
Cinnamomum *albiflorum.*
Tetranthera *Roxburghii.*
    „   *monopetala.*

Of Western Himalayan plants that have not been recorded
as natives of Kumaon, but are natives of Garhwal, there are—

Clematis *grata.*
Berberis *Lycium.*

Corydalis *crithmifolia.*
Cotoneaster *vulgaris.*

3. SIMLA.

We have applied the name (already well known to botanists)
of Simla to the whole district west of the Tons and east of
the Satlej, including Basehir, Sirmur, and numerous petty
states. It is composed principally of ranges given off from
the rapidly declining Cis-Satlej branch of the Himalaya chain,
which sweeps to the southward and westward, between the
valleys of the Satlej and Tons. The axis of this chain, at the
northern boundary of Simla, separates that province from
Kunawar, and is crossed by the Burenda and Shatul passes,
which being respectively 15,179 and 15,560 feet, indicate an
elevation of the axis scarcely lower than in Garhwal.

The plains at the foot of the Simla hills attain 1000 feet
elevation, and the outer ranges are lower than those of Garhwal and Kumaon. Rupar, close to the Satlej amongst the
outer hills, is under 1000 feet; Sabathu, a little further in, is
4200; Kassowlee 6500.

At Simla, which is situated on the main (Cis-Satlej) chain,

the elevation of the latter is 7000–8000 feet; a little further north it rises at Nagkunda to 9300, and to 10,700 at the Peak of Hattu. Chor mountain, situated on a branch of the main chain, only thirty miles from the plains, and a well known botanical habitat, is 12,100 feet, and is one of the most remarkable isolated peaks in the Himalaya. The bed of the Satlej is everywhere very low, being at Belaspur 1500, and at Rampur 3300 feet.

The flora of Simla may be considered as exceedingly well known; it presents a considerable proportion of Eastern Himalayan plants that do not appear to cross the Satlej basin, and a smaller one proportionally of western species not found in Garhwal.

Western Species.

Thalictrum *pedunculatum.*	Adonis *æstivalis.*

Eastern Species.

Clematis *nutans.*	Antidesma *paniculatum.*
Thalictrum *rostellatum.*	Betula *cylindrostachya.*
Ranunculus *diffusus.*	Alnus *Nipalensis.*
Delphinium *vestitum.*	Myrica *sapida.*
Sphærostemma *grandiflorum.*	Cupressus *torulosa.*
Stephania *rotunda.*	Potentilla *fulgens.*
Hollböllia *latifolia.*	„ *leuconotha.*
Dicentra *Roylei.*	„ *Kleiniana.*
Benthamia *fragifera.*	Sibbaldia *potentilloides.*
Daphne *papyracea.*	Sieversia *elata.*
Osyris *arborea.*	Cerasus *Puddum.*

In the tropical valley of the Satlej the vegetation resembles that of the outer hills, and dry country forms predominate, as *Colebrookia, Rœttlera,* and *Euphorbia pentagona;* whilst Bamboos, *Butea, Ægle Marmelos, Moringa pterygosperma, Capparis sepiaria,* and *Calotropis,* seem altogether absent, or are very rare.

4. Kulu.

This province consists of the mountain basin of the Beas,

and the west bank of the Satlej, and may be made to include the subtropical districts of Mandi and Suket, Nadaon and Kangra. It presents no features not common to Chamba, the next succeeding province to the north-west. Sultanpur, the capital, is 5000 feet. Kangra Fort, situated a short way within the outer ranges, is a British station, and the hills around it are extensively planted with tea. Dharmsala, above Kangra, is a sanitarium, elevated about 6000 feet. The chain bounding the Satlej on the west is considerably higher than that on its east bank, and is crossed into Suket by the Jalauri Pass, elevated 12,000 feet.

Mr. Edgeworth is the only botanist who has investigated the flora of this province, and he has (since the printing of p. 70 of this Essay) communicated a valuable collection to Sir W. Hooker's Herbarium.

5. CHAMBA.

Chamba, the next province to Kulu, is altogether like it in physical features, and consists of the mountain basin of the Ravi. It has been traversed by Dr. Thomson, who entered it from the north-west, by the Padri Pass, elevated 11,000 feet, over the chain dividing it from Jamu; thence he descended to the Ravi, in the centre of the province, where its bed is elevated less than 5000 feet; and travelling northward, left it by the Sach Pass, elevated 14,800 feet, over the range dividing it from Kishtwar.

The vegetation of Chamba appears to present few peculiarities, amongst which we may notice the appearance of *Cratægus Oxyacantha*, which here finds its eastern limit; *Litsæa consimilis, Rhododendron lepidotum,* and *Sibbaldia purpurea* have not hitherto been detected further to the west. *Fothergilla involucrata* is a curious example of a plant suddenly appearing most abundantly, and continuing so for several provinces to the westward.

6. JAMU.

Under this name we include the lower part of the Chenab valley, to the plains of the Panjab, Banahal on the southern slopes of the chain bounding Kashmir on the south, Badarwar on the confines of Chamba to the east; whilst to the north, this province passes into that of Kishtwar, which may be said to commence where the course of the Chenab changes from north-west to south-west. Though probably differing little in physical features from Chamba on the east, it is known much better, from having been traversed in several directions by botanists.

The bounding mountains of Jamu attain an average elevation of 12–14,000 feet; the Banahal Pass to Kashmir is 10,000 feet; that of Padri into Chamba has already been given as 11,000 feet; the bed of the Chenab is a little above 1000 feet near Jamu, and that town itself is 1500 feet.

The outer ranges of sandstone hills rise gradually from the plains of the Panjab (elevated 1000 feet), and are covered with a loose scrub of tropical, dry country, both eastern and western forms, as *Dodonæa, Rœttlera, Rondeletia, Phœnix sylvestris, Pinus longifolia, Solanum Jacquini,* Sissoo, *Celastrus, Zizyphus,* Mango and Pepul, *Cassia Fistula, Rhus, Salix tetrasperma, Coriaria, Bauhinia Vahlii, Euphorbia pentagona, Cocculus laurifolius.* In the temperate region, the prevalent Himalayan forms of Simla appear in much reduced numbers, with *Fothergilla, Quercus incana, Andromeda ovalifolia, Rhododendron campanulatum,* and *Sabia campanulata.* Besides these, *Quercus dilatata, Q. semecarpifolia,* and *Rhododendron arboreum,* which hardly occur further west and do not enter Kashmir, are all found in Jamu.

Of plants which probably do not occur much, if at all, further west than the Jamu hills, are—

Rhododendron *campanulatum.*	Phœnix *sylvestris.*
„ *arboreum.*	Prinsepia *utilis.*
Gualtheria *trichophylla.*	Rubus *flavus.*

Rubus *purpureus*.   Spiræa *betulæfolia*.
    „   *maculentus*.   „   *chamædrifolia*.
Potentilla *atro-sanguinea*.   „   *sorbifolia*.

Of the western forms not hitherto collected to the eastward of Jamu, are *Rubus fruticosus* and *Potentilla desertorum*.

7. RAJAORI.

Under this term we include the province of that name, and all the hill states south of Kashmir, and between the Jelam on the north-west and Jamu on the south-west; thus including the left bank of the Jelam river from where it leaves Kashmir to the plains of the Panjab.

The vegetation of the lower hills of this province has been noticed under the Panjab; that of the upper appears, so far as it is known, to be identical with that of Jamu and Chamba. *Clematis Gouriana*, which extends from Khasia, here finds its western limit.

8. KUNAWAR.

Kunawar includes the upper part of the Satlej basin, to the borders of Piti and Guge in Tibet. Its general direction is north-east and south-west; its bounding mountains are, to the south-east, the Cis-Satlej chain, and to the north-west the mountains bounding Piti. To the south-west and north-east the natural boundaries are less defined, and formed by secondary chains from the former. The province is usually divided into upper and lower Kunawar, the former approximating in climate to Piti.

The mountains which descend from the two parallel bounding chains of Kunawar to the Satlej are very lofty; they are crossed in the usual route to Tibet by the Werang Pass, 13,200, and the Runang Pass, 14,500; the passes over its southern bounding chain are the Shatul Pass, across the Cis-Satlej, leading to the Simla province, elevated 15,560; and the Kuibrang, over a more northern branch of the same, and which divides Kunawar from Tibet, is 18,300. To the north, the pass leading from Kunawar into Piti is the Hangarang, 14,800. Those to Upper Piti are much more lofty. The bed of the Satlej ascends from about 4000 feet in Lower Kunawar, to 8000 or 9000 feet at the upper extremity of the province.

As a whole the province is very dry, compared with any to the southward and eastward of it, being intermediate in this respect, as it is in geographical position, between the Tibetan and Cis-Himalayan provinces, and its flora is consequently comparatively poor in number of species. Owing to the dryness of its climate, Kunawar is sometimes selected as a retreat from the rains of Simla; and the village of Chini, elevated about 7000 feet, has thus been often visited. Plants from this province and the adjacent districts of Tibet are frequently said to be gathered in Chinese Tartary,—an unmeaning term, and one which should be disused in geographical and botanical works. Owing to the rapid transition from the climate of the humid parts of the Simla province to that of Kunawar, we have few instances to record of eastern forms finding their limits here: amongst which there are, perhaps, *Berberis aristata*, *Cassiope fastigiata*, *Potentilla fruticosa*, *P. eriocarpa*, and *P. ambigua*; and no doubt some others lurk in the more humid and shaded situations.

On the other hand, many remarkable western and Siberian forms make their appearance in Kunawar, which advance no further east. As—

Clematis *parvifolia*.	Quercus *Ilex*.
Rubus *purpureus*.	Olea *cuspidata?*
Salix *acutifolia*.	Dianthus.
Alnus *nitida*.	Paliurus *aculeatus*.
Pinus *Gerardiana*.	Eremurus *Biebersteinii*.

Whilst many species, which have been hitherto known only as natives of the dry Tibetan climate at the heads of the Himalayan rivers, become prevalent features in the flora.

The first remarkable local transition in the vegetation is

met with on the road between Chegaon and Miru, in Lower Kunawar ; but, though striking to the eye, from the prevalence of a few novel forms of plants, the total number of new species, not found commonly in Simla, amounts only to thirty or forty. Of the latter, a small-leaved ash, *Dianthus, Lychnis,* and various *Alsineæ, Artemisias* and *Leguminosæ,* contribute most to the altered character of the flora.

Of cultivated plants, the grape, apricot, all *Pomaceæ,* walnut, etc., thrive in Kunawar, and most of them better than anywhere to the eastward, but all are equally prevalent to the westward.   Their abundance, together with the beauty of the scenery of Kunawar, which is extolled by every one, the delicious climate of its almost rainless summer, and its being on the high road to Tibet, Yarkand, and Central Asia, will all contribute to render it one of the most attractive spots in our Indian possessions.

9. Lahul.

Lahul, a British province, is included by Cunningham in Tibet, from which it is however distinct in its physical features. It consists of the valleys of the head-waters of the Chenab. Of its vegetation we know very little, except from an interesting collection formed by Captain Hay, and communicated by Mr. Edgeworth, which we have not yet had time to examine. It is everywhere surrounded by lofty mountains, except towards its north-western extremity, where it is conterminous with Kishtwar. To the south it is bounded by the mountains north of Kulu, where it is crossed by the Rotang Pass, elevated 13,200 feet, an exceptional depression, the rest of the chain being very lofty.   To the west, a portion of the Himalayan axis divides it from the Tibetan province of Piti, and is crossed by the Kulzum Pass, elevated 14,850 feet; and to the north, a continuation of the same axis separates it from the Tibetan province of Zanskar, and is crossed by the Baralacha Pass, elevated 16,500 feet.

Thus hemmed in by lofty mountains, the vegetation of Lahul is probably very scanty, and nearly Tibetan in character; but pines occur even up to 11,000, and it is far more fertile than any Tibetan province. The bed of the Chenab is probably nowhere below 8500 feet elevation, and the plants must therefore be all temperate and alpine. A wild yellow Persian rose, a variety of *R. eglanteria*, here finds its eastern limit.

10. KISHTWAR.

Kishtwar includes the middle course of the Chenab valley between Lahul and Jamu. It is separated on the north from the Tibetan valleys of Zanskar and Dras by the axis of the Himalaya, which is crossed by the Umasi Pass into Zanskar, elevated 18,000 feet; and by other passes, from Wardwan into Dras, at scarcely less elevations. The district of Wardwan to the west occupies the eastern slopes of the range which separates Kishtwar from Kashmir, and is crossed by the Nabagnai Pass, of undetermined elevation, and probably by several others. To the south, Kishtwar is separated from the Chamba province by a range of 10–14,000 feet elevation, alluded to under that province. The boundary between Kishtwar and Jamu to the south-west is not defined.

The climate and vegetation of Kishtwar, like those of Kunawar, with which they are identical, are in all respects intermediate in general features between those of the plainward Himalayan provinces and of Tibet; and in more local ones between those of the provinces occupying the lower and upper course of the Chenab (Jamu and Lahul) on the one hand, and between Kashmir and these on the other. The elevation of the Chenab at about the middle of the province is from 6000 to 7000 feet, and there is hence scarcely any type of tropical vegetation, except *Paliurus, Desmodium,* and Pomegranates.

In entering Kishtwar by the Chamba province a marked change occurs in the vegetation, from the prevalence of a mixture of Kashmir and Kunawar plants which are rare or not found in the provinces skirting the plains, as a tall paniculate *Rheum,* many *Umbelliferæ, Silene inflata, Geranium,* and

2 *e*

Pteris aquilina, together with *Eremurus* in great abundance. Of other Kunawar plants are *Ephedra, Dictamnus, Rosa pimpinellæfolia, Dianthus,* and *Scutellaria orientalis.* *Pinus Gerardiana* is very common, with large walnut and other fruit-trees; and the forest vegetation resembles that of Kashmir, with the addition of *Quercus Ilex* and *Pinus Gerardiana.*

Of eastern forms, which do not, so far as we are aware, advance westward into Kashmir, there are *Clematis connata* and *Trollius acaulis.* And of Kashmir and other western forms, not hitherto collected to the eastward, there are—

Anemone *Falconeri.*   Epimedium *elatum.*
Ceratocephalus *falcatus.*   Corydalis *adiantifolia.*

11. KASHMIR.

The valley of this name consists of the upper part of the basin of the Jelam; and from its comparatively great width, level floor, abundant population, and cultivation, and from its containing by far the broadest sheets of water known anywhere within the Himalaya, it has been regarded rather as a separate country, different from the Himalaya proper, than as an integral part of that mountain mass, and one of the many series of valleys that it encloses. This erroneous impression has been much diffused from the circumstance of map-makers isolating it by a well-defined oval girdle of mountains, cut off almost entirely from the rest of the Himalaya, but which has no such independent existence. It would be out of place here to dwell upon the geological causes that have filled the Kashmir valley with deposits to the depth of many hundred feet, and which have given rise to its flat surface and its lakes, and which, if present in any of the western valleys, would render that of Kashmir less conspicuous.

Kashmir is bounded to the north by the axis of the Himalaya, which there presents a remarkable depression occupied by the Zoji Pass, elevated only 11,300 feet, and communicating with the Tibetan valley of Dras. To the south, the Pir-Panjal and Banahal ranges separate Kashmir from the provinces of Rajaori and Jamu: and the Wardwan range separates it from Kishtwar to the east. The average elevation of the main Himalayan chain north of Kashmir is about 14,000 feet; and of the Pir-Panjal, to the south of it, 12,000; its loftiest summit being 15,000. The Banahal Pass between Kashmir and Jamu is only 10,000 feet. The course of the Jelam is first from south-east to north-west, through the valley of Kashmir, when it turns south-west after leaving the Walur Lake and enters Marri. The elevation of its bed is 5300 feet at Srinagar the capital, and continues so from Islamabad to the Walur Lake, a distance of 50 miles.

Kashmir is not strictly analogous in situation or climate to Kunawar or Kishtwar, but the summer rains are so much interrupted that they can hardly be regarded as the effect of a monsoon. Kashmir contains no *Rhododendron arboreum* and no oaks, nor does it produce *Pinus Gerardiana*. Its flora is a curious mixture of the hot and dry vegetation of Afghanistan, with a few ordinary Himalayan forms on the one hand, and many Persian and Caucasian ones on the other. From its moderate elevation, and the great dryness of the atmosphere throughout the year, the summers are very hot. Rice is the staple crop, and the vine is extensively cultivated. Many of the eastern Himalayan forms which occur in Kashmir extend to Afghanistan, and some even to Persia; but their number is small when compared with those of western origin. Kashmir indeed contains many common European species, which there find their eastern limit.

Of the many western forms that inhabit the valley, the following have not been collected further east in the Himalaya, though a few probably occur in Kishtwar :—

Anemone *biflora*.	Delphinium *penicillatum*.
,, *narcissiflora*.	Nymphæa *alba*.
Ranunculus *Lingua*.	Scutellaria *galericulata*.
,, *chærophyllos*.	Lythrum *Salicaria*.
Isopyrum *thalictroides*.	Cerasus *prostrata*.

Prunus *insititia*.   Marrubium *vulgare*.
Potentilla *reptans*.   Salix *purpurea*.
  „   *grandiflora*.   „ *rubra*.
Cotoneaster *nummularia*.

Of the following list of eastern forms some may no doubt
be discovered in Marri, and even further west, in Afghanistan :—

Thalictrum *pauciflorum*.   Cotoneaster *microphylla*.
  „   *foliolosum*.   Rubus *rosæfolius*.
Anemone *rupicola*.   „   *parvifolius*.
  „   *rupestris*.   Potentilla *desertorum*.
  „   *rivularis*.   „   *argyrophylla*.
Ranunculus *hirtellus*.   Spiræa *canescens*.
Delphinium *denudatum*.   Osmothamnus *fragrans*.
  „   *incanum*.   Salix *elegans*.
  „   *ranunculifolium*.   Elæagnus *parvifolia*.
Epimedium *elatum*.   Betula *Bhojputra*.
Podophyllum *Emodi*.   Alnus *nitida*.
Euryale *ferox*.   Juniperus *recurva*.
Pyrus *variolosa*.

Kashmir affords several instances, already mentioned, of
anomalous distribution, instanced by the absence of *Andromeda ovalifolia* and *Rhododendron arboreum;* and of oaks, of
which five species occur in the adjacent provinces, namely,
Quercus Ilex, annulata, dilatata, incana, and *semecarpifolia.*
Also the appearance of *Salvinia natans,* of *Euryale ferox,* if
really wild, and *Nelumbium speciosum,* must be considered as
very singular, though the latter is found considerably further
north, on the shores of the Caspian. The bullace, *Prunus
insititia,* has been found nowhere else in a wild state, except
indeed it be a variety of *P. spinosa.* We believe also that the
cherry is truly wild in the valley, and it is abundantly cultivated in orchards. The prevalence of these, with Planes,
Lombardy Poplars, Walnuts, *Berberis vulgaris, Colchicum,
Cratægus Oxyacantha, Actæa spicata, Thalictrum minus, Alliaria officinalis,* and the great majority of the plants men-

tioned at page 109, give an eminently European cast to the whole vegetation.

In the Kashmir lakes many European forms of water-plants occur, which, from the absence of similar expanses in the temperate regions of the Himalaya, are rare or unknown elsewhere; such are *Nymphæa alba,* already mentioned, *Villarsia nymphæoides, Menyanthes trifoliata,* and *Trapa,* besides *Typha, Arundo,* and various *Potamogetons, Sium angustifolium,* several European *Menthas,* etc.

12. MARRI.

The Marri range, on the right bank of the Jelam, overhanging the platform of Rawal Pindi, is a narrow ridge separating two deep river-valleys, whose vegetation is quite tropical. On its plainward slope it produces ordinary Himalayan forms (*Rhododendron arboreum,* etc.), but the vegetation soon becomes like that of the hills of Kashmir.

The mountains of Marri properly consist of the western termination of the Himalaya (according to our definition of that chain), which sweeps round the north of Kashmir, and following the course of the Indus, turns to the southward, descending gradually into the plains of the Panjab, its most southern slopes forming the Salt range described at page 156.

Our only knowledge of the plants of Marri is derived from a very valuable collection made by Dr. Fleming, who ascended the ranges to 9700 feet. European forms abound in even a greater proportion than in Kashmir, and many Himalayan plants find there their extreme western limit; such are—

Berberis *Lycium.*	Rosa *macrophylla.*
Delphinium *saniculæforme.*	Rubus *lasiocarpus.*
Quercus *annulata.*	,, *niveus.*
,, *dilatata.*	Potentilla *Leschenaultiana.*
,, *incana.*	,, *Nipalensis.*
Pyrus *baccata.*	Spiræa *callosa.*
Cotoneaster *bacillaris.*	Machilus *odoratissimus.*

The valley of Hasora, north-west of Kashmir, is still more

arid, but not quite Tibetan, *Pinus Gerardiana* being very common. Its flora is, however, scarcely known.

Tibet.

Tibet includes the mountain valleys of the Indus and Yaru (or Brahmaputra), together with the whole axis of the Himalaya and the heads of many of the valleys which descend on the Indian side, and which are situated beyond the mass of snow throughout a great extent of the chain. Beyond the Indus and Yaru are the southern slopes of the Kouenlun, which according to our definition do not form a part of the Himalaya, but of Tibet. Politically its boundary is an irregular one, accidental circumstances having regulated the line of separation between the Indian and Tibetan states. Botanically, the boundary of Tibet is best drawn at the place where the climate becomes too arid to support such a vegetation as flourishes at equal elevations on the Indian watershed, and especially where there is a total absence of forests below 13,000 feet. The flanks of all the great Himalayan rivers, when above 13,000 feet, are, owing to the elevation, devoid of trees, whether the climate be humid or arid; but when their course is oblique, as is the case with the Satlej and the Aran, there are no trees at far lower elevations than this, and a considerable part of their upper course is through a Tibetan climate. Thus, in the valley of the Satlej the climate is too dry for trees at the junction of the Piti river, elevated 9000 feet; and the whole of Piti, as well as the upper course of the Satlej itself, forms part of Tibet. In the valleys of the Ganges and Jumna, on the other hand, whose course is perpendicular to the plains, trees ascend to 10,000 feet, and only the alpine zone is arid and hence belongs to the Tibetan Himalaya, in contradistinction to " Himalaya interior."

Tibet may be divided into two parts, one to the westward (the basins of the Indus and Satlej), the other to the eastward (those of the Yaru and Aran, and perhaps of the Monas, Subansiri, and other rivers). From the position of the Himalaya, the rain-fall is much greater at the eastern extremity of the chain than it is to the westward. Hence Western Tibet is considerably drier than Eastern Tibet; indeed, the lower part of the course of the Indus, where that river enters the Panjab plain, is situated in a rainless climate; but the lower part of the course of the Yaru, where under the name of the Dihong it joins the Brahmaputra, lies in one of the rainiest climates of the globe.

The chain of the Kouenlun, where it forms the northern boundary of Western Tibet, is not less elevated than the Himalaya, and is covered throughout a great part of its length with perpetual snow. Its axis has not been crossed by any European traveller, but has been reached by Dr. Thomson, who visited the Karakoram Pass, elevated 18,300 feet. This chain has been called the Mus-tagh, Karakoram, Hindu Kúsh, and Tsungling or Onion mountains (from the prevalence of a species of *Allium*); it is also the Belur-tagh,* which (according to Cunningham) is synonymous with "Balti mountains," and its continuation forms the Pamir range west of Yarkand. In Western Tibet, the axis of this chain is in general distant about 150 miles from the Himalaya, and the country between the two consists of a complication of ranges of lofty and rugged mountains, separated from one another by stony valleys, which on the higher parts of the courses of the rivers expand at intervals into alluvial plains.

The Indus, near its source, has an elevation of 18,000 feet, and where it debouches on the plains of the Panjab it is elevated only 1000 feet. At Le it is 10,500 feet, and at Iskardo 7200 feet. Below 10,000 feet, the summer heat, from the absence of rain, is intense, and the Tibetan flora becomes more Sindian and Persian in character. West of Kashmir and the great peak of Dayamar, the Himalaya diminishes rapidly in elevation, and allows access to the humid atmosphere, which is condensed on the first ranges of Tibet with which it comes

* The Bulut-Tag (or Cloud Mountains) of Captain H. Strachey, who confines the term to the range east of Samarkand and south of Khokand.

in contact. The Tibetan Flora of the Indus, therefore, ends a little below Iskardo, pines appearing in the district of Rondu, and throughout the valley of Hasora, which latter may hence be regarded as not Tibetan.

The mean elevation of Western Tibet exceeds that of all countries of which we have any definite knowledge, and, if not surpassed by part of Eastern Tibet, is without doubt the loftiest area of any considerable extent on the surface of the globe. Captain H. Strachey gives 15,000 feet as the approximate mean elevation; and when we consider that there are throughout Tibet many ranges of a uniform.elevation of 19–20,000 feet, and peaks innumerable of 21–25,000, as also that the very lowest level of the Indus valley (itself a mere cleft in the mountain mass) is 6000 feet, the above estimate will not be considered exaggerated. Of the passes over the main axis of the Kouenlun and Himalaya, and over their principal ramifications, far more are above than below 17,000 feet, many are 18,000, and a few 19,000; besides which many extensive areas in Guge, Nari, Nubra, Rupchu, and Zanskar, are continuously above 15,000 feet for many miles in all directions.

The climate of Western Tibet can only be approximately ascertained, no continued records of temperature, humidity of the air, or rain-fall, having ever been kept. Captain H. Strachey has however reduced all the detached observations that were procurable, and we are indebted to his valuable paper on the Physical Geography of Western Tibet* for most of the following data.

In the basin of the Indus at Le, elevated 11,800–12,000 feet, and 1300–1500 above the bed of the river, which is considerably below the mean elevation of Western Tibet, and in a sheltered locality, the mean temperature of the year is assumed to be 35°: of January 10° (variation −5° to +25°), and of July 60° (variation 50° to 70°). Constant frost sets in at that elevation early in November, and lasts till the end of February; but night-frosts continue till the middle of April,

* Read before the Royal Geographical Society, November, 1853.

and commence again in the middle of September. A rather sudden rise of temperature attends the vernal equinox, and the summer is comparatively warm, the maximum sometimes, but rarely, reaching 70°.

At 13,000 feet the mean temperature probably coincides with that of the freezing-point. At 14–15,000 feet the summer months alone are free from night-frosts, the maximum temperature is only 60° in good shade, and the winter is proportionately colder than at 12,000 feet; thaw commences at the end of April, the night-frosts are slight by the end of that month, and the mean of the day rises to 50°. At 15,500 feet it probably freezes during every night of the year. At 20,000 to 21,000 feet there is probably perpetual frost in the shade.

These numbers however give no indication of the heat to which vegetation is exposed, for, owing to the rarity of the atmosphere and cloudless skies, the sun's rays have intense power, increasing with the elevation, raising the (white glass) thermometer exposed to them sometimes upwards of 100° above the mean temperature of the air. This, combined with the fact of the temperature of the soil being always above that of the air, fully accounts for the sudden impulse given in spring to the vegetation even in the loftiest and coldest regions. The heat radiated from the naked rocks has also a very powerful effect, especially on the summer crops.

Extreme aridity is the characteristic of all Western Tibet. Rain and snow at moderate elevations are scarcely known, and have no further direct effect on vegetation than is due to the moisture of the soil produced by the melting of glaciers and snow-beds. Dew and hoar-frosts are very rare phenomena. The snow-level is nowhere below 18,000 feet; in the mountains north of the upper Indus valley it rises to 20,000.

Owing mainly to the great drought, the soil is in many places covered with an efflorescence of carbonate and other salts of soda, and salt-lakes are of frequent occurrence. Almost all the large bodies of water indeed are more or less saline, some of them intensely so, especially such as have no outlet,

2 *f*

and are hence gradually drying up. This diminution of many of the lakes is no doubt entirely attributable to a change of climate, which is extremely interesting in a botanical point of view, from its favouring the immigration of many saline types of the Caspian flora.

Where the surface is covered with salt-marshes, are found *Glaux maritima, Eurotia, Corispermum, Caroxylon, Suæda, Salsola, Chenopodium, Ambrina, Christolea, Triglochin*; and a large *Nostoc*, of a species eaten in China, floats on the surface of the pools. The carbonate of soda again appears to have no appreciable effect on the vegetation of the dry soil it encrusts; grasses, tufted *Androsaces, Astragali, Gnaphalia, Artemisiæ*, etc., being alike covered with it.

Cultivation in Tibet attains the height of 15,000 feet, and is luxuriant below 12,000 feet, barley and wheat being the grains cultivated, with rape and millet at lower levels. The indigenous vegetation is everywhere scanty. Though there is no forest, the banks of the rivers and streams are skirted by a dense scrub of bushes, chiefly *Myricaria, Hippophae, Rosa*, and *Lonicera. Populus balsamifera* and *Euphratica*, and *Juniperus excelsa* are the only trees, and these occur rarely; as does *Pinus excelsa*, which is only found towards the confines of Hasora, and can hardly be considered a Tibetan tree. *Myricaria* and *Hippophae* occasionally attain a height of twenty feet. Of cultivated trees, apricots and *Populus balsamifera* are seen up to 12,000 feet; apples, walnut, the black poplar, and *Elæagnus* up to 11,000 feet, pears to 10,000 feet, and grapes and white poplar and plane-trees to 9000 feet.

Subtropical types ascend along the course of the Indus to Rondu and Iskardo, and some of them even as far as 11,000 feet, in Nubra and Le, of which the following genera are examples :—

Capparis.	Echinops.
Peganum.	Tamarix *Gallica*.
Tribulus.	Lycium.
Sophora.	Vincetoxicum.

Plectranthus *rugosus*.
Linaria *ramosissima*.
Cyperus.
Chloris.
Cymbopogon.

Andropogon.
Eriophorum.
Saccharum.
Erianthus.

The temperate vegetation consists almost exclusively of European and Siberian types, and differs remarkably from the Himalayan in the total absence of *Rubi* and *Aconitum*. Besides the shrubs and trees mentioned above, there occur—

Salix *angustifolia*.
   ,,   *zygostemon*.
   ,,   *purpurea*.
   ,,   *acutifolia*.
   ,,   *alba*.
   ,,   *fragilis*.
Perowskia.
Ulmus *pumila*.
Populus *alba*.
   ,,   *nigra*.

Elæagnus.
Betula *Bhojputra*, var.
Loniceræ, several.
Clematis *orientalis*.
Rosa *pimpinellæfolia*.
Artemisiæ, several.
Caragana *versicolor*.
Berberis *ulicina*.
Rhamnus.
Ephedra.

The prevalent natural families are all European :—

Ranunculaceæ.
Fumariaceæ
Cruciferæ.
Alsineæ.
Leguminosæ.
Rosaceæ.
Umbelliferæ.
Saxifrageæ.
Compositæ.

Scrophulariaceæ.
Labiatæ.
Boragineæ.
Polygoneæ.
Chenopodiaceæ.
Amentaceæ.
Gramineæ.
Cyperaceæ.

The following herbaceous genera and species may be noted as often occurring :—

Ranunculus.
Anemone.
Delphinium.
Thalictrum.
Corydalis.

Hypecoum.
Draba.
Cardamine.
Matthiola.
Sisymbrium.

Stellaria.	Erigereæ.
Lychnis.	Aster.
Dianthus.	Saussurea.
Astragali, many	Gentiana.
Phaca.	Veronica *Beccabunga.*
Thermopsis.	Agrostis.
Oxytropis.	Anagallis.
Cicer.	Orobanche.
Potentilla.	Euphrasia *officinalis.*
Chamærhodos *sabulosa.*	Pedicularis.
Saxifraga.	Thymus *Serpyllum.*
Epilobeæ.	Menthæ, various.
Carum *Carui.*	Dracocephalum.
Galium *Aparine.*	Primulæ.
Tussilago *Farfara.*	Statice.
Mulgedium.	Orchis.
Tartaricum.	Herminium.
Artemisia.	Allia, several.
Allardia.	

The water-plants are *Hippuris vulgaris, Limosella aquatica, Zannichellia palustris, Ranunculus aquatilis* and *radicans, Utricularia,* and several species of *Potamogeton.*

In favourable localities a short turf covers the ground, and affords a nutritious pasturage to yaks, goats, sheep, and horses; this consists chiefly of the common Fescue grass (*Festuca ovina*) and other European species, with several species of *Stipa* and tufted *Carices.*

Owing to the great power of the sun there is scarcely any alpine vegetation, even at 15,000 feet; and above that, though plants may be gathered up to 19,000 feet, vegetation is excessively scanty, and only found by the margins of rills from melting snow. The flora of these regions includes some plants of great interest, as *Papaver nudicaule, Oxygraphis glacialis, Ranunculus hyperboreus, Taraxacum officinale, Delphinium Brunonianum, Berberis ulicina.* A small *Urtica* is everywhere common at great elevations.

The following list of genera and species that occur above

15,000 feet is of course far from complete; those with an asterisk (*) have been observed above 17,000 feet.

Corydalis *Tibetica*.
*Draba *aizoides* and others.
*Parrya.
*Cerastium.
*Lychnis.
*Thylacospermum.
*Myricaria.
*Biebersteinia *odora*.
Oxytropis *chiliophylla*.
*Astragali, several.
Thermopsis.
Potentilla *Salessovii*.
    ,,   *anserina*.
*   ,,   *Meyeri*.
*Sibbaldia *procumbens*, var.
Chamærhodos *sabulosa*.
*Saxifraga *cernua*.
*Seda.
*Saussureæ, three species.

*Aster *alpinus*.
*Artemisia.
*Leontopodium.
*Allardia.
*Pyrethrum.
Ligularia.
*Nepeta *multibracteata*.
Cynoglossum.
Lithospermum *euchromon*.
*Gymnandra.
*Primula.
Rheum.
Ephedra.
*Carices.
*Stipa.
*Lloydia *serotina*.
*Festuca *ovina*, and other
    Grasses.

Owing to the aridity of the climate all *Cryptogamiæ* are extremely rare: only three or four Ferns occur; Mosses are scarcely more common, and never fruit. A few crustaceous Lichens, on stones, and half-a-dozen *Fungi*, including several British species, have been collected.

Western Tibet is tolerably well known botanically*. It was first explored by Dr. Falconer, who visited Hasora, Dras, and Balti, and made a fine Herbarium, which is unfortunately still unexamined and undistributed, at the India House. Jacquemont visited Piti in 1830, and Dr. Royle's collectors were there also. Dr. Thomson's collections were made in Piti, Balti, Rupchu, Ladak, Zanskar, Nubra, and Dras. Captain Henry Strachey made an excellent collection in the

---

* There are a few plants in the Wallichian Herbarium, collected by Moorcroft, the first explorer in modern times of Ladak, and ticketed as from that place, but they are mostly outer Himalayan plants.

mountains round the Pangong lake, and Captain Richard Strachey and Mr. Winterbottom a very valuable one in Guge in the autumn of 1849. Mr. Lance has also sent us, through Mr. Edgeworth, a collection from Piti, Ladak, and Dras, which contains many interesting species.

Our attempts to divide Western Tibet into provinces have been attended with unusual difficulty, owing to the undefined limits of those already established, and to the fact that the natives of that country have no system of nomenclature for large areas, mountain chains, or rivers, available for our purpose. Considering how scanty the flora of Western Tibet is, not amounting perhaps to more than 500 species, and how widely the majority of these are spread, any division into provinces might perhaps have been dispensed with,—so far as the purposes of geographical distribution are concerned; but the flora of the country is far too imperfectly known in detail to warrant the assumption that particular habitats are wholly useless; and we should further be depriving future local botanists of the benefit of our local knowledge.

In the following attempt we have been guided wholly by the river systems, which enable us to divide the country into three parallel lines of provinces, that occupy (within rough limits)—1. The north slope of the Himalaya; 2. The beds of the Indus and Satlej; 3. The south slope of the Kouenlun : they are as follows :—

1. Guge, the Tibetan course of the Satlej.

2. Piti and Parang, the basins of the rivers of those names, tributaries of the Satlej.

3. Zanskar, the basin of the Zanskar river.

4. Dras, the basin of the Dras river.

5. Nari, the upper course of the Indus.

6. Ladak, the middle Tibetan course of the Indus.

7. Balti, the lower Tibetan course of the Indus and of the Shayuk rivers.

8. Nubra, the upper basins of the Nubra and Shayuk rivers, tributaries of the Indus.

1. GUGE or HUNDES is wholly under Chinese influence, and is comprised between the Himalaya and its Cis-Satlej branch. It extends from the lakes of Mansarowar and Rakastal down the course of the Satlej to Kunawar. The surface of Guge differs remarkably from the rest of Tibet in the greater extent and depth of an alluvial deposit, found elsewhere in Tibet in smaller quantity, and here forming an undulating surface, gradually declining from 15,200 feet, the level of the lakes, to 10,000 feet at the confines of Kunawar. This province, familiarly known as the plain of Tibet, and which has mainly given rise to the erroneous impression of Tibet being a steppe, plain, or table-land, is 120 miles long and 15 to 60 in breadth, and is traversed by the Satlej and its various feeders, which flow in deep narrow ravines 1000 to 3000 feet below its mean level.

The botany of Guge is scanty in the extreme; the country has been traversed by Moorcroft and Captain H. Strachey, and visited by Captain R. Strachey and Mr. Winterbottom, who collected fifty or sixty species of plants around the lakes, and calculated that not one-twentieth of its surface was covered with vegetation.

2. PITI and PARANG.—Of these two valleys, that of the Piti river is entered from Kunawar by the Hangarang Pass, elevated 14,800 feet. The Parang Pass, over the range dividing the Parang from the Piti rivers, is 18,500 feet. The lofty platform of Rupchu, which extends from the Parang Pass across the main chain of the Himalaya to the adjacent head of the Zanskar valley, and from the Chumoreri lake to the Lachalang and Tunglung Passes, is elevated 15–16,000 feet; Chumoreri lake, situated on it, being 15,200. The vegetation of the whole province is extremely scanty.

3. ZANSKAR occupies the north slope of the main Himalayan chain, parallel with Kishtwar on the south. The change in the vegetation on crossing the Umasi Pass (18,000 feet) from Kishtwar is very sudden, only two or three species found at 12–13,000 feet on the Tibetan face being identifiable with

those of Kishtwar. Padum, the capital, is 12,000 feet above
the sea; and a rich herbaceous vegetation occupies the river-
flats and ravines. The Zanskar basin is cut off from that of
the Indus by lofty ranges, and the defile through which the
Zanskar river flows to the Indus is rocky and impracticable.

4. DRAS.—This province occupies the same position relatively to Kashmir that Zanskar does to Kishtwar. The communication between Dras and Kashmir is by a remarkable
depression—the Zoji Pass, whose elevation being only 11,300
feet, gives free access to the moist winds of Kashmir, and
Dras is hence the most humid and fertile province of Tibet;
its flora approaching very closely to that of Kashmir.

The openness of the valleys of Dras, and the occurrence of
elevated plains or steppes at its north-west extremity, which
have been called the plains of Deotsu, are remarkable exceptions to the generally rugged nature of Tibet; and the fact
of Dras and Guge having both been visited and described by
European travellers before most other parts of Tibet, and their
both being so exceptionally level as compared with the rest
of that country, has materially tended to spread the erroneous impression of the whole of Tibet being a series of elevated plains.

Artemisiæ and *Umbelliferæ*, including *Prangos pabularia*,
are abundant in the Dras valley, and the prevalent *Chenopodiaceæ* of Tibet are scarce. *Vitis, Impatiens,* Black Cur-
rant, *Silene inflata, Aconitum, Hypericum, Vernonia, Juniperus, Thymus Serpyllum, Achillea Millefolia, Convallaria,* and
Tulipa, all very rare in Tibet, occur in the valley. Towards
the summit of the Pass, Dr. Thomson gathered 110 species
on the Tibet side, of which all but six or seven were Kash-
mirian.

5. NARI.—Of this province (more accurately called Nari-
Khorsum) nothing is known botanically; it is enormously
lofty, utterly barren, and almost uninhabited, except on the
lowest part of the ravine of the Indus, whose sources have
not been visited by any traveller; nor has the province been

entered except by Moorcroft : it is wholly under Chinese influence.

6. LADAK.—This province, as restricted by us, extends from Nari to Balti, a distance of 230 miles, in which the Indus descends from 14,000 feet at Demchok, to 10,500 below Le, and at 8500 enters Balti.

From Hanle, the most elevated portion of this province, to its lower end, the increase of vegetation is very gradual along the valley of the Indus. The town of Hanle (14,300 feet above the level of the sea) is situated in a very open, undulating, barren, saline plain, six to eight miles in diameter, covered with bog-soil, and bearing plants characteristic of such localities. Bushes of *Myricaria* become common at 14,000 feet, and these attain the character of small trees at 13,000; below this, Poplars, *Hippophae, Rosa,* etc. commence, and form a low brushwood. Le, the capital of the province (and of West Tibet), is 11,800 feet above the sea.

7. BALTI is a Mohamedan province, and extends from Ladak to the great bend of the Indus; it also includes the lower course of the Shayuk river, up to 10,000 feet. It is conterminous on the south with Dras and Hasora, and bounded on the north by the Kouenlun, or Mustagh. The axis of the latter is probably not less elevated than it is further east; but little is known of its slopes north of Balti, except that, owing to the damp winds finding free access by the Indus valley, they are more snowy than anywhere to the eastward.

The bed of the Indus at Tolti is elevated about 7500 feet; at Iskardo, the capital of the province, 7000; at Rondu, 6200; and at the great bend about 5000.

Throughout Balti the course of the Indus is in many places quite impracticable, from the narrowness of its defile and its rugged bounding mountains. Except in the presence of the subtropical genera mentioned at page 218, the vegetation of Balti presents little of interest. Vines abound, climbing over the poplars, and there is much cultivation in available situations.

8. NUBRA.—We have extended this province to the whole
of the south flank of the Kouenlun, from Balti to Nari; it
includes the districts of Nubra, Pangong, and Rodok, and is
comprised within the basin of the Shayuk river and its afflu-
ents, including the Pangong lakes, which have now no exit,
but which there is good evidence to prove once drained into
the Shayuk river. This is the most lofty and sterile province
of Tibet, except Nari; the axis of the Kouenlun being probably continuously upwards of 18,500 feet in elevation, and its
main ramifications being equally lofty. The valleys enclosed
between the latter extend for many miles at 16–17,000 feet,
whilst numerous peaks in all parts rise 20–23,000. The elevation of the Karakoram Pass, on the axis, is 18,300; that of
the Pangong lakes, which are very salt, 13,400 feet; and they
are surrounded by mountains of 19,000 feet. The elevation
of two of the passes over the range dividing the Indus from
the Shayuk valley, north of Le, are 17,000 and 19,000 feet.

There is little peculiarity in the vegetation of Nubra; the
plants of the lowest valleys are those of the Indus in Balti,
Populus Euphratica being plentiful. *Ulmus pumila* occurs
nowhere else in Tibet. Walnut and *Elæagnus* here find their
northern limit, and are both scarce. In respect of cultivation,
the Nubra valley is superior to any other part of Tibet of
equal elevation, being comparatively well wooded, and the
trees often affording shade, whilst green lanes blooming with
Clematis and rue, and hedges of *Hippophae* enclosing fields of
millet, wheat, buckwheat, and rape, are common around the
villages. The only peculiar plants are a curious dwarf *Berberis*
(*B. ulicina*, nob.) which grows at 14–15,000 feet, and a white-
flowered *Allium* at 11,000 feet.

EASTERN TIBET is quite unknown to us botanically and
geographically. The scanty notices published by the few travellers who have been able to penetrate into the interior of
that strictly guarded country lead to the conclusion that it
has the same general aspect as Western Tibet, as far east at
all events as Jigatzi or Teshu Lumbu and Lhassa. The oral

information of the natives of the country confirms this. We learn from Turner that showers of rain are frequent about Jigatzi during the summer months ; and as the winds in the valley of the Yaru are said to be generally east and south-east, the amount of rain-fall must increase as we descend that river, though, sheltered as it is by the Assam Himalaya and Mishmi mountains, the fall is no doubt comparatively insignificant.

Of the direction of the mountain-chain to the north of the Yaru nothing is known. The only Europeans who have visited it have been Captain Bogle in 1774, who resided at the monastery of Chammaning, in latitude $30\frac{1}{2}°$ north, when on a mission to the Supreme Pontiff; and, more latterly, Messrs. Huc and Gabet, who crossed it on their way from Kokonor to Lhassa. From the accounts of the latter travellers the country seems to be enormously elevated, and continuously so for a belt of many miles in breadth ; and to this may be added the testimony of the Tibetans themselves, and the fact of so many of the greatest rivers of Asia rising within the same area.

Dr. Hooker collected a few plants on the southern border of Tibet to the north of Sikkim, and these, amounting to only fifteen or twenty species in two days' journey, are almost identical with those from equal elevations (16–18,000 feet) in West Tibet,—a stunted *Lonicera* and *Urtica* being the prevalent species at 16,000 feet, with creeping *Carices* in the sand, and tufted plants of *Alsineæ, Draba, Androsace, Oxytropis chiliophylla, Sedum, Saxifraga,* and grasses and sedges, most of which ascend to 18,000 feet. The curious genus *Thylacospermum* forms hard, hemispherical mounds on the stony soil at these elevations, and is one of the most conspicuous features of the flora. The ground was there everywhere covered with an efflorescence of carbonate of soda, and the pools of water were full of *Ranunculus aquatilis* and *Zannichellia palustris,* also typical of similar situations in West Tibet.

In the valley of the Yaru the Dama (*Caragana versicolor*) is said to grow, and to be the only firewood ; and by the streams, in sheltered valleys, are poplars, willows, and probably ash or walnut. Where the Aran enters Nipal, at Tingri, the vegetation appears (from a small collection we have received thence) to be similar to that of Kunawar.

At Lhassa the country is open and stony, and without trees, except such as are cultivated, just as in Western Tibet. Of these, the apricot is the only one of which we have any certain knowledge. Vines have been stated to grow in the city of Lhassa (Humboldt, 'Asie Centrale'), but this has been contradicted by all our informants. Further east, in the direction of China, we learn from Huc and Gabet's Travels that the mountains are covered with forests, while towards the south-east, in the valley of the Yaru, a subtropical climate is soon reached, tea, rice, and cotton being all cultivated.

III. *Eastern India.*

The axis, or watershed, of the great meridional chain which is continuous with the Kouenlun must be sought as far north as 35° N. lat., where it penetrates between the waters of the Hoangho and those of the Yang-tse-Kiang. It is, however, probable that the watershed of the Yaru river lies considerably further south than this chain, and occupies a position nearly parallel to that river, till it reaches 28° N. lat. in 98° E. long., after which its direction is nearly north and south, and it becomes the axis of the Malayan peninsula, which separates Ava and Siam on the one hand from Yunan and Cochin-China on the other.

To the north of this chain, in Tibet, lies a vast unknown tract, in which the head-waters of the Yang-tse-Kiang perhaps ramify, as well as those of the Tsa river, which is identified by Chinese geographers with the Neay-Kiang of Cochin-China. On the southern face of the chain the Dihong, the Brahmaputra, and the Irawadi, have their sources. It may therefore be considered to be the boundary of India in this direction, as the frontiers of Ava and China run nearly along it.

The chain of mountains which separates the waters of the Brahmaputra from those of the Irawadi, branches off from this main axis at an acute angle. Its direction is south-west, and it decreases rapidly in elevation after leaving the Mishmi country, forming the Naga hills, which extend from 96½° E. long. to the sources of the Cachar and Manipur rivers, and have an average height of 6000 or 7000 feet. Here the chain bifurcates, one branch running due west as far as the great bend of the Brahmaputra, while the other runs nearly due south. The western branch, under the name of the Cachar, Jaintia, Khasia, and Garrow hills, separates the valley of Assam from that of Silhet. Its elevation varies from 4000 to 7000 feet. The other, which separates Cachar, Chittagong, and Aracan, from Ava, has been called the Aeng range; it is less known, but is in many parts probably equally elevated.

The provinces of Eastern India selected for botanical divisions are—

1. Mishmi.
2. Assam.
3. Naga and Khasia.
4. Cachar and Silhet.
5. Chittagong and Tippera.

6. Aracan.
7. Ava and Pegu.
8. Tenasserim.
9. Malayan Peninsula.

1. MISHMI.

The country between India and China to the east of Assam is as little known as any other on the globe. Between the British frontier and that of China there are interposed a number of savage tribes, constantly at war, and so extremely jealous of one another that no offers of reward have been successful in inducing them to guide travellers into the interior of their mountains, though many efforts have been made since Assam was conquered by the Indian Government during the first Burmese war. At that time (as we learn from Captain Wilcox's very interesting narrative) a corps of scientific surveyors was attached to the army in the field, in order to be ready to take every opportunity of improving our knowledge of geography. To the surveys of this corps, and in particular to Captain Wilcox himself, we are indebted for all that is known of these countries.

The Mishmi mountains, which occupy the most northerly part, are the southern and western slopes of a mass of snowy mountains which sweep round the north-west of Assam from the east bank of the Dihong to the sources of the Dihing river. The peaks of this chain are perhaps nowhere of great elevation as compared with the Himalaya, though many are covered with perpetual snow; and there are probably considerable depressions, as at the source of the true Brahmaputra, which is at the north-east angle of the chain, where the branch which runs west, and bounds Mishmi on the north, is given off. These mountains rise abruptly from the plain of Assam. They have been visited by Captain Wilcox and by Mr. Griffith, to whom we are indebted for all our information regarding their vegetation. The climate is extremely humid. The rainy season is the same as in Assam, but heavy winter rains occur, and the air is usually extremely damp.

The northern valleys of the Mishmi country appear to be included in Tibet, and from the accounts of the few travellers who have perilled their lives in attempting to ascend them, the Tibet frontier is gained in about fifteen days' march up the Brahmaputra, from the Kund or sacred pool of that river. Wilcox, indeed, approached the frontier village of Taling; and more recently a French missionary (M. Krick) reached the same spot, where he was forced to retire, owing to the jealousy of the authorities.

The flora corresponds very closely with that of Sikkim, Bhotan, and the Khasia mountains, and affords every indication of constant humidity. The mountains, up to six thousand feet, are covered with a dense tropical forest, in which *Calami, Wallichia, Areca, Caryota,* and *Arenga,* are common, with tree-ferns, *Pandanus,* and *Musa.* Oaks, chesnuts, a wild *Thea, Guttiferæ, Tiliaceæ, Verbenaceæ,* and *Araliaceæ* are characteristic trees. *Liquidambar* is also common, and parasitical *Orchideæ* and ferns are extremely abundant. A plant closely allied to *Rafflesia* (*Sapria Griffithii*), which was discovered in these mountains by Griffith, is the most remarkable form known to occur there.

The upper valley of the Brahmaputra is more open, and is richly cultivated, rice being the chief crop, and oranges the most abundant fruit-tree.

Higher up, the mountain-slopes are clad with pines of an undetermined species in great abundance. *Rhododendron arboreum* is also of frequent occurrence, and the temperate flora, so far as it is known, closely resembles that of Khasia.

The alpine flora is quite unknown; but we learn from Wilcox, who crossed a pass elevated 12,800 feet above the level of the sea[*], on his journey to the Irawadi, that stunted Rhododendrons were common, and that a species of Juniper occurred on the crest of the pass, together with *Coptis Teeta*, a remarkable Ranunculaceous genus, which is not found in the Himalaya.

Though so luxuriant and tropical, the flora of the Mishmi hills below 6000 feet elevation did not yield Griffith a rich harvest,—he did not obtain a thousand species during his residence there. These consisted chiefly of tropical orders, amongst which the following are the most numerous in species :—

Compositæ 80	Rubiaceæ 42
Gramineæ 73	Acanthaceæ 38
Labiatæ 50	Leguminosæ 31
Orchideæ 43	Cyperaceæ 22

besides 200 Ferns.

These numbers are taken from his published journals; but, from our examination of the materials from which they were computed, they must be considerably reduced, especially the Ferns.

[*] Asiat. Res. xvii. 451.

2. Assam.

The province of Assam is bounded by the Himalaya and
Mishmi mountains on the north, and by the Khasia and Naga
hills on the south.   It is a tropical valley continuous at its
western extremity with the plains of Bengal, and gradually
contracting to the eastward, till the mountains at last approach so close together that no level country remains between them.   The width of the lower valley is about thirty
miles; it is in general level, but low ranges of hills project
occasionally from both sides almost to the Brahmaputra, and
isolated hillocks occur scattered here and there over the surface.

The atmosphere is very humid, and dense fogs are frequent
in winter.   The rainy season lasts from May till October, and
the rain-fall (about eighty inches at Gowahatti), though much
less than on the mountains by which it is surrounded, is considerable.   The climate is therefore on the whole equable,
without excessive summer heat, and without great winter
cold. Lower Assam is richly cultivated, but dense forest occupies the base of the hills on either side, as well as the hillocks
which advance upon the plain.

In Upper Assam there is but little cultivation, and much
forest, which is often almost impervious from rank underwood.   Along the river the low alluvial plains, which at the
junction of the Dihong are scarcely raised 350 feet above the
level of the sea, are bare of trees, and covered with dense
grass jungle.   The mountains display a rich vegetation of the
most tropical forms which India produces.   *Anonaceæ* are
numerous, several species of *Myristiceæ* occur, and the India-
rubber fig forms large forests in some places.   *Calami* and
Plectocomia abound in the dense jungles, as well as other
rare and interesting palms, belonging to the genera *Livistonia,
Licuala, Arenga, Areca, Wallichia,* etc.   Oaks and chesnuts
are also characteristic types, as are *Guttiferæ, Ternstræmiaceæ, Magnoliaceæ, Saurauja,* and tree-ferns.

The earliest explorer of the flora of Assam was Major Jenkins, who transmitted to Sir W. Hooker very extensive collections. Wallich, Griffith, and M'Clelland visited the valley in 1835, to investigate the then recently discovered tea forests, and Griffith returned to it more than once, so that its vegetation is now well known. Mrs. Mack and Mr. Simons have also enriched the Hookerian Herbarium with many interesting Assam plants. The *Ranunculus Chinensis*, a well marked Chinese species, occurs nowhere else in India; and Griffith has pointed out a multitude of instances of similarity between the floras of these two countries, in his able Report on the cultivation of the tea-plant in the Transactions of the Agricultural Society of Calcutta. The manufacture of tea has now been carried on for some years with considerable success in Upper Assam, but the wild tea (whose abundance in the forests of some parts led to the attempt in the first instance) is no longer used for that purpose. Griffith has given a general account of the botany of the Assam valley, in his Report on the tea cultivation already alluded to; as also in his " Remarks on a collection of plants made at Sadya, in Upper Assam," published in the Calcutta Asiatic Society's Journal, and in his private journals. He mentions having collected 1500 species, and computes that the whole flora must amount to at least 6000,—an estimate which, like all such made on similar data, is greatly exaggerated, and probably doubles the actual amount.

3. Naga and Khasia Hills.

The mountain range which bounds Assam on the south is known by a great diversity of names in different parts of its course, according to the different tribes by whom it is inhabited. The only part of the range which is well explored is that called the Khasia hills, across which a good road runs, by which a communication is kept up between Silhet and Gowahatti, the capital of Assam. These mountains have been explored botanically by Wallich and Griffith, and more recently by ourselves.

The Khasia hills rise abruptly on the south from the plains of Silhet to the height of about 4000 feet, and thence more gradually to 6000 feet. The culminating point is Chillong hill, the elevation of which is about 6600 feet. Their southern slopes are exposed to the full force of the monsoon, and the rain-fall is there excessive, amounting at Churra to 500 or 600 inches annually. Further in the interior the fall is less, and it gradually diminishes in amount till the valley of Assam is entered. On the north side the slope of the mountains is less abrupt, though there too there is a sudden fall from 5000 to 2000 feet, below which level a succession of gradually lowering hills continues to the Brahmaputra.

To the westward of the Khasia hills lie the Garrows, which are lower, the maximum elevation being probably nowhere more than three or four thousand feet. To the east, beyond Jyntea or Jaintia, which is similar in general character to Khasia, and will be included by us under that designation, there appears to be a considerable depression in the range, a large river with an open valley penetrating far to the north. These hills have, however, not been explored by Europeans. To the east of Cachar again there are lofty hills, inhabited by Nagas, and also quite unexplored, except in one place, where they were crossed by Griffith in travelling from Upper Assam to the Hukum valley, on a tributary of the Irawadi.

Notwithstanding the enormous rain-fall and the great humidity of the atmosphere, the higher parts of the Khasia hills are generally bare of trees, except in ravines and occasionally on northern exposures. This remarkable peculiarity is due partly to the nature of the surface, and the free drainage, but mainly to the removal of the soil by the heavy rains, and to the furious winds which sweep over the level tops of the hills. Wherever there is shelter, trees spring up at once; and the base of the mountains, and the deep valleys which penetrate far into the interior, are clothed with dense forest.

At the base of the Khasia the vegetation is tropical, and the plants the same as those of Assam. The sheltered and well wooded dells possess a uniformly hot climate, and closely resemble similar spots on the Eastern Archipelago. *Vaccinia* are plentiful, and there are many representatives of the Malayan flora, such as *Myristica, Henslovia* (Wallich), *Polyosma, Cardiopteris, Antidesmæ, Apostasia, Cyrtosia,* and other Orchideæ, *Ternstrœmiaceæ, Sonerila, Medinilla, Erycibe, Cyrtoceras,* and *Tacca.*

Higher up, temperate climate forms become common, chiefly oaks (of which, including chesnuts, sixteen species are known), *Styrax, Magnolia, Garcinia, Sphærocarya,* and *Lauraceæ. Acanthaceæ* form a great part of the underwood, and balsams are very numerous. The open hill-sides are covered with a luxuriant herbage, remarkably rich in species; and at elevations above 5000 feet there is a remarkable predominance of northern forms, which are common on the Himalaya at greater elevations. Most of the large Himalayan genera are there represented. We find species of *Ranunculus, Anemone, Thalictrum, Delphinium, Corydalis, Geranium, Parnassia, Rubus, Potentilla, Sanguisorba, Astragalus, Saxifraga, Astilbe, Umbelliferæ, Valeriana, Senecio, Cirsium, Pedicularis, Primula, Tofieldia,* and *Iris.* Of many of the genera which abound in the temperate Himalaya there are only single species, of others there are several. *Rhododendron* is represented by several species. One of these, the common *R. arboreum,* has a very wide range in India : the others belong to the more eastern forms of the genus, and, like the species of Java, descend to very low elevations : of *Rosa* also, the only species is the Peninsular and Chinese *R. sempervirens.*

We have elsewhere (page 105) alluded to the prevalence of Chinese and Japan forms in Eastern India; many of these are Himalayan, but some are quite peculiar to the Khasia. Of these, *Pinus Sinensis, Nymphæa pygmæa, Aralia aculeata, Hamamelis Chinensis, Nepenthes phyllamphora,* and *Bowringia* of Hooker (a curious genus of ferns) are all Chinese species, which in India are almost confined to the Khasia. *Reevesia* and *Illicium* are genera confined, so far as is hitherto known, to China and the Khasia; whilst *Helwingia, Micro-ptelea, Corylopsis, Bucklandia,* and *Quercus serrata,* though all Chinese and Khasian, are also common to the Himalaya; and *Vaccinium bracteatum,* as we have elsewhere said, is found in China, the Khasia, and the Peninsula, but not in the Himalaya.

Podostemon is a remarkable genus, which is abundant in all the Khasian streams, even in the most rapid currents covering the stones in autumn with a bright green carpet. This genus is even more abundant in the Nilgiri and Ceylon streams, and also found in Mishmi, but is quite unknown in the Himalaya.

Palms are very abundant in the Khasia, though much less so than in the Malayan Peninsula and Eastern Archipelago. We collected twenty-five species, belonging to the genera *Phœnix, Licuala, Areca, Arenga, Plectocomia, Calamus, Caryota, Chamærops,* and *Wallichia.* Of these the *Chamærops* is probably identical with the Nipal and Kumaon *C. Martiana,* though not found in any intermediate part of the Himalaya. *Livistona,* which is said to occur at the northern base of the Khasia, is found no further west.

There is only one pine in the Khasia mountains, *Pinus Sinensis.* This species is not known as a native of the Himalaya, but it is not impossible that it may occur in some parts of Bhotan. It may be conjectured too that it also extends into the mountains of the eastward, but we do not yet know any details of its distribution. In the Khasia hills it is not found in the very rainy southern districts, but becomes common in the valley of the Boga Pani below Moflong, and thence extends throughout the range, and descends towards Assam. The absence of *Pinus longifolia* is curious, as there is nothing in the climate adverse to its growth; but the elevation is not sufficient to lead us to expect the occurrence of any other of the Himalayan pines, or of the subalpine plants which accompany them. The common yew is however found at 5-6000 feet, and two species of *Podocarpus* occur on the lower hills, together with *Cycas pectinata* and *Gnetum scandens,* which are abundant everywhere.

As in all very humid climates, orchids occur in very great abundance in the Khasia mountains, constituting there at least one-twelfth of the vegetation, and being by far the largest natural order of flowering plants! They are equally abundant at all elevations. Many are epiphytes, but terrestrial species are also common, both in dense woods and in open grassy places. *Scitamineæ* are very numerous. From the barrenness of the surface over a great part of the hills, grasses constitute the most prominent feature in the flora of this district, occurring gregariously in prodigious abundance. Most of the species belong to the tropical division of the order, coarse *Paniceæ* being the prevailing forms, but there are also many *Poaceæ* of European genera.

In some respects the vegetation of the Khasia approaches more closely in its features to that of the mountains of the Peninsula than of the Himalaya : this arises mainly from the form of the hills and their much less rugged outline, their valleys being more open, though with steeper flanks, and the hill-tops broader. Hence the grassy slopes being covered with clumps of shrubby vegetation, and the forest being confined to sheltered localities, are remarkable features in common with the Nilghiri, but quite foreign to the Himalaya; to which must be added a very strong resemblance in the genera and species forming the mass of the shrubby vegetation, which, though almost all Himalayan, are there less gregarious and more interspersed with large trees of different genera. These consist of :—

Rhododendron *arboreum.*	Styrax.
Pieris *ovalifolia.*	Callicarpa, several species.
Ligustrum.	Celastrus, ditto.
Eurya, two species.	Michelia, ditto.
Vaccinium *bracteatum.*	Goughia *Himalaica.*
Gaultheria, several species.	Gomphandra.
Symplocos, ditto.	Photinia, several species.

Ilex.

Eugenia.

Myrsine.

Laurineæ, various genera.

Rubiaceæ, ditto.

Compositæ, ditto.

Jasminum, ditto.

Indigofera.

Saurauja, several.

Berberis.

Casearia.

Cleyera.

Viburnum, several species.

Elæocarpus.

Elæagnus.

Turpinia.

Araliaceæ, several species.

To these must be added certain Himalayan temperate genera that are Khasian, but not Peninsular, especially oaks and chesnuts :—

Holböllia.

Manglietia.

Magnolia.

Talauma.

Spiræa.

Pyrus.

Corylopsis.

Bucklandia.

Neillia.

Pomaceæ, several.

Camellia.

Acer.

Cerasus.

Prinsepia.

Benthamia.

Leycesteria.

Itea.

Hydrangea.

Adamia.

Luculia.

Hymenopogon.

Limonia.

Wightia.

Microptelea.

Carpinus.

Helicia.

Betula.

Sabia.

Sphærostema.

Taxus.

Pinus.

Camphora.

Chamærops.

Plectocomia.

And of herbaceous forms :—

Codonopsis.

Corydalis.

Dicentra.

Panax *Pseudo-gin-seng.*

Delphinium.

Astragalus.

Astilbe.

Saxifraga.

Sanguisorba.

Lychnis.

Anisadenia.

Circæa.

Sarcopyramis.

Crawfurdia.

Primula.

Pyrola.

Monotropa.

Veronica.

Dipsacus.

Iris.

Allium.

Paris.

Polygonatum.

Of Khasian temperate forms common also to the Peninsula, but not found in the Himalaya, *Vaccinium bracteatum,* also a native of China, is almost the only example.

During our five months' residence in the Khasia we collected 2264 species of flowering-plants and nearly 200 ferns. The following natural orders are noticeable for the number of species they contain :—

Ranunculaceæ 13	Verbenaceæ 29
Menispermeæ 15	Scrophularineæ 40
Magnoliaceæ 9	Labiatæ 57
Vitaceæ 34	Cyrtandraceæ 24
Balsamineæ 22	Acanthaceæ 58
Ternstrœmiaceæ . . . 14	Asclepiadeæ 45
Aurantiaceæ 18	Polygoneæ 26
Malvaceæ ⎫	Amentaceæ 20
Byttneriaceæ ⎪	Laurineæ 24
Sterculiaceæ ⎬ . . . 37	Urticeæ 82
Tiliaceæ ⎭	Euphorbiaceæ 58
Leguminosæ 123	Gramineæ :
Rosaceæ 37	Paniceæ 122
Melastomaceæ 17	Poaceæ 42
Myrtaceæ 14	Cyperaceæ 91
Cucurbitaceæ 31	Scitamineæ 37
Umbelliferæ 19	Commelyneæ 18
Araliaceæ 30	Aroideæ ⎫ 29
Rubiaceæ 112	Orontiaceæ ⎭
Compositæ 87	Palmeæ 25
Myrsineæ 36	Orchideæ 173
Convolvulaceæ 26	

The Naga hills, to the eastward, probably exhibit a very similar vegetation to the Khasia, as their elevation is about the same. They were crossed by Griffith in the month of March, at which season vegetation at considerable elevations is nearly dormant. The greatest height attained by him was 5600 feet. He describes these hills as much more covered with forest than the Khasia,* and states that the southern slopes are moister than those to the north. As the rain-fall must be much less than it is on the southern slope of the Khasia, the greater amount of forest is probably caused by the diminished vio-

* Private Journals, p. 120.

lence of the winds, which in the Khasia sweep with tremen-
dous force over the nearly level hill-tops.

The flora of the Naga hills is only known by the few notes
published in Griffith's journals, as the collections which he
made there have not been distributed.   Except *Liquidambar*
and *Kaulfussia Assamica*, Griffith notes no plants as differing
from those of the Khasia; the general forms are therefore
certainly the same.   He especially alludes to the absence of
Coniferæ, of which however a species is said to abound on the
hills of Manipur, to the southward.   Of genera indicating
elevation, he mentions *Acer, Vaccinia, Daphne, Berberis,
Bucklandia, Crawfurdia, Viburnum,* and *Cyathea,* all equally
typical of elevation in the Khasia and Eastern Himalaya.
At lower levels, Oaks, *Gordonia, Camellia, Mesua, Bucklandia,
Magnolia, Æsculus, Pandanus, Areca, Caryota,* and tree-ferns,
are indicated as prevalent forms.

4. CACHAR AND SILHET.

The valley, or rather marshy plain of the river Súrma, which
lies to the south of the Khasia mountains, very much resembles the Assam valley in its general features.   It is an open
plain, scarcely raised above the level of the sea, which is three
hundred miles distant, and presenting here and there a few
scattered hills : below, it expands into the Jheels of Eastern
Bengal, and contracts in its upper part, as the spurs of the
Tippera and Naga hills encroach upon it, separating fertile
plains by narrow ridges covered with dense forest. The mountains which skirt this plain on the north nowhere attain an
elevation of more than 7000 feet, and those on the south are
very low and everywhere covered with dense forest.   The climate is the same as that of Bengal and Assam, but more
healthy ; the rains are heavy, the winter more mild, and the
spring moist and not hot.   The rain-fall at Silhet is very
great, more than 200 inches having been registered in one
year.   At Cachar it is equally heavy.

The vegetation of the open plains of Silhet is the same as

that of Bengal, and on the wooded hills we find a flora closely resembling that of Assam. In the moister forest, *Anonaceæ* are extremely numerous, and species of *Calamus*, tree-fern, and *Pandanus* are equally so. Oaks occur in the forests down to the level of the river Súrma, with *Camellia, Kadsura, Sabia, Rubus,* and other plants usually considered as indicating a certain degree of elevation.

The low hills which rise out of the plain in the neighbourhood of Silhet, and in several other parts of the district, are covered with brushwood, amongst which are many remarkable plants, as *Licuala peltata, Adelia castanocarpa, Trophis, Connarus, Grewia, Briedelia, Gelonium, Moacurra, Mussænda, Guettarda.* There are also some shrubs which here find their northern limit, but which are common in similar localities in Chittagong: as instances, we may mention *Dalhousiea* and *Linostoma.* In the grassy sward which covers the swampy plains interspersed among these hills, we find also *Stylidium Kunthii,* a minute annual, which is interesting as the most northerly species of the eminently Australian order to which it belongs.

Many plants from this district were communicated to Roxburgh by Mr. Smith, Judge of Silhet, Mr. Dick, and other residents, and by the Garden collectors; and are published in his ' Flora Indica.' Dr. Wallich's collectors were long at Silhet, and sent him large collections; and the authors of the present work, in the autumn of 1850, ascended the Súrma from Silhet to Silchar, and collected several hundred species.

The Jheels of Eastern Bengal are in many respects a most remarkable feature, and as they owe their origin chiefly to the excessive rain-fall of the Khasia and Silhet, and to the overflow of the Súrma, we have noticed them under this province, in preference to Bengal, in which they would otherwise have been included.

The Jheels occupy an immense area, fully 200 miles in diameter, from north-east to south-west, which is almost entirely under water throughout the rainy reason, and only par-

2 *i*

tially dry in the winter months. They extend from the very
base of the Khasia and eastern extremity of the Cachar district, southward to the Tippera hills and Sunderbunds, and
westward to the Megna and considerably beyond it, thus
forming a freshwater continuation of the Sunderbunds, and
affording a free water-communication in every direction. The
villages, and occasionally large towns, which are scattered
over the surface of the Jheels, generally occupy the banks of
the principal rivers; these have defined courses in the dry
season, their banks always being several feet higher than the
mean level of the inundated country.

Extensive sand-banks, covered in winter with a short sward
of creeping grasses and annual weeds, run along the banks of
the largest streams, and shift their position with every flood.
The remainder of the surface is occupied by grassy marshes
covered in winter with rice crops, and in summer with water,
upon which immense floating islands of matted grasses and
sedges are seen in every direction, gradually carried towards
the sea by an almost imperceptible current. The principal
floating grasses are *Oplismenus stagninus* and *Pharus aristatus*, which together form the mass of each islet; and along
with these occur *Azolla*, *Salvinia*, *Utricularia*, *Villarsia* of two
species, *Jussieua*, *Trapa*, *Pistia*, and several aquatic *Scrophularineæ*.

In shallower water, *Vallisneria*, *Hydrilla*, *Potamogeton*,
Damasonium, several *Nymphææ*, *Myriophylla*, and *Ceratopte-
res* carpet the bottom, whilst *Confervæ* and the many tribes
of fresh-water Algæ, so common in temperate latitudes, are
comparatively rare.

In the marshes the principal grasses are *Panica*, *Paspala*,
and their allies, with tall *Andropogons*, *Sacchara*, *Erianthus*,
Arundo, *Apluda*, and *Rottbœllia* in the greatest abundance.
Mixed with these are *Typha*, *Scleria* and numerous *Cyperi*,
but no large *Junci*.

On the banks of the principal streams a fringe of brushwood consists of *Stravadium*, *Tetranthera*, *Grewia*, various

Rubiaceæ, Eugenia, Gouania, and with occasionally immense quantities of *Alpinia,* more rarely *Rosa involucrata, Calamus Rotang,* and in sandy places *Tamarix.*

Convolvuli, a few *Asclepiadeæ, Cucurbitaceæ,* and all the weeds of Bengal, abound in favourable situations; and by the villages a few scattered figs, clumps of bamboo, mango, and *Areca,* are all seen, though rarely.

5. TIPPERA AND CHITTAGONG.

The valley of the Súrma is separated from that of Manipur by a meridional range of moderate elevation, which is continued to the southward, and separates Tippera, Chittagong, and Aracan from the kingdom of Ava. The nature and elevation of the axis of this range are unknown, but its ramifications extend to the sea-coast, and are separated by cultivated valleys, the direction of which is in general southwesterly or nearly due south. These ranges appear to increase in elevation as we proceed southward, but our knowledge of them is very imperfect. Blue Mountain, which lies nearly due west of Chittagong, is said to attain the considerable elevation of 8000 feet, and a peak on the same range forty miles to the south-west, in lat. 22°, is elevated (according to Wilcox's map) 3100 feet. Sitakund, thirty miles north of Chittagong, has an elevation of 1140 feet.

The provinces of Tippera and Chittagong are throughout hilly. Along the sea-coast there is in general a narrow belt of level ground, and the basins of the rivers are usually wide and well cultivated for a considerable distance inland. In the upper part of their course, however, they are hemmed in by hills, and a broad belt of impenetrable forest occupies the interior, and forms an impassable boundary between the British territories and those of Ava. The climate is similar to that of Bengal. From the proximity of the sea and the situation within the tropic, the winter is very mild, and the atmosphere always humid. The rain-fall during the monsoon is about the same as in Bengal, at least on the sea-coast and in its immediate vicinity, averaging 86 inches annually at Chittagong; on the higher ranges in the interior it is probably much more considerable. The low hills of Tippera, immediately to the south of the Súrma valley, are said to be covered to a great extent with dry bamboo jungle, extending uninterruptedly for miles and being almost uninhabited. The southern slopes may be expected to be more humid, as they are fully exposed to the rainy wind.

The vegetation of Chittagong is very similar to that of Silhet. The higher hills are covered with dense but often dry forest, and the lower ones with brushwood. Oaks (which grow down to the level of the sea), two species of nutmeg, *Dillenia pentagyna*, *Butea*, *Pongamia*, *Mesua*, *Gordonia*, *Engelhardtia*, *Henslovia*, and several *Dipterocarpi*, are conspicuous in the forests. Of the latter, *Dipterocarpus turbinatus*, which yields the well known and valuable Gurjun, or wood oil, is extremely abundant, towering over the other forest-trees. *Cycas* is common. On the drier hills we have the same shrubs which have already been enumerated as growing in similar situations in Silhet, with *Linostoma* in very great abundance, *Pterospermum*, *Dalhousiea*, *Bradleia*, *Melastoma*, *Litsæa*, *Tetranthera*, *Scepa*, *Calamus fascicularis*, *Wikstrœmia*, *Ixora*, *Adelia*, *Moacurra*, *Cæsalpinia*, *Mussœnda*, *Guettarda*, *Gelonium*, *Jasminum*, *Memecylon*, and *Congea*; and of small trees, *Ægle Marmelos*, *Amoora*, *Gaurea*, Figs, and *Micromelon*. In damp woods are many *Calami*, two *Wallichiæ*, three *Arecæ*, various *Lagerstrœmiæ*, *Meliaceæ*, many *Leguminosæ*, *Terebinthaceæ*, *Verbenaceæ*, and *Magnoliaceæ*, all growing in great luxuriance, and most of them forming gigantic forest-trees.

In consequence of the great influx of fresh water which is discharged into the Bay of Bengal by the Megna and Fenny rivers, the eastern part of that sea remains almost fresh for a very considerable distance from the shore. Even at the mouth of the Chittagong river the water is only brackish, and the maritime tropical vegetation of mangroves, and such plants, does not commence till we advance as far south as Ramri island. At the same place we find the northern limit of *Casuarina equisetifolia,* the most northerly species of the family of *Casuarineæ,* which is chiefly confined to Australia. The Indian species is extensively cultivated throughout Bengal. On the low islands along the coast the vegetation is very scanty, and chiefly consists of creeping grasses, with *Dilivaria, Excœcaria, Tamarix, Rhizophoreæ, Acrostichum aureum,* and a Composite shrub.

Our knowledge of the flora of these provinces is chiefly derived from Roxburgh's ' Flora Indica;' many of the most interesting species published there having been communicated to him from Tippera and Chittagong. Our own small collection, which was made in the months of December and January, amounts to about 600 species.

6. ARRACAN.

The province of Arracan is a narrow belt of land, 290 miles long, hemmed in between the sea and the Aeng or Youmadang range of mountains, which lies very near the coast. It is traversed from north to south by a large river, navigable for a considerable distance into the interior; and by numerous smaller rivers, all of which have tidal channels, and form a sort of delta along the coast, which is skirted by many islands. From the proximity of the mountains to the coast, and their considerable elevation, the rain-fall is very great, amounting to 160 and 180 inches annually.

The botany of Arracan is quite unknown, and the climate of the interior is very unhealthy. Along the sea-coast are forests of mangroves, and there is in all the valleys very extensive rice cultivation, the plains being inundated during the monsoon. Tobacco of superior quality is also cultivated. The mountains may be expected to produce the same plants as are found in the Malayan peninsula, to which the climate approximates very closely; they are clothed with heavy forests and bamboo jungle. The gamboge is said to be found in the

island of Cheduba, and if so, the latter is the northern limit
of that tree.

7. AVA AND PEGU.

The sources of the river Irawadi are, according to the best
authorities, between 27° and 28° of north latitude, and the
direction of its valley is nearly due north and south. The
mountains in which this immense river takes its rise probably
rival in height the Eastern Himalaya, but the meridional
ranges which bound its valley on each side do not long retain any great elevation, though they are continuously from
4000 to 8000 feet in height almost as far as the sea. The
transverse range, which separates the upper part of the western branch of the Irawadi ˙from the valley of Assam, is also
of moderate elevation, varying probably between 5000 and
6000 feet.

The slope of the valley of the Irawadi is greater than that
of the Indus or Ganges, if the estimates of elevation given by
Griffith may be relied on. The valley of Hukum is stated to
be 1000 feet above the level of the sea. The determination
however was made by boiling water, which, at such low levels,
is too fallacious a test to be depended on. The central branch
of the Irawadi, at Manchi in 27° 20′ north latitude, where it
was visited by Wilcox, has an elevation of 1800 feet[*], and
runs over a pebbly bed. Its elevation at Bhaumo, in lat.
24°, is estimated by the same authority to be about 500 feet.
The valley of the Irawadi is much less open than that of
the Ganges, being interrupted in many places by transverse
ranges. In the upper part of its course these are numerous,
and the lateral valleys they enclose are comparatively small;
but lower down there is a great expanse of level country,
though the hills occasionally attain an elevation of 3000 or
4000 feet close to the river.

The direction of the monsoon wind in the valley of the
Irawadi appears to be nearly from south to north. The

[*] As. Res. xvii. 441.

mountains to the north-east are considerably more elevated than those to the northward, over which the aerial current probably flows into the valley of Assam.

The first condensation of the moisture-laden winds takes place in the lower part of the valley, which is hemmed in by hills at the apex of its delta. Further north there are no more considerable elevations till we reach the sources of the Irawadi, so that in the central part of its course the rain-fall is comparatively small. We have therefore in Pegu a climate like that of the Gangetic delta, the rain-fall amounting at Rangoon to 85 inches; but in Ava a dry climate, like that of the Gangetic valley, or the Carnatic, prevails, with a moderate rain-fall at one season only. The upper valley is again more humid, from the loftier mountains and the more irregular surface of the country.

In the delta of the Irawadi there is a maritime vegetation of mangroves, *Sonneratia, Heritiera, Excœcaria,* and other saline plants, just as in similar salt-marshes along the coasts of the tropics. Throughout the plains of Pegu the vegetation is like that of the Gangetic delta, or the open parts of the valley of Assam. Cocoa-nut, *Corypha,* and *Borassus* are the common palms, with *Pandanus, Stravadium,* and abundance of epiphytical *Orchideæ.* On the mountains the flora is of course more varied, and is a continuation of that of Tenasserim to the south.

In Ava, with a climate and temperature very similar to that of the Carnatic, we find an almost identical vegetation. *Capparideæ* are common, with acacias, an arboreous *Euphorbia, Calotropis gigantea, Guilandina Bonduc, Zizyphus,* and *Bombax;* mangos and *Fici,* with *Borassus,* are cultivated. Teak is common on the mountains. The vegetation of the higher parts of the Irawadi is described by Griffith as very similar to that of Assam.

The valley of Manipur is drained by the most westerly tributary of the Irawadi: it is separated from Cachar by a mountain range, which is 6000 or 8000 feet high, and is pine-clad towards the summit. The valley of Hukum (or Hookhoom), which was visited by Griffith, is more open, but is surrounded on the north and east by mountains elevated 5000 and 6000 feet, and is traversed by numerous ranges of low hills.

We do not know the boundaries between the different provinces on the Irawadi, nor is it necessary for our purpose to distinguish them, as the upper country is unknown to us.

Dr. Wallich, who accompanied Mr. Crawfurd's mission to Ava soon after the close of the Burmese war in 1826, was the first botanist who explored the vegetation of the Irawadi. He ascended that river as far as the capital, and visited the mountain range bounding the Taong-dong river to the eastward, from which some of his finest plants were obtained. Mr. Griffith, in 1837, entered Ava from Assam, and descended the Irawadi to its mouth, but the collections made by him on this journey have not been distributed. Since the earlier sheets of this Introduction were printed, Dr. M'Clelland has forwarded to the Hookerian Herbarium an excellent and very valuable collection from Pegu.

8. TENASSERIM.

The province of Tenasserim is separated from Pegu by the Sitang river, and extends south to the commencement of the Malayan Peninsula, including the districts of Martaban, Tavoy, and Tenasserim. At its northern extremity, the great river of Martaban forms an extensive alluvial plain like that of Pegu, bounded to the east by mountains of considerable but unknown elevation. Elsewhere the mountains approach the coast, and are said to attain occasionally, but not continuously, an elevation of 4000 or 5000 feet. The coast is generally alluvial; tidal channels, which separate a broad and continuous belt of islands from the main, run into the interior, and the hilly tracts are covered with dense forest.

In climate Tenasserim is intermediate between Arracan and the Malayan Peninsula. The summer rains are everywhere heavy and long continued, commencing in May or the beginning of June, and lasting till November, and amounting at Tavoy to 208 inches, and at Maulmain to 175. In the more northern parts the winter is dry, the north-east wind being deprived of its moisture by high ranges of mountains. South of Tavoy the winters are more humid, and rain is of frequent occurrence at all seasons.

The vegetation of Tenasserim is a continuation of that flora which, commencing in Sikkim and Bhotan, is continued throughout the Malayan Archipelago. Oaks and *Diptero-carpi* are very common; and a pine, probably *P. Sinensis,* grows on the mountains north of Martaban. *Calami, Zalacca,* and other tropical palms, are abundant in humid jungles, and enormous bamboos in more open places. Teak is common in the interior, but has its southern limit in 15° N. lat., where the winters become too humid for its growth. The *Amherstia nobilis,* one of the most remarkable and local trees in the province, has hitherto been found only on the banks of the Salueen river; *Barclaya longifolia,* a remarkable genus of water-lilies, is confined to this province and the adjacent one of Pegu; and the *Melanorrhœa usitatissima,* or black varnish tree, abounds in many parts.

Dr. Falconer, in his able report on the teak forests of Tenasserim, gives some valuable remarks on the vegetation of the province, and the following list of prevalent timber-trees:—

Dillenia.	Elæocarpus.	Melanorrhœa.
Uvaria.	Aglaia.	Blackwellia.
Guatteria.	Heynea.	Toddalia.
Myristica.	Dipterocarpus.	Turpinia.
Cratæva.	Hopea.	Inga.
Bombax.	Vatica.	Acacia.
Sterculia.	Gordonia.	Pterocarpus.
Paritium.	Calophyllum.	Butea.
Grewia.	Garcinia.	Dalbergia.
Pterospermum.	Millingtonia.	Pongamia.

2 k

Cathartocarpus.	Diospyros.	Gynocardia.
Cassia.	Bignonia.	Trewia.
Conocarpus.	Calosanthes.	Quercus.
Lagerstrœmia.	Spathodea.	Castanea.
Jambosa.	Tetranthera.	Antidesma.
Careya.	Croton.	Ficus.
Nauclea.	Rottlera.	Artocarpus.

Martaban was visited in 1827 by Wallich, and more recently by Falconer. Mergui and Maulmain have been explored by Griffith, whose extensive collections have been distributed; and by Mr. Lobb, who has communicated some interesting plants to the Hookerian Herbarium.

9. Malayan Peninsula.

The Malayan peninsula extends from the southern extremity of Tenasserim, almost to the equator, the island of Singapur being in 1½° N. lat. Its width varies from 150 to 100 miles, and near the southern extremity it contracts to about fifty miles. A low range of hills traverses the whole length of the peninsula, rising occasionally into isolated peaks, of which the highest, Mount Ophir, near Malacca, attains 4320 feet*, but they are usually very much lower. The island of Penang is 2922 feet high.

On either side of the central axis, low ranges of hills descend towards the sea, so as to give an undulating outline to the surface. These are separated by swampy flats of considerable length, which are narrow and often under water, but there are no plains of any extent. The coast is occasionally rocky or skirted by coral reefs, at other places low and muddy. The direction of the rivers is generally at right angles to the axis. Their banks are for the most part muddy and low, and

---

* This height is taken from a paper by Logan, in the ' Journal of the Malayan Archipelago' (ii. 137). According to the same authority, Kedah peak is 3897 feet high. Mr. Logan informs us that the elevations given by Newbold for these peaks (5693 and 5705 feet) are mere guesses.

those of larger size are navigable for small vessels to a considerable distance.

The northern part of the peninsula is now subject to the kingdom of Siam, which has extended its limits to the south, so as to occupy the state of Kedah. Further south, independent Malays possess the whole of the country, except the three British settlements of Penang, Malacca, and Singapur.

From its proximity to the equator, and from the peculiarity of its shape,—a long, narrow strip of land, nearly enclosed by sea,—the Malayan Peninsula enjoys a very mild and equable climate. The monsoon winds, which are influenced by general causes at a great distance, prevail here with as much regularity as elsewhere in India, the south-west monsoon continuing while the sun is north of the equator, and the north-east monsoon from October to March, while the sun is in the southern hemisphere. Local causes, however, modify these winds very much, and regular land and sea breezes blow along the coast. Both these monsoons are rainy, as they traverse a great extent of sea, and the mountain ranges everywhere condense the vapours. The north-east monsoon is, however, more rainy than the other, because the mountains of Sumatra, which receive the first supply of moisture from the south-west monsoon, are considerably more elevated than those of the peninsula itself. The most rainy months are, therefore, from November to January, and February is the coldest month of the year. In the Straits of Malacca the rain-fall is nowhere excessive. On the hill of Penang it was in one year 116·6 inches, and on the plain at its base only 65·5 inches, while in the province of Wellesley, on the opposite coast, the amount was 79·15 inches. At Singapur the fall is 98 inches, and at Malacca the same. On the south coast of Sumatra, and on the north-eastern face of the Peninsula, the fall is probably much greater. The mean temperature of Singapur is 79·7°, and the temperature of the different months differs very little from the mean of the year.

In the equable and humid climate of Malaya, we have a

vegetation almost identical with that of Java. The surface, except where clearances have been made by man, is covered with a shady forest, rendered almost impenetrable by a dense jungle of rattan (*Calamus*), a genus which attains its maximum development in the Malayan region. Erect palms are also very numerous; chiefly of the genera *Areca, Arenga, Licuala, Cocos, Corypha,* and *Sagus.* On the coast, *Nipa* covers immense tracts. Orchids, terrestrial as well as epiphytical, *Scitamineæ, Araceæ,* and ferns, abound in the forests, which consist chiefly of gigantic *Terebinthaceæ, Sapindaceæ, Meliaceæ, Garciniaceæ, Dipterocarpeæ, Ternstræmiaceæ, Leguminosæ, Myrtaceæ, Combretaceæ, Lauraceæ,* oaks, and figs. *Dilleniaceæ,* nutmegs, *Sapotaceæ,* including *Isonandra Gutta* (the gutta-percha plant), and *Anonaceæ,* form an unusually large proportion of the flora. *Podocarpus, Dacrydium,* and *Dammara* are the only conifers, but there are several species of *Gnetum* and of *Cycas.* On the higher hills a few species of *Gaultheria, Rhododendron, Vaccinia,* and other plants of the sub-temperate zone, indicate the commencement of that rich and varied flora which covers the middle and upper parts of the mountains of Java and the Khasia, and is also found in the temperate Sikkim Himalaya.

Amongst the many rare and curious genera which occur in the forests of the Malayan Peninsula, may be mentioned *Grammatophyllum,* the most gigantic Orchid known, *Kibara,* many *Nepenthes,* several curious genera of *Aristolochiæ,* as *Thottia, Lobbia,* and *Asiphonia,* anomalous *Burmanniæ,* many *Antidesmeæ,* including *Eremostachys* and *Phytocreneæ,* as *Iodes, Cardiopteris,* and *Phytocrene* itself, many singular *Olacineæ, Santalaceæ, Loranthaceæ, Menispermeæ,* etc. The cultivated fruits are the mangosteen, durian, and nutmeg, none of which thrive elsewhere in India; with many varieties of *Citrus* and pine-apple. The littoral plants are to a great extent the same as those of Pegu and the Sunderbunds, but there are more species of mangrove and of palms. *Enhalus* and other oceanic *Cauliniæ* occur beneath high-water mark. The appearance of Australian forms in the Malay Peninsula has been alluded to at p. 103, and is shown by species of *Stylidium*, *Bæckia*, *Melaleuca*, *Casuarina*, *Leptospermum*, *Leucopogon*, *Tristania*, and *Dacrydium*. It is a remarkable fact that the teak, which abounds in some parts of Java and in the northern districts of Tenasserim, is not known to inhabit the Malayan Peninsula.

Jack was the first botanist who explored the Malayan Peninsula. Some years later, Dr. Wallich visited Penang and Singapur, where he made large collections : a part of Mr. Cuming's collection was also formed in Malaya. More recently, Griffith was for a considerable period resident at Malacca; and it is from his notes and collections that our detailed knowledge of its flora is derived. Sir W. Norris, Mr. Prince, and Dr. Oxley have also added much to our information.

IV. *Afghanistan and Beluchistan.*

The great chain of the Kouenlun, which separates the Indus and its tributaries from the Yarkand plain, is continued to the westward, under the name of the Hindu Kúsh. This chain, which has a westerly direction, with some southing, separates the basin of the Oxus on the north from that of the Kabul river, a tributary of the Indus, and from the Helmand, a river which runs towards the south-west, and is lost in the desert of Sehistan, not reaching the sea. The elevation of the chain diminishes rapidly to the westward, but few accurate determinations of its height are known. The Kalu pass, near Bamian, is 12,500 feet, and the peak of Koh-i-Baba, which rises close to it, is 17,000 feet above the level of the sea. The Erak (or Irak) pass is 12,900 feet.

From the neighbourhood of the peak of Koh-i-Baba a meridional chain runs nearly due south to the Indian Ocean, forming the watershed between the Indus on the east and the Helmand on the west. The axis of this chain passes close to Ghazni, elevated 7726 feet ; and to Quetta, 5540 feet. It

lies probably to the westward of Kelat, but our maps are not
sufficiently accurate to make its course in that direction obvious. At its point of origin this chain is more than 13,000
feet in height; where it is called the Saféd-Koh, or White
mountains, it is 14,000. Near Ghazni it is from 9000 to
10,000 feet high; and near Quetta its elevation is nearly as
great, for the peak of Chahil Tan rises to 10,500 feet. Its
eastern ramifications are high ridges which dip abruptly into
the valley of the Indus; one peak, near Dera Ismael Khan
(called Takht-i-Suliman), attains a height of 11,000 feet, and
the range south of the Kabul river rises still higher. The deceptive appearance of a chain of mountains running parallel
to and near the west bank of the Indus is given by the extremities of the eastern spurs of these ridges, and has no existence except upon our maps. To the westward, long ranges
of rugged mountains branch from it, and stretch far in a
south-west direction before they sink into the elevated table-
land of Persia. The elevation of Candahar is 3480 feet, and
that of Bamian 8500.

Excepting in the most eastern part of Hindu Kúsh, between the Kuner and the Gilgit rivers, these mountains nowhere rise to the height of perpetual snow, except on the peak
of Koh-i-Baba. Their outline is often rounded; they are in
general bare and stony, separated by wide elevated valleys,
1000 or 2000 feet below the ridges. Water being scarce,
the valleys are sterile and very rocky.

Throughout Afghanistan the climate is excessive. The
cold of the winter is intense, the spring is damp and raw, and
the summer, during which hot west winds prevail, is intensely
hot at all elevations. Winter and spring are the rainy (or
snowy) seasons, while the summer and autumn are dry. The
return upper current of moist air, which passes northward
during the prevalence of the north-east monsoon, is condensed
by the mountains, and heavy falls of snow are of frequent occurrence during winter at all elevations above 5000 feet, or a
little lower in the immediate vicinity of the Hindu Kúsh. In

the low valleys heavy rain falls at this season. Spring sets in in March in the temperate zone, and with the change of the monsoon (about the equinox or a little later) heavy rains occur, caused perhaps by the southerly direction of the monsoon wind, before the Indo-Gangetic plain becomes intensely heated, and deflects that wind into a westerly current.

The general aspect of the whole of Afghanistan is that of a desert. As the mountains rarely rise to the region of perpetual snow, water is very scarce after the termination of the spring rains; but when the country was the seat of a great empire, an energetic race of inhabitants conducted every available streamlet into artificial channels, by the help of which an extensive cultivation is still carried on in many of the valleys. Around the chief towns and many of the villages, therefore, the country is beautifully verdant. The crops are chiefly wheat and barley, even up to 10,000 feet elevation. Rice is cultivated in great quantity at Jellalabad (2000 feet), at Kabul (6400 feet), and to a considerable extent at Ghazni (7730 feet). Poplars, willows, and date-palm trees are extensively planted, as well as mulberry, walnut, apricot, apple, pear, and peach-trees, and the *Elæagnus orientalis*, which also bears an eatable fruit. The vine abounds, as in all warm and dry temperate climates.

The flora of Afghanistan is an extension of the Arabian and Persian, with a few Himalayan types. From the great solar power, and the absence of rain during summer, the heat is excessive, so that the vegetation is that of a hot, dry country. On the southern slopes of the Hindu Kúsh the great elevation of the chain produces more humidity than elsewhere in Afghanistan; and there is therefore a forest belt, which extends from 5000 to 10,000 feet. These forests are entirely confined to the mountains which rise out of the valley of Jellalabad, and do not extend further west than the 69th degree of longitude: elsewhere the country is extremely barren, and almost destitute of tree vegetation. The trees are chiefly oaks and pines. There is also a pine forest on the northern slope of the Saféd Koh range, which bounds the valley of the Kabul river on the south, it being lofty, and snow-clad almost throughout the year. The pines are *Pinus excelsa* and *Gerardiana, Abies Smithiana,* and *Cedrus Deodara*: of these the deodar appears to be the most abundant. In the temperate zone *Juniperus excelsa* is of occasional occurrence. The oak of these forests is *Quercus Ilex,* a species which extends from the south of Europe as far as Kunawar. With the oak, species of *Æsculus, Olea, Myrtus,* and *Amygdalus* occur.

In the tropical zone, which skirts the whole region, the plants are the same as those of Sind and the Panjab, which again are identical with those of tropical Arabia and of south Persia. A few scattered pistacias, with *Celtis* and *Dodonæa,* are almost the only trees; though in some valleys there are small woods of *Populus Euphratica.* The date is cultivated in Beluchistan and Southern Afghanistan up to 4500 feet, and a dwarf palm (*Chamærops Ritchieana* of Griffith, perhaps identical with the *Chamærops humilis* of Europe) occurs abundantly in many places, but with a somewhat local distribution.

Above 4000 feet, or a little higher in Beluchistan, the tropical gives place to the true oriental flora. Aromatic shrubs, chiefly *Artemisiæ* and *Labiatæ,* cover the plains, and prickly *Statice* and *Astragali* abound on the dry hills. *Cruci-feræ, Umbelliferæ, Boragineæ, Cynaraceæ,* and *Cichoraceæ* are extremely abundant, far more so than in India; with *Rosa, Lycium, Berberis,* and other Syrian shrubs. In early spring there is here, as in the Mediterranean region, an extremely luxuriant vegetation, and the genera, if not the species, are the same. *Hyacinthus, Lilium, Tulipa, Fritillaria, Narcissus, Colchicum, Ixiolirion, Anemone,* and *Delphinium* may be mentioned as instances.

In many places the soil is saline, and the *Chenopodiaceæ,* mentioned as natives of Tibet, as well as *Glaux maritima,* are abundant.

The Alpine vegetation is also a mixture of European, Siberian (and Tibetan), Oriental, and Himalayan species, with little or no peculiarity.

As instances of the Himalayan flora advancing westward beyond the Indus, we may mention the following natives of Afghanistan, none of which have hitherto been detected in Persia :—

Berberis *Asiatica.*   Loniceræ, several.
Clematis *grata.*   Impatiens, sp.
Thalictrum *pedunculatum.*   Æsculus.
Corydalis *Moorcroftiana.*   Sarcococca *pruniformis.*
Edgeworthia.   Cedrus *Deodara.*
Dalbergia *Sissoo* (cult. ?)   Pinus *longifolia.*
Mazus *rugosus ?*   „   *Gerardiana.*
Adhatoda *Vasica.*   „   *excelsa.*
Myrsine, sp.   Abies *Smithiana.*

The following have not, so far as we are aware, been found east of the Indus, nor in any part of British India :—

Delphinium *camptocarpum.*   Hypecoum *procumbens.*
Leontice *Leontopodium.*   Rosa *rubiginosa.*
Bongardia *Rauwolfii.*   Amygdalus *furcatus ?*
Glaucium *elegans.*   Ephedra *ciliata.*
   „   *corniculatum.*   Chamærops *Ritchieana.*
Rœmeria *hybrida.*   Ægilops, several species ?

Our knowledge of the botany of this province is principally due to the labours of Griffith and Stocks. Mr. Griffith accompanied the army which marched in 1838–39 from Sind, through Quetta and Candahar to Ghazni and Kabul. From Kabul he crossed the chain of the Hindu Kúsh to Bamian and Singhan, and spent some time in the Kuner valley. His collections, though formed under circumstances of great difficulty, are very good, amounting probably to about 1000 species. Dr. Stocks twice visited Beluchistan and the southern parts of Afghanistan, penetrating as far as Quetta at considerable personal hazard. Some other collections were made while the country was occupied by the British army, but we

2 *l*

have not had access to any of them. Mr. Ritchie, a Bombay officer, we believe formed a good herbarium in the mountains south of Jellalabad (the Saféd Koh), which Griffith appears to have seen, but none of the specimens have found their way into our herbaria.

EXPLANATION OF THE MAPS.

MAP I.—To face page 82 of Introductory Essay.

The Map of Isothermals for January, April, July, and October, is intended to illustrate the chapters of the Introductory Essay devoted to the Meteorology of India (page 74), and of the provinces into which we have divided that country (page 115). It is compiled (by permission) from the maps of monthly Isothermals which accompany Dove's admirable work " On the Distribution of Heat over the surface of the Globe," as translated by Colonel Sabine, and published by the British Association for the Advancement of Science.

MAP II.—To be placed at the end of the Introductory Essay.

The boundaries and names employed in the Map of India divided into Provinces, have been partially explained at page 88; it remains to add a few words on our representations of its mountain and river systems.

As regards rivers, we find these to be represented in most maps as being equally numerous, and of as great volume, in some of the most arid, as they are in the most humid provinces. This arises from the fact that the larger maps are in many cases made up from local surveys, and their component parts have hence no relative value. In an arid country like Rajwara, every streamlet carrying water for a few days in the year is of importance, and therefore mapped; whereas in Bengal, many infinitely larger perennial rivers are of no importance, and are omitted: the result is, that the two countries being brought together on a general map, appear equally well watered. We have therefore omitted in certain provinces many of the small rivers which are conspicuous in ordinary maps.

The relations of the rivers to the mountain-chains appear to us to be more or less inaccurate on our best maps of India: thus we find all the rivers on the eastern side of the peninsula of Hindostan usually represented as cutting through a coast range of hills called the Eastern Ghats; the rivers of eastern Afghanistan and Beluchistan in like manner seem to cut through a similar range parallel to the Indus; and, most extraordinary of all, the larger Himalayan rivers are made to cut through a lofty crest of that range.

The source of these errors may, we think, be traced to the neglect of a very simple law of perspective; in consequence of which, masses of mountains, of whatever configuration, resolve themselves into ranges perpendicular to the line of sight: thus, the so-called Eastern Ghats are the terminal spurs of ranges that branch off from the Peninsular chain, and which, from their number and tolerably uniform elevation and surface, form what is called the table-land of the Dekhan. The imaginary Suliman range, skirting the west bank of the Indus, is in like manner formed of the terminal spurs of ranges from a distant axis, which, with the rivers they enclose, descend at right angles to the Indus.

The Himalayan river-system is more complicated, but reducible to the same law. The great snowy peaks, as seen from the plains of India, are all thrown, by perspective, into one continuous range, and were hence originally assumed to indicate the axis of the Himalaya, and laid down as such in maps: next came the information of the natives that all the larger rivers rise behind the snowy masses; and they have consequently been represented as cutting through the supposed axis. We now know that in whatever direction the Himalaya has been explored, its axis has been found to be beyond the snowy peaks, and indicated by the river-heads. We have therefore in all cases of doubt represented the rivers as following the courses of valleys enclosed by mountains, and assumed that the geographical axis of a chain is indicated by its watershed.

We have not hesitated to contour the table-land of the Dekhan, so as approximately to represent a system of ranges descending from the meridional axis of the Peninsula to the eastern coast, and attaining an average elevation of 1500–2000 feet. We have also given to that axis itself a more interrupted and tortuous course than is usually represented; it being an error to suppose that it forms a continuous ridge of nearly uniform height parallel to the coast. Central India we have also represented as a hilly table-land, intersected by considerable valleys; of which there is ample evidence in surveys and the accounts of travellers.

For the details of the mountain systems of East Tibet there are no authorities, but we have expressed its main features,— that of an enormously elevated mountain mass. This is proved by the statements of many intelligent Tibetans, by the Chinese geographers, by the narrative of M. Huc, and by the fact of so many of the large rivers of Asia flowing from it in several directions. To omit a feature which rivals the Himalaya in dimensions, and which exercises a paramount influence over the meteorology of Eastern Asia, would deprive our map of much of the use we hope it may be of, in illustrating the relations between the vegetation and climate of India.

———————

It remains to add, that the system of spelling (which is the classical one) adopted both in the maps and the pages of our work, is rendered imperative from the fact that we hope our work may be useful to foreigners as well as to our own countrymen.

INDEX TO THE INTRODUCTORY ESSAY.

2 m

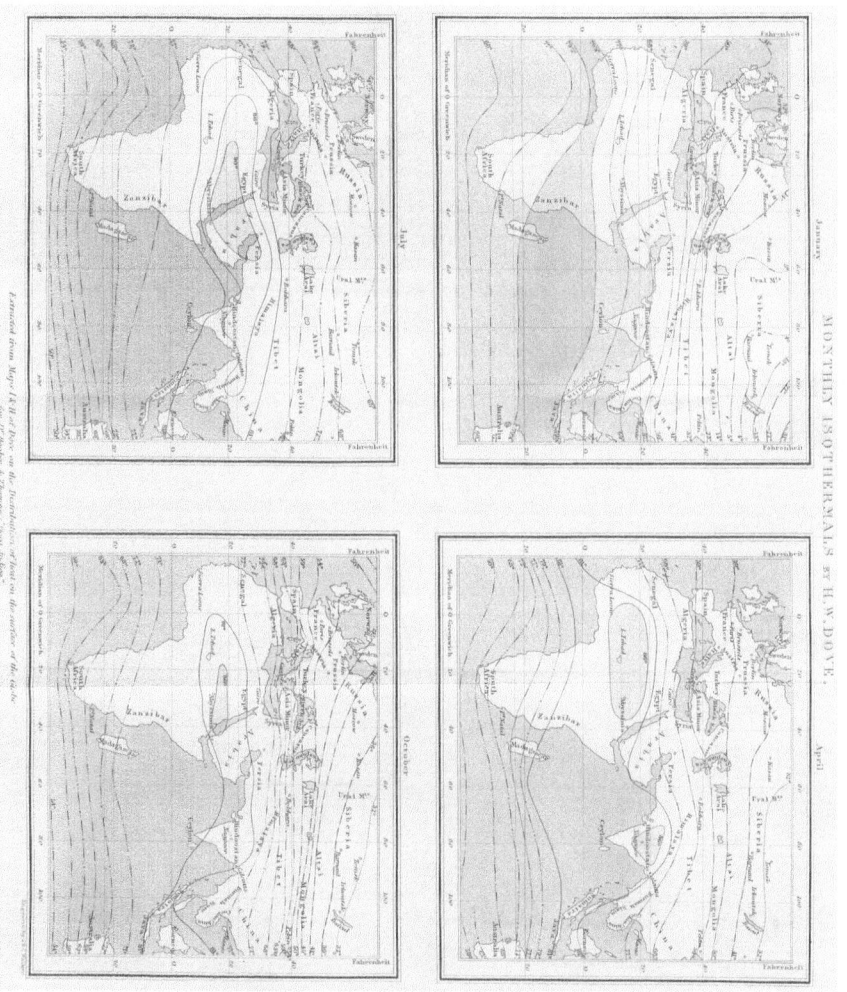

MONTHLY ISOTHERMALS by H.W. DOVE.

The material originally positioned here is too large for reproduction in this reissue. A PDF can be downloaded from the web address given on page iv of this book, by clicking on 'Resources Available'.

FLORA INDICA.

I. RANUNCULACEÆ.

Sepala 3–6, plerumque 5, hypogyna, decidua (in paucis persistentia), regularia vel varie irregularia, herbacea vel colorata. *Petala* 3–15, interdum parva, irregularia vel plane nulla. *Stamina* indefinita, antheris basifixis 2-locularibus longitudinaliter dehiscentibus. *Ovaria* plurima, secus torum elongatum vel globosum imbricata, vel uniserialia (1–5), discreta, rarissime axi subcohærentia, 1-locularia, 1- vel pluriovulata. *Carpella* achenia sicca vel folliculi, rarius baccata. *Semina* anatropa, albumine copioso, embryone minuto.

In accordance with our already stated intention to follow the arrangement of De Candolle, our Flora begins with *Ranunculaceæ*. This family was probably selected to commence the series on account of its abundance in Europe, rather than from any precise ideas of the exact degree of relationship of the different allied families. It is certainly more nearly allied to *Berberideæ* and *Papaveraceæ* than to those Orders which immediately follow it in the linear series, as is indicated by its herbaceous habit and divided leaves. *Ranunculaceæ* also exhibit a remarkable analogy or affinity to two Orders which are usually placed at a great distance from it, namely, *Umbelliferæ* and *Rosaceæ*, by means of which a transition is established between the families of *Apocarpous Thalamifloræ* and the great class of *Myrtales;* and, as we shall, under the next Order, have occasion to mention *Dilleniaceæ*, exhibit a passage to a very different series of Orders, namely, *Ternströmiaceæ* and *Ericaceæ*.

In the typical families of this class, that is to say, in *Magnoliaceæ* and *Anonaceæ*, the floral organs are (perhaps invariably) arranged in a ternary order, and in more than two rows. The closely-allied class to which *Papaveraceæ* and *Berberideæ* belong, agrees with these in respect of the multiplication of the verticils of the perianth, and partially also in the ternary arrangement of the parts of the flower, though in *Papaveraceæ* this is more generally binary. In *Dilleniaceæ*, on the contrary, the flowers are pentamerous, and the perianth in two rows. In *Ranunculaceæ* we have a complete case of transition, the arrangement being occasionally ternary, but more frequently quinary, while the petals in a considerable number of species are twice as numerous as the sepals, though it is more usual to find them equal in number. This Order and the next may therefore be considered aberrant members of the class of *Apocarpous Thalamifloræ*.

The remarkable analogy in foliage between *Ranunculaceæ* and *Umbelliferæ* was

B

first indicated by Lindley. The great sheathing base of the petiole, occasionally, though rarely, developed into a stipule, is remarkably different from anything met with in the *Anonal* families, but has much similarity to very common states of herbaceous *Saxifragaceæ* and *Rosaceæ*, where the stipules are imperfectly or not at all developed. In *Araliaceæ* we occasionally see these organs largely developed. This is also the case in most species of *Thalictrum*, in which genus also we have frequently stipellæ, analogous in position and function to those of *Leguminosæ*.

The tribe *Clematideæ*, which differs from the rest of the Order by its frutescent and generally scandent habit, and by its opposite leaves, is not equally distinct in the floral organs, which are only distinguishable from those of *Anemone* by the valvate æstivation of the calyx and the elongated stamens. The opposite leaves are, however, a very remarkable character, to which there is no approximation in any other part of the natural class, as it is usually understood, but which recurs in *Monimiaceæ*, a small Order usually placed near *Lauraceæ*, which we propose to include in this part of the series.

The position at the commencement of the linear series, which has been assigned to *Ranunculaceæ*, must not be understood to indicate that they are considered the most highly-developed family of plants, though this was in all probability assumed to be the case by the founder of our present arrangement. No part of the Dicotyledonous class presents such a well-marked superiority in organization, as to enable us to place it without hesitation at the commencement of the series. It has, however, been conjectured, with some appearance of plausibility, that those families in which special adaptation of the floral organs has attained its utmost limit, are situated at the highest part of the scale. Gamopetalous plants would therefore be more perfect than polypetalous ones, syncarpous ones more complete than apocarpous, and those with adherent sepals would take a higher place than those in which the ovary is free. It might for the same reason be conjectured that one-ovuled plants are of a superior grade to those in which many seeds are developed, especially if many carpellary leaves surround and protect a single ovule. If these views were carried out, the highest position in the vegetable kingdom would appear to belong to the family of *Loranthaceæ*, in which all these features are combined with the utmost simplicity of ovule, with a system of parasitism, and a highly abnormal mode of vegetation.

Contemplating plants from this point of view, *Ranunculaceæ* occupy a comparatively low place. In this Order all the parts of the flower exhibit the minimum amount of deviation from the ordinary type of leaf, and a most remarkable tendency to revert to it. They exhibit also a very great tendency to irregularity, not only in the assumption of abnormal forms, but also in the great variation of size of which they are susceptible in each species. These circumstances are, we think, highly confirmatory of the propriety of assigning to these plants a low place in the scale, since in all Orders in which special adaptation is carried to a high degree, the shape of the calyx-tube, as well as of the petals, and of every part of the flower, remains remarkably constant in each species. The anthers of *Ranunculaceæ* are in like manner invariably basifixed, so that the stamens do not deviate so far from the ordinary type of the leaf, as is the case in many other groups. This is also the case with the carpels, in which the analogy to leaves is very manifest, especially in the follicular species. Both stamens and carpels vary much in size in different individuals of the same species, as has already been indicated with regard to the sepals and petals. It is very important that the student should bear this fact in mind in the determination of species, undue weight having in many instances been given to the size of these organs, which frequently continue to enlarge after the expansion of the flower, in consequence of which many species have been founded on trivial and unsatisfactory characters.

It has been usual to describe the typical *Ranunculaceæ* as having extrorse anthers, but this is far from being universally, or even generally, the case, the cells being most commonly exactly lateral; it is only in the genus *Ranunculus* that the dehiscence of the anthers is evidently extrorse. This was first indicated by Asa

Gray, in the Illustrations of the Genera of Plants of the United States. We can find no difference between the anthers of *Isopyrum* or *Trollius*, both of which are considered by De Candolle to belong to *Ranunculaceæ veræ*, and those of *Actæa*, which he refers to *Ranunculaceæ spuriæ*. We have therefore followed Arnott, and Torrey and Gray, in restricting the tribe *Pæonieæ* to *Pæonia* alone. Our other tribes are like those of De Candolle.

Ranunculaceæ constitute a widely-diffused and extensive family, most abundant in the north temperate zone. The genera are well-marked, and contain mostly many species. The latter are almost always widely diffused, and very variable.

The plants of this family are in general more or less acrid; but this property exists to a very variable extent, and it is only in the genus *Aconitum* that it is so concentrated that the plants become poisonous. Few of the Indian species are officinal, though *Ranunculus sceleratus* is well known for its blistering powers, and *Coptis* is imported into Bengal from the mountains for medicinal purposes.

Tribus I. CLEMATIDEÆ.

Sepala æstivatione valvata. *Petala* nulla vel plana. *Carpella* (achenia) monosperma, semine pendulo.—Frutices *sæpissime scandentes, oppositifolii.*

1. NARAVELIA, DC.

Sepala 4–5. *Petala* 6–12, calyce longiora. *Achenia* stipiti crasso cavo insidentia, stylo barbato plumoso caudata, demum spiraliter torta. —Frutices *scandentes*, foliis *bifoliolatis*, petiolo *in cirrhum producto*.

This genus, which is scarcely distinct from *Clematis*, differing only by the conversion of the upper leaflets of the pinnate leaf into tendrils, by the presence of petals, and by the stipitate achenia, is quite tropical, growing in thickets in the hot plains of Southern India, and never rising on the mountains into the cool zone. The only species known are those described below.

1. **N. Zeylanica** (DC. Syst. i. 167; Prod. i. 10); foliolis late ovatis breviter acuminatis basi cordatis vel rotundatis subtus dense pubescentibus vel tomentosis (rarius glabratis), petalis lineari-spathulatis.—*Wall. Cat.* 4687!; *W. et A. Prod.* i. 2. Atragene Zeylanica, *L.; Roxb. Corom.* ii. *t.* 188; *Fl. Ind.* ii. 670.

HAB. Zeylania! Carnatica! Malabaria! Concan! Maisor! Dekhan! Orissa! Bengal! et secus basin Himalayæ ab Assam ad Sikkim et Nipal orientale! Ava! Malaya! in dumetis calidis præsertim montosis, sed e provinciis siccioribus extratropicalibus omnino exul.—(*v. v.*)

The leaves are generally pubescent on the under surface, but we have before us specimens from Assam and Khasia in which they are quite glabrous, as in the specimens from Prome referred to by Wight and Arnott.

2. **N. laurifolia** (Wall. Cat. 4685!); foliolis elliptico-lanceolatis acuminatis glaberrimis basi rotundatis vel subacutis, petalis anguste linearibus.—N. Finlaysoniana, *Wall. Cat.* 4686!

HAB. In Peninsula Malayana, prope Mergui, *Griffith!* et Penang, *Finlayson!*—(*v. s.*)

DISTRIB. Ins. Philippin.!

N. Finlaysoniana is a diseased state, with the achenia long, subulate, and beardless, but solid, corky, and not seed-bearing. The leaves are the same as in *N. lauri-folia.*

2. **CLEMATIS,** L.

Sepala 4–8. *Petala* nulla vel calyce breviora. *Achenia* sessilia, caudata vel rostrata.—Frutices *scandentes vel erecti.*

This very large genus, which is common in all temperate climates, is represented in the mountains of India by numerous fine species. These are all scandent, and vary much in appearance at different stages of their growth, so that their separation in the herbarium is not always easy, the specimens which are preserved there very imperfectly illustrating the character of the species and the amount of variation in the different parts of the plant. The shape and degree of division of the leaves, the size and form of the flowers, the inflorescence, and the bracts, have all been employed to yield specific characters. The mode in which the leaves are divided, whether decompound, pinnate, or ternate, appears pretty constant in each species, but the shape and degree of division of the leaflets is extremely variable. With regard to the inflorescence, the species with single-flowered peduncles seem never to pass into the paniculate group; but among the latter every amount of variation occurs, the terminal panicles in the larger species being often a mass of flowers with few and very small leaves, while those on the lower part of the stem are long-peduncled, much more leafy, and altogether quite different in aspect. The bracts too are extremely variable, and specific characters founded upon them will, we believe, be found invariably to break down when a large series of specimens are examined. Good characters are afforded by the position of the sepals in the expanded flower, but the size and pubescence of the flowers are very uncertain. The buds in all the species are at first globose, and afterwards become oval or oblong; still, within certain limits, their shape at the period of expansion of the flower appears tolerably constant. The shape and degree of hairiness of the stamens form an important character, which, however, unfortunately fails in those species which are most difficult of discrimination.

The sections into which this genus has been divided by De Candolle are on the whole natural. The majority of the Indian species belong to the subgenus *Flammula,* which is everywhere the most extensive. *Cheiropsis* is in habit undistinguishable from *Atragene;* and the amount of development of the petals or barren stamens in *Atragene* being very variable, it is not improbable that on a general revision of the genus these two groups will be united. *C. montana* has so exactly the habit of *Cheiropsis* (in which indeed De Candolle has placed it, notwithstanding its having no involucral bracts), that it appears desirable to rest the distinction of that section mainly on the one-flowered pedicels, in preference to the involucre, by doing which we include *C. montana. Bebæanthera* of Edgeworth has also so much the habit of *Cheiropsis,* that we prefer its union to that group to its retention as a distinct subgenus, especially as a transition is established between its very remarkable stamens and the ordinary form by *C. acutangula.*

The genus *Clematis* is universally diffused throughout temperate countries, if we except extratropical South America, in which its species are very rare. In Australia it is as abundant as in the Northern Hemisphere. There are no arctic or alpine species, and within the tropics they are for the most part natives of mountainous countries. Madagascar possesses a number of very remarkable forms.

In India the species of *Clematis* are most numerous in the temperate zone of the Himalaya, only two occurring in the tropical regions. Of these, one (*C. Gouriana*) is found in the open plains, but only exceptionally, as it is much more plentiful in subtropical hills. The other (*C. Cadmia*) is a rare and local plant. The species of the Eastern Himalaya and Malayan Peninsula are analogous to those of China. In the Western Himalaya and the mountains of Hindostan and the Dekhan, African types appear. In the Tibetan Himalaya a North Asiatic species is of common occurrence.

The section *Cheiropsis* is entirely confined to the cooler parts of the north temperate zone, and the Himalayan species of this section are found at greater elevations than the other species of the genus.

Sect. 1. VITICELLA, DC.—*Achenia* rostrata, ecaudata.

1. C. Cadmia (Ham. ex Wall. Cat. 4669 !); foliis ternatim decompositis integerrimis, floribus axillaribus solitariis, pedunculis medio bibracteolatis, acheniis compressis ovatis rostro recto acuminatis.—C. sulcata, *Wall. Cat.* 4667 ! Thalictrum bracteatum, *Roxb. Fl. Ind.* 2. 671 !

HAB. Bengal et Assam secus basin montium Khasia, in dumetis scandens.—(*v. s.*)

Rami elongati, graciles, profunde sulcati, glabriusculi. *Folia* glabra vel pilosula, segmentis unciam longis. *Pedunculi* medio bibracteati. *Bracteæ* sessiles, foliaceæ, foliolis sublatioribus, ovatæ, acutæ, indivisæ. *Sepala* 5–6, patentia, ½-uncialia, oblonga, acutiuscula. *Stamina* sepalis multo breviora, filamentis brevissimis planis glabratis. *Achenia* magna, adpresse sericea.

The flowers of this elegant little species are said by Roxburgh, whose description is excellent, to be very pale blue.

Sect. 2. CHEIROPSIS, DC.—*Achenia* plumoso-caudata. *Pedicelli* in axillis solitarii, vel supra ramum axillarem brevissimum dense racemosi.

2. C. acutangula (H.f. et T.); foliis pinnatim decompositis, segmentis ovato-lanceolatis basi rotundatis vel cuneatis grosse serratis, filamentis planis exacte linearibus dorso laxe pilosis, pilis supra antheras lineares introrsas fasciculum densum formantibus.

HAB. In montibus Khasia prope Molim, alt. 5000 ped., *Griffith !*—(Fl. Aug.) (*v. v.*)

Rami sparse setoso-pilosi, debiles, profunde sulcati, costis acutissimis. *Foliorum* segmenta 1–2-uncialia. *Pedicelli* foliis breviores, basi bibracteati; bracteæ subfoliaceæ, tripartitæ vel simplices. *Sepala* oblonga, acutiuscula, extus pubescentia, intus glabra. *Filamenta* membranacea, valide 1-nervia, intus glabra. *Antheræ* breves, introrse dehiscentes: *connectivum* non dilatatum. *Achenia* compressa, ovalia, sericeo-pilosa.

3. C. barbellata (Edgew. in Linn. Tr. xx. 25); foliis ternatisectis, pedicellis nudis brevibus, staminibus sepalis fere ½ brevioribus, filamentis planis lanceolatis ciliatis, antheris introrsis dorso dense pilosis.—C. Nipalensis, *Royle, Ill.* 51 ! *non alior.*

HAB. In Himalaya occidentali temperata, alt. 8–10,000 ped. Garhwal, *Royle ! Edgeworth !* Kumaon, *Strachey et Winterbottom !*—(*v. s.*)

Rami glabrati. *Caulis* teres, striatulus. *Folia* longe petiolata; *foliola* breviter petiolulata, ovata, acuminata, serrata vel grosse dentata, sæpe inciso-lobata, 1–2-uncialia. *Pedicelli* numerosi, foliis breviores. *Sepala* 4, patentia, ovato-oblonga, acuminata, purpurascentia, utrinque pubescentia, margine dense villosa, unciam longa. *Filamenta* uninervia. *Achenia* glabra.

4. **C. Nipalensis** (DC. Syst. i. 164; Prod. i. 9); foliis ternati-partitis, pedicellis brevibus supra medium involucratis, filamentis elon-gatis e basi plana filiformibus glabris.—*Wall. Cat.* 4680 ! C. montana, *Don, Prodr.* 192 ; *non alior.*

HAB. In Himalaya temperata; Garhwal ! Kumaon ! Nipal ! Bhotan ! —(Fl. Dec. Mart.) (*v. v.*)

Rami glabrati; juniores angulato-sulcati, seniores rotundati. *Folia* longe petio-lata, interdum puberula, segmentis oblongo-lanceolatis vel linearibus integerrimis vel parce dentatis lateralibus interdum trilobis vel tripartitis. *Pedicelli* numerosi, foliis breviores, supra medium bibracteati. *Bracteæ* in involucrum breve cupuliforme acute bilabiatum connatæ. *Pedicelli* supra involucrum incrassati, dense sericei. *Sepala* 4, oblonga, vix semi-uncialia, extus dense adpresse sericea, intus glabra. *Achenia* compressa, dorso gibbosa, parce sericea.

The young bud is sessile within the involucre, but the upper part of the pedicel gradually elongates, so that when the fruit is ripe the involucre is nearly in the middle.

5. **C. montana** (Ham. ex DC. Syst. i. 164); foliis ternatisectis, pedicellis nudis folia æquantibus vel superantibus, staminibus sepalis dimidio brevioribus, filamentis planis anguste ligulatis glabris.—*DC. Prod.* i. 9; *Wall. Cat.* 4681 !; *Plant. As. Rar.* iii. *t.* 217 ! C. Pun-duana, *Wall. Cat.* 4682 ! C. anemoniflora, *Don, Prod.* 192. Anemone curta, *Wall. Cat.* 4690 !

HAB. Per totam Himalayam temperatam et subalpinam, alt. 6000–12,000 ped.; exceptis jugis exterioribus Sikkim, ubi non infra 10,000 pedes occurrit; et in montibus Khasia, alt. 4–5000 ped.—(Fl. vere.) (*v. v.*)

Rami glabrati vel rarius adpresse sericei. *Folia* longe petiolata; *foliola* ovata obtusa acuta vel acuminata, grosse dentata vel lobata. *Flores* majusculi, interdum maximi, suaveolentes. *Sepala* 1–1½-uncialia, elliptica vel ovata, alba, obtusa vel acu-minata, nervosa, glabrata vel extus adpresse pilosa. *Achenia* sericea longe caudata.

The most beautiful of all the Indian species, but extremely variable. The larger-flowered forms are very sweet-scented, and when in flower, in April and May, visible from a great distance, forming dense white patches in the thickets on the hill-sides.

Sect. 3. FLAMMULA, DC.—*Achenia* plumoso-caudata. *Flores* pa-niculati (rarissime abortu subsolitarii).

§ 1. *Sepala per anthesin patentia vel reflexa.*

6. **C. smilacifolia** (Wall. in As. Researches, xiii. 414); glaber-rima, foliis maximis simplicibus late ovatis rarius ternatipartitis.—*DC. Prod.* i. 10; *Wall. Cat.* 4683 !; *Hook. Bot. Mag. t.* 4259 ! C. subpeltata, *Wall. Plant. As. Rar. t.* 20. *Cat.* 4684 ! C. Munroiana, *Wight, Ill.* i. 5. *t.* 1. C. affinis, *Wight, Ill.* i. 5. C. smilacina, *Blume, Bijd.* 1. C. glandulosa, *Blume, Bijd.* 1 ?

HAB. In provinciis humidioribus; in montibus inferioribus Sikkim !, a basi ad 5000 ped. alt.; Khasia, 2–5000 ; in montibus Concan, *Sykes !* Nilghiri, *Munro !* Travancor et Dindigal, *Wight !* Ceylon, *Thwaites !* Ava, *Wall !*—(Fl. Jun. Aug.) (*v. v.*)

DISTRIB. Java, *Blume;* Timor, *Spanoghe.*

Caules sulcati. *Folia* 5–10 unciʻas longa, 4–7 lata, late ovata, basi rotundata vel sæpius profunde cordata, valide 5–7-nervia, margine albo-cartilaginea, interdum ter-natipartita, superiora angustiora oblonga vel lanceolata, longe acuminata, basi angus-tata, integerrima vel rarius serrata seu grosse denticulata. *Paniculæ* elongatæ, ra-cemiformes vel decompositæ, pluriflorae. *Sepala* 4–5, oblonga, acuta, patentia, demum reflexa, extus fusco-tomentosa, intus glabra. *Stamina exteriora* sepalis paullo bre-viora ; *filamentis* membranaceis, late linearibus ; *antheris* brevibus, angustis ; *inte-riora* sensim breviora et angustiora, sed antheris gradatim longioribus terminata, intima fere a basi antherifera. *Achenia* ovata, compressa.

A very well-marked species, but varying like all the others, though to a less degree than most, in the shape and toothing of the leaves. The serrate-leaved forms do not constitute a distinct species, nor even variety ; specimens from the same spot, occasionally even from the same plant, presenting both forms.

7. C. triloba (Heyne in Roth nov. sp. 251, non St. Hilaire) ; mol-liter sericea, foliis parvis simplicibus vel ternatisectis ellipticis vel ovatis.—*DC. Prod.* i. 8 ; *W. et A. Prod.* i. 2.

HAB. In Peninsula, *Heyne !;* in montibus Dekhan humidioribus (in regione Mâwal dicta), *Law ! Stocks !*—(Fl. Sept.) (*v. s.*)

Caulis sulcatus. *Folia* longe petiolata, 1–2-uncialia, basi cordata vel rotundata, integerrima vel triloba, interdum ternatisecta, segmentis integris vel lobatis. *Pani-culæ* multiflorae, decomposite ramosæ, bracteis foliaceis gradatim minoribus ovatis vel ellipticis trinervibus. *Flores* albi. *Sepala* 4–6, stellatim patentia, fere uncialia, ob-longa, extus sericea. *Stamina* sepalis dimidio breviora ; *filamenta* lineari-ligulata, glabra ; *antheræ* elongatæ. *Achenia* ovata, compressa, sericeo-villosa.

Heyne's plant has hitherto been considered an obscure species, and we have seen no authentic specimen ; but the description given by Roth accords so well with the plant collected by Mr. Law and Dr. Stocks, that we have no doubt of their identity. Heyne's specimens were probably obtained from the same district, as many of his plants were, we believe, collected on the mountains near the western coast.

8. C. grata (Wall. Cat. 4668 !) ; incana, foliis pinnatipartitis seg-mentis (sæpius 5) late ovatis basi cordatis grosse inciso-dentatis utrin-que incanis vel tomentosis vel rarius superne glabrescentibus, antheris muticis.—*Wall. Plant. As. Rar. t.* 98.

HAB. In montibus Affghanistan, *Griffith !;* in Himalaya occidentali calida et temperata, alt. 2–8000 ped.; Kashmir, *Jacquemont !*; Panjab Him. ! Simla ! Kanawer ! Garhwal ! et Kumaon !—(Fl. Jul. Aug.) (*v. v.*)

DISTRIB. China borealis !

Caulis angulatus, incano-tomentosus. *Foliola* 1–2-uncialia, tripartita triloba vel incisa, sparse et adpresse pubescentia vel incano-tomentosa. *Paniculæ* decompositæ, multiflorae, ramis strictis, foliaceæ, foliis lobatis vel indivisis. *Sepala* lactea, ⅓-un-cialia, ovalia, nervosa. *Filamenta* sepalis ⅓ breviora, plana, anguste linearia, glabra. *Antheræ* breves. *Achenia* ovalia, compressa, sericea.

9. C. hedysarifolia (DC. Syst. i. 148. Prod. i. 6) ; foliis ternati-vel pinnatisectis, segmentis ovatis acuminatis coriaceis creberrime reticulatim venosis glaberrimis, filamentis ultra antheram in processum subulatum productis.—*Ker, Bot. Reg. t.* 599 !

HAB. Pegu, *Hore ! in Mus. Brit. ;* in montibus Concan, *Law !*— (*v. v.*)

Ramuli sulcati, juniores puberuli, adulti glabri. *Foliola* 2–4-uncialia, nervosa, pergamentacea, integerrima vel hinc inde grosse dentata. *Paniculæ* decompositæ, multifloræ, strictæ; *bracteæ* foliosæ, tripartitæ vel integræ, ovatæ. *Sepala* 4, ovalia, mucronata, ⅓–½-uncialia, extus dense tomentosa, intus glabra. *Stamina* sepalis ½ breviora; *filamenta* plana, glabra; *antheræ* elongatæ, connectivo longe apiculatæ. *Achenia* compressa, marginata, substipitata, pilosa.

Though very near in general appearance to some of the forms of the next species, the anthers are so peculiar that we cannot unite them. Our materials, however, are rather imperfect; but botanists in Birma and Western India will have it in their power to compare the two species in a growing state, and to decide whether the one now described, which is we think easily recognizable by its large, copiously-veined leaves and larger flowers, be distinct from all the forms of *C. Gouriana*. We have examined the original specimen of *C. hedysarifolia,* DC., in the British Museum; it is not in flower, but appears identical with the Bombay plant.

10. **C. Gouriana** (Roxb. Fl. Ind. ii. 670); glabriuscula, foliis pinnatisectis vel bipinnatisectis (rarius ternatisectis) segmentis ovatis vel oblongis acuminatis basi rotundatis vel cordatis membranaceis superne lucidis, antheris brevibus muticis.—*DC. Syst.* i. 138, *Prod.* i. 3; *Wall. Cat.* 4673!; *W. et A. Prod.* i. 2!; *Wight, Ic. t.* 933, 934, *Neilgh. Pl. t.* 12. C. cana, *Wall. Cat.* 4672!

Hab. In dumetis Indiæ tropicæ, præsertim montanæ, a Zeylania! et peninsula Malayana! ad Bengal et Assam! Behar! Dekhan! et Concan!; in montibus Khasia, et secus basin Himalayæ usque ad flumen Jelam et montes Rajaori!—(Fl. hyeme.) (*v. v.*)

Distrib. Per totam Indiam tropicam, usque ad insulas Philippinas!

Alte scandens, glaberrima, partes novellæ plerumque sericeo-pubescentes. *Folia* forma valde varia, integerrima vel grosse dentata, basi rotundata vel cordata, glaberrima sparse pilosa vel subtus ferrugineo-tomentosa. *Paniculæ* decompositæ, multifloræ, *bracteis* minutis rarius foliaceis ovatis, sæpius elongatæ folia superantes. *Flores* parvi, numerosissimi. *Sepala* ovalia, 2–3 lineas longa, extus vel margine tomentosa. *Filamenta* plana, glabra. *Antheræ* oblongæ. *Achenia* oblonga, minus compressa quam in affinibus, fusco-pilosa.

Very widely diffused throughout tropical India, in mountainous districts, climbing to a great distance over trees. A very variable plant, but not easily divisible into varieties; nor would it serve any good purpose to do so, as the broad and narrow leaved forms occur at one time with entire, at another with dentate leaves, and leaves of every shape are either glabrous or more or less pubescent underneath. The leaves vary also much in amount of division; and in a series of specimens from the Khasia mountains they are uniformly ternatipartite. These specimens, which were all collected at once, were probably elongated shoots of a luxuriant young plant flowering for the first time, as the shape of the leaflets and the inflorescence are not different from those of the ordinary forms. None of the species of continental India are liable to be confounded with *C. Gouriana*; the nearest is *C. grata*, easily distinguishable by its hoary pubescence and larger flowers. The broad-leaved tomentose forms, however, approach very near to a plant which appears to be common in Java and South China (*C. Javanica,* DC.?). The ordinary state of this species is larger-flowered than *C. Gouriana*, and the leaves, which are bipinnate, pinnate, or ternate, are soft and pubescent, without the shining surface which is characteristic of the upper surface of the leaves of that species. They are also more cut, sometimes deeply lobed, but it must be confessed that we have entire-leaved specimens before us which are scarcely to be distinguished from *C. Gouriana.* The botanists of Java or Ceylon (from which latter place two imperfect specimens in Herb. Hook. are perhaps referable to the Javanese species) will, we hope, decide a question for which unfortunately our materials are not sufficient.

11. **C. parvifolia** (Edgeworth! in Linn. Tr. xx. 25); glabra, foliis pinnatim decompositis, paniculis paucifloris (interdum unifloris), pedunculo medio bibracteato, filamentis linearibus sericeo-pilosis.—C. graveolens, *Lindl. Journ. Hort. Soc.* i. 307.

HAB. In Himalaya occidentali temperata : alt. 6–11,000 ped., *Edgeworth !* Banahal! Kanawer, *Munro ! Jacquemont !*—(Fl. Aug.) (*v. v.*)

Rami graciles, glabri vel apice tenuissime pubescentes. *Folia* decomposita, submembranacea. *Segmenta* 1–3 lineas longa rarius uncialia, indivisa vel pinnatifida, lobis lateralibus obtusis mucronatis, terminali sæpius lanceolato. *Paniculæ* pauci-floræ, foliosæ, longe pedunculatæ, vel rarius ad pedunculos unifloros medio bibracteo-latos reductæ; *bracteæ* pinnatæ vel indivisæ. *Sepala* late ovalia, obtusa, utrinque sericea. *Stamina* sepalis ⅓ breviora ; *filamenta* plana; *antheræ* muticæ. *Achenia* ovata, compressa, sericea.

This species resembles very closely in foliage *C. Flammula*, L., but differs in the large flowers; we have seen no intermediate forms.

12. **C. orientalis** (L.) ; subglabra, foliis pinnatis vel bipinnatis glaucis, filamentis planis membranaceis ciliatis apice filiformibus.

a. acutifolia ; foliolis oblongo-lanceolatis acuminatis, floribus magnis.—C. orientalis, *L. et auct.; Ledeb. Fl. Ross.* i. 3 ; *Dill. Elth. t.* 119. *f.* 145. C. tenuifolia, *Royle! Ill.* 51. C. graveolens, *Hook. Bot. Mag. t.* 4495 ! *non Lindl.* C. Ispahanica, *Boissier! Diagn.* vi. 3.

β. obtusifolia ; foliolis oblongis obtusis.—C. glauca, *Willd.; Ledeb. Fl. Ross.* i. 3.

γ. latifolia ; foliorum segmentis late ovalibus obtuse lobatis.—C. globosa, *Royle! Ill. p.* 51. C. Hysudrica, *Munro ! (in Herb. Hook.)*

HAB. In montibus Afghanistan et in Himalaya occidentali Tibetica ; *a.* in Tibet vulgaris alt. 7–14,000 ped.! *β.* in Himalaya maxime occidentali ; Kashmir, *Jacquemont!* Marri, *Fleming!* Gilgit, *Winterbottom! γ.* in montibus Piti, *Royle ! Munro !*—(Fl. Jul. Aug.) (*v. v.*)

DISTRIB. Per totam Asiam temperatam.

Glaberrima, glaucescens, rami novelli interdum sericeo-pubescentes. *Foliorum segmenta* forma valde varia, in *a* 1–3 uncias longa, oblongo-lanceolata vel linearia acuminata, integra triloba vel tripartita, integerrima vel inciso-dentata, in *β* breviora obtusiloba et apice obtusa, in *γ* omnium latissima late ovata vel deltoidea. *Paniculæ* ramosæ, multi- vel pluri-floræ, sæpius longe pedicellatæ strictæ. *Flores* straminei vel purpurascentes, in *a* 1½ unciam diametro, in cæteris var. minores. *Sepala* 4, stellatim patentia, ovata, acuminata, intus sericeo-villosa, marginibus dense tomentosis, extus glabra vel pubescentia. *Stamina* sepalis ½ breviora. *Achenia* oblonga sericeo-pilosa.

We have examined, in the Hookerian Herbarium, authentic specimens, from numerous sources, of *C. orientalis*, L., *C. glauca*, Willd., *C. longecaudata*, Ledeb., and *C. Ispahanica*, Boissier, all of which, we are quite convinced, are forms of one widely diffused and extremely variable species. The shape of the sepals, and the form and pubescence of the stamens, are the same in all the forms; but the size of the flowers, and the shape of the leaves, are very variable. In the drier parts of Tibet the variety *a* is abundant; this has usually very large flowers, but they are occasionally not larger than those of the ordinary Siberian and western forms of the species. In rich soil, and a somewhat more humid climate, the leaves become broader and more glaucous, and the flowers smaller. Cultivation appears to produce the same effect, for the figure in the Botanical Magazine, from Tibetan seed, corresponds very closely to the plate of *T. orientalis* given by Dillenius.

Clematis erecta of Linnæus, a native of the south of Europe and of western Asia, will probably also be met with in Afghanistan. It is a conspicuous plant, with a very peculiar habit for the genus; and an erect species certainly grows on the mountains between Kabul and Bamian. See Griffith, Itin. Notes, p. 307.

13. **C. nutans** (Royle! Ill. p. 51); adpresse sericea, foliis pinnatim decompositis, panicula pluriflora, filamentis sepala æquantibus e basi plana pilosa filiformibus glabris.

Hab. In Himalaya occidentali calida, alt. 2–5000 ped.: Simla! Garhwal! Kumaon!; in montibus Khasia ad Nartiang, alt. 4000 ped.!; et in monte Parasnath provinciæ Behar!—(Fl. Jul. Sept.) (*v. v.*)

Rami graciles, angulati. *Folia* pinnatim decomposita; *segmenta* ovata, oblonga vel lanceolata, 1–2-uncialia, integerrima dentata vel varie lobata. *Paniculæ* ramosæ, *bracteis* parvis. *Sepala* oblonga, acuta, extus pubescentia. *Filamenta* angustissime ligulata, basi pilosula, supra medium filiformia, glabrata. *Achenia* ovata, compressa, sericea.

Our Khasia specimens are in young fruit only, but they have quite the foliage of Dr. Royle's plant, and appear to be the wet-climate form of it, with lax, fewer-flowered panicles. The specimens from Parasnath have larger leaves than those from the Western Himalaya, the terminal leaflet being often as much as three inches in length.

14. **C. Wightiana** (Wall. Cat. 4674!); pubescens, foliis pinnatis, segmentis late ovatis, paniculis magnis decompositis, filamentis anguste ligulatis sericeo-pilosis.—*W. et A. Prod.* i. 2; *Wight! Icones, t.* 935, *Ill.* 5, *Neilg. Plants, p.* 2. *t.* 3.

Hab. In montibus Peninsulæ: Nilghiri, *Wight!* Concan, *Law!* Dekhan, *Sykes!* Orissa, in collibus secus flumen Kistna, *Wight.*

Rami validi, rotundati, adpresse pilosi. *Foliola* sæpius 5, crassa, velutino-pubescentia, 2–3 uncias longa, ovata, basi cordata, triloba, lobo medio longiore, vel 5-loba, interdum tripartita, grosse dentata, subtus dense villosa. *Paniculæ* foliis longiores; *bracteæ* foliaceæ, lobatæ vel dentatæ. *Sepala* 4, late ovata, patentia, extus villosa, ½-unciali. *Stamina* sepalis paullo breviora. *Achenia* ovata, compressa, dense sericea.

This species has the habit and general appearance of *C. grewiæflora;* it is however always smaller, and the shape of the expanded flowers is very different. In the Wallichian distribution, specimens of *C. Buchananiana,* γ *tortuosa,* have been, by some accident, mixed with this species, as in some collections (in Herb. Lindl. for instance) that plant occurs under No. 4674.

§ 2. *Sepala per anthesin erecta, apice revoluta.*

15. **C. grewiæflora** (DC. Syst. i. 140, Prod. i. 4); dense fulvo-tomentosa, foliis pinnatisectis, foliolis 3–5 crassis late cordatis 5-lobis serratis, alabastris ovalibus, filamentis e basi ligulata dense pilosa filiformibus glabris.—*Don, Prod. Nep. p.* 191; *Wall. Cat.* 4678!

Hab. In Himalaya temperata et calida, alt. 3–5000 ped.: Kumaon! Nipal! Sikkim!—(Fl. autumno.) (*v. v.*)

Rami validi, rotundati. *Foliola* 4 uncias longa et lata, superne adpresse pilosa, subtus dense tomentosa, profunde lobata. *Paniculæ* axillares, pluriflorse; *bracteæ* oppositæ vel verticillatæ, trilobæ vel grosse dentatæ. *Flores* magni, late campanulati. *Sepala* oblonga. *Filamenta* sepalis æquilonga. *Achenia* oblique ovata, sericea.

16. **C. Buchananiana** (DC. Syst. i. 140, Prod. i. 4); incana vel fusco-pilosa, foliis pinnatisectis, foliolis 5–7 fere rotundatis lobatis

grosse aristato-dentatis, sepalis oblongis, filamentis anguste ligulatis sericeo-pilosis.

α. *rugosa ;* foliis latioribus obtusiusculis rugosis reticulate venosis subtus dense albo-tomentosis, caulibus tomentosis.—C. Buchananiana, *Wall. Cat.* 4677 !; *Don, Prod.* 191.

β. *vitifolia ;* foliis tenuioribus cano-pubescentibus vel subglabris, caulibus pubescentibus.—C. vitifolia, *Wall. Cat.* 4676 !

γ. *tortuosa ;* foliis glabriusculis sericeo-venosis vel glaberrimis, caulibus laxe fulvo-pilosis.— C. tortuosa, *Wall. Cat.* 4675 !

HAB. In Himalaya temperata, alt. 5–10,000 ped.: Panjab Himalaya! Garhwal! Kumaon! Nipal! Sikkim!; et in montibus Khasia, alt. 4–6000 ped. !—α in Himalaya occidentali vulgaris, β et γ in humidioribus Himalayæ orientalis et Khasia.—(*v. v.*)

Rami validi, rotundati. *Foliola* 2–3-uncialia, basi cordata, obtusa vel acutiuscula, integra vel inæqualiter triloba. *Panicula* parce ramosa, ramis pedicellisque tomentosis. *Bracteæ* foliaceæ, interdum verticillatæ, varie incisæ. *Sepala* lineari-oblonga, dense tomentosa. *Filamenta* sepala subæquantia. *Achenia* ovata, sericea.

Though we have divided this species into three varieties, which correspond to the three species of Dr. Wallich, no great importance must be attached to this mode of grouping, as a regular gradation can be traced from the most hoary state of α, to the glabrous leaves and hairy stems of γ. The amount of variation in pubescence in this species is very remarkable; specimens which, in shape of leaves and flowers, are absolutely undistinguishable, being often quite dissimilarly clothed.

17. **C. connata** (DC. Prod. i. 4); glabra, foliis pinnatisectis, foliolis 5–7 ovatis basi cordatis, petiolis (sæpius) basi dilatatis connatis, filamentis e basi ligulata filiformibus longe pilosis.—*Wall. Cat.* 4679 ! C. venosa, *Royle, Ill.* 51. C. amplexicaulis, C. velutina, *et* C. gracilis, *Edgew.! in Linn. Tr.* xx. 24, 25.

HAB. In Himalaya temperata, alt. 6–10,000 ped.: Kishtwar! Kumaon! Nipal! Sikkim!—(Fl. autumno.) (*v. v.*)

Rami validi, sulcati, præsertim ad articulos glaucescentes. *Folia* maxima, sæpe pedalia, pinnatisecta, summa ternatisecta ; *foliola* remota, 2–5 uncias longa, 2–3 lata, acuminata, serrata vel grosse dentata, integra vel inæqualiter triloba. *Paniculæ* laxe ramosæ, rami graciles paucifloræ. *Bracteæ* (dum adsunt) foliaceæ, lanceolatæ, dentatæ vel integræ, sed sæpe minutæ. *Sepala* oblonga, acutiuscula, cano-tomentosa vel puberula, rarius glabrescentia, margine cinereo-pubescentia. *Filamenta* sepalis breviora. *Achenia* ovalia, sericeo-pilosa.

Flowers smaller than in *C. Buchananiana.* They vary much in amount of pubescence. We can find no characters of importance to separate the different forms which Mr. Edgeworth has considered as distinct. His specimens, which are now before us, certainly exhibit differences in the shape of the leaves; but we think that he has not made sufficient allowance for the great amount of variation to which all the species of this genus are subject, and feel confident that, with a suite of specimens as extensive as that which we possess, he would not even consider them as varieties.

18. **C. acuminata** (DC. Syst. i. 148, Prod. i. 6); glaberrima, petiolis basi non dilatatis, foliis ternatisectis, foliolis lucidis trinervibus ovatis acuminatis, floribus parvulis subglabris, filamentis late ligulatis longe sericeis.—*Wall. Cat.* 4670 !; *Don, Prod.* 192.

HAB. In Himalaya temperata: Kumaon! Nipal! Sikkim!—(*v. v.*)

Rami graciles, sulcati. *Folia* longe petiolata, integra vel sæpius 3-secta; *foliola* coriacea, reticulato-venosa, 3–4 uncias longa, 1½–2 lata, ovata vel ovato-lanceolata acuminata basi rotundata vel cordata, integerrima vel subserrata. *Paniculæ* decompositæ. *Bracteæ* sæpius minutæ, sed interdum foliaceæ. *Pedicelli* longi, graciles. *Alabastri* cylindracei. *Sepala* erecta, oblonga, 4–5 lineas longa, subacuta, margine præsertim puberula. *Filamenta* ad antheram usque longe sericeo-pilosa. *Achenia* sericea.

Species dubiæ et vix notæ.

1. C. *loasæfolia* (DC. Syst. i. 140, Prod. i. 4),

Described from a flowerless specimen, is indeterminable, but must be either *C. grewiæflora* or *C. Wightiana. C. loasifolia* of Don, Prod. 191, is also indeterminable without an authentic specimen, but it is probably a form of *C. Buchananiana.*

2. C. *scabiosæfolia* (DC. Syst. i. 154, Prod. i. 7); in India? (Herb. Mus. Par.)

3. C. *villosa* (DC. Syst. i. 154, Prod. i. 7); in India? (Herb. Mus. Paris.)

4. C. *comosa* (DC. Syst. i. 156, Prod. i. 8); in Ind. Or. (Herb. Mus. Paris).

This may perhaps be *C. triloba,* Heyne.

5. C. *grossa* (Wall. Cat. 4671! non Benth.)   Taong Dong, Ava.

We have examined the specimen of this plant in the Linnean Society's Herbarium. It is not in flower or fruit, and is the terminal shoot of a young plant. The leaves are 8–10 inches in length, bipinnate, the leaflets of thin texture, oblong-lanceolate, coarsely toothed or incised, and about 3 inches long. It is probably an undescribed species.

Tribus II. ANEMONEÆ.

Sepala æstivatione imbricata. *Petala* nulla vel plana. *Carpella* monosperma, semine pendulo.—Herbæ *foliis alternis.*

3. **THALICTRUM,** L.

Involucrum sub flore nullum. *Sepala* 4–5, æstivatione imbricata. *Petala* nulla. *Stamina* numerosa. *Carpella* monosperma, indehiscentia, ecaudata.—Herbæ *perennes;* caulibus *annuis;* floribus *paniculatis, albis flavis* vel *purpurascentibus.*

This is a very extensive genus, the species of which are abundant throughout the northern hemisphere and the mountains of the tropics, but which is only represented south of the tropics by one or two species at the Cape of Good Hope. All the species are subject to great variation in the size and form of the leaves, which are usually much divided. They are very widely spread over the mountainous parts of India, more especially in the Himalaya, and the Indian species seem quite as variable as those of Europe, the number of which is, we are convinced, very much over-estimated in modern systematic works. Most of the Indian species inhabit the shady mountain forests of the Himalaya during the rainy season, and are very different from those of Europe, some of them being the same as those which inhabit the Indian islands, while others will probably be found to extend into the still unknown mountain districts of West China. The alpine species however are European, and are quite as plentiful in the Himalaya as on the mountains of Europe and Siberia. It has been well observed by Seringe that characters derived from the leaves, unless supported by differences in the fruit, are insufficient to distinguish from one another the species of this difficult genus. Even the fruit seems liable to vary, though within certain limits only, and it will be found to afford the best characters for the discrimination of the species.

Sect. 1. PHYSOCARPUM, DC.—*Achenia* stipitata, lateraliter compressa, inflata vel plana.

1. **T. elegans** (Wall. Cat. 4728!); foliis pinnatim decompositis, panicula parce ramosa fere racemiformi, acheniis 6–12 longe pedicellatis obliquis obovatis membranaceis dorso et ventre alato carinatis utrinque uninerviis, carina ventrali gibbosa, stigmate incurvo sublaterali.—*Royle, Ill.* 51!; *Edgew.! in Linn. Soc. Tr.* xx. 26.

HAB. In Himalayæ zona subalpina, alt. 10–13,000 pedum : Garhwal! Kumaon! Nipal! Sikkim!—(*v. v.*)

Simplex vel subramosa, ½–1-pedalis. *Folia* 2–3 uncias longa; *foliola* minuta, rotundata, triloba vel indivisa. *Stipulæ* petiolo adhærentes, membranaceæ, fimbriatæ; *stipellæ* nullæ. *Pedicelli* subracemosi, foliis floralibus longiores, patentes. *Flores* parvi, viridi-purpurascentes. *Sepala* elliptica, trinervia. *Filamenta* filiformia; *antheræ* breves, mucronatæ.

2. **T. platycarpum** (H.f. et T.); foliis pinnatim decompositis, panicula ramosa, acheniis 4–10 breviter pedicellatis dolabriformibus stigmate recto apiculatis utrinque tricostatis.

HAB. In Tibetia occidentali : Nubra, *H. Strachey!* Hundes, *Str. et Wint.!* Milam, Kumaon, *Str. et Wint.!*—(*v. s.*)

Herba ½–1½-pedalis, ramosa, graveolens. *Folia* 2–3 uncias longa; *foliola* rotundata, triloba vel tripartita, subtus glanduloso-puberula. *Stipulæ* scariosæ; *stipellæ* nullæ. *Paniculæ* rami divaricati. *Pedicelli* fructiferi elongati. *Flores* parvi, viridescentes. *Sepala* late elliptica, membranacea. *Filamenta* filiformia; *antheræ* elongatæ, muticæ. *Achenia* glabra, oblique obovata, dorso recta, ventre valde gibbosa.—Species ut videtur rara, foliis priori, fructu sequenti affinis, ulterius examinanda.

3. **T. Chelidonii** (DC. Prod. i. 11); foliis ternatim decompositis, foliolis rotundatis crenato-lobatis basi cordatis subtus glaucis, acheniis numerosis dolabriformibus longe stipitatis utrinque trinervibus stylo recto vel incurvo apiculatis.

　　α. *reniforme ;* foliolis majoribus 1–2-uncialibus subtus valde glaucis et pulverulentis, fructu pubescente.—T. reniforme, *Wall. Cat.* 3716! T. neurocarpum, *Royle! Ill.* 51.—(*v. v.*)

　　β. *cultratum;* foliolis minoribus ¼–½-uncialibus subtus pallidis puberulis vel subglabris, fructu glabro.—T. cultratum, *Wall. Cat.* 3715! —(*v. v.*)

　　γ. *cysticarpum ;* acheniis numerosissimis obovatis forsan sterilibus longissime pedicellatis reflexis membranaceis, nervis gracillimis.—T. cysticarpum, *Wall. Cat.* 3714!—(*v. v.*)

HAB. In Himalaya temperata, alt. 6–10,000 ped. : a Kashmir! usque ad Sikkim! (7000 ad 13,000 ped.) ; et in montibus Khasia, alt. 4–6000 ped. !—(Fl. Jul. Aug.) (*v. v.*)

Herba 2–4-pedalis et ultra, laxe ramosa. *Panicula* magna, terminalis, ramis demum elongatis, racemiformibus, fere aphyllis. *Stipulæ* foliorum inferiorum maximæ, petiolis adhærentes, membranaceæ; *stipellæ* cito deciduæ. *Foliola* magnitudine valde varia, diametro ¼–1½-unciali. *Flores* longe pedicellati, majusculi, interdum maximi diametro plus quam unciali, purpurascentes, sæpe pulchre purpurei. *Sepala* elliptica, obtusa. *Filamenta* filiformia, apice vix dilatata; *antheræ* mucronatæ.

An excessively variable plant in the size of the leaves, but otherwise pretty constant to the characters above given. The powdery pubescence of the under surface of the leaf seems to be invariably present. In the variety γ, which has usually very large flowers, the fruit appears to enlarge without a corresponding development of the embryo, which, though always present, is seemingly abortive. In a very numerous suite of specimens from Sikkim not one exhibits a dilated achenium, or any indication of a perfect seed, all being quite flat. The great amount of variation in the size and shape of the fruit indicates disease, and many of the specimens are evidently in an abnormal state, having muricated or tuberculated bulbous masses, instead of branches, in their axils.

4. **T. pauciflorum** (Royle! Ill. 52); foliis biternatis subsessilibus, foliolis inciso-trilobis, floribus in panicula pauciflora longe pedicellatis, acheniis 5–15 breviter stipitatis dolabriformibus nervosis stigmate dilatato apiculatis.—T. macrostigma *et* T. secundum, *Edgeworth! Linn. Tr.* xx. 26.

HAB. In Himalaya occidentali interiore, alt. 7–13,000 ped.: Kashmir! Kishtwar! Garhwal! Kumaon!—(*v. v.*)

Herba 1-2-pedalis, glaucescens, apice tantum ramosa. *Folia* subsessilia, biternata, petiolo folioli medii elongato lateralium abbreviato; *foliola* basi rotundata vel cuneata, ½-uncialia, nervosa; floralia angustiora, pedicellis breviora. *Flores* parvi. *Sepala* ovata, acuta, trinervia. *Filamenta* filiformia; *antheræ* mucronatæ.

By an oversight, perhaps by a typographical error, Dr. Royle describes the leaves as triternate, so that it is not surprising that Mr. Edgeworth should have failed to recognize his plant in Royle's description. The leaves are uniformly biternate, and always sessile or nearly so; they are nearly uniform in size from the base to the summit of the stem.

Sect. 2. EUTHALICTRUM, DC.—*Achenia* ovali-oblonga nec compressa, stipitata vel sessilia.

§ 1. *Achenia elongata, stipitata.*

5. **T. virgatum** (H.f. et T.); subsimplex, foliis ternati-partitis subsessilibus, foliolis rotundatis lobatis rigidis, panicula terminali ramosa foliosa, acheniis plurimis breviter pedicellatis oblongis costatis.

HAB. In Himalaya orientali temperata, supra rupes humidas: Sikkim in montibus interioribus, alt. 8–10,000 ped.!—(Fl. Mai. Jun.) (*v. v.*)

Radix tuberosa; *caulis* erectus, 1–1½-pedalis, strictus, glaucus, sæpe purpurascens, simplex vel apice tantum ramosus. *Folia* ternatim partita, petiolis brevissimis; *stipulæ* minutæ, scariosæ; *foliola* ½–1 unciam longa, rigida, nervosa, rotundata, basi cuneata vel cordata, 3–5-loba, lobis integris vel obtuse tridentatis. *Flores* parvi, albi. *Sepala* elliptica, multinervia. *Filamenta* filiformia; *antheræ* muticæ. *Achenia* 10–25, utrinque acuta, stigmate sessili persistente apiculata.

6. **T. rutæfolium** (H.f. et T.); foliis pinnatim decompositis, foliolis membranaceis parvulis varie inciso-lobatis, floribus racemosis,

racemis terminalibus demum elongatis, acheniis 3–5 stipitatis arcte
reflexis oblongis incurvis stylo incurvo mucronatis obtuse costatis.
HAB. In Tibetia occidentali, alt. 10–14,000 ped.: in Nubra! et
Ladak!; et in Himalaya orientali, in regione interiore Sikkim, loco in-
certo, sed certe supra 10,000 et verosimiliter supra 12,000 ped. alt.!
—(Fl. Jul.) (*v. v.*)

Radix fusiformis, perpendicularis. *Herba* debilis, 1½–2-pedalis, glaberrima. *Folia*
inferiora longe petiolata. *Petioli* basi scarioso-stipulati, et ad ramificationes stipel-
lati, stipellis scariosis minutis cito deciduis. *Flores* parvi, pallide viridescentes.
Sepala elliptica, obtusa, multinervia. *Filamenta* filiformia; *antheræ* mucronatæ.
Pedicelli fructiferi stricti, patentes.

7. **T. pedunculatum** (Edgew. in Linn. Soc. Tr. xx. 27); foliis
triternatis, foliolis ovalibus rotundatis membranaceis obtuse lobatis,
panicula pauciflora, acheniis lineari-oblongis breviter pedicellatis valide
costatis in rostrum breve uncinatum acuminatis.

HAB. In montibus Afghanistan, *Griffith!*; in Himalaya occidentali,
alt. 6–8000 ped.: Kashmir! Banahal! Simla!—(Fl. Apr. Mai.) (*v. v.*)

Herba erecta, ramosa, bipedalis. *Folia* longe petiolata; *stipulæ* auriculares, fimbriatæ, ad basin petioli; *stipellæ* nullæ. *Petioli* partiales elongati. *Foliola* ½–1
unciam lata, rotundata vel late cuneata, triloba vel tridentata, lobis integris vel divisis. *Panicula* ramosa, pedicelli folia floralia superantes, fructiferi elongati.
Flores majusculi (½-unciales), albi. *Sepala* elliptica, obtusa. *Filamenta* filiformia;
antheræ muticæ. *Achenia* fere ¼ unciam longa, dorso subgibbosa, ventre rectiusculá,
vix subcompressa, utrinque 5-costata.

Approaches closely in general appearance to *T. orientale*, Boissier, a native of
Taurus and Asia Minor. That species, however, has sessile smaller fruit and a deciduous style.

8. **T. rostellatum** (H.f. et T.); foliis ternatim decompositis, foliolis rotundatis basi cordatis crenato-lobatis membranaceis, panicula
pauciflora foliosa, acheniis 3–5 pedicellatis lineari-oblongis striatis in
rostrum longum rectum apice uncinatum productis.

HAB. In Himalaya temperata, alt. 7–11,000 ped.: Simla, *Jacquemont!* et Sikkim interiore!—(Fl. Aug.) (*v. v.*)

Herba erecta, bipedalis, gracilis, diffusa, ramosa, radice fibrosa. *Petioli* breves,
basi stipulis reniformibus auriculati. *Stipellæ* nullæ. *Foliola* tenuiter membranacea, pallide viridia, subtus glaucescentia, ¼–½-uncialia. *Flores* in ramis fere
solitarii, parvi, albi. *Sepala* elliptica, nervosa. *Filamenta* filiformia; *antheræ*
muticæ. *Paniculæ* fructiferæ rami elongati divaricati graciles. *Rostrum* achenio
⅓ brevius.

A delicate, straggling, branched plant, approaching in the shape of its fruit to the
last species, but very different in habit, and easily distinguished by the much smaller,
long-beaked achenia.

§ 2. *Achenia breviora oblonga costata, subsessilia vel rarius (in* T.
alpino) *longe stipitata.*

9. **T. saniculæforme** (DC. Prod. i. 12); foliis ternatim decompositis, radicalibus longe petiolatis caulem sæpe superantibus, caulinis
paucis, panicula pauciflora divaricato-ramosa, pedicellis strictis elongatis, acheniis numerosis sessilibus oblongis argute costatis longe rostratis.—T. radiatum, *Royle! Ill.* 52.

HAB. In Himalaya temperata in tempore pluvioso supra arbores et rupes humidas crescens : Basehir! Garhwal, *Royle!* Nipal, *Wall.!* (in Herb. Hook.) Sikkim, in jugis interioribus, alt. 8000 ped.!—(Fl. Aug.) (*v. v.*)

Herba ½–1-pedalis, radice fibrosa, caule gracili rigido, stricte erecto, petiolis elongatis. *Stipulæ* liberæ, oblique ovales; *stipellæ* nullæ. *Foliola* rotundato-triloba, tenuia, membranacea, pallide viridia, glaberrima. *Flores* albi. *Sepala* elliptica, nervosa. *Filamenta* filiformia; *antheræ* brevissime apiculatæ. *Achenia* longe rostrata, rostro achenium æquante, apice hamato.

A curious little plant, remarkable for its very rigid habit and pseudo-parasitic mode of growth. It is more nearly allied to *T. glyphocarpum* than to any other species, but is easily distinguished by its smaller size and the very long beak of the achenium.

10. **T. glyphocarpum** (W. et A.! Prod. i. 2); foliis ternatim decompositis, panicula terminali pauciflora, filamentis clavatis, acheniis 8–15 oblongis brevissime pedicellatis valide costatis rostro brevi uncinato apiculatis.—*Wight, Ic. t.* 48.

HAB. Per totam Indiam temperatam in sylvis densis et dumetis ; in Himalaya a Simla, 6–8000 ped.! et Kumaon! usque ad Sikkim in jugis interioribus, 9–12,000 ped.! (sed nondum e Nipalia allatum); Khasia, alt. 5–6000 ped.! in monte Parasnath, prov. Behar, alt. 4000 ped.!; in montibus altioribus Peninsulæ et Zeylaniæ!—(Fl. Jul. Sept.) (*v. v.*)

Herba erecta, bipedalis et ultra, radice fibrosa. *Stipulæ* petiolo adhærentes, reniformes, membranaceæ, fimbriatæ; *stipellæ* nullæ. *Foliola* ½–1-uncialia, glaberrima, membranacea vel coriacea, rotundata, obtuse crenato-lobata vel triloba. *Panicula* divaricato-ramosa, ramis strictis subracemosis. *Antheræ* muticæ.

There are specimens of this species in the Hookerian Herbarium, from Java, collected by Mr. Lobb, so that possibly *T. Javanicum*, Blume, may be the same. The character given in the Bijdragen is, however, quite insufficient to determine whether this conjecture be well founded or the contrary ; and in any case we think the name given by Wight and Arnott, who have well characterized the species, ought to be retained.

11. **T. foliolosum** (DC. Syst. i. 175, Prod. i. 12); polygamodioicum, foliis supradecompositis, panicula ramosissima aphylla, bracteis minutis, acheniis paucis ovali-oblongis utrinque acutis argute costatis. —*Don, Prod.* 192 ; *Royle! Ill.* 51.

HAB. In Himalaya temperata ubique : occidentem versus in jugis exterioribus (5–8000), in Sikkim in interioribus tantum; et in montibus Khasia in umbrosis, alt. 4–6000 ped.!—(Fl. Aug. Sept.) (*v. v.*)

Erecta, ramosa, 3–8-pedalis. *Folia* maxima, sæpe pedalia et ultra ; *foliola* numerosissima, parva, ovalia, inciso-lobata, maxima vix pollicaria, plerumque multo minora. *Petioli* basi auriculati. *Stipellæ* nullæ. *Sepala* oblonga, obtusa, 5–7-nervia, pallide viridia vel fusco-purpurea. *Stamina* numerosa ; *filamenta* filiformia ; *antheræ* mucronatæ. *Achenia* 5–7.

The Khasia plant is very luxuriant, and generally very large-leaved.

12. **T. minus** (L.); foliis decomposite pinnatis, segmentis varie lobatis, panicula ramosa ampla, acheniis paucis oblongis vel ovalibus

utrinque acutiusculis, stylo dilatato demum deciduo apiculatis, valide costatis.

α. *vulgare;* glabrum, foliis minoribus.—T. minus, *L. et Auct.*

β. *fœtidum;* glanduloso-pubescens, foliis minoribus.—T. minus, β. glandulosum, *Koch; Led. Fl. Ross.* i. 8. T. fœtidum, *L. et Auct.* T. vaginatum, *Royle! Ill.* 52.—(v. v.)

γ. *majus;* glabrum vel glaucescens, foliis majoribus.—T. majus, *Jacq. et Auct.* T. Kemense, *Fries! Led. Fl. Ross.* i. 13. T. Maxwellii, *Royle! Ill.* 52.—(v. v.)

HAB. In Tibet occidentali vulgatissimum; et in Himalayæ occidentalis jugis interioribus, alt. 9–12,000 ped., in graminosis: Nubra! Ladak! Zanskar! Kashmir! Kishtwar! Kanawer et Piti, *Royle! etc.* Sikkim interius, alt. 11–12,000 ped.—Var. α. (forma Europæa) in India rarius occurrit in sylvis Himalayæ interioris temperatæ; β. in Tibetia vulgaris, etiam occurrit in montibus altioribus Kishtwar et Kanawer; γ. quæ in Tibet in pratis Zanskar et Piti crescit (alt. 10–11,000 ped.) vulgatior est in siccioribus Himalayæ interioris.

DISTRIB. Europa tota! Africa borealis! (et australis?); Asia temperata!

Herba 2–4-pedalis, erecta vel basi prostrata, ramosa; *radix* fibrosa. *Panicula* multiflora, fere aphylla. *Sepala* viridi-purpurascentia, elliptica, nervosa. *Antheræ* elongatæ, mucronatæ.

This species, which is extremely abundant in all parts of Europe and Siberia, is exceedingly polymorphous, and has received at the hands of European and Siberian botanists a vast number of names; while the great variation in the opinions of different authors as to the limits of the different species which they distinguish from one another is, we think, in itself sufficient to prove that the number of these has been considerably over-estimated. We have devoted much time to a careful comparison of our extensive suites of Indian specimens with the very large collection of authentically-named European and Siberian forms in the Hookerian Herbarium; and after attempting in vain to find characters sufficient to distinguish the large-leaved variety, we have felt ourselves driven to the conclusion that only one species exists in India. In this we follow Hooker and Arnott, who in the British Flora (fifth edition) have united all the European forms under *T. minus,* L. *T. saxatile,* Schleicher and DC., has been referred unhesitatingly by Planchon, in Herb. Hook., to *T. minus;* while *T. collinum,* Wallroth and Ledebour, and *T. elatum,* Murray and DC., appear to be forms of *T. majus. T. Kemense,* Fries., which is identical with *T. Maxwellii,* Royle, is distinguished by Ledebour from allied species by the presence of stipellæ at the divisions of the compound leaf. This character we have unfortunately found entirely to fail, as these organs are present or absent on different leaves of the same specimen and parts of the same leaf. The number of synonyms might be much increased if this were the proper place to do so, and if authentic specimens were available. The dingy purple hue of the densely-panicled flowers, and the long stamens, seem to characterize all the forms, but the fruit varies somewhat in length, being usually, though not invariably, thicker and shorter in the larger states.

13. **T. isopyroïdes** (C. A. Meyer in Led. Fl. Alt. ii. 346); foliis ternatim decompositis, segmentis ultimis minutis oblongis obtusis, panicula ramosa, foliis floralibus parvis sæpe bracteæformibus, acheniis 3–6 subsessilibus oblongis utrinque obtusiusculis valide costatis.— *Led. Ic. Alt. t.* 397, *Fl. Ross.* i. 7.

HAB. In montibus Beluchistan, *Stocks!* Afghanistan, *Griffith!*— (*v. s.*)

DISTRIB. Taurus! Mesopotamia! Persia! Sibiria altaica!

Herba e rhizomate horizontali erecta, glaberrima, ½–1-pedalis, subsimplex, foliis radicalibus longe petiolatis. *Stipulæ* inconspicuæ; *stipellæ* nullæ. *Foliola* profunde 3–5-partita, segmentis 1–2 lineas longis anguste oblongis vel linearibus. *Paniculæ* rami elongati, patentes, fructiferi rigidi. *Flores* parvi. *Sepala* ovalia, multinervia, ex sicco alba. *Antheræ* mucronatæ. *Achenia* stigmate persistente mucronata, lineari-oblonga.

14. **T. alpinum** (L. Spec. 767); foliis omnibus radicalibus pinnatis vel bipinnatis, scapis simplicibus racemosis, pedicellis fructiferis reflexis apice dilatatis, acheniis oblongis costatis pedicellatis.—*Ledeb. Fl. Ross.* i. 6. T. microphyllum *et* marginatum, *Royle! Ill.* 51. T. acaule, *Camb.! in Jacq. Voy. t.* 1 A. *p.* 3.

HAB. In paludosis totius Himalayæ, et Tibetiæ occidentalis, supra 10,000 ped.; in Tibet usque ad alt. 17,000 ped.—(Fl. Jun. Jul.) (*v. v.*)

DISTRIB. Europa! Asia! et America! arctica et alpina.

Herba pusilla, cæspitosa. *Foliola* rotundata, triloba vel fere tripartita, glabra, subtus glauca. *Scapi* 3–6-unciales, erecti, simpliciter racemosi. *Bracteæ* oblongæ, obtusæ, membranaceæ, infima interdum subfoliacea triloba. *Sepala* ovata, membranacea, pallide viridia, acutiuscula. *Antheræ* mucronatæ. *Achenia* striata, demum sublævia, pedicellis longitudine valde variis, achenio æqualibus vel paullo brevioribus, interdum vix ullis.

Identical with the European plant. In luxuriant specimens from Sikkim the scape is occasionally branched, and bears a small leaf at the point of ramification.

15. **T. Punduanum** (Wall. Cat. 3712!); foliis biternatis (superioribus ternatis), foliolis magnis rotundatis crenatis sub-5-lobis, panicula terminali divaricato-ramosa fere aphylla, acheniis numerosis oblongis sessilibus sulcatis stylo recto apice uncinato longe rostratis.— *Wall.! Plant. As. Rar.* ii. 26.

HAB. In Himalayæ jugis exterioribus: Kumaon (4500–7000 ped.), *Madden! Str. et Wint.!;* et in montibus Khasia, alt. 3–4000 ped., *Wall.! etc.*—(Fl. Aug. Sept.) (*v. v.*)

Herba erecta, bipedalis. *Folia* inferiora longe petiolata, biternata; media sæpe irregulariter divisa, nempe ternata, segmento terminali simplici, lateralibus ternatis vel bifoliatis. *Petioli* basi auriculati. *Stipellæ* nullæ. *Foliola* diametro 1–2-uncialia, coriacea, rigida, nervosa, supra nitida, subtus glauca, glaberrima vel tomentosa. *Sepala* 4, elliptico-oblonga, nervosa. *Filamenta* filiformia. *Antheræ* vix apiculatæ. *Achenia* glanduloso-pilosa vel glaberrima, stylo æquilongo rostrata.

It is singular that this species, which occurs in Kumaon and Khasia, should not have been met with in Nipal or Sikkim. The Kumaon specimens, which we have seen, are all glabrous; but in the Khasia mountains both states occur, that with tomentose leaves being, however, more common.

16. **T. Dalzellii** (Hook. Ic. Plant. t. 856!); foliis ternatipartitis, foliolis magnis orbiculato-reniformibus profunde cordatis crenato-lobatis vel sub-5-lobis, panicula parva, floribus ad ramorum apices glomeratis, acheniis numerosis brevissime pedicellatis oblongis sulcatis, stylo uncinato breviter rostratis.

HAB. In montibus Siadri provinciæ Concan, *Dalzell!*—(*v. s.*)

Herba rigida, pedalis. *Petioli* basi stipulis oblongis majusculis auriculati. *Stipellæ* nullæ. *Foliola* utrinque glaberrima, pallida, rigide coriacea, nervosa, diametro 1–2½-uncialia, superiora sessilia. *Sepala* elliptica, nervosa. *Filamenta* filiformia; *antheræ* muticæ.

17. **T. rotundifolium** (DC. Syst. i. 185, Prod. i. 15) ; foliis maximis simplicibus orbiculari-reniformibus inciso-lobatis et crenatis, panicula ramosa.—*Don, Prod.* 193.

HAB. Nipal, *Hamilton ! Wall.!—(v. s.)*

Herba pedalis. *Petioli* basi stipulis oblongis auriculati. *Folia* 2–4-uncialia, nitida, nervosa, subtus tomentosa. *Sepala* elliptica, obtusa. *Filamenta* filiformia.

Our specimens of this and the last species are not sufficient to enable us to decide to our own satisfaction whether or not they be distinct from one another, and even from *T. Punduanum*, to which the form of the leaflets approximates them very :losely.

4. **ANEMONE,** L.

Flores involucrati. *Sepala* 5–15, petaloidea, æstivatione imbricata. *Petala* nulla. *Achenia* mutica vel caudata, monosperma.—Herbæ *acaules*, radice *perenni.*

Chiefly a northern genus, with a few species in the mountains of South America, and several in South Africa. The Indian species are all confined to the mountains, none occurring below 5000 feet, and are mostly alpine. There is also a single species in Tasmania, and one in the mountains of the island of Sumatra.

To the sections instituted by De Candolle we have added one characterized by the small, remarkably woolly achenia. It includes *A. sylvestris*, L., *A. Virginiana*, L., and many other European and American species, and appears to form a very natural group.

Sect. 1. PULSATILLA, DC.—*Achenia* in caudas longas barbatas producta.

1. **A. Albana** (Steven in Mem. Soc. Nat. Mosq. iii. 264); foliis pinnatipartitis, pinnis profunde pinnatifidis segmentis incisis, involucri triphylli foliis basi coalitis cuneatis apice varie incisis, flore solitario campanulato nutante, sepalis late ellipticis apice reflexis.—*DC. Syst.* i. 545, *Prod.* i. 17. Pulsatilla albana, *Ledeb. Ic. Fl. Alt. t,* 109, *Fl. Ross.* i. 22. Anemone Wallichiana, *Royle! Ill.* 52.

HAB. In Tibet occidentali, alt. 12–16,000 ped.: Balti, *Winterbottom !* Zanskar ! Ladak ! Piti et Kanawer, *Jacquemont ! Royle !*—(Fl. Jun. Jul.) *(v. v.)*

DISTRIB. Armenia ! Caucasus ! Altai ! Baikal !

Dense pilosa, pilis patentibus. *Scapi* floriferi ½-pedales, fructiferi pedales. *Pedicellus* floris dense tomentosus, involucrum vix superans, fructifer elongatus. *Sepala* ¾-uncialia, intus pubescentia, extus dense sericea, pallide rubescentia. *Stamina* extima in glandulas stipitatas mutata. *Achenia* dense sericea.

Tibetan specimens agree exactly with Siberian ones in everything but the colour of the flower, in which respect they are, as it were, intermediate between the two varieties distinguished by Ledebour. Perhaps the species is only an alpine state of *A. pratensis*, which appears to differ chiefly by being larger, with larger, deeper blue flowers, a more deeply divided involucre, and more finely cut leaves.

Sect. 2. ERIOCEPHALUS.—*Achenia* lana compacta involuta, in capitulum densum globosum vel oblongum conglomerata, stylis nudis apiculata.

2. **A. biflora** (DC. Syst. i. 201, Prod. i. 19); radice tuberosa, foliis radicalibus rotundato-reniformibus tripartitis, segmentis rotundatis indivisis vel sæpius palmatim lobatis vel partitis, lobis crenato-lobatis, involucri triphylli foliis sessilibus basi cuneatis ad medium palmatim incisis, floribus in involucro 1–3.—A. Gordschakowii, *Kar. et Kir. Enum. Pl. Soony. No.* 14 ! *Ledeb. Fl. Ross.* i. 727.—*Griffith, Itin. Notes,* 349.

HAB. In Beluchistan prope Kelat, *Stocks!* Afghanistan, *Griffith!* Kashmir, alt. 6000 ped., *Jacquemont!*—(Fl. Apr. Mai.) (*v. v.*)

DISTRIB. Persia ! Sibiria altaica !

Herba florifera 2–4-uncialis, fructifera interdum 8-uncialis, glabra, vel pedicellis sparse adpresse pilosis. *Folia* crassiuscula. *Flores* pallide rubicundi, ½–1-unciales, terminalis nudus, laterales involucello diphyllo muniti. *Sepala* 5, ovalia, obtusa vel acutiuscula, adpresse pilosula, parallele nervosa, subpersistentia. *Achenia* stylo æquilongo marcescente apiculata.

The terminal flower often falls away before the lateral ones are developed, so that there are often apparently only two flowers in the involucre, each of which is involucellate. The remains of the peduncle of the first-developed flower may, however, always be discovered.

3. **A. rupicola** (Camb.! in Jacq. Voy. Bot. p. 5. t. 2); caule subterraneo horizontali, foliis longe petiolatis tripartitis, segmentis plus minus petiolatis trilobis inciso-dentatis, involucri triphylli foliis subsessilibus, floribus 1–2.

α. *sericea ;* tota molliter pilosa, alabastris et pedicellis præsertim sub flore dense tomentosis, foliis obtusius incisis.

β. *glabriuscula ;* collo, vaginis foliorum, basi involucri et pedicellis pilosis, cæterum glabra, foliis argutius incisis.

HAB. In Himalaya interiori alpina, alt. 11–15,000 ped.: Balti, *Winterbottom!* Dras! et Zanskar! Kashmir, *Jacquemont!* Kumaon, *Str. et Wint.!* Sikkim!—(Fl. Jul.) (*v. v.*)

Molliter albo-pilosa, rarius subglabrescens. *Rhizoma* sæpe elongatum, vaginis foliorum delapsorum tectum. *Involucri* folia ad medium triloba, lobis trilobis vel incisis. *Scapus* ½–1-pedalis. *Folia* 1–2-uncialia. *Flos* terminalis exinvolucellatus, lateralis involucello diphyllo munitus. *Sepala* 5, late obovata vel elliptica, obtusa vel retusa, extus molliter pubescentia 1–1½-uncialia. *Achenia* ovalia, stylo brevi subulato nudo apiculata.

This species approaches *A. sylvestris,* L., but that is always one flowered, and has the involucral leaves pedicellate.

4. **A. vitifolia** (Ham. in DC. Syst. i. 210, Prod. i. 21); foliis amplis cordatis 5-lobis subtus niveis, involucri triphylli foliis longe petiolatis foliis radicalibus conformibus, cymæ multifloræ ramis lateralibus bis terve divisis involucellatis.—*Don, Prod.* 193 ; *Lindl. Bot. Reg. t.* 1385 ! *Wall. Cat.* 4695!

HAB. Per totam Himalayam temperatam, alt. 5–8000 ped.: occidentem versus præsertim in jugis exterioribus humidioribus, in Sik-

kim, ubi in jugis exterioribus humidissimis non obvia, usque ad alt.
11,000 pedum occurrit.—(Fl. Jun.-Aug.) (*v. v.*)

Herba elata, 1-3-pedalis. *Radix* perpendicularis, cylindrica, lignosa. *Folia*
diametro 4–8-uncialia, rotundata vel ovata, late 5-loba, lobis acutis argute inciso-dentatis, superne pilis sparsis adpressis tecta vel glabra, subtus cum petiolis et scapo
niveo-floccosa rarius demum subglabra. *Cyma* decomposita, floribus gradatim evolutis, in planta juniore ramis lateralibus nondum evolutis ad speciem uni- vel pauciflora. *Sepala* 6–8, ovalia, uncialia, extus adpresse sericea, albida. *Achenia* in capitulum globosum coalita, minima, stylo nudo æquilongo apiculata.

Sect. 3. ANEMONANTHEA, DC.—*Involucra* 1-2-flora. *Pedicelli*
nudi. *Achenia* oblonga, cylindrica, angulata vel subcompressa,
parva, distincta, nec in capitulum concreta.

5. **A. Griffithii** (H.f. et T.); foliis involucralibus 3 longe petiolatis tripartitis sparse pilosis, floribus 1–2.

HAB. In Sikkim interiori in sylvis densis vallis Lachen, alt. 8–9000
ped.! et in Bhotan, *Griffith !* (No. 1720 in Herb. Hook.)—(*v. v.*)

Species *A. nemorosæ* et *A. ranunculoidi* affinis, sed specimina valde manca. Radix
et folia radicalia non suppetunt. *Scapi* 3–6-unciales. *Folia* involucri tripartita :
segmentis acutis incisis et serratis, lateralibus oblique bilobis, terminali trilobo. *Se-*
pala 6, ex sicco rubescentia, ovalia, obtusa, ½ unciam longa.

6. **A. Falconeri** (Thoms. in Hook. Ic. Plant. t. 899 !) ; foliis tripartitis, segmentis late cuneato-ovatis trilobis, involucri triphylli foliis
basi coalitis oblongis apice tridentatis vel obtusis, flore solitario, acheniis
angulatis oblongis sericeo-pilosis.

HAB. In Himalaya occidentali, alt. 6–10,000 ped., in sylvis umbrosis : Kashmir ! Kishtwar !—(Fl. Apr. Mai.) (*v. v.*)

Herba pusilla, molliter pilosa. *Rhizoma* horizontale, lignosum, fibros plurimos
emittens. *Folia* numerosa, 1-2-uncialia, longe petiolata, submembranacea, adpresse
sericea, tripartita. *Segmenta* lateralia obliqua biloba, terminale trilobum. *Petioli* piiis
longis patentibus tecti. *Scapi* 3–6-unciales, folia æquantes. *Pedicelli* florum involucro breviores vel æquales. *Flores* parvi, diametro ⅓–½-unciales. *Sepala* obovata,
obtusa. *Achenia* stylo brevissimo apiculata.
This is the plant mentioned by Falconer in the introduction to Royle's Illustrations, p. 25, as a new species of *Hepatica*. Its relationship to that genus or section is certainly very close, not only in general habit, but also in flower and fruit.
Its pedicellate flowers, however, are an obstacle to its being placed in the section (or
genus) *Hepatica*, and seem to indicate that that section is not a natural one, but
that *A. Hepatica* ought probably to find a place in the section *Anemonanthea*, along
with *A. Falconeri*.

Sect. 4. ANEMONOSPERMOS, DC.—*Involucra* pluriflora. *Pedicelli*
laterales involucellum gerentes. *Achenia* prioris.

7. **A. rupestris** (Wall. Cat. 4696 !) ; foliis trisectis, involucri trifolii foliis basi angustatis foliorum segmentis conformibus, floribus 1–3,
acheniis oblongis vix compressis glabris stylo uncinato apiculatis.

HAB. In Himalaya interiori alpina : Kashmir ad Pir Panjal jugum,
Jacquemont ! Nipal ad Gossain Than, *Wall.!* Sikkim, alt. 15,000
ped.!—(Fl. Aug.) (*v. v.*)

Radix lignosa, subhorizontalis. *Caulis* basi vestigiis petiolorum delapsorum fibrillosis involutus, multifolius. *Folia* glabra vel pilosula, 1-2-uncialia, trisecta. *Segmenta* pedicellata, tripartita vel pinnatifida, varie incisa, lobis ultimis cuneatis vel late linearibus. *Scapi* pilis patentibus hirsuti, 4-8-unciales. *Pedicellus* terminalis nudus, laterales bracteis 2 oblongis integris vel inciso-dentatis involucellati. *Sepala* 5-6, obovata, ½-uncialia, cærulescentia.

8. **A. trullifolia** (H.f. et T.); foliis late ovalibus basi rotundatis vel subcordatis trilobis, lobis obtuse tridentatis, involucri trifolii foliis oblongis subintegris vel profunde trilobis, floribus 1-3, acheniis ovalibus subcompressis rigide pilosis.

HAB. In Himalaya orientali interiori : Sikkim, alt. 11-15,000 ped.! Bhotan, *Griffith!*—(Fl. Jul.-Sept.) (*v. v.*)

Tota planta dense pilosa. *Radix* crassa, valida, sublignosa. *Folia* e collo crassissimo, reliquiis foliorum fibrosis involuto, plurima. *Petioli* late vaginantes, foliis breviores, rarius elongati. *Folia* obtusissima, subintegra vel plus minus profunde triloba, rarius pinnatisecta, 1½-3-uncialia. *Scapi* intra folia nidulantes vel elongati pedales filiformes prostrati. *Involucri* folia magnitudine valde varia. *Flos* terminalis nudus, lateralium pedicellis involucello e foliis 2 oblongis indivisis constante munitis. *Sepala* 6-8, ex sicco aurea extus pallide cærulea, obovata vel oblonga, ½-1-uncialia, extus adpresse sericea. *Achenia* parva, ovalia, subcompressa, stylo nudo apiculata, setis strigosis erectis hispidis tecta.

This species closely resembles in habit the next, but the leaves in all our specimens, which are from many localities, are very different. The achenia scarcely differ.

9. **A. obtusiloba** (Don, Prod. 194); foliis rotundatis cordatis trisectis vel tripartitis, segmentis varie lobatis, involucralibus 3 cuneato-obovatis trilobis, lobis integris vel incisis, floribus 1-3, acheniis compressis ovalibus stylo rostratis pilis rectis rigidis hirsutis.—A. Govaniana, *Wall. Cat.* 4688! A. discolor, *Royle! Ill.* 52. *t.* xi. *f.* 1. A. mollis, *Wall. Cat.* 4689! *ex parte.*

β. *glabra ;* scapo foliisque glabris vel glabrescentibus.

HAB. In Himalaya occidentali temperata et alpina, alt. 9-15,000 ped.: Marri, *Fleming!* Hasora, Balti et Kashmir, *Wint.!* Kishtwár! Sirmur, *Royle!* etc. Kanawer, *Munro!* Garhwal, *Str. et Wint.!* Nipal, *Wall.!* Sikkim! β. In alpibus Tibetiæ occidentalis, *Wint.!*—(Fl. Mai. Jun.) (*v. v.*)

Radix lignosa, crassa. *Caulis* basi reliquiis foliorum fibrosis involutus, polyphyllus. Tota planta hirsuta vel molliter pilosa, rarius glabrescens. *Folia* diametro 1-2½-uncialia, profunde cordata, segmentis basi angustatis pedicellatis vel sessilibus, latissime cuneatis varie incisis, lobis rotundatis. *Scapi* ½-1-pedales, sæpius pilis patentibus hirsuti. *Involucri* folia magnitudine valde varia. *Pedicellus* terminalis nudus, laterales (dum adsunt) bracteis 2 oblongis involucellati. *Flores* ½-1½-unciales, cærulescentes albi vel aurei, sepalis basi cæruleo-plumbeis. *Sepala* obovata vel rotundata, 5-10. *Achenii* immaturi rostrum recurvum, maturi rectum rigidum.

A very widely diffused plant in Western Himalaya, flowering on the grassy slopes of the mountains in early spring, as soon as the snow has melted. Like most very common plants, it varies a good deal in size and degree of hairiness; but these variations depend chiefly on situation, and perhaps on the age of the plant. The difference in the colour of the flowers is very remarkable, but seems quite unconnected with the variations in leaves and hairiness, as specimens of the golden-yellow and of the blue forms may be selected which are in every other respect undistinguishable. The glabrous state is a very striking variety, and we should have kept it separate had not the specimens collected in Balti and Kashmir by Mr. Winterbottom presented a series of gradations connecting it with the ordinary form of the species. The fruit is quite the same in all the varieties, the beak being hooked while young, but straight and rigid in the ripe achenium. *A. mollis,* Wall. Cat., is said to be from Khasia; but as the species is rather an alpine one in the Himalaya, and has not been found in that district by other collectors, it is probable that the specimens, which are mere fragments, have been accidentally wrongly ticketed.

10. **A. rivularis** (Ham. in DC. Syst. i. 211, Prod. i. 21, non Wall. Cat.); foliis trisectis, segmentis tripartitis vel profunde trilobis, lobis iterum trilobis irregulariter inciso-serratis, involucri 2–3-phylli foliis tripartitis, segmentis trilobis vel indivisis oblongis acuminatis argute inciso-serratis, cyma decomposita, acheniis oblongis in rostrum recurvum acuminatis.—*Don, Prod.* 193. A. hispida, *Wall. Cat.* 4694! A. Wightiana, *Wall. Cat.* 4697! *W. et A. Prod.* 3 ; *Wight, Ic. t.* 936! *Nilg. Pl. p.* 2. *t.* 4! *Hook. Ic. Plant. t.* 176! *Lindl. Bot. Reg.* 1842, *t.* 8! A. dubia, *Wall. Cat.* 4698 (*fide W. et A.*) ; *W. et A. Prod.* 3. A. geraniifolia, *Wall. Cat.* 4693!

HAB. Ladak: in aquosis infra 10,000 rara!; ubique in Himalaya temperata, alt. 5–10,000 ped., in graminosis humidis et secus vias: in Sikkim ubi ad alt. 13,000 ped. ascendit in jugis interioribus tantum ; in montibus Khasia!; in peninsulæ australis et Zeylaniæ montibus temperatis!—(Fl. per totam æstatem.) (*v. v.*)

Sericeo-pilosa, 1–3-pedalis. *Radix* crassa, lignosa. *Folia* 2–6-uncialia, majora longissime petiolata, circumscriptione rotundata vel reniformia, profunde cordata. *Involucri* folia sæpe 3–5-uncialia, subsessilia vel alato-petiolata, segmentis ultimis oblongis vel lanceolatis. *Inflorescentia* decomposite cymosa. *Involucelli* folia sæpius bina, bipartita, segmentis lanceolatis vel linearibus parallele nervosis, inciso-serratis. *Sepala* 5–8, ovalia, obtusa, extus sericea, ½–⅔-uncialia, intus alba, extus cærulescentia. *Achenia* ½-uncialia.

This species, which grows at a lower elevation than any other, is the only one which extends within the tropics, and, indeed, except *A. elongata,* Don, which is found in Khasia, the only species yet known in India out of the Himalaya. Notwithstanding the formidable array of synonyms which we have brought together, it is by no means a variable plant, except in size. A monstrous state, in which the flower is converted into a leafy umbel, sometimes six inches in diameter, is common in northern India. The original specimen of *A. dubia,* in the Linnæan Society's herbarium, belongs by some accident to *A. nemorosa,* or some closely allied plant. There can, however, be no doubt that that synonym is correctly referred here, as we have the authority of Wight (Nilg. Plant. p. 2) for uniting it to *A. Wightiana.*

Sect. 5. OMALOCARPUS, DC.—*Achenia* ovalia, valde compressa.

11. **A. demissa** (H.f. et T.) ; foliis trisectis, segmentis petiolatis late cuneato-obovatis varie incisis, scapis prostratis, involucri foliis 3 sessilibus trilobis, floribus 1–6 exinvolucellatis, acheniis glabris stylo deflexo apiculatis.

HAB. In Himalayæ orientalis provincia Sikkim, alt. 13–16,000 ped.! —(Fl. Jul.) (*v. v.*)

Villosa vel rarius glabrata. *Radix* crassa, sublignosa, collo incrassato polyphyllo. *Folia* uncialia, petiolo longitudine vario. *Scapi* 3–12-unciales. *Pedicelli* floriferi breviusculi, fructiferi sæpe elongati. *Sepala* ovalia, ¼–½-uncialia, cærulescentia. *Ovaria* glabra. *Achenia* late ovalia, ¼ unciam longa.

12. **A. narcissiflora** (L. Sp. 763) ; foliis palmatim 5-sectis, segmentis cuneatis profunde incisis, laciniis angustis, involucralibus 3–5
tridentatis vel incisis, floribus umbellatis, acheniis ovalibus late alatis
stylo obliquo rostratis.—*DC. Syst.* i. 212, *Prod.* i. 21 ; *Ledeb. Fl. Ross.*
i. 18. A. umbellata, *Willd.; DC. Syst.* i. 213, *Prod.* i. 22 ; *Deless. Ic.
sel.* i. *t.* 18.

HAB. In montibus Kashmir borealis versus Gares, *Winterbottom!*—
(Fl. Jul.) (*v. s.*)

DISTRIB. In alpibus Europæ mediæ et australis, Asiæ temperatæ
et Americæ bor. occid.

Planta pedalis, sericea vel glabriuscula. *Folia* longe petiolata, diametro 2-polli-
caria, ad basin usque secta, segmentis inciso-lobatis linearibus vel oblongis obtusis
vel acutis, plerumque adpresse sericeis. *Pedunculi* nudi, abbreviati, uniflori, rarius
elongati. *Sepala* 5, elliptica, ½–¾ poll. longa.

The specimen of this species, which was collected by Mr. Winterbottom, is in no
way different from some forms of the European plant, which is evidently very variable. The lobes of the leaves are, however, less deeply cut, and the teeth not so
narrow, as in the ordinary state. It is very silky, and in so far belongs to the var.
villosissima of DC., or the var. δ, Led. l. c.

13. **A. polyanthes** (Don, Prod. 194); foliis reniformi-cordatis
5–7-lobis, segmentis trilobis grosse crenatis, involucri foliis 3–5 trilobis, lobis varie inciso-crenatis, pedunculis subquinis unifloris vel umbellatim multifloris interdum decompositis, involucellis incisis vel trilobis, acheniis ovalibus stylo subulato oblique rostratis late alatis.—A.
longiscapa, *Wall. Cat.* 4691! A. villosa, *Royle! Ill.* 52. A. obtusiloba, *Lindl. Bot. Reg.* 1844, *t.* 65, *et* A. Govaniana, *Ib. p.* 45, *non
Wall. nec Don.* A. scaposa, *Edgew! Linn. Tr.* xx. 27.

HAB. In Himalaya interiori, alt. 10–12,000 ped.: Kishtwar! Kanawer! Garhwal! Kumaon! Nipal! Sikkim!—(Fl. Jun. Jul.) (*v. v.*)

Planta 1–2-pedalis, sericeo-pilosa, pilis plerumque patentibus. *Folia* sæpius
longe petiolata, 3–6 uncias lata, ultra medium vel fere ad basin lobata. *Scapi* validi,
erecti, foliis longiores. *Involucri* folia magnitudine valde varia, 1–2-uncialia.
Pedunculi nunc uniflori, involucro breviores vel longiores, nunc elongati umbellatim
decompositi. *Sepala* 4–5, ½–¾-uncialia, alba, extus adpresse sericea vel subglabra.
Ovaria glabra vel parce pilosa. *Achenia* ¼ unciam longa.

This is much larger and stouter than the last species, to which, however, it is so
nearly allied that future observations may render it necessary to unite them. The
umbels of *A. narcissiflora* are occasionally compound, and those of the present
species are not always so : so that the shape of the leaves is the only valid mark of
distinction, and that, as we know, is a very variable character in this genus. The
discovery by Mr. Winterbottom of *A. narcissiflora*, L., in Kashmir, throws still more
doubt on the distinctness of this species.

14. **A. tetrasepala** (Royle! Ill. 53); foliis reniformi- vel rotun-
dato-cordatis 5-lobis vel 5-partitis, segmentis plerumque acutis integris vel trilobis argute subduplicato-serratis, involucro sæpius maximo
3–4-phyllo, foliis basi angustatis late obovato-cuneatis trilobis, lobis
argute dentatis, pedunculis umbellatim decompositis, acheniis oblongis
late alatis stylo arcte deflexo terminatis.

HAB. In Himalaya occidentali, alt. 8–11,000 ped.: Marri, *Fleming!*
Kashmir, *Royle! Winterbottom!* Kishtwar!—(*v. v.*)

Herba 1–2-pedalis, glabra vel sublanata. *Folia* coriacea, 3–10 uncias lata, sæpius
longe petiolata, utrinque glabra vel subtus adpresse sericea. *Scapi* folia superantes,
glabri vel patentim pilosi. *Involucri* folia 1–4 uncias longa. *Involucelli* foliola
obovata, varie incisa vel lineari-oblonga, indivisa. *Flores* sæpius plurimi. *Sepala*
4–5, obovata vel orbicularia, glabra, ½–¾-uncialia, alba.

This, which is one of the largest and most robust species of the genus, is in general character very closely allied to the preceding, from which it chiefly differs in being less hairy, with larger leaves, the segments of which are acute and sharply toothed, and not, as in *A. polyanthes*, cut into blunt serratures. The involucre is generally very much larger in the present species, but we have seen specimens in which it is very small. In the only specimen which we possess with ripe fruit the achenium has the style so much inflexed as to be closely adpressed to the fruit, but this character may not be constant.

15. **A. elongata** (Don, Prod. 194) ; foliis tripartitis, segmentis obovato-cuneatis acutis grosse inciso-serratis, involucri parvi foliis tribus, pedunculis 3–5 unifloris di-trichotomisve, involucelli foliolis parvis, acheniis paucis (1–3) ovalibus vel orbiculatis subobliquis anguste alatis stylo brevi recto rostratis.—*A.* rivularis, *Wall. Cat.* 4692! *non Ham. nec alior.*

HAB. In Himalaya temperata : Garhwal, *Str. et Wint. No.* 5 ! Nipal, *Wall.!;* et in montibus Khasia prope Nonkrim, alt. 5000 ped.!

Radix fusiformis, perpendicularis. *Caulis* erectus, glabratus vel tenuiter pubescens. *Folia* longe petiolata, 2–5 uncias lata, tripartita, segmento medio trilobo lateralibus bilobis. *Scapi* valde elongati, 1–3-pedales, laxiflori, involucris pro planta parvis. *Pedunculi* dichotomi vel imperfecte umbellati (fructiferi elongati), terminalis exinvolucellatus, laterales involucellum parvum 1–3-folium gerentes, simplices vel umbellati. *Umbellulæ* radii pauci. *Sepala* ½-uncialia, alba.

Remarkable for its much elongated stems and scapes. The inflorescence is intermediate between umbellate and cymose, the central terminal flower being usually distinct and solitary, though occasionally all the peduncles are similarly umbellate. In the latter case, however, the central flower may have withered or been abortive.

5. **ADONIS,** L.

Sepala 5–8. *Petala* 8–16, fovea nectarifera nulla. *Achenia* plurima, angulosa, ecaudata, *stylo* recto vel recurvo apiculata.—Herbæ *caulescentes, foliis multifidis.*

This genus contains two very natural groups, *Adonia* and *Consoligo*, each of which has a representative in the Indian Flora. The species of *Adonia* are annual, and usually occur in corn-fields. They are very closely allied to one another, if, indeed, they be not all forms of one or at most two very variable species. The section *Consoligo* comprises a few perennial-rooted plants, which are natives of mountainous districts of south Europe and temperate Asia.

1. **A. æstivalis** (L. spec. 772); annua, caule folioso, petalis planis expansis, acheniis angulatis rugosis stylo subrecto apiculatis in spicam oblongam dispositis.—*DC. Syst.* i. 224, *Prod.* i. 24 ; *Ledeb. Fl. Ross.* i. 23 ; *W. et A. Prod.* i. 3 ; *Royle, Ill.* 53 ? A. Inglisii, *Royle! Ill.* 53.

E

HAB. Inter segetes in montibus Indiæ bor.: Beluchistan, *Stocks!*
Afghanistan, *Griffith!* Kashmir, *Winterbottom!* Kanawer, *Royle! etc.*
Sirmur ad Kotgarh, *Str. et Wint.!*; et in montibus Nilghiri, *Wight*,
sed fide cl. Munro in hortis tantum.—(Fl. Mai.–Jul.) (*v. v.*)

Herba erecta, 1–2-pedalis, simplex vel superne ramosa, glabra vel tenuiter pubescens. *Folia* 1–3-uncialia, pinnatim decomposita, segmentis anguste linearibus.
Flores ad ramorum apices solitarii, diametro ½–1½-unciales, coccinei, petalis basi
atropurpureis. *Sepala* petalis ½ breviora. *Achenia* late ovalia, angularia, superne
prope apicem tuberculata, et basi dente acuto quasi calcarata.
The Indian plant agrees perfectly with European and Siberian specimens. The
achenia vary a good deal in shape, and do not, we fear, afford good characters, though
many of the species described by European botanists seem to have no other distinguishing marks. The broad petals and globose flowers of *A. autumnalis* serve to
distinguish the typical form of that species from the ordinary state of *A. æstivalis*,
but intermediate forms are common. Dr. Royle's description seems partly taken
from *A. autumnalis*, of which we have seen no Indian specimens, those in Herb.
Royle being *A. æstivalis*.

2. **A. Pyrenaica** (DC. Prod. i. 25); radice perenni, foliis radicalibus longe petiolatis multifidis caulinis subsessilibus, ramis unifloris,
acheniis stylo uncinatim recurvo apiculatis in capitulo ovali vel subgloboso dispositis.—*Deless. Ic. sel.* i. *t.* 21.

HAB. In montibus Kashmir, *Jacquemont! Winterbottom!* et in Tibet
occid. prov. Guge, *Str. et Wint.!* (*Trollius, No.* 3.)—(Fl. Jun. Jul.)
(*v. s.*)

DISTRIB. In mont. Pyrenæis! necnon in Apenninis et Hungaria,
DC.

Radix valida, fusiformis, subhorizontalis, collo squamis magnis membranaceis
vaginantibus involuta. *Caules* e collo plures vel solitarii, ½–1½-pedalis, basi plerumque
nudi, superne foliosi. *Folia radicalia* longe petiolata, caulem floriferum sæpe fere
æquantia, cito marcescentia, decomposite pinnatisecta, segmentis ultimis anguste
linearibus. *Flores* majusculi, aurei. *Sepala* 7–8, obovata, pallida. *Petala* 12–15,
obovato-cuneata, obtusa, 1–1½-pollicaria, sepalis subduplo longiora. *Achenia* magna,
angulata, glabra, in capitulum densum aggregata.
Our Indian specimens are in flower only, and we had considered them at one time
a distinct species. A more careful examination, however, has shown us that the
characters on which we relied are of no value, and that our plant is in no way distinguishable from that of Western Europe. *A. vernalis*, L., chiefly differs in the
absence of radical leaves, for the floral characters are by no means constant. It is
very remarkable that the Himalayan plant should be the same as that of Western
Europe, and different from that of the Caucasus and Siberia.

6. CALLIANTHEMUM, C. A. Meyer.

Sepala 5, decidua. *Petala* 5–15, ungue fovea nectarifera impressa.
Achenia subglobosa, stylo brevi apiculata. *Semen* pendulum.—Herbæ
alpestres, caulescentes vel acaules, radice *perennante,* foliis *decompositis,*
floribus *albis.*

The only other known species of this genus, *C. rutæfolium*, C. A. Meyer, is a
native of the alps of Europe and Siberia.

1. **C. pimpinelloides** (Royle! Ill. 45); acaulis, foliis bipinnati-

sectis, scapis 1-floris.—Ranunculus pimpinelloides, *Don! in Royle, Ill.*
53. C. Cachemiricum, *Camb.! in Jacq. Voy. Bot. p.* 5. *t.* 3.

HAB. In montibus Himalayæ interioris, alt. 9–13,000 ped.: Kashmir, *Falconer, Jacquemont! Winterbottom!* Kanawer, *Royle!* Kumaon, *Str. et Wint.!* Sikkim?—(Fl. vere.) (*v. v.*)

Herba pusilla, 2–4-uncialis, glabra; *radice* fibrosa, collo squamis involuto. *Folia* sæpius longe petiolata, bipinnatisecta, segmentis rotundatis bis ternatim sectis. *Scapi* folia æquantes, 1-flori. *Flores* diametro unciales. *Sepala* herbacea, late elliptica. *Petala* 8–12, sepalis subtriplo longiora, 6–8 lineas longa, oblongo-cuneata, retusa, fovea parva. *Achenia* ovalia, utrinque obtusa, vix compressa, rugosa, stylo brevi apiculata.

Our specimens from Sikkim are unfortunately so imperfect that their identity with the plant of the Western Himalaya is very doubtful. They are in fruit only, and have larger and less divided leaf-segments, but are not otherwise distinguishable.

Tribus III. RANUNCULEÆ.

Sepala æstivatione imbricata. *Petala* plana. *Carpella* monosperma, semine erecto.—Herbæ *foliis alternis.*

7. OXYGRAPHIS, Bunge.

Sepala 5, persistentia. *Petala* 10–15, fovea nectarifera impressa. *Achenia* in capitulum globosum collecta, membranacea, stylo subulato rostrata. *Semen* erectum.—Herbæ *alpestres acaules*, radice *perennante*, foliis *integris*, floribus *aureis*.

This genus is remarkable in the Order for its persistent sepals, which afford a curious analogy with *Nuphar*. The only known species are those described below.

1. **O. glacialis** (Bunge, Enum. Pl. Alt. 35); foliis ovalibus integerrimis crenatisve obtusis.—*Ledeb. Fl. Ross.* i. 47. Ficaria glacialis, *Fisch. in DC. Prod.* i. 305. Ranunculus Kamtschaticus, *DC. Prod.* i. 43, *fide Ledeb.*

HAB. In Himalayæ interioris summis alpibus: Kumaon, *Str. et Wint.!* Sikkim, alt. 16–18,000 ped.!—(Fl. Jul. Aug.) (*v. v.*)

DISTRIB. Sibiria altaica! Davuria! Kamtschatka?

Herba acaulis, 1–4-uncialis, glabra. *Radix* fibrosa. *Folia* crassiuscula, ½–1½ uncias longa, ⅓–1 unc. lata, petiolo subæquilongo, basi vaginante. *Scapus* solitarius, erectus, 1-florus. *Sepala* late elliptica, obtusa, post anthesin aucta. *Petala* 12–15, anguste oblonga, 4 lineas longa, sepalis duplo longiora, infra foveam callo transversali instructa. *Achenia* numerosa, stylo subulato recto terminata, in capitulum globosum collecta.

2. **O. polypetala** (H.f. et T.); foliis rotundato-subreniformibus crenato-lobatis.—Ranunculus polypetalus, *Royle! Ill.* 54. *t.* 11. *f.* 2. Callianthemum Endlicheri, *Walp. Rep.*

HAB. In Himalaya occidentali interiori, alt. 12–15,000 ped.: Zanskar! Sirmur, *Royle!* Kanawer, *Munro!* Kumaon, *Str. et Wint.!*—(Fl. vere.) (*v. v.*)

Herba pusilla, cæspitosa, acaulis, radicibus fibrosis. *Petioli* 1–3-unciales. *Folia* diametro ⅓–1-uncialia, basi cordata, profunde crenato-lobata, vel triloba, lobis crenatis. *Scapi* 2–4-unciales, debiles, 1-flori. *Flores O. glacialis* sed paullo majores, diametro unciales. *Petala* oblonga, spathulata, fovea nectarifera parva ecallosa. *Achenia* prioris.

8. CERATOCEPHALUS, *Mœnch.*

Sepala 5, decidua. *Petala* 5, fovea nectarifera impressa. *Achenia* supra receptaculum spicata, basi utrinque gibba, apice longe rostrata. *Semen* erectum.—Herbæ *annuæ acaules*, floribus *flavis*.

A genus consisting of one very variable species, which is a native of the Mediterranean region of Europe and the corresponding climates of Asia. As a genus it is not sufficiently distinct from *Ranunculus*, with which it is connected by means of *R. oxyspermus*, Willd., and *R. orientalis*, L., which have long-beaked fruit. When the family is again monographized it will probably be reduced, but the sections of *Ranunculus* will at the same time require a thorough revision.

1. **C. falcatus** (Pers. Syn. 341).—*DC. Prod.* i. 26; *Ledeb. Fl. Ross.* i. 26.   C. Orthoceras, *DC. Prod.* i. 26; *Deless. Ic. Select.* i. *t.* 23; *Led. Fl. Ross.* i. 26.   Ranunculus falcatus, *L. Sp.* 781.; *Schlecht. Anim. Ran.* 6.

HAB. In graminosis siccis montium Indiæ boreali-occidentalis : Beluchistan ! Afghanistan ! Kashmir ! Kishtwar !—(Fl. vere.) (*v. v.*)

DISTRIB. Europa austr.! Asia temperata !

Herba pusilla, tenuiter tomentosa vel rarius glabra. *Petioli* sursum dilatati. *Folia* ternatisecta, segmentis linearibus sæpe bifidis, interdum pinnatisecta. *Scapi* plures, 1-flori, 1–2-unciales. *Flores* 2–3 lineas lati. *Sepala* 5, oblonga, plurinervia. *Petala* æquilonga, obovata, trinervia, flava, fovea nectarifera minuta. *Achenia* in spicam oblongam fere uncialem disposita, rostro recto vel falcato.

We have examined a great number of authentic specimens of the two species usually distinguished, from all the countries in which they occur, and find the shape and size of the beak of the fruit very variable, as is also the amount of development of the crest on its dorsum. We have, therefore, no hesitation in adopting Schlechtendal's opinion, and uniting the two supposed species. All the forms occur in Indian specimens, and it is not uncommon to find on the same individual both glabrous and hairy spikes.

9. RANUNCULUS, L.

Sepala 3–5, decidua. *Petala* 5–15, basi fovea nectarifera impressa. *Achenia* in spicam vel capitulum collecta, stylo brevi apiculata. *Semen* erectum.—Herbæ *annuæ vel perennes, sæpius caulescentes*, floribus *albis vel flavis*.

This very large genus has representatives in all parts of the globe. The tropical species are very few, and chiefly marsh-plants; but in all parts of the temperate zone, and at considerable elevations in the torrid zone, its species are numerous, some growing in water or in marshy places, others in pastures or in woods, while many of the smaller kinds are found to extend into the arctic zone, or to rise on the mountains to the uppermost limits of vegetation. Being in general widely diffused, and capable of existing under very different circumstances, the species are extremely variable, and in consequence very difficult of determination and definition; the shape of the leaves in particular varies much. In the great majority the leaves are palmately divided into lobes, and the amount of division seems (as was long ago pointed out by Seringe) to vary indefinitely. To such an extent, indeed, does the variation extend, that occasionally species very dissimilar in fruit are in a flowering state absolutely undistinguishable. This tendency to mutability of form is unfortunately not confined to the leaves, but extends to the size and degree of branching of the stem, to the size of the flowers, to the shape of the head of fruit and of the individual carpels, and to the amount of pubescence; and in consequence the genus is in a state of complete chaos, the descriptions given in books being quite insufficient for the determination of the species. Very frequently the diagnoses of the same plant given by different authors are quite irreconcilable, and the most different species are occasionally found in herbaria under the same name. A careful examination of extensive suites of specimens from all parts of the world has convinced us that no single character, except the colour of the flowers, is to be relied upon absolutely. The shape of the leaves is the least constant of all, and in four-fifths of the genus is undefinable in words; and even the shape of the style or beak of the achenia, which seems to be mainly relied on as a character, will, unless used with great caution, lead to very erroneous conclusions, as straight and curved styles may be seen on the same specimen, frequently even in the same capitulum. Nothing is more common than to find in botanical works that a newly-described species is "facile distinctus" by a certain character, which, if an extensive series of specimens be examined, will be found to be no character at all. At the same time we seek in vain in such works for any recognition of the great amount of variation to which the different organs are subject, though the fact must be familiar to all careful observers of nature. And yet with this mass of ill-assorted descriptions in books, new species are almost daily being added to the list, not a few being described without a knowledge of the ripe fruit. We believe that no greater boon could be conferred upon science than a careful series of observations on the amount of variation to which cultivated specimens of any common *Ranunculus* are liable during a series of years.

Sect. 1. BATRACHIUM, DC.—*Carpella* transverse rugosa. *Flores* albi, petalorum ungue flavo.

1. **R. aquatilis** (L. Sp. 781); fluitans, foliis submersis capillaceomultifidis, emersis (dum adsunt) rotundato-reniformibus.—*DC. Prod.* i. 26; *Don, in Royle, Ill.* 54; *Schlecht. Anim. Ran.* 7; *Ledeb. Fl. Ross.* i. 27; *Torrey et Gray, Fl. N. Am.* i. 15. R. divaricatus *et* fluitans, *Ledeb. l. c.* R. peucedanifolius, *All.; Schlecht. Anim. l. c.* R. Pantothrix *et* fluviatilis, *Auct.*

HAB. Beluchistan! Afghanistan! Kashmir! Ladak usque ad 14,500 ped. alt.! Panjab Himalaya, *Jacquemont!* Kumaon, alt. 5–12,000 ped.! in Tibetia Sikkimensi, alt. 17,000 ped.!; in India calida rarissima: ad Saharunpur in planitie Gangetica superiore, *Royle!*—(Fl. per totam æstatem.) (*v. v.*)

DISTRIB. Europa! usque ad Islandiam! Asia temperata usque ad Chinam! Tasmania! Abyssinia! Algeria! Teneriffa! America borealis temperata usque ad mare arcticum!

Herba aquatilis, in lacubus et aquis lente fluentibus fluitans, radicibus fibrosis. *Caules* sæpius elongati, graciles. *Folia submersa* petiolata, rariusve sessilia, 1–3 pollices longa, circumscriptione rotundata, dissecta, segmentis capillaceis; *emersa* (in speciminibus Indicis adhuc non observata) rotundato-reniformia, inciso-crenata, triloba vel tripartita. *Pedunculi* oppositifolii, 1-flori. *Flores* magnitudine valde varii, diametro $\frac{1}{4}$–$1\frac{1}{2}$-pollicares. *Achenia* in capitulum globosum collecta, ovali-oblonga, vix compressa.

The Indian forms, so far as hitherto observed, belong to the state called *Panto-thrix*, in which the leaves are all submerged and divided into capillary segments, but not so much elongated as in *R. peucedanifolius*, All., which Schlechtendal considers the only distinct species. This plant is not very common in India, lakes and tranquilly flowing streams being of rare occurrence in the exterior Himalaya, though frequent in the inner parts of the chain and in Tibet, where, accordingly, our plant is very generally diffused. In the plains it is confined to the extreme north, where it will probably be found skirting the base of the Himalaya in all parts of the Panjab. It is usual to divide this species into several, characterized by the absence or presence of the reniform leaves, and by variations in the size and shape of the multifid ones, as well as by the hairiness or smoothness of the plant and carpels; but we quite agree with Seringe, that all these forms are states of one very variable species, to which we are quite willing, with that very accurate observer, to unite *R. hederaceus* and *R. tripartitus.*

Sect. 2. HECATONIA, DC.—*Carpella* lævia vel minute punctulata. *Flores* (in Indicis) flavi.

§ 1. *Folia (radicalia saltem) indivisa (in* C. pulchello *interdum tri-loba).*

2. **R. Lingua** (L. Sp. 773); foliis basi semiamplexicaulibus lanceolatis, floribus magnis 5-petalis.—*DC. Syst.* i. 246, *Prod.* i. 32; *Hook. Fl. Lond. t.* 171; *Ledeb. Fl. Ross.* i. 31; *Torrey et Gray, Fl. N. Am.* i. 16.

HAB. In aquosis Kashmir, *Jacquemont!*—(*v. s.*)
DISTRIB. Europa; Asia temp.! America temp.!

Herba erecta, 2–4-pedalis, perennis, glabra vel adpresse pubescens. *Folia* lineari-lanceolata, 4–8 poll. longa, nervosa, integra vel remote denticulata, inferiora lamina abortiva ad vaginas amplexicaules reducta. *Flores* diam. bipollicares. *Sepala* orbicularia, margine membranacea, puberula. *Achenia* subcompressa, glabra, rostro rectiusculo.

3. **R. reniformis** (Wall. Cat. 4709!); caule erecto, foliis late ovatis orbicularibusve basi cordatis vel truncatis grosse dentatis, petalis 12–15 obovatis.—*W. et A. Prod.* i. 3; *Wight! Ill.* i. 5. *t.* 2, *Ic. t.* 75.

HAB. In montibus Peninsulæ australis altioribus!—(*v. s.*)

Herba erecta, spithamæa vel 1–2-pedalis, ramosa, pluriflora, pilis laxis parce setosa. *Rhizoma* horizontale. *Folia radicalia* crassa, sparse setosa vel glabra, forma varia, obtusa, diametro 1–3-pollicaria. *Folia caulina* pauca, infimum lanceolatum serratum in petiolum attenuatum, superiora linearia minuta. *Flores* diametro fere unciales. *Achenia* in capitulum globosum collecta, ovali-oblonga, tumida, stylo recto abrupte apiculata.

4. **R. sagittifolius** (Hook. Ic. Plant. t. 173!); caule erecto, foliis oblongis cordato-sagittatis crenatis, petalis 5 fere orbicularibus.—R. hastatus, *Walker, ex Wight, Ill.* i. 5.

HAB. In Zeylaniæ montibus, alt. 6–8000 ped., *Walker! Gardner!*

Herba e rhizomate horizontali erecta, 1–2-pedalis. *Caulis* glaber, superne paniculatus. *Petioli* laxe pilosi. *Folia radicalia* 1–3 uncias longa, ½–1½ lata, obtusa (rariusve acuta), sagittata, auriculis rotundatis; *caulina* oblonga lanceolatave, inciso-serrata vel pinnatifida; *suprema* lineari-oblonga. *Flores* diam. pollicares. *Achenia* prioris.

Chiefly distinguished from the last by the number of petals, for the leaves are pro-

bably very variable in both. *R. Javanicus,* Bl. (Hook. in Lond. Journ. Bot. vii.
t. 17), is closely allied to both, but is more glabrous than either, with procumbent or
sarmentose stems, and leaf-opposed, one-flowered pedicels. The achenia are the same
in all three.

5. **R. pulchellus** (C. A. Meyer, in Led. Fl. Alt. ii. 333); caule
erecto, foliis radicalibus ovali-oblongis indivisis vel ad medium usque
trilobis.—*Ledeb. Ic. Alt. t.* 111! *Fl. Ross.* i. 33. R. longicaulis, *C. A.
Meyer, l. c.; Ledeb. Ic. Alt. t.* 117. R. salsuginosus, *Wall. Cat.* 4708!
non Pall. R. flammula, *Don! in Royle Ill.* 53. R. membranaceus,
Royle! Ill. 53. R. nephelogenes, *Edgew.! in Linn. Tr.* xx. 28. Ranunculus, *No.* 18, 19, *Str. et Wint. Herb.!*

HAB. Afghanistan, *Grif.!* in Tibet occ. alt. 10–18,000 ped., ubique
vulgatissimus!; Kanawer! et in Sikkim int. alt. 14–18,000 ped.!—(*v. v.*)

DISTRIB. Sibiria altaica! et Baikalensis! Mongolia Chinensis!

Herba erecta, simplex vel sæpius parce ramosa, 1–12-uncialis, *ramis* elongatis
subaphyllis apice 1-floris. *Folia radicalia* lanceolata, oblonga vel late ovalia, obtusa
vel acuta, nervosa, indivisa vel grosse dentata vel ad medium usque 3–7-loba, lobis
oblongis basi non angustatis; *caulina inferiora* petiolata, lanceolata, indivisa vel trifida, *suprema* sessilia, linearia vel trisecta. *Pedunculi* elongati, sulcati, pubescentes.
Flores solitarii, diametro ½-pollicares. *Sepala* patentia, elliptica, membranacea, dorso
pilosa puberulave, apice sæpe nigricantia. *Petala* late obovata, sepalis dimidio longiora. *Achenia* in capitulum ovale vel oblongum collecta, numerosa, parva, glabra,
ovalia, vix compressa, stylo subrecto compresso fere æquilongo apiculata.

Planta polymorpha: variat—1. simplex vel ramosa; 2. foliis omnibus indivisis,
vel radicalibus indivisis caulinis trifidis vel trisectis, vel foliis omnibus trifidis; 3.
glabra vel pubescens, interdum adpresse sericea. Hæ formæ ut varietates non dis-
tinguendæ, quum formæ innumeræ intermediæ occurrunt. In forma sericea (*R.
membranaceus,* Royle) folia radicalia interdum glaberrima sunt, omnino ut in planta
typica.

6. **R. lobatus** (Jacquem. mss. Camb.! in Jacq. Voy. Bot. p. 5.
t. 1 B); caule diffuso non stolonifero, foliis radicalibus rotundatis crenato-lobatis.—R. salsuginosus, *Don! in Royle Ill.* 53.

HAB. In Himalayæ interioris alpibus, alt. 12–16,000 ped.: Zanskar!
Ladak! Piti! Kanawer! Hundes! Kumaon! Sikkim!—(*v. v.*)

Herba 2–5-pollicaris, glabra vel puberula, 1-pauciflora. *Folia radicalia* ½–1-pol-
licaria, rotundata vel reniformia, coriacea, basi cordata vel cuneata, apice obtusa, crenato-dentata; *caulina* tridentata, sæpe fasciculata. *Flores* ⅔–¾ poll. diam. *Sepala*
ovalia. *Petala* duplo longiora, late obovata, emarginata vel rotundata. *Achenia* in
capitulum ovatum collecta, obovata vel subglobosa, vix compressa, stylo longo recto
apiculata.

Intermediate between *R. Cymbalariæ* and *R. pulchellus,* but differing from both
in habit and in its large flowers.

7. **R. Cymbalariæ** (Pursh, Fl. Bor. Am. ii. 392); stoloniferus,
foliis rotundatis vel oblongis varie lobatis, scapis 1-paucifloris.—*DC.
Prod.* i. 33; *Schlecht. Anim.* 22; *Ledeb. Fl. Ross.* i. 34; *Hook. Fl.
Bor. Am.* i. 11; *Torrey et Gray, Fl. Bor. Am.* i. 17. R. salsuginosus,
Pall.; DC. Prod. i. 33; *Schlecht. Anim.* 22. R. plantaginifolius,*Murr.;
Ledeb. Fl. Ross.* i. 33. R. halophilus, *Schlecht. Anim.* 23. *t.* iv. *f.* 1. R.
tridentatus, *H. B. K.*

a. major, foliis orbicularibus inciso-crenatis, acheniis longioribus in capitulum oblongum collectis.

β. *alpinus* (Hook.); minor, foliis ellipticis vel oblongis apice tridentatis, acheniis latioribus brevioribus in capitulum globosum collectis.
HAB. In Tibet occ. ubique, inter Iskardo, alt. 7000 ped.! et Hundes! (*a.* rarior, in paludosis 10–12,000 ped.; β. vulgatissimus, usque ad 17,000 ped. alt. adscendens) et in Sikkim interioris alpibus, alt. 11,500–14,000 ped.!—(*v. v.*)
DISTRIB. Sibiria! Persia! America bor. in planitie a Novo Eboraco in montibus a Mexico usque ad mare arcticum! in Americæ austr. temperatæ et tropicæ alpibus!

Herba parva (sæpe pusilla), stolonifera, ad nodos radicans et foliosa. *Folia* forma valde varia, orbicularia elliptica vel oblonga, regulariter inciso-crenata, vel apice tantum triloba, basi rotundata vel cordata, ¼–1 poll. longa. *Scapi* folia æquantes vel longiores, 1–6- (vel rarissime 12-) unciales, 1–pauciflori, aphylli, vel ad ramificationes bracteas lineares incisasve gerentes. *Flores* ¼–½-unciales, flavidi. *Sepala* 5, ovalia, patentia vel reflexa, membranacea, subcolorata. *Petala* 5–8, anguste obovata. *Achenia* numerosa, obovata, compressa, dorso gibbosa, stylo brevi uncinato vel recto apiculata, utrinque longitudinaliter tricostata.

This, though a very variable plant in form of leaf and in size, is well characterized by the longitudinally ribbed fruit. It is extremely widely diffused, and as all the forms have a wide extension, there can, we think, be no doubt, notwithstanding slight differences in the shape of the leaves and fruit, that only one species exists.

§ 2. *Folia omnia secta, caulis radicans.* (R. diffusus, *DC.*, in quo caulis ad nodos radicans cum affinibus, in § 4 quærendus.)

8. **R. radicans** (C. A. Meyer in Ledeb. Fl. Alt. ii. 316); prostratus, radicans, foliis reniformibus 3–5-lobis, floribus oppositifoliis, acheniis numerosis subglobosis.—*Ledeb. Ic. Alt. t.* 116, *Fl. Ross.* i. 34. R. natans, *C. A. Mey. in Led. Ic. Alt. t.* 114, *et Fl. Ross.* i. 34.
HAB. In Tibet occidentali alpino: Ladak 14–16,000 ped., *H. Strachey;* Hundes, *Str. et Wint.!*—(*v. v.*)
DISTRIB. Sibiria, *Ledeb.!*

Herba prostrata, glabra, in paludosis radicans, et radices plurimas albas fibrillosas emittens, interdum fluitans. *Folia* ad nodos in ramulo abbreviato axillari plura, reniformia, pollicaria, 3–5-loba vel 3–5-fida, lobis rotundatis vel crenatis. *Flores* parvi, diametro ⅓-unciales, oppositifolii vel subterminales, longe pedicellati. *Sepala* reflexa. *Petala* late obovata, fere rotundata, sepalis vix longiora. *Achenia* plurima, parva, in capitulum fere ⅓-unciale globosum collecta, vix compressa, stylo brevissimo mucronata.

The two species distinguished by Meyer and Ledebour differ in nothing but the degree of division of the leaves.

9. **R. hyperboreus** (Rottb. Act. Hafn. x. 458. t. 4. f. 16); pusillus, stoloniferus vel erectus, foliis 3–5-fidis, floribus solitariis, acheniis numerosis parvis subglobosis.—*DC. Prod.* i. 35; *Ledeb. Fl. Ross.* i. 35; *Torr. et Gray, Fl. N. Am.* i. 20. R. pygmæus, *Wahl. Fl. Lapp.* 157. *t.* 8. *f.* 1; *DC. Prod.* i. 35; *Ledeb. Fl. Ross.* i. 36; *Torrey et Gray, l. c.*
HAB. In summis alpibus: Ladak, *H. Strachey!* Kanawer, *Jacquemont!* et Sikkim, alt. 15–17,000 ped.!—(*v. v.*)

DISTRIB. Europa! Asia! et America arctica!

Herba pusilla, erecta vel longe stolonifera et ad nodos radicans. *Folia radicalia* circumscriptione orbicularia, diam. 2–5 lin., 3–5-fida vel -partita, lobis oblongis vel cuneatis sæpius integris. *Caules* 1–2-pollicares, folia 1–2 sessilia triloba vel tripartita gerentes. *Flores* solitarii, 2–3 lineas diam. *Sepala* reflexa. *Petala* sepalis paullo longiora, obovata. *Achenia* in capitulum parvum globosum collecta, late ovalia vel fere globosa, vix compressa, stylo brevi recto vel reflexo apiculata.

Ledebour admits two varieties of *R. hyperboreus*, differing in the straight or hooked style; and *R. pygmæus* only differs from the latter in the want of stolones. In Sikkim both the erect and the stoloniferous states occur, and specimens from that province are identical in every respect with those of northern Europe. The heads of fruit of this species are a good deal like those of *R. radicans*, but smaller.

§ 3. *Folia secta, caulis erectus; achenia ovalia, subcompressa, lateribus convexis.* (R. pulchellus, *C. A. Meyer*, foliis trifidis vel integris, in § 1 quærendus.)

10. **R. Chærophyllos** (L. Sp. 780); foliis trisectis, lobis plerumque linearibus, scapo uni- vel paucifloro, acheniis in spicam oblongam dispositis.—*DC. Prod.* i. 27.

β; foliis primordialibus integris late ovalibus grosse dentatis.

HAB. β. In Himalayæ maxime occidentalis montibus: Balti, alt. 12,000 ped., *Winterbottom!*

DISTRIB. Europa australis! Asia Minor! Persia!

Herba erecta, 6–12-pollicaris, subvillosa, radice bulbosa fibros crassos emittente. *Folia radicalia* 1–3-uncialia, trisecta, segmentis tripartitis et varie incisis, lobis linearibus; *caulina* pauca, tripartita vel linearia. *Flores* flavi, pollicares. *Sepala* oblonga, patentia. *Petala* duplo longiora, late obovata. *Achenia* numerosa, ovalia, compressa, in stylum longum rectum sensim attenuata.

11. **R. cæspitosus** (Wall. Cat. 4701!); foliis radicalibus reniformibus pedatim multipartitis, sepalis patentibus, acheniis in capitulum oblongum dispositis subglobosis.—R. pedatifidus, *Ledeb. Fl. Ross.* i. 732; *non Smith in Rees' Cycl.*

HAB. In Tibetia occidentali, et in alpibus Himalayæ interioris, alt. 11–16,000 ped.: Nubra! Ladak! Zanskar! Kanawer! Kumaon! Nipal! Sikkim!—(Fl. Jun.–Aug.) (*v. v.*)

DISTRIB. Asia et America temperata et arctica!

Herba erecta vel diffusa, pilosa, caule ramoso 3–18-pollicari. *Folia radicalia* rotundata vel reniformia, diam. ½–1 poll. pedatim 7–multifida vel -partita, segmentis rotundatis oblongis vel linearibus; *caulina* subsessilia, inferiora pedatim multipartita, superiora 5–3-partita, segmentis omnium linearibus. *Caules* sæpe plures, ramosi, graciles, ramis elongatis folia 1–2 parva gerentibus, apice 1-floris. *Flores* ⅓–⅔-pollicares. *Sepala* elliptica, sericeo-pilosa. *Petala* oblonga vel obovata. *Achenia* numerosa, parva, ovali-subglobosa, vix compressa, stylo recto apiculata, glabra.

This elegant species agrees so well with the description of *R. amœnus*, Ledeb., which is universally considered to be the Siberian form of *R. affinis*, R. Br., that it is difficult to consider it different, especially as there are specimens referred by botanists to that species, not only of Siberian origin, but also from arctic America, which are undistinguishable from the Indian plant described above. *R. Dahuricus*, Turcz. mss. (which is quoted by Ledebour as a synonym of his *R. pedatifidus*), is certainly the same as the Indian plant, if the specimen in the Hookerian Herbarium may be relied upon as authentic; and it differs from the usual Siberian states of *R. affinis* chiefly in the very small glabrous achenia, which are exactly the same as those of *R. pulchellus*, to which plant the present bears a striking resemblance in general habit, notwithstanding the great difference in the shape of the leaves. We retain it as distinct from *R. affinis*, not only on account of a certain difference of habit, but because, if united to it, it will be necessary to reduce the next species also. We have ascertained, by an inspection of the original specimens, that *R. pedatifidus* of Smith is the same as *R. amœnus*, Ledeb.

12. **R. hirtellus** (Royle! Ill. 53); foliis radicalibus reniformibus 3-partitis vel segmentis lateralibus ad basin fere fissis pedatim 5-partitis, sepalis adpressis, acheniis in capitulum ovatum vel oblongum collectis obovatis compressis utrinque convexis.—R. lætus, *Wall. Cat.* 4702 *C. ex parte.* R. attenuatus, R. nervosus, *et* R. Choorensis, *Royle! Ill.* 53.

HAB. In sylvis umbrosis Himalayæ occidentalis temperatæ, alt. 7–12,000 ped.: a Kashmir ad Kumaon: et in pratis subalpinis et alpinis usque ad 14,000 ped. Forma parviflora latifolia in Himalaya exteriori vulgaris, angustifolia in sylvis interioribus, humilis foliis multifidis in alpibus.—(Fl. per totam æstatem.) (*v. v.*)

Herba perennis, pluricaulis, adpresse pubescens vel glabrescens, radice fibrillosa. *Caules* e collo plures, ½–1½-pedales, sæpius elongati, basi nudi, apice ramosi. *Folia radicalia* sæpe longe petiolata, circumscriptione rotundata vel reniformia, tenuia, nervosa, pilis sericeis adpressis utrinque vestita, rarius glabrescentia, diametro 1–3-pollicaria, ad basin fere tripartita, segmentis lateralibus bilobis oblique ovalibus, terminali late cuneato-trilobo, omnibus argute inciso-dentatis; seu pedatim 5-partita, lobis oblongis indivisis, apice tridentatis vel acute palmatim inciso-multifidis. *Folia caulina* palmatim 3–5-partita, segmentis cuneatis trifidis vel tridentatis 1–1½-pollicaribus. *Flores* ⅓–⅔-pollicares, in spec. alpinis majusculi. *Sepala* ovalia, adpressa, pilosa. *Petala* obovata, sepalis duplo longiora. *Achenia* in capitulum ovale collecta, obovata, subcompressa, immarginata, rugosula, dense pilosa vel glabra.

Variat—1. grandiflorus et parviflorus; 2. foliorum lobis integris, inciso-dentatis vel palmatim partitis; 3. acheniis tomentosis vel glaberrimis; 4. caulibus elongatis erectis multifloris, vel abbreviatis diffusis 1-pauciflores.

After a careful comparison of very extensive suites of specimens of the numerous forms which we have here united under one name, with previously-described European species, we have been unable to identify our plant with any, though certain states of this very variable plant certainly approach very near to some forms of both *R. auricomus* and *R. affinis*. That the hairy and glabrous fruited states of the Indian plant belong to one species we do not in the least doubt, these variations bearing no definite relation to the differences in the shape of the leaves. We believe therefore that all the forms which occur in the wooded region of the Himalaya are referable to one very variable species. With regard to the alpine forms we are less certain, as our specimens, though numerous, are destitute of good fruit, without which it is impossible satisfactorily to determine the affinities in this very difficult genus.

13. **R. auricomus** (L. Sp. 775); foliis radicalibus rotundato-reniformibus vel tripartitis crenatis, caulinis digitatim partitis, segmentis linearibus integris vel inciso-serratis, acheniis velutinis in capitulum globosum collectis orbicularibus subcompressis anguste marginatis stylo uncinato apiculatis.—*DC. Syst.* i. 266, *Prod.* i. 34; *Ledebour, Fl. Ross.* i. 38; *Torrey et Gray, Fl. N. Am.* i. 17. R. cassubicus, *L.; DC. Syst. et Prod. l. c.; Ledeb. l. c.*

HAB. In montibus Afghanistan, *Griffith!* (ex spec. imperfect.)

DISTRIB. Europa omnis! Asia temperata! Grœnlandia.

Herba erecta, ramosa, ½–1-pedalis, glabra vel puberula, multiflora. *Folia radicalia* plerumque indivisa inciso-crenata, rarius tripartita, interdum abortiva.

We have introduced this species into our list on the authority of some very imperfect specimens collected by Mr. Griffith in Afghanistan. We have done so mainly for the purpose of calling the attention of travellers in the Himalaya to this species, in order that they may institute a search for it in the woods of the temperate region of these mountains in early spring. It is our impression, from a survey of numerous specimens of this and closely allied species, that *R. pedatifidus* or *affinis*, and all its forms, as well as *R. abortivus*, of America, are only varieties of *R. auricomus*, and that *R. polyrhizus*, Stev., is the alpine state of the same plant. *R. montanus*, L., seems a very imperfectly-defined plant, consisting of the dwarf states of *R. acris* and *bulbosus*, and occasionally also of *R. auricomus*. The main distinctions between *R. auricomus* and *R. hirtellus* are the undivided radical leaves, the large size of the achenia of the former, and their forming a globose capitulum; and an examination of the degree of variation of the Indian species in these points would probably throw much light on a very difficult question.

14. **R. nivalis** (L. Sp. 778); caulibus 1-floris, foliis radicalibus reniformibus 5–7-partitis, caulinis sessilibus 3–5-partitis, sepalis ellipticis dorso dense fusco-villosis.—*DC. Syst.* i. 273, *Prod.* i. 35; *R. Br. in Parry's 1st Voy. App.* 264; *Ledeb. Fl. Ross.* i. 36; *Torrey et Gray, Fl. N. Am.* i. 20.

HAB. In Himalaya alpina: Sikkim, alt. 15,000 ped.! (Tankra Pass). —(*v. v.*)

DISTRIB. Europa, Asia, et America arctica! in America in montibus scopulosis ad lat. 52° descendit.

Radix crassa, perpendicularis. *Planta* pusilla, 1–2-pollicaris. *Folia radicalia* pedatim 5–7-partita, semipollicaria, segmentis obovatis vel oblongis, lateralibus trilobis. *Caulis* superne villosus, plurifoliatus. *Folia caulina* basi late membranaceo-dilatata, conformia, supremum sessile 3–5-partitum. *Sepala* elliptica. *Petala* obovata, sepalis vix longiora. *Achenia* non suppetunt.

This little plant, which is unfortunately not in a sufficiently advanced state for accurate determination, may be referred provisionally to *R. nivalis*, L., to which *R. Altaicus* of Laxmann, and *R. Eschscholtzii* of Schlechtendal, should apparently be united. The above description refers to the Indian plant only.

15. **R. sceleratus** (L. Sp. 776); glaber, foliis radicalibus reniformibus tripartitis, sepalis reflexis, acheniis in capitulum oblongum congestis obovatis non compressis.—*DC. Prod.* i. 34; *Don, Prod.* 195; *Royle! Ill.* 53; *Ledeb. Fl. Ross.* i. 45; *Torrey et Gray, Fl. N. Am.* i. 19. R. Indicus! *Roxb. Fl. Ind.* ii. 671; *Wall. Cat.* 4699! R. carnosus, *Wall. in Hb.* 1824. Hecatonia palustris, *Lour. Fl. Coch. Chin.* 371.

HAB. In Indiæ borealis planitie ubique in arenosis prope aquam, secus Indum, Gangem, et Brahmaputra flumina, et in Himalaya occidentali subtropica, sed vix supra 5000 ped. alt., a Kashmir! ad Kumaon!; et in Malwa! ad ripas fluminis Nerbada, *Rottler in Hb. Royle!* (in Peninsula deest.)—(Fl. Febr. Mart.) (*v. v.*)

DISTRIB. Europa tota, Asia temperata, China (*Loureiro*), Africa borealis, America temp. usque ad lat. 67°.

Herba annua, erecta, 1–3-pedalis, glabra vel apice summo interdum subpuberula, caule carnoso sæpius ramosissimo. *Folia radicalia* reniformia, 1–3 uncias lata, ad basin fere tripartita, segmentis obovato-cuneatis trifidis et obtuse incisis; *caulina* (exceptis infimis) sessilia, tripartita, segmentis anguste oblongis 1–2 uncias longis inciso-pinnatifidis vel tridentatis. *Flores* diam. $\frac{1}{2}$–$\frac{1}{3}$-poll. *Sepala* reflexa, petala oblonga æquantia. *Achenia* in capitulum 3–4 lineas longum collecta, numerosa, minuta, oblique obovata, obtusa vel stylo brevissimo apiculata. *Receptaculum* oblongum, pilosum.

§ 4. *Folia secta. Caulis erectus, rarius prostratus. Achenia plano-compressa.*

16. **R. diffusus** (DC. Prod. i. 38); prostratus vel diffusus, pilis patentibus hirsutus, foliis tripartitis, pedunculis unifloris oppositifoliis, floribus parvis, sepalis patentibus, acheniis in capitulum globosum collectis punctatis, receptaculo parvo piloso.—*Don, Prod.* 195. R. Napaulensis, *DC. Prod.* i. 39. R. trilobatus, *Don, Prod.* 194. R. geranioides, *Blume, Bijd.* 2. R. hydrocotyloides, *Wall. Cat.* 4703! R. mollis, *Wall. Cat.* 4704!; *Don! in Royle Ill.* 53. R. obtectus, *Wall. Cat.* 4705!

HAB. In Himalaya temperata, alt. 6–8000 ped.: Simla! Garhwal! Kumaon! Nipal! Sikkim, 9–11,000 ped.!—(Fl. per totam æstatem.) (*v. v.*)

Caules pilis patentibus albis vel rufis longe pilosi, interdum basi glabrescentes, prostrati, ad nodos interdum radicantes, vel apice adscendentes. *Folia* 1–3-pollicaria, utrinque molliter pilosa, radicalia primordialia in adulto sæpe evanida, trisecta, foliolis trilobis, cætera et caulina profunde triloba, lobis inciso-dentatis. *Pedunculi* oppositifolii et terminales, 1-flori, foliis æquilongi vel duplo longiores. *Flores* parvi, diam. $\frac{1}{2}$–$\frac{2}{3}$ poll. *Sepala* oblonga, pilosa. *Petala* obovata, sepalis subduplo longiora. *Achenia* ovata, margine acuta, lateribus utrinque intra marginem 1-costatis, in stylum sensim attenuata.

The position of the costa or elevated ridge of the disk of the carpels, in this and allied species, varies very much, being sometimes close to the margin, at other times distant from it.

17. **R. subpinnatus** (W. et A. Prod. i. 4); diffusus, pilis patentibus hirsutus, foliis trisectis, segmentis petiolatis, pedunculis unifloris oppositifoliis, floribus magnis, sepalis patentibus, acheniis in capitulum globosum collectis punctatis, receptaculo parvo piloso.—*Wight! Ic. t.* 49.

HAB. In montibus Nilghiri, *Wight!*—(*v. s.*)

Rhizoma horizontale. *Caules* elongati, prostrati vel rarius adscendentes, patentim pilosi. · *Folia* ternatim rarius quinatim pinnatisecta; *foliola* molliter pilosa, longe petiolata, late cordata, tripartita, segmentis profunde incisis, floralia subsessilia tripartita. *Flores* oppositifolii et terminales, ultrapollicares, longe pedunculati. *Sepala* villosa. *Petala* late obovata. *Achenia* in capitulum 3 lineas longum collecta, marginata, 2 lin. longa, carinata, stylo recto vel recurvo compresso apiculata.

Very like the last, but the leaves are much more divided, and the flowers a good deal larger. When carefully examined in their native country, connecting links will probably be found. Both require comparison with *R. repens*, L., which is a widely diffused plant, and varies very much.

18. **R. lætus** (Wall. Cat. 4702! excl. lit. C. partim); erectus, adpresse pilosus, foliis tripartitis, panicula multiflora, sepalis patentibus, acheniis in capitulum globosum congestis epunctatis, receptaculo ob-

longo glabro.—*Royle! Ill.* 53. R. distans, *Royle! ib.* R. brevirostris,
Edgew. in Linn. Tr. xx. 28 ? R. riparius, *Edgew. ib.?*
    Hab. In Himalaya temperata vulgaris, alt. 3–10,000 ped.! in Sikkim
in jugis interioribus !—(Fl. per totam æstatem.) (*v. v.*)

   Rhizoma horizontale, vel radix descendens ·fusiformis. *Caulis* erectus, ramosus,
1–2-pedalis, multiflorus, adpresse albo-pilosus, pilis rarius prope basin caulis subpaten-
tibus. *Folia radicalia* tripartita, supra sparse pilosa, infra adpresse sericea, diam.
2–4-pollicaria; segmenta late ovalia, basi cuneata, rarius in petiolum angustata, in
lobos plures argute dentatos grosse incisa; *caulina* sessilia, tripartita, segmentis oblongis grosse incisis. *Panicula* divaricato-ramosa, multiflora. *Flores* diam. pollicares. *Sepala* ovalia, extus villosa. *Petala* fere orbicularia, basi cuneata, sepalis
duplo longiora. *Achenia* ovalia, in capitulum diam. 3 lin. collecta, lævia, marginata,
in stylum brevem acutum basi latum compressum sensim attenuata.
   This species has the habit and general appearance of *R. acris, nemorosus, lanugi-
nosus,* etc., but we have not been able to identify it with any of them, though we
must confess that the characters by which it is distinguished from all of these are of
the smallest possible importance, as being derived from the achenia, which vary to a
very great degree. Many specimens of these European species can be selected from
among the great numbers now before us, which, without fruit, are undistinguishable
from the Indian plant; and there is a specimen in the Hookerian Herbarium from
Fries, marked *R. sylvaticus,* which, with widely different foliage, has exactly the
same beak as the ordinary state of the Indian plant. There is no doubt that the
leaves vary extremely in all these species; and if the characters derived from the
achenia be found insufficient, which we believe will be the case, we fear that many of
the supposed species now distinguished by authors, and the present among the number, must be reduced to *R. acris,* L. We have only seen very imperfect specimens
of Mr. Edgeworth's plants, but we believe them to be rather abnormal states than
distinct species. The alpine one closely resembles some of our own specimens, and
R. riparius seems only a mountain plant, casually carried down to the plains.

   19. **R. bulbosus** (L. Sp. 778); caule erecto, basi bulboso adpresse piloso, foliis ternatim pinnatisectis, panicula multiflora, sepalis
reflexis, acheniis in capitulum globosum collectis impunctatis, receptaculo oblongo glabro.—*DC. Prod.* i. 41; *Royle! Ill.* 53; *Ledeb. Fl. Ross.*
i. 44.
   Hab. In Himalaya occ. temp.: Kanawer, *Royle!*—(*v. s.*)
   Distrib. Europa tota et Asia occidentalis ! in Americam temperatam, ex Torrey et Gray, ex Europa introducta.

   Caulis pedalis, ramosus. *Folia* trisecta, segmentis profunde trifidis grosse incisis;
caulina tripartita, segmentis linearibus pinnatifido-lobatis. *Flores* ¾–1-pollicares.
Sepala ovata, pilosa. *Petala* late obovata. *Achenia* in capitulum diam. 3-lineare
congesta, ovalia, marginata, in stylum brevem late triangularem acutum sensim attenuata.
   We do not feel at all certain that this plant has not been introduced through some
mistake among Dr. Royle's Indian plants. It is certainly not common in the Himalaya, as it has not been found by any of the recent travellers in these mountains.

   20. **R. fibrosus** (Wall. Cat. 4706!); caule erecto patentim hispido, foliis ternatim pinnatisectis, segmentis ad basin usque partitis,
panicula multiflora, sepalis reflexis, acheniis in capitulum magnum globosum collectis marginatis punctatis, receptaculo subgloboso sericeo-
piloso.

HAB. In Nipalia, *Wall.!* et in oryzetis et paludosis montium Khasia, alt. 2–6000 ped.!—(Fl. Jul.–Sept.) (*v. v.*)

Caulis sesqui- vel bipedalis, e basi prostrata interdum ad nodos radicans erectus. *Radix* dense fibrillosa. *Folia radicalia* longe petiolata (cum petiolo spithamæa et ultra, majora fere *Heraclei*), adpresse pilosa, ternatisecta; foliolis longe petiolatis, terminali tripartito, lateralibus bipartitis, segmentis omnibus lobatis, lobis ultimis 1–2-pollicaribus oblongis acutis inciso-serratis foliorum superiorum lineari-oblongis. *Flores* paniculati, diam. pollicares. *Petala* obovata, sepalis duplo longiora. *Achenia* diametro ½ poll., ovata, plano-compressa, margine acuta, intra marginem linea elevata marginata, in stylum rectum compressum sensim attenuata.

21. **R. Chinensis** (Bunge, Mem. Sav. étr. St. Petersb. ii. 76); caule erecto hirsuto, foliis trisectis, segmentis bi-trisectis, calyce reflexo, acheniis in capitulum oblongum collectis dorso tricostatis punctatis, receptaculo elongato oblongo piloso.

HAB. In paludosis provinciæ Assam, *Jenkins! Griffith! Masters! Simons!*—(*v. s.*)

DISTRIB. China borealis, *Bunge!*

Radix fibrosa. *Caulis* erectus, ramosus, 2–3-pedalis, cum petiolis pilis rufis adpresse hispidissimus. *Folia radicalia* 3–6-pollicaria (petiolis 6–12-poll.), adpresse pilosa, 2–4-tim ternatisecta, segmentis plerumque petiolatis palmatim incisis, ultimis cuneato-oblongis grosse dentatis vel inciso-serratis. *Flores* terminales, paniculati, ¾-pollicares. *Sepala* hirsuta. *Achenia* in capitulum oblongum ¾ pollicem longum congesta, costis lateralibus valde prominentibus, dorso fere truncata, tricostata.

Sect. 3. ECHINELLA, DC.—*Carpella* echinata vel tuberculata. *Flores* flavi.

22. **R. flaccidus** (H.f. et T.); caule prostrato filiformi, foliis rotundatis crenato-lobatis, acheniis tuberculatis.

HAB. In Himalayæ temperatæ paludosis: Kumaon, alt. 10,000 ped., *Str. et Wint. No. 2!* Sikkim, alt. 9–10,000 ped.! Bhotan, *Griffith!*—(Fl. Jun. Jul.) (*v. v.*)

Herba pusilla. *Caulis* glaber, 3–6-pollicaris, interdum ad nodos radicans. *Folia* longe petiolata (petiolo 1–2-poll.), diam. 2–4-lin., rotundata, basi cordata vel cuneata, 3–5-loba vel crenata, glabra. *Pedunculi* 1-flori, oppositifolii, strigoso-pilosi. *Flores* (²⁄₁₀ poll. diam.) minuti, flavi. *Sepala* ovalia, reflexa. *Petala* vix majora, obovata. *Achenia* 6–12, in capitulum globosum collecta, ovalia, vix compressa, pubescentia, stylo brevi reflexo apiculata.

23. **R. Wallichianus** (W. et A. Prod. i. 4); caule prostrato, foliis ternatim pinnatipartitis, acheniis compressis marginatis punctatis et tuberculatis.—*Wight, Ic. t.* 937! *Nilg. Pl. t.* 5. R. pinnatus, *Wight, Ill.* i. 6.

HAB. In montibus temperatis Zeylaniæ! et Malabariæ austr.!

Herba perennis. *Caulis* prostratus, ad nodos radicans, pilis patentibus hirsutus, rarius glabriusculus. *Folia radicalia* longe petiolata, pilosa, segm. ½–1 poll. longis ovatis trifidis et grosse dentatis; *caulina* tripartita. *Pedunculi* oppositifolii, 1-flori. *Flores* ½-pollicares. *Sepala* ovata, reflexa. *Petala* anguste obovata, sepalis duplo longiora. *Achenia* plano-compressa, orbicularia.

Dr. Wight has identified the Nilghiri plant with that of Ceylon, and applies to both the name of *R. pinnatus*, Poiret, founded on a plant said to occur in Ceylon and

South Africa.   Specimens of a South African species, in Herb. Hook., have pinnated leaves, which the present species has not, and are therefore probably the true *R. pinnatus*, Poiret.

24. **R. muricatus** (L. Sp. 780); foliis rotundatis trifidis, acheniis plano-compressis aculeis rigidis vel tuberculis obtusis asperis.—*DC. Prod.* i. 42; *Ledeb. Fl. Ross.* i. 47; *Torrey et Gray, Fl. N. Am.* i. 24. HAB. In Indiæ borealis planitie et montibus, usque ad alt. 5–6000 ped.: Beluchistan! Afghanistan! Panjab! Peshawer! Kashmir! Kishtwar! DISTRIB. Europa media et australis! Asia Minor! Persia! Africa bor.! ins. Atlant.! America bor.! a Virginia ad Louisianam,.America austr. temp.!—(*v. v.*)

Herba annua, erecta vel diffusa, glabra vel pilis sparsis patentibus hirsuta. *Folia* 1–2-pollicaria, profunde trifida, segmentis grosse inciso-crenatis, superiora basi cuneata triloba. *Pedunculi* angulati, oppositifolii, uniflori vel terminales paniculati. *Flores* ⅓–½-pollicares. *Sepala* ovata, reflexa. *Petala* paullo longiora, obovata. *Achenia* in capitulum magnum globosum collecta, long ⅓ poll., ovalia, marginata, rostro recto compresso utrinque costato apice uncinato terminata, rarissime inermia.

25. **R. arvensis** (L. Sp. 780); foliis radicalibus obovatis apice 3–5-dentatis, caulinis 3-partitis, acheniis paucis plano-compressis undique aculeis rectis vel tuberculis irregularibus tectis.—*DC. Prod.* i. 41; *Led. Fl. Ross.* i. 46; *Wall. Cat.* 4700! *Royle, Ill.* 53! R. tuberculatus, *DC. Prod.* i. 41.

HAB. Inter segetes Beluchistan! et Afghanistan! et Himalayæ occid. temp. a Kashmir! ad Kumaon!—(*v. v.*)

DISTRIB. Europa media et austr.! Madeira! Asia Minor! Sibiria occ.!

Herba annua, erecta, glabra, ramosissima, multiflora. *Folia radicalia* 1–2-pollicaria; *caulina* inferiora, petiolata, tripartita, segmentis anguste elongato-cuneatis 1–2-pollicaribus apice tridentatis, superiora sessilia multifida segmentis linearibus. *Flores* diam. semipollicares. *Petala* obovata, sepalis subduplo longiora. *Achenia* 5–10, ⅓ poll. longa, oblique obovata, stylo longo rectiusculo mucronata.

Ledebour distinguishes the form with tuberculated (not echinated) achenia as a variety, but both states occasionally occur on the same individual.

Species dubia vel vix nota.

1. R. *vestitus* (Wall. Cat. 4707).

The specimens in the Linnean Society's Herbarium consist of a few long-petioled radical leaves, without stem or flowers, and are quite undeterminable.

Tribus IV. HELLEBOREÆ.

Sepala colorata, æstivatione imbricata. *Petala* plana vel irregularia, rarius nulla. *Carpella* follicularia, polysperma.—Herbæ *foliis alternis*.

10. **CALTHA,** L.

Sepala 5 vel plura, regularia, colorata. *Petala* nulla. *Ovaria* secus suturam ventralem per totam longitudinem ovuligera. *Folliculi* 5–30. —Herbæ *perennes*, floribus *flavis vel albis*.

This genus consists of a few species natives of marshes in the arctic and temperate regions in both hemispheres, with one species in the mountains of tropical America.

1. **C. palustris** (L. Sp. 784); caule erecto vel adscendente (interdum ad nodos radicante), foliis orbicularibus vel reniformibus.—*DC. Prod.* i. 44; *Ledeb. Fl. Ross.* i. 48; *Torr. et Gray, Fl. N. Am.* i. 26. C. Himalensis, *Don, Prod.* 195; *Royle! Ill.* 54. C. Govaniana, *Wall. Cat.* 4710! *Royle! Ill.* 54. C. paniculata, *Wall. Cat.* 4711!

β. *alba*.—C. alba, *Jacquem. Mss.; Camb. in Jacq. Voy. Bot.* 6. *t.* 4.

HAB In paludosis Himalayæ interioris temperatæ et subalpinæ, alt. 8–10,000 ped.: a Kashmir! ad Nipal!—(Fl. Jun.) (*v. v.*)

DISTRIB. Europa, Asia, et America temperata! Japonia, *Siebold*.

Herba sæpius erecta, 1–2-pedalis. *Folia radicalia* longe petiolata, rotundata vel reniformia, interdum subdeltoidea, basi profunde cordata, lobis divaricatis, eleganter crenato-dentata, rarius integra, diametro 3–5-pollicaria; *caulina* petiolata vel subsessilia, petiolis basi auriculis rotundatis membranaceis stipulatis. *Caules* ramosi, multiflori, rarius humiles, subsimplices, 1-flori. *Flores* paniculati, diam. 1–2-pollicares, aurei (in var. β albi). *Sepala* ovalia vel obovata. *Folliculi* oblongi, juniores acuti, maturi sæpius truncati rarius subacuti, coriacei, stylo brevi apiculato, 5–10, in speciminibus Indicis interdum 20.

The Indian plant is identical with that of Europe, and varies in the same way in the degree of dentation and in size, being very luxuriant at moderate elevations, and becoming small and stunted at its highest level. The white-flowered variety is a remarkable one, but it is undistinguishable in the herbarium.

2. **C. scaposa** (H.f. et T.); acaulis, multiscapa, foliis ovali-oblongis.

HAB. In alpibus Sikkim int., alt. 15–17,000 ped. in paludosis.—(Fl. Jul.) (*v. v.*)

Radix crassa, fibrosa. *Folia* omnia radicalia, longe petiolata, ovali-oblonga, basi profunde cordata, coriacea, integerrima vel repando-crenata, 1–1½-pollicaria. *Scapi* 3–6-pollicares, nudi, 1-flori. *Flores* diametro pollicares et ultra, aurei. *Sepala* 5, obovata, tarde decidua, interdum sub fructu maturo persistentia. *Folliculi* 8–30, chartacei, stipitati, erecto-patentes vel recurvi, semipollicares, lineari-oblongi, stylo subulato apiculati.

11. CALATHODES, H.f. et T.

Sepala 5, ovalia, æstivatione imbricata, colorata. *Petala* 0. *Stamina* indefinita; *filamenta* filiformia; *antheræ* lineari-oblongæ, adnatæ, loculis marginalibus lateraliter dehiscentibus. *Ovaria* 10 vel plura, extus basi gibba, oblonga, rostrata. *Ovula* 8–10 prope basin ovarii, placentis 2 nerviformibus intramarginalibus prope suturam ventralem sitis inserta, horizontalia, rhaphe inferiori. *Styli* retrorsum uncinati, superne stigmatosi.—Herba *perennis, erecta*, Trollii *facie;* foliis *palmatim sectis;* floribus *flavis*.

This is a very remarkable plant, which has the flower of *Caltha* with the divided leaves of *Trollius*. The habit is so different from that of *Caltha* that the two could scarcely be united, even if the floral organs were the same in all respects; whilst the insertion of the ovules in *Calathodes* is so remarkable, that no doubt can exist as to the propriety of distinguishing it generically.

1. C. palmata (H.f. et T.).

HAB. In graminosis Sikkim, alt. 10,000 ped.!—(Fl. Jun.) (*v. v.*)

Herba erecta, 1¼–2-pedalis, simplex vel parce ramosa, glabra. *Folia* diam. 2–4-poll., longe petiolata; radicalia cito marcescentia; caulina superne numerosa, palmatim trisecta, segmentis basi cuneatis, profunde trilobis, lobis argute incisis; auriculæ stipulares membranaceæ dilatatæ, petiolo adhærentes, pollicares. *Flores* terminales, solitarii, aurei, diam. fere bipollicares. *Fructus* ignotus.

12. TROLLIUS, L.

Sepala 5 vel plura, regularia, colorata. *Petala* 5 vel plura, unguiculata, lamina plana, basi fovea nectarifera impressa. *Folliculi* 5 vel plures.—Herbæ *erectæ, perennes,* floribus *flavis.*

A very small genus, containing a few variable species, all natives of the north temperate or arctic zone. In the polysepalous species the flower has a globose shape, and is very different in appearance from that of the species which have only five sepals.

1. T. pumilus (Don, Prod. 195); caule subnudo unifloro, foliis coriaceis aristato-dentatis, petalis longiuscule unguiculatis.

HAB. In Himalayæ interioris alpibus: Kumaon, alt. 14,000 ped., *Str. et Wint.*, No. 2! Nipal, *Wall. ex Don.* Sikkim, alt. 15–17,000 ped.!—(Fl. Jun. Jul.) (*v. v.*)

Caulis basi fibrillis involutus. *Folia* radicalia longe petiolata, glabra, fere rotundata, 1–2-pollicaria, rigide coriacea, palmatim 5-partita, segmentis cuneato-obovatis acutis trilobis argute dentatis vel incisis. *Scapi* erecti vel adscendentes, nudi vel 1–3-foliati, spithamæi-pedalesve. *Flores* 1–1¼-pollicares. *Sepala* 5–6, rotundata, apice erosa. *Petala* 10–12, filamentis æquilonga, ungue cylindrico, lamina cuneato-oblonga obtusa apice incrassata 3-nervi, basi foveolata et saccata. *Folliculi* 5 vel plures, late oblongi, transverse nervosi, stylo abrupte rostrato.

2. T. acaulis (Lindl. in Bot. Reg. 1842, Misc. 56; *Ib.* 1843. *t.* 32); caule humili superne folioso, foliis 5-partitis argute incisis, petalis anguste cuneatis, ovariis subulatis.—T. pumilus, *Royle! Ill.* 54.

HAB. In Himalayæ occidentalis interioris alpibus, alt. 11–13,000 ped.: Kishtwar! Kumaon, *Str. et Wint.* No. 1!—(Fl. Jun.) (*v. v.*)

Radix fibrosa. *Caulis* 2–8-pollicaris, erectus, foliosus, basi squamis apice interdum folium parvum gerentibus involutus. *Squamæ* membranaceæ, nervosæ, acutæ, extus fibris plurimis (reliquiis squamarum anni præteriti) circumdatæ. *Folia radicalia* interdum serotina circumscriptione orbicularia palmatim 5-partita, segmentis oblongo-lanceolatis argute incisis; *caulina* breviter petiolata, petioli basi stipulis magnis tenuissime membranaceis auriculati. *Flores* magni, bipollicares. *Sepala* 7, late ovalia, obtusa, pollicaria, aurea. *Petala* 14, 3 lineas longa, staminibus paullo breviora, ungue brevissimo, lamina lineari- vel cuneato-oblonga, obtusa, basi foveolata. *Ovaria* elongata, in stylum longum subrecurvum angustata.

Dr. Royle, who obtained this plant from the mountains south of Kashmir, considered it to be the same as the species previously described by Don, but the character given in the Prod. Fl. Nep. applies evidently to our first species. *T. acaulis* is very near *T. Americanus*, though apparently quite distinct in the shape of its ovaries, which are narrow, and taper into the long subulate style, but in *T. Americanus* are much shorter and truncate. The latter is also in general a much taller plant, but some small states of it very much resemble in appearance *T. acaulis.*

13. **COPTIS,** Salisb.

Sepala 5–6, regularia, colorata. *Petala* unguiculata, cucullata vel linearia, non foveolata. *Folliculi* longe stipitati, stellatim patentes.— Herbæ *rhizomate horizontali perennantes;* foliis *ternatim sectis;* scapis *paucifloris;* floribus *albidis.*

The few known species of this genus are confined to the colder parts of the north temperate zone, one species being European and Siberian, while the rest are confined to North America. The Indian species is scarcely known, although its root is apparently much esteemed as a drug by the inhabitants of the mountains east of Assam, in which it is indigenous, and whence it is exported to Bengal. It is very bitter.

1. **C. Teeta** (Wall. Tr. Med. et Phys. Soc. Calc. viii. 347, et in Linnæa xii. 227); foliis trisectis, segmentis lobato-pinnatifidis, scapo paucifloro, bracteis foliaceis lobato-tripartitis.—*Griffith Journ.* 37.

HAB. In montibus Mishmi, in zona temperata, *Wall. Griff.*

Radix subcarnosa, multiceps, fibrillosa, iutus luteo-aurea. *Folia* glabra, rigida, circumscriptione ovato-cordata, attenuato-acuminata, 4-poll.; *segmenta* lateralia bipollicaria, petiolo ½-poll., semicordata, extus fere bipartita, terminale duplo majus utrinque attenuatum, omnia inciso-pinnatifida. *Scapus* gracilis, folia æquans; flores vix ultra 3, parvi, alterni, pedunculati. *Sepala* oblongo-lanceolata, acuta, semipollicaria, fugacissima. *Petala* lineari-ligulata, obtusa, sepalis triplo breviora. *Folliculi* plures.

Our description is condensed from that of Wallich, as we have not had an opportunity of seeing this rare plant.

14. **ISOPYRUM,** L.

Sepala 5–6, regularia, colorata. *Petala* unguiculata vel sessilia, lamina cucullata vel planiuscula, non foveolata. *Folliculi* 2–20.— Herbæ *annuæ vel perennes;* foliis *ternatim sectis;* floribus *albis.*

The species of this genus are natives of shady woods or of mountain-rocks in the north temperate zone. The alpine species have a very peculiar habit, but the caulescent ones resemble *Thalictrum.*

1. **I. adiantifolium** (H.f. et T.); caule folioso, foliis caulinis oppositis, petalis longe unguiculatis, lamina parva rotundata subbiloba, ovariis 3.

HAB. In Himalayæ orientalis sylvis humidissimis: Sikkim prope Dorjiling, alt. 7500 ped.!—(Fl. April.) (*v. v.*)

Rhizoma horizontale, squamis rotundatis concavis tectum. *Caulis* erectus, 3–6-pollicaris, simplex vel dichotome ramosus, basi nudus, superne foliosus. *Folia radicalia* longe petiolata, petiolis basi auriculatis, trisecta. *Foliola* longe petiolulata, petiolis partialibus basi membranaceis stipellatis, terminale indivisum, lateralia in segmenta 5–7 petiolata subdichotome secta; *segmenta* omnia tenuissima late cuneata vel fere rotundata, 3–4 lineas longa, antice inciso-crenata, subtus glauca. *Folia caulina* opposita vel quaternatim verticillata, petiolis basi stipulis membranaceis rotundatis auriculatis, majora foliis radicalibus conformia. *Flores* diametro semipollicares. *Sepala* ovalia, obtusa. *Petala* minuta, longe unguiculata, lamina rotundata, plana vel saccata, biloba vel eroso-dentata. *Filamenta* filiformia, petalis duplo longiora, sepalis ½ breviora. *Antheræ* ovales.

2. **I. thalictroides** (L. Sp. 783); caule folioso, foliis caulinis alternis, petalis breviter stipitatis cucullatis, ovariis 2–4.—*DC. Prod.* i. 48; *Led. Fl. Ross.* i. 53. I. anemonoides, *Kar. et Kir. Enum. Pl. Soong.* 55; *Led. Fl. Ross.* i. 735.

HAB. In Himalaya occidentali ad portum Garés inter Kashmir et Balti, alt. 10,000 ped., *Winterbottom!*—(*v. s.*)

DISTRIB. In montibus Pyrenæis, Sabaudia, Carniolia! Borussia! Polonia, Lithuania! Sibiria Altaica!

Rhizoma horizontale, fibrillosum vel squamis vestitum. *Caules* 4–8-pollicares, folia paullo superantes. *Folia radicalia* longe petiolata, 2–3-ternata, segmento terminali trilobo, lateralibus bilobis; *caulina* biternata, suprema ternata vel simplicia. *Flores* pauci, diam. ½–⅔-poll. *Sepala* ovalia, obtusa. *Petala* stipiti filiformi insidentia, ovalia, cucullata, obtusa.

Mr. Winterbottom's plant is identical with specimens of *I. anemonoides* of Karelin and Kirilow; but the characters by which these botanists distinguish that species from *I. Thalictroides* are, we fear, not of sufficient importance. The petals in the European plant vary much in shape, and those of Carniolian specimens in Herb. Hook., collected by Mr. Bentham, are the same as those of the Altai plant, while in others every possible degree of intermediate form may be observed. The scaly elongated root is therefore the only constant distinguishing character, and that depends in all probability on the age of the plant, or on some other accidental circumstance.

3. **I. grandiflorum** (Fisch. in DC. Prod. i. 48); subacaule, scapis unifloris opposite bibracteolatis, ovariis 3–7.—*Ledeb. Fl. Ross.* i. 53; *Wall. Cat.* 9123! (spec. aphylla valde imperfecta); *Royle! Ill.* 54. *t.* i. *f.* 3. I. microphyllum, *Royle! Ill.* 54. *t.* i. *f.* 4. Aquilegia anemonoides, *Willd.; DC. Prod.* i. 51 (*indic. Ledeb.*).

HAB. In Himalaya occidentali alpina et Tibetica, alt. 13–17,000 ped.: Dras! Kanawer! Hundes, *Str. et Wint.!* Garhwal! Kumaon! —(*v. v.*)

DISTRIB. Sibiria Altaica et Baikalensis!

Radix perpendicularis, lignosa, fusiformis. *Caules* cæspitosi, inferne petiolis induratis foliorum delapsorum basi dilatatis vaginantibus exasperati. *Folia* longe petiolata, petiolis basi auriculato-stipulatis, 2–4-ternatisecta, segmentis ultimis oblongis vel obovato-cuneatis obtuse incisis. *Scapi* aphylli, 1–4-pollicares, versus apicem bracteas 2 oppositas lanceolatas vel lineares, rarius ternatisectas, basi in auriculas magnas membranaceas dilatatas, gerentes. *Flores* diametro ultra-pollicares. *Sepala* late ovalia, obtusa. *Petala* obovata vel obovato-oblonga, basi saccata, magnitudine varia, obtuse bifida, retusa obtusave, trinervia.

The amount of division of the leaves varies just as much in Siberian as it does in Himalayan specimens. We have therefore reduced *I. microphyllum* of Royle, which is not otherwise distinct. The petals are always, we believe, bifid in the Siberian plant; generally entire, but sometimes retuse or emarginate in the Himalayan one.

15. **AQUILEGIA**, L.

Sepala 5, regularia, colorata. *Petala* 5, infundibuliformia, deorsum in calcar producta. *Stamina* interiora sterilia, membranacea. *Folliculi* 5 vel plures.—Herbæ *perennes;* foliis *ternatim sectis;* floribus *magnis, versicoloribus.*

The genus *Aquilegia* is limited to a few species, all natives of the north temperate zone. In Europe, Northern Asia, and North America, they are common in mountain woods and pastures, rising into the alpine region. In India the genus occurs only in the Western Himalaya and in the mountains of Tibet, to which at least five species have been considered peculiar. We have, however, found that all the European species distinguished by Linnæus and subsequent authors occur in the Himalaya. It has therefore been necessary to submit all these to a critical examination, the result of which has very unexpectedly been that all the European and many of the Siberian forms generally recognized belong to one very variable species. We do not include *A. parviflora*, Ledeb., which, judging from the figure and description, and from a single specimen, is very distinct; nor *A. Canadensis*, L., to which *A. Sibirica*, Lam., *A. atropurpurea*, DC., *A. Davurica*, Patr., *A. formosa*, Fisch., and perhaps *A. lactiflora*, Kar. et Kir., ought probably to be referred as synonyms. This species is universally recognized as distinct by American botanists, and appears readily distinguished by the exserted stamens, the shape of the petals, and the small, straight, inflated, and suddenly contracted spurs. *A. cærulea*, Torrey, or *A. leptoceras*, Nutt. non Fischer, is probably a large-flowered form of the same species; and even *A. hybrida*, Sims (Bot. Mag. t. 1221), which has hooked spurs, retains the other characteristics of *A. Canadensis*. None of these varieties have been found in the Himalaya.

We feel that it is difficult to explain briefly, and at the same time clearly, the grounds on which we have come to the conclusion that all the synonyms quoted below must be considered states of one very variable species. Our Indian specimens are numerous, and exhibit many different forms, which it was not difficult to throw into tolerably well marked groups by their general appearance, with the exception of a few intermediate specimens. On comparing them with the general herbarium, it was at once apparent that these groups corresponded pretty closely to the commonly recognized species of authors, so that our course appeared easy. As soon, however, as we attempted to frame diagnoses which should be applicable not only to the Indian plants, but to those of Europe with which we had identified them, we found that the great amount of variation to which this genus is subject interposed insuperable difficulties.

Authors have availed themselves of four classes of characters to distinguish from one another the species of *Aquilegia*. 1. The shape of the floral organs. 2. The nature and degree of pubescence. 3. The height of the stem, the number of its leaves, and the amount of ramification. 4. The degree of division of the leaves, and the stalked or sessile leaflets. Linnæus described only two European species, *A. alpina*, with straight spurs, and *A. vulgaris*, with hooked spurs; but subsequent authors consider the pubescence a prominent character, as may be seen by the names *viscosa*, *glandulosa*, *pubiflora*. Both De Candolle and Treviranus, however, have long ago admitted the inefficacy of this character, and stated their belief that *A. viscosa* cannot be distinguished from *A. vulgaris;* and though systematists in general have not followed their example, that is only because the wish to make species prevails over the authority of scientific inquirers,—we cannot say over their example, since both the above-named authors, while stating their opinion that the species are not distinct, have kept them separate.

The shape of the leaves, though not noticed by Linnæus, has been relied upon by De Candolle and others, for the separation of *A. alpina* from *A. vulgaris*. The result has been that specimens which would otherwise be referred, from the shape and size of the flowers, to *A. alpina*, have been separated from it by more recent authors under various names, because the leaves were less deeply cut. Species have even been distinguished by the leaflets being sessile or stalked. We do not repeat the remarks which we have already so frequently had occasion to make regarding the great degree of variation to which the foliage of *Ranunculaceæ*, and indeed of all cut-leaved families, is subject: an examination of any large collection of specimens, or a careful observation of nature, ought to convince every one of the little confidence which is to be placed in such distinctions, and no argument will have any weight with those who attach specific value to trifling variations.

To the size and degree of branching, which too often constitute the only distinctions between alpine and lowland plants, and which have, in the genus *Aquilegia* as well as elsewhere, been used as specific characters, it is not possible to attach much weight.　It will be found that size is not accompanied by any constant characters, but that in this genus, as in most or all of those which are common in alpine regions, every variety has its dwarf and tall state.　It is well known to gardeners that the species of *Aquilegia* do not retain their stature in cultivation for any length of time, but that they become by degrees tall and luxuriant, and totally unlike their original condition.　All the more luxuriant states of *A. vulgaris*, indeed, are probably of garden origin, as the wild species in all mountain countries attain no great size.

It is, however, upon the shape of the floral organs and the size of the flowers that specific characters are in general founded.　The colour of the flowers has also occasionally been employed as an auxiliary; but little stress having been laid upon it, we need only remark that in gardens every colour is common, and that changes in that respect are known to be produced by artificial circumstances.　The colour of the anthers, which has occasionally been relied on, seems to depend in a great measure upon the depth of colour of the perianthial leaves, being yellow when they are white or pale, and leaden or bluish when they are dark.

We cannot find in the published descriptions of this genus that any of the European species have any smell.　There can be no doubt, however, that the more alpine Himalayan forms growing in dry places are sweet-scented, and that they even retain their agreeable odour when raised from seed in gardens in this country.　At the same time, these sweet-smelling forms are in no way distinguishable from European specimens of *A. viscosa* and *A. Pyrenaica*, and the odour seems to depend on the development of the viscid glands so abundant in such states.

In passing in review the floral organs of the supposed species here reduced to *A. vulgaris*, it may be remarked in the first place that, including the straight or hooked spurs, all the characters derived from them are those of degree only.　The sepals vary from long acuminate to quite obtuse, and their size is equally variable, as is also that of the flowers.　In structure there is not even a shadow of a difference, and characters derived from proportion allow of the separation of any number of species.　The shape of the inner abortive scariose filaments, the parastemones of Reichenbach, has been relied upon by that author; but they appear to vary very very much, and not to be deserving of any attention.

If the straight and hooked spurs were a constant character, it would form an admirable specific distinction.　Unfortunately this is by no means the case, as may at once be seen by examining the diagnoses of authors, in which the words subincurved, slightly curved, etc., are of common occurrence.　On many specimens too, perfectly straight and much incurved spurs may be met with on one plant.　Some curious instances of the difficulties which beset authors who attempt to retain all the species usually distinguished may be mentioned.　Reichenbach quotes *A. alpina*, DC. (characterized by straight spurs, *apice subincurva*), under his *A. Sternbergii*, to which he ascribes much incurved spurs (*valde incurvata*).　He is mainly led to this by the figure of Delessert, who undoubtedly ought to be supposed to know the plant intended by De Candolle, but who represents a specimen with a much hooked spur, without any indication that the diagnosis of the species does not correspond with his figure.　In like manner Lindley figures (Bot. Reg. 1847. t. 64) *A. leptoceras*, Fisch., raised from seeds sent by the author himself to the Horticultural Society.　The plate represents a much incurved spur, while the accompanying description, copied from Fischer, dwells especially upon the perfectly straight spur as the distinguishing mark of the species.

However paradoxical the views which we have thus expressed may appear to those who, on the authority of European systematists, have been in the habit of viewing *Aquilegia* as a very large genus, containing upwards of thirty well-marked species, we trust that they will not be rejected without due consideration, and a careful study of large series of specimens, both dried and in a botanical garden. Many botanists to whom we have shown the extensive series of specimens in the Hookerian Herbarium, have been surprised at the amount of variation exhibited, and at the intricate interlacing, so to speak, of the different characters by which their separation into well characterized species is rendered impossible. We have devoted weeks to the study of these plants, in the hope of arriving at some definite results, and we shall be only too happy to have characters pointed out to us on which confidence may be placed. Those at present given in books fail completely in practice.

1. **A. vulgaris** (L. Sp. 752); foliis biternatim sectis, lobis varie incisis, sessilibus vel petiolatis, sepalis genitalia superantibus, petalorum calcaribus sensim attenuatis.

a. normalis; subglabra vel pubescens, sepalis ovatis obtusiusculis, calcaribus petalorum laminas subæquantibus validis.—A. vulgaris, *DC. Syst.* i. 334, *Prod.* i. 50; *Ledeb. Fl. Ross.* i. 55; *Reich. Ic. Germ. t.* 114. A. atrata, *Koch, Fl. Germ.* A. Sternbergii, *Reich. Ic. Fl. Germ. t.* 118. A. Pyrenaica, *Reich. Ic. Germ. t.* 117 (*non aliorum*).

β. *viscosa;* glandulosa, pubescens, floribus ut in *a.*—A. viscosa, *Gouan; DC. Syst.* i. 335, *Prod.* i. 50.

γ. *pubiflora;* molliter pubescens, sepalis ovato-lanceolatis sæpe attenuatis, calcaribus abbreviatis incurvis.—A. pubiflora, *Wall. Cat.* 4714!; *Royle! Ill.* 54. A. nigricans, *Reich. Ic. Germ. t.* 115. A. viscosa, *Reich. ib. t.* 116.

δ. *grandiflora;* pubescens vel glandulosa, sepalis ovalibus acutiusculis vel obtusis, calcaribus crassis rectis vel incurvis, foliis plerumque profunde incisis.—A. alpina, *L. Sp.* 752; *DC. Syst.* i. 336, *Prod.* i. 50; *Deless. Ic. sel.* i. *t.* 48. A. glandulosa, *Fisch.; DC. Prod.* i. 50; *Sweet, Fl. Gard.* i. *t.* 55; *Ledeb. Fl. Ross.* i. 56. A. jucunda, *F. et M. Ind. Hort. Pet.; Ledeb. Fl. Ross.* i. 736.

ε. *Pyrenaica;* molliter pubescens vel glandulosa, sepalis ovatis acutis, calcaribus elongatis gracilibus rectis vel incurvis.—A. Pyrenaica, *DC. Syst.* i. 337, *Prod.* i. 50. A. leptoceras, *Lindl. Bot. Reg.* 1847, *t.* 64. A. Kanawerensis, *Camb. in Jacq. Voy. Bot. t.* 5; *Hook. Bot. Mag.* 4693. A. glandulosa, *Knowles et Westcott, Bot. Cab.* i. *t.* 10. A. glauca, *Lindl. Bot. Reg.* xiii. *t.* 46. A. fragrans, *Benth. in the Botanist,* iv. *t.* 181. A. Moorcroftiana, *Wall. Cat.* 4713!; *Royle! Ill.* 54. A. Olympica, *Boissier! Ann. Sc. Nat.* xvi. 360.

HAB. In Himalaya occidentali temperata et alpina. *a.* In jugis interioribus! β. In Himalaya Tibetica! γ. In jugis exterioribus vulgaris, a Kumaon! ad Kashmir!; in interioribus in *a.* transiens. δ. In montibus Balti, *Winterbottom !* Kumaon, *Wall. !* ε. in Himalayæ alpibus et per Tibetiam occidentalem inter 10–14,000 ped. alt. vulgaris! DISTRIB. Europa et Asia temp.

It must not be supposed that the varieties indicated above are in any way constant. On the contrary, it will be found that they pass into one another in every possible way, and the botanist must expect to find many specimens which he will have great difficulty in referring to any of the forms characterized.

1. The variety α is that commonly cultivated in gardens in England, where it is often very luxuriant, with large leaves, the lobes of which are little divided. In a wild state it is seldom glabrous, and is very variable in size. In India it is less common than some other forms, but specimens of γ and ε are often barely, if at all, distinguishable from α.

2. Viscid specimens from Zanskar and Piti are identical with an authentic specimen of *A. viscosa*, Gouan, in Herb. Hook.

3. The *Aquilegia pubiflora* of Wall., which is common in the rainy Himalaya from Kumaon westward as far as Kashmir, was long considered by us as having claims to specific distinction. A more careful study of the genus has, however, shown us that there are no characters in the leaves which can be relied upon, and that the elongated or acuminate sepals, which we had regarded as a sufficient character, occur equally in European specimens, which are only distinguished from *A. vulgaris* and *A. viscosa* by those botanists in whose opinion every trifling variation of aspect affords specific characters. There is no doubt that the ordinary state of this variety is quite distinct in appearance from the common alpine state of *A. vulgaris;* but not only do the specimens from the interior of the mountains gradually obliterate these differences, but many specimens from the outer hills, where it cannot be supposed that two species grow, differ in their large leaves as well as in the large flowers and broader sepals, from the normal state of *A. pubiflora.*

4. δ is a very remarkable form, but it is perhaps less deserving of being distinguished as a variety than any other, as the monstrous or considerably enlarged flowers on which its main character depends, occur in very different localities, and with every variety of leaf and size. Some of the states of this plant, when the spur is abnormal, and the sepals are much enlarged and obtuse, are very remarkable, and at first sight have the appearance of being specifically distinct. They must, however, be regarded rather as monsters than as anything else. Some specimens from Mr. Winterbottom must be mentioned, as being included under this variety, lest it should be supposed that we consider them as distinct. These seem to be identical, in flower at least, with *A. jucunda*, Fischer.

5. The smaller forms of *A. Pyrenaica*, DC., with a slender, perfectly straight spur, and viscid pubescence, are readily distinguished from the ordinary form of *A. vulgaris;* but unfortunately they pass, in every country in which they occur, by a series of imperceptible gradations, into *A. vulgaris.* The stem becomes tall and branched, and softly pubescent, the spur becomes much curved, and the flowers much larger. Visiani has well pointed out, in the 'Flora Dalmatica,' the uncertainty of the spur as a character, and has stated his conviction that *A. Pyrenaica*, with a straight spur, is not distinct from *A. viscosa.* Nor are the rigid glaucous leaves a sufficient character to distinguish certain states of this variety from the remainder.

16. **DELPHINIUM**, L.

Sepala 5, posticum calcaratum. *Petala* 4 (interdum in unum calcaratum coalita); duo postica basi calcarata, 2 antica unguiculata. *Ovaria* 1–7.—Herbæ *annuæ vel perennes*, caulibus *erectis.* Folia *palmatim lobata.* Flores *conspicui, cærulei vel purpurascentes.*

The species of this genus are all natives of the north temperate zone, growing either in woods or in the grassy pastures of mountainous districts, or in dry, stony, and desert hot places. None have been found in the mountains of the Indian Archipelago. They are all extremely variable, and the genus is in consequence, if possible, in still greater confusion than the other genera of *Ranunculaceæ.* The mode of ramification and denseness of the racemes, the shape and size of the flowers, and the length of the spur, seem to vary almost as much as the shape of the leaves; and we have not been successful in our attempt to arrive at definite ideas regarding the limits of European and North Asiatic species, from the materials at our disposal.

We have therefore avoided, in all doubtful cases, identifying the Indian species with those of other countries, though we think it probable that, on a revision of the genus with good materials, several of them will be found to agree with already known species. We are satisfied indeed that the greater part of the described forms will then be reduced to a few common types.

The Indian *Delphinia* are mostly Himalayan, but one or two Persian forms extend as far east as the Panjab. One has been found in tropical India, but not further south than the northern part of the Siadri mountains, whence it extends westwards to Abyssinia. One of the Himalayan species inhabits a very humid climate, and extends into the higher parts of the Khasia hills, but in general these plants seem to prefer the drier climates of western India. Several species are alpine, and these form a very peculiar group, with large half-closed membranous flowers, remarkable in general for their strong odour of musk, whence the mountaineers erroneously suppose that the musk-deer feeds on them, and thereby communicates the peculiar odour to its glandular secretion.

Sect. 1. CONSOLIDA, DC.—*Ovarium* 1. *Petala* in unum calcaratum coalita.—Species *annuæ*. [We omit *Delphinium Ajacis*, L. (Wall. Cat. 4720!; W. et A. Prod. i. 4. *D. pauciflorum*, W. et A. Prod. i. 4; vix Don) which is only cultivated in India.]

1. **D. camptocarpum** (Fisch. et Mey. in Led. Fl. Ross. i. 58, 37); caule rigido ramosissimo, foliis trisectis, caulinis sessilibus, segmentis fere integris.—D. Persicum *et* Aucheri, *Boissier!* Ann. Sc. Nat. xvi. 362.

HAB. In montibus Beluchistan supra 5000 ped., *Stocks!* Afghanistan, *Griff.* No. 1376!—(*v. s.*)

DISTRIB. Persia! Asia Caspica!

Radix annua, perpendicularis. *Caulis* pedalis vel ½-pedalis, ramosissimus, ramis rigidis divaricatis foliosis, glaber vel incano-puberulus, vel superne breviter viscoso-pilosus. *Folia* trisecta: segmenta radicalium tripartita vel trifida, caulinorum sessilium oblongo-cuneata pollicaria tridentata integrave. *Flores* longe racemosi, violacei. *Pedicelli* patentes stricti, prope basin bracteolis 2–3 minutis linearibus muniti. *Sepala* oblonga, obtusa, ⅓-pollicaria, posticum in calcar cylindricum obtusum rectum adscendens ⅔-pollicare productum. *Petala* in corollam monopetalam 4-nervem calcaratam coalita, calcar cum calcare sepali postici subconforme et in eo inclusum. *Folliculus* 1 subrecurvus, obtusus, stylo persistente coronatus, transversim rugosus, glaber vel hirsutus.

A plant with a very well marked habit, but varying much in degree of pubescence, and to a certain extent in the length of the fruit. It seems to be widely distributed in the hot deserts of western Asia.

Sect. 2. DELPHINASTRUM, DC.—*Ovaria* 3–7. *Petala* 4, postica calcarata, antica pilosa vel barbata.

2. **D. penicillatum** (Boissier, Ann. Sc. Nat. xvi. 369); caule paucifoliato, foliis 5-partitis, segmentis inciso-pinnatifidis lobis linearibus, racemis elongatis multifloris subsimplicibus, pedicellis flore parvo brevioribus, calcare apice dilatato obtusissimo incurvo.

HAB. In montibus aridis Indiæ bor. occ.: Beluchistan, *Stocks, No.* 994! Panjab prope Rawal Pindi, *Vicary!* Marri, *Fleming!* Kashmir prope Baramula, *Wint.!* Banahal!—(Fl. Apr. Mai.) (*v. v.*)

DISTRIB. Persia.

Radix lignosa, descendens. *Caulis* erectus, 1–1½-pedalis, fere nudus, tomento brevi glanduloso denso tectus, rarius glaber. *Folia* utrinque sparse pilosa, circumscriptione reniformia. *Bracteæ* et bracteolæ minutæ, lineares. *Sepala* pallide purpurascentia, anguste obovata, ½-pollicaria, extus pilosula, posticum calcare breviore vel subæquilongo apice inflato gibboso munitum. *Petala postica* ⅔ poll. longa, antice obtuse angulata, lamina tenui obtuse biloba, calcare incurvo obtuso apice ampliato; *antica* ½-pollicaria, lamina alte biloba, utrinque dense et longe pilosa. *Ovaria* 3, pilosa, rarius glabra.

The shape of the leaves, and of the anterior petals, which are very large and covered with long shaggy hairs, in combination with the very short pedicels, seem to be the best characters of this species. The pubescence is entirely wanting in Major Vicary's specimens, which are rather imperfect, with scarcely any leaves, but identical in flowers with those from other localities. The spur is singularly variable, being sometimes only slightly incurved, at other times doubled up, so that the apex almost touches the base of the sepals.

3. **D. saniculæfolium** (Boissier! Diagn. Or. vi. 6); caule paucifoliato ramoso, foliis tripartitis segmentis cuneato-ovatis trilobis, racemis elongatis strictis, pedicellis flores parvos vix superantibus, calcare recto sepala æquante.

HAB. Afghanistan, *Griffith, No.* 1373! Panjab, in montosis prope Indum flumen, *Fleming!*—(*v. s.*)

DISTRIB. Persia!

Radix lignosa, elongata, descendens. *Caulis* erectus, glaber vel adpresse puberulus, parce ramosus, ramis rigidis vimineis divaricatis. *Folia* coriacea, adpresse cinereopuberula vel tomentosa, circumscriptione rotundata, 1½–3-poll., lobis incisis; *caulina* trisecta. *Racemi* multiflori, pedicellis flores æquantibus vel paullo longioribus. *Bracteæ* et bracteolæ 1–2 lineas longæ, lineares. *Flores* pallide cærulescentes. *Sepala* vix ½ poll. longa, extus puberula, oblonga, calcare recto æquilongo. *Petala posteriora* antice obliqua, obtuse angulata, apice bidentata, calcare subulato recto; *antica* biloba, pilosa. *Folliculi* 3, glabri vel puberuli, inflati (in Indicis speciminibus non visi).

It is quite possible that this may be only a form of the next, from which it chiefly differs in the size of the flowers, and somewhat also in aspect. Several more of Boissier's species may in all probability be referred to this or to the preceding, but the specimens before us are too imperfect to enable us to state decidedly to which they belong; nor are the diagnoses, which rest on very trivial characters, sufficient to settle the matter.

4. **D. denudatum** (Wall. Cat. 4719!); caule paucifoliato ramoso, foliis palmatim 5–7-partitis segmentis inciso-lobatis, racemis divaricatoramosis laxis, floribus magnis longe pedicellatis, calcare recto sepala æquante.—D. pauciflorum, *Royle! Ill.* 55 (*vix Don, Prod.*).

HAB. In Himalaya occidentali temperata in graminosis calidis: a Kashmir ad Baramula, *Wint.!* usque ad Kumaon!—(Fl. Apr. Jun.) (*v. v.*)

Caulis erectus, glaber vel apicem versus puberulus. *Folia radicalia* longe petiolata, 2–6-pollicaria, 5–7-partita, segmentis ovalibus basi angustatis cuneatis incisobipinnatifidis, lobis oblongis vel lineari-oblongis; *caulina* 3–5-secta, superiora sessilia, segmentis lineari-pinnatifidis linearibusve. *Pedicelli* lineari-bracteolati. *Flores* ultrapollicares, pallide cærulei. *Sepala* ¾-pollicaria, cum calcare æquilongo extus subflavida. *Petala postica* obtuse calcarata, antice obliqua, obtuse angulata, apicem ver-

H

sus angustata et bidentata; *antica* profunde biloba, utrinque longe pilosa. *Folliculi* 3, pilosuli vel subglabri.

A very common plant in the outer mountains of the Western Himalaya, varying much in size, but in general readily recognizable by its few-flowered, much-branched stems. *D. pauciflorum* of Don is probably correctly referred by Wight and Arnott to their plant of the same name, which is apparently a state of *D. consolida;* but Don's description is not certainly referable to any known plant, for though the greater part of it can apply only to *D. consolida,* the petals are those of a plant of the section *Delphinastrum.*

5. D. dasycaulon (Fresen. Mus. Senkenb. ii. 272); caule ramoso paucifoliato, foliis radicalibus amplis rotundato-reniformibus late 5-lobis, lobis trilobis et grosse incisis, caulinis 5-partitis segmentis argute incisis, racemis laxis elongatis, sepalis extus incano-tomentosis, calcare conico subrecurvo duplo longioribus.—*Walp. Rep.* i. 52.

HAB. In summis montibus Dekhan occidentalis prope Júnir (Jooneer), *Stocks! Gibson!*—(Fl. Aug. Sept.) (*v. s.*)

DISTRIB. Abyssinia, *Schimper!*

Caulis erectus, 1½-3-pedalis, pilis incanis vel fulvis villosus vel tomentosus. *Folia radicalia* numerosa, plerumque longe petiolata, diam. 3–6-pollicaria, lobis late trapezoideis, utrinque pubescentia, sericea vel villosa, subtus pallida et conspicue reticulatim nervosa; *caulina* ad basin secta, segmentis linearibus incisis; *floralia* indivisa, linearia. *Pedicelli* flores æquantes vel duplo superantes, tomentosi, bracteolis 2 alternis subulatis. *Flores* læte cærulei. *Sepala* ¾-pollicaria, versus apicem macula pallida dense pilosa notata. *Petala postica* cartilaginea, calcare subulato recto, antice obliqua, angustata, acuta vel bidentata; *antica* biloba, pilosa. *Folliculi* 3, recti, ¾ poll. longi, tomentosi.

We can find no difference between Dr. Stocks' specimens and those distributed by Schimper, except that the latter are more villous, and want the radical leaves. As Fresenius in his diagnosis describes the leaves as quinquepartite, the same deficiency probably exists in all the specimens collected by Schimper. As a species *D. dasycaulon* seems very distinct, and its occurrence in Western India is very interesting as a proof of the affinity which exists between the flora of that country and that of Western Africa. Many more instances of this will be met with in the course of our work.

6. D. incanum (Royle! Ill. 55); caule folioso, foliis tripartitis segmentis lineari-multifidis, racemis elongatis multifloris, pedicellis flores majusculos æquantibus vel superantibus, calcare recto sepalis longiore.

HAB. In Himalaya interiori occidentali, alt. 6–8000 ped.: Kashmir! Kanawer!—(Fl. Aug. Sept.) (*v. v.*)

Radix lignosa, cylindrica vel tuberosa, perpendicularis. *Caulis* strictus, erectus, bipedalis et ultra, striatus, sæpe angulatus, incanus vel subtomentosus, basi interdum glabrescens. *Folia* petiolata vel subsessilia, petiolis basi dilatatis. *Inflorescentia* subsimplex vel rarius paniculata, pedicellis bracteolis pluribus linearibus munitis. *Flores* læte cærulei. *Sepala* ovalia, ⅔-pollicaria, incana. *Petala postica* antice obtuse angulata, bidentata, calcare subulato; *antica* bifida, pilosa. *Folliculi* 3, ½-pollicares, brevissime tomentosi.

A handsome, tall, large-flowered species, strikingly like some forms of *D. grandiflorum,* L., but with bifid (not entire) anterior petals. The petals seem to be invariably entire in that species, and they are always bifid in the Indian plant, except in some specimens (unfortunately flowers only, without leaves,) from the mountains of Tibet behind East Nipal, in which they are very slightly emarginate. It is, nevertheless, extremely probable that our species is not distinct from *D. grandiflorum,* which seems to be very widely distributed.

7. **D. cæruleum** (Jacquem.! ex Camb. in Jacq. Voy. Bot. p. 7. t. 6); caule folioso ramoso, foliis palmatim 5–7-partitis segmentis obovato-cuneatis inciso-lobatis lobis obtusis, racemis patulis paucifloris laxis, floribus magnis longe pedicellatis, calcare recto sepalis longiore, ovariis 5.

HAB. In Himalaya interiori alpina : Garhwal, alt. 14–15,000 ped., *Str. et Wint.!* Sikkim, alt. 14–17,000 ped.!—(*v. v.*)

Caulis sæpe divaricato-ramosissimus, 3–6-pollicaris vel pedalis, incano-tomentosus, superne laxe sericeo-pilosus. *Folia* ½–1¼ poll. lata, rotundata, segmentis obovato-oblongis, superiorum linearibus. *Flores* pallide cærulei, pedicellis pluribracteatis. *Sepala* ¾ poll. longa, ovalia, obtusa, extus pubescentia. *Petala posteriora* antice rotundata, vix obliqua, calcare subulato ; *antica* lamina obcordata vel obovata, utrinque parce pilosa. *Folliculi* 5, pubescentes vel pilosi, ½-pollicares.

Seemingly very distinct from the last, not only in general habit but in floral characters. It is, however, compared by Cambessèdes with *D. grandiflorum*, L., but distinguished by the smaller flowers, by the more velvety pubescence, and particularly by the number of ovaria. It is also not unlike some small states of the American *D. azureum*, Mich., and *D. pauciflorum*, Nutt.

8. **D. ranunculifolium** (Wall. Cat. 4716 !); caule elato folioso, foliis palmatim 5-lobis, lobis cuneato-ovatis inciso-lobatis, racemis elongatis multifloris, pedicellis flores magnos excedentibus, calcare recto sepalis æquilongo.—D. pyramidale, *Royle! Ill.* 56.

β. *incisum;* foliis palmatim 5-partitis.—D. incisum, *Wall. Cat.* 4717 !

HAB. In Himalaya occidentali interiori : Pir Panjal, Kashmir, *Royle!* Kanawer, alt. 11,000 ped., *Jacquemont!* Garhwal et Kumaon, *Blinkworth!*—(*v. s.*)

Caulis erectus, 2–3-pedalis et ultra, pilis patentibus hirsutus. *Folia* superne pilosula, subtus dense tomentosa, rotundata, basi cordata, diam. 3–4-poll., petiolis æquilongis, ad medium 5-fida, lobis trifidis et inciso-dentatis ; *superiora* tripartita ; *floralia* lanceolata, pedicellis 1–2-pollicaribus tomentosis breviora. *Bracteolæ* 2, sæpe flori adpressæ. *Flores* (ex sicco) sordide cærulescentes. *Sepala* extus pilosa, ¾-poll., ovalia, obtusa. *Petala postica* antice recta acuta, atropurpurea, calcare subulato ; *antica* lamina bifida pilosa. *Folliculi* 3 pubescentes.

9. **D. altissimum** (Wall.! Plant. Asiat. Rar. ii. t. 128); caule ramoso paucifoliato, foliis reniformibus subtus albidis palmatim 5–7-fidis, lobis argute incisis, racemis laxis paucifloris, floribus magnis longe pedicellatis, calcare sepalis longiore longe subulato incurvo.—*Wall. Cat.* 4718 ! *Griff.! Itin. Notes, p.* 54: *No.* 827.

HAB. In Nipalia, *Wall.!;* in montibus Khasia, alt. 5–6000 ped., *Griffith!*—(Fl. autumno.) (*v. v.*)

Radix fusiformis. *Caulis* gracilis, 2–4-pedalis, pilis patentibus vel subreflexis hirsutus, rarius basi glabrescens. *Folia radicalia* longissime petiolata, pet. 6–12-pollicari, diam. 3–6-pollicaria, utrinque sparse pubescentia vel glabriuscula, 5–7-loba, lobis late cuneatis trilobis et argute dentatis : *floralia* subsessilia, triloba, suprema linearia bracteæformia. *Flores* fœtidi, violacei. *Sepala* ovalia, obtusa, extus pilosa, ¾–1-pollicaria. *Petala postica* calcare subulato, antice oblique angulata, bidentata, atropurpurea ; *antica* biloba, pilosa. *Folliculi* 3, pubescentes.

Dr. Wallich's plant is only known to us by an imperfect specimen in the herbarium of the Linnean Society, and by the figure quoted. The Khasia plant is remarkable for the extremely disagreeable odour of the flowers.

10. **D. vestitum** (Wall. Cat. 4715 !) ; caule hispido paucifoliato, foliis reniformibus palmatim 5-fidis, lobis cuneato-ovatis grosse incisodentatis, racemis elongatis strictis multifloris, pedicellis flores magnos superantibus, calcare incurvo sepalis æquilongo.—*Royle! Ill.* 55. D. rectivenium, *Royle! Ill.* 56.

Hab. In Himalaya temperata et subalpina, alt. 8–12,000 ped.: Simla! Kanawer! Garhwal! Kumaon! Nipal!—(Fl. Aug. Sept.) (*v. v.*)

Caulis erectus, 2–3-pedalis, pilis rigidis subreflexis dense hispido-pilosus. *Folia radicalia* petiolo 6–12-poll., utrinque hispida, diam. 4–5-pollicaria; *caulina* 1–2, subsessilia, floralia bracteæformia triloba vel lanceolata. *Inflorescentia* subramosa vel simplex, racemis sæpe pedalibus. *Bracteolæ* membranaceæ, suboppositæ, lanceolatæ, flori non adpressæ. *Sepala* ¾-poll., extus pilosa, membranacea, fere rotundata. *Petala postica* antice obtuse angulata, bidentata, calcare subulato; *antica* biloba, dorso parce pilosa. *Folliculi* 3, pilosi.

11. **D. Kashmirianum** (Royle! Ill. 53. t. 12); caule folioso subsimplici, foliis reniformibus palmatim 5-lobis, lobis inciso-dentatis, racemis paucifloris corymbosis, floribus longe pedicellatis, calcare recto saccato obtuso sepalis paullo breviore, ovariis 3–7.—D. Jacquemontianum, *Camb.! in Jacquem. Voy. Bot.* viii. *t.* 7.

Hab. In alpibus Himalayæ occ. Tibeticæ, alt. 11–16,000 ped.: Gilgit, *Wint.!;* Dras! Suru, *Lance!* Kanawer, *Jacquemont!* Kumaon, *Str. et Wint.!*—(*v. v.*)

Caulis erectus, 1–2-pedalis, glabriusculus vel pilis mollibus patentibus sparsis pilosus, plurifoliatus. *Folia* longe petiolata, 3–4-pollicaria, pubescentia vel pilosa, orbiculari-reniformia, lobis grosse et sæpius argute incisis; *caulina* superiora sessilia tripartita, floralia lanceolata. *Racemus* simplex vel ramosus, ramis elongatis apice paucifloris. *Pedicelli* bracteolas plures lanceolatas gerentes. *Flores* cærulei. *Sepala* pollicaria, extus dense pilosa, late ovalia, membranacea, nervosa. *Petala postica* antice obliqua et obtusa, angulata, apice biloba, atropurpurea, calcare subulato uncinato; *antica* lamina profunde bifida, dorso aureo-pilosa. *Folliculi* 3–7, pubescentes, ⅔ poll. longi.

12. **D. viscosum** (H.f. et T.); caule ramoso paucifoliato, foliis reniformibus palmatim 5–7-fidis, lobis grosse et obtuse crenatis, ramis elongatis plurifloris, floribus longe pedicellatis, calcare cylindrico incurvo sepalis æquilongo, ovariis 3 glabris.

Hab. In Himalayæ alpibus interioribus: Sikkim, alt. 15–16,000 ped.!—(Fl. Aug. Sept.) (*v. v.*)

Caulis erectus, bipedalis, pilis fulvis patentibus brevissimis tectus, paniculatim ramosus. *Folia* breviter petiolata, in caule 2–3, secus nervos parce pilosa, cæterum glabra, 3–4-pollicaria, crenato-lobata, lobis glandula apiculatis; *superiora* parva, triloba vel tripartita. *Pedicelli* bracteolas 2–3-lineares flori non adpressas gerentes. *Flores* purpureo-cærulei. *Sepala* late ovalia, subacuta, ¾-poll., membranacea, nervosa, extus pilosa. *Petala postica* atropurpurea, calcare subulato incurvo, lamina antice oblique obtuse angulata et angustata, integra vel crenulata; *antica* utrinque albo-pilosa, obtuse biloba. *Folliculi* 3, ½-poll., glabri, sutura ventrali longe ciliati.

This species appears distinct both in habit and characters, but our specimens are not numerous, and future observers may discover connecting links. We do not re-collect that it has any smell, nor do we find any note indicating that it is a musky plant, like the last and all the following species, from which, if inodorous, it is pro-bably quite distinct.

13. **D. moschatum** (Munro, mss.); caule folioso ramoso, foliis re-niformibus palmatim 5-fidis lobis inciso-crenatis, ramis multifloris, calcare saccato conico obtuso sepalis ½ breviore, ovariis 3 tomentosis.

HAB. In Himalaya int. occ. Tibetica, alt. 12–14,000 ped.: Kanawer, *Munro!* Hundes, *Str. et Wint. No.* 8 !—(*v. s.*)

Caulis erectus, 3–5-pedalis, ramosissimus, foliosus, glaber, apice viscoso-puberulus. *Folia* glabra, longe petiolata, 3–4-pollicaria, petiolis 6–8-poll. basi dilatatis, ½-5-fida, inciso-crenata, dentibus glandula apiculatis; *floralia* tripartita, summa oblonga. *Pa-nicula* divaricato-ramosa, multiflora. *Flores* pollicares, pallide cærulei. *Sepala* fere rotundata, membranacea, nervosa. *Petala postica* calcare subulato incurvo, glabrius-cula, lamina atropurpurea antice obtuse angulata apice bifida; *antica* lamina utrin-que pilosa profunde biloba. *Folliculi* 3-pollicares, tomento brevi fulvo pilosa.

Chiefly distinguished from the last by being much more robust and leafy, and much less hairy, by having a strong musky smell, by the bifid posterior petals, and by the larger hairy fruit. Both species, however, are imperfectly known, and, like all this group, require careful examination and comparison in the living state.

14. **D. glaciale** (H.f. et T.); caule simplici folioso, foliis reni-formibus tripartitis segmentis late cuneatis palmatim multifidis lobis linearibus, racemo corymboso, calcare saccato conico obtuso sepalis breviore, ovariis 4–5.

HAB. In Himalaya orient. interiori: Sikkim, alt. 16–18,000 ped.!— (Fl. Aug. Sept.) (*v. v.*)

Herba 3–6-pollicaris, tota pilis glandulosis patentibus hirsuta, et moschum putri-dum redolens. *Petioli* inferiores elongati, basi vaginantes. *Folia* diametro bipolli-caria. *Pedicelli* exteriores elongáti. *Bracteolæ* plures, alternæ, linearilobæ vel line-ares, suprema a flore remota. *Flores* inflato-subglobosi, maximi, pallide cærulei. *Sepala* membranacea, nervosa, ultra poll. longa, extus laxe pilosa, fere orbicularia, calcare ⅓-⅔-pollicari, obtusissimo. *Petala postica* lamina apice vix obliqua, biden-tata, atropurpurea, calcare subulato subincurvo; *antica* lamina dorso pilosa semibifida. *Folliculi* ½ poll. longi.

15. **D. Brunonianum** (Royle! Ill. 56); caule simplici folioso, fo-liis reniformibus semiquinquefidis, lobis cuneato-ovalibus grosse inciso-dentatis, floribus corymbosis, calcare late saccato conico obtuso, ovariis 5–6.

HAB. In Tibetia occidentali, in summis alpibus, alt. 14–18,000 ped.: Nubra! Ladak! Hangarang!—(Fl. Aug. Sept.) (*v. v.*)

Herba moschata. *Caulis* erectus, 6–8-pollicaris, rarius pedalis, viscoso-puberulus vel tomentosus. *Petioli* inferiores 3–5-pollicares, basi vaginantes. *Folia* adpresse pubescentia, 3–4-poll., dentibus apice glandulosis; *floralia* inferiora tripartita, su-prema lanceolata. *Pedicelli* erectiusculi, corymbosi, nudi, apicem versus bibracteo-lati, bracteolis calyci adpressis. *Flores* pallide cærulei. *Sepala* fere orbicularia, pol-licaria, membranacea, nervosa, calcare ½-poll. *Petala postica* lamina pallida vix obliqua, obovato-spathulata, biloba, calcare cylindrico incurvo obtuso piloso; *antica* utrinque pilosa, lamina bipartita. *Folliculi* 7 lin. longi, viscoso-puberuli.

This, which is the most northern of the musky group, is distinguished from all the others by the hairy pale-coloured posterior petals, a character which appears constant in a considerable series of specimens. It differs a good deal in habit from all the other species except the last, the leaves of which are very different, so that we do not hesitate to keep it distinct. At the same time we readily admit that it is quite possible that more extended observation will show that the characters derived from the follicles and petals are of less importance than we at present believe, in which case several of the species above described must necessarily be reduced.

17. ACONITUM, L.

Nirbisia, *G. Don, Syst. Gardening,* i. 63; Calthæ sp., *Ham. in Edin. Journ. Sc.* i. 249.

Sepala 5, inæqualia; supremum (*cassis*) convexum vel fornicatum, cætera plana. *Petala* 2 superiora intra cassidem abscondita, unguiculata, apice in saccum (*cucullum*) forma varium expansa, cætera minima vel abortiva. *Ovaria* 3–6.—Herbæ *perennes, erectæ,* foliis *palmatisectis.* Flores *ochroleuci, violacei, vel sæpius cærulei.*

This genus is entirely confined to the northern hemisphere, the species being chiefly European and north Asiatic. A few only are American. Some inhabit woods, others mountain pastures, and the latter are often very alpine. The Indian species are all temperate Himalayan, and occur in every part of that chain in nearly equal proportions, but most abundantly perhaps to the eastward in the humid parts of Nipal and Sikkim, where they grow in very wet places, generally near streams. Four of the Himalayan species are endemic, but three are common to these mountains and Europe. Of these, two inhabit the forest region, but one (the common *A. Napellus*) is in India always alpine, and confined to the driest regions in the interior.

There appears to be no necessity for following Reichenbach into the critical details by which he has illustrated this Protean genus, as most botanists appear convinced that he has enormously over-estimated the number of species. Most Aconites grow with great luxuriance in rich soil, and have besides been very extensively cultivated : they therefore vary much in luxuriance, and in the size of the flowers. The shape of the sepals and petals is also far from constant, and it is upon slight differences in these that Reichenbach relies for the discrimination of his species. These differences are merely of degree, and are so trifling, that an examination of the plates of his monograph of the genus, will, we think, satisfy most persons, that at least three-fourths of his species are mere varieties. In this opinion we are supported by the authority of Seringe, who seems to have studied the genus with great care in the mountains of Switzerland, as well as in a state of cultivation, and whose testimony to the great amount of variation in all parts of the flower is quite in accordance with what we have observed in the Indian species. The characters of the species are difficult to express in words with precision, as they are chiefly derived from variations in the shape of the posterior sepal or helmet, and of the petals, which are very irregular.

The roots of certain species of this genus constitute the celebrated *Bikh* poison of the Himalaya. The result of our inquiries into this interesting subject has been, that no individual species is particularly prized, but that several yield this virulent poison. The degree of virulence varies greatly according to the soil, exposure, climate, and altitude, at which the plant grows,—to such a degree indeed, that we have grounds for believing that the same species which is violently deleterious in humid shaded localities, is all but inert in drier, loftier, colder, and more sunny places. That this is no anomaly in the vegetable kingdom is notorious to persons familiar with the influence of external causes on the development of medicinal properties in the Hemp and Poppy. So far as our experience goes, *A. Napellus, ferox, palmatum,* and *luridum* are all extensively used as *Bikh*, and are indiscriminately called by that name throughout the Himalaya. We have not detected any characters by which the dried roots of these species can be specifically recognized, nor do we believe that any such exist. Their form and size seem to depend on local circumstances, and their colour on the mode of drying. With regard to native information, upon which so much stress is laid, our experience has proved it to be utterly worthless so far as regards the discrimination of the species of Aconite; even the most intelligent hill-men have no exact knowledge on the subject.

1. **A. Lycoctonum** (L. Sp. 753); foliis palmatis, racemis laxis paniculatis, floribus violaceis vel ochroleucis, casside conica vel cylindracea, petalorum ungue recto filiformi, calcare elongato cylindrico uncinato vel contorto, folliculis 3 divaricatis, seminibus transversim plicato-rugosis.—*DC. Prod.* i. 57; *Ledeb. Fl. Ross.* i. 66. A. læve, *Royle! Ill.* 56.

Hab. In Himalaya occidentali temperata, alt. 7–10,000 ped.: Kashmir! Chamba! Kanawer! Kumaon!—(Fl. Aug. Sept.) (*v. v.*)

Distrib. Europa! (excl. Britannia); Asia temperata!

Herba elata, foliosa, sæpius glabra, sed interdum pubescens vel etiam tomentosa, paniculatim ramosa. *Folia* diam. 6–10-poll., rotundato-reniformia, ultra medium palmatim 7–9-fida, lobis cuneato-ovatis trilobis et argute incisis; superiora sessilia, 5–3-partita, segmentis oblongis grosse incisis vel indivisis. *Racemi* elongati, axillares, laterales et terminales, puberuli vel tomentosi; bracteis lineari-lanceolatis minutis. *Flores* flavidi, ochroleuci vel pallide violacei, magnitudine valde varii, puberuli. *Cassis* dorso cylindracea vel subconica, antice in rostrum breve porrecta. *Petala* longe unguiculata, ungue filiformi erecto, apice in saccum dilatato, calcare cylindrico recurvo uncinato vel subcontorto apice obtuso, labello oblongo emarginato. *Ovaria* 3, glabra vel pubescentia. *Folliculi* ½–⅔-poll.

A well marked and widely diffused species, varying much in the size and shape of the helmet, and in the degree of curvature of the spur of the petals, which is either at once abruptly reflexed and convolute, or straight with a recurved tip. The latter shape is that assigned by authors to *A. orientale*, Miller, or *A. ochroleucum*, Willd., while the former is ascribed to the true *A. Lycoctonum*. In both, we find the shape of the spur very variable, as also in the Indian plant, which is identical in general appearance with the northern forms of *A. Lycoctonum*, except that it is usually somewhat smaller-flowered.

2. **A. luridum** (H.f. et T.); foliis palmatim 5-fidis, racemo laxiusculo simplici, floribus sordide rubicundis, casside postice gibbosa hemisphærica antice late et obtuse rostrata, petalorum ungue erecto brevi lato, cucullo horizontali maximo, calcare brevi lato obtusissimo, folliculis 3–5 erectis, seminibus triquetris lævibus.

Hab. In Sikkim interiori, alt. 14,000 ped.! (ad Tankra et Chola).—(Fl. Aug.) (*v. v.*)

Radix fusiformis. *Caulis* erectus, 2–3-pedalis, paucifoliatus, puberulus. *Folia radicalia* longe petiolata, pet. fere pedalibus, utrinque adpresse puberula, ultra medium 5-fida, lobis cuneato-ovatis grosse crenato-dentatis; *caulina* 5–3-partita, argute inciso-dentata; *floralia* bracteæformia, tridentata, lanceolata vel linearia. *Racemus* simplex, ½–1-pedalis. *Pedicelli* bracteis et floribus plerumque breviores, inferiores interdum remoti, elongati, 2–3-bracteolati. *Sepala* fulvo-tomentosa. *Cassis* long. ¾ poll., gibbere ⅓ poll. alto fere hemisphærico. *Petala* ungue erecto ⅓ poll. lato canaliculato, labello apicem versus porrecto, emarginato. *Ovaria* glabra vel pilosa.

This species is very distinct from any hitherto described in the form of the helmet, which is dome-shaped behind, with a very broad high arched projection. The petals resemble a hammer with a very short handle.

3. **A. palmatum** (Don, Prod. 196); panicula pauciflora, floribus viridi-cæruleis, casside convexa fornicata, petalorum ungue anguste lineari incurvo apice in cucullum globosum ecalcaratum dilatato, folliculis 5 erectis glabris, seminibus transversim plicato-rugosis.—*Wall.* *Cat.* 4723! *Royle, Ill.* 56.

HAB. In Himalaya temperata: Nipal ad Gossain Than, *Wall.!* Sikkim, in monte Tonglo, alt. 10,000 ped.! (Fl. Jun. Jul.)—(*v. v.*)

Caulis erectus, simplex, 2–3-pedalis, foliosus, glaber, panicula subramosa pauciflora. *Folia* longe petiolata, glabra, diam. 4–6-poll., circumscriptione rotundata, basi cordata, sinu latissimo, ultra medium palmatim 5-fida, lobis cuneato-ovatis grosse inciso-lobatis; *floralia* similia, minora. *Flores* in ramis elongatis subsolitarii, fere pollicares. *Cassis* convexa, altitudine longitudinem excedente, breviter rostrata. *Folliculi* 1–1½-pollicares.

4. **A. variegatum** (L. Sp. 751); ramis flexuosis, racemis laxis paucifloris, floribus viridescentibus vel cæruleis, casside altissime fornicata, petalorum ungue recto, calcare adscendente reflexo, ovariis 5, seminibus transversim plicatis.—*DC. Prod.* i. 59; *Ledeb. Fl. Ross.* i. 68.

HAB. In Himalaya orientali: Sikkim in valle Lachung, alt. 9000 ped.!—(Fl. Sept.) (*v. v.*)

DISTRIB. Europ. austr.! Caucasus!

Herba gracilis, debilis, ramis elongatis flexuosis imo scandentibus glabris vel adpresse puberulis. *Folia* 2–4-pollicaria, profunde palmatim 5-fida vel -partita, segmentis cuneato-ovatis inciso-dentatis; *floralia* similia sed minora et altius incisa. *Panicula* divaricato-ramosa. *Flores* ⅔–1-pollicares. *Cassis* antice abrupte rostrata. *Ovaria* glabra.

The occurrence of a species only known as a native of Europe and western Asia in the interior of Sikkim, without any known intermediate station, is very remarkable; but the Sikkim specimens agree so exactly with the species to which we have referred them, that the identity of the two cannot be doubted. Our specimens, however, are not numerous, and are in flower only. They were all obtained in one locality, and the species was not observed in any other part of Sikkim. This may, however, perhaps be ascribed to its slender subscandent habit, rather than to its rarity. It will probably be found to be a native of Western China, and to extend thence to Eastern Siberia, where a very similar species (*A. volubile,* Pall.) appears to be common.

5. **A. ferox** (Wall. in Ser. Mus. Helv. i. 160, non Plant. As. Rar. t. 41); foliis ovalibus 5-fidis, racemo terminali multifloro basi composito, floribus sordide cæruleis, casside alte fornicata acute et breviter rostrata, petalorum ungue incurvo-filiformi, calcare recurvo obtuso, folliculis 5 erectis pubescentibus, seminibus triquetris dorso transversim membranaceo-plicatis.—*DC. Prod.* i. 64; *Wall. Cat.* 4721, B! C! D! (*non* A, nec *Plant. Asiat. Rar.* t. 41 quoad iconem). A. virosum, *Don, Prod.* 196.

HAB. In Himalaya interiori temperata, alt. 10–14,000 ped.: Garhwal! Kumaon! Nipal! Sikkim! (Fl. Jul. Aug.)—(*v. v.*)

Caulis erectus, 3–6-pedalis, foliosus, molliter pubescens. *Folia* circumscriptione ovalia, basi cordata, utrinque pubescentia et subtus ad nervos pilosa, rarius subglabra,

4–5-pollicaria, palmatim 5-fida, lobis ovalibus vel oblongis basi cuneatis, superne pinnatifide incisis et grosse dentatis; *floralia* conformia, minora, suprema trifida vel lineari-lanceolata. *Racemus* terminalis, sæpe pedalis, basi compositus; *pedicelli* longi, floriferi patentes, fructiferi patentes vel erecti, bracteolis pluribus alternis muniti, apice dilatati. *Flores* ultra-pollicares. *Sepala* extus fulvo-pubescentia. *Petala* ungue longo incurvo superne in saccum magnum inflatum dilatata, calcare recurvo obtuso, labello elongato oblongo apice retuso. *Filamenta* pilosa. *Ovaria* plerumque 5, dense villosa.

As this is the best known and most extensively distributed *Bikh*, or poisonous Aconite, of the Himalaya, it appears desirable to retain for it Dr. Wallich's original name, notwithstanding that he has confused with it certain states of the next species, one of which he has figured in the 'Plantæ Asiaticæ Rariores.' As the descriptions given by Seringe, De Candolle, and Don apply chiefly, if not entirely, to this species, there is much less inconvenience in retaining than there would be in changing the name.

6. **A. Napellus** (L. Sp. 751); foliis multifidis, racemo denso vel laxo terminali interdum basi composito (in alpinis paucifloro), floribus cæruleis, casside hemisphærica sensim in rostrum breve producta, petalorum ungue incurvo filiformi, calcare brevi obtuso interdum brevissimo, folliculis 3–5 erectis (in planta Indica 5 tomentosis), seminibus triquetris lævibus.—*Seringe, Mus. Helv.* i. 162; *DC. Prod.* i. 62; *Torrey et Gray, Fl. N. Am.* i. 34; *Ledeb. Fl. Ross.* i. 69. A. dissectum, *Don, Prod.* 197; *Wall. Cat.* 4724! *Royle! Ill.* 56. A. ferox, *Wall. Cat.* 4721 A! (*non* B, C, D), *Plant. As. Rar. t.* 41. A. delphinifolium, *Reich.; Ledeb. Fl. Ross.* i. 70. A. multifidum, *Royle! Ill.* 56.

Planta polymorpha; formæ Indicæ sequentes :—

1. Caule erecto basi glabro superne tomentoso, foliis palmatim partitis, segmentis inciso-pinnatifidis, lobis linearibus, racemo simplici denso vel laxiusculo.—A. dissectum, *Don.* A. ferox, *Wall. l. c.*

2. Caule humili diffuso basi glabro superne pubescente, foliis lineari-multifidis, racemo paucifloro, floribus longe pedicellatis.—A. multifidum, *Royle!*

3. Caule humili foliisque adpresse puberulis, foliis rotundato-reniformibus palmatim 5-lobis, lobis obtuse inciso-crenatis.—A. delphinifolium γ, *Ledeb.* A. rotundifolium, *Kar. et Kir. in Led. Fl. Ross.* i. 740?

HAB. In Himalaya interiori alpina, alt. 10–16,000 ped.: a Gilgit, *Winterbottom!* usque ad Nipal, *Wall.!* et Sikkim!—(Fl. Jul.–Sept.) (*v. v.*)

DISTRIB. Europa australis! Asia et America temperata et arctica!

1. *Caulis* erectus, 2–3-pedalis, simplex, foliosus, glaber, apice tomento brevi fulvo pilosus. *Folia* glabra, 3–6-pollicaria, trisecta vel pedatim 5-partita, segmentis inciso-pinnatifidis, lobis divaricatis linearibus acutis. *Racemus* terminalis, densus vel (in spec. et ic. Wallichianis) laxus, pedicellis stricto-erectis. *Bracteæ* trifidæ vel lanceolatæ. *Flores* ¾–1-poll. *Sepala* extus puberula. *Petala* ungue longo incurvo, sacco parvo, calcare brevi obtuso subrecurvo, labello inflexo æquilongo.

2. *Caulis* diffusus, ½–1-pedalis. *Folia radicalia* 1–2-pollicaria, rotundato-reniformia, ad basin 5-partita, lobis inciso-multifidis; *caulina* pauca, lineari-multifida; *floralia* trifida vel linearia. *Flores* ultra-pollicares, in summo caule 3–5, cærulei;

I

petioli elongati, sæpe bipollicares. *Cassis* elongata, minus fornicata, quam in forma typica. *Petala* longe unguiculata, calcare brevissimo obtuso, labellum non æquans.

3. *Caulis* adscendens vel prostratus, 3–12-pollicaris. *Folia radicalia* numerosa, longe petiolata, reniformia, diam. 1–2-poll., ultra medium 5-fida, segmentis rotundato-trilobis vel obtuse tridentatis; *caulina* sessilia, palmatim 5-partita. *Flores* fere prioris, sepalis sæpe longius persistentibus.

This is at once the most widely diffused and the most variable Aconite, being extremely abundant in temperate Europe, Asia, and America, in mountain pastures, and ascending into the alpine region. In America and Asia it is found abundantly, even on the borders of the arctic zone. At low elevations it is very luxuriant, and as it grows generally in rich soil near villages or the huts of the mountain shepherds, it sports to a great extent. At high elevations it becomes very small, and assumes many forms, which, considered *per se*, would at once be regarded as specifically distinct, but which, when traced by the assistance of numerous suites of specimens, are found to present no well-defined characters.

To the Indian botanist who has not had an opportunity of observing the amount of variation to which this species is subject in different parts of the world, or of studying extensive suites of specimens in a dried state, the association of all these varied forms under one specific name will doubtless appear at first sight very surprising. The Himalayan forms, however, are quite similar to those of other countries. The smaller alpine states are the same as those of Siberia and North America, and some of the larger specimens are strikingly like Pyrenean and Spanish specimens, which exhibit a very peculiar facies, but which even Boissier does not consider specifically distinct.

Though this plant yields a part of the Bikh poison of the Himalaya, yet we are informed by our friend Colonel Munro that the roots of the alpine form are eaten by the hill-men of Kanawer as a pleasant tonic, under the same name (*Atees*) as those of the next species.

7. **A. heterophyllum** (Wall. Cat. 4722!); foliis vix lobatis, racemo multifloro simplici, floribus ochroleucis vel cæruleis, petalorum cucullo ecalcarato, folliculis 5 erectis, seminibus argute triquetris lævibus.—*Royle! Ill.* 56. *t.* 13. A. cordatum, *Royle! Ill.* 56. A. Atees, *Royle, Journ. As. Soc.* i. 459 (ex ipso auctore).

HAB. In Himalaya occidentali temperata, alt. 8–13,000 ped.: Dras et Kashmir! Simla! Kumaon!—(Fl. Sept. Oct.) (*v. v.*)

Radix fusiformis, perpendicularis. *Caulis* erectus, foliosus, simplex vel ramosus, 1–3-pedalis, glaber, superne velutino-pubescens. *Folia* radicalia petiolata, rotundato-reniformia vel cordata, obscure 5-loba, grosse duplicato inciso-crenata vel dentata glabra, caulina late cordata, brevissime petiolata vel amplexicaulia, floralia oblonga vel lanceolata. *Racemi* laterales et terminales, multiflori, laxi vel densi. *Pedicelli* erecti, floribus æquales vel longiores. *Bracteolæ* 2–3, submembranaceæ, ovatæ vel oblongæ, alternæ. *Flores* ultra-pollicares, ochroleuci, purpureo-venosi, vel læte cærulei. *Sepala* extus puberula; *cassis* convexa, navicularis; lateralia oblique ovalia, antica sinuosa, lanceolata. *Petala* ungue late lineari subincurvo, apice in cucullum subglobosum obtusum inflatum ecalcaratum dilatata. *Ovaria* 5, pubescentia. *Folliculi* ¾ poll. longi, puberuli, erecti.

According to Dr. Royle, the roots of this plant are employed in Indian Materia Medica as a tonic, under the name of Atees.

18. **CIMICIFUGA,** L.

Sepala 4–5, regularia, elliptica. *Petala* 3–5, rarius nulla, forma varia. *Ovaria* 1–8. *Folliculi* totidem. *Semina* ala scariosa lacera circumdata.—Herbæ *perennes*, foliis *bi-tri-ternatim sectis*, floribus *racemosis*.

One East Europe and Siberian and two or three North American species constitute the whole of the genus, which is distinguished from *Actæa* by the dehiscent fruit only, as in one of the American species the ovary is solitary.

1. **C. fœtida** (L. Syst. Nat. ed. 12. 659); foliolis ovatis lanceolatisve, *petalis* 2–4 emarginatis vel bifidis, ovariis 4–8.—*Ledeb. Fl. Ross.* i. 72. C. frigida, *Royle! Ill.* 57. Actæa Cimicifuga, *L.; DC. Prod.* i. 64. A. frigida, *Wall. Cat.* 4725! Actinospora frigida, *Fisch. et Meyer.*

HAB. In sylvis Himalayæ temperatæ, alt. 7–12,000 ped.: Kashmir, *Jacquemont! Royle!* Nipal, *Wall.!* Sikkim! Bhotan, *Griffith!*—(Fl. Jul.) (*v. v.*)

DISTRIB. Europa orient.! et Sibiria!

Herba elata, foliosa, subglabra, apice ferrugineo-tomentosa. *Folia* ternatim vel quinatim 2-3-pinnatisecta, foliolis 1½-3-poll. subtus ad nervos pubescentibus vel subglabris grosse inciso-serratis. *Racemi* simplices vel paniculam simpliciter ramosam elongatam sæpe pedalem formantes. *Flores* parvi, flavescentes. *Petala* forma valde varia, subsaccata, et fere integra, vel planiuscula ½-biloba, lobis apice incrassatis. *Folliculi* ½-pollicares, breviter vel longe pedicellati.

The form of the petals varies much, as well as the length of the pedicel of the fruit and the shape of the leaflets; nor can we find any character to distinguish the Indian plant from the common North Asiatic species. *C. Americana* is also very closely allied, but differs in having much more elongated racemes and longer palercoloured seeds.

19. **ACTÆA**, L.

Sepala 4–5, regularia, elliptica. *Petala* oblonga vel linearia, 4–5 vel plura. *Ovarium* solitarium, oblongum, stigmate sessili peltato. *Fructus* indehiscens, baccatus, polyspermus.—Herbæ *perennes*, foliis *bi-tri-ternatim sectis*, floribus *albidis racemosis*.

Two species, one common in the temperate parts of the northern hemisphere and in the Himalaya, the other confined to America (and perhaps not really distinct), constitute the whole of this genus.

1. **A. spicata** (L. Sp. 722); foliolis ovato- vel oblongo-lanceolatis inciso-serratis, racemo simplici, pedicellis filiformibus.—*DC. Prod.* i. 65; *Ledeb. Fl. Ross.* i. 71. A. brachypetala, *DC. Prod.* i. 65 (*excl. var. δ*). A. rubra, *Bigelow; Torrey et Gray, Fl. N. Am.* i. 35. A. arguta, *Nuttall; Torrey et Gray! l. c.* A. acuminata, *Wall. Cat.* 4726! *Royle! Ill.* 57.

HAB. In Himalayæ temperatæ sylvis: Marri, *Fleming!* Kashmir! Kumaon! Bhotan, *Griffith!*—(Fl. Mai. Jun.) (*v. v.*)

DISTRIB. Europa! Asia! et America! temp.

Caulis erectus, bipedalis, basi squamosus, aphyllus. *Folia* pedalia, decomposita, foliolis 1½-2½-pollicaribus. *Racemus* terminalis, 1-3-pollicaris. *Baccæ* ellipticæ vel subglobosæ.

The thick fleshy peduncles and petioles are probably sufficient to distinguish *A. alba* of Bigelow; but the other supposed species are unquestionably identical, the colour of the fruit alone appearing to vary.

Tribus V. PÆONIEÆ.

Sepala persistentia, herbacea, æstivatione imbricata. *Petala* plana.
Antheræ demum tortiles. *Ovaria* 2–5, multiovulata, disco carnoso
cincta. *Folliculi* totidem. *Semina* magna, albumine carnoso.

20. PÆONIA, L.

Sepala 5, persistentia, herbacea. *Petala* 5–10. *Ovaria* 2–5. *Folliculi* ovati, polyspermi. *Semina* subglobosa.—Herbæ *erectæ,* foliis
pinnatim decompositis, floribus *conspicuis solitariis purpureis vel albis.*

Natives of Europe and Northern Asia, and of North America west of the Rocky
Mountains, growing in mountain woods and pastures. The species have been much
cultivated, and floricultural botanists have devoted much labour to the discrimination
of many supposed species, chiefly of garden origin. Specimens in herbaria are often
very imperfect, but we observe that authors readily admit glabrous-leaved varieties
of the pubescent-leaved species (a character on which De Candolle divides the genus
into two sections), and that glabrous and densely tomentose fruited plants are not
considered specifically distinguishable. We are inclined to believe that all the erect-
fruited herbaceous forms belong to one species, as we can see no distinctions in the
shape of the leaves sufficient to distinguish from one another *P. officinalis* and *P.
peregrina.* We have not had an opportunity of seeing fruiting specimens of *P. ano-
mala,* L., the common Siberian species, which is undistinguishable in the herbarium
in a flowering state, but is characterized by the spreading carpels.

1. **P. officinalis** (L. Sp. 747) ; herbacea, foliis biternatim sectis,
foliolis incisis, lobis oblongis lanceolatisve acutis, folliculis 1–3 erectis
tomentosis vel glabris.—*DC. Syst.* i. 339, *Prod.* i. 65. P. peregrina,
DC. Prod. i. 66. P. intermedia, *C. A. Meyer in Led. Fl. Alt.* ii. 277 ;
Led. Fl. Ross. i. 74. P. Emodi, *Wall. Cat.* 4727 ! *Royle ! Ill.* 57.

HAB. In Himalaya occidentali temperata interiori, alt. 5–10,000
ped.: a Kashmir ! ad Kumaon !—(Fl. Mai.) (*v. v.*)
DISTRIB. Europa australis ! Sibiria Altaica !

Herba erecta, 1–2-pedalis, glabra. *Folia* 6–12-poll., subtus pallida, glabra vel
pubescentia. *Flores* albi rosei vel purpurei (in Indicis albi), extus bracteis 2–3
calyci adpressis foliaceis lanceolatis. *Sepala* ovalia vel rotundata, 2–3 ext. apice in
appendicem foliaceam lanceolatam expansa. *Discus* parvus annularis ovaria cingens.
Ovaria dense tomentosa vel glabra.

Himalayan specimens of this species are not distinguishable from those of Europe
and Siberia. The pubescence of the fruit is of no value as a character, for among
the few Indian specimens in fruit before us, that of Jacquemont is quite glabrous,
while Major Madden's is densely strigose.

II. DILLENIACEÆ.

Sepala 5, persistentia, æstivatione imbricata (quincuncialia), rarius
pluriserialia. *Petala* 5, decidua, æstivatione imbricata, unum sæpe
exterius. *Stamina* hypogyna, pluriserialia, indefinita. *Antheræ* basifixæ, introrsæ vel laterales, rarius extrorsæ, biloculares, longitudinaliter

vel poris 2 apicalibus dehiscentes. *Ovaria* discreta vel in axi mediante columna centrali cohærentia, unilocularia (rarius solitaria) intus vel basi ovulifera. *Styli* discreti, terminales. *Carpella* dehiscentia folliculária, vel subbaccata indehiscentia. *Semina* 1 vel plura, arillata, testa crustacea granulata vel cancellata, rhaphi brevi, amphitropa. *Embryo* minutus, hilo proximus. *Albumen* carnosum.—Arbores, frutices, *vel* herbæ, *interdum scandentes.* Folia *alterna, decidua vel persistentia.* Petioli *basi dilatati vel stipulis adnatis cito deciduis muniti.*

This Order, which is usually placed next to *Ranunculaceæ*, is undoubtedly very nearly allied to that Order, as well as to *Magnoliaceæ*, but has also a marked relationship to some Orders which are generally placed at a considerable distance, being connected both with *Ternströmiaceæ* and *Ericaceæ* by means of *Saurauja*, which, though referred by most botanists to the former of those Orders, is by Lindley and Planchon considered an undoubted member of the present family. We shall have an opportunity of entering fully into the question of these curious affinities when we describe the genus *Saurauja*, which, on account of its syncarpous fruit, we propose to place in or near *Ternströmiaceæ*.

The species of *Dilleniaceæ* are either tropical or Australian. The latter country probably contains the largest part of the Order, in the shape of small herbaceous plants or under-shrubs, sometimes climbing. No species are found beyond the tropics in America, but in the eastern hemisphere a few stragglers extend as far as the base of the Himalaya, and into southern China.

Tribus I. DELIMEÆ, DC.

Stamina superne dilatata, *antheris* remotis obliquis divaricatis.

1. **DELIMA,** L., DC.

Trachytella, *DC. Syst.* i. 410, *Prod.* i. 70.   Leontoglossum, *Hance in Walp. Ann.* ii. 18. iii. 812.

Flores hermaphroditi. *Sepala* 5. *Petala* 4–5. *Stamina* indefinita. *Ovarium* solitarium, depresso-subglobosum, in stylum subulatum attenuatum; *ovula* 2–3 e basi adscendentia. *Folliculus* ovalis, angulo interiori dehiscens. *Semen* solitarium, arillo cupuliformi denticulato cinctum.—*Frutex scandens sarmentosus,* foliis *asperrimis,* floribus *in paniculas terminales dispositis albis.*

Tropical climbing shrubs, all but one American.

1. **D. sarmentosa** (L. Sp. 736).—*Burm. Fl. Ind.* 122. *t.* 37.*f.* 1; *DC. Prod.* i. 69; *Wall. Cat.* 6632! *Bot. Mag.* 3058! *Hook. et Arn. Bot. Beech.; Benth. Kew Journ. Bot.* iii. 256. D. intermedia, *Blume Bijd.* 4; *Hassk. Pl. Jav. Rar.* 176. Actæa aspera, *Lour. Fl. Coch. ed. Willd.* i. 405. Tetracera sarmentosa, *Willd.; Roxb. Fl. Ind.* ii. 645. Trachytella Actæa, *DC. Prod.* i. 70. Leontoglossum scabrum, *Hance, l. c.*

α. *glabra;* fructu glabro.

β. *hebecarpa;* fructu piloso.—D. hebecarpa, *DC. Syst.* i. 407, *Prod.* i. 70; *Deless. Ic. Sel. t.* 72; *Wall. Cat.* 6633!

HAB. In Zeylania! Malaya! Ava! Chittagong! Silhet! et Assam!—
(*v. v.*)

DISTRIB. Java! Ins. Philippin.! China australis!

Folia obovata, ovali-oblonga vel late lanceolata, obtusa vel acuta, 2–5 poll. longa, 1–2 lata, nervosa, nervis superne exaratis subtus prominulis parallelis numerosis, scaberrima et utrinque cum ramis parce adpresse pilosa, integra subcrenata vel serrata, serraturis mucronatis. *Panicula* divaricato-ramosa, adpresse pilosa vel tomentosa, multiflora, hinc inde foliosa. *Flores* diametro ¼–½-poll. *Sepala* reflexa.

A very variable and widely diffused plant. The hairy-fruited variety is well marked, but does not appear to possess any other character, nor to be distinguished in any way in distribution or range. The shape of the leaves is very variable, those on barren shoots and on young plants being larger, more hairy, and more conspicuously serrated. According to Planchon (in Herb. Hook.) the Linnæan specimen of *D. sarmentosa* has hairy fruit. As the glabrous-fruited variety is common in China, there can be no reasonable doubt that Loureiro's *Actæa* is correctly referred here. His *Calligonum* (l. c. p. 418, *Trachytella Calligonum*, DC. Syst., Prod. i. 70) is more doubtful, being described *fructu gemino polyspermo*, which agrees with no Indian *Dilleniacea*. Indeed Loureiro, in comparing his plant with *Delima sarmentosa* of Burmann, expressly says that the fruit is very different. The *Piripu* of Rheede (Hort. Mal. vii. t. 54), which is referred by Willdenow to *D. sarmentosa*, but which De Candolle rejects under the name of *D. Piripu*, as too different in appearance to be considered to belong to the present species, appears to us to be *Polygonum Chinense*, L. If Rheede's figure be compared with that of Burmann (Fl. Ind. t. 30. f. 3) the resemblance will, we think, strike every one. Rheede's description is, however, very defective.

2. TETRACERA, L.

Flores hermaphroditi vel abortu polygami. *Sepala* 4–6 ; *petala* totidem. *Stamina* indefinita. *Ovaria* 3–5, ovulis pluribus biserialibus. *Folliculi* totidem, crustacei, nitidi, angulo interiori dehiscentes. *Semina* 1–5, arillata.—*Frutices scandentes, vel rarius* arbores. Paniculæ *terminales, vel ramulos axillares foliosos terminantes.*

Natives of tropical Asia, Africa, and America. Continental India north of Malaya possesses only two species, but in the islands of the Indian Archipelago they appear more numerous, and some of those described by Blume from Java may yet be found in the Malayan Peninsula. *T. Heyneana* (Wall. Cat. 6630!) is (as has been noted by M. Planchon in Herb. Hook.) a Euphorbiaceous plant.

1. **T. lævis** (Vahl, Symb. iii. 71); foliis glaberrimis superne nitidis nervis distantibus, sepalis intus sericeo-pilosis, folliculis 1–2-spermis.—*Wall. Cat.* 6627 ! *DC. Syst.* i. 402, *Prod.* i. 68. T. Malabarica, *Lam. Ill. t.* 485. *f.* 1? T. Rheedei, *DC. Syst.* i. 402, *Prod.* i. 68 ; *W. et A. Prod.* i. 5 ; *Wight, Ic. t.* 70 ; *Rheed. Mal.* v. *t.* 8.

HAB. Zeylania! Malabaria! Concan, *Graham.*—(*v. s.*)

Frutex scandens, ramis rigidis angulatis, cortice nitido cinereo. *Folia* oblonga vel lanceolata, utrinque angustata vel apice abrupte acuminata, integerrima vel remote et inconspicue paucidentata, 3–5 poll. longa, 1–2 lata, nervis primordialibus utrinque 5–6. *Paniculæ* pauci- vel multifloræ, glabræ, rarius parce strigoso-puberulæ. *Bracteolæ* ad ramificationes parvæ, lineares. *Sepala* ciliata, extus glabra. *Semina* 1–2, nitida, arillo inciso-lacero, segmentis late lanceolatis.

The principal characters by which this species is distinguished from the next are its greater smoothness and narrower nearly entire leaves, and the number of seeds, in *T. lævis* 1–2, in *T. Assa* generally more. We have not a sufficient number of specimens to enable us to judge of the validity of these differences, but in any case Vahl's description applies to the present and not to the next species, as has been correctly observed by Wallich. Wight and Arnott, however, have referred it to *T. Assa*, considering that it must be the same as *T. Malabarica* of Lamarck, from whom he received his specimens. This is not quite conclusive, because Lamarck may have had both species before him. His figure certainly resembles *T. Assa*, but he represents only one to two seeds, and his description, in many points, seems applicable to the peninsular plant. He states that his specimens were from Sonnerat, who may have communicated to him both species, as he collected in the Malayan Archipelago, as well as on the continent of India.

2. **T. Assa** (DC. Syst. i. 402, Prod. i. 68); foliis superne glabris subtus præsertim ad nervos adpresse pilosis nervis approximatis, sepalis utrinque glabris ciliatis, folliculis 3–5-spermis.—*W. et A. Prod.* i. 5 (*in adnot.*); *Wall. Cat.* 6629. T. Malabarica, *Lam. Ill. t.* 485, *f.* 1? T. dichotoma, *Bl. Bijd.* 3. T. trigyna, *Roxb. Fl. Ind.* ii. 645.

HAB. In Chittagong! et in Peninsula Malayana: ad Penang! Malacca! Singapur!—(*v. v.*)

DISTRIB. Java! Ins. Philipp.!

Frutex scandens, cortice fusco vel pallido. *Rami* novelli strigoso-pilosi, rarius glabrescentes. *Folia* oblonga, utrinque acuta, remote serrato-dentata, 2½–4 poll. longa, 1–2 lata, nervis primordialibus 8–12. *Paniculæ* 3–12-floræ, strigoso-pilosæ. *Semina* atra, nitida. *Arillus* fimbriato-lacerus, segmentis filiformibus semen superantibus.

DC. says that there is no arillus, but his specimens probably had only abortive seeds. Roxburgh describes it as orange-coloured wool.

3. **T. Euryandra** (Vahl, Symb. iii. 71); foliis ovalibus vel oblongis crasse coriaceis supra lucidis subtus scabridis demum glabris, sepalis ovalibus extus pubescentibus.—*DC. Syst.* i. 402, *Prod.* i. 68; *Roxb. Fl. Ind.* ii. 646. T. lucida, *Wall. Cat.* 6631!

HAB. In insula Singapur, *Wall.!*—(*v. s.*)

DISTRIB. In Moluccis et Nova Caledonia.

Ramuli volubiles, glabri, juniores pilis scabris asperi; partes novellæ cinereo-tomentosæ. *Folia* 3 poll. longa, 1¾–2 lata, petiolo ¼-poll., integerrima vel remote denticulata, subtus pallida, secus nervos pubescentia. *Paniculæ* terminales, foliosæ, multifloræ. *Sepala* 5. *Petala* 3, oblonga. *Folliculi* 3, ovati, læves. *Semina* 2, nigra, arillo amplo laciniato cincta.

We have not seen specimens of the New Caledonian plant, so that the identification is a little doubtful. The description given by DC., however, agrees very well with Dr. Wallich's plant, which is unfortunately only in bud. We have therefore taken the character of the flower and fruit from DC., but the remainder of the description represents the Indian plant.

4. **T. macrophylla** (Wall. Cat. 6628!); foliis obovato-oblongis utrinque scabris, panicula elongata multiflora, sepalis oblongis nervosis extus scabridis pubescentibus, folliculis monospermis.

HAB. In Singapur, *Wall.!*—(*v. s.*)

Frutex verosimiliter scandens. *Ramuli* læves, tomento scabrido fulvo pubescentes. *Folia* 5–8 poll. longa, 2½–4½ lata, petiolo ¼-poll., obtusa vel subtruncata, interdum obtuse acuminata, margine subsinuata. *Panicula* fructigera elongata, fere pedalis: *ramuli* 2–3-pollicares, multiflori. *Sepala* ⅔ poll. longa. *Folliculi* 1–4, sepala æquantes, obovato-oblongi, argute rostrati, brunnei, nitidi. *Semen* solitarium, atratum, arillo cupuliformi fimbriato obliquo in sicco albido involutum.

Tribus II. DILLENIEÆ, DC.

Filamenta apice non dilatata; *antheræ* lineares, elongatæ.

3. ACROTREMA, Jack.

Sepala et *petala* 5. *Stamina* 15 vel plura, pluriserialia. *Filamenta* libera, erecta. *Antheræ* poris 2 terminalibus dehiscentes. *Ovaria* 3, discreta, 2- vel multiovulata. *Carpella* intus dehiscentia; *semina* membranaceo-arillata, testa crustacea cancellata.—Herbæ *perennantes subacaules*, foliis *magnis*, floribus *racemosis flavis*.

This genus is remarkable as being the only herbaceous one among the tropical section of this family. It is very peculiar in habit, consisting of low, large-leaved, almost stemless plants, which are natives of the southern parts of both peninsulas.

§ 1. *Folliculis polyspermis.*

1. **A. Arnottianum** (Wight! Ill. i. 9, t. 3); foliis obovatis basi late cordatis, racemis bracteis distichis ovatis dense imbricatis, pedicellis elongatis patentim pilosis.—A. costatum, *Wall. Cat.* 1117 B! A. Wightianum, *Wall. Cat.* 3669! (*non W. et A.*)

HAB. Malabar et Courtalam, *Wight!*—(*v. s.*)

Rhizoma decumbens, lignosum, fibrillos crassos emittens. *Folia* e collo plura, obovata, argute dentata, basi cordata, 6–12 poll. longa, 3–4 lata, longe ciliata et utrinque præsertim supra nervos pilis laxis tecta, cæterum glabra. *Petioli* 1–3-pollicares, late alati, vaginantes. *Racemi* axillares vel in axilla folii delapsi, 2–4-pollicares, bracteis ovatis, integris vel bilobis, membranaceis, fuscis, laxe pilosis demum fere glabris dense imbricati. *Pedicelli* 3–4-pollicares, cum calycibus laxe hirsuti. *Sepala* 5 lineas longa.

2. **A. uniflorum** (Hook.! Ic. Plant. t. 157); foliis obovato-oblongis basi angustatis rotundatis, racemis brevissimis, bracteis lanceolatis dense imbricatis, pedicellis elongatis adpresse pilosis.

HAB. In Zeylaniæ montosis!—(*v. s.*)

Rhizoma horizontale, elongatum, lignosum. *Caulis* abbreviatus, foliosus. *Folia* 4–8 poll. longa, 1–3 lata, denticulata, superne inter nervos et versus marginem longe pilosa, cæterum glabra vel scabra, subtus pallida, secus nervos adpresse pilosa, repando-denticulata, juniora subplicata. *Petioli* 1–2-poll., anguste marginati, basi vaginantes. *Racemi* laterales in axillis foliorum superiorum brevissimi, simplices vel a basi ramosi, bracteis oblongis vel lanceolatis integris vel bidentatis tecti, dense adpresse tomentosi. *Pedicelli* 1–2-pollicares, cum calycibus pilis adpressis hirsuti. *Flores* iis *A. Arnottiani* dimidio minores.

A good deal like the last, but easily distinguishable by its less membranous leaves, which are more narrowed at the base, by its much shorter racemes, which are often quite concealed by the sheathing bases of the leaves, and by the much smaller flowers. There is an extensive series of specimens of this species in the Hookerian Herbarium, from which we learn that it flowers in its first year, and that in young plants, before the rhizoma is developed, the leaves are considerably smaller, proportionally narrower, more rugose, and sometimes bullate.

3. **A. lanceolatum** (Hook. Ic. Plant. sub t. 157); foliis anguste lanceolatis acutis sinuato-dentatis distanter nervosis superne glabris nitidis subtus ad nervos adpresse pilosis.

HAB. In Zeylaniæ montibus temperatis, *Wight! Thwaites!*—(*v. s.*)

Folia anguste spathulata vel fere linearia, 3–6 poll. longa, ½ poll. lata, superne argute sinuato-dentata, dentibus glandula apiculatis basin versus angustata dentibus obtusioribus: *petiolis* abbreviatis, alatis. *Inflorescentia A. uniflori.* *Pedicelli* 1–2-pollicares, laxe patentim pilosi.

This species is only known from a few very imperfect specimens in the Hookerian Herbarium: these appear to be young plants, the rhizoma being scarcely developed. There are traces of an inflorescence like that of the last species, but no flowers in a state fit for examination.

§ 2. *Folliculis 1–2-spermis.*

4. **A. costatum** (Jack, Mal. Misc., et in Hook. Bot. Misc. ii. 82); foliis obovatis basi sagittatis, racemis scapiformibus erectis laxis, bracteis lanceolatis non imbricatis, floribus breviter pedicellatis.—*Wall. Cat.* 1117 A! A. Wightianum, *W. et A. Prod.* i. 6; *Wight! Ill.* i. 9.

HAB. In Travancor, *Wight!;* Malaya, ad Penang et Singapur, *Jack! Wall.!*—(*v. s.*)

Rhizoma lignosum, subhorizontale. *Folia* 4–6 pollices longa, ½–2 lata, dentato-serrata, scabra, superne secus costam inter nervos et versus margines molliter pilosa, subtus pallida, secus nervos adpresse-pilosa. *Petioli* brevissimi, auriculati, vaginantes. *Scapi* (cum pedicellis et calycibus) patentim pilosi, supra medium floriferi; *pedicelli* bracteis duplo longiores. *Sepala* ½-pollicaria. *Stamina* 15. *Ovaria* biovulata, ovulis axi insertis collateralibus adscendentibus.

The materials at the disposal of Wight and Arnott at the time of the publication of the Prodromus were so imperfect that they did not discover that their specimens belonged to two different species. One specimen (belonging to the present species), which had good flowers and fruit, was employed for the analysis of the flowers given in the Prodromus, but all the others belonged to *A. Arnottianum,* which alone occurs in the Wallichian Herbarium, under the name of *A. Wightianum.* Dr. Wight had, however, retained in his own collection the specimen of *A. costatum,* along with one of those of *A. Arnottianum;* and when he had occasion to revert to the subject for the 'Illustrations,' having acquired additional materials, he detected the differences, which he has clearly indicated in that work. Dr. Wight has also pointed out the probable identity of the *A. Wightianum* of the Prodromus with *A. costatum,* Jack; and after a comparison of the solitary specimen from Travancor in the Wightian Herbarium, with those of Jack and Wallich, we can find no differences. As the description of *A. Wightianum* in W. A. Prod., which must be considered the authority for the species, agrees in all essential points with *A. costatum,* the former name must necessarily be suppressed.

4. **SCHUMACHERIA,** Vahl, Arnott.

Sepala 5. *Petala* 5. *Stamina* indefinita, unilateralia, monadelpha, pluriserialia, filamentis in columnam brevem oblique cylindricam coalitis. *Antheræ* subsessiles, lineari-oblongæ, obtusæ, apiculatæ, biloculares; loculis lateraliter dehiscentibus. *Ovaria* 3, discreta, dense pilosa, uni-

K

ovulata. *Styli* filiformi-subulati. *Carpella* indehiscentia. *Semen* erectum, subglobosum, basi arillatum, testa crustacea.—Frutices *scandentes*, ramis *rigidis flexuosis*, foliis *coriaceis conspicue penninerviis*, spicis *axillaribus vel terminalibus paniculatis*, floribus *sessilibus secundis bibracteolatis.*

An obscure and imperfectly described genus of Vahl's, identified and characterized by Arnott in 1834. It appears to have more affinity with the Australian forms of the Order than any other tropical genus has. The species are all natives of Ceylon, and require study in their native country to determine the amount of variation to which they are subject.

1. **S. angustifolia** (H.f. et T.); foliis oblongo-lanceolatis longe acuminatis serratis, spicis axillaribus foliis brevioribus.

HAB. In Zeylania, *Walker! Gardner!* etc.—(*v. s.*)

Rami juniores sericeo-incani. *Folia* basi rotundata vel angustata, superne glabra nitida, subtus præsertim ad nervos adpresse pilosa, 4–6 poll. longa, 1–1¼ lata. *Petioli* basi vaginantes, ½-pollicares. *Paniculæ* 1–2-pollicares, simplices vel ramosi.

2. **S. alnifolia** (H.f. et T.); foliis late ovalibus utrinque obtusissimis sinuato-crenatis, spicis axillaribus ramosis foliis dimidio brevioribus.

HAB. In Zeylaniæ montibus, *Gardner!*—(*v. s.*)

Rami pilis adpressis scabri. *Folia* crassa, coriacea, 4–6 poll. longa, 3–4 lata, superne glabra, subtus pubescentia.

3. **S. castaneæfolia** (Vahl, Act. Hafn. vi. 122); foliis late oblongis crenatis, panicula terminali divaricato-ramosa multiflora.—*Arn. in Edin. N. Phil. Journ.* xvi. 315; *Wight! Ill.* i. 9. *t. 4.*

a. Vahlii (Arn. l. c.); foliis utrinque acutis.

β. Grahamii (Arn. l. c.); foliis utrinque vel basi rotundatis.

HAB. In Zeylania!—(*v. s.*)

Rami juniores incani. *Folia* oblonga, forma admodum varia, breviter petiolata, 4–6 poll. longa, 2½–3½ lata; inferiora multo majora, interdum fere pedalia, crassa, superne glabra, subtus ad nervos puberula. *Panicula* interdum foliosa. *Flores* magnitudine varii.

5. **WORMIA,** Rottb.

Capellia, Blume.

Sepala 5, coriacea. *Petala* 5. *Stamina* indefinita. *Antheræ* basifixæ, lineares, apice poris dehiscentes, omnes conformes, vel interiores elongatæ, patentim recurvæ. *Ovaria* 5–10, multiovulata, axi vix cohærentia, stylis longis subulatis terminata. *Carpella* demum ad suturam ventralem, dehiscentia. *Semina* arillata, testa crustacea.—Arbores *interdum excelsæ*, floribus *conspicuis, flavis*, foliis *penninerviis*, stipulis *petiolo adnatis, cito deciduis vel rarius persistentibus.*

This genus was founded by Rottböll in 1783, on a Ceylon plant. Decandolle, in the Systema, united with this the *Lenidia* of Poiret,-founded on a Madagascar plant, and added the *Dillenia dentata* of Thunberg as a third species, but with well-founded doubts as to the propriety of distinguishing it from Rottböll's plant. The fourth and last species in DC. Syst. Veg. is a New Holland plant, *D. alata* of Brown. Of this we have seen no authentic specimen; but a plant from Cape York collected by Mr. Macgillivray during the voyage of the 'Rattlesnake' agrees perfectly with the description given by DC. It has, however, the inner stamens elongated and re-curved, as in many *Dilleniæ*, and in the genus *Capellia* of Blume, which is otherwise, both in habit and character, identical with *Wormia*. It appears therefore unadvisable to retain this genus *Capellia*. Jack had, indeed, some time before the publication of the Bijdragen, described Blume's *Capellia* as a genuine *Wormia*, adopting the same view of the limits of the genus that we now do, and making its characters depend chiefly on the dehiscent fruit. We have, however, availed ourselves of the character indicated by Blume, as a means of dividing the genus *Wormia*. We have included provisionally in the genus several species which have hitherto been referred to *Dillenia*, but which have so entirely the aspect of the known species of *Wormia*, that in all probability they will be found, when better known, to be members of it. Except one Madagascar and one New Holland plant, the species are all tropical Indian.

Sect. 1. CAPELLIA, Blume.—*Stamina* interiora longiora, patentim recurva.

1. **W. excelsa** (Jack, Mal. Misc. et in Hook. Comp. Bot. Mag. i. 221); foliis ovalibus acutis denticulatis, petiolis late marginatis, ra-cemis oppositifoliis elongatis multifloris.—Capellia multiflora, *Blume, Bijdr.* 5.—*Cuming, No.* 2358!

HAB. In Peninsula Malayana ad Malacca! et Singapur!—(*v. s.*)
DISTRIB. Java, *Blume!*

Arbor excelsa. *Folia* 4–12-pollicaria, in petiolum 1–2-pollicarem late marginatum sensim angustata, glabra, subtus ad nervos sparse pilosa, juniora cum pedunculis et omnibus partibus novellis sæpe floccoso-tomentosa. *Racemi* simplices, rarius sub-ramosi, folia æquantes vel superantes; bracteæ lanceolatæ, cito deciduæ. *Pedicelli* alterni, apice clavati, ½–1 poll. longi. *Sepala* late ovalia, glabra. *Petala* late obo-vata, bipollicaria. *Stamina* exteriora flava, interiora purpurea. *Folliculi* 6–8, li-neari-oblongi.

2. **W. oblonga** (Wall. Cat. 951!); foliis ovali-oblongis integris vel obscure crenatis, petiolis non marginatis, pedunculis oppositifoliis 2–5-floris paniculam terminalem formantibus.

HAB. In Peninsula Malayana: Penang, *Wall.!* Malacca, *Griffith!*— (*v. s.*)

Arbor? *Folia* 4–8-poll., petiolo 1½-pollicari gracili basi vix dilatato, glabra vel subtus ad nervos puberula. *Racemi* folia superantes, flexuosi, puberuli; pedi-celli clavati, apice pilosi. *Sepala* crassa, coriacea, fere orbicularia, extus adpresse sericeo-pilosa. *Petala* late obovata, ultra 2 poll. longa. *Ovaria* 8–10, multiovulata.

Sect. 2. EUWORMIA.—*Stamina* æquilonga.

3. **W. triquetra** (Rottb. Nov. Act. Hafn. ii. 532. t. 3); foliis late ovalibus subtruncatis grosse repando-dentatis vel sinuatis, petiolis (nisi supremis) non marginatis, racemis 5–6-floris folia subæquantibus. —*DC. Prod.* i. 75. W. dentata, *DC. Prod.* i. 75; *W. et A. Prod.* i. 7 *in adnot.* Dillenia dentata, *Thunb. in Linn. Tr.* i. 201. *t.* 20.

HAB. In Zeylania!—(*v. s.*)

Arbor. *Rami* glabri, partes novellæ interdum tenuissime incanæ, cito glabre-scentes. *Folia* glabra, coriacea, 5–8 pollices longa, 4–5 lata, basi rotundata, versus apicem profundius sinuata. *Petioli* non dilatati, superne canaliculati, 2–3-pollicares. *Racemi* 6-pollicares; pedicelli superne clavati, glabri, pollicares. *Sepala* ovalia, extus adpresse sericea. *Petala* 1¼-pollicaria, obovata. *Ovaria* 5.

In this species, as in *W. Madagascariensis* and *alata*, the lateral foliaceous processes on the petioles break off spontaneously, leaving a circular scar on the stem. On the petioles of the youngest leaves they are occasionally somewhat persistent, and at the apex of the branch there is a sheathing bract, (evidently a dilated petiole, as it occasionally has the lamina developed,) in which the terminal bud is enclosed.

4. **W. pulchella** (Jack, Mal. Misc. et in Hook. Comp. Bot. Mag. i. 221); foliis obovatis obtuse mucronatis integerrimis, petiolis non marginatis, pedunculis unifloris.

HAB. In Malaya ad Malacca, *Griff.!*—(*v. s.*)

DISTRIB. Sumatra, *Jack*.

Arbor humilis. *Folia* 4–5-pollicaria, petiolo fere pollicari, glabra, coriacea. *Pedunculi* oppositifolii, 1½–2-pollicares, ebracteati. *Sepala* late ovalia, glabra, pollicaria. *Folliculi* 5. *Semina* pauca, arillo rubro pulposo.

Griffith's specimens are very imperfect, but correspond in everything with Jack's description. They are not in fruit; we have therefore taken the character of the seeds from the Comp. Bot. Mag. Jack describes the peduncles as axillary, but in Griffith's specimens they are evidently leaf-opposed, and only appear axillary in consequence of several leaves growing close together towards the extremity of the branchlet.

5. **W. retusa** (H.f. et T.); foliis obovatis sinuato-dentatis subtruncatis et retusis, pedunculis oppositifoliis 1–3-floris.—Dill. retusa, *Thunb. in Linn. Tr.* i. 200. *t.* 19; *Lam. Ill. t.* 492. *f.* 2; *DC. Prod.* i. 76; *Wall. Cat.* 6625! *W. et A.! Prod.* i. 6.

HAB. In Zeylania, *Thunberg.*—(*v. s.*)

Arbor, ramis glabris, junioribus puberulis. *Folia* 4–6-pollicaria, petiolo 1–1½-poll. *Sepala* ovalia, glabra. *Petala* obovata, pollicaria. *Ovaria* 5–6.

6. **W. bracteata** (H.f. et T.); foliis ovalibus vel obovatis crenatis, pedunculis plurifloris folia non æquantibus, pedicellis bibracteolatis.— Dillenia bracteata, *Wight! Ic. t.* 358.

HAB. In montibus provinciæ Maisor, in regione "Balaghat" dicta. —*Wight!* (*v. s.*)

Arbor, cortice cinereo rugoso, ramulis partibusque novellis sericeis. *Folia* ad apices ramorum conferta, 3–6 poll. longa, 1½–3 lata, nervis obliquis crebris parallelis, superne lucida glabra, subtus pallida, adpresse pubescentia, demum fere glabra. *Racemi* oppositifolii; *bracteæ* obovato-spathulatæ. *Sepala* ovalia, dorso sericea. *Petala* obovata, 1¼ poll. longa. *Ovaria* 5. *Folliculi* totidem, membranacei. *Semina* obovata, arillo parvo carnoso.

This appears a very distinct species. Probably Roxburgh's *Dillenia repanda* is the same, but his description is so imperfect that the point cannot be determined with certainty without the inspection of specimens. The locality assigned by Roxburgh to his plant is Hindostan,—that is to say, probably the mountains of Behar, the vegetation of which is analogous to that of the hilly country of Maisor and the Dekhan.

7. **W. integra** (H.f. et T.); foliis obovatis obtusis subintegris, pedunculis subsolitariis.—Dillenia integra, *Thunberg in Linn. Tr.* i. 199. *t.* 18; *Lam. Ill. t.* 492. *f.* 1; *DC. Prod.* i. 76.

Hab. In Zeylania, *Thunberg.*

Arbor. Folia integerrima vel a medio ad apicem obscure serrulata, subspitha-mæa, palmam lata. *Petioli* villosi, pollicares, canaliculati. *Sepala* oblonga. *Petala* obovata, 1¼-pollicaria (ex icone Thunberg.).

We have not seen anything like this from Ceylon; in general character it approaches very closely to *W. oblonga,* Wall., but that belongs to the first section.

6. DILLENIA, L.

Sepala et *petala* 5. *Filamenta* filiformia, pluriserialia; *antheræ* lineares, exteriores erectæ introrsæ, interiores recurvæ extrorsæ. *Carpella* 5–20, indehiscentia, cum axi centrali in pseudo-baccam calyce persistente involutam cohærentia. *Semina* in pulpa gelatinosa nidulantia, exarillata.—Arbores, foliis *penninerviis, sæpe maximis,* floribus *conspicuis, albis vel flavis.*

The species of this genus are all Indian, and inhabit the dense tropical forests among the mountains. One species skirts the base of the Himalaya to 28° N. lat. Most of them flower before the expansion of the leaves, which are generally of great size, and vary a good deal in shape. On this account the species are very difficult of discrimination, and it is possible that we may have reduced their number too much. We believe, however, that it is much more advantageous to science to limit our lists to the species which are well known than to establish new species on insufficient grounds; and we must leave to botanists in India who may have an opportunity of observing these trees in their native forests or in cultivation, the task of ascertaining the degree of variation to which they are subject, especially in size and shape of leaves, and in the length of the petiole.

Sect. 1. EUDILLENIA.—*Flores* albi. *Semina* margine pilosa.

1. **D. speciosa** (Thunb. Linn. Tr. i. 200); foliis petiolatis oblongis vel lanceolatis acutis argute serratis, floribus coætaneis solitariis maximis, carpellis viginti polyspermis.—*Sm. Exot. Bot. t.* 2, 3; *DC. Prod.* i. 76; *Ham. in Linn. Tr.* xv. 99; *Roxb. Fl. Ind.* ii. 650; *Wall. Cat.* 943 *excl. C! W. et A.! Prod.* i. 5; *Wight! Ic. t.* 823. D. elliptica, *Thunb. Linn. Tr.* i. 200; *DC. Prod.* i. 76. D. Indica, *L. Sp.* 745. Syalita, *Rheed. Mal.* iii. *t.* 38, 39.

Hab. In sylvis densis in regionibus montosis: Zeylania! Malabar, *Rheed., Wight!* Concan, *Graham;* Orissa, *Roxb.;* Behar, *M'Clelland!* secus basin Himalayæ a Nipalia, *Wall.!* ad Assam! in Silhet! Chittagong! Ava! et per totam peninsulam Malayanam, *Griffith!*—(Fl. Jun. Jul.) (*v. v.*)

Distrib. Per totam Indiam tropicam.

Arbor mediocris, late comosa. *Folia* oblonga vel lanceolato-oblonga (arborum juniorum basin versus longe angustata), acuta vel abrupte acuminata, 8–10 poll. longa, 3–4 lata, petiolo 1–2-poll., superne glabra, subtus ad nervos cum petiolis subpilosa. *Flores* diametro 9-pollicares (*Roxb.*). *Sepala* fere rotundata, crassissima, glabra. *Petala* late obovata. *Antheræ* lineares, exteriores erectæ, interiores flavæ, patentes. *Fructus* calyce aucto inclusus, depresse subglobosus, diam. 3-poll., loculis carnosis circa axin crassum carnosum dense verticillatis, ibique inter se et cum placenta spongiosa partim cohærentibus, cæterum liberis; seminibus in axi numerosis compressis margine pilis simplicibus inarticulatis villosis, testa crassa granulata.

A widely distributed plant, which is also much cultivated in the hotter parts of India as an ornamental tree. It is, we think, doubtful whether the *Songium* of Rumphius be meant for this species. It is at any rate so totally unlike, that it is not desirable to quote it.

Sect. 2. COLBERTIA, Salisb., DC.—*Flores* flavi. *Semina* glabra.

2. **D. ovata** (Wall. Cat. 945!); foliis petiolatis ovatis margine denticulatis superne glabriusculis vel ad nervos puberulis subtus cum petiolis fusco-tomentosis, pedunculis coætaneis unifloris oppositifoliis.

HAB. In insula Penang, *Porter!*—(*v. s.*)

Folia 8 poll. longa, 5 lata, petiolo 1–1½-pollicari. *Flos* (a spec. discretus super eandem chartam affixus) majusculus. *Sepala* ovata, crasse coriacea, extus pube-scentia, sesquipollicaria.

This is seemingly very distinct from any other known species, but the specimen in the Wallichian Herbarium is very imperfect.

3. **D. aurea** (Sm. Exot. Bot. ii. t. 92, 93); foliis petiolatis ovato-oblongis vel obovatis remote crenato-denticulatis supra glabris subtus molliter pubescentibus, floribus ante folia enatis ramulos laterales breves terminantibus solitariis aureis.—*DC. Prod.* i. 76; *Ham. in Linn. Tr.* xv. 101; *Wall. Cat.* 6624! D. ornata, *Wall. Plant. As. Rar.* i. 20. *t.* 23, *Cat.* 947! Colbertia obovata, *Blume?*

HAB. In sylvis densis secus basin Himalayæ Nipalensis, *Ham.!* et in Ava in provincia Martaban, secus ripas fluminum Attran et Saluen, *Wall.!*—(Fl. vere) (*v. s.*)

DISTRIB. Java?

Arbor excelsa, ramis cinereis. *Folia* approximata, pedalia, petiolo pollicari. *Flores* magni, speciosi, odorati (diam. 3–4-poll., *Wall.*, iis *D. speciosæ* paullo minores, ex *Ham.*). *Pedunculus* in ramo brevi terminalis, pollicaris, crassus, bracteis aliquot parvis ovatis valde deciduis prope basin munitus. *Sepala* glauca, dorso villis longis sericeis cito deciduis vestita. *Petala* obovata, bipollicaria. *Ovaria* 8–12. *Fructus* (cum calyce) magnit. Pomi minoris. *Semina* plura, glabra.

There can be little doubt that the descriptions of Smith (or Hardwicke), Hamilton, and Wallich, are all referable to one species, which will probably be found to extend throughout the jungles along the base of the eastern Himalaya, and of the central axis of the Malayan Peninsula. In the figure in 'Exotic Botany' the styles are not well represented, but this is probably a mistake of the artist.

4. **D. scabrella** (Roxb. Hort. Beng. 43, Fl. Ind. ii. 643); foliis petiolatis ovali-oblongis denticulatis utrinque scabrido-pilosis subtus pallidis, floribus ante folia enatis secus ramos ad cicatrices foliorum delapsorum fasciculatis, pedicellis 2–3-bracteolatis, carpellis 5–7.— *Wall. Pl. As. Rar.* i. 20. *t.* 22, *Cat.* 944 A! *et* B! (*excl. folio magno, quod verosimiliter ad* D. auream *referendum est*). D. pilosa, *Ham.!* in *Linn. Tr.* xv. 102, *non Roxb.* Colbertia scabrella, *Don, Prod. Nep.* 226.

HAB. In sylvis densis Assam! et Silhet!—(*v. v.*)

Arbor 30–40-pedalis, ramosa. *Folia* 6–10 poll. longa, 4–6 lata; petiolo ½–1-poll., pubescente, basi dilatato, semiamplexicauli. *Flores* suaveolentes, diam. sesqui-pollicares; pedicelli sæpius terni, tuberculo insidentes, bipollicares; bracteolæ oblongæ, sparsæ vel oppositæ, interdum persistentes.　*Carpella* 5–7, circa axin carnosum verticillata, semina pauca gerentia.　*Testa* glabra, granulata.

5. **D. floribunda** (H.f. et T.); foliis late ovalibus petiolatis margine fere integris supra glabris subtus ad nervos adpresse puberulis, floribus ante folia enatis secus ramos supra tuberculos parvos umbellatis, pedicellis ebracteolatis.—Colbertia floribunda, *Wall. Cat.* 950 !

HAB. Martaban, in sylvis ad ripas Saluen fluminis, *Wall.!*—(*v.s.*)

Arbor. Folia 1½-pedalia, 10 poll. lata, coriacea, petiolo glabro tripollicari. *Flores* iis *D. pentagynæ* similes.

Two leaves and a truncheon of wood, bearing a few half-withered flowers, constitute all that is known of this plant.

6. **D. pentagyna** (Roxb. Cor. Pl. i. t. 20); foliis petiolatis vel subsessilibus oblongo-lanceolatis acutis basi longe angustatis denticulatis vel subrepandis adultis glabris junioribus utrinque subpilosis, floribus ante folia enatis secus ramos quasi in axillis foliorum delapsorum fasciculatis, pedicellis ebracteatis.—*W. et A.! Prod.* i. 5 ; *Ham.! in Linn. Tr.* xv. 100 ; *Roxb. Fl. Ind.* ii. 652 ; *Grah. Cat. Bomb.* 2.　D. augusta *et* pilosa, *Roxb. Fl. Ind.* ii. 652.　Colbertia Coromandeliana, *DC. Prod.* i. 75 ; *Wall. Cat.* 949 !　C. augusta, *Wall. Cat.* 948 !

HAB. In sylvis densis ad radices montium : Malabar ! Concan ! Dekhan ! Orissa ! Behar ! Malaya ! Ava ! Chittagong ! et secus basin Himalayæ ab Assam ad prov." Oude" dictam !—(Fl. Apr.)　(*v. v.*)

Arbor mediocris, late comosa.　*Folia* maxima, 1–2-pedalia (in arboribus junioribus interdum 4–5-pedalia), subtus pallida, adulta coriacea, glabra vel subtus puberula, juniora membranacea pilosa vel sericea.　*Petioli* 1–4-pollicares, marginati, basi dilatati, semiamplexicaules.　*Flores* super tuberculos paucos umbellati, diametro pollicares ; pedicelli 5–6, 1–2-poll.　*Sepala* ovata.　*Petala* oblonga.　*Stamina* 10 int. cæteris longiora.　*Ovaria* 5 ; *semina* 1–2, cæteris abortientibus.

Seemingly a widely distributed tree, very variable in the shape of its leaves.　The two supposed species distinguished by Roxburgh have never been seen in flower. We should, however, perhaps have kept *D. pilosa* provisionally distinct, on account of its sessile leaves, had it not been that Wight's specimens of *D. pentagyna* exhibit that character very markedly, and are nevertheless regarded by him, we believe justly, as only a state of *D. pentagyna.* These trees are well worthy the attention of Indian botanists, as it is only in that country that it can be finally decided whether several species be confounded under this name.

7. **D. grandifolia** (Wall. Cat. 946 !); foliis petiolatis anguste oblongis grosse inciso-dentatis utrinque pubescentibus costa subtus petiolis et caule furfuraceo-tomentosis.

HAB. Penang, *Wall.!*—(*v. s.*)

The specimen of this plant in the Wallichian Herbarium at the Linnean Society consists of two leaves, both imperfect towards the apex.　One of these is young ; the other was probably at least two feet long, as the portion preserved measures twenty-two inches.　There are no flowers nor fruit, but the tomentum of the stems and petioles renders it probable that the species is distinct from the last.　Wall. Cat. No. 943 C. is, we think, a leaf of the same species.

In the Hookerian Herbarium there is a specimen distributed as *D. grandifolia,* Wall., which is either a new species or a remarkable form of one of those described above.　Its leaves, which though young appear nearly fully developed, are ovate or

somewhat obovate, sharply denticulate, 10 inches long by 5 broad, with a petiole an
inch long. They are slightly silky above, but probably become glabrous with age, and
are pubescent below, especially on the nerves. The flowers, which unfortunately are
in a very imperfect state, seem the same as in *D. pentagyna.* For the present it appears sufficient to call attention to this plant; we really know nothing of the variations and mode of growth of these trees, and to found species on single specimens,
especially where the flowers and leaves are detached, as is almost always the case in
this genus, would lead to irremediable confusion.

III. MAGNOLIACEÆ.

Flores hermaphroditi, rarissime unisexuales.   *Sepala* et *petala* hypogyna, ternatim (rarissime quinatim) pluriserialia, æstivatione imbricata,
decidua.   *Stamina* indefinita, circa torum cylindricum inserta, libera;
antheræ basifixæ, loculis linearibus lateraliter vel introrse dehiscentibus.
Ovaria plurima (rarissime pauca aut solitaria), discreta sive lateraliter
inter se cohærentia, uniserialiter verticillata vel sæpius supra torum
elongatum spicatim disposita, unilocularia.   *Ovula* in sutura ventrali 2
vel plura, rarius e basi adscendentia solitaria, anatropa.   *Embryo* in
basi albuminis copiosi oleosi non ruminati, minutus, hilo proximus.—
Arbores *vel* frutices *sæpe aromaticæ;* foliis *alternis simplicibus integerrimis,* stipulis *lateralibus petiolo adnatis cito deciduis, rarius nullis.*

In this family the petals are always imbricated in more than one row, or, in other
words, the perianth always consists of more than two series. The sepals are often
identical in texture and appearance with the petals, but sometimes they are readily
distinguishable from them, and they are then usually three in number. In the Indian
species this ternary arrangement occurs more or less distinctly in all the species
which we have had an opportunity of examining; but other authors describe the
perianth of some species as pentamerous.

We follow the usual course in including *Wintereæ* as a tribe of *Magnoliaceæ.* The
absence of stipules, however, is so very marked a character, in an Order in which
these organs are so constantly and conspicuously present, that it may be questioned
whether it would not be more advisable to separate them. This is, however, a matter of little consequence, till the systematic value of natural groups is better established, as their position would in any case remain the same, their affinity being much
greater with *Magnolia* and its allies than with any other group.

The stipulation of *Magnoliaceæ* is very peculiar. In the leaf-bud each scale is
composed of a pair of stipules at first united throughout their whole length, but latterly more or less split. From the dorsum of the scale, at a distance below the apex,
which varies in each species, rises the rudimentary leaf, which is longitudinally folded
inwards in vernation. In the outermost scale of the bud the foliaceous portion of the
leaf is usually very small, and falls away at a very early period, leaving a distinct
cicatrix at the top of the very evident petiole, along which the two stipules, which
are united to form the scale, are adherent. After the development of the branch, the
stipules remain at first adherent to the petiole on each side, but very soon wither and
fall off, leaving an elongated cicatrix on the petiole, which varies in proportional
length to the petiole on the different species.

In the flower-bud the spathes are exactly analogous to the scales of the leaf-bud;
but the tendency to development is in reverse order: the innermost, which is adpressed to the flower, rarely shows any tendency to leaf-development, but splits to
the base before falling off; while in the outer spathes the petiole is generally distinct,
with a scar at its apex marking the spot from which the rudimentary leaf has fallen

away. In many species, indeed, this leaf is occasionally developed, and in some it is normally so.

The nature of the integuments of the seed in *Magnoliaceæ* has generally been misunderstood, except by Gärtner, whose account is quite accurate. The true structure has recently been pointed out by Asa Gray (Genera of N. Am. Plants, i. 61). The outermost coat, which is fleshy, and often of a bright scarlet colour, has generally been considered an arillus; it has, however, been traced by Asa Gray to the primine of the ovule, and correctly regarded as testa. It is traversed in its whole length from the hilum to the chalaza at the opposite end of the seed by the rhaphe. The inner crustaceous coat, usually considered as testa, is conspicuously marked at the end most remote from the hilum by the chalaza. A third coat may be distinguished, consisting of a very delicate membrane, which adheres pretty firmly to the albumen.

Dr. Wallich appears to have made a curious mistake as to the position of the embryo, unless indeed (in the Tent. Fl. Nap. p. 4) for ' umbilicus *internus*' we ought to read ' *externus*,' in which case his view would be the same as that suggested by Blume (Fl. Javæ, p. 9), that the true hilum is where the brittle seed-coat is inserted into the fleshy one,—a view which is manifestly only tenable on the supposition that the latter is arillus.

The lateral position of the rhaphe with respect to the ovule and seed is worthy of note. It is well represented by Mr. Sprague in the plates of Asa Gray's work just quoted, but is not noticed in the text.

The plants of this family are all more or less aromatic, and their flowers have often an extremely powerful perfume. The Himalayan species are large trees, and yield valuable timber. The bark of many of the American species possesses bitter and tonic qualities, but none of those of India are known to do so. In the tribe *Illicieæ* these tonic and aromatic properties are more marked; but their presence in the whole Order is indicated by the transparent dots of the flower, and by the glandular markings of the woody tissue.

The species of *Magnoliaceæ* are chiefly natives of mountainous countries. They are probably more abundant in Western China, in eastern continental India, and in the Indian Archipelago, than in any other part of the world. Many species occur in the more humid parts of the temperate Himalaya, but one only extends as far west as Kumaon. The western peninsula produces only two species, and Ceylon not more than one. From China several extend to Japan. North America, excluding Mexico, which seems to contain several species of this family, produces eight species. A few are natives of the West Indies and the mountainous parts of tropical South America. In Africa they appear to be entirely wanting.

Tribus I. WINTEREÆ, R. Br.

Ovaria simplici serie verticillata vel solitaria. *Stipulæ* nullæ.

1. ILLICIUM, L.

Florès hermaphroditi. *Sepala* et *petala* 12–36, multiserialia. *Stamina* numerosa, antheris adnatis. *Ovaria* 6–15, stylo subulato intus stigmatoso apiculata. *Ovula* solitaria, e basi loculi adscendentia.— Frutices *sempervirentes aromatici;* foliis *integerrimis, glabris, ad ramorum apices confertis;* floribus *axillaribus, solitariis vel ternis, flavidis vel purpurascentibus.*

Two species of this genus are natives of the warmer parts of the eastern United States, one inhabits Japan, and one Southern China. The Indian species will probably also be found to extend into the interior of Southern China. The fruit of the

Chinese species is largely exported to India and Europe under the name of Star-Anise.

1. **I. Griffithii** (H.f. et T.); foliis ellipticis vel lanceolatis utrinque acutis sæpe apice acuminatis, sepalis et petalis circa 24, staminibus totidem, carpellis 12–15 superne rostratis.—*Griffith! Itin. Notes*, 38, 80.

HAB. In montibus Khasia, in sylvis densis humidis, alt. 4–5000 ped., *Griffith!*—(Fl. vere.) (*v. v.*)

Frutex 10–15-pedalis, cortice griseo rugoso, ramis junioribus angulatis. *Gemmæ* squamis numerosis imbricatis involutæ. *Folia* nitida, (sicca) læte viridia, subtus fusco-lutea, 2–4 pollices longa, 1–2 lata. *Sepala* rotundata, subciliata. *Petala* exteriora late ovalia, sepalis majora, ½ pollicem longa, interiora gradatim minora et angustiora. *Filamenta* lata, plana; *antheræ* ovali-oblongæ, introrsæ. *Carpella* carnosa, endocarpio crasso coriaceo, in fructum superne planiusculum subumbilicatum 1⅓ poll. latum ⅔ poll. altum congesta, sed inter se non cohærentia, dorso convexa, superne in rostrum erectum vel subincurvum subulatum producta, superne inter rostrum et axin dehiscentia. *Semen* solitarium, testa nitida luteo-fusca, rhaphe superiore.

Though the species of *Illicium* are all very much alike in habit and in the shape of the leaves, they appear to possess sufficient marks of distinction in the flower and fruit. *I. Griffithii* is readily distinguished from the Chinese and Japanese species by the more numerous and strongly-beaked carpels. The flowers resemble those of *I. parviflorum*, but the petals are much more numerous; their colour is unknown. All parts of the plant are aromatic, even in the dried state; the fruit has not, either when fresh or dried, at all the smell of anise, but possesses a faint agreeable odour like that of the leaves and wood. It is rather a local plant in the Khasia hills. Griffith found it at Mamloo, near Churra, and it occurs also in the deep valley of the Kala-pani.

Tribus II. MAGNOLIEÆ, DC.

Ovaria secus torum elongatum spicata. *Stipulæ* conspicuæ.

2. **TALAUMA**, Juss.

Sepala 3. *Petala* 6 vel plura. *Gynophorum* sessile. *Ovaria* biovulata. *Carpella* lignosa, in fructum strobiliformem coalita, irregulariter et quasi circumscisse dehiscentia. *Semina* in foveolis receptaculi centralis persistentis pendula.—Arbores *vel* frutices, floribus *terminalibus solitariis.*

A very distinct genus, easily recognized when in fruit by the peculiar dehiscence of the carpels, and by the seeds adhering to the persistent axis after the separation of the greater part of each carpel. In this genus, as well as in *Magnolia* and *Michelia*, the cord by which the seeds are suspended is composed of a mass of highly elastic spiral vessels, which are capable of extension by the weight of the seed, and yet quite strong enough to support its weight for a considerable time. The seeds of *Talauma*, therefore, remain suspended to the woody central axis long after the carpels have fallen away. The species are all tropical or subtropical, and appear to be about equally numerous in the Old and New World. The Asiatic species hitherto described are four in number, all natives of Java and the islands of the Archipelago. One of these only, so far as we know, extends into the Malayan peninsula, but two very fine new species have been obtained from the mountainous countries north of Bengal. In the Madras peninsula and Ceylon this genus is wanting.

1. **T. Hodgsoni** (H.f. et T.); foliis obovato-oblongis, fructu

magno, carpellis subtetragonis argute rostratis diametro transversali longitudinale excedente, rhachi profunde excavata, foveolis rotundatis. HAB. In sylvis densis Sikkim exterioris subtropici, alt. 3–5000 ped.! —(Fl. Aprili.) (*v. v.*)

Arbor excelsa. *Rami* glabri, apice glaucescentes. *Folia* coriacea, margine sub-sinuata, basin versus angustata, apice obtusa vel acuminata, utrinque glabra, (in sicco) conspicue et crebre reticulata, 8–20 poll. longa, 4–9 lata, petiolo bipollicari. *Areola stipularis* petiolum fere æquans. *Flores* terminales, solitarii, pedunculo brevi crasso 1–2-annulato suffulti. *Spathæ* valde deciduæ. *Sepala* 3, ovalia, crassa, 3½ polli-caria, extus herbacea, apice et marginibus roseis. *Petala* 6, ovalia, albida, 3 inte-riora minora. *Ovaria* in capitulum ovatum stamina longe superans collecta, stylis reflexis squarrosa. *Fructus* ovalis, 4–6 pollices longus. *Carpella* diam. transv. fere pollicari, longit. 1½-pollicari, dorso irregulariter tuberculata, angulo superiore in rostrum producto, serie inferiore basi etiam rostrata.

We gladly avail ourselves of this fine plant, to commemorate the eminent services of our friend B. H. Hodgson, Esq., of Dorjiling, to whose exertions the Natural History of the Himalaya is so much indebted.

2. **T. Rabaniana** (H.f. et T.); foliis lanceolatis, fructu magno, carpellis elongatis obtuse rostratis diametro longitudinali transversale excedente, rhachi leviter excavata, foveolis verticaliter elongatis sub-tetragonis.

HAB. In montibus Khasia, in sylvis densis prope Nunklow!—(Fl. vere.) (*v. v.*)

Arbor excelsa, ramis glabris. *Folia* coriacea, utrinque glaberrima, (in sicco) con-spicue reticulata, 8–12 poll. longa, 2–4 lata, petiolo pollicari. *Areola stipularis* petiolo ⅓ brevior. *Pedunculi* terminales, solitarii, 1–2-annulati. *Flores* ignoti. *Fructus* ovalis, 4–6 pollices longus. *Carpella* irregulariter obovata, dorso pustulis minutis tuberculata, diam. transv. fere pollicari, longitudinali 1½-pollicari.

We found this large tree bearing ripe fruit in the month of October. As *T. Hodg-soni*, which ripens its fruit at the same season, flowers in early spring, this species will probably be found to do so too. We propose to dedicate this and another species of the Order to Lieutenants Cave and Raban, of the Silhet Light Infantry, to whom we are under great obligations for assistance in forwarding our pursuits while in the Khasia mountains, and whose gardens at Churra show what skill and perseverance may accomplish in overcoming the obstacles which an ungenial climate opposes to horticulture.

3. **T. mutabilis** (Blume, Fl. Jav. Magn. 35. t. 10, 11, 12 B); foliis ovalibus vel lanceolatis utrinque acutis, fructu parvo, carpellis mucrone brevi crasso recurvo instructis.—Manglietia Candollei, *Wall. Cat.* 6497! *non Blume.*

a; foliis ovalibus utrinque acutis subtus adpresse pilosis.

β; foliis ovali-oblongis acuminatis subtus puberulis.

γ; foliis oblongis vel lanceolatis fere glabris.

HAB. Penang, *Wall.!* Moulmein, *T. Lobb!*—(*v. s.*)

DISTRIB. Java, *Blume!*

Frutex 6–10-pedalis, cortice lævi fusco. *Folia* 6–12 poll. longa, 2–4 lata, forma admodum varia, basi acuta, apice longe acuminata, coriacea, nitida, venosa, superne glabra. *Petioli* 1–2-pollicares, basi incrassati, cylindrici, subgeniculati. *Flores* ter-minales, solitarii, pedunculo crasso annulato sericeo vel subvilloso suffulti. *Sepala* 3, crassa, late ovalia, convexa, viridescentia, 1–2 poll. longa. *Petala* 6, subæquilonga, obovata, alba. *Stamina* petalis plus triplo breviora. *Ovaria* 9–12. *Carpella* in strobilum ovalem compactum bipollicarem coalita, dorso gibba, confertim tuberculata, sordide viridia, crasse coriacea, vix lignosa. *Foveolæ* rhachidis scrobiculatæ.

Wallich's specimen in the Linn. Soc. Herb. has no flower; but the terminal peduncle from which it has fallen away is present, and the leaves agree with Mr. Lobb's specimen, which again we have been able to identify with the variety γ of Blume, from whom, in consequence of the paucity of our own materials, we have taken our diagnosis and description.

3. MANGLIETIA, Blume.

Sepala 3. *Petala* 6 vel plura. *Gynophorum* sessile. *Ovaria* 6- vel pluri-ovulata. *Carpella* sublignosa, inter se in fructum ovalem vel oblongum cohærentia, demum soluta, et medio dorso longitudinaliter dehiscentia.—Arbores *excelsæ*, floribus *terminalibus*.

This genus may be readily known, when in fruit, by the somewhat fleshy carpels cohering into a solid fruit. When in flower it is only to be distinguished from *Magnolia* and *Talauma* by the more numerous ovules. *Michelia* is in most cases readily distinguished by the numerous axillary flowers and the stipes of the gynophore. The species of *Manglietia* are all Asiatic; and one Javanese species, with the two described below, constitute all that is known of the genus.

1. **M. insignis** (Bl. Fl. Jav. Magn. 23); gemmis apicem versus fulvo-villosis, foliis lanceolatis, fructu oblongo purpureo.—Magnolia insignis, *Wall.! Tent. Fl. Nap. t.* 1, *Plant. Asiat. Rar.* ii. *t.* 182, *Cat.* 973!

HAB. In Nipalia, alt. 6–10,000 ped., *Wall.!; in montibus Khasia*, alt. 3–6000 ped.!—(Fl. vere.) (*v. v.*)

Arbor excelsa, ramis glabris rugosis crebre transverse annulatis. *Folia* coriacea, lanceolata vel oblongo-lanceolata, acuta vel acuminata, utrinque glaberrima, superne nitida, subtus pallida, (in sicco) crebre reticulata, 4–8 poll. longa, 1–2½ lata, petiolo vix pollicari; areola stipularis ⅓–½ petioli æquans. *Pedunculus* terminalis, brevis, crassus. *Flores* suaveolentes, ex albo rosei. *Alabastri* ovato-oblongi, bipollicares, spatha 1 subrotundata membranacea caduca involuti. *Sepala* 3, rubescentia, oblonga, obtusa, 3-pollicaria. *Petala* 9, forma varia, interiora sensim minora. *Carpella* purpurea, in conum oblongum 3–4-pollicarem dense compacta, axin versus cuneata, dorso (siccitate) tuberculato-rugosa. *Semina* 3–6.

We collected this species plentifully in the forests of the Khasia range, but unfortunately in fruit only. Our description of the flower is therefore entirely derived from Wallich. The species appears to vary much in the shape of the leaves, and we are not quite satisfied that all our Khasia specimens belong to one species. We can divide them easily into two sets, one with broad elliptic lanceolate very coriaceous leaves, the other with narrower, much larger, and thinner leaves. Both states, however, occur among Dr. Wallich's Nipal specimens.

2. **M. Caveana** (H.f. et T.); foliis obovato-oblongis obtusis apice breviter mucronatis vel obtuse acuminatis, fructu ovali vel subgloboso.

HAB. In montibus Khasia, alt. 2–3000 ped.!—(*v. v.*)

Arbor excelsa, cortice cinereo, ramulis crassiusculis rugosis glabris. *Folia* versus ramorum apices approximata, oblonga, apice rotundata et in acumen breve obtusum vel acutum producta, 8–10 poll. longa, 3–4 lata, petiolo 2-pollicari, coriacea, subtus glauca, utrinque (sicca) conspicue reticulato-venosa. *Areola* stipularis petioli ⅔ long. æquans. *Pedunculus* terminalis, solitarius, 1–2-pollicaris, glaber. *Carpella* in fructum 3–5-pollicarem coalita, dorso rotundata, siccitate tuberculis parvis albidis verruculosa. *Semina* 2–6.

Nearly allied to Blume's *M. glauca*, but apparently quite distinct. We have, however, seen no specimen of the Javanese plant, and know the Khasia species only in fruit. The origin of the specific name has been already given, under *Talauma Rabaniana*, at p. 75.

4. MAGNOLIA, L.

Sepala 3. *Petala* 6–12. *Gynophorum* sessile. *Ovaria* biovulata. *Carpella* coriacea, inter se libera, imbricato-spicata, dorso longitudinaliter dehiscentia.—Arbores *vel* frutices, floribus *terminalibus*.

The terminal flowers, the more densely spiked carpels, and the definite ovules, in general suffice to distinguish *Magnolia* from *Michelia*. There is, however, no broad line of distinction between the two, some *Micheliæ*, as we shall immediately see, being as it were intermediate. *Magnolia* is the least tropical genus of the Order. It is best known as an American genus, six species being described from the United States. There are, however, several Japanese and Chinese species; and the Himalayan ones which we are about to describe appear normal members of the genus.

1. **M. Campbellii** (H.f. et T.); foliis ovalibus vel ovatis utrinque glaberrimis vel subtus albo-sericeis, floribus ante folia enatis maximis, spathis dense fusco-pilosis, petalis 9–12, carpellis obtusis.—Magnolia, *Griffith! Itin. Notes*, 152.

HAB. In sylvis densis Himalayæ orientalis exterioris, alt. 8–10,000 ped.: Sikkim! Bhotan!—(Fl. Aprili.) (*v. v.*)

Arbor excelsa, interdum 150-pedalis, trunco erecto, ramis tortis patentibus, cortice pallido rugoso. *Folia* ovalia ovata vel oblonga, interdum anguste obovata, acute vel abrupte breviter acuminata, basi subcordata vel rotundata, interdum obliqua, 4–12 poll. longa, 2–6 lata, petiolo pollicari, tenuia, submembranacea, superne glaberrima, secus nervos (in sicco) glaucescentia, subtus glaberrima vel secus costam et nervos sericea, rarius tota superficie adpresse sericea, juniora dense tomentosa. *Areola stipularis* brevissima. *Flores* diametro 6–10-pollicares, pulcherrimi, suaveolentes, rosei vel rarius albi; *spathæ* 2 vel plures, late ovatæ, extus fusco-pilosæ, exteriores plerumque foliiferæ, intima flori adpressa. *Sepala* et *petala* conformia, 12–15, late ovalia, 3–5-pollicaria, 4–5 interiora minora. *Carpella* in spicam cylindricam 6–8-pollicarem approximata, ovalia, obtusa. *Semina* 1–2, testa aurantiaca.

This superb species, which is so conspicuous a feature in the scenery of Sikkim, will aptly record the services of Dr. Campbell, Resident at Dorjiling, in connection with the rise and progress of that important place, and also his many contributions to our knowledge of the geography and productions of the Himalaya. It flowers in the month of April, when quite leafless. The shape and clothing of the leaves varies more than is usual in the genus; on very young trees the leaves are quite glabrous, and much more membranous than on the adult plant.

2. **M. globosa** (H.f. et T.); foliis membranaceis ovatis superne glabris subtus præsertim ad nervos fusco-tomentosis, floribus coætaneis, petalis 6, carpellis breviter apiculatis.

HAB. In Sikkim interiori temperato, alt. 9–10,000 ped.!—(Fl. Jun.) (*v. v.*)

Arbor 40-pedalis. *Ramuli* adulti glabri, cortice lævi stramineo, juniores fuscotomentosi. *Folia* 5–9 poll. longa, 3–6 lata, petiolo 1–1½-pollicari, ovata acuta vel obtusiuscula, cum mucrone brevi, superne nitida, subglabra, subtus pallida, glaucescentia, ad nervos præsertim fusco-tomentosa, juniora dense tomentosa. *Areola sti-pularis* petiolum fere æquans. *Pedunculi* terminales, solitarii, dense tomentosi. *Alabastri* globosi, 1½-pollicares, bractea spathacea ovata purpurea involuti. *Flores* globosi, nivei, suaveolentes. *Sepala* 3, fere tripollicaria, late obovata. *Petala* 6, late obovata, interiora minora. *Carpella* in spicam oblongam 2–3-pollicarem congesta, breviter apiculata. *Semina* 1–2.

This is the species which attains the greatest elevation, and penetrates furthest into the interior of the Himalaya. It seems nearly allied to *M. conspicua* of Japan, a species now common in our gardens, and will, in all probability, prove equally hardy.

3. **M. sphenocarpa** (Roxb. Cor. iii. t. 266); foliis oblongis glabris, floribus coætaneis, spathis cinereo-incanis, petalis 6, carpellis longe rostratis.—*Wall. Cat.* 975 !—Liriodendron grandiflorum, *Roxb. Fl. Ind.* ii. 65. Michelia macrophylla, *Don, Prod. Nep.* 226. Talauma Roxburghii, *G. Don, Gen. Syst.* i. 85.

HAB. In montibus subtropicis Bengaliæ orientalis prope Chittagong, *Roxb.;* in mont. Khasia, alt. 2–3000 ped., *Roxb. Wall.!;* in Nipal, *Wall.!*—(Fl. vere.) (*v. v.*)

Arbor mediocris, ramosa. *Rami* tuberculis crebris notati, adulti glabri, juniores cum omnibus partibus novellis cinereo-incani vel subtomentosi. *Medulla* septata. *Folia* oblonga, versus basin angustata, obtusa vel vix acuta, coriacea, utrinque glaberrima aut subtus minutissime puberula, 8–16 poll. longa, 3–6 lata, petiolo 1–2-pollicari. *Nervi* subtus validi, obliqui, paralleli. *Pedunculi* validi, terminales, solitarii, incano-tomentosi, anuulis plurimis approximatis notati. *Flores* magni, albi, suaveolentes, spathis pluribus cito deciduis involuti. *Sepala* 3, extus herbacea. *Petala* 6, alba, ovalia, crassa, carnosa, margine undulata. *Ovaria* plurima in conum imbricata, rostro ensiformi villoso. *Carpella* in strobilum cylindricum 8–12 (vel ex Roxb. 16) pollices longum dense imbricata, extus tuberculata, rostro ultrapollicari ruguloso lateraliter compresso.

The very coriaceous leaves and the long-beaked fruit remove this species to a considerable distance from the other Himalayan species. On this account Dr. Wallich has, in his Catalogue, proposed to constitute of it a new genus (*Sphenocarpus*), but it seems to us to possess no characters of sufficient importance to make it desirable to separate it.

5. **MICHELIA,** L.

Sepala et *petala* plerumque conformia et concolora, 9–21. *Gynophorum* stipitatum. *Ovaria* 2–6- vel pluri-ovulata. *Carpella* coriacea, laxe spicata, sæpe subremota, dorso longitudinaliter dehiscentia.—Arbores *sæpe excelsæ,* floribus (*excepta* M. Cathcartii) *axillaribus.*

The laxly spiked carpels, numerous ovules, and axillary flowers, in general sufficiently characterize this genus. One or other of these characters, however, occasionally fails us, and the stalked gynophore or torus alone remains; and by that character, in combination with most of those just enumerated, the genus may with certainty be known. Thus, though *M. Punduana* and *Nilagirica* have not more than two ovules, and would thus technically be referable to *Magnolia,* yet their axillary flowers and distant carpels sufficiently distinguish them from that genus. The most anomalous species is *M. Cathcartii,* which has terminal flowers, and more densely imbricated carpels than are usually seen in *Michelia.* Its numerous ovules and stipitate gynophore, however, prevent its being referred to *Magnolia,* and its general habit seems to demand its admission among the *Micheliæ.* This genus is entirely Indian. Two species are natives of the mountains of the Madras Peninsula, and one of Ceylon. In the shady forests of the eastern Himalaya five species form a prominent feature in the vegetation of the temperate zone, at elevations between five and eight thousand feet. They are, however, impatient of drought, and one only extends as far west as Kumaon. In the Khasia hills and the Malayan peniusula other species occur; and the latter, when we become better acquainted with the vegetation of its mountains, may be expected to yield many species. The genus is common in Java and the islands of the Eastern Archipelago.

§ 1. *Floribus terminalibus.*

1. **M. Cathcartii** (H.f. et T.); foliis oblongo-lanceolatis acuminatis utrinque secus costam pilosis cæterum glabris ætate glabrescentibus, floribus albis, sepalis cum petalis novem, staminibus gynœcium fere superantibus, carpellis dense spicatis.

HAB. In sylvis temperatis Sikkim exterioris, alt. 5–6000 ped.!— (Fl. Aprili.) (*v. v.*)

Arbor excelsa, cortice griseo. *Ramuli* pubescentes, novelli cum gemmis dense sericei. *Folia* tenuiter coriacea, subtus pallida, (in sicco) conspicue reticulatim nervosa, 4 poll. longa, 1½ lata, petiolo ½-poll. *Areola stipularis* petiolum fere æquans. *Flores* solitarii, terminales vel gemmæ laterales evolutione (interdum per florationem, serius semper) ad speciem laterales. *Pedunculus* pollicaris. *Spatha* 1, calyci approximata, elliptica, mucrone piloso apiculata. *Flores* albi, diam. 3–4-pollicares. *Sepala* et *petala* triserialia, oblonga, interiora sensim paullo minora. *Stamina* petalis interioribus vix breviora. *Antheræ* lineares, introrsæ, connectivo obtuse mucronato apiculata. *Carpella* secus rhachin 2–4-pollicarem spicata. *Semina* 1–4.

§ 2. *Floribus axillaribus; ovulis 3 vel pluribus.*

2. **M. Champaca** (L. Sp. 756); foliis ovato-lanceolatis basi acutis apice acuminatis sæpe longe angustatis superne glabris subtus plus minus puberulis ætate glabrescentibus, floribus flavis, sepalis cum petalis 15–20 interioribus multo angustioribus.—*DC. Syst.* i. 447, *Prod.* i. 79; *Wall. Cat.* 969! (*excl.* K); *Roxb. Fl. Ind.* ii. 656; *W. et A.! Prod.* i. 6; *Wight, Ill.* i. 13; *Blume, Bijdr.* 7, *Fl. Jav. Magn.* 9. *t.* 1. M. rufinervis, *DC. Syst.* i. 449, *Prod.* i. 79; *Bl. Bijdr.* 8. M. Doltsopa, *Ham. in DC. Syst.* i. 448, *Prod.* i. 79; *Don, Prod. Nep.* 226; *Wall.! Tent. Fl. Nap.* 7. *t.* 3, *Cat.* 971!; *Wight! Ill.* i. 13. M. aurantiaca, *Wall. Cat.* 6492! *Pl. As. Rar.* ii. *t.* 147; *Wight, Ill.* i. 13. M. pubinervia, *Bl. Fl. Javæ Magn.* 14. *t.* 4. M. Rheedei, *Wight! Ill.* i. 14. *t.* 5. *f.* 6 (*fructus maturus tantum*); *Rheed. Mal.* i. *t.* 19; *Rumph. Amb.* ii. *t.* 67.

HAB. In Himalaya temperata: Kumaon et Nipal, *Wall.!;* in sylvis Pegu et Tenasserim, *Wall.!;* et in montibus temperatis peninsulæ australis ad Nilghiri et Courtalam, alt. 3–5000 ped., *Wight!*—(Fl. vere.) (*v. v.*)

DISTRIB. In Java sylvestris (*Blume*), et per totam Indiam tropicam necnon in calidis totius orbis culta.

Arbor magna, umbrosa, culta plerumque mediocris. *Ramuli* cinerei, calloso-punctati, glabri, juniores (cum omnibus partibus novellis) pubescentes vel cinereo- aut fusco-sericei. *Folia* 8–10 poll. longa, 2¼–4 lata, petiolo 1–1½-poll., superne nitida, subtus pallida, puberula vel pubescentia, ætate sæpe glabrescentia. *Areola stipularis* paullo ultra medium petiolum extensa. *Alabastri* breviter pedicellati, cinereo- vel fusco-sericei. *Flores* flavi vel aurantiaci, suaveolentes. *Sepala* et *petala* 15–20, 1½-2-pollicaria, exteriora oblonga cuneata acutiuscula, interiora multo angustiora lineari-oblonga acuta. *Carpella* in spicam 3–4-pollicarem congesta, subsessilia.

The *Champaca* of Rheede and Rumphius, adopted by Linnæus and all following authors, and universally recognized, notwithstanding the brevity of the original description of Linnæus, is only known as a cultivated tree. Indigenous trees, however, have been described by Wallich, Blume, and Wight, from the regions investigated and illustrated with so much success by these botanists, which very closely resemble the cultivated tree, differing only, it appears to us, in such characters as are chiefly affected by cultivation. In all, the flowers have the same structure, and the leaves the same shape and degree of variation. The pubescence is much more considerable in the wild plants described by Wallich and Blume than in the cultivated *Champaca;* and though Wight describes his *M. Rheedei* as glabrous, his specimens were in fruit only, whilst flowering ones in our possession from the same localities are quite as pubescent as *M. Doltsopa* from Nipal. Blume has recognized the affinity of his *M. pubinervia* with *M. Doltsopa,* Wall., while at the same time he fully admits its close affinity to the cultivated *Champaca* of Java, by relying on characters for its separation which are of very subordinate importance. For these reasons, after a very careful examination of all the specimens to which we have access, we have convinced ourselves that all the synonyms adduced above are referable to one species. *M. rufinervis* of De Candolle (not of Blume) is a cultivated Mauritius plant; a specimen in Herb. Hook., which agrees exactly with the description, is a luxuriant young shoot, with copious brown silky pubescence, but with leaves like those of *M. Champaca.* De Candolle's specimens were also without flowers, and probably of the same age. It is more difficult to decide whether the *Doltsopa* of De Candolle and Don be the same as that of Wallich, as the descriptions given by the two former authors of *M. Doltsopa* and *M. Kisopa* are very brief, and so obscure that they cannot be referred with certainty to either species, but partake of the characters of both. In these circumstances, as the original specimens are not available, having been dispersed with the Lambertian Herbarium, we have thought it advisable to follow Wallich in the use of the names *Doltsopa* and *Kisopa,* considering him in fact as the authority for the species, which he was the first to characterize in a satisfactory manner.

3. **M. excelsa** (Blume, Fl. Javæ Magn. 9, in adnot.); foliis oblongis vel oblongo-lanceolatis acutis superne glabris subtus fusco-sericeis ætate glabrescentibus, floribus albis, sepalis cum petalis 12.— *Wall. Cat.* 6494! *Wight, Ill.* i. 14. Magnolia excelsa, *Wall.! Tent. Fl. Nap.* 5. *t.* 2.

HAB. In Himalaya orientali temperata, alt. 6–8000 ped.: Nipal, *Wall.!* Sikkim! Bhotan, *Griffith!* et in Khasia, alt. 5000 ped., *Simons!* —(Fl. vere.) (*v. v.*)

Arbor excelsa, ramosa. *Ramuli* rugosi, grisei, punctis callosis conspersi. *Gemmæ* fusco-pubescentes. *Folia* coriacea, acuta vel acuminata, superne nitida, subtus (juniora dense, seniora sparse) tomento brevi adpresso cinnamomeo sericea, rarius subglabrescentia, 5–8 poll. longa, 2–3 lata, petiolo pollicari. *Areola stipularis* paullo ultra medium petiolum extensa. *Alabastri* subsessiles, dense fusco-tomentosi, bipollicares et ultra, spathis pluribus deciduis involuti. *Sepala* 3, obovata, coriacea. *Petala* 9–10, anguste obovata, interiora sensim angustiora et breviora. *Carpella* secus rhachin 4–8-pollicarem laxe disposita, subsessilia, ½-pollicaria. *Semina* 1–4.

4. **M. lanuginosa** (Wall.! Tent. Fl. Nap. 8. t. 5); foliis oblongis vel lanceolatis superne nitidis glabris subtus dense cinereo-tomentosis, floribus albis, sepalis petaliscum 18.—*Wall. Cat.* 6493! *Wight, Ill.* i. 14. M. velutina, *DC. Prod.* i. 79.

HAB. In Himalaya orientali temperata, alt. 5–7000 ped.: Nipal!
Sikkim! Bhotan!; et in Khasia, *T. Lobb!—(v. v.)*

Arbor excelsa, cortice rugoso fusco. *Ramuli* juniores griseo-tomentosi, novelli
cum gemmis et petiolis dense stramineo- vel cinereo-tomentosi. *Folia* lanceolata vel
oblonga, superne in sicco tenuissime reticulata, 6–10 poll. longa, 2–3½ lata. *Pedun-culi* breves alabastrique dense tomentosi. *Sepala* et *petala* 1½–2 poll. longa, exteriora anguste obovato-oblonga, obtusa, interiora paullo angustiora mucronata vel
acuta. *Ovaria* cum gynophoro dense tomentosa, stylo filiformi in sicco nigro glabro.
Carpella in spica 4–5-pollicari discreta, pedicellata, obovata, pollicaria et ultra, verrucosa. *Semina* 1–3.

Wallich states that this species flowers in spring; in Sikkim, however, it does not
flower till August and September, nor does it in that country attain the great size to
which it grows in Nipal.

5. **M. Kisopa** (Ham. in DC. Syst. i. 448); foliis lanceolatis vel
oblongo-lanceolatis utrinque glabris basi acutis apice acutis vel acuminatis, floribus dilute flavis, sepalis et petalis 12 anguste obovatis
interioribus vix minoribus.—*DC. Prod.* i. 79; *Wall.! Tent. Fl. Nap.* 8.
t. 4; Don, Prod. Nep. 226; *Wall. Cat.* 970! *Wight, Ill.* i. 13.

HAB. In Himalaya temperata, alt. 5–7000 ped.: a Kumaon! ad
Nipaliam maxime orientalem!—(Fl. vere.) *(v. v.)*

Arbor excelsa, cortice rugoso cinereo, partibus novellis gemmisque cinereo-sericeis.
Folia coriacea, in sicco nervis crebris reticulata, superne nitida, subtus pallida; juniora subtus adpresse incana, 5–6 poll. longa, 1½–2 lata; petiolo pollicari. *Cicatrix*
stipularis ultra medium petiolum extensa. *Alabastri* ½ poll. longi, cinereo-sericei,
spathis 2, quarum exterior cito decidua, involuti. *Flores* odore debili, brevissime
pedunculati. *Petala* pollicaria. *Ovula* 5–6 vel plura. *Carpella* in spicam 3–4-
pollicarem disposita, compressa, rotundata, vix ½-pollicaria. *Semina* 3–4.

§ 3. *Floribus axillaribus; ovulis duobus superpositis.*

6. **M. oblonga** (Wall. Cat. 972!); foliis obovato-oblongis basi
angustatis apice obtuse acuminatis utrinque glaberrimis, (floribus albidis?), sepalis et petalis 12 exterioribus anguste obovatis interioribus
lanceolatis.—*M.* lactea, *Ham. mss.; Wall. Cat.* 6491!

HAB. In sylvis secus basin montium Khasia, *Ham.! Wall.!—(v. s.)*

Ramuli minutissime tuberculati; partes novellæ glabræ. *Folia* in sicco crebre
reticulata, 4–6 pollices longa, 2½ lata, superne nitida, subtus pallida vel glaucescentia.
Flores axillares, brevissime pedicellati, spathis pluribus involuti. *Alabastri* oblongi,
elongati, glabri, fusci. *Petala* sesquipollicaria. *Fructus* non visus.

7. **M. Punduana** (H.f. et T.); foliis oblongis basi obtusis vel
acutis apice abrupte acuminatis utrinque glabris, (floribus albis?), pe-
rianthii foliis 9 obovato-cuneatis exterioribus obtusis interioribus mucronatis.—Liriodendron liliifera, *Roxb. Fl. Ind.* ii. 654. Magnolia
Punduana, *Wall. Cat.* 974!

HAB. In montibus Khasia, alt. 3–5000 ped.!—(Fl. Nov.) *(v. v.)*

Arbor excelsa, cortice rugoso fusco, ramorum juniorum lævi viridi. *Partes no-*
vellæ fusco-sericeæ. *Folia* tenuiter coriacea, laxe reticulata, subtus pallida, 4–6-
pollicaria, petiolo vix pollicari. *Cicatrix stipularis* petiolum longitudine fere æquans.
Alabastri ovati, fere pollicares, breviter pedunculati, fusco-sericei. *Spathæ* 2, in-
volucrantes, exterior citissime decidua, profunde bifida, ad bifurcationem apiculata,

M

sed nunquam foliifera. *Stamina* gynœcium æquantia, connectivo apice longe subulato. *Carpella* in spicam oblongam 3-4-pollicarem laxe congesta, approximata, ½-
pollicaria, compressa, rotundata. *Semina* 1-2.

8. **M. Nilagirica** (Zenker, Plant. Ind. t. 20); foliis ellipticis utrinque acutis vel ovalibus obtuse acuminatis utrinque glabris vel subtus
secus costam pubescentibus, floribus albis, petalis cum sepalis 12 exterioribus obovatis interioribus oblongo-lanceolatis acutis.—*Wight! Ill.*
i. 14, *Icon*. t. 938! *Spic. Neilgh*. t. 6.   M. Pulneyensis, *Wight! Ill*. i.
14. *t*. 5, *excl. f*. 5 *et* 6.   M. Champaca, *Wall. Cat*. 969 K! (*nec aliæ*
lit.)   M. ovalifolia, *Wight! Ill*. i. 13.

β. *Walkeri;* arbuscula, foliis oblongis vel lanceolatis plerumque
subtus glaucescentibus 2-3 poll. longis, floribus minoribus.—M. Walkeri *et* M. glauca, *Wight, Ill*. i. 13.

HAB. In montibus altioribus peninsulæ australis, alt. 6-8000 ped.,
Wight! et in summis montibus Zeylaniæ, *Walker!* etc.—(*v. s.*)

Arbor magnitudine varia, plerumque excelsa, in Zeylania interdum fruticosa;
partes novellæ sericeo-villosæ. *Folia* forma valde varia, 3-5 pollices longa, 1½-2
lata, petiolo ⅔-poll. *Cicatrix stipularis* dimidium petiolum æquans. *Alabastri*
1-1½-pollicares, cum pedunculo longitudine vario dense fusco-sericei (in β cinereo-
sericei). *Spathæ* 2. *Carpella* in spicam 2-3-pollicarem disposita, subcompressa, rotundata. *Semina* plerumque solitaria.

Our variety β (from Ceylon) has at first sight so very different an aspect from the
peninsular plant, that we can scarcely persuade ourselves that it is not distinct.   We
have, however, failed to discover satisfactory characters to distinguish these plants in
the dried state; but botanists who have an opportunity of observing the living plant
may perhaps be more successful.   The Ceylon plant, of which we have seen a
rather extensive suite of specimens, varies much in the size of the flower and in the
shape of the leaves; and the small lanceolate-leaved states appear to pass insensibly
into a plant with oval leaves, which, though usually more coriaceous, are sometimes
quite undistinguishable from those of the typical *M. Nilagirica*.   These small states,
which have sometimes nine instead of twelve petals, seem in many of our specimens
to be diseased, the flowers being unusually small, the stamens few and abbreviated,
and the young carpels abnormally swollen, as if punctured by an insect, and apparently abortive.   Perhaps, therefore, it will be found that the broad-leaved arborescent state is the normal form in Ceylon as well as in the peninsula, and that the
lanceolate-leaved state is an accidental variety.   *M. glauca* of Wight is certainly
only an abnormal form, with broadly obovate leaves, for the glaucous hue of the
under surface is not confined to specimens with that form of leaf, but is seen equally
in the oval and lanceolate-leaved plants, and is often observed on the same specimen
with leaves not at all glaucous below.

IV. SCHIZANDRACEÆ.

Flores unisexuales.   *Sepala* et *petala* hypogyna ternatim vel quinatim pluriserialia, æstivatione imbricata.   *Stamina* definita vel indefinita,
toro depresso vel conico inserta.   *Filamenta* libera vel plus minus
coalita.   *Antheræ* adnatæ, biloculares, plerumque varie heteromorphæ.
Ovaria indefinita, in capitulum oblongum vel subglobosum coalita.
Ovula in sutura ventrali 2-3, amphitropa vel fere campylotropa.   *Baccæ*
dissepimento spurio transverse bi- (rarius tri-) loculares, dispermæ.

Semina superposita, reniformia, in pulpa nidulantia ; testa lævis, crustacea ; albumen copiosum, oleosum ; embryo minutissimus.—Frutices *scandentes, volubiles, glaberrimi,* ramulis *elongatis,* junioribus *basi squamis gemmæ persistentibus stipatis,* foliis *integris integerrimis vel dentatis,* floribus *plus minus conspicue pellucide punctatis.*

We have only been deterred from following Asa Gray in considering this small group as a section of *Magnoliaceæ,* by the unisexual flowers and marked difference in habit, and in particular by the frequently toothed leaves. Its position is undoubtedly in the immediate neighbourhood of *Magnoliaceæ,* between that Order and *Anonaceæ,* to certain genera of which (especially *Stelechocarpus*) the aspect of the flowers, and the occasionally truncal inflorescence, indicate a certain degree of approach.

The family is a very small one. One species inhabits damp woods in the southern United States of America, and the remainder the Indo-Chinese region, from Japan to the Malayan Archipelago, Ceylon and Malabar, and the Himalaya. The leaves and flowers are mucilaginous, the fruit and seeds faintly aromatic, and the woody fibre exhibits glandular disks similar to those of *Illicium* and *Drimys.*

The structure of the andrœcium, which is the most conspicuous character of the plants of this Order, is nevertheless only of importance for the distinctiou of species, as those plants which are most closely allied, differ very remarkably from one another in the degree of combination of the filaments. *Schizandra,* with five monadelpho·is stamens, is, however, a good genus. The shape of the fruit, on the contrary, is, we think, a natural character, dividing this small Order into two well-marked groups, which, in accordance with the views of Blume in his monograph of the Javanese species, we regard as of generic value. Of these, *Kadsura,* with globose fruit, contains the original species of Japan, and several others ; while *Sphærostema,* with the baccate carpels arranged on an elongated torus, extends from the Western Himalaya tu Java.

1. KADSURA, Juss.

Sarcocarpon, *Blume.*

Sepala 3. *Petala* 6–9. *Stamina* 15 vel plura. *Filamenta* discreta vel in globum coalita. *Ovaria* numerosa. *Stylus* obconicus, lateralis. *Carpella* baccata, inter se libera, capitulum globosum formantia.— Frutices *scandentes, mucilaginosi,* floribus *albis vel rubescentibus.*

1. **K. Roxburghiana** (Arn. in Jard. Mag. Zool. Bot. ii. 546) ; foliis ovatis vel oblongis carnosulis, filamentis monadelphis, ovariis biovulatis.—Kadsura Japonica, *Wall. Tent. Nap.* 12 (*non Juss. nec alior.*), *Cat.* 4987 A ! B *partim!* (*specim. dextr.*) 4985 B ! Uvaria heteroclita, *Roxb. Fl. Ind.* ii. 663.

HAB. In Assam ! et Silhet ! ; in montibus Khasia a basi ad altitudinem 5000 ped.! ; et in vallibus calidioribus Sikkim !—(Fl. Mai. Jun.) (*v. v.*)

Frutex alte scandens, trunco diametro pollicari et ultra. *Cortex* rugosus. *Ramuli* læves, annulati, basi interdum squamis stipati. *Folia* cum caule articulata, acuta vel acuminata, integerrima vel remote et obscure denticulata, 3–6 poll. longa, 1¼–3 lata, petiolo ½-poll. *Pedunculi* petiolum duplo superantes, crassiusculi, basi squamulis gemmaceis persistentibus suffulti, et infra medium bracteas 4–6 ovatas minutas gerentes. *Flores* diametro semipollicares. *Sepala* rotundata. *Petala* rotundata, convexa, carnosula, interiora minora. *Filamenta* basi in columnam centralem cylindricam coalita ; exteriora pauca, superne breviter libera, crassa, cylindrica, apice in connectivum carnosum late cuneatum subtruncatum dilatata ; superiora usque ad anthercas coalita. *Antheræ* lincari-oblongæ, connectivo lateraliter adnatæ, longitudinaliter dehiscentes. *Baccæ* in capitulum globosum diam. 1-2-poll. congestæ, cuneato-subglobosæ, coccineæ, pisi vel fabæ minoris magn. *Semina* 1-2.

The sweet but flavourless fruit of this species is eaten by the inhabitants of Sikkim and Eastern Bengal. Though confounded by Wallich with *K. Japonica*, it appears to be quite distinct from the plant figured and described by Siebold and Zuccarini. We have not seen Japanese specimens, but a specimen in the Hookerian Herbarium, brought from Hongkong by Major Champion, and referred by Bentham without hesitation to *K. Japonica*, has larger flowers on very short pedicels, which are more covered with bracts than those of the Indian plant. The leaves are also thicker and firmer, scarcely toothed, and longer-petioled. The leaves vary much in shape in all the species, and, as is often the case among scandent plants, the foliage of the long suckers is very different from that of the lateral shoots of the second year.

2. **K. Wightiana** (Arn. l. c. ii. 546); foliis late ovalibus obtuse acuminatis basi cuneatis, filamentis discretis, ovariis triovulatis.—*Wight*, *Cat. No.* 2478.

HAB. In Zeylania, alt. 2-3000 ped., *Walker!*; Malabar, *Wight.*— (*v. s.*)

Frutex scandens, glaber, cortice rugoso fusco. *Ramuli* abbreviati (an semper ?). *Folia* basi cuneata, integra vel vix denticulata, subtus pallida, 2-3 poll. longa, $\frac{1}{2}$-2 lata, petiolo $\frac{1}{2}$-poll. *Pedunculi* axillares, validi, petiolum vix superantes, bracteis pluribus squamæformibus ovatis deciduis. *Sepala* inæqualia, parva. *Petala* 9, ovalia, obtusa, iut. minora. *Baccæ* prioris.

This appears to be a more rigid shrub than the last, with smaller and broader leaves, and short, thick, woody branches. We have not seen the male plant. It is worthy of note that *K. Japonica* is said by Siebold and Zuccarini to have also occasionally three ovules and seeds.

3. **K. scandens** (Blume? Fl. Jav. Schiz. p. 9. t. 1).

A specimen of a *Kadsura* in the Benthamian Herbarium, collected by Griffith at Malacca, is very distinct from either of the former species, and probably belongs to *K. scandens ;* but as it consists of a single leaf attached to the stem, and a few male flowers, we do not feel justified in appending a description of that plant, especially as Blume's figure and description of the *andrœcium* are unsatisfactory, and also not easily reconciled with what we see in the single flower which we have been able to examine. The leaves of *K. scandens* (and of our plant) are ovate or ovato-oblong and acuminate, quite entire, glabrous, 4-6 inches long, and 2-4 broad, with a petiole 1-2 inches long. The flowers are axillary and solitary, and the pedicel is shorter than the petiole. Blume further describes the stamens as free on the cylindrical torus, with the connective extending beyond the anther into a fleshy gibbous process. This does not seem to be the case in the specimen from Malacca, but the flower has been so much compressed that we cannot determine the structure with anything like accuracy. *K. scandens* is further readily recognizable by the shape of the carpels, which are terminated by an obtuse hooked mucro.

2. SPHÆROSTEMA, Bl.

Sepala 3. *Petala* 6-9. *Stamina* 15 vel plura, monadelpha. *Carpella* globosa, secus torum cylindricum spicata.—*Frutices scandentes, volubiles, glaberrimi,* floribus *albis, flavidis vel rubescentibus.*

§ 1. *Filamentis basi monadelphis, apice liberis.*

1. **S. grandiflorum** (Bl. Fl. Jav. Schiz. 17); foliis ovato- vel ob-

longo-lanceolatis acuminatis basi acutis remote denticulatis, pedunculis laxis elongatis, toro fructus elongato crasso carnoso.—Kadsura grandiflora, *Wall. Tent. Nap. p.* 10. *t.* 14, *Cat.* 4985 A *partim! (spec. dextrum) (non* B *nec* C).

HAB. In Himalaya temperata, alt. 7–10,000 ped. : a Simla ! ad Bhotan !—(Fl. Mai. Jun.) (*v. v.*)

Ramuli graciles, cortice fusco. *Folia* 3–6 poll. longa, 1–2 poll. lata, petiolo 1–1½-poll., subcarnosa, supra lucida, subtus pallida. *Pedunculi* axillares, 1–2-pollicares, basi squamosi, cæterum nudi. *Flores* diam. ultrapollicares, penduli, suaveolentes, albi flavidi vel rosei. *Petala* rotundata vel late ovalia, interiora sensim minora. *Filamenta* indefinita, superne libera, cylindrica. *Antheræ* ovales, connectivo crasso, loculis discretis connectivo lateraliter insertis lineari-oblongis subextrorse longitudinaliter dehiscentibus. *Torus* fructus cylindricus, 6–9ˌpoll. longus, incrassatus, carnosus, rubescens. *Baccæ* globosæ, coccineæ, pisi magn., superne lineola brevi notata ; *testa* seminis crustacea, minute punctulata.

2. **S. elongatum** (Bl. Fl. Jav. Schiz. 17. t. 5) ; foliis ovatis acutis vel acuminatis basi cuneatis, pedunculis elongatis filiformibus, toro fructus vix carnoso brevi.—Sphærostema grandiflorum, *Wall. Cat.* 4985 A *partim! (spec. sinistrum)* C !

HAB. In Himalaya orientali temperata : Nipal, *Wall.!* Sikkim ! ; et in mont. Khasia, alt. 5–6000 ped.!—(Fl. per tot. æst.) (*v. v.*)
DISTRIB. Java.

Rami fusci, rugosi, verruculosi. *Ramuli* læves, glaucescentes, basi squamis persistentibus stipati. *Folia* sæpe longe acuminata, subtus pallida vel glauca, 3–4 poll. longa, 1½–2 lata, petiolo 1–1½-poll., rubescente. *Pedunculi* axillares, sæpe prope basin ramuli ad axillas foliorum delapsorum plures, pseudo-fastigiati, petiolos duplo superantes, basi squamulis 1–2 subulatis muniti, cæterum nudi vel interdum medio unibracteolati. *Flores* diametro ¾-poll., flavidi. *Sepala* parva, inæqualia. *Petala* plerumque sex, ovata, carnosula, margine membranacea, interiora majora. *Stamina* prioris vel *Kadsuræ Roxburghianæ*. *Torus* fructus 2–3-pollicaris. *Baccæ* grani piperis magnitudine, substipitatæ, globosæ, superne cicatrice lineari longiuscula notatæ.

It is a striking proof of the difficulty of distinguishing the plants of this family in a dried state, that Dr. Wallich has confounded this species with the preceding, from which it differs in many important particulars. We were fortunate enough to find it abundantly in Khasia, as well as in Sikkim, where it grows at a lower level than *S. grandiflorum.* We refer our plant without hesitation to the species figured and described by Blume, notwithstanding the absence of the bractlet on the pedicel in all our specimens, because it agrees in all other essential particulars, and one of the pedicels in the plate is represented as without a bractlet.

§ 2. *Filamentis in globum coalitis ; antheris circiter* 15, *alveolis andrœcii longitudinaliter adnatis, bilocularibus, longitudinaliter dehiscentibus.*

3. **S. propinquum** (Bl. Fl. Jav. Schiz. 16) ; foliis ovato-lanceolatis basi rotundatis vel cuneatis apice longe acuminatis, pedunculis petiolos subæquantibus, toro fructus elongato parum incrassato.— *Wall. Cat.* 4986 ! 4987 *B ! (spec. sinistr.)* Kadsura propinqua, *Wall. Tent. Nap. p.* 11. *t.* 15. S. pyrifolium, *Blume, Fl. Jav. Schiz. p.* 16. *t.* 4?

HAB. In Himalaya exteriori temperata, alt. 4–6000 ped. : Kumaon, *Str. et Wint.!* Nipal, *Wall.!*—(*v. s.*)

Rami glaberrimi. *Folia* serrata, denticulata, carnosula, 3–5 poll. longa, 1¼ lata, petiolo ½-poll. *Pedunculi* solitarii vel subfasciculati, basi squamulis pluribus suffulti, medio bracteola 1 semiamplexicauli persistente. *Alabastri* globosi, pisi magn. *Sepala* ovata, inæqualia. *Petala* 6, fere rotundata, coriacea. *Baccæ S. grandi-flori*, sed minores, secus torum sexpollicarem spicatæ.

4. **S. axillare** (Bl. Bijdr. 22, Fl. Jav. Schiz. 14. t. 3); foliis lanceolatis longe acuminatis basi rotundatis vel cuneatis, pedunculis plerumque brevissimis, toro fructus filiformi abbreviato.

HAB. In mont. Khasia, alt. 4–5000 ped.!—(Fl. per tot. æst.) (*v. v.*) DISTRIB. Java.

Ramuli angulati, rufescentes, glabri. *Folia* coriacea, superne nitida, subtus pallida, margine integerrima vel distanter denticulata, 3 poll. longa, ½–¾ poll. lata, petiolo ½-poll. *Pedunculi* axillares, petiolo breviores, sæpe brevissimi, squamis rotundatis imbricatis scariosis tecti. *Flores* coccinei vel lutescentes, diametro ½-poll. *Sepala* rotundata, parva. *Petala* 9, triserialia, ovato-rotundata, interiora multa minora. *Torus fructus* 1–2-pollicaris. *Baccæ* numerosæ, substipitatæ, globosæ. *Semina* 2, vel abortu solitaria.

V. ANONACEÆ.

Flores hermaphroditi, rarius unisexuales. *Sepala* 3, hypogyna, æstivatione plerumque valvata, basi sæpe coalita. *Petala* serie duplici 6, æstivatione valvata vel imbricata, rarissime serie interiore deficiente sepalis numero æqualia. *Stamina* indefinita, multiserialia, rarius subdefinita; plerumque numerosissima, dense conferta. *Filamenta* abbreviata. *Antheræ* biloculares, connectivo lato superne producto sublateraliter vel extrorse adnatæ, loculis remotis vel contiguis, sublateraliter vel extrorse dehiscentibus. *Ovaria* plurima, rarius definita, rarissime solitaria, 1-locularia, supra torum convexum vel concavum sessilia, interdum inter se subcohærentia. *Ovula* solitaria vel bina e basi erecta, vel in sutura ventrali 1 vel plura, vel indefinita, in *Monodora* parieti undique inserta, anatropa. *Stigmata* terminalia, libera vel inter se subcohærentia. *Carpella* sessilia aut stipitata, libera vel in fructum multilocularem coalita, sicca vel pulposa, indehiscentia, rarius follicularia. *Semina* solitaria vel numerosa. *Albumen* copiosum, ruminatum. *Embryo* minutus.—Arbores *vel* frutices *sæpe scandentes vel sarmentosi, plerumque aromatici ;* foliis *alternis integerrimis exstipulatis,* floribus *terminalibus vel axillaribus, solitariis vel varie congestis.*

This large and very natural Order is readily distinguishable from its near allies by a combination of well marked characters. The ternary arrangement of the parts of the perianth, the small, closely packed, extrorse, almost sessile anthers, the numerous small ovaries, the distinct often stipitate fruits seated on a rounded torus, and the ruminated albumen, characterize all the typical species, though one or other of these characters is occasionally absent, or unavailing as a distinction. The ruminated albumen, though universal in the Order, occurs also in *Myristicaceæ,* and to a small extent in a few genera of *Menispermaceæ.* The ternary arrangement of the flower is also universal, but is met with in many of the neighbouring families. The sepals always form a single verticil; and the petals, which never exceed six in number (in two rows), are in a few instances reduced to a single row by the suppression of the inner series. In *Magnoliaceæ* they are generally much more numerous. The anthers are always more or less extrorse, but the number of stamens is far from constant, being in many genera reduced to 18, 15, 12, 9, and even as low as 6. The ovaries are occasionally subdefinite, or even solitary, and the carpels are sometimes dehiscent. The valvate æstivation of the petals, which, when present, is the most conspicuous character of the Order, is wanting in the Section *Uvarieæ*.

The state of this comparatively little known Order is still very unsatisfactory, notwithstanding that it has received the attention of many of the principal botanists of the day, nor is it to be expected that the tribes and genera can be established on a proper basis, till the species have been much more carefully and completely examined than their very imperfect condition in herbaria has hitherto permitted them to be. Their study, indeed, even under the most favourable circumstances, presents great difficulties to the student of dried plants, from the minute size of the stamens and ovaries, and from the bad state of preservation in which the flowers occur in herbaria. Though the flowers are often large, they are generally more or less fleshy, and in drying become much flattened and distorted, so that the restoration of the natural state is almost impossible. The determination of the number of ovules is, in particular, a very difficult matter, as the minute ovaries are always much compressed; and their walls are so brittle, that the dissection necessary for the isolation of the ovules can only be effected by much patience, and with an abundance of materials.

The number of species of *Anonaceæ* known to the older botanists was too small to permit of any great progress being made by them towards the proper circumscription of the genera. These were first accurately defined, and the species carefully described, by Dunal, in a monograph of the Order, published in 1817. At that time only 103 species of the Order were known, most of them very imperfectly. Of these scanty materials M. Dunal has certainly made much; and his work, which has formed the foundation of all that has since been done, has been well characterized by M. Alph. De Candolle as being a monument of talent and sagacity, considering the period when it appeared. The 'Systema' and 'Prodromus' of De Candolle contain no additions to the labours of Dunal, who had at his command all De Candolle's materials; and since that period the Order has not been treated generally, except by M. Alph. De Candolle, in a memoir in the fifth volume of the Geneva Transactions, in which the additions to the Order, up to the year 1832, are reviewed. The number of known species is there stated at 204.

Much attention has, however, been directed to the definition and arrangement of the genera of *Anonaceæ*, in all the works which have been published of late on tropical botany; and so many remarkable forms have been figured, that much greater facilities are now afforded for the correct appreciation of affinities, than were available to the older botanists. The works of St. Hilaire, Martius, and Richard, on American Botany, and the 'Flora Javæ' of Blume, have all contributed much to our knowledge of the Order. The careful analyses and excellent descriptions of the Eastern forms in the last-mentioned work, in particular, have been of the greatest service to us.

From the time when the number and position of the ovules was first indicated by Brown as an important character in *Anonaceæ*, in his remarks when founding the genus *Artabotrys*, in the 'Botanical Register,' this character has been generally employed, not only for the distinction of genera, but also for the formation of the primary divisions of the Order. But though the number and position of the ovules is nearly constant in each species, and therefore constitute most important characters for the distinction of genera, the higher groups thus characterized appear to us unnatural, and we therefore think it desirable to employ other characters for their circumscription. Five aberrant tribes appear to be at once distinguishable by well marked and easily recognizable characters.

The first of these, which may be called *Uvarieæ*, from its principal genus, has its petals imbricated in æstivation. This important character was first indicated by Bentham, in the Niger Flora.

A second tribe, which we propose to call *Mitrephoreæ*, has been indicated by Mr. Bennett, in his valuable remarks under *Saccopetalum* in Horsfield's ' Plantæ Javanicæ Rariores.' It comprises a number of genera, in which the inner petals are more or less unguiculate at the base. In *Orophea* and *Mitrephora*, which may be considered the typical genera of this tribe, the claw is long and slender; but in others, which appear to form a transition to the typical genera of *Anonaceæ*, it is very short and much broader.

A third aberrant tribe has also been indicated by Mr. Bennett; it comprises the genera *Alphonsea, Saccopetalum*, and *Miliusa*. Mr. Bennett has characterized this tribe by the small size of the outer petals, and by their similarity to the calyx; but in the genus *Alphonsea*, which evidently forms a part of it, this character is not present, while it occurs in *Phæanthus*, which cannot be separated far from *Guatteria*, as well as in some other species not naturally allied to the genera above mentioned. The true character of the tribe, we think, lies in the shape and structure of the anthers, which, instead of being densely wedged together as in the other tribes, are broadly oval or oblong, with large short cells, and a small terminal apiculus of connectivum. These anthers rise above one another in a laxly imbricated manner, so that the greater part of each is exposed; whereas the normal stamen of the Order is erect and columnar, with the dilated process of the connective alone visible, while the linear anther-cells are completely concealed.

Monodoreæ, which we propose to regard as a fourth tribe, contains only a single species, characterized by the distribution of the ovules over the whole surface of the solitary ovary. This very remarkable structure, which is very rare among plants, occurs in the nearly allied family *Lardizabaleæ*, to which this tribe exhibits an interesting transition. It is found also, curiously enough, in the apocarpous monocotyledonous Order *Butomaceæ*.

The remarkable Australian genus *Eupomatia*, described by Mr. Brown in 'Flinders' Voyage,' and referred by him without doubt to *Anonaceæ*, cannot surely be separated from the remainder of the Order, but forms a fifth aberrant tribe, the well known characters of which it is unnecessary to repeat here. We believe that this interesting plant has not been found by any botanist but its illustrious discoverer, and though it has been introduced into our conservatories, it has never flowered there.

In the remainder of the Order the perianth is valvate in æstivation, the petals are never unguiculate, the anthers are numerous and densely packed, and the ovules are either erect from the base of the ovary, or arise from the ventral suture. This combination of characters, marking the typical *Anonaceæ*, is present in about one-half of the Indian species, and in a much larger proportion of those of America. Among these, *Anoneæ*, with the ovaries cohering together in the flower, and afterwards developed into a compound fruit, form a well-marked tribe. The remaining genera we propose to divide into two tribes, *Xylopieæ*, with thick fleshy inner petals, which are triquetrous, except at the base, and *Guatterieæ*, with coriaceous inner petals, not materially different in shape or texture from the outer ones. These tribes appear to us very natural; but they pass by such insensible gradations into one another, that the limit between the two is quite arbitrary.

In the formation and circumscription of the genera, it has been our aim in the first instance to bring together those species which possess a similar habit, and which appear to us to form natural groups, and to select as generic distinctions such characters as are common to the species thus associated. This has led us to study with care the relative importance of the floral organs, and we have in consequence made considerable alterations in the limits of the genera. We cannot expect that the conclusions at which we have arrived will be final, as our attention has been confined almost entirely to the Asiatic forms; but it may be serviceable to the future monographist of this difficult Order, to state the degree of value we are disposed to attach to each character.

The ovaries of *Anonaceæ* are generally very numerous and small, and closely packed together. In *Uvaria* they are columnar, and quite straight, and grooved along the inner face; but generally they are rounded on the back, and oblong in shape. They are usually very hairy, but sometimes perfectly glabrous. This character, though constant in each species, is of no avail for the distinction of genera. The style is invariably terminal, and is either continuous with the ovary, and undistinguishable from it except by the absence of a cavity, or separates by a joint. In the latter case the mass of styles often coheres together by means of a viscid or gelatinous fluid. The style is usually grooved on the inner face, and is stigmatic over its whole surface, and often covered with papilli. Sometimes it is short and capitate, more generally oblong, and occasionally elongate and subulate. With occasional exceptions, which will be noticed under the genera in which they occur, these characters seem constant. The number of ovaries is of less value. In *Xylopia*, a very natural genus, they vary from one to ten, and in *Orophea* from three to fifteen. In *Asimina* and other genera their number is equally uncertain.

The number and position of the ovules are of great importance as generic characters. When solitary, the ovule is either erect from the base of the cell, as in *Unona* and *Guatteria*, or attached to the ventral suture, as in *Ellipeia* and in some *Miliusæ*. In *Artabotrys* and *Anaxagorea* there are always two collateral ovules, erect from the base of the cell. When the ovules are definite, and attached to the ventral suture, their number seems less constant. Thus, in *Unona* they vary from two to eight, but are nearly constant in each species. In *Miliusa* they vary from one to two, and in *Xylopia* from two to six. In *Polyalthia* and *Phæanthus* there are two superposed ovules inserted very near the base of the cell, one of which seems occasionally absent, in which case *Polyalthia* is with difficulty distinguishable from *Guatteria*. When the ovules are numerous they are arranged more or less distinctly in two rows, and are closely packed together: they are then occasionally subdefinite, especially where the ovary is very short, but this is in no case a character of generic value. The section *Kentia* of *Melodorum*, where they are reduced to two, is the only very marked exception to the importance of the difference between definite and indefinite ovules in the Order.

The shape of the stamens forms a very important character in *Anonaceæ*, whenever it deviates from the ordinary type. This type, which depends mainly upon the great compression of the anther, is nearly sessile, cuneate, tetragonal, with two dorsal cells almost in contact with one another, and the connective produced beyond the anthers into a depressed rounded head. More rarely the cells are distant, and almost lateral. The process of the connective is, however, in some genera elongated, and not at all depressed or truncate. In one section of *Uvaria* the anthers are flat and almost foliaceous; and in the whole group of *Saccopetaleæ* they are ovoid, with a scarcely conspicuous process of the connectivum. When the stamens are definite in number they are very irregular in shape, but usually trapezoidal, with a thick fleshy connective and small dorsal anther-cells.

The torus varies remarkably in amount of development. Where the number of ovaries and stamens is definite, it is very small; but in general it is large and conspicuous, being sometimes cylindrical and elongated, as in *Ranunculus* or *Magnolia*, but more generally conical, somewhat after the fashion of *Rubus*, or broadly cylindrical and truncated. It is not unfrequently slightly concave in the centre; and this concavity becomes extreme in *Xylopia*, where the stamens are borne on the outside of the torus, which completely encloses the ovaries. The modifications of this organ are very constant, but not always sufficiently capable of definition to render them available to the systematist.

The shape of the petals has been much neglected in the formation of genera. Blume, however, has employed it as a sectional character in *Uvaria*, under which genus he has united most of the many-ovuled *Anonaceæ*, and also in *Polyalthia*, in which he includes many of those with two ovules. The sections thus formed are highly natural, as the species included in them agree very closely in habit; and we have accordingly raised them to the rank of genera, following an indica-

N

tion given by Blume himself. Throughout the Order the shape of the petals appears to afford characters of great importance, and the facility with which it can be determined makes it of great practical utility. The particular modifications are readily recognized, and have for the most part been already indicated; others will be specially noticed under the different genera.

As an accessory character the inflorescence is deserving of attention, since it will often be found that its different modifications correspond with generic groups. The inflorescence of *Anonaceæ* is generally definite and terminal, but very often, by the continuance of the growth of the axillary bud, the flowers become leaf-opposed. Frequently the leaves on the flower-bearing branch are reduced to mere bracts or scales, in which case we have axillary cymes. These are occasionally so far reduced as to bear only one flower, with several empty bractlets at the base of the peduncle; but truly axillary and solitary flowers are very rare.

The nature of the fruit appears to bear less relation to the natural groups than any of the characters enumerated above. The number of ovaries which ripen their seeds, and the number of seeds which are developed, vary much. Many-seeded fruits occur in the same capitulum with one-seeded ones in many *Uvariæ* and *Melodora*. In *Unona* the many-jointed pods are frequently reduced to one joint. Occasionally (as in *Guatteria*) the shape of the seed and the nature of the testa afford good characters, but the fruit of many species being yet unknown, the universality of this character is still doubtful. The dry and fleshy fruit is also a very uncertain character, as the endocarp appears to remain long dry, and at last suddenly to become pulpy : this we have observed in several genera. We have therefore made no use of characters derived from the fruit, except for the purpose of distinguishing species.

In distribution *Anonaceæ* are one of the most tropical Orders. The most northern species known is *Asimina pygmæa*, which is found on the southern shores of Lake Erie, in North America. In South America they do not extend beyond 32° S. In Africa some occur at Natal, but none in the Cape district. In the Mediterranean province and throughout Europe they are unknown. In China a few occur as far north as Hongkong, but none in North China or Japan. In India only one species extends to 30° N., and in Australia one only is known further south than Moreton Bay, namely *Eupomatia*, which is a native of New South Wales.

So many *Anonaceæ* are still undescribed, and the materials which exist in herbaria are still so imperfect, that the number of species cannot be definitely estimated. A conjectural estimate may, however, be formed. We have described 123 species. Blume has enumerated 31 from Java alone; and from the materials we have seen, we think we may safely assume that the Malayan Archipelago contains at least as many as continental India. In Australia they are probably much less numerous, the climate of that country being very much drier : several very interesting forms have, however, been brought from the northern and eastern coasts of that continent, and their number will probably be hereafter considerably increased. On the whole, we may assume the number of eastern species to be about 250. For America we may perhaps allow an equal number, as Von Martius has enumerated 97 species in the Brazilian flora, and they are very numerous in equatorial America. From Africa few are as yet known, but, as has been pointed out by Bentham, they bear a very large proportion to the whole amount of the flora of western tropical Africa, and they extend throughout the whole of the continent as far as Abyssinia, Madagascar, and Natal; their number may therefore be guessed at 100; which would make the total number of species in the Order 600.

In India the *Anonaceæ* are most abundant in the Malayan peninsula, from which 55 are known. Ceylon has about half that number, of which all but three are different from those of Malaya. They exhibit a marked preference for the humid provinces, and are almost entirely wanting in the drier ones. The number lessens as we proceed northward, but they are still numerous in the forests at the base of the Khasia mountains and in the Assam valley. Further west they rapidly diminish in number, though a few creep along the base of the Himalaya as far as Nipal. The forms characteristic of Ceylon and Malabar extend north along the chain of the Ghats to Concan, Kandesh, and even the mountains of Orissa, and in greatly diminished numbers to the hills of Behar, whence a single species reaches the base of the Himalaya in Garhwal at Dehra.

The forms characteristic of the Madras and Malayan peninsulas respectively are scarcely intermingled in any part of these regions, the number common to the two being only six, of which three are common to Ceylon and Malaya, and three are found in Khasia or the eastern Himalaya, as well as in Behar.

Few *Anonaceæ* rise to any height on the mountains, as might indeed be expected from the tropical character of the Order. In Ceylon they are found up to 6000 feet, and in Khasia up to 5000. In Brazil, according to St. Hilaire, their greatest elevation is 4000 feet. As about a fourth part of the Indian species are scandent, it is curious that no scandent species has yet been described from America. It may also be observed that in America one-ovuled species predominate, whereas in Asia the majority are many-ovuled. In India the species of *Anonaceæ* generally inhabit dense forests, and no representatives occur of the many shrubby species which in Brazil clothe the *campos*, or open grassy plains.

In addition to the published materials regarding this Order, we have had access to a number of drawings and descriptions of Penang species made many years ago by Sir W. Hooker. We have also found in the Hookerian Herbarium many useful remarks by M. Planchon, who appears to have studied the Order with care: these are chiefly identifications of species; but several of our new genera have also been indicated by him, though without any characters being given.

CONSPECTUS TRIBUUM.

A. Carpella in fructum multilocularem coalita　.　. ANONEÆ.
B. Carpella discreta.
　　a. Petala æst. imbricata .　.　.　.　.　.　.　. UVARIEÆ.
　　b. Petala æst. valvata.
　　　　α. Stamina laxe imbricata, antheris in flore
　　　　　　conspicuis .　.　.　.　.　.　.　.　.　.　. SACCOPETALEÆ.
　　　　β. Stamina densissime conferta, invicem antheras occultantia.
　　　　　　i. Petala interiora unguiculata　.　.　. MITREPHOREÆ.
　　　　　　ii. Petala interiora haud unguiculata.
　　　　　　　　1. Pet. int. incrassata triquetra　.　. XYLOPIEÆ.
　　　　　　　　2. Pet. int. exterioribus subconformia vel basi tantum excavata .　. GUATTERIEÆ.

In *Uvarieis* et *Mitrephoreis* paucis stamina definita occurrunt.

CONSPECTUS GENERUM.

I. UVARIEÆ.—Petala æstivatione imbricata.
　　A. Stamina definita (12–21) .　.　.　.　.　.　. 1. *Sageræa.*
　　B. Stamina indefinita.
　　　　i. Ovula numerosa.
　　　　　　a. Flores dioici. Stamina in toro cylindrico breviter cuneata .　.　.　. 2. *Stelechocarpus.*
　　　　　　b. Flores hermaphroditi. Stamina plano-compressa, in toro planiusculo disposita .　.　.　.　.　.　.　.　. 3. *Uvaria.*
　　　　ii. Ovulum solitarium in sutura ventrali　. 4. *Ellipeia.*

II. MITREPHOREÆ.—Petala interiora unguiculata.
  A. Ovulum solitarium e basi erectum; sta-
    mina definita (12–21) . . . . 5. *Popowia.*
  B. Ovula 2 prope basin ovari superposita;
    stamina indefinita . . . . . . 6. *Goniothalamus.*
  C. Ovula in sutura ventrali 2–6, stamina de-
    finita . . . . 7. *Orophea.*
  D. Ovula indefinita, stamina indefinita . . . 8. *Mitrephora.*
III. ANONEÆ.—Carpella in fructum multilocularem
  cohærentia . . . . 9. *Anona.*
IV. XYLOPIEÆ.—Petala interiora incrassata, triquetra.
  A. Torus conicus; antheræ connectivi processu
    oblongo apiculatæ . . . . 10. *Melodorum.*
  B. Torus planiusculus . . . . 11. *Habzelia.*
  C. Torus excavatus; antheræ truncato-capita-
    tæ; stigmata elongata . . . . . . 12. *Xylopia.*
V. GUATTERIEÆ.—Petala planiuscula vel basi tan-
  tum excavata.
  A. Petala basi circa ovaria constricta, lamina
    erecta vel patente plana.
    a. Ovarium 1, ovula in sut. ventr. plura . 13. *Cyathocalyx.*
    b. Ovaria plura, ovula 2 e basi erecta . 14. *Artabotrys.*
  B. Petala plana, coriacea vel tenuia.
    a. Ovula indefinita, biserialia . . . . 15. *Cananga.*
    b. Ovula definita.
      a. Petala subconformia.
        1. Ovula 2–8, in sutura ventrali re-
          gulariter disposita . . . . . 16. *Unona.*
        2. Ovula 2, prope basin superposita 17. *Polyalthia.*
        3. Ovulum 1, e basi erectum . . 18. *Guatteria.*
        4. Ovula 2, e basi erecta . . . . 19. *Anaxagorea.*
      b. Petala interiora crassa, exterioribus
        minora, conniventia . . . . . 20. *Oxymitra.*
      c. Petala exteriora minuta, sepalis con-
        formia . . . . 21. *Phæanthus.*
VI. SACCOPETALEÆ.—Stamina laxe imbricata.
  A. Petala exteriora minuta, sepalis subcon-
    formia.
    a. Ovula in sutura ventrali 1–2 . . . 22. *Miliusa.*
    b. Ovula indefinita . . . . . . 23. *Saccopetalum.*
  B. Petala subæqualia . . . . 24. *Alphonsea.*

Tribus I. UVARIEÆ.

Petala plano-convexa, coriacea, obtusa, æstivatione imbricata. *Stamina* indefinita, dense conferta, rarius (in *Sageræa*) definita.

This tribe is, we think, a very natural one. The scandent habit of most *Uvarieæ*
indicates an approach to *Schizandraceæ*, which is confirmed by the imbrication of the

petals, as well as by the occasional separation of the sexes, and the tendency to repand leaves. The truncal inflorescence of *Stelechocarpus* recurs in *Kadsura cauliflora*, Blume.

The American genus *Asimina*, in which the petals are only very slightly imbricated, approaches *Unona* by their thin, almost membranous texture, and by its stamens and torus.

1. SAGERÆA, Dalzell in Hook. Kew Journ. iii. 307.

Flores hermaphroditi vel unisexuales. *Sepala* 3, rotundata, æst. imbricata. *Petala* 6, biserialia, orbicularia, carnosula, concava, æst. imbricata. *Stamina* 12–21, abbreviato-cuneata, carnosa, truncata, dorso antherifera ; antheræ biloculares, loculis oblongis longitudinaliter dehiscentibus. *Torus* planus. *Ovaria* definita 3–6, lineari-oblonga ; ovula circiter 10, suturæ ventrali inserta, biserialia.—Arbores, foliis *coriaceis lucidis glaberrimis*, floribus *axillaribus fasciculatis.*

When the ovaries are three in number they alternate with the sepals. The stamens closely resemble those of *Bocagea* (among *Saccopetaleæ ?*) and of *Orophea* among *Mitrephoreæ;* but there is too little resemblance in other respects among those genera which have subdefinite stamens, to render it advisable to form of them a distinct section, as has been done by Blume and Endlicher. No species of *Sageræa* are known save those described below; but *Guatteria polita*, Wall. Cat. 6450, from Tenasserim, which has no flowers nor fruit, has the habit of the genus, and is probably a congener, if indeed it be not referable to one of the species described below.

1. **S. laurina** (Dalz.! l. c.) ; foliis lineari-oblongis, pedicellis 1–5 basi squamulosis, floribus hermaphroditis 12-andris, sepalis glabris, carpellis globosis.— Guatteria laurifolia, *Graham, Cat. Bomb. p.* 4.

HAB. In sylvis Concan utriusque !—(Fl. Oct. Nov.) (*v. s.*)

Arbor mediocris, elegans, Lauri facie. *Ramuli* rugosi, nigricantes, glabri. *Folia* basi rotundata vel acutiuscula, apice angustata, 5–7 poll. longa, 1½–2 lata, petiolo ½-poll., tenuiter coriacea, rigida, nervis crebre reticulatis. *Pedicelli* ½-pollicares, bracteola in medio pedicello ovali vel rotundata. *Flores* albi. *Petala* fere semipollicaria. *Antheræ* exteriores interdum anantheræ. *Carpella* globosa, glabra, circiter sexsperma.

This tree is said by Mr. Dalzell to yield valuable timber of a reddish colour. It will probably be found to be also a native of Malabar.

2. **S. elliptica** (H.f. et T.) ; foliis lineari-oblongis, floribus axillaribus vel secus ramos crassiores solitariis aut fasciculatis, pedicellis brevissimis basi squamulosis, floribus dioicis 12-andris, sepalis ciliatis, carpellis obovatis.—Uvaria elliptica, *Alph. DC. Mém.* 27 ; *Wall. Cat.* 6470 !

HAB. In prov. Tenasserim ad Tavoy, *Wall.!*—(*v.s.*)

Arbor excelsa, cortice ramulorum albido vel griseo lævi glabro. *Folia* 10 poll. longa, fere 3 lata, pet. ¼-poll., basi obtusa, apice obtusa vel acuta, crasse coriacea, nervis vix conspicuis. *Bracteola* in medio pedicello rotundata. *Sepala* margine membranacea. *Petala* late ovalia, margine ciliata, ¼ poll. longa. *Carpella* immatura obovato-oblonga.

3. **S. Thwaitesii** (H.f. et T.) ; foliis auguste oblongis, floribus secus ramos crassiores fasciculatis icosandris hermaphroditis.

HAB. In Zeylania, *Thwaites!* (No. 2702.)—(*v. s.*)

Arbor. Cortex ramulorum rugulosus, atrofuscus. *Folia* 8–12 poll. longa, 3–4½ lata, petiolo ½–⅔-poll., coriacea, glaberrima, lucida, subtus pallidiora. *Pedicelli* 2–4, basi bracteati, squamulis minutis, pollicares. *Sepala* abbreviata, obtusa. *Petala* rotundata, exteriora majora. *Stamina* toro planiusculo inserta, compressa, late cuneiformia, irregularia, plana, truncata, dorso antherifera. *Ovaria* 3, subglobosa, irregulariter angulata, parce strigosa, stigmate depresso coronata.

This species is a good deal like the last, but the flowers are larger and hermaphrodite. The stamens are also more numerous; and as these characters are usually constant in the Order, there can be little doubt that the two are specifically distinct. We have only seen one specimen and a single flower, and the fruit is unknown.

2. **STELECHOCARPUS,** Blume.

Uvariæ sectio, *Blume, Fl. Jav. Anon.* 13.

Flores dioici, fœminei majores. *Sepala* 3, rotundata. *Petala* 6, ovalia vel rotundata, æqualia, æst. imbricata. *Stamina* indefinita, secus torum anguste conicum dense imbricata, breviter cuneata, connectivo ultra antherarum loculos extrorsos contiguos truncato capitato. *Ovaria* numerosa, torum hemisphæricum obtegentia, oblique ovalia, ovulis in axi 6–8. *Stigma* sessile, depressum, radiatum. *Carpella* magna, globosa, polysperma.—Arbor, foliis *coriaceis lucidis*, venis *arcuatis distantibus subtus prominulis*, inflorescentia *supra ramos fasciculata.*

This genus seems to have no very close affinity with any other in the Order. The rounded imbricated petals constitute an approach to *Uvaria*, but the ovaries are widely different, as well as the whole habit. It approaches *Schizandreæ* in its diœcious flowers, which are rare in the Order, but occur in several very distant parts of it, and do not appear to be of much moment in deciding affinity. The habit and foliage, as remarked by Blume, are a good deal like those of some *Magnoliaceæ* (*Talauma pumila* for example), but the truncal inflorescence, and the aspect of the flowers, recall that of some species of *Schizandreæ*. These, however, are perhaps distant or fanciful analogies, of no real value. We retain the name adopted by Blume for the section of *Uvaria*, to which he refers his plant; but the other species, *U. reticulata*, Blume, must, according to our views of affinity, be excluded. Of that plant the male flower only is known, so that its position cannot be indicated with certainty. If the female flower presents no obstacle, it may form part of the genus *Mitrephora*, notwithstanding its dioicality; but if it differs, it must form a new genus close to it, and to *Orophea*, Bl.

1. **S. Burahol** (Blume, Fl. Jav. Anon. 48. t. 23, 25 C, sub Uvaria); foliis oblongo-lanceolatis utrinque acutis, floribus fœmineis longe pedicellatis, carpellis globosis breviter pedicellatis.

HAB. In peninsula Malayana ad Singapur, *Lobb!*—(*v. s.*)
DISTRIB. Java, *Blume.*

Arbor excelsa. *Ramuli* nigricantes, rugulosi, glabri. *Folia* sæpe acuminata, coriacea, rigida, 5–8 poll. longa, 1½–3 lata, petiolo ⅓-pollicari, utrinque glaberrima, venulis (in sicco) conspicue reticulatis. *Flores* secus truncum et ramos in tuberculis lignosis bracteis squamæformibus dense imbricatis onustis pubescentibus fasciculati; *masculi* fœmineis multo minores, pedicellis ½–1 poll. longis ebracteatis pubescentibus, sepalis minutis, petalis ⅛ poll. longis; *fœminei* triplo majores, pedicellis bipollicaribus validis apice subclavatis rugosis costatis infra medium bracteatis, petalis latioribus. *Ovaria* sericea, obliqua, dorso superne gibbosa. *Carpella* pauca, toro globoso insidentia, sesquipollicaria, baccata, aromatica. *Semina* 4-6, ovalia, subcompressa, subrugosa, castanea, margine elevato cincta, triserialia.

Our specimens being in flower only, we have derived our character of the fruit from Blume's detailed description. There are specimens in our own Indian collections of a tree from the forests north of Chittagong, which, though in leaf only, appear to belong to this species.

3. UVARIA, L.

Sepala 3, æstivatione valvata, lata, basi sæpe coalita. *Petala* 6, rotundata, ovalia, vel oblonga, æst. biserialiter imbricantia, plano-convexa, basi interdum plus minus coalita. *Stamina* indefinita, multiserialia, plano-compressa, oblonga vel lineari-oblonga, antherarum loculis remotis dorsalibus linearibus, connectivo in processum oblongum subfoliaceum vel truncatum et abbreviatum producto. *Torus* parum elevatus, truncatus, pubescens, inter ovaria sæpe dense tomentosus. *Ovaria* indefinita, recta, lineari-oblonga, angulata, intus sulcata, pubescentia, stylo continuo apice truncato, marginibus involutis, succum gelatinosum effudente coronata ; *ovula* indefinita, biserialia. *Carpella* polysperma, forma valde varia, interdum abortu meio- vel monosperma. —Frutices *scandentes vel saltem sarmentosi, pube vel tomento stellato,* inflorescentia *plerumque oppositifolia, rarissime axillari.*

Notwithstanding the exclusion of many species, this genus still remains a very extensive one. The species appear to be all scandent, and they are entirely confined to the Old World, through which they are widely distributed, from western Africa to the Philippine Islands. *Uvaria Brasiliana* of Von Martius, with an arillus and dehiscing fruit, and stamens like those of *Anona,* certainly does not belong to the genus. It ought probably to be associated with *Asimina* or *Porcelia,* as has been suggested by Asa Gray.

The principal characters of the genus *Uvaria,* as now limited, are the equal petals, imbricate in æstivation, and the narrow, linear, cylindrical ovaries, perfectly straight, with a very short style, which is marked at the apex with a horse-shoe-like impression, continuous with the ventral groove of the ovary. The ovules are always numerous, and the carpels always (except by abortion, and that not typically, but casually) numerous, or at least scarcely definite.

The genus divides itself naturally into two sections, characterized by very different forms of stamen. In one of these, containing the majority of the species, the stamens are flattened, and the outer series generally very thin, and sometimes barren, or without anthers. In the other, which contains *U. Zeylanica,* L., the original species of the genus (to which, therefore, if division be carried further, the name must attach), the stamens are narrower and truncate at the apex. This is, however, only a question of degree, the outer stamens, even in this section, being terminated by a projection of the connectivum.

The petals are occasionally united at the base in *U. Narum* and other species, in which case they form a single verticil, like the tubular perianth of most monocotyledonous plants, though belonging to two distinct series, alternating with one another.

Sect. 1. MACRANTHI.—*Connectivum* in processum magnum subfoliaceum productum. *Antheræ* loculi remoti. *Stamina* exteriora tenuia, subfoliacea, interdum ananthera.

1. **U. purpurea** (Bl. Bijdr. 11, Fl. Javæ Anon. 13. t. 1 et 13 A) ; foliis cuneato-oblongis vel oblongo-lanceolatis basi angustatis corda-

tis, pedunculis unifloris, bracteis 2 magnis rotundatis submembranaceis nervosis ante florationem alabastrum involventibus, petalis ovalioblongis, carpellis baccatis oblongo-cylindricis dorso bicostatis longe pedicellatis.— *Wall. Cat.* 6485 ! (*excl.* E *et* G). U. grandiflora, *Wall. Pl. As. Rar.* ii. *t.* 121 ; *Roxb. Fl. Ind.* ii. 665 ; *W. et A. Prod.* i. 9 ; *Alph. DC. Mém.* 29. U. platypetala, *Champ.; Benth. in Hook. Kew Journ.* iii. 257. Unona grandiflora, *DC. Prod.* i. 90.

HAB. Pegu, *Wall.!;* et in Penins. Malay.: ad Penang, *Phillips!* Malacca, *Griff.!* et Singapur, *Lobb !—(v. s.)*

DISTRIB. Sumatra ! Java ! Ins. Philippin. (*Cuming,* 1380 !) Hongkong !

Frutex scandens, sarmentosus. *Ramuli* distichi, nigricantes, rugosuli, juniores pilis stellatis fusco-tomentosi ; partes novellæ omnes dense fusco-tomentosæ. *Folia* acuta vel acuminata, rarius obtusa, 6–12 poll. longa, 2–4 lata, petiolo ¼–⅓-poll., coriacea, supra sparse stellato-pubescentia, nervo medio piloso, demum glabrescentia, nitida, nervosa, subtus pilis fulvis stellatis dense tomentosa, plerumque margine undulato-repanda. *Pedunculi* extra-alares, sæpe oppositifolii, pollicares, subclavati. *Bracteæ* tomentosæ, deciduæ. *Alabastri* globosi. *Sepala* extus tomentosa, late ovata, obtusa, nervosa. *Flores* purpurei, diametro 3-pollicares, suaveolentes. *Petala* 1–1½-pollicaria, sub lente pubescentia, interiora angustiora, basi angustata. *Baccæ* 1–4 poll. longæ, leviter torulosæ, oblique acutiusculæ, flavæ, tomentosæ, dorso costis 2 prominentibus distantibus notatæ, pedicello 1–1½-pollicari suffultæ. *Semina* partitionibus cellulosis separata.

De Candolle, who described this plant in the Prodromus, from specimens brought to Europe by Leschenault, gives Bengal as the locality ; but these specimens were doubtless collected in the Calcutta garden, where the plant has long been cultivated. We have not seen the fruit, which, however, is described and figured by Wallich.

2. **U. ferruginea** (Ham. mss.) ; foliis obovatis vel oblongis, pedunculis oppositifoliis unifloris medio unibracteatis, petalis ovato-oblongis cinereo-tomentosis.

HAB. In prov. Ava ad Meaday, *Hamilton!—(v. s. in Herb. Mus. Brit.*)

Frutex scandens. *Ramuli* elongati, læves, fulvo-tomentosi, paucifoliati. *Folia* distantia, supra adpresse pubescentia, subtus laxe stellatim pubescentia, ad costam et nervos fulvo-tomentosa, 4–6 poll. longa, 2½–3½ lata, petiolo ½-poll. *Pedunculi* solitarii, dense ferrugineo-tomentosi, ½–⅔ poll. longi, medio bracteam lanceolatam semipollicarem gerentes. *Sepala* lata-ovata, ⅓ poll. longa ; *petala* ¾ poll. longa.

There is only one specimen of this plant in the British Museum, but it seems so unlike anything else we have seen that we have no hesitation in describing it as distinct. As we have not examined the flowers, we have only the general appearance to guide us in referring it to *Uvaria.* It is evidently a climber, and the specimen exhibits a less woody appearance than is usual in the Order, being an elongated, soft shoot, with few and distant leaves.

3. **U. Hamiltonii** (H.f. et T.) ; foliis obovato-oblongis superne angustatis et in acumen gracile productis membranaceis utrinque pubescentibus, pedunculis oppositifoliis 1–2-floris, bracteola parva in medio pedicello, petalis late obovatis, carpellis longe pedicellatis ovalisubglobosis tomentosis.—U. purpurea, *Wall. Cat.* 6485 E !

HAB. In montibus Behar prope Monghir, *Hamilton !;* et secus basin Himalayæ orientalis : in prov. Sikkim ! et Assam !—(*v. v. fruct.*)

Frutex alte scandens. *Ramuli* nigricantes, juniores fulvo-tomentosi. *Folia* 5-8 poll. longa, 2-3½ lata, petiolo ½-poll., interdum ovalia vel lineari-oblonga, basi angustata, rotundata vel subcordata, utrinque secus nervos (subtus densius) molliter pubescentia; pilis stellatis. *Pedunculi* abbreviati, vix ¼-pollicares; *pedicelli* dense ferrugineo-tomentosi, 1-1½-pollicares, medio bracteolam ovalem vel oblongam gerentes. *Flores* magni. *Calyx* rotundatus, ½-pollicaris, fulvo-tomentosus, nervosus. *Petala* 6, pollicaria, fere rotundata, obtusa, utriuque dense tomentosa. *Stamina, ovaria,* et *torus U. purpureæ. Torus* fructus dilatatus, subglobosus. *Carpella* ¾ poll. longa, pedicello pollicari suffulta, carnosa, (sapore subdulci,) cinereo-tomentosa. *Semina* biserialia, oblonga, compressa; testa fusca, nitida, lævi.

Apparently a very distinct species. The fruit when dried is very like that of *U. rufa,* Blume, which is represented with slightly projecting seeds. This seems, however, in our species to be the result of drying, and we have no notes of the appearance in a fresh state.

4. **U. semecarpifolia** (H.f. et T.); foliis oblongis vel obovato-oblongis obtusis cum mucrone brevi basi subcordatis, pedunculis oppositifoliis abbreviatis 3-6-floris, petalis ovalibus cinereo-incanis, carpellis ovali-subglobosis fulvo-tomentosis lævibus breviter pedicellatis.

HAB. In Zeylania, *Walker! Thwaites!;* in Malaya ad Malacca, *Griffith!—(v. s.)*

Frutex scandens, ramulis fulvo-tomentosis. *Folia* 6-10 poll. longa, 2½-4½ lata, petiolo ¼-poll., coriacea, rigida, supra (præsertim secus costam et nervos) puberula, demum glabra, subtus pilis stellatis pubescentia. *Pedunculi* fulvo- vel cinereo-tomentosi, vix semipollicares, bracteis pluribus rotundis muniti; pedicelli vix ½-pollicares, medio bracteolati. *Flores* in specimine nondum aperti. *Fructus* subglobosi, diametro fere pollicares, pedicello ⅛-pollicari. *Semina* biserialia, 8-10, lævia, testa nitida, fusca.

5. **U. macrophylla** (Roxb. Fl. Ind. ii. 663); foliis oblongis vel obovato-oblongis abrupte acuminatis supra glabriusculis subtus fusco-tomentosis, pedunculis oppositifoliis plurifloris, petalis ovalibus, carpellis numerosis subsessilibus ovato-oblongis glabris.—*Wall. Plant. As. Rar. t.* 122, *Cat.* 6487! (excl. F. quoad sp. fructiferum). U. cordata, *Wall. Cat.* 6486! U. rufescens, *Alph. DC. Mém.* 26 (*excl. descr. fructus*). Guatteria cordata, *Dunal, Anon.* 129. *t.* 30; *DC. Syst.* i. 505, *Prod.* i. 93.

HAB. In sylvis montanis Silhet et Chittagong, *Roxb.;* in Ava! Tenasserim! Malaya!—(*v. s.*)

DISTRIB. Java.

Frutex scandens. *Ramuli* ferrugineo-tomentosi. *Folia* coriacea vel juniora membranacea, ovalia vel late oblonga, basi cordata, 6-12 poll. longa, 3-6 lata, petiolo ¼-½-pollicari, supra secus costam pubescentia, et tota superficie sub lente stellatim puberula, subtus pilis stellatis tomentosa. *Pedunculi* dense cinereo- vel fulvo-tomentosi, pollicares, bracteis pluribus ovalibus tomentosis. *Pedicelli* pollicares, versus medium bracteolam gerentes. *Alabastri* globosi, dense cinerei. *Sepala* ad medium coalita. *Petala* rubescentia, ⅔-pollicaria, basi plus minus coalita. *Stamina* exteriora late linearia, ananthera, truncata. *Carpella* 15-20, toro subgloboso inserta, ovali-oblonga vel subglobosa, 1-1½ poll. longa, interdum subtorulosa, baccantia, stipite vix lineam longo crasso insidentia. *Semina* biserialia.

From their large size, and the consequent imperfection of the specimens, many of the closely allied species of this genus are not readily distinguishable without ripe fruit ; or at least the suites of specimens available in herbaria are not sufficient to enable a correct conclusion to be drawn as to the extent to which the leaves vary. Wallich's figure and description are the authority for this species, and there are good specimens of it in the Linnæan herbarium, both in flower and fruit. We have also before us numerous flowering specimens from all parts of Trans-Gangetic India, but no fruit, except on the Wallichian specimens. There is, however, a very similar species from the Philippines (Cuming, 751), which has long-pedicelled glabrous carpels ; and as this is undistinguishable save by the fruit, it is quite possible that some of our specimens may belong to it. Others are probably referable to *U. semecarpifolia*, or to a third species, as there are considerable differences in the form and texture of the leaves, some being membranous and some rigidly coriaceous : this, however, may depend on age. *U. littoralis* and *ovalifolia* of Blume are also evidently very nearly allied, but we do not venture to unite them without seeing specimens.

In the Wallichian collection at the Linnean Society there is a specimen in fruit under the letter F, which certainly does not belong to this species, though we think the larger leaves on the same sheet do. In this the carpels are globose, rugulose, covered with brown tomentum, and more than half an inch in diameter, with a pedicel more than an inch long. The leaves on the specimen are elliptic-obovate, subcordate at base, five inches long by three broad, stellato-pubescent below, but they are not sufficient to identify it with any of the species here described. We are, however, inclined to believe that M. Alph. De Candolle must have received a similar fruiting specimen along with his specimen, because we cannot doubt (notwithstanding the discrepancies in his character of the fruit) that his *U. rufescens* is Wallich's *U. macrophylla*. This appears evident when the localities of *U. rufescens*, given by De Candolle, are compared with those in Wall. Cat., in which *U. rufescens* is not referred to.

6. **U. dulcis** (Dunal, Anon. 90, t. 13) ; foliis ovalibus vel oblongis supra puberulis subtus dense furfuraceo-tomentosis, pedunculis abbreviatis subumbellatim 1–4-floris, petalis oblongis basi coalitis.—*DC. Syst.* i. 483, *Prod.* i. 88 ; *Spr. Syst.* ii. 639.   U. Javana, *Dun. Anon.* 91. *t.* 14 ; *DC. Syst.* i. 483, *Prod.* i. 88.

HAB. In peninsula Malayana ad Malacca, *Griff.!*—(*v. s.*)
DISTRIB. Java, *Bl.!*

Frutex alte scandens. *Ramuli* atro-fusci, juniores stellato-tomentosi. *Folia* obtusa vel acutiuscula vel abrupte acuminata, basi rotundata vel retusa, margine subrepanda, 3–5 poll. longa, ½–2 lata, petiolo tomentoso 2–3 lineas longo, coriacea, supra pilis minutis stellatis vel simplicibus sub lente tantum conspicuis tecta, subtus venosa. *Pedunculi* lignosi, ¼–½-pollicares, bracteis pluribus ovatis parvis tomentosis, pedicelli 1–4, ¼–1 poll. longi, medio bracteolati, bracteolis bracteis similibus. *Alabastri* globosi, dense cinereo-tomentosi. *Flores* odorati. *Sepala* late ovalia, basi concreta, obtusiuscula. *Petala* patentia, fere pollicaria, utrinque tomentosa. *Stamina* abbreviata, ext. sterilia.

Sufficiently distinct from *U. macrophylla* in the smaller size of the leaves, and in the shape and aspect of the flowers. It is near *U. rufa*, Blume (a species which has not been found within our limits), but that is more frequently one-flowered, and the petals are a good deal smaller and broader. *U. microcarpa*, Champion (from Hongkong), is also closely allied, but nearer to *U. rufa*, from which it differs by the smooth not transversely sulcate carpels, and by the large flowers, which are not distinguishable from those of *U. macrophylla*, Roxb. Blume distinguishes *U. Javana* from *U. dulcis* by the stellate, not simple, hairs of the upper surface of the leaves. He seems, however, to trust entirely to Dunal's figure, without having seen specimens of *U. dulcis*. In our specimens we see simple and stellate hairs intermixed.

Uvaria.] FLORA INDICA. 99

7. **U. sphenocarpa** (H.f. et T.); foliis brevissime petiolatis anguste obovatis vel cuneato-oblongis acuminatis supra minute scabris subtus pilis fulvis stellato-tomentosis, pedunculis unifloris, bracteis rotundatis imbricatis squamæformibus, petalis ovalibus, carpellis sessilibus obovato-cuneiformibus tuberculatis.

HAB. In insulæ Zeylaniæ montosis, *Walker!* *Champion!* *Thwaites!*
—(*v. s.*)

Frutex verosimiliter scandens. *Ramuli* elongati, flexuosi, graciles, nigricantes: juniores fulvo-tomentosi. *Folia* brevissime petiolata, basi rotundata vel retusa, sensim vel abrupte in acumen longum gracile attenuata, 3–5 poll. longa, 1¼–2¼ lata, petiolo 1–2 lineas longo, supra atro-viridia, sicca nigricantia, sub lente stellato-pilosa. *Pedunculi* oppositifolii, semipollicares, tomentosi. *Flores* parvi, diametro vix semipollicares. *Alabastri* globosi. *Sepala* dense fulvo-tomentosa, in cyathum obtuse trilobum coalita, in fructu subpersistentia. *Petala* ovalia, obtusa, cinereo-tomentosa. *Torus* fructus depresso-globosus. *Carpella* 8–10 vel pauciora, vertice rotundata, pilis fulvis rigidis dense tomentosa, supra medium grosse et irregulariter tuberculata, pollicaria. *Semina* dissepimentis cellulosis separata, oblique biserialia.

This is a remarkable species, readily known by its very peculiar fruit. We have seen only one expanded flower, and have therefore not examined the ovaries. The habit and characters, however, leave no doubt as to the genus to which it ought to be referred.

8. **U. hirsuta** (Jack, Mal. Misc. et in Hook. Bot. Misc. ii. 87); foliis oblongis apice plerumque longe acuminatis breviter petiolatis supra longe et laxe pilosis subtus densius hirsutis, pedunculis unifloris supra basin unibracteatis rarius bifloris, petalis ovalibus puberulis, carpellis oblongis hirsutis longe pedicellatis.—*Blume, Fl. Jav. Anon.* 22. *t.* 5; *Wall. Cat.* 6458! (excl. C, quæ planta stipulata, forsan *Dipterocarpi* species). U. pilosa, *Roxb.! Fl. Ind.* ii. 665. U. velutina, *Bl. Bijdr.* 13, *non Roxb.* U. trichomalla, *Bl. Fl. Jav. Anon.* 42. *t.* 18.

HAB. In Penins. Malayana ad Penang, *Jack, Wall.!* Singapur, *Lobb!*
—(*v. s.*)
DISTRIB. Java, *Bl.*

Frutex sarmentosus. *Ramuli* cinerei vel nigricantes, rugulosi, juniores pilosi; partes novellæ pilis fulvis patentibus laxe hirsutæ. *Folia* basi subangustata, rotundata vel emarginata, 5–6 poll. longa, 1¾–2½ lata, petiolo vix 2 lineas longo, supra pilis longis plerumque simplicibus, subtus pilis stellatis hirsuta, coriacea, nervis obliquis prominentibus. *Pedunculi* extra-alares, laxe hirsuti, uniflori et paullo supra basin articulati, ibique plerumque bractea oblonga ½ poll. longa decidua pilosa muniti, vel abbreviati, apice biflori, pedicellis pollicaribus supra medium bracteolam similem gerentes. *Alabastri* globosi, laxe pilosi, diam. ⅔-pollicares. *Sepala* subrotundata, obtusa, reflexa, membranacea, nervosa, extus dense pilosa, petala fere æquantia. *Petala* sanguinea, puberula, fere ½-pollicaria. *Stamina* fere ad apicem antherifera, connectivi processu abbreviato obtuso. *Ovaria* dense fulvo-pilosa. *Torus* fructus incrassatus, globosus, diam. pollicaris. *Carpella* 10–20, oblonga vel obovato-oblonga, obtusa vel mucronata, interdum subtorulosa, dorso obscure carinata, 1–1½-poll., pedicello æquilongo, rarius seminibus pluribus abortivis abbreviata, subglobosa.

Some of the specimens in the Wallichian Herbarium are identical with *U. trichomalla* of Blume, which is no way different from the ordinary form of the species. *U. hirsuta*, Blume, is rather more softly hairy, but the floral characters present no differences of importance.

9. **U. bracteata** (Roxb. Fl. Ind. ii. 660); foliis oblongis vel ob-ovato-oblongis tenuiter coriaceis subtus præsertim secus nervos sparse puberulis, pedunculis oppositifoliis plerumque bifloris, bractea ad bifur-cationem ovali petiolata foliacea decidua, bracteolis ovalibus sessilibus, petalis conniventibus, carpellis oblongis obtusis subsessilibus.—*Wall. Cat.* 6468! U. Gomeziana, *A. DC. Mém.* 27; *Wall. Cat.* 6459!

HAB. Silhet, *Roxb.!* Tenasserim, *Wall.!*—(Fl. Maii, Fr. Sept.)— (*v. s.*)

Frutex alte scandens. *Ramuli* elongati, cinerei vel nigricantes, juniores puberuli; partes novellæ tomentosæ. *Folia* acuta vel breviter acuminata, basi angustata, ro-tundata, supra nitida, glabra, secus costam et petiolum pubescentia, demum glabrata, subtus pallida, 4–7 poll. longa, 1¾–2½ lata, petiolo ⅓–½-poll. *Pedunculi* ½-polli-cares, pubescentes, bractea semipollicari. *Flores* pallide flavescentes, nutantes, dia-metro vix ½-pollicares. *Sepala* reniformi-rotundata, pubescentia, ad medium coa-lita, undulata. *Petala* pubescentia, lacera, interiora basi subsaccata, ⅓-poll. *Sta-mina* anguste linearia; *connectivi* processus oblongus, carnosus. *Torus* fructus in-crassatus, ¾-poll. *Carpella* numerosa, baccantia, juniora tomentosa, matura pube-rula, 1–2-pollicaria. *Semina* pauca, biserialia, ovalia, compressa, ⅔-poll., hilo magno depresso.

The flowers of this species remain so long connivent, that the petals probably drop off without expanding. It appears very distinct from all but the next following species, the foliage of which is very similar.

10. **U. Lobbiana** (H.f. et T.); foliis oblongis vel obovato-ob-longis obtusis et obtuse mucronatis coriaceis subtus sub lente furfura-ceis, pedunculis 2–4-floris ad bifurcationem bracteam rotundatam cori-aceam amplexicaulem gerentibus, pedicellis sub flore ample bracteolatis, petalis patentibus ovalibus.

HAB. In Penins. Malayana ad Malacca, *Griff!* Singapur, *Lobb!*— (*v. s.*)

Frutex scandens, *U. bracteatæ* similis. *Ramuli* cinerei, juniores squamulis minutis stellatim radiantibus scabri; partes novellæ dense stellato-furfuraceæ. *Folia* 4–7 poll. longa, 1¾–3 lata, petiolo ¼–½-poll., rigida, supra nitida, glabra, vel secus costam pubescentia, subtus glabriuscula vel pilis minutis stellatis furfuraceis tecta. *Pedun-culi* cum bracteis et alabastris dense albido-furfuracei; bracteæ et bracteolæ crassæ, semipollicares. *Alabastri* globosi. *Flores* diametro pollicares. *Sepala* rotundato-reniformia, ultra medium in cyathum trilobum margine crispato-undulatum coalita. *Petala* coriacea, utrinque verrucosa, ovalia, obtusa; interiora paullo minora et angus-tiora. *Stamina* late lineari-oblonga, apice truncata, exteriora ananthera, processu connectivi quadrato plano.—In Herb. Benthamiano videmus ramulum fructiferum aphyllum a Griffithio in Malacca lectum, et cum foliis hujus speciei distributum, in quo *carpella* plus quam viginti, pedicellis 2–3-pollicaribus stellato-pubescentibus angu-latis suffulta, globosa vel ovalia, obliqua, ¼–1 poll. longa, tuberculis parvis verrucosa, fulvo-tomentosa.

Though very closely resembling the last species in leaves and habit, this seems to differ in many important points. The leaves are much thicker and firmer, with more transverse nerves; the flowers are larger, much more tomentose, with very different bracts; and the calyx is remarkable, being patent and undulated like a ruff or frill, even in very young buds. The stamens, too, are very different, and exhibit to perfection the peculiar foliaceous flattened form characteristic of the section. The outer series are sterile; on the next the anthers are very short, and these gradually lengthen from without inwards, the process of the connective at the same time becoming thicker. The fruit, described from Mr. Bentham's Herbarium, closely resembles that in the Linnean Soc. Collection, under 6487 F.　See our remarks under *U. macrophylla.*

11. **U. subrepanda** (Wall. Cat. 6483 !); foliis oblongis vel obovato-oblongis, pedunculis axillaribus solitariis pluribracteatis, petalis anguste oblongis.

HAB. In Peninsula Malayana ad Singapur, *Wall./—(v. s.)*

Frutex (forsan scandens), ramulis gracillimis, junioribus cum omnibus partibus novellis pubescentibus. *Folia* membranacea, acuta, superne glabra, secus costam pubescentia, subtus puberula, secus nervos pubescentia, demum glabrescentia. *Pedunculi* pollicares, graciles, stellato-puberuli, medio et basin versus bracteolis parvis cucullatis muniti.

This species is very imperfectly known, the Wallichian specimens being few and in flower only. Its axillary flowers seem to distinguish it from all its allies.

Sect. 2. NARUM.—*Stamina* apice truncata, connectivo ultra antheras vix producto.

12. **U. macropoda** (H.f. et T.); foliis coriaceis oblongis vel lanceolatis utrinque glabris, floribus terminalibus solitariis, alabastro subgloboso longe mucronato granulato, petalis ovali-oblongis, carpellis longissime pedicellatis oblongis argute tricostatis.

HAB. In Zeylaniæ montibus, *Walker / Thwaites!—(v. s.)*

Frutex scandens. *Ramuli* cinerei, rugulosi, glabri; partes novellæ pilis stellatis paucis adpressis sparsis sub lente scabridæ. *Folia* plerumque in acumen angustata, rigida, supra nitida, subtus· pallidiora, 3–6 poll. longa, 1–2 lata, petiolo ¼-poll. *Pedunculi* in ramulis terminales, solitarii, ½–1-pollicares, pilis stellatis furfuracei. *Sepala* extus tuberculata, rotundata, ½–⅔ poll. longa, abrupte in mucronem ¼-poll. angustata, intus dense tomentosa. *Petala* fere pollicaria, utrinque dense furfuraceotomentosa. *Stamina* et *ovaria U. Nari.* *Torus* fructus incrassatus, subglobosus. *Carpella* 15–30, pedicellis 3–6-pollicaribus filiformibus superne clavatis argute triquetris suffulta, mucronata, 1–2-pollicaria, glabra, granulosa.

13. **U. lurida** (H.f. et T.); foliis coriaceis oblongo-lanceolatis utrinque glabris vel subtus minute puberulis, floribus subsolitariis terminalibus, alabastris obtusis granulosis, petalis late obovatis obtusis.

β. *macrophylla ;* foliis 6–9 poll. longis 2–2¾ latis, floribus minoribus sæpe oppositifoliis, carpellis junioribus oblongis subtorulosis, pedicellis 2–3-pollicaribus angulatis.—*Wall. Cat.* 6473 C !

HAB. In montibus Khasia versus Assam, alt. 2000 ped.!　β. Peninsula, *Herb. Madr. in Wall. Cat.!* in montibus Concan austr., *Dalzell!—*(Fl. Nov.) (*v. v.*)

Frutex alte scandens. *Ramuli* grisei vel nigricantes, graciles, rugulosi, glabri; partes novellæ pilis stellatis tomentosæ. *Folia* oblongo-lanceolata, basi rotundata, supra nitida, subtus pallidiora, rigida, juniora plerumque sub lente pilis minutis sparsis, ad costam nervosque densioribus et stellatis puberula, 3–5 poll. longa, 1–2 lata, petiolo ½-poll. stellato-puberulo. *Flores* ad ramulorum apices solitarii vel bini, luride purpurei, diametro bipollicares. *Pedicelli* pollicares, pilis stellatis dense furfuracei, superne subclavati. *Alabastri* depresso-subglobosi, tuberculis parvulis granuloso-asperi. *Sepala* valvata, suturis ante dehiscentiam indistinctis, late ovata, intus dense furfuracea. *Petala* pollicaria, basi plus minus in unguiculum contracta, unguiculis basi cohærentibus. *Fructus* ignoti.

Very closely allied to *U. Narum,* but not so near in general appearance to that species as is *U. macropoda,* which is, however, very distinct in fruit. The variety β corresponds very closely in foliage with specimens from the Philippines (Cuming, No. 1607, 1729); but these again have the flowers of the Khasia plant, and many oblong, glabrous, slightly granular carpels, nearly 1¼ inch long, on stalks the same length, and with numerous flat round seeds in two rows. The buds of this and the preceding species are remarkable for the strong union of the calyx-lobes, the lines of separation of which are not distinguishable till they are about to dehisce.

14. **U. Narum** (Wall. Cat. 6473 A! B!); foliis oblongo-lanceolatis vel lineari-oblongis utrinque glabris, floribus terminalibus subsolitariis, alabastris lævibus obtusis, petalis obovato-oblongis, carpellis ovoideis obtusis longiuscule pedicellatis.—*W. et A. Prod.* i. 9; *Wight, Ill.* i. *t.* 6. U. Zeylanica, *Lam. non L.* Unona Narum, *Dun. Anon.* 99; *DC. Syst.* i. 486, *Prod.* i. 89.

HAB. In Zeylania! Malabaria! Carnatica! Maisor! Concan austr., *Dalzell!—(v. s.)*

Frutex scandens. *Ramuli* glabri, nigricantes, rugulosi; partes novellæ vix subfurfuraceæ. *Folia* acuta vel breviter acuminata, supra lucida, 3–6 poll. longa, 1¼–1¾ lata, petiolo glabro ¼-poll. *Flores* rubescentes, diametro 1–1½-poll. *Pedicelli* filiformes, glabri, 1–1½-poll. *Sepala* rotundato-ovata, glabra vel versus marginem adpresse stellato-tomentosa. *Petala* basi plus minus coalita, utrinque fulvo-tomentosa. *Torus* fructus depresso-globosus. *Carpella* 30–40, pedicello 1-2-pollicari, glabra, obscure torulosa. *Semina* 4–5, ovalia, compressa, hilo magno terminali, interdum 1-2 tantum et tunc carpella subglobosa.

This appears to be a widely-diffused plant in southern India, but it is possible that more than one species are still confounded under it. Wight figures the seeds as forming two rows; but in all the specimens which we have seen, they occupy the whole breadth of the seed, and are consequently in one row. These doubtful points can only be settled by careful study of the plant in a living state. The amount of variation in the shape of the petals appears considerable, and in several specimens from Ceylon they are united beyond the middle, and are occasionally increased in number to seven or eight.

15. **U. Zeylanica** (L. Sp. ii. 756); foliis (parvis) ellipticis vel lanceolatis acuminatis glabris brevissime petiolatis, pedunculis solitariis terminalibus vel oppositifoliis, petalis ovato-oblongis, carpellis ovoideis subsessilibus fulvo-incanis.—*Dun. Anon.* 88; *DC. Syst.* i. 481, *Prod.* i. 88. U. lutea, *Wall. Cat.* 6462! *non Roxb.* U. Heyneana, *W. et A. Prod.* i. 8, *non Wall.* U. coriacea, *Vahl, Symb.* iii. 72. Guatteria Malabarica, *Dun. Anon.* 134. G. montana, *DC. Syst.* i. 508, *Prod.* i. 94; *Rheed. Mal.* v. *t.* 17.

HAB. In sylvis Zeylaniæ! Malabariæ et Travancor!—(v. s.)

Frutex alte scandens, ramosissimus, dense foliosus. *Ramuli* graciles, rigidi, cortice cinereo ruguloso, adulti glabri, juniores adpresse tomentosi. *Folia* basi acuta, apice plerumque longe angustata, 2½–3¼ poll. longa, ¾–1¼ lata, petiolo vix 1/10-poll., crasse coriacea, rigida, atro-viridia, subtus (sicca) lutescentia, petiolis et costa subtus subpuberulis, nervis inconspicuis. *Pedunculi* ramulos terminantes, vel ramulo excurrente laterales, vix ½-pollicares, tomentosi, bracteolis 2–3 parvis oblongis prope basin muniti. *Alabastri* subglobosi, pisi magnitudine. *Flores* rubescentes (ex Burmannio punicei), diametro pollicares. *Sepala* ovata, membranacea, nervosa. *Petala* extus pubescentia, intus glabra. *Stamina* brevia, late oblongo-cuneata. *Torus* fructus globosus, parvus. *Carpella* 4–12, pulposa, utrinque obtusa, apice brevissime mucronulata, ovoidea vel (seminibus plerisque abortientibus) globosa ½-pollicaria, pedicello brevissimo vix lineam longo. *Semina* 3–6.

There has been a good deal of confusion with respect to this species, partly caused by Wallich's having mistaken it for *U. lutea*, Roxb. (*Alphonsea lutea*), and partly by Wight and Arnott having accidentally reversed the labels of this plant and *U. Heyneana*, Wall. (*Orophea Heyneana*), after comparing their collection with the Wallichian Herbarium, or perhaps rather owing to the accidental shifting of the labels of these two plants in Dr. Arnott's herbarium.

16. **U. micrantha** (H.f. et T.); foliis (parvis) oblongo-lanceolatis brevissime petiolatis obtuse acuminatis superne secus costam pubescentibus, pedunculis oppositifoliis vel terminalibus paucifloris bracteatis, floribus parvis, petalis fere rotundatis pubescentibus, carpellis ovalibus vel globosis glabris pedicellatis.—Guatteria micrantha, *A. DC. Mém.* 42; *Wall. Cat.* 6449! Polyalthia fruticans, *A. DC. Mém.* 42; *Wall. Cat.* 6430! Uvaria elegans, *Wall. Cat.* 6474 B! (*non* A).

HAB. Ava! Tenasserim! Malaya!—(*v. s.*)

Frutex verosimiliter scandens. *Ramuli* graciles, cortice nigricante punctulis albis consperso, juniores cinereo-incani, partes novellæ fulvo-tomentosæ. *Folia* nitida, tenuiter coriacea, rigida, nervosa, præter costam superne pilosam glaberrima, juniora subtus puberula, 2–3½ poll. longa, ¾–1⅓ lata, petiolo pubescente vix $\frac{1}{10}$ poll. longo. *Pedunculi* ¼–½-pollicares, 1–3-flori, fusco-tomentosi. *Bracteæ* 2–3, rotundatæ vel oblongæ, tomentosæ, parvæ. *Flores* vix ½ poll. diametro, albi (ex scheda Wall.). *Sepala* rotundata, extus pubescentia, glanduloso-punctata, in fructu persistentia. *Carpella* 15–20, glabra, granulosa, ¼–⅓ poll. longa, pedicello ¾-pollicari oblique inserta. *Semina* 1–3.

17. **U. parviflora** (H.f. et T.); foliis oblongis acuminatis basi plerumque acutis membranaceis, floribus extra-alaribus lateralibus solitariis vel cymosis minutis, pedicellis medio 1-bracteolatis.

HAB. In penins. Malayana ad Penang, *Phillips!*—(*v. s.*)

Frutex scandens. *Ramuli* graciles, glabri, cortice nigricante rugoso; gemmæ tomentosæ. *Folia* 4–6 poll. longa, 1½–2½ lata, petiolo ⅓-poll., tenuia, reticulato-nervosa, pellucido-punctata. *Pedunculi* abbreviati, pluribracteati, 1–4-flori, bracteis squamæformibus, pedicelli ⅓–½ poll. longi, puberuli, medio bracteolam lineari-oblongam gerentes. *Alabastri* globosi. *Sepala* pubescentia, rotundata.

Uvaria Hasseltii, Blume, Anon. 46. t. 21, is so closely allied to this, that we had almost united them; in that, however, the petioles are scarcely a line long, so that for the present they must be kept distinct. *U. Hasseltii* is known in fruit only. The carpels are three and shortly pedicellate.

Species dubia.

18 **U. sclerocarpa** (Alph. DC. Mém. 27); foliis ovalibus glabriusculis basi subciliatis, pedunculis axillaribus, toro fructigero capitato, carpellis (immaturis) ovoideo-acutis longe pedicellatis coriaceis, seminibus paucis oblique jacentibus planiusculis.—*A. DC.*; *Wall. Cat.* 6461!

HAB. Tenasserim prope Moulmein, *Wall.!*—(*v. s.*)

Ramuli glabri, rugulosi. *Folia* 4–5 poll. longa, 2–2½ lata, petiolo basi articulato. *Flores* ignoti. *Torus* fructus basi pubescens, globosus. *Carpella* usque ad 12, glabra, atro-fusca, ovoidea, vix acuta, ½–⅔ poll. longa, pedicello paullo longiore, crassa, indurata.

The fruit in the specimens which we have seen is far from ripe. It is perhaps a species of *Saccopetalum.*

4. **ELLIPEIA,** H.f. et T.

Sepala 3, parva. *Petala* 6, rotundata, obtusissima, æstivatione imbricata, interiora exterioribus minora, basi angustata. *Torus* convexiusculus. *Stamina* indefinita, linearia, connectivo truncato ultra antheras parallelas producto. *Ovaria* numerosa, strigosa, oblonga. *Ovulum* 1, suturæ ventrali supra medium insertum, globosum. *Stylus* oblongus, pubescens. *Carpella* monosperma, obliqua.—Frutex *forsan scandens,* floribus *paniculatis.*

This is a very remarkable plant, which cannot well be associated with any of the existing genera of *Anonaceæ.* The ovarium resembles that of *Melodorum* or *Mitrephora,* but the imbricated petals and the very different stamens forbid its union with either. The single ovule attached to the ventral suture has few parallels in the Order. From its decidedly imbricated petals, it belongs undoubtedly to the tribe *Uvarieæ,* in which it will be readily distinguished by the ovary and style, and the one-seeded carpels, which are curiously oblique, as in some species of the genus *Miliusa.* In both the cause is probably the same, the development of a single ovule attached to the ventral suture, not to the base of the cell, which is the usual position in the Order. (Name from ελλιπης, *defective.*)

1. **E. cuneifolia** (H.f. et T.) ; foliis anguste obovato-oblongis abrupte acuminatis basin versus cuneato-angustatis obtusis vel subcordatis, floribus in panicula laxa terminali dispositis.

HAB. Malaya prope Malacca, *Griffith!*—(*v. s.*)

Ramuli ferrugineo-velutini. *Folia* 6–8 poll. longa, 2–3 lata, petiolo vix ¼-poll., basin versus longe angustata, coriacea, rigida, supra lucida et præter costam tomentosam glabra, subtus adpresse fulvo-tomentosa, nervis obliquis parallelis numerosis conspicuis. *Panicula* terminalis, ramosa, multiflora, plerumque aphylla. *Flores* dense tomentosi, bractea rotundata concava calyci adpressa. *Sepala* rotundata, bracteam æquantia. *Petala* exteriora coriacea, convexa, utrinque fulvo-tomentosa, ¾-pollicaria, interiora multo minora. *Carpella* oblonga, ⅔ poll. longa, adpresse tomentosa, pedicello 1¼-pollicari, oblique inserta, infra medium mucrone parvo apiculatum.

Tribus II. MITREPHOREÆ.

Petala æstivatione valvata ; interiora basi unguiculata. *Stamina* dense conferta, rarius definita.

The genera which are associated in this tribe are all well marked by habit and characters, except *Popowia,* which is so imperfectly known that its position must still be considered doubtful.

5. **POPOWIA,** Endl.

Oropheæ species, *Blume, Fl. Jav.*

Sepala 3, ovata. *Petala* 6 ; exteriora minora, ovata, sepalis paullo majora ; interiora crassa, ovata, concava, apiculo inflexo, basi late unguiculata, æst. valvata. *Stamina* numero subdefinita, 12–21, cuneata, truncato-capitata, antherarum loculis dorsalibus oblongis discretis. *Ovaria* 5–7, ovali-oblonga, strigoso-pilosa, stylo magno obovato verruculoso recurvo. *Ovulum* e basi erectum, solitarium (vel 2 parietina). *Carpella* monosperma.—Arbores, foliis *parvis, nervis distantibus obliquis inconspicuis,* floribus *minutis oppositifoliis.*

This genus, which was established by Endlicher for the reception of *Bocagea pisocarpa* of Blume, appears to be the proper place for the Wallichian species which we here refer to it, notwithstanding some discrepancies in the structure of the ovary between it and Blume's plant. Our specimens are so imperfect that we have been able to examine very few ovaries, but in every case we found the ovules solitary and erect. The petals being very different from those of *Orophea,* it does not seem desirable to unite our plant to that genus, though probably, unless the genus *Popowia* had been already established, we should have put it there till its structure was better known. The flowers are small, and the petals are only very slightly unguiculate at the base, so that the genus is intermediate between *Mitrephoreæ* and *Guatterieæ.* A plant from Natal, in South Africa, and another from North Australia, collected by Armstrong, seem to be referable to the same genus. The imperfectly known *Uvaria Vogelii,* Hook. fil., from the Quorra, in West Africa, is perhaps also a congener.

1. **P. ramosissima** (H.f. et T.) ; foliis ovatis vel oblongis subtus secus nervos tomentosis, staminibus 18.—Guatteria? ramosissima, *Wall.* *Cat.* 7294 !

HAB. In peninsula-Malayana? *Wall.!* loco speciali omisso, sed cum pluribus plantis e Penang longe post cæteras Anonaceas Herbarii Wallichiani distributa.—(*v. s.*)

Arbor ramosissima. *Ramuli* rugulosi, cortice nigricante glabro ; juniores laxe ferrugineo-tomentosi. *Folia* basi rotundata, acuta vel acuminata, 3–4 poll. longa, 1¼–1¾ lata, petiolo vix ¹⁄₁₀ poll. longo, tomentoso, tenuiter coriacea, opaca, utrinque glabra, præter costam nervosque subtus pubescentes. *Pedunculi* oppositifolii, filiformes, vix ¼ poll. longi, tomentosi, uniflori, medio unibracteati. *Alabastri* rufopilosi. *Ovaria* 5, dense aureo-strigosa ; *ovulum* solitarium e basi erectum. *Fructus* ignotus.

6. **GONIOTHALAMUS,** Blume.

Polyalthia, § Goniothalamus, *Bl. Fl. Jav.*

Sepala 3, plerumque magna. *Petala* 6, æstivatione valvata ; exteriora plana, crasse coriacea, ovata, oblonga vel elongata ; interiora late unguiculata, crasse coriacea, laminis incurvis in mitram conicam arcte cohærentibus. *Stamina* indefinita, connectivo ultra antheras lineari-oblongas discrete biloculares in processum ovalem vel capitatum producto. *Torus* parum elevatus, truncatus, medio sæpe excavatus. *Ovaria* indefinita (rarius subdefinita), lineari-oblonga, strigoso-pilosa, biovulata. *Ovula* axi paullo supra basin inserta, superposita, in mucilagine nidulantia. *Stylus* oblongus vel sæpius elongatus, intus sulcatus. *Carpella* oblonga, semine solitario fere *Guatteriæ.*—Arbores *parvæ vel* frutices, foliis *supra nitidis, nervis obliquis parallelis distantibus non prominentibus, venulis prope marginem arcuatis conspicuis junctis,* pedunculis *axillaribus vel supra-axillaribus unifloris.*

P

This genus was established as a section of *Polyalthia* by Blume, who, however, only described one species. It is so well marked, both in characters and habit, that we have no hesitation in regarding it as a distinct genus. The thick, strictly valvate, and broadly-clawed inner petals, closely connivent into a mitriform cap, occur in no other genus. The nearest approach to this structure is found in *Oxymitra* among *Guattterieæ;* but there the inner petals are not unguiculate. The species of *Goniothalamus* appear for the most part to be undershrubs, rarely rising to the size of trees. The leaves of many are very thick and coriaceous ; but the thinner-leaved species and the young leaves of the others are pellucid-dotted. The nervation is peculiar, the principal veins being connected by loops, which often form a very conspicuous intramarginal nerve. Many of the species are unfortunately very imperfectly known, and we are by no means satisfied with the diagnoses given, our materials not being sufficient to enable us to form an opinion of the amount of constancy of the characters on which we have relied. We therefore recommend a careful study of the floral organs of these plants to those botanists who may have an opportunity of observing them in a living state. *Goniothalamus* appears to be entirely an Asiatic genus. Several species occur among Cuming's Philippine plants, and others will probably yet be met with in the Malayan Archipelago.

1. **G. Wightii** (H.f. et T.); foliis lanceolatis subtus pallidis glabriusculis, pedunculis axillaribus solitariis.—*Wall. Cat.* 9009!

HAB. In montibus Travancor ad Courtalam, *Wight!*—(*v. s.*)

Arbor? *Ramuli* graciles, nigricantes, rugulosi, glabri; partes novellæ fusco-pubescentes. *Folia* utrinque acuta, 3–5 poll. longa, ¾–1¼ lata, petiolo 2–3 lineas longo, tenuiter coriacea, subtus pallida (in sicco flavescentia), glabra vel juniora sparse puberula, minutissime pellucido-punctata. *Pedunculi* ½–1-pollicares, subclavati, basi bracteis pluribus oblongis minutis puberulis distiche imbricatis muniti, supra medium bracteola parva rotundata amplectente decidua. *Sepala* ovata, vix acuta, extus puberula, fere ½ poll. longa. *Petala exteriora* ¾-poll., ovalia vel ovata, obtusa, breviter et late unguiculata, basi intus areola oblonga notata, utrinque fusco-sericea; *interiora* ⅓ poll. longa, ovata, late unguiculata, in mitram coalita, lamina late trapezoidea, acuta, extus fusco-sericea, intus apicem versus sericea, cæterum glabra. *Torus* truncatus. *Ovaria* dense aureo-strigosa, stylo oblongo apice dilatato compresso retuso dimidio longiora. *Carpella* calyce persistente suffulta, pedicello 2 lineas longo stipitata, oblonga, ⅔-pollicaria, atro-fusca, glabra.

The style of this species is shorter and broader than that of those described below. In this respect it agrees with *G. macrophyllus,* Blume, the original species of the genus.

2. **G. salicinus** (H.f. et T.); foliis anguste lanceolatis basi acutis apice in acumen plerumque obtusum longe angustatis, floribus paullo supra-axillaribus semipollicaribus, petalis exterioribus angustis lineari-bus tomentosis.

HAB. In Zeylania ad montem "Adam's Peak" dictum, *Walker!*— (Fl. Mart.).—(*v. s.*)

Ramuli graciles, foliosi, cortice ruguloso nigricante; partes novellæ fusco-tomentosæ. *Folia* 3–4½ poll. longa, ½–1 lata, supra glabra, subtus sub lente sparse pubescentia; petioli vix ¼-poll., pilis atro-fuscis strigosi, demum glabrescentes. *Pedicelli* ¼ poll. longi, solitarii, pilis atro-fuscis strigosi, a basi ad medium bracteis 3–4 minutis oblongis acuminatis muniti. *Sepala* ovata, acuminata, dense strigosa, ¼-poll. *Petala* exteriora ½-poll., lineari-triangularia, interiora dimidio breviora. *Ovaria* pauca, 7–10, dense fusco-strigosa; *stylo* subulato, æquilongo.

3. **G. Thwaitesii** (H.f. et T.); foliis oblongis breviter et obtuse acuminatis, pedunculis axillaribus petiolos triplo superantibus, floribus

ultrapollicaribus, petalis exterioribus ovato-lanceolatis glabris basi in unguem brevem latum angustatis.

HAB. In sylvis Zeylaniæ, alt. 2–3000 ped.! Travancor ad Courtalam, *Wight !—(v. s.)*

Ramuli foliosi, glabri, cortice nigricante ruguloso; gemmæ fusco-pubescentes. *Folia* $3\frac{1}{2}$–6 poll. longa, $1\frac{1}{4}$–$2\frac{1}{4}$ lata, pet. $\frac{1}{4}$–$\frac{1}{2}$-poll., rigida, coriacea, utrinque glaberrima, supra lucida, subtus pallida, marginibus in sicco recurvis. *Pedunculi* $\frac{3}{4}$–1-pollicares, apice subclavati, ima basi bracteis pluribus minutis squamæformibus muniti. *Sepala* lata, ovata, vix acuta, basi coalita, coriacea, in fructu persistentia. *Petala* ext. crasse coriacea, $1\frac{1}{4}$–$1\frac{1}{2}$ poll. longa, apice obtusiuscula, glabriuscula, subgranulosa, ungue basi areola depressa oblonga fusco-pubescente notato, int. in mitram ovatam acutam coalita. *Ovaria* lineari-oblonga, strigosa, stylo subulato paullo longiore. *Torus* planus. *Carpella* numerosa vel abortu pauca, brevissime pedicellata, paullo ultra $\frac{1}{2}$ poll. longa, ovalia, utrinque obtusa. *Semen* 1, conforme, testa tenui, papyracea, lævi.

4. **G. Gardneri** (H.f. et T.); foliis anguste oblongo-lanceolatis basi acutis apice obtusis vel obtuse et breviter acuminatis, pedunculis axillaribus vel paullo supra-axillaribus petiolos vix superantibus, floribus sesquipollicaribus, petalis exterioribus oblongo-lanceolatis glabris basi vix unguiculatis.

HAB. In sylvis Zeylaniæ, alt. 2–3000 ped., *Walker ! Gardner !—* *(v. s.)*

Ramuli prioris. *Folia* 5–8 poll. longa, 1–2 lata, petiolo $\frac{1}{4}$-poll., crasse coriacea, supra nitida, subtus pallida, nervi crebriores quam in priore specie. *Pedunculi* $\frac{1}{4}$–$\frac{1}{2}$ poll. longi, squamis distichis bracteata. *Sepala* basi cordata, submembranacea, in sicco nervosa. *Petala* ext. $1\frac{3}{4}$-poll., glabriuscula, basi areola triquetra pubescente notata. *Ovaria* adpresse pilosa, stylo longe subulato triplo longiore superata.

Though certainly close to the last species, this appears sufficiently distinct. Its leaves are longer and narrower, and its flowers larger, than those of *G. Thwaitesii.* The fruit is unknown.

5. **G. Malayanus** (H.f. et T.); foliis lineari-oblongis longe acuminatis utrinque glaberrimis, pedunculis petiolos parum superantibus, floribus pollicaribus, petalis exterioribus oblongo-lanceolatis pubescentibus.

HAB. In Malaya ad Malacca, *Griffith !—(v. s.)*

Ramuli elongati, glabri, cortice ruguloso albo ; partes novellæ fusco-tomentosulæ, cito glabrescentes. *Folia* basi acutiuscula vel rotundata, 6–9 poll. longa, $1\frac{3}{4}$–3 lata, petiolo $\frac{1}{3}$-poll., supra lucida, subtus pallidiora. *Pedicelli* axillares, $\frac{1}{4}$–$\frac{3}{4}$-poll., fulvo-tomentosi, basi distiche bracteolati, cæterum nudi. *Sepala* ovata, acuta, tomentosa. *Petala exteriora* 1–$1\frac{1}{4}$-pollicaria, late unguiculata, dorso linea longitudinali subcarinata, tenuissime fusco-sericea, basi areola lata glabra notata ; *interiora* in mitram extus dense albido-sericeam vix semipollicem altam coalita. *Ovaria* dense aureo-sericea, stylo æquilongo subulato terminata. *Torus* fructus globosus, tomentosus. *Carpella* pollicaria, elongato-oblonga vel cylindrica, interdum medio parum constricta, apiculata, pedicello $\frac{1}{4}$ poll. longo suffulta, atro-fusca, granulosa, pilis paucis sparsis aureo-sericeis vestita, demum glabrescentia.

This species seems identical in foliage with *G. giganteus*, but is very distinct in the size and structure of the flower.

6. **G. cardiopetalus** (H.f. et T.) ; foliis obovato-oblongis vel lineari-oblongis basi acutis apice abrupte acuminatis margine undulatis, pedunculis supra-axillaribus 1–3 verticaliter uniseriatis petiolo bre-

vioribus, floribus ¾-pollicaribus, petalis ext. ovalibus obtusis.—Polyalthia cardiopetala, *Dalz. in Hook. Kew Journ. Bot.* ii. 39. Uvaria obovata, *Heyne ex Wall. Cat.* 6471!

HAB. In montibus Canara, *Rottler* (in Hb. Royle)! *Heyne! Dalz. Gibson!*—(Fl. Apr.) (*v. s.*)

Arbor parva. *Ramuli* glabri, cortice fusco ruguloso. *Gemmæ* adpresse fulvopubescentes. *Folia* 6-9 poll. longa, 2-3 lata, petiolo ½-poll. *Pedunculi* basi pluribracteati. *Sepala* reniformi-rotundata, brevissima, apiculata, velutino-puberula. *Petala exteriora* crasse coriacea, utrinque adpresse tomentosa; *interiora* ⅓ breviora, in mitram ovalem obtusiusculum coalita. *Ovaria* lineari-oblonga, stylo æquilongo. *Fructus* ignotus.

Our description is taken from Heyne's specimens in Wallich's collection in the museum of the Linnean Society, and from a specimen just received from Dr. Gibson. Dalzell's description quite corresponds, so that we have no doubt of the identity of the two.

7. **G. sesquipedalis** (H.f. et T.); foliis lanceolatis vel lineari-oblongis obtusis abrupte et obtuse acuminatis, pedunculis supra-axillaribus petiolo brevioribus, floribus vix ¾-pollicaribus, petalis exterioribus oblongis longe acuminatis.—Guatteria sesquipedalis, *Wall. Plant. As. Rar.* iii. *t.* 266! *Cat.* 6446! G. macrophylla, *A. DC. Mém.* 42, *non Blume; Wall. Cat.* 6451!

HAB. In montibus Khasia a basi ad alt. 4000 ped.! in prov. Silhet! et Tenasserim, *Wall.!*—(Fl. Apr. Mai.) (*v. v.*)

Frutex 2-4-pedalis, subsimplex, erectus, cortice griseo vel nigricante rugoso glabro. *Folia* 9-13 poll. longa, 2-3¼ lata, petiolo ¾-poll., coriacea, glaberrima, supra nitida, subtus pallida, marginibus in sicco recurvis, minutissime pellucido-punctata. *Pedunculi* ¼-½-pollicares, basi bracteis pluribus minutis squamæformibus distichis muniti. *Sepala* ovata, glabriuscula, ½-poll., in fructu persistentia. *Petala exteriora* vix puberula, intus obscure carinata, basi macula oblonga tomentosa notata; *interiora* subæquilonga, in mitram elongatam apice attenuatam coalita, extus pubescentia, intus dense fulvo-sericea. *Torus* truncatus. *Ovaria* linearia, dense aureo-strigosa, intus sulcata, stylo cylindrico recurvo æquilongo. *Carpella* 8-10 vel abortu plerumque pauciora (3-4), pedicello vix lineam longo suffulta, ⅔-poll., ovalia, apice mucronata, glabra, minute granulata.

8. **G. Simonsii** (H.f. et T.); foliis lineari-oblongis vel anguste obovato-oblongis basi acutis apice longe acuminatis subtus puberulis, pedunculis axillaribus petiolo brevioribus, floribus ultrapollicaribus, petalis exterioribus oblongo-lanceolatis.

HAB. In montibus Khasia, alt. 2-3000 ped., *Simons!*—(Fl. Jun.) (*v. v.*)

Arbor parva, vix 20-pedalis, erecta, parum ramosa. *Rami* elongati, rugosi, cortice cinereo; *ramuli* lævigati, cum omnibus partibus novellis dense ferrugineo-tomentosi. *Folia* 9-15 poll. longa, 2¾-4¼ lata, petiolo ½-pollicari, apice in acumen angustum fere lineare ½-1 poll. longum, apice obtusum, subito angustata, tenuiter coriacea, minute pellucido-punctata, supra glabra, nitida, subtus pallidiora, secus petiolum et nervos ferrugineo-tomentosa. *Venulæ* arcuatæ, in nervum submarginalem conspicuum coalitæ. *Pedunculi* plerumque ad axillas foliorum delapsorum secus ramos nudos dispositi, vix ⅛ poll. longi, basi bracteolis oblongis vel ovatis squamæformibus distichis muniti. *Sepala* late ovata, acuta, nervosa, pubescentia, ½ poll. longa, in fructu persistentia. *Petala exteriora* apice obtusiuscula, crasse coriacea, dense pubescentia, basi vix angustata, ibique intus areola lata notata, 1¼ poll. longa; *interiora* in mitram ½-poll. altam dense tomentosam coalita. *Torus* fructus dilatatus, depresso-globosus, diam. ⅔-poll. *Carpella* non visa.

There is in the Hookerian Herbarium a single flower of this very fine species, from which we have not ventured to remove the inner petals, so as to expose the stamens and ovaria. The petals, however, sufficiently indicate that it belongs to this genus, independently of the habit and nervation, which are markedly those of *Goniothalamus.* The ferruginous tomentum of the under surface of the leaves, and the strong marginal nerve, make this a very distinct species.

9. **G. giganteus** (H.f. et T.); foliis oblongis vel lineari-oblongis basi acutis apice longe et obtuse acuminatis, pedunculis petiolos longe superantibus, floribus maximis, petalis exterioribus ovatis basi unguiculatis.—*Uvaria gigantea, Wall. Cat.* 6469 A! *et* B! (*partim.*)

HAB. In Penins. Malayanæ sylvis vulgaris, *Wall.! Griff.!—(v. s.)*

Arbor? *Rami* elongati, *stricti*, cortice albo ruguloso; *ramuli* graciles, foliosi, glabri; *gemmæ* fulvo-tomentosæ. *Folia* 6–10 poll. longa, 1¾–3 lata, petiolo ⅓-poll., coriacea, rigida, supra atro-viridia, lucida, subtus pallida, cum petiolis sub lente tenuissime adpresse puberula, demum glabrata; costa argute carinata, scabrida. *Pedunculi* plerumque in axillis foliorum delapsorum positi, penduli, 1–1½-pollicares, fusco-pubescentes, apice subclavati, basi bracteolis paucis squamæformibus muniti. *Alabastri* aureo-sericei. *Sepala* e basi lata ovata, obtusiuscula, utrinque adpresse tomentosa, fere ⅔ poll. longa. *Petala exteriora* basi in unguem subcontracta, tenuia, foliacea, plana vel margine undulata, 4 poll. longa, 2 lata, utrinque pubescentia, basi intus aureo-sericea; *interiora* dense sericea, in mitram ovalem acutam ⅔ poll. altam coalita. *Torus* planus, parum elevatus, medio excavatus. *Ovaria* lineari-oblonga; stylo filiformi dimidio longiore apice subclavato. *Fructus* ignotus.

The flowers of this species are larger than those of any other with which we are acquainted, and the petals appear to increase considerably in size after expansion. The measurements given above are those of the largest petals we have seen. There are a good many specimens of Cuming's from the Philippines in various states, which are undistinguishable from the present species in shape and size of leaves, but with certain differences in the flowers, the constancy of which will require further confirmation. One of these has rigid peduncles and a glabrous calyx, while another seems to have much smaller flowers. All these, however, are in a very imperfect state, nor is *G. giganteus* itself sufficiently well known as to the amount of variation to which its flowers are liable.

10. **G. Walkeri** (H.f. et T.); foliis elongatis lineari-oblongis basi acutis apice breviter et obtuse acuminatis, pedunculis axillaribus unifloris brevissimis.

HAB. In Zeylania, *Walker!—(v. s.)*

Arbor? *Ramuli* elongati, validi, cortice griseo rugoso glabro. *Partes novellæ* vix puberulæ. *Folia* (etiam sicca) aromatica plerumque basi longe attenuata, tenuiter pellucido-punctata, 8–13 poll. longa, 2–3 lata, petiolo vix ½-pollicari, rigida, utrinque glaberrima, supra nitida, subtus pallida, nervis inconspicuis. *Sepala* in fructu persistentia, late ovata, acuta, nervosa, ¼-poll. *Torus* parum incrassatus, subglobosus. *Carpella* numerosa, ovali-oblonga, mucronata, glabra, ½-poll., pedicello vix lineam longo suffulta. *Semen* erectum, solitarium.

This unfortunately very imperfectly known plant has many points of resemblance with *G. macrophyllus* of Blume, the original species of the genus. It differs, however, considerably in the shape of the leaves, in the length of the peduncles, and in the position of the flowers, all characters of too great importance to permit of our combining the two. In *G. macrophyllus*, Blume, the flowers are about an inch long. That species differs somewhat from the rest of the genus in the shorter,

broader style, which is not more than half the length of the ovary; but this character cannot be considered of much importance, in the absence of other differences.
Blume does not represent his species with looping nerves; but authentic specimens
communicated by himself show them to be so, and to be dotted, like all the other
thin-leaved species of *Goniothalamus*.

11. **G. Griffithii** (H.f. et T.); foliis oblongis obtuse acuminatis
basi acutis, pedunculis axillaribus solitariis, sepalis obtusis, petalis longe
acuminatis.

HAB. In Mergui, *Griffith!*—(*v. s.* in Hb. Wight.)

Ramuli rugosi, grisei vel nigricantes. *Folia* 6–8 poll. longa, 2½–3 lata, petiolo
½-poll., tenuiter coriacea, in sicco nervosa, glaberrima, nitida, subtus pallida, pellucido-punctata. *Pedunculi* petiolum æquantes, basi distiche squamigeri, deflexi. *Sepala* basi coalita, fere rotundata, obtusa, in sicco nervosa, ⅔ poll. longa, puberula.
Petala exteriora bipollicaria, oblongo-lanceolata, longe attenuata, basi parum contracta, crasse coriacea, glabra; *interiora* in mitram ½-poll. altam coalita, parce strigoso-pubescentia. *Stamina* lineari-oblonga, ultra antheras in processum carnosum
acutum producta. *Ovaria* strigoso-pilosa, stylo longe subulato terminata.

Very near *G. macrophylla*, Blume, but with flowers twice as large, and a different
style. It is, however, described from a single specimen, and, as we have already
said, our materials are not sufficient to enable us to ascertain the value of characters
in this genus.

7. **OROPHEA,** Blume.

Bocagea, *Bl. Fl. Jav. non St. Hil.*

Sepala 3. *Petala* 6, æst. valvata; exteriora ovalia, interiora unguiculata, laminis in mitram cohærentibus. *Stamina* definita, 6–12, toro
vix convexo inserta, carnosa, ovalia, dorso antheram bilocularem gerentia. *Ovaria* 3–15, oblonga vel obovata; *ovula* in sutura ventrali 2–4.
Stigma sessile, capitatum vel oblongum.—Arbores *vel* frutices, foliis
parvis, floribus *axillaribus fasciculatis vel cymosis mediocribus vel parvis*.

This genus, which was originally instituted by Blume in the Bijdragen, was afterwards reduced by him to *Bocagea*, St. Hilaire. M. Alph. De Candolle and Mr.
Bennett have, however, both objected to this, and stated their conviction that there
are too many important differences between the two genera to justify their union.
The long-clawed inner petals, usually more delicate in texture than is common in the
Order, distinguish it from all the genera except *Mitrephora*, from which it may at
once be known by the definite stamens. The reduced number of stamens is the chief
resemblance between *Orophea* and *Bocagea*, which have no close agreement in habit
or inflorescence. The stamina of *Orophea* are in structure more like those of *Saccopetaleæ* than the more ordinary state of these organs in *Anonaceæ*; but we do not
place sufficient reliance upon this character to induce us to refer the genus to that
section of the Order, because the majority of characters appear to indicate the propriety of associating it with *Mitrephoreæ*. In this group, however, it certainly forms
the transition to *Saccopetaleæ*, standing as it were on the border between the two
tribes. *Bocagea*, with small inner petals, not contracted at the base, appears to belong
to *Saccopetaleæ*. The species of *Orophea* are all Asiatic, and are confined to the
most tropical provinces. In the Western Peninsula they do not occur north of Malabar. One only is found in Ava, but several inhabit the Malayan Peninsula, and extend thence to the Malayan Archipelago, which appears to possess many species.

1. **O. Heyneana** (H.f. et T.); foliis coriaceis ovato-lanceolatis obtusis glaberrimis, pedunculis abbreviatis oppositifoliis 1–2-floris, sta-

minibus 12, ovariis 6–9 dense strigosis quadriovulatis.—Uvaria Heyneana, *Wall. Cat.* 6463! U. lutea, *Wight, Cat. No.* 31 b! U. lutea, β,
W. et A. Prod. i. 8 (*non Roxb. nec Wall. Cat.*).
HAB. In Zeylania, *Thwaites!* in montibus Courtalam, *Wight!*—
(*v. s.*)

Arbor humilis. *Ramuli* graciles, cortice cinereo rugoso; adulti glabri; juniores
cum gemmis tomento fusco pubescentes. *Folia* basi rotundata, apice sæpe longe
angustata, 2–4 poll. longa, 1–1½ lata, rigida, utrinque glaberrima, supra lucida, subtus pallida, nervis obliquis, venulis (in sicco) creberrimis reticulatis. *Pedunculi* vix
¼ poll. longi, tomentosuli; bracteæ minutæ, distichæ, alternæ, rotundatæ. *Sepala*
rotundata, extus puberula, in fructu decidua. *Petala exteriora* plana, membranacea,
nervosa, ⅔ poll. longa, cuneato-lanceolata, apice longe acuminata, utrinque puberula,
parallele nervosa; *interiora* trapezoidea, acuta, ungue fere ½ poll. longo, extus parce,
intus dense villosa. *Torus* dense strigosus. *Stamina* omnia fertilia, late cuneata.
Carpella 4–8, stellatim patentia, pedicello brevissimo suffulta, ovoidea vel subglobosa,
utrinque obtusa, ½ poll. longa, fusco- vel cinereo-incana. *Semina* 1–3.
We have already referred (under *Uvaria Zeylanica*, L.) to the mistake into which
Wight and Arnott have fallen with respect to the synonymy of this species. The
description of *U. lutea*, W. et A., being partly taken from the present plant and
partly from Roxburgh's plate and description of the true *U. lutea*, does not apply
precisely to either.

2. O. uniflora (H.f. et T.); foliis ellipticis obtuse acuminatis glabris, pedunculis axillaribus abbreviatis unifloris gracilibus, staminibus
12 biserialibus, ovariis 6 late ovalibus biovulatis.
HAB. In montibus Travancor prope Courtalam, *Wight!*—(*v. s.*)

Ramuli graciles, rugulosi, cortice fusco, glabri; partes novellæ vix puberulæ. *Folia*
tenuiter coriacea, elliptica vel oblongo-lanceolata, basi acuta, 1½–2½ poll. longa, ¾–1
lata, petiolo vix lineam longo. *Pedunculus* ¼–½ poll. longus, basi squamulis pluribus
bracteatus, superne nudus vel squamula 1 minutissima. *Sepala* rotundata, ciliata.
Petala exteriora 2 lineas longa, rotundata, membranacea, glabriuscula; *interiora* trapezoidea, obtusa vel acutiuscula, ungue petala ext. æquante. *Stamina* fere rotundata. *Stigmata* lineari-oblonga, stellatim patentia, ovariis longiora. *Carpella* globosa, atro-fusca, glabra, semipollicaria, pedicello lineam longo suffulta.

3. O. Zeylanica (H.f. et T.); foliis ovali-oblongis obtuse acuminatis, pedunculis axillaribus solitariis vel fasciculatis 1–4-floris fusco-
pubescentibus, staminibus 6, ovariis circa 15 biovulatis.
HAB. In Zeylania, *Thwaites!* Canara, *Stocks!*—(*v. s.*)

Frutex? ramosissimus; ramulorum cortice griseo ruguloso; partes novellæ fusco-
pubescentes. *Folia* tenuiter coriacea, 2–3½ poll. longa, 1–1¾ lata, petiolo ⅛ poll.
longo; juniora subtus puberula et pilis albidis ciliata, demum glabrata. *Pedunculi*
longitudine valde varii, graciles, sæpius pluriflori, bracteis minutis ad basin pedicel-
lorum. *Flores* ¼-poll. diametro. *Sepala* orbicularia, tomentosa. *Petala exteriora*
rotundata, venosa, pubescentia, margine incana; *interiora* trapezoidea, apice incrassata, glabra, margine pubescentia. *Ovaria* glabra, obovata. *Stigma* capitatum.
Carpella globosa, baccata, lævia, glabra, diam. ⅓-poll., pedicello vix lineam longo
suffulta.

4. O. polycarpa (Alph. DC. Mém. 39); foliis elliptico-lanceolatis, pedunculis axillaribus filiformibus 1–3-floris glabris, staminibus 6
uniserialibus?, ovariis 9–12 glabris.—*Wall. Cat.* 6431!
HAB. In Martabania secus ripas fluminis Saluen, *Wall.!*—(*v. s.*)

Ramuli grisei, vix pubescentes, juniores fusco-pubescentes. *Folia* plerumque longe attenuata, obtusa, basi acuta, membranacea, glaberrima, 4–5 poll. longa, 1½–1¾ lata, petiolo sesquilineam longo. *Pedunculi* gracillimi, 1–2 poll. longi. *Sepala* ovata acuta, ciliata. *Petala exteriora* rotundata, patentia, ciliata; *interiora* duplo majora, margine et apice pubescentia.

We have not had an opportunity of examining a flower, and have therefore derived our description of the perianth, etc., from Alph. De Candolle. The flowers appear to be very small.

5. **O. acuminata** (Alph. DC. Mém. 39); foliis oblongo-lanceolatis ad nervos subtus velutinis, pedunculis filiformibus 1–3-floris pubescentibus, staminibus 6 uniserialibus?, ovariis 6 dense strigosis bi-ovulatis.—*Wall. Cat.* 6432!

HAB. In Tenasserim prope Tavoy, *Wall.!—(v. s.)*

Ramuli graciles, rugulosi, nigri, juniores dense tomentosi. *Folia* plerumque longe attenuata, membranacea, supra glabra, subtus præsertim secus costam puberula vel velutina, nervis validis obliquis, venulis transversis parallelis (ut in *Oxymitra*) conspicuis, 4–5 poll. louga, 1–1½ lata, petiolo 1–2 lineas longo. *Pedunculi* basi bracteis pluribus subulatis pilosis muniti, ultrapollicares. *Sepala* minuta, ovato-lanccolata, dense pilosa. *Petala exteriora* minuta, membranacea, ovalia, pilosa; *interiora* longe unguiculata, iis *O. Zeylanicæ* conformia.

6. **O.? obliqua** (H.f. et T.); foliis oblongis vel lanceolatis acutis rigide coriaceis basi inæqualibus utrinque glaberrimis, floribus terminalibus 1–3 fasciculatis, carpellis ovalibus.

HAB. In Zeylania, *Gardner!* prope Galle, *Champion!—(v. s.)*

Arbor? *Ramuli* læves, glaberrimi, atro-fusci. *Folia* brevissime pedicellata, basi inæquilatera, nempe uno latere rotundata, altero acuta vel acuminata, 4–5 poll. longa, 1¼–2 lata, petiolo vix lineam longo, supra lucida, subtus pallida; margines in sicco recurvi. *Flores* minuti, glabri, brevissime pedicellati. (Ex. icone Champion, stamina 6, ovaria 3.) *Pedunculus* fructus incrassatus, clavatus. *Carpella* diam. ½-poll., atro-fusca, glabra, lævia. *Semina* 2, rotundata, subcompressa; testa nitida, brunnea, scrobiculata.

This is a very remarkable plant, which, without a knowledge of the structure of the flowers, we are induced to refer to *Orophea*, from a certain general resemblance, especially in the obliquity of the leaves, to *O. latifolia*, Blume, the flower of which is also scarcely known. The shape of the fruit differs too much to permit of the two being considered the same species : but perhaps, when the flowers of both are known, they will be found to be congeners, and to be deserving of being generically separated from *Orophea*. Mr. Thwaites' No. 2612, according to a fragment just received, is a different but closely allied species.

8. **MITREPHORA,** Blume.

Uvaria, § Mitrephoræ, *Bl. Fl. Jav. Anon.*

Sepala 3, rotundata. *Petala* 6, æstivatione biseriatim valvata ; *exteriora* ovata, nervis subconspicuis ; *interiora* basi unguiculata, lamina fornicata. *Torus* depresse conicus, subtruncatus, medio parum excavatus, pilosus. *Stamina* numerosa, oblongo-cuneata, antheris dorsalibus remote bilocularibus, connectivo truncato-capitato. *Ovaria* oblonga, glabra. *Ovula* in axi biserialia, numerosa. *Stylus* oblongus, intus sulcatus.—Arbores *sæpe excelsæ*, foliis *coriaceis, nervis crebris parallelis conspicuis.*

Acting on the suggestion of Alph. De Candolle and Bennett, this very natural group, which was separated by Blume from the remainder of the many-ovuled *Uvariæ* as a section, is now constituted a distinct genus. It is closely allied in floral characters to *Orophea*, but the indefinite stamens and numerous ovules at once distinguish it. The only known species of the genus besides the following are those described by Blume from Java.

1. **M. tomentosa** (H.f. et T.); foliis ovato- vel oblongo-lanceolatis subtus fulvo-tomentosis, pedunculis oppositifoliis abbreviatis paucifloris, carpellis subglobosis dense tomentosis longe pedicellatis.

HAB. In prov. Assam, *Jenkins! Masters! Simons!*; et Chittagong! —(*v. s.*)

Arbor. Ramuli validi, cortice cinereo rugoso, punctis depressis conspersi, puberuli; juniores cum omnibus partibus novellis fulvo-tomentosi. *Folia* obtusa, acuta vel acuminata, basi rotundata, subcoriacea, superne secus costam pilosa, cæterum glabra, nitida, subtus cum petiolo pilis asperis fulvis tomentosa, nervis prominentibus, obliquis, parallelis, versus basin folii magis approximatis, 3–6 poll. longa, 1½–3 lata, pet. vix ¼-poll. *Pedunculi* tomentosi, 2–4-flori. *Bracteæ* late rotundatæ, amplexicaules, crassæ, tomentosæ, deciduæ. *Pedicelli* ¼–½-poll., supra medium bracteolati. *Alabastri* dense fulvo-tomentosi. *Sepala* late ovata, acuta. *Petala exteriora* ovata, acuta, fere pollicaria, intus vix pubescentia, parallele nervosa; *interiora* late unguiculata, lamina late ovata. *Ovaria* glabra. *Torus* fructus depressoglobosus, tomentosus, diam. ½-poll. *Carpella* 10–20, late ovoidea vel subglobosa, pollicaria, pedicello 1–1½-poll. suffulta, granuloso-tuberculata.

2. **M. obtusa** (Blume, Fl. Jav. Anon. 32. t. 10 et 14 C. sub Uvaria); foliis ovatis vel oblongis subtus adpresse pubescentibus, pedunculis oppositifoliis vel terminalibus pollicaribus, pedicellis elongatis gracilibus, carpellis oblongis velutinis longe pedicellatis.—Uvaria obtusa, *Bl. Bijdr.* 13 ; *Wall. Cat.* 6484 !

HAB. In Peninsula Malayana ad Penang, *Wall.!*—(*v. s.*)
DISTRIB. Java.

Arbor procera, ramosissima. *Ramuli* divaricati, rigidi, cortice rugoso, sæpe transverse fisso, nigricante, adulti glabri, juniores cum omnibus partibus novellis fusco-tomentosi. *Folia* magnitudine et forma valde varia, interdum fere rotundata vel elongato-oblonga, obtusa vel acuta, vel breviter et obtuse acuminata, basi rotundata, 2–5 poll. longa, 1–2¼ lata, petiolo fusco-tomentoso ¼–⅓-poll., rigide coriacea, superne nitida, præter costam præsertim basin versus fulvo-pubescentem glabra, subtus pallidiora, venis obliquis prominulis. *Pedunculi* ferrugineo-tomentosi ; bracteis pluribus (4–5) alternis distichis deciduis lineam longis. *Pedicelli* 1–2-pollicares, supra medium bracteolam convexam rotundatam minutam gerentes. *Alabastri* globosi, subtrigoni, dense tomentosi. *Petala exteriora* ovalia, ¼-pollicaria, extus adpresse velutina, intus parce pubescentia, parallele nervosa, flavescentia, purpureo-striata ; *interiora* ungue filiformi, lamina ovali obtusa extus pubescente, intus dense tomentosa. *Torus* fructus incrassatus. *Carpella* 7–15, subcarnosa, ¾-pollicaria, utrinque obtusa, oblonga, vel seminibus pluribus abortivis subglobosa, pedicello 1–1½ poll. longo suffultis, tomento brevi fuscescente velutina. *Semina* 4–5.

All the specimens we have seen from the Malayan Peninsula are in flower only, and our description of the fruit is copied from Blume. In the absence of fruit the identification is somewhat doubtful, but no difference can be detected except that they are larger-leaved than Blume's figure and specimens, and we learn from that author that the leaves are exceedingly variable in size. Cuming's No. 1135 is a very nearly allied species, but distinguished by its carpels being nearly sessile, and covered with lax ferruginous tomentum.

Q

Species dubia, floribus vix notis.

3. **M.? excelsa** (H.f. et T.); foliis rigide coriaceis obovato-oblongis abrupte acuminatis basi cordatis subtus dense fulvo-furfuraceis, cymis abbreviatis axillaribus 2–3-floris, bracteolis rotundatis imbricatis, floribus parvulis.—Uvaria excelsa, *Wall. Cat.* 6477 !

HAB. In penins. Malayana ad Penang, *Wall.!—(v. s.)*

Arbor (ex scheda Wallichiaua) excelsa. *Ramuli* rugulosi, cortice atro-fusco, juniores puberuli; partes novellæ stellato-tomentosæ. *Folia* supra nitida, sub lente minute squamulosa, subtus oblique nervosa, 6–8 poll. longa, 2–3¾ lata, petiolo ¼-poll. *Cymæ* vix ¼ poll. longæ, tomentosæ. *Flores* albi. *Alabastri* globosi, vix ¼-pollicares. *Sepala* orbicularia, extus dense fulvo-villosa, basi subcohærentia. *Petala* (quantum ex alabastro juniore judicare licet) *exteriora* crasse coriacea, rotundata, acutiuscula, intus subcarinata, utrinque adpresse tomentosa; *interiora* æst. valvata, ovata, crassissime coriacea.

The specimens of this plant distributed by Wallich are very imperfect, nor are those in the Linnean Society's Herbarium sufficiently good to enable its genus to be determined with certainty. It would perhaps have been better to have left it for the present in *Uvaria*, where it was placed by Wallich; but the arborescent habit is not consistent with that genus, and the petals appear to be decidedly valvate.

Tribus III. ANONEÆ.

Petala æstivatione valvata, haud unguiculata. *Stamina* indefinita. *Carpella* in fructum multilocularem coalita.

The cohesion of the ovaries and carpels at once distinguishes this tribe from all the others. In floral characters it approaches *Melodorum* and *Artabotrys*, some species of *Rollinia* in especial bearing much resemblance to those of the latter genus. All the species are uniovulate, and the whole tribe is American, except a few species which have been naturalized in the Old World. *Lobocarpus*, W. et A. (Prod. i. 7), which, from the characters assigned, would belong to this tribe, is founded on imperfect specimens of an Euphorbiaceous plant closely allied to *Bradleia*, in which the very immature fruit is terminated by a thick, erect style, slightly lobed at the apex.

9. ANONA, L.

Sepala 3, minuta, basi coalita. *Petala* 6, æst. biseriatim valvata; *exteriora* carnosa, triquetra, basi excavata, vel tota concava. *Stamina* indefinita; connectivo ultra antherarum loculos lineares extrorsos contiguos in processum ovalem producto. *Torus* hemisphæricus. *Ovaria* numerosa, subcoalita, stylo oblongo terminata. *Ovula* solitaria, erecta. *Carpella* numerosa, in fructum multilocularem carnosum ovalem vel rotundatum coalita. *Semina* in loculis solitaria, erecta, testa lævi nitida.—Arbores *vel* frutices *Americani*, pedunculis *terminalibus vel oppositifoliis.*

This is a very extensive genus, which contains the well-known tropical fruits, the Custard Apple, Soursop, Bullock's-heart, etc. All the species are natives of South America or the West Indies; but as two are extensively cultivated in India, and are often found in a more or less naturalized state, it is desirable to include them in our Flora. As it is not necessary to study an American genus for the sake of two naturalized plants, our diagnoses are taken from Von Martius's elaborate monograph. Both species belong to his section *Atta.**

* Sect. ATTA.—*Petala* interiora minima, squamæformia, interdum plane deficientia.

1. **A. squamosa** (L. Sp. 757); foliis membranaceis, junioribus pubescentibus subtus glaucis oblongo-lanceolatis acuminatis acutis vel obtusis basi acutiusculis, pedunculis unifloris subsolitariis, fructu ovato-globoso vel conico, areolis convexo-prominentibus viridi-flavis vel glaucescentibus.—*Roxb. Fl. Ind.* ii. 657 ; *DC. Syst.* i. 472, *Prod.* i. 85 ; *Bl. Fl. Jav. Anon.* 107. *t.* 53 B (*fructus*); *Wall. Cat.* 6490 ! ; *W. et A. Prod.* i. 7 ; *Bot. Mag. t.* 3095 ; *Martius, Fl. Bras. Anon.* 14. *t.* 5. *f.* 1 (*fructus*).—*Rheed. Mal.* iii. *t.* 29 ; *Rumph. Amb.* i. *t.* 46.

HAB. In hortis ubique culta et sæpe in dumetis subspontanea, præsertim in provinciis australioribus.

DISTRIB. In insulis Antillis indigena ; per totum orbem tropicum culta.

The leaves of this species are smaller than those of the next, and more frequently obtuse than acute. When in fruit the two are readily distinguished.

2. **A. reticulata** (L. Sp. 757); foliis membranaceis subtus scabriusculis oblongis aut oblongo-lanceolatis acuminatis basique acutiusculis, pedunculis lateralibus 2–4 confertis, fructu cordato-ovato obtuso cortice scabriusculo crasso fusco vel subrubello leviter pentagono-reticulato, areolis planiusculis, seminibus nigricantibus.—*Roxb. Fl. Ind.* iii. 657; *DC. Syst.* i. 473, *Prod.* i. 85 ; *Bl. Fl. Jav. Anon.* 108 ; *Wall. Cat.* 6489 ! ; *W. et A. Prod.* i. 7 ; *Bot. Mag. t.* 2911, 2912 ; *Mart. Flor. Bras. Anon.* 15.—*Rheed. Mal.* iii. *t.* 30, 31. ʼ

HAB. In hortis culta, hinc inde subspontanea.

DISTRIB. In insulis Antillis indigena ; per totum orbem tropicum culta.

Rheede, Roxburgh, Blume, Wallich, and Wight unite in bearing testimony that no species but the two above described are in common cultivation in India, so that *A. Asiatica*, L., which is referred conjecturally by Brown to *A. muricata*, L., is rightly considered by Martius a spurious species, made by mixing the characters of *A. squamosa* and *reticulata*, to one or other of which species it is generally referred.

Tribus IV. XYLOPIEÆ.

Petala æstivatione valvata, haud unguiculata ; interiora difformia, triquetra. *Stamina* indefinita. *Carpella* discreta, interdum definita.

10. **MELODORUM,** Dunal.

Unona, § Melodorum, *Dunal, DC.* Uvaria, § Melodoræ, *Blume (excl. sp.).* Polyalthia, § Kentia, *Blume.*

Sepala 3, parva, basi plus minus coälita. *Petala* 6, æst. biseriatim valvata ; exteriora convexa, interiora superne triquetra. *Stamina* indefinita, multiserialia, connectivo ultra antherarum loculos lineares extrorsos contiguos in processum ovalem vel oblongum carnosum productо. *Torus* convexo-conicus. *Ovaria* numerosa, oblonga, secus suturam ventralem multi- vel pluri-ovulata, rarius biovulata. *Styli* oblongi.—*Frutices plerumque scandentes,* foliorum nervis *subtus conspicuis obliquis rectis vel vix incurvis parallelis,* inflorescentia *terminali vel oppositifolia,* alabastris *triquetris tomentosis.*

This genus corresponds with the section *Melodorum* of *Unona*, as left by Dunal (judging from the characters, not from the species included), and with the *Melodoræ* division of *Uvaria* of Blume in the Fl. Javæ, excluding, however, almost all Dunal's species, and a few of those included by Blume, which do not appear naturally allied to the majority. Loureiro's *Melodorum* is different, as we have determined by an inspection of the materials in the British Museum. In that collection there is an authentic specimen of *M. fruticosum*, Lour., which is an undescribed plant, of doubtful affinity, as we have not examined the flower, but certainly not belonging to this genus. It has no fruit. There is no authentic specimen of *M. arboreum*, Lour., but it is described as a large tree, and is perhaps a *Mitrephora*. A specimen from Sir George Staunton, which is so named, is an *Uvaria*, nearly allied to, if not identical with, *U. microcarpa*, Champion. This, however, does not accord with Loureiro's description. Notwithstanding the exclusion of both Loureiro's species, it appears desirable to retain the name for the group to which it was applied by Dunal and Blume, whose works and plants are well known to botanists, rather than to substitute a new one. Loureiro's plants will probably both be found to belong to well-known genera. At all events, his descriptions are not sufficient to identify the species nor to distinguish the genus: it would therefore, we think, be manifestly unjust to Dunal and Blume not to retain their name.

As defined above, the genus is a very natural one, well marked by the triquetrous buds. The thick, firm, fleshy petals are strictly valvate in æstivation, and the inner ones are concave near the base only, while towards the apex they are triquetrous and acutely carinate internally, so that the two inner faces rest against the corresponding ones of the next petals, exactly as in *Anona*. The numerous stamens, with linear, parallel, approximate anther-cells, are terminated (generally) by a fleshy process of the connectivum, which is analogous to that of *Anona*, but often much more developed. The conical torus and oblong styles, much slenderer than the ovary, are also important characters. The sepals are often persistent in the fruit, and the species are all scandent.

The generic character might be made still more definite, by introducing the number of ovules, which is in general great, were it not that there are several species in which they are reduced to two. The type of these aberrant species is *Polyalthia Kentii*, Blume (*Melodorum Kentii*, H.f. et T.), a plant which has not hitherto been found within our limits, but which so closely resembles *M. elegans*, H.f. et T., a many-ovuled species, that the two are undistinguishable when placed together, except by an examination of the flowers. In consequence of this close resemblance, which extends to all parts of the flower, we think it better to retain *M. Kentii* and *M. pisocarpum* in *Melodorum*, than to institute a new genus which is not indicated by habit. Indeed, the number of ovules is in this case of less importance than other characters, because *M. elegans* and *M. Kentii* agree in so many points that they form a natural section of the genus, characterized by the peculiar thickened petals, the glabrous, glandular dotted ovaries, and pitted seeds.

Besides the species described below, one or two of which have already been figured by Blume, several exist in herbaria from the Philippine Islands. The genus is, however, so far as is known, entirely Asiatic, no Australian, African, or American species being known. It is still more remarkable that no species occur in Ceylon, or in the Madras Peninsula, or anywhere west of the Ganges; though in Malaya they are very abundant, and many species extend along the coast of Arracan and Chittagong to Silhet and Khasia, and one or two to the base of the Himalaya, where they are found as far west as Sikkim.

Sect. 1. EUMELODORUM.—*Petala* exteriora anguste marginata. *Ovaria* strigoso-pilosa. *Ovula* numerosa. *Semina* non scrobiculata.

1. **M. rubiginosum** (H.f. et T.); foliis oblongis obtusis vel acu-

tis supra sparse puberulis secus costam et venas villosis subtus pube
minuta fulva tomentosis, panicula terminali pauciflora, pedicellis pollicaribus, floribus magnis, carpellis ovoideis pedicello æquilongo suffultis,
seminibus numerosis oblongis lævibus.—Uvaria rubiginosa, *Alph. DC.*
Mém. 26; *Wall. Cat.* 6465!   U. nervosa, *Wall. Cat.* 6479!

HAB. In sylvis Silhet! Chittagong! et Tenasserim, *Wall.!*—(*v. v.*)

Frutex scandens, ramulis nigricantibus vel griseis rugulosis pubescentibus; partes
novellæ dense fulvo-tomentosæ. *Folia* utrinque obtusa, basi interdum subcordata,
apice acuta, rarius in acumen breve subulatum angustata, tenuiter coriacea, 3–6 poll.
longa, 2–3 lata, majora pedalia, 5 poll. lata, petiolo ¾-poll., supra opaca; nervi sub-
curvati, subtus prominuli. *Flores* axillares, solitarii vel plerumque in paniculam
terminalem 5–6-floram congesti, dense fulvo-tomentosi. *Pedicelli* pollicares et ultra,
infra medium bracteolas 1–2 parvas ovatas gerentes. *Sepala* lata, vix lineam longa.
Petala exteriora ultrapollicaria, extus tomentosa, intus cinereo-incana, ovali-oblonga;
interiora paullo breviora, angustiora, e basi ovali concava, in rostrum longe trique-
trum producta, dorso cinerea, intus subglabra. *Torus* inter stamina glaber, inter
ovaria dense strigosus. *Processus* connectivi oblongus. *Ovaria* dense strigosa,
stigmate puberulo. *Carpella* 5–10 vel plura, fulvo-tomentosa, pollicaria. *Semina*
biserialia, nitida, septis tenuibus separata; testa atro-fusca.

2. **M. latifolium** (Dunal, sub Unona); foliis ovalibus vel oblongis
supra pubescentibus secus costam tomentosis subtus cum petiolo dense
fulvo-tomentosis, racemis lateralibus et terminalibus laxis folio dimidio brevioribus, carpellis numerosis subglobosis, seminibus compressis
transverse rugulosis.—Unona latifolia, *Dunal, Anon.* 115; *DC. Syst.*
i. 497, *Prod.* i. 91. Uvaria latifolia, *Bl. Fl. Jav. Anon.* 37. *t.* 15, 25 A.
U. longifolia, *Bl. Bijd.* 13.—*Wall. Cat.* 9411!

HAB. In Peninsula Malayana ad Malacca, *Griffith!*—(*v. s.*)
DISTRIB. Java, ins. Molucc., ins. Philippin. (Cuming, 1548.)

Frutex scandens vel sarmentosus. *Ramuli* læves, fulvo-tomentosi. *Folia* remo-
tiuscula, 4–10 poll. longa, 2–4 lata, pet. ½–⅓-poll., utrinque rotundata vel basi subcordata et apice emarginata, subcoriacea, opaca. *Racemi* ramulos sæpe terminantes,
subpaniculati, 4–5 vel rarius pluriflori. *Pedicelli* solitarii vel fasciculati, bracteis
parvis ovatis vel oblongis suffulti, infra medium bracteolati. *Alabastri* ¾-poll., dense
tomentosi. *Sepala* ovata, acutiuscula, vix 2 lin. longa. *Petala exteriora* ovato-
oblonga, extus tomentosa; *interiora* ⅓ breviora, glabra, oblongo-acuminata, infra
medium concava, superne triquetra. *Stamina* et *ovaria* prioris. *Carpella* 8–15,
supra torum incrassatum subglobosum umbellata, tota tomento ochraceo tecta, obtusa vel mucronata, pericarpio subbaccato post maturitatem atro-purpureo. *Semina*
4–8.

We have not seen fruiting specimens of this fine species, and have in consequence
followed Blume in that part of our description. The species, though confined in continental India to the extreme southern part of the Malayan Peninsula, seems abundant to the eastward. Cuming's Philippine specimens are slightly different, in
having more numerous flowers, forming a sort of panicle beyond the leaves.

3. **M. lanuginosum** (H.f. et T.); foliis lanceolatis vel oblongo-
lanceolatis basi rotundatis supra præter costam fulvo-pubescentem glabris nitidis· subtus cum petiolo dense fulvo-lanuginosis, floribus paucis
terminalibus vel oppositifoliis magnis dense lanuginosis.—Uvaria lanuginosa, *Wall. Cat.* 6454!

HAB. Malaya ad Penang! et Singapur!—(*v. s.*)

Frutex sarmentosus, ramulis dense fulvo-tomentosis. *Folia* sæpius longe angustata, basi rotundata, 6–8 poll. longa, 1¾–2¾ lata, petiolo ⅔-poll., inferiora in quovis ramulo minora oblonga obtusa. *Flores* 2–4 in racemo abbreviato congesti; bracteæ minutæ, deciduæ. *Pedicelli* ½–1-pollicares, prope basin bracteolam ovalem semiamplectentem ¼ poll. longam gerentes. *Sepala* ovata, ½–⅔ poll. longa, fulvo-tomentosa, obscure nervosa, basi connata. *Petala exteriora* 1½–1¾-poll., e basi lata oblonga, obtusa, crasse coriacea; *interiora* vix breviora, utrinque incana, basi concava, glabrata, superne elongato-triquetra. *Stamina* et *ovaria* priorum, sed stylus glaber. *Carpella* (immatura) 15–20, pedicello æquilongo suffulta, anguste oblonga, obtuse rostrata, fulvo-tomentosa.

4. **M. manubriatum** (H.f. et T.); foliis oblongis vel oblongo-lanceolatis rigidis supra glaberrimis lucidis secus costam fulvo-pubescentibus subtus dense fulvo-tomentosis, floribus subternis fasciculatis mediocribus, carpellis longe pedicellatis oblique ovalibus obtusis fulvo-tomentosis.—Uvaria manubriata, *Wall. Cat.* 6456!; *Cuming, No.* 2339!

HAB. In Malaya vulgaris!—(*v. s.*)

Frutex scandens, ramosus, foliosus. *Ramuli* nigricantes vel grisei, rugulosi, sparse verrucosi, glabrescentes; juniores cum omnibus partibus novellis dense fulvo-tomentosi. *Folia* basi rotundata vel subcordata, apice acuta vel sensim acuminata, 3–5 poll. longa, 1¼–1¾ lata, petiolo ¼–⅓-poll. *Flores* oppositifolii vel ad apices ramorum fasciculati. *Pedicelli* tomentosi, ½–¾-pollicares, infra medium bracteolam ovatam minutam gerentes. *Alabastri* dense tomentosi. *Sepala* ovata, acuta, ¼-poll. *Petala exteriora* e basi ovali-oblonga obtusa, extus tomentosa, intus basi glabra, superne incana; *interiora* paullo breviora, ovato-lanceolata, supra medium triquetra, extus incana, intus glabra, granulosa. *Stamina* elongata, connectivi processu subgloboso. *Ovaria* dense aureo-strigosa. *Stylus* oblongus, pilosus. *Carpella* immatura, 6–7, ovali-oblonga, obtusa, obliqua, intus gibbosa, dense fulvo-tomentosa; pedicelli tomentosi, 2-pollicares, basi dilatati, toro coadunati et persistentes. *Semina* 5–6.

This species in general appearance closely resembles the last, but it is smaller in all its parts, the flowers in particular being very considerably less. The fruit is very peculiar, and unlike that of any other species.

5. **M. Wallichii** (H.f. et T.); foliis anguste oblongo-lanceolatis plerumque longe acuminatis basi rotundatis distanter nervosis superne glaberrimis subtus pallidis et sub lente præsertim secus costam et venulas pube minutissima adpresse puberulis, pedicellis fasciculatis paucis, floribus mediocribus, carpellis longe pedicellatis.—Uvaria bicolor, *Wall. Cat.* 6466! *non Roxb.*

HAB. In montibus Silhet! et Khasia! subtropicis.—(*v. v.*)

Frutex alte scandens. *Ramuli* cortice cinereo ruguloso, juniores pubescentes; partes novellæ aureo-sericeæ. *Folia* 4–6 poll. longa, 1¼–2 lata, petiolo ⅓-poll., tenuiter coriacea. *Nervi* valde obliqui, quam in cæteris speciebus remotiores. *Pedunculi* oppositifolii, brevissimi; *pedicelli* 1–4, ½-pollicares, tomentosi, basi bracteolis ovatis parvis muniti. *Sepala* extus tomentosa, intus glabra, granulata. *Petala* ovato-lanceolata, ⅔-poll.; exteriora extus fulvo-tomentosa, interiora glabra. *Stamina* et *ovaria U. verrucosi.* *Carpella* immatura dense tomentosa, ovalia, ¼-pollicaria, pedicellis pollicaribus angulatis tomentosis suffulta, toro globoso insidentia.

Dr. Wallich, who found this species in the Calcutta Botanic Garden, and also received it from Silhet, has considered it to be *U. bicolor* of Roxb.; and it is probable that both species may have been so marked in the Calcutta garden, and perhaps confounded by Roxburgh. This is the more likely, because there are two small fragments of the true *M. bicolor* glued on the same sheet with *M. Wallichii* in the Linn. Soc. Herb., and several more of the same among the duplicates of that collection, though that plant does not elsewhere occur in the Wallichian collection. The only tangible part of Roxburgh's description is the globose fruit, and that is only applicable to the species called by us *M. bicolor*. This we have compared with an authentic specimen from Roxburgh in the British Museum, in flower and fruit. We only know the fruit of the present species by a single specimen collected by us in Khasia. It is not unlike that of *M. verrucosum* in shape, but less wrinkled; but it is far from ripe.

6. **M. verrucosum** (H.f. et T.); foliis oblongis vel lanceolatis crebre nervosis supra nitidis pube minuta puberulis secus costam subtomentosis subtus fulvo-pubescentibus secus nervos sericeo-villosis, floribus fasciculatis vel subracemosis plerumque terminalibus, alabastris latis obtuse triquetris, carpellis longe pedicellatis obliquis rostratis valde rugosis.

HAB. In mont. Khasia, alt. 4–5000 ped.—(Fl. Jul., Fr. Oct.) (*v. v.*)

Frutex alte scandens, ramosus, densè foliosus. *Ramuli* grisei vel rufescentes, lenticellis albidis creberrimis notati, rugulosi, puberuli, demum glabrescentes, juniores cum omnibus partibus novellis fulvo-tomentosi. *Folia* 3–5 poll. longa, 1¼–2 lata, petiolo pollicari, basi rotundata vel acutiuscula, apice obtusa vel acuta. *Flores* ad apices ramulorum 1–5, solitarii, fasciculati, vel racemulum pauciflorum foliosum formantes, floribus inferioribus oppositifoliis. *Pedicelli* fulvo-tomentosi, pollicares, basi pluribracteati, et versus medium bracteolas 1–2 ovales amplectentes 2–3 lineas longas gerentes. *Alabastri* quam in cæteris speciebus breviores, latiores, dense tomentosi. *Sepala* ¼-poll., late ovata, costata, in fructu subpersistentia. *Petala exteriora* late ovata, extus dense tomentosa, intus cinereo-incana, medio linea vix elevata obscure carinata, ⅔–¾ poll. longa; *interiora* paullo breviora, triangulari-oblonga, intus concava, glabriuscula, impressionibus staminum areolata, superne longe triquetra. *Processus* connectivi magnus, ovalis. *Ovaria* dense fusco-pilosa. *Torus* fructus globosus. *Carpella* pulposa, subglobosa, vel late ovoidea, bipollicaria, obliqua, obtuse mucronata, irregulariter verrucoso-tuberculata, tomentosula, suffulta pedicellis 1½–3-pollicaribus crassis clavatis longitudinaliter striatis. *Semina* biserialia, partitionibus cellulosis separata, ¾–1 poll. longa, oblonga, compressa, irregulariter transverse rugosa, margine toto annulo lato prominente tuberculato versus hilum lævigato medio secus longitudinem profunde sulcato circumdata.

This species is certainly very near the last, but it differs, we think, essentially in the leaves being much more tomentose beneath, with more numerous transverse nerves, and especially by the very broad short flowers. We are, however, unacquainted with the fruit of *M. Wallichii*, except in a very young state. If the species be distinct, which we believe them to be from their general aspect, no doubt good characters will be found in the fruit.

7. **M. bicolor** (H.f. et T.); foliis oblongis vel oblongo-lanceolatis tenuiter coriaceis crebre nervosis supra præter costam fulvo-pubescentem glabris subtus cum petiolo pilis cinereis vel fulvis longe et adpresse sericeis, floribus fasciculatis extra-axillaribus majusculis, carpellis globosis breviter pedicellatis tomentosis.—*Uvaria bicolor, Roxb. Fl. Ind.* ii. 662 (*non Wall.*).

HAB. In provinciis Silhet! et Assam! et secus basin Himalayæ in montibus inferioribus Sikkim!—(*v. v.*)

Frutex scandens. *Ramuli* elongati, foliosi, cortice cinereo, rugulosi, pubescentes, juniores, ut omnes partes novellæ, fulvo-tomentosi. *Folia* 4–10 poll. longa, $1\frac{3}{4}$–$3\frac{1}{4}$ lata, petiolo $\frac{1}{4}$–$\frac{1}{3}$-poll., basi obtusa, apice obtusa vel acuta vel longe acuminata. *Pedicelli* 1–3, dense tomentosi, 1–$1\frac{1}{2}$ poll. longi, basi bracteati et paullo supra basin bracteola parva ovata amplectente muniti. *Flores* fere pollicares, laxe lanuginosi. *Sepala* ovata. *Petala exteriora* ovato-lanceolata; *interiora* paullo breviora, apice breviter triquetra, utrinque glabra, granulata. *Stamina* et *ovaria M. verrucosi*. *Carpella* diametro pollicaria, pedicello crasso, $\frac{1}{4}$–$\frac{1}{3}$-poll. *Semina* circa 8, biserialia, partitionibus cellulosis separata, atro-fusca, nitida, lævia, immarginata.

This is evidently the most common species throughout the mountainous countries north and east of Bengal, as we have before us specimens from all the collectors who have visited these countries. We found it abundantly in the wooded districts of Silhet and Cachar; but, as we did not meet with it in the Khasia hills, except at the very base, it is probably confined to the lowest levels, while the two last species occur at considerable elevations. The present species is, as we have mentioned under *M. Wallichii*, the same with a specimen from Roxburgh in the British Museum, and, as it has globose fruit, there can be no reasonable doubt that it is that described in the Flora Indica. It varies more than most species in the shape and size of the leaves; in one specimen in fruit they are only from two to three inches long.

8. **M. fulgens** (H.f. et T.); foliis ovato- vel oblongo-lanceolatis basi rotundatis apice in acumen longe attenuatum productis rigide coriaceis supra parce puberulis subtus cum petiolo pube brevissima molliter fusco-sericeis, floribus parvis tomentosis axillaribus vel ramulos terminantibus subracemosis.—Uvaria fulgens, *Wall. Cat.* 6482! Myristica Finlaysoniana, *Wall. Cat.* 6793!

HAB. In peninsula Malayana: ad Malacca, *Griffith!* et Singapur, *Wall.!*—(*v. s.*)

DISTRIB. Ins. Philippin. (*Cuming*, 2340!)

Frutex verosimiliter scandens. *Ramuli* graciles, flexuosi, lævigati, juniores pube brevissima cinerea vel flavescente incani. *Folia* 3–4 poll. longa, $1\frac{1}{4}$–$1\frac{1}{2}$ lata, petiolo $\frac{1}{4}$–$\frac{1}{3}$-poll., supra pilis rigidis adpressis paucis demum evanidis vel secus costam tantum superstitibus puberula. *Flores* plerumque versus apices ramulorum congesti, sæpe foliis ramuli plerisque abortivis vel deciduis racemos 4–5-floros simulantes. *Pedicelli* $\frac{1}{4}$–$\frac{1}{2}$-pollicares, tomentosi, bracteis lineari-subulatis suffulti, bracteolis 2–3 squamæformibus rotundatis muniti. *Sepala* rotundata, extus tomentosa, intus glabra, acutiuscula, basi coalita. *Petala exteriora* ovata, crassa, semipollicaria, extus aureo-sericea, intus cinereo-incana, macula basilari ovali glabra (atro-purpurea); *interiora* fere dimidio breviora, oblonga, acutiuscula, granulata, glabra, dorso convexa, intus concava, apice tantum triquetra. *Stamina* et ovaria sequentis.

The specimens in the Wallichian herbarium, under the number quoted above, are in a very imperfect state, and the leaves appear to belong to several very distinct species. A portion of them, however, certainly belong to this.

9. **M. Griffithii** (H.f. et T.); foliis anguste oblongis basi rotundatis apice obtusiusculis emarginatis rigide coriaceis supra præter costam pubescentem glabris subtus cum petiolis rufo-pubescentibus, pedunculis oppositifoliis vel in ramulo brevi axillari unifoliato subterminalibus brevissimis, pedicellis fasciculatis, floribus parvis.

HAB. In prov. Tenasserim ad Mergui, *Griffith*, 790!—(*v. s.*)

Frutex alte scandens. *Ramuli* nigricantes, rugulosi, puberuli, lenticellis flavidis punctati, juniores cum omnibus partibus novellis rufo-tomentosi. *Folia* 3–5 poll. longa, $1\frac{1}{4}$–$1\frac{3}{4}$ lata, petiolo $\frac{1}{3}$-poll. *Pedunculi* abbreviati, sæpe vix lineam longi. *Pedicelli* 2–5 vel plures, $\frac{1}{4}$–$\frac{1}{3}$ pollicem longi, tomentosi, basi bractea squamæformi suffulti, medio bracteola amplectente munita. *Sepala* ovalia, obtusa, tomentosa. *Petala exteriora* oblonga, obtusa, extus tomentosa, intus subglabra; *interiora* ¼ breviora, glabra, granulosa, dorso prope apicem tomentosa, acuta, apice triquetra. *Stamina* processu connectivi elongato late ovato terminata. *Ovaria* dense aureo-strigosa. *Fructus* ignotus.

We have seen only two specimens of this species, one in the Hookerian Herbarium, the other in that of Dr. Wight. The short axillary branches, bearing at their apex one leaf with a cyme of flowers and a terminal bud, are apparently peculiar to this species. The upper cymes are sessile and leaf-opposed. It should be compared with *Uvaria sphærocarpa*, Blume, with which it may be identical.

10. **M. polyanthum** (H.f. et T.); foliis lanceolatis vel oblongis basi acutis et in petiolum subdecurrentibus apice acutis vel acuminatis rigide coriaceis supra præter costam pubescentem glabris subtus pallidis pube molli brevissima ope lentis tantum conspicua incanis, floribus parvis in cymas oppositifolias congestis.—Uvaria polyantha, *Wall. Cat.* 6467!

HAB. Assam et Khasia, *Griffith!* Silhet, *Wall.!—(v. s.)*

Frutex verosimiliter scandens. *Ramuli* flexuosi, glabri, cortice nigricante, ruguloso, lenticellis albis consperso; gemmæ fusco-pubescentes. *Folia* forma et magnitudine admodum varia, late lanceolata vel oblongo-lanceolata vel lineari-oblonga, alia 3–4 poll. longa, 1¼–1½ lata, alia 6–8 poll. longa, 2–3 lata. *Petioli* glabri, supra profunde sulcati. *Cymæ* numerosæ, oppositifoliæ. *Pedunculi* tomentosi, abbreviati, 1–2 lineas longi, interdum vix ulli. *Flores* 3–7. *Pedicelli* brevissimi, basi bracteis oblongis minutis suffulti, medio unibracteolati.' *Sepala* ovato-rotundata, obtusiuscula. *Petala exteriora* ovalia, ¼ poll. longa, extus tomentosa, intus margine et sub apice puberula; *interiora* conformia, fere dimidio minora, crassa, extus incana, intus glabra et scrobiculata, apice breviter triquetra. *Stamina* connectivi processu cuneato-oblongo apiculata. *Ovaria* strigosa. *Fructus* ignotus.

11. **M. rufinerve** (H.f. et T.); foliis oblongis basi rotundatis apice acutis vel obtusis utrinque glaberrimis subtus pallidis, floribus oppositifoliis fasciculatis.

HAB. In provincia Silhet!—(v. v.)

Frutex alte scandens, trunco diametro 3–4-pollicari; cortice nigerrimo, ruguloso, ramulorum lævi glabro. *Gemmæ* aureo-pubescentes. *Folia* 6–10 poll. longa, 2½–4 lata, petiolo ⅔-pollicari, cylindrico, vix sulcato, tenuiter coriacea, costa nervisque parallelis, subtus in sicco rufescentibus. *Fasciculi* florum sessiles vel brevissime pedunculati, 2–7-flori. *Pedicelli* (forsan nondum plene evoluti) ¼-pollicares, aureosericei, infra medium bracteola minuta squamæformi. *Alabastri* minuti, vix lineam longi, triquetri, sericei. *Sepala* basi coalita, extus pubescentia. *Stamina* processu connectivi ovali apiculata. *Ovaria* multiovulata. *Torus* convexo-conicus.

This is a very distinct species from any with which we are acquainted, but unfortunately all our specimens have only very young buds. It will however be readily recognized by the large size of the leaves, and probably by its small flowers; but from the peculiar mode of development of the flowers in this Order, it is by no means easy to satisfy oneself how nearly they have attained their full size. We found it in dense forests on the banks of the Soorma river, between Silhet and Cachar, in Nov. 1850.

12. **M. prismaticum** (H.f. et T.); foliis ovalibus vel oblongis utrinque rotundatis apice in acumen breve gracile subito. productis rigide coriaceis utrinque glaberrimis supra lucidis subtus pallidis (in sicco rufescentibus), floribus magnis axillaribus solitariis breviter pe-

R

dunculatis, alabastris elongato-triquetris, carpellis glabris ovoideis.—
Uvaria prismatica, *Wall. Cat.* 6455 !

HAB. In Malaya ad Penang, *Wall.!* Malacca, *Cuming*, 2344 ! Singapur, *Wall.!*—(*v. s.*)

Frutex scandens, ramis validis rugosis nigricantibus. *Ramuli* glabri; partes novellæ fulvo-puberulæ. *Folia* 5–9 poll. longa, 2½–4 poll. lata, petiolo ¾-pollicari; inferiora in ramulo sæpe abbreviata, 2½ poll. tantum longa. *Pedunculi* petiolo paullo breviores, validi, infra medium bracteis pluribus squamæformibus cito deciduis tecti, superne nudi. *Sepala* in cyathum late campanulatum obscure trilobum demum explanatum vix ⅓ poll. diametro coalita, extus tomentosula, in fructu persistentia. *Petala exteriora* elongato-subulata, extus fulvo-incana, intus cinerea, basi concava, superne plana triquetra, ultra pollices 2 longa; *interiora* ⅓–½ poll. longa, ovata, acuta, vix puberula, intus concava, apice tantum triquetra. *Stamina* elongato-linearia, connectivi processu brevi ex ovali subgloboso apiculata. *Ovaria* strigosa, multiovulata. *Carpella* (in specimine uno visa) 4–5, sesquipollicaria, ovoidea, utrinque obtusa, glabra, granulosa, nigricantia, antice costata, costa superne evidentiore. *Semina* biserialia, septis cellulosis separata, compressa, lævia, colore helvolo, hilo terminali maximo.

This fine species deviates somewhat from the ordinary form of the genus, in the less divided calyx, the elongated flower, and the great excess of size of the outer petals. The staminal process, too, is shorter than it generally is in the genus, but still sufficiently distinct, and in all other characters it quite agrees with the other species.

Sect. 2. KENTIA, *Bl. in Fl. Javæ*, *Anon. p.* 71. (Polyalthiæ *sectio.*)
—*Petala* exteriora marginibus valde incrassatis latissimis, infra medium tantum excavata. *Ovaria* glabra, pellucido-glandulosa. *Ovula* numerosa vel definita. *Semina* scrobiculata.

13. **M. elegans** (H.f. et T.); foliis oblongo-lanceolatis in acumen gracile obtusum angustatis mucronatis basi plerumque rotundatis rigide coriaceis subtus adpresse fulvo-pubescentibus, pedunculis axillaribus solitariis petiolum paullo superantibus, floribus parvis, carpellis parvis subglobosis glabris pedicello æquilongo suffultis.—Uvaria elegans, *Wall. Cat.* 6474 A! (*non* B).

HAB. Malaya ad Penang, *Wall.!*—(*v. s.*)

Frutex scandens, ramosissimus. *Ramuli* graciles, flexuosi, rugulosi, nigricantes, punctis callosis flavidis conspersi, glabri (juniores puberuli); gemmæ fusco-sericeæ. *Folia* longiuscule petiolata, 3–4 poll. longa, 1–1½ lata, petiolo ½-poll., subtus pallida; in sicco flavescentia, supra pube rara sparsa sub lente tantum conspicua tecta. *Pedunculi* graciles, puberuli, bracteolis 2–3 minutis squamæformibus prope basin muniti. *Alabastri* ovato-triquetri, fusco-sericei. *Sepala* minuta, ovata, basi cohærentia, in fructu persistentia. *Petala exteriora* ⅓-pollicaria, ovata, obtusa, extus fulvo-sericea, intus cinereo-incana, basi excavata, rubra, linea longitudinali elevata notata, marginibus superne latis planis; *interiora* plus quam dimidio minora, ovata, crassa, extus convexo-carinata, incana, intus basi profunde concava, circa marginem superiorem excavationis longe pilosa, superne triquetra, a medio ad apicem incana. *Stamina* processu connectivi brevi crasso ovali apiculata. *Ovaria* glabra, oblonga, pellucide glandulosa, stylo oblongo brevi. *Ovula* in axi 8–10, biserialia, in pulpo nidulantia. *Carpella* ½ poll. longa, late ovoidea, granulosa. *Semina* 2–6, compressa, obovata, atro-fusca, nitida, biserialia, septis cellulosis separata, scrobiculata.

This elegant little species resembles so closely in general appearance *M. Kentii* (*Polyalthia Kentii*, Bl.), that we had without hesitation referred it to that species, till the examination of the flower showed remarkable differences. The petals are different in shape,, and the ovules are undoubtedly numerous, and not, as in *M. Kentii*, reduced to two. Blume describes *P. Kentii* as a tree; but his own specimens, we think, indicate that it is, like the present plant, a climber.

14. **M. pisocarpum** (H.f. et T.); foliis ellipticis vel ovalibus basi rotundatis vel acutiusculis apice angustatis obtusis sæpius emarginatis rigide coriaceis supra glabris subtus pallidis pube minutissima incanis, floribus axillaribus solitariis, alabastris triquetro-súbglobosis, carpellis pisiformibus.

HAB. In sylvis prope Malacca, *Griffith !—(v. s.)*

Frutex verosimiliter scandens, ramosissimus. *Ramuli* foliosi, nigricantes, vix rugulosi, glabri, juniores incano-puberuli; gemmæ aureo-sericeæ. *Folia* 1¾–3 poll. longa, 1¼–1¾ lata, pet. ⅓-poll., in sicco creberrime reticulata, nervis parallelis leviter arcuatis inconspicuis. *Pedunculi* ¼–½-pollicares, basi articulati, bracteis 2 minutis squamæformibus deciduis muniti. *Sepala* ovalia, basi coalita, coriacea, granulosa, glabra, in fructu persistentia. *Petala exteriora* late ovalia, fere rotundata, extus aureo-sericea, intus cinereo-incana, macula basilari (ex sicco atro-purpurea) glabra, ovata; *interiora* dimidio breviora, anguste oblonga, dorso sericea, carinato-convexa, intus basi excavata, glabra, supra medium triquetra. *Stamina* connectivi processu subgloboso carnoso terminata. *Ovaria* oblonga, ovulis in axi 2 superpositis. *Styli* oblongi, ovarium fere æquantes. *Carpella* toro capitato glabro umbellatim inserta, 4–8 vel plura, ovalia vel globosa, 2–3 lineas longa, pedicello æquilongo, glabra, granulosa. *Semina* 1–2, superposita, plano-convexa, rotundata, regulariter foveolata, atro-fusca, margine rhaphe lata percursa.

This species is very closely allied to *M. Kentii*, Blume, but the broader emarginate leaves and the smaller flowers sufficiently distinguish it. The terminal process of the stamens, too, is different.

11. **HABZELIA,** Alph. DC.

Sepala 3, acuta, basi connata. *Petala* 6, biserialia, æstivatione valvata, e basi inflata intus concava, elongato-linearia, apice triquetra; interiora paullo minora. *Torus* planus. *Stamina* linearia, connectivo ultra antheras dorsales biloculares in processum oblongum producto. *Ovaria* indefinita, oblonga, in conum conniventia, dense strigosa, multiovulata, stylo æquilongo subulato recto terminata. *Carpella* elongata, cylindrica, numerosa. *Semina* oblonga, non compressa, septis cellulosis separata.—Arbores, foliis *coriaceis, nervis obliquis parallelis falcatis subtus prominulis,* floribus *axillaribus elongatis triquetris.*

This genus was instituted by M. Alph. De Candolle for the reception of the *Unona Æthiopica* of authors, which, however, as we learn from Richard (Fl. Cubæ), has the hollowed torus of *Xylopia*, and is therefore perhaps more properly referable to that genus. Our materials do not enable us to determine this point; but the two species described below present all the characters assigned by De Candolle to *Habzelia*, and one of them has a fruit very like the Guinea pepper of commerce. They much resemble *Habzelia Æthiopica* of Alph. De Candolle in general appearance, so far as we can judge from a very imperfect specimen of that species, and recede considerably from the majority of the species of *Xylopia*. The stamens of our Indian species are remarkable for the curious structure of the anthercells, which are hairy, and divided by a series of transverse partitions, which give them an appearance of being jointed.

1. **H. ferruginea** (H.f. et T.); foliis lineari-oblongis acutis basi rotundatis subtus fusco-pubescentibus, floribus axillaribus solitariis.

HAB. In penins. Malayana ad Malacca, *Griffith !—(v. s.)*

Arbor. Ramuli vix rugulosi, cortice fusco, pubescentes, demum glabrati, juniores cum omnibus partibus novellis dense ferrugineo-tomentosi. *Folia* 5–7 poll. longa, 1½–2 lata, petiolo vix 2 lineas longo, coriacea, rigida, supra glabra, lucida, subtus pallida, fusca, petiolis nervisque tomentosis. *Pedunculi* ½ poll. longi, prope basin squamulis minutis bracteolati, tomentosi. *Sepala* ovata. *Petala* 1–1½-pollicaria, fulvo-tomentosa. *Ovaria* denseferrugineo-strigosa. *Carpella* 2 poll. longa, lævia, glabra. *Semina* oblonga, testa spongiosa, uniserialia.

We have only seen two or three detached carpels, and have not been able to find any arillus.

2. **H. oxyantha** (H.f. et T.) ; foliis ovalibus vel oblongis abrupte acuminatis utrinque glaberrimis, floribus axillaribus fasciculatis.—Uvaria oxyantha, *Wall. Cat.* 6478 !

HAB. In penins. Malayana ad Singapur, *Wall.!—(v. s.* in Herb. Linn. Soc.)

Arbor. Ramuli validi, rugosi, cortice fusco, glabri, juniores puberuli. *Folia* 7–8 poll. longa, 2½–3½ lata, pet. ½-poll., coriacea, supra nitida, subtus glauca. *Pedunculi* petiolum paullo superantes, adpresse puberuli. *Sepala* connata, late ovata, demum revoluta, dorso puberula. *Petala* griseo-puberula, 1¼–1½ poll. longa, siccitate dorso costata. *Stamina* et *ovaria* prioris.

12. **XYLOPIA,** L.

Cœlocline, *Alph. DC. Mém.* Patonia, *Wight, Ill.* i. 18.

Sepala 3, basi (sæpe alte) connata. *Petala* 6, æstivatione biseriatim valvata, elongata, subæquilonga, crasse coriacea; exteriora concava, marginibus planis; interiora basi tantum excavata, superne triquetra. *Torus* conicus, interne excavatus et ovaria includens, externe stamina gerens. *Stamina* indefinita, oblonga, antherarum loculis dorsalibus remotis, connectivo truncato capitato. *Ovaria* definita, 1–5, sericea, intra torum abscondita, oblonga vel ovalia, stylis elongatis exsertis in conum conniventibus apice clavatis. *Ovula* 2–6, secus suturam ventralem horizontalia.—Arbores *forsan humiles, foliosæ, ramosæ,* foliis *coriaceis lucidis,* floribus *axillaribus solitariis vel fasciculatis,* alabastris *longis triquetris.*

The genus *Xylopia* was originally founded by Linnæus, but its characters were remodelled by St. Hilaire, who first established it firmly, as we now recognize it, distinguishing it by the shape of the petals, the position of the ovules, the peculiar hollow torus, and the dehiscence of the fruit. Von Martius distinguishes it by the same characters, while A. Richard, in the 'Fl. Cubæ,' depending principally on the torus, omits all mention of the dehiscence of the fruit, and unites with it Alph. De Candolle's *Cœlocline* and *Habzelia.* The latter genus we have already alluded to. *Cœlocline,* which has the same torus and ovary as *Xylopia,* appears distinguished by less important characters, as the dehiscence of the fruit sometimes seems to occur at a very late period, and the presence of arillus is perhaps of no great moment in the Order. Of the Indian species described below, the only one which is known in fruit has the arillus of *Xylopia;* we are therefore disposed for the present to follow A. Richard in retaining that genus entire, as all the species are very similar in habit, and the flowers of the Asiatic ones are in no way distinguishable from those of the American ones. If the fruit of any of the Indian plants described below be found to differ from that of the typical species of the genus, the difference will in all probability be regarded as of no more than sectional value. *Xylopia* is very closely allied to *Melodorum*, but is readily known by the erect habit, the peculiar torus, and the truncate stamens. Its petals only differ by being more elongated. *Anona*, which is also like it in flower, is distinguishable by a multitude of characters. Many of the species, on a casual inspection, so much resemble the genus *Diospyros*, that the two genera are often intermixed in herbaria.

Xylopia is widely distributed throughout tropical regions, for, though probably most abundant in America, a number of species are known from West Africa, and A. Richard mentions one as a native of Mauritius. Blume describes none from any part of the Malayan Archipelago, nor have we seen any among the collections of Cuming. At present therefore it would appear that in India they are confined to Ceylon and the Malayan Peninsula.

1. **X. Malayana** (H.f. et T.); foliis oblongis basi acutis obtuse acuminatis glaberrimis, inflorescentia axillari subtriflora, pedicellis medio unibracteatis, ovariis 5–7.

HAB. Malacca, *Griff.!*—(*v. s.*)

Arbor. *Ramuli* graciles, glabri, cortice fusco ruguloso, juniores vix puberuli. *Folia* 3–5 poll. longa, 1½–2 lata, petiolo ¼-poll., coriacea, firma, laxe reticulato-venosa, areolis magnis, supra lucida, subtus (in sicco) brunnea. *Pedunculi* vix lineam longi, subtriflori, pubescentes; *pedicelli* vix longiores. *Alabastri* strigosi, pubescentes, vix semipollicares, argute triquetri. *Sepala* majuscula, acuta. *Petala* ext. summo apice tàntum triquetra. *Ovaria* dense et longe albo-pilosa. *Ovula* 2.

2. **X. parvifolia** (H.f. et T.); foliis oblongo-lanceolatis acuminatis basi acutis utrinque glabris, inflorescentia axillari 3–5-flora, pedicellis bracteolis pluribus rotundatis imbricatis tectis, ovariis 5.—Patonia parvifolia, *Wight, Ill.* 19!

HAB. In Zeylania, *Walker!*—(*v. s.*)

Arbor vel *frutex* floribundus. *Ramuli* fulvo-pubescentes, demum glabrati, fusci, albo-punctati. *Folia* 2–3 poll. longa, ¾–1¼ lata, petiolo ⅓-poll., crasse coriacea, venulis creberrimis reticulatis notata, supra nitida, subtus pallida. *Fasciculi* florum subsessiles; *pedicelli* brevissimi; *bracteola* suprema calyci adpressa, rotunda vel reniformis. *Alabastri* ½ poll. longi, fusco-sericei. *Sepala* ad medium coalita, acuta. *Petala* utrinque pubescentia. *Ovula* 4–6.
Patonia Walkeri, Wight! Ill. i. 19, is a species of *Diospyros.*

3. **X. nigricans** (H.f. et T.); foliis ellipticis vel lanceolatis obtuse acuminatis glaberrimis, floribus axillaribus solitariis vel ternis, bracteolis 1–2 minutis, ovariis 5.

HAB. Ceylon, *Thwaites!* No. 615, 1038.—(*v. s.*)

Frutex. *Ramuli* graciles, foliosi, cortice ruguloso, albido, glabro; partes novellæ puberulæ. *Folia* basi acuta, in sicco atro-viridia, tenuiter coriacea, forsan undulata, subtus pallidiora, 3 poll. longa, 1¼ lata, petiolo ⅙-poll. *Pedicelli* vix ½-pollicares, graciles, bracteolis deciduis. *Alabastri* ⅓-pollicares. *Flores* fere *X. parvifoliæ*, sed petala exteriora fere ad apicem excavata, mucrone triquetro abbreviato. *Ovula* circa 4.
We learn from a memorandum by Mr. Thwaites, that the carpels of this species dehisce when ripe, along the suture, and expose the seeds nestling in red pulp.

4. **X. caudata** (H.f. et T.); foliis oblongo-lanceolatis longe et obtuse acuminatis mucronulatis subtus sericeo-incanis, floribus minutis solitariis vel fasciculatis, ovariis 2.—Guatteria? caudata, *Wall. Cat.* 6452!

HAB. In Malaya: Singapur, *Wall.!* Malacca, *Griffith!*—(*v. s.*)

Frutex ramosissimus. *Ramuli* stricti, gracillimi, foliosi, cinerei, rugosuli, glabri; juniores pubescentes; partes novellæ albo-sericeæ. *Folia* oblonga vel lanceolata, in acumen longissimum obtusum angustata, $1\frac{3}{4}$–$2\frac{1}{4}$ poll. longa, $\frac{3}{8}$–1 poll. lata, petiolo vix 2-lineari, tenuiter coriacea, supra (præsertim secus costam) minute puberula, demum glabrata, subtus sericea. *Pedicelli* axillares, brevissimi, basi bracteis minutis squamæformibus muniti, bracteola 1 rotundata calyci adpressa. *Sepala* basi coalita, extus pubescentia. *Petala* utrinque sericea, subæquilonga, $\frac{1}{4}$-pollicaria; *exteriora* oblonga, concava, obtusa; *interiora* basi tantum concava, superne elongata triquetra. *Ovaria* 2, ovalia, longe albo-sericea, in sutura ventrali biovulata. *Styli* elongati, clavati, e toro concavo longe exserti. *Carpella* 1–2, vix $\frac{1}{2}$-pollicaria, breviter pedicellata, ovalia, utrinque pubescentia (indehiscentia?). *Semina* 2, superposita, septo membranaceo separata.

5. **X. Championii** (H.f. et T.); foliis ellipticis glabris subtus sub lente sericeo-puberulis, pedunculis axillaribus solitariis abbreviatis, alabastris oblongis obtusis fusco-sericeis, petalis exterioribus latiusculis obtusis, ovario solitario quadriovulato.

HAB. In Zeylania, *Gardner!* *Champion!*—(*v. s.*)

Arbor. *Ramuli* graciles, cortice nigricante, vix rugosulo, glabro; juniores et gemmæ fusco-sericeæ. *Folia* elliptica vel late lanceolata, basi acuta, apice acuminata, 3–$4\frac{1}{2}$ poll. longa, 1–$1\frac{3}{4}$ lata, petiolo vix $\frac{1}{2}$-pollicari, tenuiter coriacea, firma, pellucidopunctata; nervi obliqui, inconspicui, venulis creberrimis reticulatis. *Pedunculi* vix 2 lineas longi, fusco-sericei, infra medium bracteas 2–3 minutas squamæformes amplectentes gerentes. *Sepala* in cyathum trilobum acutilobum coalita. *Petala exteriora* $\frac{1}{2}$-$\frac{3}{5}$-pollicaria, crassa, concava, late oblonga; *interiora* $\frac{1}{3}$ breviora, basi cuneata, concava, superne late triquetra. *Ovarium* longe pilosum. *Fructus* (ex icone cl. Champion) brevissime pedicellatus, oblongus, $1\frac{3}{4}$-pollicaris, forsan follicularis. *Semina* 4, ex icone videntur arillata.

Tribus V. GUATTERIEÆ.

Petala omnia conformia, haud unguiculata, plerumque plana, in paucis basi concava, superne plana vel irregularia, æstivatione valvata vel subaperta. *Stamina* indefinita, dense conferta. *Carpella* discreta, interdum solitaria.

In this tribe, exceptin the two first genera, which are anomalous, the petals are quite flat, and generally eathery or membranous in texture. In this respect they differ essentially from *Anoneæ* and *Xylopieæ*, and approach *Uvarieæ*, from which they are only distinguished by the valvate (not imbricate) æstivation of the petals. In some of the genera the petals are only valvate in the very young bud, separating from one another at an early period. The two first genera are in many respects intermediate between *Xylopieæ* and *Guatterieæ*, but as their petals are uniform in shape, and the inner ones, though concave at the base, are not triquetrous above, we refer them to the present tribe.

13. **CYATHOCALYX**, Champion.

Sepala 3, in cyathum tridentatum coalita. *Petala* 6, æst. biseriatim valvata, basi concava, circa genitalia constricta, superne plana, coriacea. *Torus* depresso-conicus, concavus. *Stamina* indefinita, anguste cuneata, apice truncata, dorso antherifera. *Ovarium* unicum, toro concavo partim inclusum, oblongum, superne angustatum. *Stigma* magnum, pel-

tatum, rotundatum.   *Ovula* in sutura ventrali indefinita, biserialia.—
Arbores, foliis *glabris lucidis,* floribus *terminalibus et oppositifoliis, solitariis vel fasciculatis.*

This very remarkable genus is clearly but briefly defined by Major Champion in
the Hookerian Herbarium. The flowers are intermediate between those of *Xylopieæ*
and those of *Guatteriæ.* The torus is that of *Xylopia,* but the petals are flatter
than in *Habzelia,* and very like those of some species of *Artabotrys.* The stamens
are more like those of *Guatteriæ,* and the single ovary with many ovules is very
anomalous in the Order. It will be interesting to determine on the living plant the
position of the solitary carpel with respect to the floral envelopes; our specimens
are not sufficiently numerous to enable us to ascertain this point. One species only is
known, which appears to be a native both of Ceylon and the Malayan coast.

1. **C. Zeylanicus** (Champion, mss. in Herb. Hook.); foliis oblongo-lanceolatis, pedunculis 1–3 fere pollicaribus, petalis lineari-oblongis.

HAB. In Zeylania, alt. 1–3000 pedum, *Walker!* *Gardner!* ad Han-
tani, Kandy, et Narawelle Galle, *Champion!; *et in Tenasserim ad
Mergui, *Griffith!* (No. 1032; specimen fructigerum, sed quantum e
specimine suppetente judicandum a planta Zeylanica nullo modo dis-
tinguendum).

Arbor. Ramuli læviusculi, atro-fusci, adulti glabri, juniores aureo-pubescentes.
Folia 6–10 poll. longa, 2–3 lata, petiolo ½-pollicari, utrinque glaberrima, acuminata,
basi acuta, coriacea, nervis obliquis, incurvis, venulis crebre reticulatis. *Sepala* in
cyathum truncatum ¼ poll. altum ½ poll. diam. coalita, dentibus 5 rotundatis obtusis
remotis, fusco-sericea. *Petala* 2¼ poll. longa, fusco-sericea, obtusa, ½–⅔ poll. lata.
Ovarium obscure costatum, glabrum. *Fructus* (ex spec. Griffithiano, in Herb. Wight)
late ovalis, utrinque obtusissimus, sesquipollicem longus, plus quam pollicem latus,
subtorulosus, glaber, aromaticus. *Semina* biserialia, 8–10. *Testa* lævis. *Albumen*
ruminatum.

14. **ARTABOTRYS,** R. Br.

Sepala 3, basi cohærentia. *Petala* 6, æst. biseriatim valvata, basi
concava, et circa genitalia constricta, sursum patentia, forma varia.
Stamina indefinita, oblonga vel cuneata, connectivo superne truncato
plano. *Antheræ* dorsales, loculis remotis. *Torus* plano-convexus.
Ovaria indefinita vel subdefinita (5–30), ovalia vel oblonga, stylis ovalibus vel lineari-oblongis plerumque reflexis terminata. *Ovula* 2, e
basi erecta, collateralia. *Carpella* forma varia. *Semina* 1–2, magna.—
Frutices *sarmentosi vel scandentes,* foliis *lucidis,* floribus *interdum suaveolentibus solitariis vel fasciculatis,* pedunculis *lignosis uncinatim re-
trofractis.*

This genus, which was first characterized by Mr. Brown in the ʻ Botanical Register,ʼ
is readily known by its peculiar habit, and by its floral characters. The uncinate
woody peduncles have no parallel in the Order. The position of the genus in the
family is somewhat doubtful, but, on the whole, its nearest affinities appear to be with
the tribe in which we have placed it. The concave base of the petals resembles that
of *Xylopia* and *Habzelia,* but they are all similar in shape, and the upper part is
usually quite flat. In one species, however, it is cylindrical or clavate, and in another triquetrous,—a circumstance which has been overlooked in the preparation of the
analytical table of the genera. *Artabotrys* is for the most part an Asiatic genus, one

species only being known from tropical West Africa, and none from America. It occurs in about equal proportions in continental India and the Eastern Archipelago, and one extends into south China. One species is very extensively cultivated as an ornamental shrub in gardens throughout the East.

1. **A. odoratissimus** (R. Br. in Bot. Reg. t. 423, non Blume); foliis oblongo-lanceolatis glabris utrinque acutis, pedunculis 1–2-floris, petalis fere æquilatis, lamina plana oblongo-lanceolata, ovariis paucis glabris, carpellis oblongis obtuse acuminatis.—*Wall. Cat.* 6415!; *W. et A. Prod.* i. 10.   A. hamatus, *Blume, Fl. Jav. Anon.* 60. *t.* 29, 31 *C.* Anona hamata, *Dun. Anon.* 106. *t.* 27; *DC. Syst.* i. 491, *Prod.* i. 90. Uvaria odoratissima (*et* U. uncata), *Roxb. Fl. Ind.* ii. 666!

HAB. In Zeylania et Malaya forsan indigenus, in hortis sæpissime cultus.—(Fl. per totum annum.) (*v. v. cult.*)

DISTRIB. Java, *Blume.*

Frutex elegans, ramulis sarmentosis, cortice albido-punctato.   *Folia* 2–8 poll. longa, 1–2 lata, in acumen longum obtusiusculum producta, tenuiter coriacea.   *Petala* 1–1¾ poll. longa, fusco-sericea, demum glabra.   *Carpella* 2–2½ poll. longa.

We are obliged to differ in opinion from Blume as to the plant originally described by Brown as *A. odoratissimus*, which appears to us not to be the species figured in the 'Flora Javæ' under that name, but that which Blume has figured and described as *A. hamatus.* As the plant figured in the 'Botanical Register' was introduced into England from Calcutta, it must necessarily be that cultivated in the garden there.   Now we find no indication in Roxburgh or Wallich of the cultivation of any species but one in continental India; Wight and Arnott notice only one from the Peninsula, and we find only one in the Hookerian Herbarium from Ceylon. Blume's *A. odoratissimus* is therefore entirely an eastern form, of which we shall subjoin a description*, as it will probably occur in the Malayan Peninsula, as well as in the Archipelago. Mr. Brown is, we believe, right in referring all the species enumerated by Dunal and DC. to that now described.

2. **A. Zeylanicus** (H.f. et T.); foliis oblongis acutis vel obtuse acuminatis utrinque glabris, pedunculis plurifloris, petalis fere æquilatis, lamina plana lanceolata dense fulvo-tomentosa, ovariis numerosis tomentosis, carpellis obovatis mucronatis.

HAB. In Zeylaniæ sylvis, *Walker! Champion! Thwaites!*—(*v. s.*)

Frutex alte scandens, cortice ruguloso, brunneo, glabro, ramulorum juniorum fusco-pubescente.   *Folia* lanceolata vel oblonga, acuminata, subtus pallida, nervis crebris

---

* A. Blumei (H.f. et T.); foliis oblongis obtuse acuminatis, pedunculis unifloris, petalis exterioribus cæteris duplo latioribus lamina ovata, ovariis 8–10 villosulis, carpellis ovalibus vel subglobosis abrupte mucronatis.—A. odoratissimus, *Blume, Fl. Jav. Anon.* 59. *t.* 28, 31 *B, excl. syn.*   A. hamatus, *Benth. in Hook. Kew Journ.* iii. 257.

HAB. In Java, *Blume;* in insula Hongkong, *Champion!*—(*v. s.*)

Frutex elegans, densus, glaber, ramulis sarmentosis, cortice cinereo vel atro-fusco. *Folia* 3–8 poll. longa, 1¼–3 lata, utrinque glabra, nitida, coriacea.   *Sepala* ovata, acuta, parce strigosa.   *Petala* fusco-sericea; *exteriora* ⅔ poll. longa, lamina ovali-oblonga, obtusa; *interiora* lineari-oblonga, paullo breviora.   *Carpella* glabra, 1–1½-pollicaria, obtuse rostrata.

This appears to differ from *A. odoratissimus* in the shape and texture of the leaves, and in the shorter and broader petals, and more rounded fruit.

reticulatis, 4–6 poll. longa, $1\frac{1}{2}$–$2\frac{1}{2}$ lata, petiolo vix 2 lineas longo. *Pedunculi* oppositifolii, validi, lignosi (rarius elongati, foliosi, vix hamati). *Pedicelli* $\frac{1}{2}$-pollicares, fulvo-pubescentes, bractea oblonga suffulti et basi bracteolis pluribus distichis squamæformibus muniti. *Sepala* acuminata, extus dense fusco-tomentosa. *Petala* cinnamomea, tomentosa; *exteriora* $1\frac{1}{4}$ poll. longa, crasse coriacea, medio dorso obscure carinata; *interiora* pollicem longa, subconformia. *Ovaria* numerosa, dense villosa, stylo subulato apiculata. *Torus* vix convexus, in fructu subglobosus, fusco-tomentosus, cicatricibus pluribus magnis notatus. *Carpella* granulata, strigoso-tomentosa, $\frac{2}{3}$–1 poll. longa.

3. **A. caudatus** (Wall. Cat. 6417!); foliis oblongo-lanceolatis basi acutis in acumen longum obtusum abrupte acuminatis utrinque glaberrimis, petalorum lamina plana e basi quadrata anguste lineari.

HAB. In montibus Silhet, *Wall.!—(v. s. in Herb. Soc. Linn.*)

Frutex alte scandens. *Ramuli* rugulosi, cortice atro-fusco; partes novellæ fusco-puberulæ. *Folia* 3–5 poll. longa, $1\frac{1}{2}$–2 lata, petiolo bilineari, tenuiter coriacea, utrinque lucida. *Pedunculi* normales, pluriflori. *Sepala* acuta. *Petala* $1\frac{1}{2}$ poll. longa, fusco-strigosa, inter se fere conformia; interiora paullo angustiora.

4. **A. Burmannicus** (Alph. DC. Mém. 36); foliis oblongis in acumen gracile productis subtus rufo-hirsutulis, petalorum laminis triquetris filiformibus subulatis cinereo-pubescentibus.—*Wall. Cat.* 6418!

HAB. Ava, *Wall.!* Mergui, *Griffith!—(v. s.*)

Frutex scandens, ramulis striatulis pubescentibus, cortice fusco; partes novellæ dense fusco-tomentosæ. *Folia* oblonga, utrinque acutiuscula, vel lauceolata, acumine obtuso, membranacea, 3–7 poll. longa, 1–$2\frac{1}{2}$ lata, petiolo vix 2 lineas longo, supra præter costam pubescentem glabra, subtus pilis brevibus laxis præsertim secus costam nervosque hirsuta. *Pedunculi* oblique oppositifolii, rufo-pubescentes, plerumque uniflori. *Petala* fere ut in *A. suaveolente*, sed lamina triquetra. *Carpella* obovato-oblonga, glabriuscula, atro-fusca, $\frac{3}{4}$ poll. longa.

5. **A. suaveolens** (Blume, Fl. Javæ, Anon. 62. t. 30, 31 D); foliis oblongo-lanceolatis acuminatis basi acutis utrinque glabris, petalorum laminis elongatis cylindricis, carpellis oblongis.—*Wall. Cat.* 6416! Unona suaveolens, *Blume, Bijdr.* 17.—*Rumph. Amb. v. t.* 14.

HAB. In sylvis densis Silhet, *Wall.!;* in Malaya ad Penang, *Wall.!* et Malacca, *Griff.!—(v. v.*)

DISTRIB. Per totum archipelagum Malayanum et insulas Moluccas et Philippinas, *Blume, Cuming, etc.*

Frutex alte scandens. *Ramuli* rugosi, striati, atro-fusci, glabri, juniores lævigati, pilis sericeis puberuli. *Folia* tenuiter coriacea, lucida, glaberrima, subtus pallidiora, secus costam pubescentia, demum glabra, juniora adpresse sericea, 3–5 poll. longa, 1–$1\frac{1}{2}$ lata, petiolo vix $\frac{3}{4}$-poll. *Pedunculi* validi, lignosi, uncinati, versus apicem strigoso-pilosi, multiflori. *Flores* in fasciculos plurifloros congesti, bracteis subulatis cito deciduis suffulti, flavidi, suaveolentes, $\frac{1}{2}$ poll. longi.

15. **CANANGA,** Rumph. (non Aublet).

Sepala 3. *Petala* 6, æst. aperta, biserialia, longa, linearia, æqualia. *Stamina* numerosa, linearia, connectivo ultra antheras dorsales in processum carnosum ovatum acutum producto. *Torus* convexiusculus, medio subconcavus. *Ovaria* oblonga, in stylum anguste oblongum

s

sensim attenuata. *Ovula* numerosa, biserialia. *Stigmata* subcapitata, ope gelatinis inter se subcoalita.—Arbor *excelsa, floribunda.*

In habit and general appearance this genus closely resembles *Unona*, but the indefinite ovules prevent its being referred to that genus. The peculiar stamens and seeds are in themselves, we think, sufficient to justify us in distinguishing it. The seeds are pitted like those of the section *Kentia* of *Melodorum*, and of some *Cucurbitaceæ;* and the inner surface of the brownish-yellow, brittle testa is covered with sharp tubercles, which penetrate into the albumen, taking the place of the flat plates which are found in the rest of the Order. *Cananga* of Aublet is not distinguishable from *Guatteria,* which is not to be regretted, as the name was incorrectly applied to an American group.

1. **C. odorata** (H.f. et T.); foliis ovato-oblongis longe attenuatis plerumque obliquis margine undulatis, pedunculis axillaribus 2–4-floris. —Uvaria odorata, *Lam. Ill. t.* 495. *f.* 1; *Roxb. Fl. Ind.* ii. 661!; *Wall. Cat.* 6457!; *W. et A. Prod.* i. 8; *Bl. Bijd.* 14, *Fl. Jav. Anon.* 29. *t.* 9, 14 *B.* Unona odorata, *Dun. Anon.* 108; *DC. Syst.* i. 492, *Prod.* i. 90. Uvaria Cananga, *Vahl.* U. farcta, *Wall. Cat.* 6460! U. axillaris, *Roxb. Fl. Ind.* ii. 667. U. Gærtneri, *Dunal, Anon.* 89; *DC. Syst.* i. 482, *Prod.* i. 88. Unona leptopetala, *Dun. Anon.* 114; *DC. Syst.* i. 496, *Prod.* i. 91; *Deless. Ic. Sel. t.* 88. U. velutina, *Bl. Fl. Jav. Anon.* 31. *non Dunal, nec Roxb.—Gært. Fr.* ii. *t.* 114. *f.* 2.

HAB. Ava, *Wall.!;* Tenasserim, *Griff.!* in tropicis utriusque orbis frequentissime culta.—(*v. s.*)

DISTRIB. Java, *Blume !* Ins. Philip., *Cuming !*

Arbor excelsa. *Ramuli* validi, cortice fusco albido-punctato, glabri, juniores puberuli; partes novellæ cinereæ. *Folia* basi rotundata, apice in acumen longum plerumque obliquum attenuata, 5–8 poll. longa, 2–3 lata, petiolo semipollicari, tenuia, nervosa, supra glabra, subtus præsertim ad nervos tenuissime puberula, demum glabrata. *Pedunculi* axillares, vel sæpius ad axillas foliorum delapsorum, ½-pollicares, interdum abbreviati, in axilla sæpe plures. *Pedicelli* pollicares, cinereo-incani, ad apices pedunculorum subumbellati, bracteis minutis squamosis, vel una interdum foliacea, bracteola 1 versus medium pedicellum oblonga. *Sepala* rotundata, acutiuscula, cinereo-tomentosa, decidua. *Petala* in alabastro juniori dense sericeo-villosa, demum elongata, fere tripollicaria, basi ⅓ poll. lata. *Torus fructus* dilatatus, cylindricus. *Carpella* numerosa, pulposa, pedicellis ultrapollicaribus suffulta, ovalia vel obovata, glabra, nigra, ⅔ pollicem longa. *Semina* pulpo immersa, biserialia, numerosa, plano-compressa, obovata, pallida, badia, irregulariter scrobiculata.

This species, which is very generally cultivated throughout tropical India as an ornamental tree, does not appear to be a native of Bengal or Madras, though it is certainly indigenous to the eastward.

16. **UNONA**, L.

Sepala 3. *Petala* 6, biserialia, æstivatione valvata, tenuiter coriacea, elongata, rarius 3, serie interiore suppressa. *Stamina* numerosa, tetragono-oblonga, connectivo ultra antheras dorsales subdistantes oblongas vel lineari-oblongas in processum subglobosum vel truncatum producta. *Torus* parum elevatus, apice truncatus, planus vel aliquantulum excavatus, inter stamina glaber, inter ovaria pilosus. *Ovaria* indefinita, oblonga, strigoso-pilosa. *Ovula* in axi superposita, 2–7, adscendentia, uniserialia. *Stylus* ovalis vel oblongus, recurvus, interne per totam longitudinem sulcatus.   *Carpella* indefinita.—Arbores *erectæ vel* frutices *scandentes,* floribus *majusculis axillaribus vel extra-axillaribus plerumque solitariis.*

If we except the last section, the species of this genus are readily known by their elongated fruit, separated by constrictions into one-seeded joints. This structure occurs in no other genus of the Order. We have, however, abstained from introducing it into the generic character, because we are unwilling for the present to separate from the genus several species in which the fruit is unknown, and one at least in which it is not jointed. All these species agree with the more typical *Unonæ,* in the thin, more or less elongated petals, in the shape of the ovary and style, and in the ovules being definite in number, and inserted into the ventral suture in a single row. These characters appear to us constant, and they are, we believe, sufficient to characterize the genus, without its being necessary to have recourse to the fruit. One or two species are scandent, while the majority are erect; but there is a great similarity in general aspect in all. The young leaves and petals are always pelluciddotted. When the ovules are reduced to two, the genus approaches very close to *Polyalthia,* but is readily distinguished by the position of the ovules. That genus is also well marked by the nervation of the leaves, which is peculiar, and very different from that found in *Unona.* The section *Dasymaschalon* is remarkable for the entire suppression of the inner petals, but its habit is quite that of the typical *Unonæ;* and the other characters (especially the fruit) are so identical, that it does not appear to us advisable to separate it. *Unona* is entirely an Asiatic genus, nor do we know any species in addition to those described below, except *U. virgata,* Blume, which appears to be referable to our section *Pseudo-Unona.*

Sect. 1. DESMOS.—*Petala* 6. *Carpella* inter semina constricta.

1. **U. dumosa** (Roxb. Fl. Ind. ii. 670); scandens, foliis obovatis vel ovalibus basi cordatis supra glabris subtus dense tomentosis, pedunculis extra-alaribus gracilibus pendulis, petalis obovatis spathulatis apice angustatis, carpellis 2–3-articulatis.—*Wall. Cat.* 6429!

HAB. In provincia Silhet, *Roxburgh, Wallich!*—(Fl. Apr., fr. Oct.) (*v. s.*)

Frutex dumosus, scandens, ramis griseis, rugosis, junioribus fulvo-pubescentibus; partes novellæ dense fulvo-tomentosæ. *Folia* obtusa vel acuta, 3–5 poll. longa, 1½–2¼ lata, petiola vix ¼-poll., juniora utrinque pubescentia. *Pedunculi* sæpe oppositifolii, 1–1½-poll., tomentosi, supra medium vel prope basin bracteola 1 ovata tomentosa ¼-poll. longa munita. *Sepala* late ovalia vel fere rotundata, basi cordata, acuta, tenuia, nervosa, glanduloso-punctata, utrinque sericea, semipollicem longa. *Petala* tenuia, nervosa, glanduloso-punctata, utrinque adpressa, pubescentia; *exteriora* fere tripollicaria, 1–1⅓ poll. lata, obtusiuscula, basi in unguem latum angustata; *interiora* paullo breviora et angustiora. *Torus* medio depressus. *Ovaria* triovulata. *Stigma* breve, obovatum.

The fruit is only known from Roxburgh's description, unless *Uvaria heterocarpa,* Bl. Fl. Jav. Anon. 41. t. 17. belongs to this species, which, from the general resemblance, is probably the case. I hesitate, however, to quote that species, because it has stellate hairs, which I have not found in the plant now described; they are, however, very densely compacted, and may occasionally be stellated. Blume's plant is from Java, but of doubtful locality. The carpels are pubescent, and have from one to two joints, of the size of a pea, the terminal one mucronate.

2. **U. Dunalii** (Wall. Cat. 6425!); scandens, foliis oblongis vel oblongo-lanceolatis submembranaceis utrinque glabris vel subtus sparse

at adpresse pubescentibus, pedunculis axillaribus vel terminalibus, petalis e basi lata lanceolatis, carpellis 1–3-articulatis.

HAB. Concan, *Stocks!*; in sylvis Chittagong ad montem Sitakund! et in peninsula Malayana ad Penang, *Wall.!*—(*v. v.*)

Frutex scandens, cortice griseo rugoso. *Ramuli* elongati, graciles, atro-fusci, glabri, punctulis albis conspersi; partes novellæ pubescentes. *Folia* obtusa vel acuminata, subtus pallida, 3–4½ poll. longa, 1¼–1¾ lata, petiolo pubescente, ¼-poll. *Pedunculi* ½–1¼-pollicares, pubescentes, infra medium bracteolis 1–3 squamæformibus minutis muniti. *Flores* pallide flavescentes, suaveolentes. *Sepala* lata, ovata, ⅓-poll., acutiuscula. *Petala* nervosa, subglabra; *exteriora* 1½ poll. longa, ½ poll. lata; *interiora* paullo minora et angustiora. *Torus* parum elevatus, vix excavatus, inter ovaria longe et dense strigosus. *Ovula* 4–6. *Pedunculus* fructus interdum elongatus, 4-poll. *Torus* globosus, pisiformis. *Carpella* 10–15, pedicello ¼–½-pollicari adpresse fulvo-strigoso. *Articuli* 1–3, sæpe solitarii, rugulosi, pubescentes vel glabri, ovales, ultimus acutus vel mucronatus.

3. **U. Zeylanica** (H.f. et T.); foliis elongato-lanceolatis submembranaceis utrinque glabris vel subtus vix puberulis, pedunculis axillaribus brevibus, petalis oblongo-lanceolatis acutis, articulis fructus 1–4.

HAB. In Zeylania, alt. 2–3000 ped.!—(*v. s.*)

Arbor? *Ramuli* elongati, graciles, grisei vel nigricantes, rugulosi, glabri; partes novellæ pubescentes. *Folia* basi acuta, apice acuminata, 5–8 poll. longa, 1¼–2 lata, plerumque glabra, sed interdum subtus oculo armato pilis sparsis puberula, superne nitida, subtus pallida, glaucescentia; petioli ⅙-poll., glabri. *Pedunculi* solitarii vel bini, interdum supra-axillares, graciles, ½–¾-poll., basi squamellati, cæterum nudi, glabri. *Sepala* oblonga-lanceolata, extus puberula, 2 lin. longa. *Petala* coriacea, glabra; *exteriora* pollicaria, ¼ poll. lata; *interiora* ⅔-poll., angustiora. *Stamina* latissima, abbreviata. *Ovula* 2–4, in gelatine immersa. *Torus* superne parum excavatus, in fructu globosus. *Carpella* 10–20, (immatura) pedicello 2 lineas longo. *Articuli* 1–4, glabri, rugulosi, ultimus apiculatus.

The joints of the fruit are less markedly distinct than usual, but that is probably only because they are immature.

4. **U. Lawii** (H.f. et T.); foliis oblongo-lanceolatis tenuiter coriaceis supra sub lente sparse puberulis (demum glabratis) subtus glaucis pubescentibus, pedunculis suboppositifoliis gracilibus, petalis anguste linearibus, carpellis 1–3-articulatis.

HAB. In sylvis Malabar, *Wight!* Concan, *Law!*—(*v. s.*)

Ramuli graciles, foliosi, rugosi, grisei vel nigricantes, verruculosi, juniores pubescentes; partes novellæ sericeæ. *Folia* basi rotundata, apice acuminata, acuta aut obtusiuscula, 2½–4½ poll. longa, ¾–1½ poll. lata, petiolo ⅓-poll., pubescente. *Pedunculi* pubescentes, pollicares, infra medium bracteam 1 ovatam sæpe acuminatam 1–2 lineas longam gerentes. *Sepala* ovato-lanceolata, extus tenuiter tomentosa, ⅓-poll. *Petala exteriora* 2½ poll. longa, vix ⅛ poll. lata, adpresse pubescentia, e basi rotundata, linearia, tenuiter coriacea; *interiora* 1¼–1¾ poll. longa, sericeo-pubescentia, e basi elliptica intus tuberculata rugosa anguste linearia. *Ovaria* 2–3-ovulata. *Torus* convexus, medio depresso-concavus. *Carpella* indefinita, pedicello ¾-pollicari suffulta, articulis 1–3 ovali-oblongis nigricantibus, ultima mucronata.

This plant, which is usually smaller and narrower-leaved than any of the states of *U. discolor*, in general appearance closely resembles that species, but is readily distinguished by the narrow petals.

5. **U. discolor** (Vahl, Symb. ii. 63. t. 36); foliis oblongis oblon-

go-lanceolatis vel lanceolatis basi rotundatis vel cordatis rarius acutis apice plerumque acutis supra glaberrimis nitidis subtus glaucis glabris vel sparse pubescentibus, pedunculis extra-axillaribus, petalis e basi lata lanceolatis apice obtusiusculis sericeis vel subglabris, carpellis 1–6-articulatis.—*Dunal, Anon.* 111; *DC. Syst.* i. 494, *Prod.* i. 90; *Alph. DC. Mém.* 28; *Wall. Cat.* 6420 ! *excl.* B, E, F; *Roxb. Fl. Ind.* ii. 669; *Bl.! Fl. Javæ, Anon.* 53; *W. et A.! Prod.* i. 9. U. Chinensis, *DC. Syst.* i. 495, *Prod.* i. 90. U. Amherstiana, *A. DC. Mém.* 28; *Wall. Cat.* 6424 ! U. lævigata, *Wall. Cat.* 6428 ! U. biglandulosa, *Bl. Bijdr.* U. undulata, *Wall.! Plant. As. Rar.* iii. *t.* 265. U. Roxburghiana, *Wall. Cat.* 6423 B ! (non A). U. Lessertiana, *Dun. Anon.* 107. *t.* 26; *DC. Syst.* i. 492, *Prod.* i. 90. Uvaria cordifolia, *Roxb. Fl. Ind.* ii. 662 ? Desmos Chinensis, *Lour.!*

α. *pubiflora;* foliis late lineari-oblongis acuminatis sæpe 5–7-poll. basi cordatis, floribus sericeis. (U. discolor, *Auct.*)

β. *lævigata;* foliis oblongis vel lanceolatis plerumque 3–4-pollicaribus basi rotundatis, floribus glabrescentibus. (U. Chinensis, *Auct.* U. undulata, *Wall.*)

γ. *pubescens;* foliis subtus dense pubescentibus secus costam tomentosis.

δ. *latifolia;* foliis late ovalibus. (U. discolor, β bracteata, *Blume, Fl. Jav. Anon. t.* 26, 31 A *quoad folia.*)

HAB. Per totam Indiam australiorem et humidiorem in sylvis tropicis : Malaya ! Tenasserim ! Ava ! Chittagong ! Sikkim ! secus basin Himalayæ ; Concan ! Orissa ! Carnatica ; Zeylania !—(*v. v.*)

DISTRIB. Java, China austr.

Frutex vel arbor mediocris. *Rami* divaricati, nigro-fusci, tuberculis albis conspersi; partes novellæ pube aureo-fuscescente subsericeæ. *Folia* in sicco sæpe nigricantia, 2–8 poll. (plerumque 4–5) longa, 1–2½ lata, petiolo vix ¼-poll. *Pedunculi* graciles, 1–2 poll. longi, infra medium bracteolam oblongam vel lanceolatam 1–3 lineas longam (rarius foliaceam 1–2-pollicarem) deciduam gerentes. *Flores* solitarii, nutantes, sordide virentes, demum flavescentes vel ochroleuci. *Sepala* basi vix coalita, membranacea, punctis glandulosis conspersa, sericeo-pubescentia vel subglabra, ovato-lanceolata, acuta, semipollicaria. *Petala* demum 2-pollicaria et ultra, interiora angustiora et plerumque paullo breviora. *Stamina* oblonga, antherarum loculis inæqualibus, interioribus brevioribus, connectivo in processum ovalem producto. *Torus* depressus, medio aliquot excavatus. *Ovaria* 5–6-ovulata. *Pedunculus fructifer* sæpe incrassatus. *Torus* incrassatus, globosus. *Carpella* plurima, pedicello ⅛–1-pollicem longo suffulta. *Articuli* late ovales, 1–6, pisi magnitudine, utrinque obtusi, ultimus apiculatus, rugulosi, demum baccati, viridi-purpurascentes.

We have described at considerable length this very variable plant, in order to bring to notice as far as possible the various forms which it assumes. It has been well pointed out by Blume that the most different forms of leaves occur on the same tree, and often on the same specimen. The various degrees of pubescence of the flower seem somewhat more constant, the glabrous state being that common in China, while the pubescent-flowered form is that generally found in India. Wallich, however, does not hint that his *U. lævigata* is of Chinese origin. The very pubescent state γ, from the base of the Sikkim Himalaya, is only known to us in fruit, and the flowers may possibly prove it to be a distinct species. The variety δ is a very remarkable one, but the leaves are not always of that extreme width, but pass by insensible gradations into the ordinary state. In the specimen figured by Blume the

broad leaves occur with a much enlarged bractlet on the peduncle; but in a specimen from Griffith (from Malacca) this is not the case.

6. **U. Desmos** (Dunal, Anon. 112); arborea? foliis oblongis vel
lineari-oblongis basi rotundatis apice acutis vel acuminatis subcoriaceis
rigidis subtus pubescentibus, pedunculis extra-alaribus elongatis gracillimis nutantibus, petalis ovato-lanceolatis, carpellis 1–5-articulatis.
—U. Cochin-Chinensis, *DC. Syst.* i. 495, *Prod.* i. 91; *Alph. DC. Mém.*
28. U. pedunculosa, *Alph. DC. Mém.* 28 ; *Wall. Cat.* 6422 ! U. discolor, *Wall. Cat.* 6420 E ! F ! Desmos Cochinchinensis, *Lour.!*

Hab. Ava ! Tenasserim ! et Malaya !—(*v. s.*)

Ramuli rugulosi, atro-fusci, maculis pallidis notati, juniores cum omnibus partibus novellis pilis rigidis puberuli. *Folia* plerumque acuta, inferiora in ramulo
sæpe obtusa, supra nitida glabra, vel juniora sub lente sparse puberula præsertim
secus costam, subtus pilis adpressis in sicco fuscescentibus pubescentia, 6–8 poll.
longa, 1½–3 lata ; petioli ⅓-poll., pubescentes, demum glabrati. *Pedunculi* 4–8-pol-
licares, supra medium bracteola parva oblonga vel lineari cito decidua muniti, apice
subclavati. *Sepala* ovato-lanceolata, acuminata, fere ⅔-pollicaria. *Petala* adpresse
pubescentia, juniora dense aureo-sericea. *Torus* supra leviter excavatus. *Ovaria*
ovali-oblonga, 4–6-ovulata. *Carpella* toro parvo insidentia, pedicellis ½–1-pollicari-
bus suffulta ; articuli pisiformes, subglobosi, strigosi vel subglabri, ultimus apiculatus.

Unona fulva, Wall. Cat. 6427 ! which has no flowers or fruit, is probably referable either to the present species or to *U. Dasymaschala.*

Sect. 2. DASYMASCHALON.—*Petala* 3, uniserialia, interioribus
plane deficientibus. *Carpella* inter semina constricta.

7. **U. longiflora** (Roxb. Fl. Ind. ii. 668); fruticosa, foliis longe
petiolatis oblongis vel lineari-oblongis magnis membranaceis utrinque
glaberrimis supra nitidis subtus glaucis, pedunculis axillaribus unifloris,
petalis longissimis lineari-lanceolatis, carpellis 1–4-articulatis.—*Wall.
Cat.* 6419 !

Hab. Assam ! Khasia infra 3000 ped.! Silhet ! Chittagong !—(Fl.
Apr., Mai.; fr. Oct.) (*v. v.*)

Frutex ramosus. *Ramuli* elongati, grisei, rugulosi, juniores læves, pallidi vel (in
sicco) flavescentes, glaberrimi, partes novellæ glaberrimæ ; gemmæ axillares sub-
puberulæ. *Folia* acuta vel acuminata, inferiora in ramulo interdum obtusa, 6–15
poll. longa, 2–4 lata, petiolo incrassato cylindrico ½–⅔-poll., punctis pellucidis creberrimis notata. *Pedunculi* basin versus pluribracteati, bracteis linearibus minutis
fulvo-pubescentibus rarius flores abortivos in axilla foventibus, superne graciles, apice
subclavati, longitudine valde varii, nunc pollicares, plerumque elongati, 4–10-polli-
cares. *Alabastri* juniores fulvo-pubescentes, serius glabri, elongati, 4–6 pollices
longi. *Sepala* 3, rotundato-reniformia, mucronata, extus pubescentia, vix 2 lineas
longa. *Petala* 3, quorum 2 sæpissime in unum coalita, plano-convexa, crasse coriacea, apice contorta, lævia, glabra, basi ½–⅔ poll. lata, extus aurantiaca vel ex au-
rantiaco viridescentia, interne flavida. *Ovaria* 2–4-ovuláta. *Stigmata* pubescentia,
recurva. *Torus* apice truncatus. *Carpella* numerosa, toro dilatato cylindrico inserta,
pedicello ¼–1-pollicari suffulta, rugulosa, juniora pilis adpressis subpubescentia, demum glabra ; articuli 1–4, oblongi vel ovales, ⅓–⅔ poll. longi, terminalis apiculatus.

This magnificent species varies remarkably in the shape of the joints of the fruit,
which are sometimes oval, while at other times they are linear-oblong. We cannot,
however, find that these differences bear any definite relation to the length of the

flower-stalk, which is also a somewhat variable character, and we have no doubt that all the forms are referable to one species.

8. **U. Dasymaschala** (Bl. Fl. Jav. Anon. 55. t. 27); arborea? foliis brevissime petiolatis obovato-oblongis basi cordatis subtus glaucis, pedunculis axillaribus gracilibus, petalis lineari-lanceolatis, carpellis 1–7-articulatis.—*Alph. DC. Mém.* 28; *Wall. Cat.* 6421! U. discolor, *Wall. Cat.* 6420 B! U. Alphonsii, *Wall. Cat.* 6426!

    a. Blumei; ramulis glabris, foliis subtus glabris vel sub lente sparse puberulis.

    β. *Wallichii;* ramulis fulvo-tomentosis, foliis subtus dense pubescentibus siccis purpureo-glaucis.

HAB. Ava, Tenasserim, penins. Malay., *Wall.!*—(*v. s.*)
DISTRIB. Java.

Ramuli nigricantes, glabri, ad axillas foliorum et in omnibus partibus novellis pubescentes (in β dense fusco-tomentosi). *Folia* apice plerumque acuta vel acuminata, inferiora in ramulo sæpe oblonga, obtusa, 5–9 poll. longa, 2–4 lata, coriacea, opaca, supra glabra. *Petiolus* incrassatus, lineam longus. *Pedunculi* penduli, juxta basin bracteola 1 minima lineari munita. *Sepala* ¼-poll., late ovata, mucronata. *Petala* plana, coriacea, fere 3 uncias longa, juniora extus pubescentia, intus longitudinaliter carinata. *Torus* convexo-truncatus. *Ovaria* 6–7-ovulata. *Carpella* pedicello ¼-pollicari stipitata; articulis oblongis strigoso-pilosis, demum glabrescentibus, ultimo apiculato.

The young petals of this species are distinctly carinate on the inner surface, and therefore deviate a little from the ordinary structure of *Unona,* and approach somewhat to the outer series of the genus *Goniothalamus.* When fully developed, however, they are very like those of the last species; and the stamens, torus, ovary, and fruit are precisely those of *Unona.* We have not seen enough of specimens in good state to enable us to say with certainty that there is only one species; but the general habit of both varieties is so much alike, that we believe the differences will not be found of specific importance when the flowers of both are better known.

Sect. 3. PSEUDO-UNONA.—*Petala* 6. *Carpella* inter semina non constricta.

9. **U. pannosa** (Dalz. in Hook. Kew Misc. iii. 207); foliis ovato-lanceolatis obtuse acuminatis, floribus axillaribus subsessilibus, petalis oblongo-lanceolatis villosis, carpellis ovoideis laxe pilosis.—Uvaria mollis, *Wall. Cat.* 6475!

HAB. In montibus Concan, *Dalzell!* Malabar, *Wight!*—(Fl. Aug. Oct.) (*v. s.* in Herb. Linn. Soc. et in Herb. Wight.)

Arbor. Ramuli rugulosi, cortice griseo, juniores pilis fuscis patentibus dense pubescentes vel tomentosi. *Folia* pallida, 2½–4 poll. longa, ¾–1½ lata, petiolo vix 2 lineas longo, pellucido-punctata, coriacea, supra glaberrima, subtus parce pubescentia, ad costam nervosque velutina, demum glabrata; nervi obliqui, remoti. *Sepala* extus villosa, ovata, acuta, 3 lin. longa. *Petala* 1–2-pollicaria, basi unguiculata, subæquilonga; interiora paullo angustiora. *Stamina* brevia, cuneata; processus connectivi capitatus, subtruncatus. *Torus* elevatus, convexus, dense aureo-strigosus. *Ovaria* 8–12, dense et longe aureo-strigosa, in stylum brevem angustata. *Stigma* depresso-capitatum, pilosulum. *Ovula* in axi 2–3. *Torus* fructus tomentosus. *Carpella* subsena vel abortu pauciora, ovalia, utrinque obtusa, brevissime pedicellata, ⅔ poll. longa. *Semina* 1–3, magna; testa nitida, lævi.

We have not seen a specimen from Dalzell, but, from the description, we have no doubt that his plant is the same as that of Wallich, whose specimens are partly communicated by Wight (in fruit), and partly by Heyne (in flower), but in both cases without special locality.

In Dr. Wight's Herbarium there is a specimen in fruit of a species nearly allied to *U. pannosa*, Dalz., and bearing at the same time a very close resemblance in foliage and general habit to *Unona virgata*, Blume, Bijdr. (*Uvaria virgata*, Bl. Fl. Jav. Anon. t. 19 et 25 B.) The fruit of Blume's plant is however very different. Dr. Wight's specimen, which was gathered at Quilon, in Malabar, in October, 1835, has oblique oblongo-lanceolate leaves, acute at the base, and long acuminate, 5–8 inches long and 2–2½ broad, thin and membranous, with oblique distant nerves, prominent below, and united into loops a long way within the margin. The petioles are scarcely ¼ inch in length, thickened and cylindric, and the leaves are glabrous above and very slightly downy on the midrib below. The specimen bears one fruit, supported on a pedicel little more than a line in length in the axil of the lowest leaf. Two carpels remain, and the scars of two more are visible; they are oblong and obtusely mucronate, ⅝ of an inch in length, with a pedicel a line long, two-seeded, and slightly constricted in the middle between the seeds, yellowish-brown, slightly pubescent and granular.

Species ob fructum ignotum dubiæ.

10. **U. præcox** (H.f. et T.); foliis lanceolatis acuminatis, floribus in axillis foliorum delapsorum cum ramulo solitariis longiuscule pedicellatis nutantibus, petalis linearibus elongatis glabriusculis.

Hab. Assam, *Simons!*—Fl. Febr. (*v. s.*)

Arbor forsan humilis. *Ramuli* rugulosi, cortice griseo, in specimine suppetente florido foliis adultis orbati, floribus una cum foliis novellis e gemmis axillaribus evolutis. *Folia* (novella) tenuia, incano-puberula, 2–3 poll. longa, ¾–1¼ lata, petiolo vix ¼-poll. *Pedunculi* graciles, pollicares, cum ramulo tuberculo axillari piloso inserti. *Sepala* lineari-oblonga, acuta, membranacea, ⅜ poll. longa. *Petala* 2½–3-pollicaria, tenuissima. *Stamina* truncato-capitata. *Ovaria* glabra; stylo oblongo, piloso. *Ovula* in sutura ventrali 2–3.

A very singular species, of which we have before us several specimens collected by Mr. Simons, all in good flower, and covered with young shoots and scarcely-expanded leaves. It seems to be a soft-wooded plant, and the branches have the appearance of being jointed, from the peculiar development of the young shoots. These are much smaller than the *pulvinar*, or tubercle, from which they spring, and the flower-stalk is inserted into it exterior to the branchlet, both being immersed in short rigid hairs. As there are no adult leaves on our specimens, it is doubtless a deciduous-leaved plant.

11. **U. stenopetala** (H.f. et T.); foliis brevissime petiolatis obovato-lanceolatis vel lineari-oblongis basin versus angustatis basi obtusis et oblique emarginatis, floribus secus ramos crassiores dense fasciculatis, petalis angustissime linearibus elongatis.

Hab. In prov. Tenasserim ad Moulmein, *Lobb!*—(*v. s.*)

Distrib. Java, *Lobb!*

Arbor? *Ramuli* graciles, juniores fusco-pubescentes. *Folia* acuminata, 4–6 poll. longa, 1¼–1¾ lata, petiolo vix lineam longo, tenuiter coriacea, supra glabra, subtus pallida, secus costam pubescentia. *Pedunculi* in massam lignosam varie ramosam coaliti, squamulis vestiti; pedicelli pubescentes, ¼–½ poll. longi, basi bracteolas 1–2 lanceolatas gerentes. *Sepala* basi subconnata, longe angustata, ⅛ poll. longa, ciliata, extus pubescentia. *Petala* 2–3-pollicaria, vix lineam lata, extus sericeo-puberula; *exteriora* basi parum dilatata, connata, obtuse carinata; *interiora* paullo angustiora, basi remota. *Ovaria* 4–7, villosa, oblonga. *Ovula* 5, horizontalia.

In the absence of fruit, this species and the next cannot be better placed than in the genus *Unona*. They are probably congeners of *Unona pannosa*, Dalzell, and of *Uvaria virgata*, Blume.

12. **U. cauliflora** (H.f. et T.); foliis lineari-oblongis vel lineari-lanceolatis brevissime petiolatis basi rotundatis obtusis, floribus secus ramos fasciculatis, sepalis dense pilosis, petalis linearibus sericeis.

HAB. In peninsula Malayana ad Singapur, *Lobb!*—(*v. s.*)

Arbor?　Ramuli dense fusco-tomentosi. *Folia* 5–6 poll. longa, 1¼–1¾ lata, petiolo vix lineam longo, incrassato, tomentoso, tenuiter coriacea, supra lucida, subtus (secus costam densius) fusco-pubescentia, nervis falcatis. *Flores* in fasciculo pauciores quam in præcedente. *Pedicelli* pedunculo abbreviato lignoso ramoso inserti, ¼ poll. longi, tomentosi. *Sepala* ovato-lanceolata, ½-pollicaria. *Petala* 2½ poll. longa, ¼ poll. lata, (ex sicco) læte rubra. *Stamina* indefinita, breviter cuneata, truncata, antherarum loculis discretis. *Ovaria* ovalia, dense albo-strigosa. *Ovula* in sutura ventrali 3–4.

17. **POLYALTHIA,** Blume.

Polyalthia, § 1, *Blume, Fl. Jav. Anon.* 70.

Sepala 3. *Petala* 6, biserialia, ovata vel elongata, coriacea, plano-convexa, æst. valvata. *Stamina* indefinita, connectivo truncato capitato; antheris lineari-oblongis dorsalibus. *Torus* apice truncatus, planus. *Ovaria* indefinita, oblonga, in sutura ventrali prope basin biovulata. *Stylus* oblongus. *Carpella Guatteriæ.*—Arbores, foliis *coriaceis, nervis obliquis distantibus apice arcuatim connexis inconspicuis*, floribus *axillaribus vel extra-alaribus*.

The genus *Polyalthia*, as originally instituted by Blume, included four very distinct groups, all of which were clearly distinguished by that author as sections. Blume made the character of the genus to rest mainly on the two-ovuled ovary. This indeed is the principal point of resemblance between the different groups which he brought together under this genus, while they possess, it appears to us, too many and important points of distinction to permit of their being associated together. We have therefore considered each of Blume's sections as a distinct genus, except *Kentia*, which is so closely allied to our genus *Melodorum* that it does not appear necessary to retain it as a genus. As Blume has foreseen the probability of this being done, and has given to each section a name, indicating at the same time to which he desired the generic name to be attached, we have of course made no alteration in that respect. The true *Polyalthia*, in the restricted sense, as characterized above, is much more closely allied to *Guatteria* than to any other genus, the flowers being in no respect different, except by the increased number and different position of the ovules. The species have, however, a peculiar habit, not like that of the majority of *Guatteriæ*, the very short-petioled leaves giving them a peculiar facies. *Guatteria suberosa*, however, approaches the genus *Polyalthia* in this respect very closely, and forms a direct transition from the one genus to the other. We learn from Blume that species of *Polyalthia* are numerous in the Malayan Archipelago, while within our limits they are entirely confined to the Malayan peninsula. Besides Blume's species and those described below, we have before us several species from the Philippines, collected by Cuming. We have also seen an imperfect specimen from Ceylon, in Dr. Wight's Herbarium, which resembles *P. obliqua*, but has oblong-lanceolate leaves more membranous than those of that species, and long-pedicelled flowers opposite the leaves, not axillary as in *P. obliqua*: it is probably a very distinct species.

T

1. P. cinnamomea (H.f. et T.); foliis lanceolatis basi angustatis cordatis brevissime petiolatis, pedunculis extra-alaribus unifloris, petalis oblongis acutis extus sericeis, carpellis globosis breviter pedicellatis fusco-tomentosis.—Guatteria cinnamomea, *Wall. Cat.* 6444 ! G. multinervis, *Wall. Cat.* 6445 !

HAB. In peninsula Malayana ad Penang et Singapur !—(*v. s.*)

Arbor, ramis cinereis rugosis. *Ramuli* fulvo-tomentosi. *Folia* 6–10 poll. longa, 2–3½ lata, petiolo 2–3 lineas longo, pubescente, incrassato, tenuiter coriacea, superne glabra, nitida, secus costam puberula, subtus pubescentia, secus costam nervosque tomentosa. *Pedunculi* solitarii vel bini, fulvo-tomentosi, ½-pollicares, basi bracteolis 2–3-linearibus parvis muniti. *Sepala* rotundata, acuta, pubescentia. *Petala* æqui longa, ⅔-pollicaria; exteriora paullo latiora, crasse coriacea. *Ovaria* dense strigosa. *Torus fructus* incrassatus. *Carpella* fere pollicaria, pedicello ⅓ poll. longo, tomento denso longo intertexto fusco vestita. *Semen* erectum, conforme; testa pericarpio coadunata, endospermio nitido tenui.

2. P. obliqua (H.f. et T.); foliis subsessilibus lineari-oblongis obtuse acuminatis basi parum angustatis oblique cordatis, pedunculis axillaribus unifloris, petalis oblongis obtusis extus sericeis, carpellis globosis pedicellatis.

HAB. In peninsula Malayana ad Malacca, *Griffith* !—(*v. s.*)
DISTRIB. Borneo, *Low* !

Arbor. Ramuli viminei, foliosi, glabri, cortice lævigato nitido fuscescente betulino, pustulis albidis minutis crebris tuberculato; partes novellæ adpresse pubescentes. *Folia* 4–6 poll. longa, 1–1¾ lata, petiolo vix lineam longo, rigide coriacea, glaberrima, utrinque lucida, subtus pallidiora. *Pedunculi* ½–1 poll. longi. *Sepala* rotundata, pubescentia. *Petala* ¼ poll. longa, crasse coriacea, æqualia. *Torus* truncatus. *Ovaria* strigosa, subdefinita. *Carpella* atro-fusca, glabra, granulosa, pisi majoris magnitudine, pedicello ½ poll. longo.

3. P. cauliflora (H.f. et T.); foliis breviter petiolatis lanceolatis basi vix acutis apice obtuse acuminatis, pedunculis axillaribus fasciculatis elongatis, petalis linearibus extus strigoso-villosis.—Uvaria cauliflora, *Wall. Cat.* 6476 !

HAB. In peninsula Malayana ad Singapur, *Wall.*!—(*v. s. in Herb. Linn. Soc.*)

Arbor ramosissima. *Ramuli* graciles, virgati, glabri, cortice cinereo nitido vix ruguloso, lenticellis minutis sparsis tuberculato; partes novellæ tomentosæ. *Folia* 3–6 poll. longa, 1–1¾ lata, petiolo vix 2 lineas longo, coriacea, firma, supra glabra, nitida, subtus secus costam nervosque pubescentia, reticulato-nervosa. *Pedicelli* in axillis foliorum delapsorum subterni, filiformes, apice subclavati, 1½–2 poll. longi, pubescentes, ima basi bracteolis paucis squamæformibus muniti. *Sepala* ovata, acutiuscula, extus adpresse pilosa. *Petala* pollicaria, æquilonga; exteriora paullo latiora. *Torus* cylindricus, truncatus. *Ovaria* strigoso-pilosa. *Fructus* ignotus.

18. **GUATTERIA,** Ruiz et Pavon.

Sepala 3, rotundata vel ovata, parva. *Petala* 6, æstivatione biseriatim valvata, plana, ovata oblonga vel linearia. *Stamina* indefinita, late cuneata; connectivo truncato, capitato; antherarum loculis dorsalibus remotis. *Ovaria* numerosa, oblonga, ovulo 1 e basi erecto. *Stylus* oblongus, basi intus sulcatus. *Torus* parum elevatus, plano-convexus, interdum medio excavatus. *Carpella* sicca, pericarpio tenui sæpe fragili. *Semen* erectum.—Arbores *sæpe proceræ, vel* frutices (*interdum scandentes ?*) *habitu variæ*, foliis *oblique nervosis*, inflorescentia *axillari vel oppositifolia.*

We retain the genus *Guatteria* nearly as left by Blume and Martius. It is still very extensive, and perhaps not quite natural, though we have not been able to find any good characters for subdividing it. The greater number of the species are American.

1. **G. longifolia** (Wall. Cat. 6442 !) ; foliis e basi lata longissime angustatis, floribus versus apicem pedunculi axillaris racemosis, petalis elongatis, carpellis ovoideis.—*W. et A.! Prod.* i. 10 ; *Wight, Ic. t.* 1. Uvaria longifolia, *Lam.; Roxb.! Fl. Ind.* ii. 664. Unona longifolia, *Dun. Anon.* 109 ; *DC. Syst.* i. 492, *Prod.* i. 90.

HAB. In sylvis Zeylaniæ ! et Tanjor, *Wight!* per totam Indiam tropicam frequentissime culta.—(*v. v. cult.*)

Arbor procera, elegans, ramis adscendentibus. *Ramuli* graciles, glabri; gemmæ pubescentes. *Folia* basi plerumque in petiolum angustata, rarius rotundata, submembranacea, margine undulata, 5–8 poll. longa vel interdum fere pedalia, 1–2 rarius 3 poll. lata, pet. ¼–½-poll.; utrinque glaberrima, lucida, pellucido-punctata, nervis obliquis parallelis, venulis crebre reticulatis. *Pedunculi* ad axillas foliorum delapsorum, breves, ½ poll. longi vel plerumque multo breviores, interdum vix ulli, solitarii vel bini (et tunc 1 sessilis), cinereo-incani, basi bracteolis minutis squamæformibus muniti. *Pedicelli* plurimi, secus pedunculi apicem dense racemosi, bracteis parvis suffulti, elongati, graciles, 1–2-pollicares. *Sepala* extus puberula. *Petala* ¾-poll., e basi lata subulata, parallele nervosa. *Torus fructus* pubescens. *Carpella* plerumque pauca, ¾ poll. longa, pedicello ½-poll., glabra, fusca, vix granulata.

This well known and very ornamental tree is commonly planted along roads in Bengal and throughout the southern parts of India, but scarcely at all beyond the tropics, the winters of the northern parts of Hindostan being probably too cold for it. Roxburgh did not know its native country, and we learn from Blume that it is not a native of Java. It appears, however, to be really indigenous in Ceylon, and in the southern part of the Madras Peninsula.

2. **G. bifaria** (Alph. DC. Mém. 41) ; foliis ellipticis acuminatis basi acutis punctatis superne glabris subtus pubescentibus, pedunculis axillaribus nudis 1-floris, petalis lineari-lanceolatis, carpellis longe stipitatis ovoideis.—*Wall. Cat.* 6447 !

HAB. In Ava circa Prome, *Wall.!*—(*v. s. sine flore.*)

This species is said to differ from the next only by the linear petals. The specimens in the Linnean Society's herbarium, which are the only ones we have seen, are unfortunately not in flower. We cannot, however, distinguish them in any way from *G. cerasoides ;* our diagnosis is, therefore, taken verbatim from Alph. De Candolle.

3. **G. cerasoides** (Dunal, Anon. 28) ; foliis lanceolatis vel oblongo-lanceolatis subtus pubescentibus, pedicellis 1–3 ad apicem pedunculi axillaris tuberculiformis, petalis ovato-oblongis, carpellis ovoideis apiculatis longe pedicellatis.—*DC. Syst.* i. 503, *Prod.* i. 93 ; *Wall. Cat.* 6436 ! *W. et A.! Prod.* i. 10. Uvaria cerasoides, *Roxb.! Cor.* i. t. 33, *Fl. Ind.* ii. 666.

HAB. In montibus tropicis et subtropicis Bahar, *Ham.!* Orissa,

Roxb.! Dekhan! Maisor! Courtalam, *Wight!* et forsan totius Carnaticæ.—(*v. s.*)

Arbor. Ramuli cinerei, rugosi, glabri, juniores cum omnibus partibus novellis laxe tomentosi. *Folia* valde approximata, basi rotundata vel parum angustata, 3–8 poll. longa, 1–2 lata, pet. vix 2-lin., tenuia, submembranacea, supra glabra, nitida, secus costam pubescentia, subtus laxe pubescentia, pube secus costam et nervos petiolumque densiore. *Pedicelli* axillares, ½–1 poll. longi, graciles, solitarii vel 2–3 e tuberculis axillaribus lignosis interdum in pedunculum vix lineam longum elongatis orientes, pubescentes, basi squamis aliquot suffulti et hinc inde bracteis 2–3 foliaceis usque ad ½ poll. longis ovatis interdum plane deficientibus muniti. *Sepala* pubescentia, ⅓ poll. longa. *Petala* vix longiora, crasse carnosa, adpresse puberula. *Ovaria* fusco-strigosa. *Torus fructus* parum dilatatus, pubescens. *Carpella* numerosa, sparse puberula, nigra, ¼ poll. longa, pedicello duplo longiore.

That portion of Wall. Cat. 6436 D, which is marked as having been collected by Hamilton at Goalpara, in Eastern Bengal, close to Assam, is *Hyalostemma*. We have therefore omitted that locality, as the tree does not appear to occur on the east of the Ganges, but to be confined to the drier regions of Behar, and the eastern part of the Madras Peninsula. It remains, however, to be determined whether or not the species be not also a native of similar climates in the drier parts of Ava, if, as is probable, the last species consist of the leaves of this species with the flowers of some other (probably *G. Simiarum*).

4. **G. suberosa** (Dun. Anon. 128); foliis brevissime petiolatis oblongis subtus puberulis, pedicellis plerumque solitariis infra medium unibracteatis, petalis ovalibus, carpellis globosis breviter pedicellatis. —*DC. Syst.* i. 504, *Prod.* i. 93; *Wall. Cat.* 6437!; *W. et A. Prod.* i. 10. Uvaria suberosa, *Roxb. Cor.* i. *t.* 34, *Fl. Ind.* ii. 667.

HAB. In Zeylania! Carnatica! Orissa! Bahar! Bengal! Assam! Tenasserim!—(Fl. Apr. Mai.) (*v. v.*)

DISTRIB. Ins. Philipp. (*Cuming*, 1051! 1191!); an vere indigena? (In Java non indicatur a Blume.)

Frutex vel arbor parva. *Rami* fusco-cinerei, rugosi, pallide tuberculati, cortice sæpe incrassato suberoso, adulti glabri, juniores laxe pubescentes; gemmæ sericeæ. *Folia* utrinque obtusa vel rarius basi parum angustata et apice acuta, interdum basi obliqua, tenuia, fere membranacea, margine undulata, supra glabra, subtus plus minus pubescentia, demum glabrata, 2½–5 poll. longa, 1–1¾ lata, pet. 1–2 lineas longo. *Pedunculi* axillares, brevissimi. *Pedicelli* plerumque solitarii (rarius 2), ½–1-pollicares, graciles, apice subclavati, pubescentes, infra medium bracteolam lineari-subulatam gerentes. *Sepala* pubescentia, in fructu subpersistentia, parva. *Petala* ovalia, obtusa, pubescentia; interiora ⅓ poll. longa, exterioribus fere duplo longiora. *Ovaria* fusco-strigosa. *Torus fructus* parvus, globosus, tomentosus. *Carpella* numerosa, globosa, mucronulata, intus sulcata, subsericea, demum glabrescentia, pisum parvum magnit. æquantia, pedicello æquilongo.

5. **G. persicæfolia** (H.f. et T.); foliis lanceolatis acuminatis basi angustatis obliquis subtus sparse puberulis, pedicellis 2–3 fasciculatis pedunculo supra-axillari brevissimo suffultis, petalis fere rotundatis exterioribus minoribus, carpellis globosis.

HAB. In Zeylania ad Narawelle, *Champion!*—(Fl. Apr.) (*v. s.*)

Frutex. Ramuli grisei, rugulosi, dense foliosi, glabri, juniores fusco-pubescentes. *Folia* plerumque in acumen longum gracile attenuata, 2–4 poll. longa, ⅔–1¼ lata, petiolo vix lineam longo, coriacea, subtus pallida, nervis inconspicuis intra marginem arcuatis. *Pedunculus* vix lineam longus. *Pedicelli* ¼–½ poll. longi, fusco-pubescentes. *Flores* parvi. *Sepala* ovata, acuta, strigoso-tomentosa. *Petala* crasse coriacea, strigoso-pubescentia; *exteriora* sepalis duplo majora, rotundata, abrupte acuminata; *interiora* ext. fere duplo majora, ¼ poll. longa, rotundata, acutiuscula. *Ovaria* dense strigosa. *Torus fructus* parvus. *Carpella* 10 vel plura, pedicello vix lineam longo suffulta, pisi magnitudine, granulata, vix puberula.

6. **G. Corinti** (Dun. Anon. 134); foliis ovatis vel oblongis rarius lanceolatis lucidis coriaceis utrinque præter costam puberulam glaberrimis, pedicellis axillaribus solitariis, petalis ovali-oblongis, carpellis ovoideis granulosis strigoso-pubescentibus.—*DC. Syst.* i. 507, *Prod.* i. 94; *W. et A. Prod.* i. 10; *Wight, Ill. t.* 398. G. acutiflora, *Wall. Cat.* 6438 ! (*excl. D*).

HAB. In Zeylania ! Malabar ! Tanjor ! et Courtalam !—(*v. s.*)

Frutex (scandens, ex Wight) ramosus, foliosus. *Ramuli* cinerei vel nigricantes, cortice rugoso glabro, juniores puberuli; gemmæ strigoso-tomentosæ. *Folia* acuminata, basi acuta vel rotundata, 2–4 poll. longa, 1–2 lata, pet. 2 lin. longo, pellucido-punctata, supra atro-viridia, nervis obliquis, venulis conspicue reticulatis, subtus pallidiora; petiolo et costa utrinque pilis adpressis sparsis puberulis, demum glabratis. *Pedicelli* filiformes, 1–2-pollicares, strigoso-pubescentes, basi et infra medium bracteola minuta squamæformi rarius foliosa muniti. *Sepala* rotundata, acutiuscula, extus tomentosa. *Petala* pubescentia; interiora paullo majora, ½–⅔ poll. longa. *Ovaria* strigoso-pilosa. *Torus fructus* non dilatatus. *Carpella* 5–15, fere ¼ poll. longa, pedicello æquilongo strigoso.

Apparently a very variable plant in form of leaf; but all the specimens before us certainly belong to one species. Wight figures the fruit as globose, but in most of the specimens before us it is a little ovoid. *G. sempervirens*, Dunal, and *G. acutiflora*, Dunal, founded entirely on Rheede's figures (Hort. Mal. v. t. 16, 18), appear to us to differ in no character of importance from t. 14 of the same volume, which is considered to represent *G. Corinti*, Dun. The shape of the petals varies from acute to obtuse, and the flowers vary a good deal in size. *G. sempervirens* is said by Rheede to be common in Malabar, and *G. acutiflora* to grow in mountainous places; and it is not likely that no trace of these species (if distinct) should be found in Dr. Wight's extensive collections. There can at least be no doubt that *G. acutiflora*, which is the only one of the three figured in flower, is only the flowering state of one of the other two.

7. **G. Jenkinsii** (H.f. et T.); foliis oblongo-lanceolatis utrinque glaberrimis, pedicellis axillaribus plerumque solitariis, petalis (magnis) ovato-lanceolatis, carpellis oblongis pedicellatis.

HAB. Assam, *Jenkins!* Silhet! Malacca, *Griff.!*—(*v. v.*)

Arbor (forsan scandens) ramosissima. *Ramuli* stricti, graciles, glabri, cortice nigricante vel griseo ruguloso, juniores fusco-pubescentes. *Folia* acuta vel acuminata, basi rotundata vel acutiuscula, 4–7 poll. longa, 1½– fere 3 lata, petiolo ½-poll., tenuiter coriacea, lucida, subtus pallida, nervis obliquis remotis parallelis, venulis conspicue reticulatis. *Pedicelli* (rarius bini) ¾–1 poll. longi, stricti, fulvo-tomentosi, basi bracteolis 2–3 minutis squamæformibus stipati, medio et prope apicem bracteolam rotundatam vel oblongam deciduam gerentes. *Sepala* ¼ poll. longa, ovalia, extus tomentosa. *Petala* basi angustata, plana, tenuiter coriacea; ext. 1⅔, int. 1¼ poll. longa (⅓ poll. lata); juniora cinereo-incana, demum puberula tantum. *Ovaria* strigoso-pilosa. *Torus fructus* globosus, tomentosus. *Carpella* numerosa, oblonga, apiculata, ⅔-pollicaria, pedicello æquilongo, glabra, minute granulosa.

8. **G. coffeoides** (Thwaites, mss.); foliis lanceolatis vel oblongo-lanceolatis utrinque glaberrimis, pedicellis axillaribus solitariis, petalis

lanceolatis, carpellis ovoideis obtuse rostratis in pedicellum æquilongum
attenuatis.

HAB. In Zeylania, *Thwaites!* (No. 2503.)—(*v. s.*)

Arbor, cortice cinereo rugoso glabro; partes novellæ cinereo-puberulæ. *Folia*
basi acuta, vel rarius rotundata, longe attenuata, tenuia, pellucide punctata, margine
undulata, oblique nervosa, 4–7 poll. longa, 1–2½ lata, petiolo ¼-pollicari, *Pedicelli*
axillares vel ad axillas foliorum delapsorum, pollicares, adpresse puberuli, basi articulati, bracteis 2–3 squamæformibus deciduis muniti. *Sepala* fere rotundata. *Petala* coriacea, e basi lata lanceolata obtusiuscula pollicaria, interiora parum longiora.
Torus fructus globosus, tomentosus. *Carpella* granulosa, cinereo-puberula, ½ poll.
longa.

There is only one expanded flower on the specimen before us: this we have not
examined, but, as a memorandum on the accompanying ticket informs us that the
ovules are solitary, we refer this plant without hesitation to *Guatteria*.

9. **G. fragrans** (Dalzell in Hook. Kew Misc. iii. 206); foliis oblongo-lanceolatis ovalibus vel ovatis interdum obovatis valide costatis,
pedunculis axillaribus decompòsitis multifloris, petalis anguste linearibus, carpellis magnis ovoideis cinereo-incanis longe pedicellatis.

HAB. Concan, *Dalzell! Law!* Malabar, *Wight!*—(*v. s.*)

Arbor. *Ramuli* cortice griseo rugoso, lenticellis albidis sæpe notati, glabri; gemmæ subtomentosæ. *Folia* sæpe obliqua, basi rotundata, apice plerumque obtusa,
mucronata, sed interdum acuminata, 4–9 poll. longa, 2–5 lata, petiolis ½–⅔-pollicari-
bus, membranacea, utrinque glabra, nitida, juniora secus costam nervosque puberula,
costa subtus sæpe tuberculata. *Pedunculi* ad axillas foliorum delapsorum, secus
ramulos crassiores siti, abbreviati, validi, vix pollicares. *Pedicelli* filiformes, pollicares, incano-puberuli, infra medium bracteola caduca rotundata muniti. *Sepala*
minuta, rotundata, extus pubescentia, apice recurva. *Petala* 1½ poll. longa, 2 lin.
lata, longe attenuata, subæqualia, incano-puberula. *Pedicelli fructus* incrassati,
lignosi. *Torus* dilatatus, depresso-globosus, diam. ½-poll., fulvo-tomentosus. *Carpella* 10–20, obliqua, dorso gibbosa, 1–1½ poll. longa, pedicello æquilongo, apice obtusa et obtuse mucronata. *Semen* conforme. *Testa* cum exocarpio fragili arcte
coalita. *Endospermium* albidum, nitidum, papyraceum, transverse fibrosum.

10. **G. Simiarum** (Ham. ex Wall. Cat. 6440!); foliis ovalibus
vel ovatis valide costatis, pedicellis axillaribus 2–3 fasciculatis nudis,
petalis lineari-oblongis obtusis interioribus longioribus.

HAB. In Silhet, *Wall.!* Assam inf., *Ham.!*—(*v. s. ex Hort. Calc.*)

Arbor, cortice cinereo glabro, gemmis pubescentibus. *Folia G. fragrantis*, sed
minora, utrinque glaberrima. *Pedicelli* graciles, pollicares, incano-puberuli. *Petala*
glabra, interiora pollicaria, exteriora ⅓ breviora.

G. lateriflora, Blume, is evidently closely allied to this species as well as to the
preceding. All three seem to vary much in the size and shape of the leaves, but to
be readily distinguishable by the inflorescence.

11. **G. membranacea** (Alph. DC. Mém. 41); foliis oblongo-
lanceolatis acuminatis nervo centrali superne velutino subtus piloso,
pedicellis axillaribus brevibus bractea ovata acuta stipatis, carpellis
brevistipitatis paucis ovoideis velutinis.

HAB. In prov. Tenasserim et Tavoy, *Wall. ex A. DC. l. c.*

We cannot identify this description with anything in Wall. Cat. from Tavoy, so
that probably the only specimens of this species known are at Geneva.

12. **G. nitida** (Alph. DC. Mém. 41); foliis (magnis) oblongis vel lineari-oblongis obtusis vel acutis utrinque glabris, pedicellis axillaribus solitariis medio bracteolatis, petalis ovalibus adpresse tomentosis, carpellis magnis subglobosis pedicellatis.—*Wall. Cat.* 6439!

HAB. In prov. Tenasserim ad Tavoy, *Wall.!* in penins. Malayana ad insulam Singapur, *Lobb.*—(*v. s.*)

Arbor. Ramuli nigricantes, rugulosi; partes novellæ pubescentes. *Folia* 8–12 poll. longa, 3–5 lata, petiolo ½-poll. incrassato, supra nitida, subtus pallida, nervis conspicuis parallelis obliquis rectiusculis. *Pedicellus* pollicaris, prope basin articulatus et verosimiliter bractea munitus (e cicatrice in specimine conspicua), medio bracteola ovali semi-amplexicauli munitus. *Sepala* ad medium coalita, rotundata, obtusissime pubescentia, subciliata, ½ poll. longa. *Petala* crassa, intus ferrugineo-velutina, fere pollicaria, interiora paullo minora. *Ovaria* lineari-tetragona.

13. **G. biglandulosa** (Blume, Fl. Jav. Anon. 102. t. 51); foliis oblongis plerumque acuminatis supra glabris subtus glaucis tenuissime adpresse sericeis, pedunculis unifloris extra-alaribus, carpellis oblongis pedicellatis.

HAB. In Malaya ad Malacca, *Griffith!*—(*v. s.*)

DISTRIB. Java, *Spanoghe ex Blume.*

Caulis (ex Spanoghe in Bl. *l. c.*) fruticosus. *Ramuli* nigricantes, rugulosi, juniores pube sericea fulva vestiti, demum glabrescentes. *Folia* 4–6 poll. longa, 1½–3 lata, petiolo incrassato sericeo vix semipollicari, basi rotundata vel acutiuscula, marginibus prope petiolum glanduloso-incrassatis, recurvis; nervi validi, leviter incurvi; venatio *Oxymitræ. Flores* ignoti. *Pedunculi fructigeri* pollicares. *Carpella* oblonga, apiculata, atro-fusca, granulosa, sparse puberula, demum glabrata.

Species dubiæ.

14. **G. costata** (H.f. et T.); foliis oblongis vel elliptico-lanceolatis acutis vel acuminatis subtus præsertim secus costam sparse puberulis, pedicellis extra-alaribus solitariis abbreviatis.—Uvaria costata, *Wall. Cat.* 6480!

HAB. In Ava ad flumen Attran, *Wall.!*—(*v. s. sp. imperfect.*)

Arbor? Ramuli cortice nigricante vel cinereo ruguloso, puberuli, demum glabrati, juniores fulvo-tomentosi. *Folia* rigida, coriacea, 5–6-pollicaria, 1½–2½ lata, petiolo ½-poll., basi acuta, supra saturate-viridia, subtus argentea, nervis obliquis rectiusculis parallelis prominentibus, venulis transversis subconspicuis (ut in *Oxymitra*). *Pedicelli* ¼-pollicares. *Sepala* in fructu persistentia, ovata, basi subcoalita. *Carpella* non adsunt.

Though the specimens of this plant in the Linnean Society's Herbarium exhibit neither fruit nor flowers, we have thought it right to give a description of the species, which is not nearly related to any other with which we are acquainted within our limits, except the following, to which it is apparently very closely allied, though quite distinct. It is also, to all appearance, very near *G. macrophylla*, Blume (Fl. Javæ), from which it differs chiefly by its smaller leaves, which are more silvery and less hairy beneath. Its evident close relationship to that species induces us to transfer it from *Uvaria* to *Guatteria*, a step which we should otherwise have hesitated to take till better materials were available.

15. **G.? pallida** (H.f. et T.); foliis oblongo-lanceolatis acuminatis basi rotundatis, floribus monoicis secus ramos subsessilibus fasciculatis minutis, fasciculis oppositifoliis, carpellis oblongis breviter pedicellatis.

HAB. In provincia Silhet secus basin montium Khasia!—(*v. v.*)

Frutex divaricato-ramosus. *Ramuli* rugulosi, nigricantes, glabri; partes novellæ vix puberulæ. *Folia* 6–8 poll. longa, 1¾–2½ lata, petiolo ¼-poll. incrassato granuloso, tenuiter coriacea, pallide viridia, utrinque glaberrima, subtus micantia et fere argentea, nervis arcuatis remotis, venulis reticulatis inconspicuis *Flores* supra tubercula lignosa fasciculati; bracteolis squamæformibus. *Alabastri* minuti, globosi, vix ½ lineam diametro. *Bractea* 1 rotundata calyci adpressa. *Sepala* æstivatione imbricata, dorso strigoso-pilosa, rotundata, ciliata. *Petala* 6. MAS.:—*Stamina* supra columnam centralem conicam sessilia, indefinita, apice truncato-capitata. FŒM.:— *Ovaria* numerosa, toro cylindrico inserta, basi staminibus effœtis suffulta, oblonga, dense sericea, stigmate parvo subsessili. *Ovula* 2? e basi erecta. *Carpella* numerosa, ¾-pollicaria, pedicello ⅓-poll. puberulo suffulta, granulosa, glabra.

In this very curious plant we have a combination of characters not very usual in the Order. We have unfortunately only been able to examine a few very young buds, and have no more at our command, so that no confidence is to be placed in the details of the flower, and in particular in the shape of the petals, which often vary much after expansion. In general habit, as well as in the unisexual flowers, our species so closely resembles *Uvaria reticulata*, Blume, Fl. Javæ, t. 24, of which the male plant only is known, that we have little doubt the two are congeners, so that the inner petals of our plant will probably prove to be clawed and valvate. That species, however, as we have elsewhere mentioned, forms the type of a new genus: but it appears desirable not to give a generic name till it can be accompanied by a definition.

19. **ANAXAGOREA,** St. Hilaire.

Sepala 3, basi coalita. *Petala* 6, ovalia vel oblonga, subæqualia, æst. biseriatim valvata. *Stamina* indefinita. *Antheræ* lineares, extrorsæ vel sublaterales, connectivo apiculatæ, interiores interdum deformatæ, filamentis filiformibus elongatis suffultæ. *Torus* vix convexus. *Ovaria* subdefinita, basi solida, ovalia vel oblonga. *Ovula* 2, collateralia, e basi loculi erecta. *Stylus* subglobosus vel oblongus. *Carpella* pedicello elongato clavato suffulta, valvis 2 dehiscentia. *Semina* 2, erecta, nitida, exarillata.—*Arbores*, floribus *inconspicuis e viridi albicantibus*.

This is a very curious genus, deviating remarkably from the ordinary type of *Anonaceæ*. Several South American species have been described by St. Hilaire and Martius. To these Blume has added one from Java, which is evidently a congener, though it differs in some trifling particulars from the others. We have now the pleasure of indicating a second Asiatic species, which presents too many deviations from Blume's description to be identifiable with *A. Javanica*.

1. A. Zeylanica (H.f. et T.); foliis oblongis vel lineari-oblongis abrupte et obtuse acuminatis basi acutis, floribus solitariis oppositifoliis, staminibus omnibus conformibus, ovariis 1–4 ovali-oblongis.

HAB. In Zeylania, *Walker! Champion!*—(*v. s.*)
DISTRIB. Ins. Philippin., *Cuming,* 831! *Lobb,* 457!

Ramuli rugulosi, scabridi, cortice cinereo, etiam juniores glaberrimi. *Folia* membranacea, utrinque glaberrima, subtus pallidiora, secus costam scabrida, 3–6 poll. longa, 1¼–1¾ lata; petiolo ¼-poll.; nervi recti, obliqui, intra marginem in arcus continuos anastomosantes. *Pedicelli* abbreviati, plerumque petiolis vix longiores, bracteis 2 amplexicaulibus, una prope basin, altera versus medium, muniti. *Flores* diametro ¾-poll. *Sepala* late ovalia. *Petala* ovalia, exteriora crasse coriacea, interiora tenuiora. *Antheræ* processu brevissimo connectivi apiculatæ, fere sessiles. *Stylus* ovali-oblongus. *Carpella* spathulata, stipite late compresso, sesquipollicem longa, mucronata.

This species differs from *A. Javanica* in several important points. In that species the ovaries are 8–10, globose, broader than and very distinct from their pedicel, and terminated by a depressed, almost globose style. The abnormal inner anthers of that species are also wanting in the *A. Zeylanica*, which, however, agrees in every respect with Cuming's and Lobb's specimens from the Philippiues. Cuming's No. 496 l, also from Luzon, is not certainly identical with our plant. We have not examined its flowers, but, according to a note by M. Planchon in Herb. Hook., the anthers are different from those of the Java species. The leaves of *A. Zeylanica* are somewhat like those of *A. prinoides*, St. Hil.; but in that species they taper gradually, and are not abruptly acuminated. The fruit is also very oblique, and the dorsal margin is bent at a right angle, almost semispathulate, while in *A. Zeylanica* it is nearly regular in outline, both margins being alike.

20. **OXYMITRA,** Blume.

Polyalthiæ § Oxymitra, *Bl. Fl. Jav. Anon.*

Sepala 3. *Petala* 6, biseriatim valvata; exteriora multo majora, elongata, plana, tenuia; interiora ovata vel oblonga, conniventia, basi interdum angustata. *Stamina* lineari-oblonga, apice truncato-capitata, antherarum loculis linearibus dorsalibus discretis. *Torus* alte conicus vel subtruncatus. *Ovaria* oblonga, dense strigosa; ovula in sutura ventrali prope basin 1–2 funiculo elongato adscendentia. *Stylus* obovatus. *Carpella* monosperma.—Frutices *scandentes*, pedunculis *extra-alaribus unifloris*, nervis *foliorum obliquis validis remotiusculis, nervulis crebris subparallelis transversis connexis.*

This genus, one of those united by Blume under *Polyalthia*, but distinguished as a section under the name we have adopted, appears, so far as our limited materials enable us to judge, a very natural one. The species are all very similar in habit and general appearance, and are especially alike in the peculiar nervation, the numerous cross nerves between the distant principal ones being strongly marked. In the two species of which we have been able to examine the ovaries, we find one ovule rising from very near the base, and supported by a longish funiculus, with the nucleus projecting beyond the exostome; but Blume describes and figures *U. cuneiformis* as two-ovuled. We have, therefore, here, as well as in *Phæanthus*, one- and two-ovuled species in the same genus: and perhaps it will be found that the number varies in the same species, but, from the extreme difficulty of examining the ovaries of these plants after drying, this can only be determined by investigating the recent plant. The genus is in floral characters very close in the natural series to *Unona*, differing little from the section *Dasymaschalon*, except in number and position of ovules, and in the presence of small inner petals.

In addition to the four described below, Cuming's No. 1896, from the Philippines, and *O. cuneiformis*, Blume, are the only species referable to *Oxymitra* which have come under our notice.

1. **O. latifolia** (H.f. et T.) ; foliis ovalibus vel obovatis obtusissimis basi cordatis, pedunculis medio unibracteatis.

HAB. In ins. Penang, *Phillips!—(v. s. in Herb. Hook.)*

Frutex scandens. *Ramuli* validi, elongati, cortice nigricante ruguloso, punctis minutis albis conspersi, juniores dense ferrugineo-tomentosi. *Folia* 6–12 poll. longa, 4–7 poll. lata, petiolo ½-poll., coriacea, rigida, superne glaberrima, nitida, subtus dense pubescentia, secus costam nervosque ferrugineo-tomentosa, nervulis seconda-

U

riis valde conspicuis. *Pedicelli* ¼–½-pollicares, medio bracteam parvam oblongam gerentes. *Sepala* rotundata, basi coalita. *Petala exteriora* tomentosa, oblongo-lanceolata, apicem versus angustata, verosimiliter 1½-pollicaria, sed in specimine imperfecta; *interiora* late ovata, vix ½ poll. longa. *Ovaria* dense aureo-pilosa.

2. **O. unonæfolia** (H.f. et T.); foliis oblongis in acumen gracile abrupte productis subtus valde glaucis utrinque glaberrimis, carpellis breve pedicellatis oblongis apiculatis.—Guatteria unonæfolia, *Alph. DC. Mém.* 41; *Wall. Cat.* 6435 !—(*v. s. in Herb. Linn. Soc.*)

HAB. In provincia Tenasserim ad Tavoy, *Wall.!*—(*v. s.*)

Frutex scandens. *Ramuli* nigricantes, glabri, rugulosi ; partes novellæ aureo-pubescentes. *Folia* basi rotundata vel acutiuscula, apice longe acuminata, 5–7 poll. longa, 2–3 lata, petiolo vix ¼-poll., submembranacea, supra nitida. *Pedunculi fructiferi* ½ poll. longi. *Sepala* in fructu persistentia, late ovata, extus fusco-pubescentia. *Carpella* 5–6, oblonga, utrinque obtusa cum mucrone, glabra, granulosa, atro-fusca, pedicellis pubescentibus 0·2 lineam longis suffulta.

This plant, which is in fruit only, with one very young bud, is referred here, from its close resemblance to *Polyalthia* (*Oxymitra*) *cuneiformis*, Blume, the original species of the genus, and to the next species.

3. **O. glauca** (H.f. et T.); foliis oblongis acutis vel obtusis utrinque glaberrimis subtus valde glaucis, carpellis ovoideis utrinque obtusis.

HAB. In ins. Penang, penins. Malayanæ, *Phillips!*—(*v. s.*)

Frutex verosimiliter scandens. *Ramuli* glabri, graciles, nigricantes ; gemmæ puberulæ. *Folia* 3–5 poll. longa, circa 2 poll. lata, petiolo ½-poll., basi rotundata. *Pedunculi fructiferi* pollicares, graciles. *Carpella* numerosa, atro-fusca, pedicello æquilongo suffulta, ⅓ poll. longa, glabra.

Like the last, this is only known in fruit, but it seems quite distinct.

4. **O. fornicata** (H.f. et T.); foliis lanceolatis vel oblongo-lanceolatis subtus glaucis cinereo-pubescentibus, pedunculis medio bracteatis, petalis oblongo-lanceolatis nervosis, carpellis anguste oblongis mucronatis.—Uvaria fornicata, *Roxb. Fl. Ind.* ii. 662! U. Roxburghiana, *Wall. Cat.* 6423 A! (*nec* B).

HAB. In Assam, *Simons!* Silhet, *Roxb.!* Mergui, *Griffith!*—(Fl. Mai.) (*v. s.*)

Frutex scandens. *Ramuli* graciles, nigricantes, vix rugulosi, juniores puberuli ; partes novellæ fusco-tomentosæ. *Folia* basi parum angustata, sed rotundata, acuminata, 4–6 poll. longa, 1–1¾ lata, pet. ½-poll.; tenuiter coriacea, rigida, supra glabra vel juniora sparse puberula, nervis nervulisque subtus conspicuis. *Pedunculi* ½-pollicares, subclavati, fusco-tomentosi, bractea oblonga. *Sepala* ovato-lanceolata, ½–¾ poll. longa, nervosa, tenuia, adpresse pubescentia. *Petala exteriora* bipollicaria, utrinque adpresse pubescentia, tenuia ; *interiora* ½-pollicaria, ovata, acuta, conniventia, basi concava, marginibus superne latis planis. *Carpella* adpresse fulvo-tomentosa, ⅔ poll. longa, pedicello brevi crasso (vix lineam longo) suffulta.

The only fruiting specimen of this species which we have seen is in the British Museum.

21. **PHÆANTHUS**, H.f. et T.

Sepala 3, elongato-triangularia, parva. *Petala* 6 ; exteriora parva, sepalis conformia et æqualia ; interiora multo majora, crasse coriacea, plana, æstivatione valvata. *Stamina* linearia, connectivo truncato-capitato, antheris lineari-oblongis dorsalibus remotis. *Ovaria* lineari-oblonga, 8–12 vel plura, ovulis solitariis vel binis, suturæ ventrali infra medium insertis, adscendentibus. *Stylus* oblongus, intus sulcatus.— Arbor? floribus *extra-alaribus.*

The plant on which we have founded this genus seems widely diffused throughout the Eastern Archipelago. It is closely allied in characters to *Guatteria* and *Oxymitra*, but the large size of the inner petals, which are as much developed as in *Saccopetalum*, render it necessary to constitute it a distinct genus. The small size of the outer petals indicates an approach to *Miliuseæ*, but its stamens are quite those of the normal division of the Order. The ovule in our Indian species appears always solitary, but in a second species, collected by Cuming in the Philippines (No. 525, 1084), we several times found a second ovule. In several species of *Polyalthia* and *Oxymitra*, the ovules have occasionally appeared to us solitary; but it is so difficult to feel certain of the non-presence of ovules in the ovaries of dried specimens of this tribe of plants, that this point must remain ˙doubtful till these species are examined in a growing state. (Name from φαιος, *brown*, and ανθος.)

1. **P. nutans** (H.f. et T.) ; foliis oblongis vel lanceolatis subtus pubescentibus, floribus longe pedicellatis, petalis interioribus ovato-oblongis longitudinaliter costatis —Uvaria nutans, *Wall. Cat.* 6481 ! U. tripetala, *Roxb. Fl. Ind.* ii. 667. U. ophthalmica, *Roxb. mss. in Herb. Linn. Soc.!; Don. Gen. Syst.* i. 93.

HAB. In penins. Malayana ad Malacca, *Griff !* et Singapur, *Wall.!* —(*v. s.*)

DISTRIB. Ins. Moluccanæ.

Ramuli cinerei, rugosi, glabri, juniores cum partibus novellis ferrugineo-tomentosi. *Folia* basi acutiuscula, apice acuta vel acuminata, rarius obtuse acuminata, 4–6 poll. longa, 1½–2¼ lata, pet. ½-poll., tenuiter coriacea, supra præter costam pubescentem glabra, subtus (juniora densius) pubescentia, petiolo et costa tomentosis. *Pedunculi* oppositifolii vel sæpe in ramulo infrafoliacei, ferrugineo-tomentosi, 1–2-pollicares, graciles, apice subclavati, prope basin articulati, ibique bracteas plures lineari-lanceolatas parvas gerentes, in axillis interdum alabastros minutos rarius (vel nunquam) evolutos foventes. *Petala exteriora* ¼-poll.; *interiora* ¾–1 poll. longa, convexo-plana, utrinque adpresse pubescentia, sulcis 5 profundis exarata. *Ovaria* strigoso-pilosa.

The species from the Philippines has much smaller sepals, and the petals are longer and narrower, and not ribbed. Its leaves also are more glabrous. One of Cuming's specimens is in fruit. The carpels are oblong-apiculate, ½ inch long, with a pedicel rather longer than themselves, and one-seeded.

Tribus V. MILIUSEÆ.

Petala æstivatione valvata, haud unguiculata. *Stamina* laxe imbricata; *antheræ* late ovales.

This little tribe deviates considerably from the remainder of the Order in the structure of the anthers. Its species are all Indian, unless *Anona tenuiflora* of Martius, which we only know by the figure given by that author, be a genuine member of it. Of that species only the male plant is known.

22. **MILIUSA,** Lesch., Alph. DC.

Hyalostemma, *Wall., Lindley.*

Flores dioici vel hermaphroditi. *Sepala* 3, minuta. *Petala* 6; exteriora minuta, sepalis fere conformia; interiora multo majora, tenuiter coriacea, æstivatione valvata, serius cohærentia. *Stamina* indefinita, toro cylindrico inserta, in dioicis totum torum tegentia, in hermaphroditis pluriserialiter circa ovaria imbricata. *Antheræ* extrorsæ, ovales, subdidymæ, biloculares, connectivo vix apiculatæ. *Ovaria* lineari-oblonga, stylo oblongo terminata. *Ovula* 1–2, rarius plura; suturæ ventrali inserta.—Arbores *mediocres vel humiles*, pedunculis *axillaribus, solitariis vel fasciculatis, rarius extra-alaribus.*

Hyalostemma being only distinguished from *Miliusa* by the number of ovules, and that not constantly, as *H. Wallichiana* and *macrocarpa* have not unfrequently two ovules, we have united the two genera. In one species Wight indicates the occasional presence of three and even four ovules. The original species of *Miliusa* is hermaphrodite, while Wallich's *Hyalostemma* is diœcious; but as other species are polygamous, we cannot regard this character as of generic importance.

1. **M. montana** (Gardner, mss. in Herb. Wight et Hook.); foliis ovato- vel oblongo-lanceolatis plerumque acutis glabriusculis, pedunculis solitariis flores hermaphroditos æquantibus vel paullo superantibus, sepalis petalisque exterioribus glabriusculis ciliatis, toro strigoso-piloso. —Guatteria montana, *Moon Cat. ex Gardn. et Wight.*

β. *major;* foliis duplo majoribus, carpellis globosis sessilibus glabris atro-fuscis minute granulatis pisi magnitudine.—*Wall. Cat.* 6433 *C.?*

Hab. In Zeylaniæ montosis, alt. 2–3000 ped., ut videtur vulgaris, *Moon aliique.* β. In Zeylania, Malabar, et Maisor, *Wight!*—(*v. s.*)

Frutex ramosissimus, ramis strictis, dense foliosis. *Cortex* cinereus, rugosus, verruculosus, glaber; partes novellæ fusco-pubescentes. *Folia* ovata, forma et magnitudine valde varia, interdum pauca ad ramulorum basin interdum ovalia, vel subglobosa, obtusissima, in var. α 1–2 poll. longa, ½–¾ lata, petiolo brevissimo, utrinque glabra vel juniora subtus puberula. *Pedunculi* basi pluribracteati, cæterum nudi, graciles. *Sepala* acuta, petalis exterioribus dimidio breviora. *Flores* ⅔ poll. longi. *Petala* interiora coriacea, nervosa, acuta, glabriuscula. *Ovaria* subglabra, late ovalia, toro circa basin ovariorum albo-piloso cæterum glabro strigoso-piloso inserta. *Ovula* 2. *Stigma* ovale.

The smallest of all the species. It is probably a wiry, rigid shrub, and may readily be known by being much more glabrous than either of the two following. The var. β appears to be only distinguished by the larger size of the leaves; but as the fruit of the smaller one is unknown, it is possible that the two may be very distinct.

2. **M. Indica** (Lesch. in A. DC. Mém. p. 36); ramis tomentosis, foliis ellipticis plerumque obtusis subtus pubescentibus, pedunculis abbreviatis, floribus hermaphroditis, sepalis petalisque exterioribus incanis, petalis interioribus ovatis pilis stellatis pubescentibus, carpellis incano-pubescentibus sessilibus.—*W. et A.! Prod.* i. 10. *Wall. Cat.* 6433 A! B!

Hab. In montibus Travancor ad Courtalam, *Lesch., Wight!*—(Fl. Apr.) (*v. s.*)

Frutex foliosus; ramulis gracilibus, flexuosis, fusco-tomentosis, demum glabrescentibus, et tunc cortice cinereo ruguloso-pustulato. *Folia* tenuiter coriacea, basi rotundata, elliptico- vel oblongo-lanceolata vel elongato-lanceolata, rarius ovata vel ovalia, superne lucida, glabra; juniora supra pilis paucis sparsis stellatis munita, secus costam pubescentia, subtus pubescentia, secus costam fere tomentosa, 2–2½ poll. longa, ¾–1¾ lata, petiolo vix lineam longo. *Flores* ¾ poll. longi. *Petala exteriora* sepalis æquilonga, duplo latiora; *interiora* ovata, nervosa, pubescentia. *Torus* dense strigosus. *Ovaria* dense albo-pilosa, lineari-oblonga, plerumque biovulata. *Stigmata* oblonga. *Carpella* numerosa, fere sessilia, subglobosa, pisi magnitudine, apiculata, dense cinereo-tomentosa.

Perhaps there is a glabrous-fruited species undistinguishable in leaves from this, but it is more probable that all the glabrous and sessile-fruited forms are referable to *M. montana.*

3. **M. Zeylanica** (Gardner in Herb. Wight); ramulis pubescentibus, foliis oblongo-lanceolatis subtus adpresse puberulis, pedunculis abbreviatis, floribus hermaphroditis, sepalis petalisque ext. pubescentibus, petalis interioribus lineari-oblongis.

HAB. In Zeylania, alt. 2–3000 ped., *Walker! Gardner!* etc.—(*v. s.*)

Arbor parva. *Ramuli* fusco-grisei, rugosi, tuberculati, juniores cum omnibus partibus novellis fusco-tomentosi. *Folia* coriacea, rigida, opaca, obtusa acuta vel acuminata, basi rotundata, et sæpe obliqua, 2–4 poll. longa, 1–1½ lata, petiolo bilineari, plus vel minus tomentoso, demum utrinque glabrata. *Pedunculi* floribus dimidio breviores, tomentosi, basi pluribracteati. *Sepala* et *petala exteriora* fere æqualia, oblonga, vix acuta, pilosa, molliter ciliata. *Petala* interiora quam in cæteris speciebus angustiora, adpresse puberula. *Torus* strigoso-pilosus. *Ovaria* numerosa, pilis rigidis strigosa, late ovalia, stigmate ovali terminata. *Ovula* plerumque 2. *Fructus* ignotus.

4. **M. Wightiana** (H.f. et T.); ramulis glabriusculis, foliis anguste oblongo-lanceolatis vel lineari-oblongis obtuse acuminatis basi acutis utrinque glabris, pedunculis floribus duplo longioribus gracilibus, floribus polygamo-dioicis?, carpellis globosis granulatis glabris pedicello æquilongo suffultis.

HAB. In montibus humidioribus peninsulæ australis prope Courtalam, *Wight!*—(Fl. Jul. Aug.) (*v. s.*)

Arbor parva. *Ramuli* cinerei, rugulosi, glabri, tuberculis crebris pustulati. *Folia* tenuiter coriacea, lucida, 3–4 poll. longa, ¾–1 lata, petiolo brevissimo glabro. *Pedunculi* ¾–1 poll. longi, vix puberuli, basi squamis paucis remotiusculis muniti. *Flores* magnitudine eorum *M. montanæ. Sepala* et *petala exteriora* minuta, reflexa. *Petala interiora* oblonga, extus puberula. *Torus* glaber. *Stamina* (in flore uno) uniserialia, longiora et majora quam in cæteris speciebus. *Ovaria* numerosa, glabra, ovali-oblonga, ovulo 1, suturæ ventrali inserto. *Stylus* oblongus.

Of this very distinct species there are several specimens in Dr. Wight's herbarium, but only one or two flowers. It is intermediate in many respects between the original species of *Miliusa* and *Hyalostemma.*

5. **M. Wallichiana** (H.f. et T.); foliis elongato-oblongis longe acuminatis glabris lucidis, pedunculis elongatis bracteis pluribus linearisubulatis sparsis munitis, floribus polygamo-dioicis, sepalis petalisque exterioribus subæqualibus longe linearibus pubescentibus, toro fructifero apice tantum carpella gerente breviter strigoso, carpellis numerosis transverse ovalibus, seminibus 1–2.

HAB. In sylvis Assam! Silhet et Cachar! et in montibus Khasia a basi ad altitudinem 4000 pedum usque.—(Fl. Sept.–Nov.) (*v. v.*)

Arbor parva, cortice cinereo lenticellis crebris ruguloso, sæpe aliquot suberoso;

partes novellæ sparse puberulæ, plerumque cito glabrescentes. *Folia* basi parum obliqua, rotundata, in acumen longum attenuata, 3–6 poll. longa, ¾–1¾ lata, petiolo vix 2 lineas longo, glabro vel puberulo, supra lucida, interdum secus nervos puberula, subtus pallida, fere glabra. *Pedunculi* solitarii vel gemini, axillares vel paullo supraaxillares et terminales, puberuli vel glabri, 1–1½ poll. longi, graciles, substricti. *Stamina* in flore fœmineo pauca effœta, uniserialia, 4–10 (an semper?). *Sepala* ¼-poll. *Petala* interiora fere pollicaria, ovata, obtusa, nervosa, secus margines incana. *Stamina* in flore masculo capitulum globosum formantia, toro oblongo superposita. *Ovaria* oblonga, glabra, stylo oblongo apiculata. *Carpella* fabæ minoris magnitudine, basi subumbilicata, pedicello æquilongo suffulta.

The characters of this species are less accurately defined than we could wish, from the imperfection of the materials available for its discrimination from the next. We have very few flowering specimens of *M. Roxburghiana*, and those imperfect; but the present species we have abundantly, both with male and female flowers, and we believe we are right in identifying the fruiting specimens with them. That there are two very well-marked species, readily distinguishable by the fruit, cannot be doubted, but careful observations are still required to discriminate the two in a flowering state. Both are common in Khasia and Silhet. *Guatteria umbilicata*, Dunal, 135. t. 33, DC. Syst. i. 508, Prod. i. 94, certainly belongs to the genus *Miliusa*, and is perhaps referable to this species.

6. **M. Roxburghiana** (H.f. et T.); foliis oblongis vel longe oblongo-lanceolatis abrupte acuminatis subtus molliter tomentosis vel pubescentibus, pedunculis pubescentibus flori æquilongis, bracteis 2–3 lineari-subulatis munitis, floribus dioicis, carpellis torum tomentosum ovalem omnino tegentibus pedicello longo suffultis oblongis glabris granulosis.—Hyalostemma Roxburghianum, *Wall. Cat.* 6434! Guatteria globosa, *Alph. DC. Mém.* 43; *Wall. Cat.* 6448!

HAB. In sylvis tropicis et subtropicis Sikkim! Assam! Khasia! Silhet! Chittagong! Tenasserim!—(*v. v.*)

Arbor parva, cortice griseo ruguloso, pubescente, ramorum juniorum fusco-tomentoso. *Folia* forma et magnitudine sicut indumento valde varia, 3–7 poll. longa, 1–2 lata, petiolo vix lineam longo. *Flores* axillares, sæpe fasciculati. *Carpella* pedicellis ¾ poll. longis suffulta, subobliqua. *Semina* 1–2, dum 2 plano-convexa.

This is in general readily known from the last by the much more pubescent leaves and branches, but these are sometimes, though rarely, almost glabrous.

7. **M. macrocarpa** (H.f. et T.); foliis oblongis vel lanceolatis basi acutis longe acuminatis glabris, floribus ignotis, carpellis numerosis oblongis glabris longe pedicellatis, seminibus 1–2.

HAB. In sylvis densis Sikkim, alt. 5–6000 ped.!—(*v. v.*)

Arbor verosimiliter parva. *Ramuli* grisei vel nigricantes, glabri; partes novellæ puberulæ. *Folia* 5–9 poll. longa, 1½–2¼ lata, petiolo 2 lineas longo, supra lucida, subtus pallida, tenuia, fere membranacea. *Pedicelli fructus* stricti, superne incrassati, bipollicares. *Torus* fusco-strigosus. *Carpella* ¾ poll. longa, utrinque obtusa, pedicellis compressis æquilongis suffulta. *Semina* solitaria vel bina, et tunc carpella duplo majora.

Though our specimens are only in fruit, they belong evidently to a very distinct species, which we have no hesitation in considering as congeneric with those species described above. The carpels must be considered as transversely oblong, as they are attached to the pedicel obliquely by one end; and when there are two seeds the second is superposed, and both lie transversely, as is the usual arrangement in the Order.

8. **M. velutina** (H.f. et T.); foliis ovatis vel oblongis basi cordatis utrinque velutino-tomentosis, pedunculis extra-alaribus, pedicellis elongatis dense tomentosis ebracteatis, floribus (hermaphroditis) extus dense tomentosis, carpellis pubescentibus breviter pedicellatis.—Uvaria velutina, *Dunal, Anon.* 91; *DC. Syst.* i. 484, *Prod.* i. 88. Uvaria villosa, *Roxb. Fl. Ind.* ii. 664. Guatteria velutina, *Alph. DC. Mém.* 42; *Wall. Cat.* 6441!

HAB. In sylvis siccis Bengaliæ occidentalis et Bahar secus basin montium, *Roxb.! Ham.!;* Garhwal in sylvis prope Kheri secus basin Himalayæ copiose, *Edgeworth!;* et in Ava ad Taong Dong et secus ripas fluminis Atran, *Wall.!*—(Fl. Mart. Apr.) (*v. s.*)

Arbor, cortice pallido ruguloso. *Ramuli* dense tomentosi. *Folia* brevissime pedicellata, parva, forma valde varia, acuminata, acuta vel obtusa, 3–6 poll. longa, 1¾–4 lata, petiolo ⅒-pollicari, subtus densius velutina. *Pedunculi* ramosi. *Pedicelli* 3–6, 2–4 pollices longi, graciles. *Sepala* ovata, dense tomentosa. *Petala exteriora* sepalis conformia; *interiora* late ovata, ¼–⅓ poll. longa, extus dense tomentosa, intus atro-fusca, subglabra. *Ovaria* velutina. *Carpella* baccata, nigricartia, puberula, sicca semipollicaria, pedicello ¼-poll. suffulta. *Semina* 1–2.

Roxburgh says that the fruit much resembles black cherries.

23. SACCOPETALUM, Bennett.

Sepala 3, parva. *Petala* 6; exteriora sepalis fere conformia; interiora multo majora, velutina, marginibus cohærentia, denique libera, basi saccata. *Torus* subglobosus. *Stamina* indefinita, multiserialia; antheræ subsessiles, dorsales, biloculares, connectivo apiculatæ. *Ovaria* ovali-oblonga. *Ovula* in axi biserialia, 6 vel plura.—*Arbores late comosæ, deciduæ*, floribus *ante vel cum foliis novellis nascentibus, majusculis.*

The genus *Saccopetalum* was established by Mr. Bennett in Horsfield's ' Plantæ Javanicæ Rariores' for a Javanese plant, a native of the southern coast of that island. In that work Mr. Bennett clearly indicated the close relationship of *Saccopetalum* to *Miliusa* and *Hyalostemma*, and pointed out that these three genera formed a well marked section of the family, to which he omitted to give a name. Mr. Bennett at the same time indicated *Mitrephoreæ* as the connecting link by which the *Saccopetaleæ* pass into the more ordinary state of the family.

1. **S. longiflorum** (H.f. et T.); foliis ovali-oblongis vel oblongo-lanceolatis acuminatis supra glabris subtus adpresse puberulis, floribus breviter pedicellatis, petalis elongatis, stylis elongatis filiformibus apice subclavatis.—*Wall. Cat.* 6443!

HAB. In Bengalia superiori ad Purneah, *Hamilton, ex Wall. Cat.;* verosimiliter in sylvis densis Terai dictis secus basin Himalayæ.— (*v. s.*)

Arbor forsan excelsa. *Ramuli* rugulosi, glabri, cortice cinereo vel albido, lenticellis minutis asperato; partes novellæ pubescentes. *Folia* 8 poll. longa, 3 lata, petiolo ⅓-pollicari, tenuiter coriacea. *Flores* ad axillas foliorum delapsorum solitarii, pedicellis ¼-pollicaribus incano-pubescentibus. *Sepala* oblonga, 2 lin. longa. *Petala* interiora oblongo-lanceolata, obtusa, 1⅓ poll. longa, tenuia, puberula, basi subsaccata. *Torus* dense albo-villosus. *Ovaria* albo-strigosa, numerosa. *Ovula* circa 10. *Styli* ovario fere duplo longiores, gummi exudantes. *Carpella* subdefinita (in spec. Herb.

Bentham 7), toro globoso inserta, subglobosa, diam.1-1¼-poll., pedicello ¾-poll. crasso antice sulcato suffulta, glabra, minutissime granulosa, atro-fusca.

We find in Wallich's Catalogue two barbarous names for this plant, one being Roxburgh's manuscript name, the other that of Hamilton. As neither has been published, and we cannot tell which ought to have the preference, we have adopted neither, gladly availing ourselves of any legitimate excuse for escaping the employment of such extremely local and probably quite erroneous names.

2. **S. tomentosum** (H.f. et T.) ; foliis ovalibus vel ovato-oblongis acutis utrinque pubescentibus vel (junioribus saltem) subtus tomentosis, pedicellis elongatis gracilibus, petalis oblongis obtusis, stylis ovalibus.—Uvaria tomentosa, *Roxb. Cor.* i. *t.* 35, *Fl. Ind.* ii. 667 ; *Dun. Anon.* 90 ; *DC. Syst.* i. 483, *Prod.* i. 88 ; *Wall. Cat.* 6472 ! ; *W. et A. Prod.* i. 8 ; *Graham, Cat. Bomb.*

Hab. In montibus Coimbator, *Wight !* Concan, *Graham, Law !* Orissa, *Roxburgh !* Bahar, *Hamilton ;* et in Nipalia centrali in sylvis secus basin Himalayæ prope Gorakpur, *Hamilton !*—(Fl. Apr. Mai.) (*v. s.*)

Arbor. Ramuli rugulosi, cortice griseo, juniores fulvo-tomentosi. *Folia* 4–6 poll. longa, 2½–3 lata, petiolo ¼-pollicari, basi rotundata vel cordata, tenuiter coriacea, opaca, subtus pallida. *Pedunculi* oppositifolii, abbreviati, 1–2 lineas longi, 1-2-flori. *Pedicelli* 2–3 poll. longi. *Sepala* lineari-oblonga, 2 lin. longa. *Petala* interiora oblonga, obtusa, extus puberula, intus incano-tomentosa. *Torus* dense albo-villosus. *Ovaria* late ovalia, 4–6-ovulata. *Carpella* 5–15, subglobosa, pollicaria, pedicello ½–⅓-pollicari suffulta, dense fulvo-tomentosa. *Semina* 3–4, in pulpa nidulantia.

24. ALPHONSEA, H.f. et T.

Sepala 3, rotundata. *Petala* 6, subæqualia, æst. biseriatim valvata, ovata. *Torus* alte hemisphæricus. *Stamina* indefinite pluriserialia. *Ovaria* 1–10 (rarissime solitaria), plerumque conum ultra stamina longe eminentem formantia, lineari-oblonga, subangulata, strigoso-pubescentia, ovulis in sutura ventrali indefinitis biserialibus. *Stylus* oblongus vel depressus.—Arbores *interdum proceræ,* foliis *crasse coriaceis nitidis,* floribus *in fasciculos oppositifolios congestis, parvis.*

With the stamens of *Saccopetalum,* this genus has the petals equal in size, or nearly so, the outer being usually somewhat larger. The general appearance is very distinct from that of *Saccopetalum,* the leaves being very thick and coriaceous, and quite glabrous, except when very young. We are happy to have this opportunity of associating the name of M. Alphonse De Candolle with an Order which he has so ably investigated, by dedicating to him this very distinct genus.

1. **A. ventricosa** (H.f. et T.); foliis oblongis vel lineari-oblongis, pedicellis longiusculis medio bracteolatis, carpellis longe pedicellatis.— Uvaria ventricosa, *Roxb. ! Fl. Ind.* ii. 658 ; *Wall. Cat.* 6453 !

Hab. In sylvis prov. Chittagong !—(*v. v.*)

Arbor excelsa. *Ramuli* grisei, rugosi, glabri, juniores cum omnibus partibus novellis ferrugineo-tomentosi. *Folia* basi rotundata, apice in acumen longum obtusiusculum sensim angustata, 6–10 poll. longa, 1½–4 lata, petiolo vix ¼-poll., coriacea, firma, (in sicco) venulis creberrimis reticulatis notata, supra lucida, præter costam pubescentem glabra, subtus (in sicco) pallidiora, lutescentia, petiolo et costa pubescentibus demum glabratis. *Pedunculi* brevissimi, plerumque vix ulli; *pedicelli* numerosi, ½–1 poll. longi, tomentosi, basi bractea ovata parva et medio bracteola squamæformi munita. *Flores* e viridescente albi, odorati. *Sepala* basi coalita, extus pubescentia. *Petala* utrinque fulvo-tomentosa, subtrapezoidea, ¾ poll. longa, basi saccata. *Ovaria* 10; *stylo* oblongo *Carpella* 4–8, baccata, ovoidea vel subglobosa, tomentosa, sesquipollicaria, pedicello pollicari. *Semina* 4–8, biserialia.

2. **A. lutea** (H.f. et T.); foliis ovali-oblongis, pedicellis abbreviatis, carpellis lævibus breve pedicellatis.—Uvaria lutea, *Roxb.! Cor.* ii. *t.* 36, *Fl. Ind.* ii. 666 (*non Wall. nec W. et A.*) A. Russelii, *Wall. Cat.* 6464!

HAB. In montibus Orissa, *Roxburgh! Russell!* Silhet, *Wall.!* Ava, *Wall.!*—(*v. s.*)

Arbor excelsa. *Ramuli* foliosi, glabri, cortice griseo ruguloso. *Gemmæ* fulvo-tomentosæ. *Folia* basi rotundata, obtuse acuminata, coriacea, utrinque glabra, supra nitida, subtus pallida et (in sicco) reticulato-venulosa, juniora subtus secus costam puberula, 3–4 poll. longa, 1½–2½ lata, petiolo vix ¼-poll. *Flores* in fasciculos oppositifolios multifloros sessiles vel brevissime pedunculatos congesti. *Pedicelli* abbreviati, fulvo-tomentosi. *Sepala* rotundata, decidua. *Petala* vix ½-pollicaria, ovata, exteriora paullo majora. *Stamina* prioris. *Ovaria* circa 10, strigoso-pilosa, in conum conniventia, stylo ovali. *Torus fructus* dilatatus, globosus, cicatricibus magnis rotundatis uniserialibus lateralibus notatus. *Carpella* 4–6, late ovalia, utrinque obtusa, brevissime pedicellata, 1–1½ poll. longa, lævia, fulvo-puberula, pulposa, matura læte flava. *Semina* 3–6.

3. **A. Zeylanica** (H.f. et T.); foliis lanceolatis vel elliptico-lanceolatis, floribus in fasciculis paucifloris congestis, carpellis verrucosis pedicellatis.—Uvaria lutea *a, W. et A. Prod.* i. 8 (*excl. syn.*); *Wight, Cat.* 28! Guattaria acutiflora, *Wall. Cat.* 6438 D!

HAB. In Zeylaniæ montosis, alt. 2–3000 ped., *Walker! Gardner! Thwaites!* et in montibus Travancor ad Courtalam, *Wight!*—(*v. s.*)

Arbor ramosissima, foliosa. *Ramuli* graciles, virgati, cortice cinereo glabro ruguloso, juniores puberuli; gemmæ fusco-sericeæ. *Folia* basi acuta, sæpius longissime acuminata, apice plerumque obtusa, 2¼–4 poll. longa, ⅔–1¼ lata, pet. gracili ¼-poll., coriacea, utrinque glabra, supra nitida, petiolo et costa subtus pubescentibus, demum glabratis, nervis inconspicuis parallelis, venulis crebre reticulatis. *Pedunculi* extra-alares, interdum oppositifolii, vix ½-pollicares, tomentosi. *Pedicelli* 1–3, semipollicares, pubescentes, basi bractea ovata minuta suffulti, cæterum nudi. *Petala* ⅔-poll., extus puberula. *Ovaria* 5, stylo depresso. *Torus fructus* magnus, globosus. *Carpella* subglobosa, toro lateraliter inserta, 1¼ poll. diam., pedicello crasso ¼–½-poll. oblique inserta, irregulariter tuberculata, fulvo-tomentosa (pulposa?). *Semina* 4–6, uniserialia, plano-compressa, septis cellulosis separata, rotundata, lævinscula, cinerea, hilo oblongo conspicuo.

VI. MYRISTICACEÆ.

Flores dioici. *Sepala* 2–4 (plerumque 3), hypogyna, basi et sæpe alte coalita, coriacea, æstivatione valvata. *Petala* nulla. *Stamina* 3–18, monadelpha; antheris ovalibus vel linearibus, extrorse bilocularibus, longitudinaliter dehiscentibus, columnæ varie adnatis. *Ovarium* plerumque solitarium (rarissime 2 discreta), liberum, centrale, subglobosum, ovulo 1 erecto anatropo; stigmate subsessili, lobato vel depresso-capitato. *Fructus* bivalvis, monospermus. *Semen* erectum, arillo involutum.

Testa carnosa, tunica interior dura, fragilis. *Albumen* ruminatum, se-
baceum. *Embryo* prope hilum minutus, cotyledonibus divaricatis planis
vel undulatis, radicula infera.—Arbores (*rarius* frutices) *tropicæ sæpe
excelsæ, plus minus aromaticæ, succo acrido sæpius rubicundo scatentes,*
foliis *alternis integerrimis distichis paralleli-nerviis, (junioribus saltem)
pellucido-punctatis,* floribus *inconspicuis sæpe minimis, in axillis glomeratis vel paniculatis.*

This small Order is well known, from its containing the tree which yields the
Nutmeg of commerce; and most of the species possess similar aromatic qualities,
though occasionally these are very faint, and in some instances confined to the arillus
(in the officinal nutmeg called *Mace*), or to the fleshy part of the fruit. Several species are said to be employed in India to adulterate the true nutmeg, and in America
one or more yield when fresh a tolerable substitute for that valuable spice, though
their aromatic qualities are unfortunately not permanent. We follow Endlicher in
bringing this Order into contact with *Anonaceæ,* to which it appears in most respects
closely allied. The habit, alternate dotted leaves, valvate æstivation, extrorse anthers, apocarpous ovaries, ruminated albumen, and minute embryo, are the chief
points of resemblance. The young leaves of nutmegs are in vernation quite like
the leaves of *Mitrephora,* and an arillus is sometimes present in *Anonaceæ,* in which
Order the flowers are also occasionally unisexual. On this last character so much
stress is laid by Lindley, that he removes *Myristiceæ* (associated with *Menispermeæ*
and *Lardizabaleæ*) to a considerable distance from the Ranal alliance, although he
fully recognizes their relationship to *Anonaceæ,* and transfers *Hyalostemma* from
that Order to this, on a mistaken supposition that it is apetalous; in order apparently not to invalidate this mark of distinction. We do not overlook the important
points of affinity which exist between *Myristicaceæ* and *Monimiaceæ* and *Atherospermeæ,* which Orders are also included by Lindley in his group *Menispermales.*
These are certainly entitled to great weight, especially that of the apocarpous fruit,
which removes those Orders far from *Lauraceæ.* The opposite leaves, however,
distinguish them from all the Ranal alliance except *Clematideæ.*

Myristicaceæ differ remarkably from *Anonaceæ* in the great development of the
arillus. The hilum is generally large and quite basal, and the arillus springing from
its margin envelopes the whole of the seed. The arillus has, at the same time, an
organic connection with the tissues around the micropyle, and in the common nutmeg it is perforated opposite that aperture by a small slit, which is usually quite
exterior to the cicatrix of the hilum; hence the arillus of the nutmeg has been
regarded by Planchon as an arillode, and its connection with the hilum is supposed
by that author to be spurious. The vascular tissue of the arillus might be expected
to throw light upon the origin of that body; but we find it to rise all round the
basal cicatrix, which includes not only the hilum, but an areola around the micropyle, to which the arillus is firmly attached. The examination of the fully-formed
arillus, therefore, leads us to infer that it is developed at once from the hilum and
the margin of the micropyle; but this must remain doubtful till the development of
the ovule and its nucleus has been carefully studied. The arillus is generally fleshy,
but sometimes thin and very membranous; and although usually divided towards the
apex into long linear lobes, which in the cultivated nutmeg and some other species
are very deep, in a part of the Order it is quite entire, and scarcely perceptibly perforated at the apex. Towards the base it contains a good deal of vascular tissue, the
vessels being spirally marked, but not unrollable. The cellular tissue is dense, and
in each cell there is an opaque yellowish mass, which nearly fills it, and which is
probably the seat of the aroma. The seed has three coats; of these, the outer or
testa is fleshy (as in *Magnoliaceæ*), and very thin on the sides, but thicker at the base
and apex. It is traversed on one side by a rhaphe, formed of numerous vascular
cords passing from the hilum to the chalaza, where it divides into many branches,
which ramify to a great extent over the inner surface of the testa. The chalaza is

often not quite terminal, and the seed is then slightly oblique, the dorsum or nourhaphal surface being the most convex. The middle coat is hard, woody, and brittle, and consists of a single layer of prismatic cells radiating from within outwards. The innermost coat, which is probably the nucleus, is thin and fleshy, and gives off the numerous irregularly branching, much anastomosing plates which divide the albumen. These are largest near the chalaza, from which they appear to spring when the seed is cut vertically. The albumen is composed of hexagonal cellular tissue, with thin transparent walls, each cell enclosing an opaque mass which nearly fills its cavity. The cells of the processes of the endosperm are smaller and darker coloured than those of the albumen.

This is a very tropical Order, usually forming handsome trees, often with a lofty, undivided trunk, and horizontal, more or less verticillate branches. In India none of the species are known further north than 26° N. lat., on the northern face of the Khasia hills. From America only thirteen are described by Mr. Bentham, in a general enumeration of the American species, so that the Order is eminently Indian. The species are probably most numerous in the eastern part of the Malayan Archipelago. A few are found in tropical New Holland, but none, so far as is known, in China. From Africa no species have been described, but in the British Museum there are two specimens marked "*Myristica?*" One of these, from Cape Coast, collected by Brass, is a subscandent stipulate plant, apparently belonging to *Malvaceæ* or *Euphorbiaceæ*, but the other (brought by Afzelius from Sierra Leone) is in fruit, and, judging from the general aspect, probably belongs to this Order.

As Nutmegs are generally lofty trees, inhabiting dense forests, and are almost invariably diœcious, many of the species are very imperfectly known; most frequently one sex only exists in herbaria, or, if the flowers of both sexes be known, the fruit is perhaps a desideratum. Great caution is necessary in identifying fruiting and flowering specimens gathered at different times. Of many of the species we have only seen single specimens, and have no means of determining the amount of variation to which they are subject. We have also had few opportunities of observing this family in a living state, but we think it probable that the shape of the leaves will be found to vary very much, and that it ought to be used with great caution as a specific character. For these reasons we attach but little importance to the diagnoses and descriptions here given. In one or two cases only have our materials been sufficient to enable us to offer an opinion on the limits of species; in general we have been obliged to content ourselves with describing as accurately as possible the individual specimen before us.

Myristica? sesquipedalis, Wall. Cat. 6809! is, as Dr. Wallich himself suspected, a laurel, as is also *M. glaucescens*, Wall. Cat. 6790! *M. Finlaysoniana*, Wall. Cat. 6793! is a species of *Melodorum.* Wall. Cat. 9017, referred doubtfully to *Myristica*, must also be excluded, as it certainly does not belong here.

1. **MYRISTICA,** L.

Character ordinis.

There are no doubt several very distinct genera among Nutmegs, but the structure (especially that of the female flowers) is so very little known, that the time has not yet come for establishing these genera on a secure basis. We therefore follow Blume and Bentham in retaining the genus *Myristica* for the present entire, and in dividing it into sections according to the modifications of the andrœcium.

Sect. 1. KNEMA.—*Calyx* trilobus vel tripartitus, extus tomentosus. *Columna* staminea apice in discum margine antheriferum dilatata. *Stigma* dilatatum, margine pluridentatum. *Cotyledones* planæ.— Flores *fastigiati ad apicem pedunculi axillaris abbreviati.*

The species of *Knema* form, on the whole, a well-marked group, distinguishable

at first sight by their rigid leaves and peculiar inflorescence. The staminal disc is
generally toothed on the margin, each tooth bearing an anther, which spreads out
horizontally ; but in *Myristica laurina,* Blume, which nevertheless is a genuine species of the section, the anthers are vertical and sessile on the pyriform disc, almost
as in *M. Irya,* Gærtner.

1. **M. Hookeriana** (Wall. Cat. 6802 A ! non B) ; ramulis densissime floccosis, foliis maximis anguste obovato-oblongis vel lineari-
oblongis basin versus paullo angustatis basi subcordatis, disco antherifero 18-dentato, fructu maximo dense floccoso.

HAB. In peninsula Malayana : in ins. Penang, *Wall.!*—(*v. s.*)

Arbor verosimiliter elata. *Ramulorum* cortex facile separabilis, tomento floccoso
densissimo tectus, ut omnes partes novellæ. *Folia* juniora floccosa, cito glabrescentia,
rigide coriacea, subtus glauca, fere bipedalia, 4–8 poll. lata, acuta vel breviter acuminata, petiolis dense floccosis $\frac{3}{4}$-pollicaribus. *Florum situs* e specimine Herb. Wall.
(in Mus. Soc. Linn.) unico florido insectorum morsu abnormi, non rite determinan-
dus, verosimiliter ut in cæteris speciebus axillaris. *Calyx* majusculus, $\frac{1}{3}$-poll., obo-
vato-globosus, $\frac{1}{2}$-trilobus, extus laxe tomentosus. *Columna staminea* striata, apice
parum concava, ramis brevissimis dentiformibus. *Fructus* oblongus, $2\frac{1}{2}$-poll., valvis
crassissimis ; arillus carnosulus, lobatus. *Semen* oblongum, pollicare, (ex sicco) atro-
fuscum. *Chalaza* obliqua.

The flowers of this superb species are unfortunately imperfectly known ; but it
certainly belongs to *Knema,* and will be easily recognized by the extremely dense
pubescence of all its parts except the leaves. Wall. Cat. 6802 B seems to belong
to *Laurineæ.*

2. **M. longifolia** (Wall. Cat. 6801 !) ; ramulis furfuraceis, foliis
magnis oblongo-lanceolatis basi cordatis rarius rotundatis apice angustatis obtusis, floribus axillaribus dense fastigiatis subsessilibus, disco
antherifero concavo 12–18-lobo, fructu oblongo tomentoso.—*Bl. Rumph.*
i. 188. M. linifolia, *Roxb. Fl. Ind.* iii. 847 ?

HAB. In montibus Khasia, *Wall.!* Chittagong ! peninsula Malayana,
Griffith !—(Fl. Jan.) (*v. v.*)

Arbor excelsa, ramulis validis elongatis, cortice laxo subpapyraceo cinereo vel fusco ;
partes novellæ dense tomentosæ. *Folia* forma et magnitudine admodum varia, in-
terdum lineari-oblonga vel obovato-oblonga, $\frac{1}{2}$–$1\frac{1}{2}$-pedalia, 2–6 poll. lata, petiolo $\frac{3}{4}$–1-
pollicari, utrinque glabra, supra nitida, subtus pallida et sæpissime glaucescentia,
costa basi furfuracea. *Pedunculi* brevissimi, lignosi, bracteis imbricatis rotundatis
onusti. *Pedicelli* brevissimi vel flores subæquantes, medio unibracteolati. *Calyx*
dense furfuraceus ; maris globosus, demum pyriformis, $\frac{1}{4}$-pollicaris ; fœm. ovalis. *Discus antheriferus* obscure triqueter, 15–18-lobus, ramis horizontalibus subtus anthe-
riferis. *Fructus* bipollicaris, rectus, utrinque obtusus. *Semen* erectum, chalaza terminali ; *arillus* pallidus, tenuis, profunde lobatus ; *albumen* album.

The shape of the leaves, the size and shape of the (male) flowers, and the number
of disc-lobes, appear to vary much in this species, of which we have a great many
specimens before us from many different localities. These are, however, chiefly
male plants ; and possibly characters may be afforded by the female flowers and fruit
for the discrimination of more than one species ; at any rate, a careful study of the
whole genus, in a living state, is necessary before the species can be considered as
established on a satisfactory basis.

3. **M. erratica** (Hf. et T.) ; ramulis tenuiter furfuraceis, foliis
anguste lanceolatis vel late linearibus apice attenuatis acutis basi acutis, floribus ad apicem pedunculi axillaris paucis, disco plano 12-lobo,

fructu late ovali vel subgloboso tomento brevissimo incano, arillo crasse carnoso.

HAB. In montibus Khasia, alt. 2–3000 ped.!—(*v. v.*)

Arbor excelsa, trunco apice horizontaliter et subverticillatim ramoso. *Ramuli* elongati, foliosi, rugulosi, adulti cinerei, glabrati. *Folia* 6–10 poll. longa, 1¼–2½ lata, pet. ⅔-poll., supra glabra, lucida, subtus glauca, secus costam et nervos cinereo-furfuracea. *Pedunculi* axillares, vix ½-pollicares, rigidi, lignosi; pedicelli eorum apicem versus inter bracteas squamæformes fasciculati, graciles, ¼–½ poll. longi, cinereo-incani, supra medium bracteolam latissimam amplectentem minutam gerentes. *Alabastri* masculi subglobosi, obtuse triquetri. *Fructus* plerumque solitarii, pedicello crasso, ovales, minute fusco-puberuli, 2-pollicares. *Valvæ* crassæ. *Arillus* crassissimus, carnosus, demum coccineus, æqualis, apice lacerus. *Nucleus* oblongus, lævis, obliquus, chalaza infra-apicali.

This species (the most northern of all Nutmegs) is readily distinguished from *M. longifolia* by the much fewer and smaller flowers on longer stalks, by the flat-topped disc (if that character is to be relied on), and especially by the very different fruit. The leaves also are narrower, much thinner in texture, and not cordate at the base. The leaves of both, however, vary a good deal, and some of the smaller ones of *M. longifolia* are not easily distinguished from those of the present species.

4. **M. attenuata** (Wall. Cat. 6791!); ramulis tenuiter furfuraceis, foliis oblongo-lanceolatis longe attenuatis basi acutis vel rotundatis subtus glaucis, pedunculis axillaribus paucifloris, disco plano 12-lobo, fructu ovali vel oblongo fulvo-tomentoso, arillo tenuissimo.—M. amygalina, *Graham, Cat. Bomb., non Wall.*

HAB. In montibus Concan, *Dalzell! Law!*—(*v. s.*)

Arbor forsan excelsa. *Ramuli* elongati, graciles, adulti glabrati, cortice nitido ruguloso. *Folia* secus costam et nervos subfurfuracea, demum glabra, 5–8 poll. longa, 1¾–2½-lata, petiolo ½-poll. *Pedunculi* axillares, 1–2 lineas longi, apice bracteati, 3–4-flori. *Pedicelli* petiolos æquantes vel paullo superantes, filiformes, apice subclavati, prope florum unibracteolati. *Flores* subglobosi, laxe tomentosi. *Fructus* sesquipollicares. *Arillus* apice lobatus. *Semen* oblongum; chalaza unilaterali, a vertice ¼ pcll. distante.

Our specimens from Law and Dalzell agree very well with those of Heyne. The leaves have a tendency to vary in shape, and are sometimes quite rounded at the base. The fruit is broader in the Wallichian specimens than in the others, but the seed is the same in both.

5. **M. glaucescens** (H.f. et T.); ramulis glabrescentibus, foliis oblongo-lanceolatis basi plerumque rotundatis supra glaberrimis subtus glaucis, pedunculis brevissimis paucifloris, disco plano 10–15-radiato, fructu ovali-oblongo ferrugineo-pulverulento, arillo tenui.—M. Sumatrana, *Bl. Rumph.* i. 187. Knema glaucescens, *Jack. in Mal. Misc. et in Hook. Comp. Bot. Mag.* i. 149, *non Wall.*

HAB. In penins. Malayana ad Malacca, *Griff.! Cuming,* 2315!—(*v. s.*)

DISTRIB. Sumatra, *Jack;* ins. Philipp., *Cuming,* 1042! 1309!

Arbor ramulis rugulosis; partes novellæ furfuraceæ, cito glabrescentes. *Folia* 5–8 poll. longa, 1½–2½ lata, petiolo ½-poll., juniora secus costam et nervos furfuracea, cito glabrescentia. *Pedicelli* graciles, petiolum æquantes, medio bracteolati. *Flores* ferrugineo-tomentosi, fœminei ovali-oblongi, masculi subglobosi. *Fructus*, secundum Jack, l. c., parvus, olivæ magnitudinem non attinens, ex spec. in Herb. Bentham asservato subglobosus, cerasi magnitudine, valvis tenuibus, arillo indiviso.

Specimens from Malacca in the Hookerian Herbarium appear to us to be identical with the plant described by Jack, and with the numbers of Cuming quoted above. If we are correct in these identifications, the species must be widely distributed. In our description we have chiefly followed Jack.

6. **M. gibbosa** (H.f. et T.); ramulis glabriusculis, foliis anguste lanceolatis acuminatis basi acutis subtus pallidis (in sicco flavescentibus), fructu oblongo tomentoso obliquo hinc gibbo, arillo tenui.

HAB. In mont. Khasia!—(*v. v.*)

Arbor, ramulis gracilibus foliosis, cortice griseo ruguloso; partes novellæ tenerrime furfuraceæ, cito glabrescentes. *Folia* glaberrima, coriacea, supra nitida, subtus pallida, secus costam subfurfuracea, cito glabrescentia, 4–7 poll. longa, 1–1$\frac{3}{4}$ lata, pet. $\frac{2}{3}$–$\frac{3}{4}$ poll. *Pedicellus fructus* $\frac{1}{2}$-pollicaris (pedunculo brevissimo). *Fructus valvæ* crasse coriaceæ, intus castaneæ, nervosæ, rugulosæ. *Arillus* apice tantum sublacerus. *Semen* obliquum, chalaza longe infra-apicali.

This species, which appears very distinct both in leaves and fruit, was obtained by us in the Khasia hills, with ripe fruit, in the month of June. We are inclined to associate with our Khasia plant a single specimen of a male plant in flower, collected by Griffith at Mergui, on the Tenasserim coast, which is evidently quite unlike all the states of *M. corticosa*, and which cannot be confounded with any other species. The leaves of this Mergui plant are identical with those of *M. gibbosa;* and if the two be the same, the male flowers of that plant may be thus described:—*Pedunculi* axillares, validi, $\frac{1}{2}$ pollicem longi, lignosi, nudi, apice in ramos plures abbreviatos dense squamosos divisi. *Pedicelli* plures, $\frac{1}{4}$–$\frac{1}{2}$-pollicares, medio bracteolam minutam amplectentem gerentes. *Calyx* ovalis vel campanulatus, $\frac{1}{2}$-trilobus, ferrugineopubescens. *Columna* staminea superne subclavata. *Antheræ* 12, suberectæ, disco breviter lobato adnatæ.

7. **M. corticosa** (H.f. et T.); ramulis glabratis, foliis anguste lanceolatis utrinque acutis rariusve obtusis subtus glaucis, pedunculis axillaribus abbreviatis plurifloris, disco plano 8–12-lobo, fructu ovali pulverulento, arillo carnoso.—Knema corticosa, *Lour. Fl. Coch.* 742! Myristica globularia, *Lam.* M. glauca, *Bl. Bijdr.* 576, *Rumph.* i. 182. *t.* 60. M. lanceolata, *Wall. Cat.* 6794! M. missionis, *Wall. Cat.* 6788! M. angustifolia, *Roxb. Fl. Ind.* iii. 847!

HAB. Chittagong! Tenasserim! et in penins. Malayana!—(Fl. Jan. Apr.) (*v. v.*)

DISTRIB. Java! Cochin China! Borneo!

Arbor sæpe excelsa. *Ramuli* cortice rubicundo vel fusco-cinereo, glaberrimi; partes novellæ vix subpulverulentæ. *Folia* elongato-lanceolata, interdum sed rarissime obtusa, 4–7 poll. longa, 1–1$\frac{1}{2}$ vel rarius 2 poll. lata, petiolo $\frac{1}{2}$–$\frac{2}{3}$-poll., rigide coriacea, glaberrima, venulis in sicco creberrimis reticulatis. *Pedunculi* 1–3 lineas longi, areolis squamarum delapsarum notati. *Pedicelli* graciles, supra medium bracteolati. *Flores* tenuissime tomentosi. *Arillus* puniceus, multipartitus. *Chalaza* obliqua.

Our Chittagong plant is undoubtedly the same with that from Singapur in Herb. Wallich, and we can in no way distinguish *M. missionis*, Wall., which was perhaps, therefore, obtained by König from the Straits, and was not a native of the Madras Peninsula. Our specimens quite agree with Blume's figure and detailed description, and with Loureiro's specimens in the British Museum. We adopt Loureiro's name, because we have no doubt that, on a general revision of the Order, his genus *Knema* will be kept up.

8. **M. intermedia** (Bl. Rumph. i. 187); ramulis glabriusculis,

foliis rigidis anguste lanceolatis utrinque acutis, floribus ferrugineo-tomentosis in pedunculo axillari pluribus, disco sub-15-lobo superne in
processum oblongum obtusum vel subconicum producto.

HAB. Malacca, *Griff.!* Singapur, *Lobb!—(v. s.)*
DISTRIB. Java, *Bl.*

Arbor. Ramuli fusco-cinerei. *Folia* glaberrima, fere *M. corticosæ. Pedunculi*
lignosi, vix 2 lineas longi. *Pedicelli* 5–10, crassiusculi, supra medium bracteolati.
Flores dense ferrugineo-tomentosi. *Columna* staminea brevis, disco late peltato.
Fructus (ex Blume) pyriformes, ochraceo-pulverulenti.

With quite the habit and general appearance of the last species, this has a very
different anther-disc, which, instead of being concave or flat, is elevated into a large
obtuse beak. Blume's description is very short, and we are therefore in some doubt
as to our being justified in considering his plant the same as ours. He says that it
differs from his *M. glauca* (our *M. corticosa*) chiefly in having "connectivum latius
atque apice subdilatato-obtusatum." In a specimen from Griffith in Herb. Hook.
the common peduncle, apparently from disease, and probably from insect puncture,
is converted into a globular woody mass, entirely covered with long brown hairs.

9. **M. furfuracea** (H.f. et T.); ramulis petiolisque dense furfuraceo-tomentosis, foliis anguste vel lineari-oblongis rigidis utrinque glaberrimis, pedunculis axillaribus brevissimis, floribus (fœmineis) pyrifor-
mibus dense tomentosis.—Knema glaucescens, *Wall. Cat.* 6810! *non
Jack.*

HAB. In penins. Malayana ad Penang, *Wall.!—(v. s.)*

Arbor parva, a prioribus ramulis dense tomentosis facile distincta. *Folia* supra
lucida, utrinque acuta, 4–7 poll. longa, 1– fere 2 lata, glaberrima, subtus ferruginea.
Flores ¼-pollicares. *Fructus* immaturi subglobosi, dense tomentosi.
The female plant only is known of this species, which appears sufficiently distinct
from all its congeners.

Sect. 2. IRYA.—*Calyx* 2–3-lobus, glaber. *Columna* staminea pyriformis, apice concava, margine antherifera. *Stigma* depressum,
integrum vel vix lobatum. *Fructus* globosus.—Flores *minuti,
in paniculis axillaribus ramosissimis dispositi*, bracteis *ovatis
glanduloso-punctatis citissime deciduis involuti.*

The staminal column of this section is not very different from that of some species
of *Knema*, but the inflorescence and habit are very different, approaching closely to
some of the American species, among which *M. Hostmanni*, Benth., seems to have a
very similar andrœcium. *M. micrantha*, Wall. Cat. 6807! from Finlayson's Herbarium, and therefore probably from Siam, which is beyond our limits, also belongs to
this section. It is larger-flowered than *M. Irya*, Gært., and the authers in the specimen before us are more numerous, so that it is perhaps a distinct species.

10. **M. Irya** (Gærtn. Fr. i. 195. t. 41); foliis oblongis acuminatis.
—M. Javanica, *Bl. Bijdr.* 576, *Rumph.* i. 190. t. 62. M. sphæro-
carpa, *Wall. Plant. As. Rar.* i. t. 89, *Cat.* 6796! M. exaltata, *Wall.
Cat.* 6804 *partim (quod ad specimina florifera ad Moulmein lecta!)*

HAB. Ava ad Martaban et Moulmein, *Wall.!; et Zeylania!—(v. s.)*
DISTRIB. Java, *Blume.* Borneo, *Low!*

Arbor sæpe procera, ramosa, cortice fusco ruguloso glabro; partes novellæ tenuissime puberulæ. *Folia* forma varia, oblonga vel lineari-oblonga, basi rotundata aut
obtusiuscula vel interdum acutiuscula, utrinque glabra, 5–10 poll. longa, 1½–2½ lata,

pet. $\frac{1}{3}$–$\frac{2}{3}$-poll. *Paniculæ* ad axillas foliorum delapsorum, ramosissimæ, 3–6-pollicares, floribus minutis glabris. *Antheræ* 6–8. *Fructus* diametro fere pollicaris, extus granuloso-tuberculatus, glaber, valvis crassis coriaceis intus glabris rufescentibus. *Arillus* tenuis, carnosulus, coccineus vel aurantiacus, completus. *Nucleus* globosus, hilo parvo.

Sect. 3. PYRRHOSA.—*Calyx* 2–4-lobus, glaber. *Columna* staminea depressa, subglobosa, apice concava, tota antherifera.— Flores *in paniculis axillaribus dispositi.*

Blume appears to have brought together under the name of *Pyrrhosa* all the paniculate Nutmegs; but the species included by him in the group by no means agree in floral characters. As defined above, however, it constitutes a very natural group. *M. (P.) Javanica,* Blume, belongs to our last section, and *M. (P.) Horsfieldii,* Blume, though anomalous as to inflorescence, seems rather referable to *Eumyristica.*

11. M. crassifolia (H.f. et T.); ramulis validis, foliis ovalibus vel oblongis utrinque obtusis, antheris 4 sessilibus ovalibus in columnam depresse globosam coalitis.

HAB. In peninsula Malayana ad Malacca, *Griff.!—(v. s.)*

Arbor, cortice fusco rugoso glabro. *Folia* crasse coriacea, nervis obliquis rectis apice tantum curvulis, subdistantibus, 5–9 poll. longa, 3–4½ lata, petiolo ¾-poll. *Paniculæ* axillares, ramosissimæ, pubescentes, 4–6 poll. longæ, bracteis ad ramos primordiales subpersistentibus ovalibus. *Flores* minuti, ad apices ramulorum fasciculati; calyx plerumque bivalvis.

12. M. amygdalina (Wall. Plant. As. Rar. t. 90, Cat. 6797!); foliis oblongo-lanceolatis vel cuneato-oblongis utrinque acutis, paniculis ramosissimis puberulis, floribus brevi-pedicellatis, fructu oblongo, arillo apice lacero.—M. exaltata, *Wall. Cat.* 6804! (*excl. spec. ad Moulmein lectis supra ad M. Irya relatis.*)

HAB. In Ava inf., *Wall.!;* Malacca, *Griff.!—(v. s.)*
DISTRIB. Sumatra, *Marsden!* (*in Herb. Hook.*)

Arbor procera. Ramuli teretes, glabri, cortice rugoso cinereo. *Folia* 5–8 poll. longa, 1¾–2¼ lata, petiolo ½-poll., basi plerumque longe attenuata, utrinque glabra, subtus ferruginea. *Paniculæ* 3–5-pollicares, plerumque ad axillas foliorum delapsorum sitæ, ad ramificationes bracteas ovatas acuminatas cito deciduas gerentes. *Calyx* globosus, plerumque trilobus. *Columna* staminea glabra, 3–4-loba, antheris linearibus apice incurvis. *Fructus* paniculati, breviter pedunculati, glauci, amygdali magnitudine, hinc sulco profundo exsculpti; valvæ crassæ. *Arillus* aurantiacus, tenuis, carnosus. *Chalaza* infra-apicalis.

The flowers of this species are as small as those of *M. Irya,* and very numerous, and the panicles are elongated, with long, much-spreading branches.

13. M. globularia (Bl. Rumph. i. 191. t. 64. f. 2, non Lam.); foliis ellipticis vel lanceolatis utrinque glabris, paniculis ferrugineo-tomentosis, floribus minutis globosis ad apices ramulorum fasciculatis brevissime pedicellatis, fructu ovoideo parvo.

HAB. In penins. Malayana ad Malacca, *Griff.!—(v. s.)*

Arbor mediocris. Ramuli fusco-cinerei, glabri; partes novellæ ferrugineo-puberulæ. *Folia* utrinque acuta, apice plerumque acuminata, 4–5 poll. longa, 1¼– fere 2 lata, petiolo ¾-poll., coriacea, supra nitida, subtus ferruginea, nervis subdistantibus parallelis ultra medium minus conspicuis curvatis. *Paniculæ* ad axillas foliorum delapsorum 2–3-pollicares. *Calyx* 3–4-lobus. *Antheræ* in globum 3–4-lobum coalitæ. *Fructus* (ex Blume, *l. c.*) ½-pollicaris.

14. **M. Wallichii** (H.f. et T.) ; foliis anguste oblongis acutis basi rotundatis coriaceis supra lucidis subtus pallidis (in sicco fulvis) præsertim secus costam et nervos tomentosulis, paniculis masculis elongatis ferrugineo-tomentosis (pedicellis florum brevissimis), fœmineis brevioribus incrassatis, fructu ovali.—M. Horsfieldii, *Wall. Cat.* 6806 ! (*non Blume*).
Hab. In penins. Malayana ad Malacca, *Griffith !* et Singapur, *Wall.!* —(*v. s.*)

Arbor præcedenti simillima, sed partes novellæ dense tomentosæ, et *folia* majora 6–9 poll. longa, 2½–3 lata, pet. ⅔-poll. *Nervi* crebriores, subtus validi, versus marginem arcuati. *Flores* aliquanto majores, calyce 3–4-lobo, glabro. *Flores* fœminei pedunculo simplici vel parum ramoso abbreviato (fructifero incrassato lignoso) inserti, calyce globoso late 3-dentato. *Ovarium* globosum, tomentosum. *Stigma* sessile, globosum, capitatum. *Fructus* junior glaber, glaucus, calyce persistente profunde tripartito, segmehtis ovatis fere 2 lineas longis glabris.
The female flowers are described from a specimen collected by Lobb at Singapur, which has all the appearance of belonging to this species; and the young fruit is iu the Wallichian Herbarium, but detached.

15. **M. tomentosa** (H.f. et T.); foliis obovatis vel cuneato-oblongis basi attenuatis subacutis subtus pilis stellatis laxe tomentosis, paniculis laxe fusco-pilosis ramosis.—Myristicea? *Wall. Cat.* 9025 !
Hab. In penins. Malayana ad Penang, *Wall.!*—(*v. s.*)

Arbor cortice cinereo, ramulis cum omnibus partibus novellis ferrugineo tomentosis, demum glabratis. *Folia* submembranacea, 5–8 poll. longa, 2–4 lata, petiolo ½-poll., obtusa vel acuta, nervis obliquis subtus prominentibus, infra tenuissime puberula subtus præsertim ad uervos fulvo-tomentosa. *Paniculæ* 2–3-pollicares, axillares, ante expansionem bracteis dense tomentosis involutæ. *Calyx* glaucus, 3–4-fidus.

16. **M. glabra** (Bl. Bijdr. 576, Rumph. i. 191. t. 64. f. 1); foliis lanceolatis vel obovato-oblongis acutis basi longe attenuatis utrinque glabris subtus pallidis vel ferrugineis, paniculis axillaribus ramosis, floribus glabris ovali-subglobosis, pedicellis calycem subæquantibus, fructu ovali-oblongo, arillo subcompleto.—M. integra, *Wall. Cat.* 6799 ! M. floribunda, *Wall. Cat.* 6805 !
Hab. In montibus Silhet, *Wall.!* et in penins. Malayana ad, Singapur, *Wall.!*—(*v. v.*)
Distrib. Java, *Bl.*

Arbor, ramulis rugulosis, etiam junioribus glabris; gemmæ vix puberulæ. *Folia* 5–7 poll. longa, 1½–3 lata, pet. ⅔–½-poll. *Paniculæ* sæpe elongatæ, 2–5-pollicares, ramis pollicaribus. *Flores* quam in præcedentibus majores, ovales, fere lineam longi, plerumque trilobi. *Antheræ* in massam ovalem coalitæ. *Fructus* ovalis, ultra-pollicaris, hinc infra medium sulco exsculptus, valvis crassissimis ; arillus tenuiter carnosus, coccineus. *Testa* cinerea ; chalaza lateralis vix supra medium posita.
This is larger-flowered than *M. amygdalina,* the only species with which it is liable to be confounded, and it seems quite distinct, though, as in most of the allied species, only the male flowers and fruit are known. The fruit is very like that of *M. amygdalina,* with which it agrees in having a deeply excavated mark on one suture near the base. The undivided arillus, however, will readily distinguish it, unless that character be found to be a variable one.

17. **M. Farquhariana** (Wall. Cat. 6798) ; foliis anguste oblongis acuminatis basi acutis rigide coriaceis utrinque glabris supra lucidis

Y

subtus glaucis, paniculis ramosis, floribus majusculis ad apices ramorum brevium umbellatis, pedicello calycem ovalem superante, fructu ovali, arillo fere ad basin lacero.

HAB. In Malaya ad Malacca, *Griff.!* et Singapur, *Wall.!;* et in montibus Kúrg, *Hohenacker,* No. 541!—(*v. s.*)

DISTRIB. Ins. Philipp., *Cuming,* 901!

Arbor, ramulis rubescentibus vix rugulosis glabris; gemmæ vix puberulæ. *Folia* 4–8 poll. longa, 1¼–2 lata, petiolo ½-poll., marginibus (in sicco) recurvis. *Paniculæ* ad axillas foliorum delapsorum 2–3-pollicares, ramis elongatis vel abbreviatis. *Flores* in umbellas subsessiles vel breviter pedicellatas 5–10-floras congesti. *Calyx* ⅛ poll. longus, 3–4-lobus, glanduloso-punctatus. *Antheræ* 8, biloculares, apice ultra connectivum productæ, liberæ, incurvæ, in columnam oblongam subsessilem coalitæ. *Flores fœminei* ignoti. *Fructus* in paniculis ½–1-pollicaribus lignosis parum ramosis subracemosis pauci (interdum subsolitarii), ovoidei, glabri, ¾ poll. longi. *Arillus* profunde lacerus. *Chalaza* subterminalis.

The elongated calyx and anther-column of this species indicate an approach to *Eumyristica,* and the inflorescence is more simple than in the majority of *Pyrrhosæ.* The anthers are, however, distinctly involute, and lobed at the apex.

Sect. 4. EUMYRISTICA.—*Calyx* ovalis vel elongatus, 3–4-lobus. *Antheræ* in columnam cylindricam basi nudam coalitæ. *Inflorescentia* varia.

18. **M. superba** (Hf. et T.); foliis lanceolatis utrinque acutis subtus tomento fulvo furfuraceis, paniculis furfuraceis elongatis ramosis, floribus majusculis ad apices ramulorum umbellatis.

HAB. In insula Penang, *Phillips!*—(*v. s.*)

Arbor forsan elata, cortice ruguloso pallide fusco albo-punctato; ramuli tomentosi. *Folia* 12–18 poll. longa, 3–7 lata, pet. ¾-poll., firma, supra glabra, subtus laxe tomentosa (margine in sicco recurva), nervis conspicuis numerosis obliquis rectiusculis prope marginem sursum arcuatis. *Paniculæ* in axillis foliorum delapsorum sitæ, 4–6 poll. longæ, divaricato-ramosæ, ad ramificationes bracteis parvis rotundatis demum induratis muniti. *Flores* in umbella 3–7, ⅓ poll. longi, pedicello æquilongo suffulti. *Calyx* glaber, ovalis, apice 3–4-lobus. *Antheræ* in columnam ovalem subsessilem coalitæ, elongato-lineares, circa 18, apice longitudine subinæquales, columna connectivo haud apiculata.

There is no apiculus of connective beyond the anthers, but as the mass of the column narrows into a rounded point, some of the anthers are continued to the apex, while the others stop abruptly a little lower.

19. **M. elliptica** (Wall. Cat. 6798 A!); foliis tenuibus lanceolatis vel oblongo-lanceolatis utrinque acutis glabris subtus glaucis, pedunculo supra-axillari brevi paucifloro, floribus fasciculatis pedicellatis.

HAB. In Malaya ad Penang et Singapur, *Wall.!*—(*v. s.*)

Arbor, ramulis fuscis lævibus glabris. *Folia* 6–10 poll. longa, 2¼–3½ poll. lata, petiolo ½-pollicari. *Pedunculus* ½-pollicaris. *Bractea* rotundata, calyci adpressa. *Calyx* masculus oblongus, subinflatus, tridentatus, ⅓ poll. longus. *Columna* staminea breviter pedicellata, cylindrica, connectivo apiculata. *Calyx* fœmineus urceolatus. *Ovarium* aureo-strigosum, intra calycem stipitatum, oblongum, superne attenuatum, sulcatum. *Stigma* oblongum, hinc canaliculatum. *Fructus* oblongus, 2½ poll. longus, glaber, valvis crassis carnosis. *Arillus* apice lobatus. *Chalaza* subapicalis.

We have described this plant from imperfect materials, but it seems very distinct from any of our other species, as well as from those described by Blume. The *B.*

of Wallich's Catalogue, from which the fruit is described, has a mark of doubt appended to it, and its leaves are too imperfect to enable us to determine whether or not it be identical with the flowering specimens.

20. **M. laurifolia** (H.f. et T.); foliis ovali-oblongis vel oblongis glabris, pedunculis masculis abbreviatis crassis apice plurifloris, fructu ovali tomentoso.

HAB. In Zeylania, *Walker! Gardner! Thwaites!*—(*v. s.*)

Arbor, cortice griseo vel rufescente rugoso. *Ramuli* lævigati, glabri; partes novellæ vix puberulæ. *Folia* utrinque obtusa vel acutiuscula, supra lucida, subtus pallida, conspicue oblique nervosa, 6–9 poll. longa, 2–4 lata, petiolo ultra-pollicari. *Pedunculi* masc. axillares, plerumque bini, ¼–½ poll. longi, lignosi, cicatricibus bractearum crebre notati. *Flores* pedicellis vix lineam longis suffulti, dense tomentosi, ovales, pedicellis duplo longiores, bractea rotundata calyci adpressa suffulti. *Calyx* breviter 3-lobus. *Columna staminea* pedicellata, apice processu connectivi apiculata. *Antheræ* circa 9. *Fructus* solitarius, ovalis, in spec. immaturis dense tomentosus, pedicello ¼–½ poll. suffultus.

21. **M. obtusifolia** (Wall. Cat. 6808!); foliis obovato-oblongis rigidis supra glabris nitidis subtus ferrugineo-glaucis sparse puberulis, paniculis fructiferis parce ramosis fusco-tomentosis.

HAB. In Malaya ad Singapur, *Wall.!*—(*v. s. in Herb. Linn. Soc.*)

Folia 6–12 poll. longa, 3–4½ lata, petiolo 1–1½-pollicari, nervis obliquis distantibus rectis apice arcuatis subtus prominentibus notata. *Flores* ignoti. *Paniculæ* fructiferæ 1½ poll. longæ. *Fructus* ⅔-pollicares, oblongi, tomentosi. *Semen* subglobosum, arillo subcompleto, testa tenui.

22. **M. Malabarica** (Lam. Act. Paris. 1788. p. 162); foliis anguste oblongis vel elliptico-lanceolatis, inflorescentia mascula axillari dichotome cymosa multiflora, floribus laxe umbellatis, fructu oblongo fulvo-incano.—*Blume, Rumph.* i. 185. M. dactyloides, *Wall. Cat.* 6786! (*vix Gærtner*). M. notha, *Wall. Cat.* 6787!—*Rheede, Hort. Mal.* iv. *t.* 5.

HAB. In sylvis densis Malabariæ et Concan, *Sykes! Dalzell! Law!* —(Fl. Nov. Feb.) (*v. s.*)

Arbor excelsa; ramuli glabri, cortice rubescente læviusculo. *Folia* superne augustata, apice obtusa, basi acuta vel rarius rotundata, 4–8 poll. longa, 1¼–4 lata, petiolo ½–1-pollicari, glaberrima, subtus glauca. *Cymæ* masculæ 1–3-pollicares; ramuli oppositi, apice umbellatim pluriflori. *Pedicelli* ¼–½ poll. longi, graciles. *Alabastri* fere globosi. *Calyx* inflatus, late ovalis, trilobus, pubescens, basi bractea amplectente adpressa latissima munita. *Antheræ* 15 in columnam solidam oblongam pedicello brevi crasso pubescente suffultam coalitæ, connectivo apice apiculatæ. *Flores fœminei* ad apicem pedunculi axillaris umbellati, pauci (in spec. 2); *masculis* majores, urceolati. *Ovarium* dense tomentosum. *Fructus* 2½–3 pollices longus, oblongus, pericarpio bivalvi crasso carnoso. *Arillus* ex rubro flavescens, irregulariter fissus, lacunosus, lobis versus apicem in conum seminis apici insidentem contortuplicatis. *Semen* ovoideum, erectum, utrinque obtusum, subobliquum, ventre planiusculum, dorso convexius, tegmine lignoso tenui, fragili, arilli pressione irregulariter sulcato, medio latere chalaza notato. *Rhaphe* a hilo ad chalazam linearis.

According to Rheede, the pericarp is acid and astringent, with a disagreeable smell. The aril is less agreeably flavoured than true mace, and the nut has scarcely any taste or smell.

23. **M. Horsfieldii** (Bl. Bijd. 577, Rumph. i. 192); foliis ovato-oblongis acuminatis subtus stellato-pubescentibus, floribus dense glo-

merulato-paniculatis.—M. Iryaghedhi, *Wight, Icones, t.* 1857. *Gærtn.*
i. *t.* 41, *ex parte.* M. Iriagedi, *Spr. Syst. Veg.* iii. 65. M. ferruginea,
Wall. Cat. 6803 ! Horsfieldia odorata, *Willd. Sp.* iv. 872.

HAB. In Zeylaniæ sylvis !—(*v. s.*)

Arbor excelsa, cortice nigro-fusco striatulo glabro, ramulorum juniorum dense tomentoso; partes novellæ floccoso-tomentosæ. *Folia* 6–9 poll. longa, 2¼–3½ poll.
lata, petiolo ½–1-pollicari, basi subcordata vel rotundata vel interdum subacuta, margine recurva, glabra, læte viridia, ochreo-tomentosa. *Paniculæ* axillares; masculæ
3–6 poll. longæ, ramosæ, ramis alternis, capitula 3–5 subsessilia gerentes; fœmineæ
plerumque multo breviores, simplices, dense furfuraceo-tomentosæ. *Flores* suaveolentes; *masculi* dense glomerati, sessiles, obconici, mutua pressione angulati, calyce
3–4-dentato.   *Antheræ* 6 in columnam gracilem clavatam apice connectivo haud
apiculatam coalitæ.   *Flores fœminei* laxiores, subsessiles, basi turgidi.   *Ovarium*
tomentosum, stigmate sessili indiviso. *Fructus* ovoidei, ferrugineo-tomentosi. *Arillus* carnosus, completus, indivisus.

Blume has very properly rejected the barbarous name employed by Gærtner,
which is only doubtfully referable to the present species, as the synonyms quoted belong elsewhere. As *M. Horsfieldii* is stated by Blume to be only known in a cultivated state in Java, *M. ferruginea*, Wall., is probably also cultivated at Singapur.
It is, according to Blume, closely allied to the Madagascar species, *M. Madagasca-
rensis* and *M. acuminata*, Lam.

VII. MONIMIACEÆ.

Flores unisexuales, rarius hermaphroditi. *Sepala* basi plus minus
coalita. *Petala* nulla vel sepalis alterna interdum pluriserialia, æstivatione imbricata. *Stamina* perigyna, definita et uniserialia, vel sæpius
indefinita et calycis tubo inserta; basi plerumque glandulis stipata.
Antheræ biloculares. *Ovaria* indefinita, unilocularia, ovulo solitario
pendulo anatropo. *Drupæ* siccæ; *semen* pendulum; *albumen* carnosum; *embryo* minutus hilo versus; *radicula* supera; *cotyledones* divaricatæ.—Arbores *vel* frutices, foliis *oppositis, exstipulatis, integris, den-
talis vel integerrimis*, inflorescentia *cymosa axillari vel terminali.*

The genus *Hortonia* furnishes precisely the information required to settle definitely the position of the Order to which it belongs, for it cannot be doubted that it
is a genuine Monimiaceous plant, notwithstanding its hermaphrodite flowers, numerous petals imbricated in several rows, and definite stamens. The opposite exstipulate leaves, slightly perigynous stamens furnished with glands, solitary pendulous
anatropous ovules, and, above all, the peculiar character of the fruit and embryo of
Hortonia, agree so precisely with the Order, that its right to a place there cannot
be disputed.

Monimiaceæ being generally apetalous, have sometimes been considered achlamy-
deous and involucrate; but the regularly imbricated perianth of *Hortonia* is opposed
to this view of their structure, which had already indeed been rendered improbable
by the regular alternation of the inner series of segments with the outer in *Boldoa*.

If the presence of a perianth be admitted, the place of *Monimiaceæ* is necessarily
among apocarpous orders; and its minute embryo, with divaricating cotyledons in
copious albumen, bring it naturally into the great class upon which we are now engaged, notwithstanding the more or less perigynous insertion of the stamens of the
greater number of genera, and the opposite leaves, which indeed occur likewise in
Clematideæ. The glandular appendages of the filaments, and the valvular dehiscence
of the anthers of *Atherospermeæ* (which must share the position of *Monimiaceæ*) do

not require us to look to *Laurineæ* for the allies of these plants, as they are present also in *Berberideæ.*

The divarication of the cotyledons is a curious character, the physiological import of which it is not easy to determine. It is of frequent but by no means universal occurrence in the great class of plants to which we have referred *Monimiaceæ.* Among *Ranunculaceæ* it occurs in some species of *Clematis*, as was first indicated by De Candolle. Among *Magnoliaceæ* also it is common, and it occurs in all *Myristicaceæ*, and in the whole of the first tribe of *Menispermaceæ.* In *Hortonia* and *Hedycarya* the embryo is included within the albumen, a portion of which penetrates between the divaricating cotyledons. In *Hortonia* the central albumen is very distinct from that near the surface, being paler in colour as well as laxer in texture ; and a longitudinal section of the seed shows that the line of demarcation between the two is continuous with the apex of the cotyledons, and that the exterior albumen is perisperm, while that between the cotyledons, and continuous thence almost to the base of the seed, is perisperm. Evident traces of the embryo-sac may be seen covering the embryo, which occupies a small cavity in the perisperm. It is, however, evanescent below.

In *Boldoa* the embryo is figured by Lindley as being altogether exterior to the albumen ; but we find the structure even more anomalous than he describes it, although his analysis of the seed of that curious plant is, as might be expected, quite accurate. Dr. Lindley describes a thick fleshy testa and spreading cotyledons resting on the albumen. We find a thin brown coat, not readily separable from the albumen, and traversed by a broad rhaphe, which terminates in a thickened large circular chalaza like that of *Hortonia.* Within this coat there is a fleshy layer of considerable thickness, and the ovate widely-divaricating cotyledons rest upon another fleshy mass, which is everywhere readily separable from the outer, except sometimes at the base, and is undoubtedly albumen, and no doubt endosperm, that is to say, developed in the embryo-sac. It will be seen that this structure only differs from that of *Hortonia* by the larger size and greater divarication of the cotyledons, and by the increased mass and more complete separation of the inner albumen from the outer fleshy layer. We think that the same explanation will apply to both genera, and that the fleshy coat of *Boldoa* is perisperm. That it cannot be testa, as Lindley supposes, is, we think, proved by the position of the chalaza exterior to it.

The nearest allies of *Monimiaceæ* in the class to which we propose to refer them, are, we think, *Schizandraceæ.* We are led to this conclusion by the principle long ago laid down by Mr. Brown, that the most perfect species of a group ought to be kept in view in determining the affinities of the whole. In the present family, *Hortonia*, which is hermaphrodite and petaliferous, appears to claim the highest place, and the resemblance of its flower-buds to those of *Kadsura* or *Illicium* must strike every one. The ovaries and style are also very like those of *Kadsura*, while the oily albumen and the embryo are quite Magnoliaceous. At the same time *Monimiaceæ* form undoubtedly a very distinct family, not closely allied to any other, but presenting evident relations to all the Orders of the class. It is worthy of note that Dr. Wight, in founding the genus *Hortonia*, referred it to *Schizandraceæ*, an Order with which he was only acquainted by means of books.

Monimiaceæ are a very small Order, and are almost entirely confined to the southern hemisphere, our Indian species and a few which inhabit Mexico and Panama being the only exceptions. Tropical South America is the great centre of the Order, whence it extends south through Peru to Chili. In Africa several are natives of Madagascar and the islands of Mauritius and Bourbon, but none have yet been obtained from the continent. Australia and New Zealand also contain a few species.

1. **KIBARA,** Endlicher.

Brongniartia, *Blume, non Kunth ;* Sciadicarpus, *Hasskarl.*

Flores diclines. *Calyx* turbinatus, basi bibracteolatus, ore squamis (ex Blume 4, secundum Hasskarl pluribus) bi-triserialibus conniventibus subclauso.   MASC. *Stamina* 5–7 ; filamenta brevia ; antheræ basifixæ. FŒM. *Ovaria* indefinita, pyramidata.   *Stigma* sessile, obtusum.   *Drupæ* siccæ, stipitatæ, calyci demum fisso et reflexo indurato insidentes.   *Albumen* carnosum.—Arbor *procera*, foliis *inconspicue repando-dentatis*, inflorescentia *axillari racemosa*, floribus *minutis aurantiacis.*

This genus seems abundantly distinct by the definite stamens, and, according to Hasskarl, by the scales of the involucre in several rows. It wants the staminal glauds of *Hortonia.* Our materials do not enable us to add anything to the information given regarding it by Blume and Hasskarl. The latter tells us that the cotyledons are very small and adpressed. Only one species is known.

1. **K. coriacea** (Endl. Gen. 314).—Brongniartia coriacea, *Bl. Bijdr.* 436.   Sciadicarpus Brongniartii, *Hassk. Plant. Jav. Rar.* 209.
HAB. Ad Malacca, *Griffith!*—(*v. s.*)
DISTRIB. Java, *Blume.*

Arbor excelsa; rami crassi; teretes, ad nodos subcompressi ; ramuli complanati, puberuli. *Folia* ampla, coriacea, ovato-oblonga, acuta vel acuminata, 5–10 poll. longa, 2½–5 poll. lata, petiolo ½–1-pollicari, versus apicem repanda aut subserrata, supra glaberrima, subtus ad nervum medium puberula. *Cymæ* axillares, petiolos superautes. *Calyx* subglobosus, vix ¼-pollicaris. *Filamenta* parva, obovata, complanata; *antheræ* minutæ. *Drupæ* 1–15 calyci indurato reflexo margine truncato insidentes, oblongæ, ½-pollicares, pedunculo ¼-pollicari suffultæ.
The details given above are chiefly taken from Hasskarl's descriptiou, as we have only seen one very imperfect fruiting specimen, bearing a single drupe.

2. **HORTONIA,** Wight.

Flores hermaphroditi.   *Petala* (cum *sepalis*) circa 30, multiserialia, æstivatione imbricata, basi subcohærentia, exteriora carnosa rotundata, interiora sensim longiora et tenuiora, intima ligulata acuta.   *Stamina* 7–10, ad marginem tori explanati uniserialia.   *Filamenta* cylindrica, basi glandulis 2 magnis carnosis cucullatis extus stipata.   *Antheræ* extrorsæ, adnatæ, late ovales, biloculares, longitudinaliter dehiscentes.   *Ovaria* indefinita (15–20), oblongo-subulata, stigmate sessili dilatato erecto acuto.   *Drupæ* dense glomeratæ, siccæ, ovoideæ, lateraliter compressæ.   *Putamen* læve, fragile.   *Semen* pendulum ; *testa* tenuis ; *chalaza* magna, basilaris ; *rhaphe* marginalis conspicua.   *Albumen* oleosum ; *embryo* minutus, in albumine inclusus, cotyledonibus ovalibus obtusis divaricatis, radicula supera.—Frutices *glabri*, foliis *integerrimis*, inflorescentia *axillari cymosa*, floribus *pallide flavis.*

1. **H. floribunda** (Wight ex Arn. in Jard. Mag. Zool. Bot. ii. 545).
Var. *a. acuminata ;* foliis oblongo-lanceolatis acutis vel longe acuminatis subtus pallidis.—H. floribunda, *Wight, Ic. t.* 1997. H. acuminata, *Wight, Ic. t.* 1998, *fig. dext.*
Var. *β. ovalifolia ;* foliis ovalibus obtusiusculis crassioribus margine reflexis subtus lucidis.—H. ovalifolia, *Wight, Ic. t.* 1998. *fig. sin.*
HAB. In Zeylaniæ sylvis, alt. 4–6000 ped., *Walker! Gardner! Wight! Thwaites!*—(*v. s.*)

Frutex magnus, ramulis glabris, ad nodos compressis; partes novellæ pilis stellatis subfurfuraceæ.   *Folia* 3-6 poll. longa, $\frac{3}{4}$-2$\frac{1}{2}$ lata, petiolo $\frac{1}{4}$-$\frac{2}{3}$-pollicari.   *Cymæ* foliis breviores, 1-3-pollicares, pluri- vel pauciflori.   *Drupæ* in sicco acutæ, $\frac{1}{2}$-pollicares, pedicello brevi suffultæ.

Dr. Wight distinguished three species by the inflorescence and shape of the leaves; but these characters appear too variable to be relied on, several specimens now before us being quite intermediate in shape of leaf.   The number of flowers is certainly not at all to be depended upon.

VIII. MENISPERMACEÆ.

Flores abortu unisexuales, plerumque dioici.   *Sepala* 4-12, plerumque 6 biserialia, rarius 4, rarissime 5 uniserialia, interdum multiserialia, discreta vel rarissime gamosepala, æst. imbricata, rarissime valvata.   *Petala* 6, serie duplici imbricata, rarius 4 vel nulla, interdum basi gamopetala (rarissime 5).   *Stamina* petala numero æquantia et iis opposita, vel plura (9–18) rarissimo 3; *filamenta* libera, vel in columnam centralem cylindricam aut globosam coalita (in *Odontocarya* biserialia, petalorum numero dupla); *antheræ* valde variæ, ad apicem filamenti adnatæ et tunc extrorse vel lateraliter vel introrse imo transverse dehiscentes, seu circa discum peltatum horizontaliter dispositæ, seu supra globum irregulariter sitæ; in floribus fœmineis effœtæ vel nullæ.   *Ovaria* plerumque 3 (petalis exterioribus opposita), rarius solitaria, interdum 6 vel plura, toro inserta vel gynophoro brevi suffulta, uniovulata.   *Ovula* amphitropa, suturæ ventrali peltatim affixa, rarissime anatropa, tegumento unico, micropylo superiore, chalaza basin ovarii spectante.   *Styli* terminales, subulati vel depressi, interdum 3–5-lobi.   *Carpella* drupacea, styli cicatrice terminali vel sæpius basilari notata; *putamen* lignosum vel fere osseum, obscure bivalve, per ovarii maturationem in unaquaque fere specie modo diverso deformatum.   *Semina* putaminis cavitati plerumque valde irregulari conformia, hippocrepice curvata, vel uncinata vel circa processum internum putaminis conchæformia, rarissime recta.   *Testa* tenuissime membranacea.   *Albumen* copiosum vel parcum, oleosum, æquabile vel membranæ nucleariæ laminis tenuibus transversis ruminatum, interdum nullum.   *Embryo* in speciebus exalbuminosis crassus carnosus cotyledonibus amygdalinis, in albuminosis centralis vel parum excentricus; *radicula* semper superior et ad styli cicatricem spectans, sed in seminibus hippocrepicis fere basilaris, cylindrica.   *Cotyledones* valde variæ, sæpe foliaceæ divaricatæ et in loculis separatis albuminis inclusæ, plerumque normaliter appositæ, lineari-oblongæ vel semicylindricæ, radiculam diametro non superantes, in exalbuminosis crassæ amygdalinæ.—Frutices *scandentes vel sarmentosi,* foliis *exstipulatis alternis plerumque palminerviis et sæpe peltatis,* petiolis *basi (et interdum etiam apice) pseudo-articulatis et sæpius basi vel utrinque incrassatis.*   Flores *inconspicui, plerumque minuti, paniculati, racemosi vel cymosi, rarissime in axillis solitarii.*

As now correctly limited by the exclusion of *Lardizabalaceæ,* and of a number of genera which were only referred hither because their structure was quite unknown,

Menispermaceæ constitute a very natural Order. The scandent habit, alternate, exstipulate, palminerved leaves, petioles dilated and jointed at the base, minute, unisexual, thalamiflorous flowers, arranged in a ternary order, in at least three and usually four rows, and imbricated in æstivatiou, the small scale-like petals, definite stamens, definite apocarpous ovaries, solitary amphitropal ovules and fleshy drupes, characterize all the typical species, and form a combination of characters which is to be met with nowhere else. Of these the alternate exstipulate leaves, petioles not sheathing at the base, minute thalamiflorous flowers, solitary ovules, and peculiar drupaceous fruit, are constant, but all the others are subject to exception.

The scandent habit is almost universal in the Order, but it is absent in *Cocculus laurifolius*, which is a small, erect (or somewhat sarmentose) tree.

The peculiar structure of the petiole so common in *Menispermaceæ* can only be compared to that of a few *Euphorbiaceæ*. The petiole is generally elongated and cylindrical, without any marked groove in front, and has the appearance of being articulated with the stem, but the leaves are seldom very deciduous. The joint is sometimes a little above the base, so that a small projection is left on the branch after the leaf falls away. Above the base, and sometimes also near the leaf, the petiole is generally thickened, but contracts suddenly. The thickened portion of the petiole is often weaker in texture than the remainder, and exhibits a tendency to that twisting which is characteristic of the petioles of *Clematideæ*. In most species of *Cocculus* the petiole is short, not dilated at either extremity, and scarcely articulated.

The leaves of *Menispermaceæ* vary much in shape and texture. The most common shape is broad cordate, or nearly round ; they are often peltate, but this mode of attachment to the petiole is frequently present and absent in the same species, and occurs at times in young individuals, even when absent in the adult plant. Many, however, have elongated leaves. The palmate arrangement of the nerves is not confined to the peltate and broad cordate-leaved species, those with elongated leaves being always three-nerved at the base. The leaves often present a great variety of form, size, and texture in the same individual, so that copious suites of specimens are necessary for the proper illustration of each species.

The flowers are almost always unisexual, but Mr. Miers mentions the occurrence of hermaphrodite flowers in *Tiliacora* and *Odontocarya*. In the female flower imperfect stamens are usually present, and in the male more or less distinct traces of the gynœcium are usually found, except in the tribe *Cissampelideæ*, and other monadelphous genera, in which the staminal column occupies the centre of the flower.

The ternary arrangement of the parts of the perianth is of very general occurrence. The most remarkable exception is met with in Mr. Miers's genus *Odontocarya*, in which he describes the calyx and corolla as forming each a single verticil of five leaves. *Odontocarya* appears, however, from Mr. Miers' analysis, to be a genuine menispermaceous plant, the embryo having the laterally divaricating cotyledons of the tribe *Tinosporeæ*. In a part of the *Cissampelideæ* the segments of the perianth are arranged in a binary (or more rarely quaternary) order, and in the same tribe they are not unfrequently combined into a gamopetalous calyx and corolla. The solitary sepal of *Cissampelos* (usually called petal) in the axil of the bract is evidently formed of two combined sepals.

The number of verticils of the perianth is normally four, and they are usually sufficiently distinguishable into calyx and corolla, the latter being much the smallest, so that the petals were often described by the older botanists as nectaries or scales. Occasionally the petals are reduced to a single verticil of three, or entirely suppressed ; sometimes also the sepals are increased by the addition of one or more verticils, or of a number of irregularly imbricated bracts. In *Coscinium* the petals are larger than the sepals.

The imbricated æstivation of the perianth is not without exception, as in *Tiliacora* and several *Limaciæ* the inner sepals are valvate. This has been pointed out in *Tiliacora* by Mr. Miers, and in *Limacia* by Dr. Asa Gray. The petals of *Coscinium* are in like manner very slightly imbricated.

The stamens are normally free and definite, one being placed opposite each petal, so that they form two verticils. In *Limacia triandra* they are reduced to three; in another species of the same genus their number is nine; and in *Menispermum* and *Calycocarpum* they are indefinite. In *Pycnarrhena, Chasmanthera,* and *Abuta* (of Pöppig) the filaments are partially monadelphous; and in *Parabæna, Aspidocarya,* and the whole of the tribe *Cissampelideæ,* they are combined into a central column, bearing on its apex a flat peltate disc, antheriferous round the margin. In *Anamirta* the structure is still more complex, the anthers being united into a globose mass.

The ovaries are sometimes seated directly on the torus, but not unfrequently they are supported by a distinct gynophore, which becomes very conspicuous as the fruit advances to maturity. Their number is usually three; but in the tribe *Cissampelideæ* they are always solitary, and in *Cocculus ovalifolius* and *Coscinium* there are generally six. In *Tiliacora* they are indefinite in number.

The ovary of *Menispermaceæ* is generally oval or oblong, straight on the ventral suture, and rounded on the back, with a terminal style. The ovules are solitary and peltate, and inserted at or below the middle of the ventral suture, with the micropyle invariably superior, and the chalaza at the broad end of the ovule, which is nearest the base of the ovary. In *Aspidocarya,* and an undetermined species nearly allied to it, in which the seed is pendulous and anatropous, the ovule is probably attached near the apex of the ovary; but nevertheless the micropyle and foramen have the same position as in the rest of the Order.

During the ripening of the fruit great changes take place in the structure of the ovary. The dorsum grows more rapidly than the ventral part, so that the style or its cicatrix, which is terminal in the ovary, is in the ripe fruit more or less lateral, and in a large part of the Order is situated close to the base of the carpel. While this irregular development of the parietes of the ovary is proceeding, the inner wall gradually hardens into a more or less woody putamen, sometimes very thick and almost bony, at other times thin and brittle, and variously tuberculated. At the same time the podosperm lengthens as the hilum of the seed is carried by the increasing curvature of the walls of the ovary further and further from the base of the fruit; while the putamen, which thus becomes as it were doubled upon itself, invests it with a bony sheath, which takes a great diversity of form in different parts of the Order.

Mr. Griffith* has thrown out a conjecture that the woody or bony portion of the fruit is not putamen, but testa. This view receives some support from the fact that only one very delicate coat can be detected on the seed, and from the peculiar mode in which the bony coat adapts itself to the shape of the seed; but it is not borne out by a study of the development of the ovule, which we have been able to trace so satisfactorily as to ascertain beyond a doubt that this coat belongs to the ovary, and not to the ovule.

The form of the embryo is very different in different tribes of the Order. Except in *Aspidocarya* it is always more or less curved; and in the greater part of the Order, where the style-scar is situated near the base of the fruit, the radicle, which always points towards it, is brought almost into contact with the base of the fruit and the chalazal extremity of the seed. In the division *Heterocilineæ* the cotyledons are foliaceous and very thin, and (usually laterally) divaricated, so as to occupy distinct cavities in the albumen. The seed is therefore broad, and, but for the peculiar mode of growth of the putamen, would be quite flat, as it is in the genus *Aspidocarya.* This, however, causes it to assume a globular shape; but it is hollow within, and moulds itself on an internal process of the putamen, which Mr. Miers has called *condyle.*† In the remainder of the Order the narrow, strap-shaped or hemi-

---

* Itinerary Notes, p. 165.

† We have not adopted this term, partly because it does not represent an organ or structure analogous to that so called in osteology, and partly because we hesitate to apply specific terms to modifications of structure which are confined to small

spherical cotyledons have the ordinary position. The seed has therefore an elongated cylindrical shape, and is curved like a horse-shoe or siphon, the hilum occupying the bottom of the concavity. The putamen then forms a bony sheath, which closely invests the seed; but the concavity of the horse-shoe, along which the nutritive vessels run from the base of the fruit, is filled up by one or more bony plates, variously perforated, and sometimes enclosing empty spaces.

The albumen of *Menispermaceæ* varies as much as the form of the embryo. It is generally only present in small quantity, and in the tribe *Pachygoneæ* it is entirely wanting. Most commonly it is fleshy and homogeneous; but in several genera not otherwise very closely allied, namely in *Tinospora*, *Abuta*, and *Tiliacora*, it is very oily, and ruminated by means of transverse membranous plates. In *Anamirta* it contains small granular masses of a different texture from the greater portion; and finally, in *Coscinium* it is irregularly ruminated by cellular plates springing from the hilum, the nature of which has not been accurately determined.

The close relationship of *Menispermaceæ* to the great class of *Apocarpous Thalamiflora*, in which they are generally placed, may be considered well established; as the ingenious arguments by which Dr. Lindley attempts to maintain his opinion that they are more nearly related to apetalous orders have been well answered by M. Decaisne; who has shown, as we think, successfully, that neither the structure of the wood (to which we shall advert more particularly further on) nor the unisexual flowers, are to be relied on as indications of affinity.

To all the Orders of this great class, *Menispermaceæ* present more or less affinity by means of aberrant species, though the typical forms collectively possess such a peculiar habit as to make them a very natural family. With *Anonaceæ* they are connected by means of the genera with ruminated albumen; with *Myristicaceæ* through *Coscinium*; with *Lardizabalaceæ* and *Berberideæ* through *Burasaia*; to *Ranunculaceæ* they are only allied through *Berberideæ*, and to *Magnoliaceæ* through *Schizandraceæ*. *Dilleniaceæ* are the most distant, lying at the opposite extremity of the class, so as to form a passage to a very different series of Orders.

Menispermaceæ agree with *Berberideæ* in the structure and number of the parts of the perianth, in the usually definite stamens, in the solitary ovarium of *Cissampelideæ*, and in the comparatively large embryo, differing, however, in many important points.

From *Lardizabalaceæ*, which they approach very closely in the number of parts and in the diœcious flowers, *Menispermaceæ* are readily distinguished by their solitary ovules. The remarkable position of the indefinite ovules of all the genera of *Lardizabalaceæ* except *Decaisnea*, the anatropous seeds, and the minute embryo, are other important distinctions. The compound leaves of *Lardizabalaceæ* exist in *Burasaia*, which was by Decaisne doubtfully referred to that Order, but which Mr. Miers places in *Menispermaceæ*. Its structure appears to us to be quite intermediate between the two; but though the anatropous ovules are anomalous in *Menispermaceæ*, the seed has, according to Thouars, the divaricating cotyledons of the tribe *Tinosporeæ*. The genus *Lardizabala* has amphitropous seeds, excavated on one side, so as to resemble those of *Tinospora*, but their minute embryo is not Menispermeous.

Anonaceæ, which in general are so very distinct in habit and characters, are yet immediately connected with *Menispermaceæ* by those genera which have definite stamens, as well as by the remarkable occurrence of ruminated albumen in several genera of *Menispermaceæ*. The abnormal genus *Pycnarrhena* approaches in habit to such aberrant *Anonaceæ* as *Stelechocarpus* and *Guatteria pallida*.

Schizandraceæ form the link which connects *Menispermaceæ* with *Magnoliaceæ*; but the relationship is not very near, except by means of *Sabia*, which is very closely allied to both Orders; and by the amphitropous ovules of both.

Notwithstanding the close relationship which is now fully established as existing

---

Natural Orders, and are not of universal occurrence in them. Mr. Miers' condyle we shall call *processus internus putaminis*, and we shall designate it as *condyliformis*, when (as in *Tiliacora*) it resembles a condyle in form.

between *Menispermaceæ* and the Orders just mentioned, it cannot be denied that the large size of the embryo, and the small quantity of albumen, are very abnormal in the class to which they belong, and indicate that their true position is at one extremity of this class, and that, as in the case of *Dilleniaceæ*, they form a passage from it to another part of the vegetable kingdom. In fact, we think that the relationship of *Menispermaceæ* to the Malval alliance, in which we include *Euphorbiaceæ*, is unmistakable. A. St. Hilaire has already indicated the resemblance in the andrœcium to *Phyllanthus*, and indicated the connection which is established between *Menispermaceæ* and *Malvaceæ* by means of *Euphorbiaceæ;* and De Candolle has noticed an approach in the same parts to *Sterculiaceæ*.

The relationship which exists between *Menispermaceæ* and *Euphorbiaceæ* appears to us to be too close to be merely regarded as one of analogy. We do not attach much weight to the unisexuality of both Orders, nor can we adduce the scandent habit of *Plukenetia, Dalechampia, Pterococcus, Tragia,* and other *Euphorbiaceæ*, as a very important resemblance. The peltate leaves of species of *Mappa, Jatropha,* and many other *Euphorbiaceæ*, and the pseudo-articulation of the leaves of *Cicca, Conceveiba, Cleidion,* and others, may also be regarded as distant resemblances. It is the close agreement in structure both of the male and female flowers of many of the trimerous genera of *Euphorbiaceæ* to those of *Menispermaceæ* which we are disposed to regard as important. The stamens of *Euphorbiaceæ* are so often identical with those of *Menispermaceæ*, that it is needless to enumerate instances, which occur as well among the genera with free stamens as among those in which the stamina are united into a central column. The ovaries of the two Orders, again, are in many instances undistinguishable, except by their being united in the one and free in the other; and the mode of division of the styles of *Euphorbiaceæ* is repeated in some genera of *Menispermaceæ*. If to this we add the Euphorbiaceous male flower of Mr. Miers' genus *Odontocarya*, the peltate ovules of *Glochidion* and allied genera, the loculicidal dehiscence of the putamen, which is always more or less evidently present in *Menispermaceæ*, the frequently curved embryo of *Euphorbiaceæ*, and the peculiar structure of the cocci of *Phyllanthus*, as figured by Jussieu, with cavities like those so characteristic of *Menispermaceæ*, we have a series of resemblances which cannot be neglected.

In the structure of their stems *Menispermaceæ* almost invariably depart from the ordinary type of exogenous vegetation, and there are few or no natural orders of Dicotyledonous plants of equal number of species in which this departure is so great and so uniform.

The greatest differences of opinion have existed amongst botanists as to the value of the characters derived from a study of the vegetative organs, and especially the axis of Exogens, in a systematic and physiological point of view; the more theoretical observers have predicted far too much from the inquiry, the purely systematical have too often neglected it. Those who have combined a sufficiently extensive knowledge of systematic and physiological botany have for the most part considered the structure of the wood to be of very subordinate value: we ourselves adopt this view, from the writings of Brown, Adrien de Jussieu, and Decaisne, with whose observations our experience entirely coincides; and we would (with Decaisne) recommend a careful study of *Menispermaceæ*, and a comparison of the woods of the different genera one with another, and with other plants, as strongly corroborative of this opinion. In a systematic point of view, however, the wood often becomes a safe guide to the affinities of a plant when the organs of vegetation and reproduction are arrested in development, or defeat our attempts at analysis; on the other hand, in a physiological point of view, the structure of the common axis rather tends to confound our preconceived ideas of the necessary adaptation of structures to particular functions, and of these functions being indicated by structure. Without presuming to say that no relation exists between the habit of plants and their wood, or their wood and floral organs, we may affirm that we have never been able to detect any, though we have studied the subject in the forests of the most favourable localities. One broad fact has indeed been generally recognized, that most climbing plants have abnormal woods, but there are few natural orders of any extent amongst which scandent genera are not to be found; these have often no further relation to one another than their scandent habit, and the woods of nearly-allied species often differ essentially; add to these, the fact that the wood of erect Exogens sometimes presents as great anomalies as that of scandent ones, and even in some cases imitates the latter, and the value of the fact in its broadest aspect is considerably diminished.

Menispermaceæ exhibit very unequally, but always more or less, certain features common to most scandent plants; as, a spongy stem, abundance of cellular tissue, and of sclerogen cells in it, laxity of pleurenchyma, irregular development of woody bundles and liber, absence of rings of annual increase, wood often wholly composed of dotted, scalariform, or pitted vessels, always accompanied by many of very large diameter, and lastly, great anomalies in the structure of the pith.

Such characters are more or less common to the allies of *Menispermaceæ*, as *Kadsuraceæ, Anonaceæ*, and *Clematideæ*, and also to plants having no direct affinity with these or with one another, as *Phytocreneæ, Nepentheæ, Chloranthaceæ, Malpighiaceæ*, some *Santalaceæ, Balanophoreæ, Piperaceæ, Combretaceæ, Verbenaceæ, Vitis*, etc., and some scandent *Leguminosæ* and *Compositæ*. Amongst all these the only recognizable relation between function and structure is, perhaps, the fact that the lax tissues and abundance of large air-vessels in the wood, ensure a free circulation of fluids and gases through vessels which, by reason of the many convolutions and contortions to which they are subjected, are peculiarly exposed to constriction.

The prevalence of these peculiarities in Menisperms suggests three subjects of inquiry:—1. Do they indicate a high or low position of *Menispermaceæ*, amongst Exogens? 2. Do they indicate a transition to Endogens? 3. Do they betray any affinity with other natural orders placed at a distance in our systems?

1. At the outset of the first of these questions, we are met by the inquiry, what constitutes perfection and imperfection in wood structure, and indeed in the Vegetable Kingdom generally? Under the notes that are appended to *Ranunculaceæ*, will be found some on comparative complexity in the floral organs, which are applicable to *Menispermaceæ*, and which argue their belonging to a low type. But, by a parity of reasoning, the same arguments applied to the wood of this Order may by some be assumed to indicate a highly developed type. In illustration of this, we may remark that there is much more complexity in the construction of a three years old stem of *Coscinium*, than in *Magnolia*, or most other Exogens of the same age; for whereas there is in most ordinary Exogens an annual repetition of parenchyma and pleurenchyma, with few large vessels, but without change in relative position, and with little variation in the structure of the component parts of each year's growth, we have in *Menispermaceæ* many structurally different forms of cellular and vascular tissue annually developed in the stem, besides liber-bundles, and further, in some, a double system of Exogenous bundles of wood and of liber is developed, wholly independently of those first deposited.

It may be argued, that the great prevalence of parenchyma, and constant irregularities in the development of the various vascular tissues, denote imperfection; when it will be answered, that during several years the growth of *Menispermaceæ* is always normally Exogenous, that the simplest theoretical plan upon which this could be continued would be by the annual repetition of the same, and that a deviation from this type and arrangement implies a modification of structure for another and higher function; in short, that, in the vegetable as in the animal kingdom, specialization and complexity of organs for the performance of special functions implies relative elevation in the scale. It is true that we may not be able to recognize the function, but in this, as in all similar cases, we must assume that when a structure is fully developed, it implies the existence of a function in either a latent or active condition.

Decaisne, in his admirable essay on *Lardizabaleæ*, has thrown great light upon the structure of Menispermeous wood, and treated the whole subject, in its many bearings, in a most masterly manner; he indeed was the first to show the relations between the ages of the particular organs and some of the abnormal characters they present; and the mere fact of there being in some cases definite periods for the formation and suppression of the liber, indicates the existence of functions that will one day find expression as natural laws　In pursuance of Decaisne's investigations, we would recommend the study of the anatomy of the internodes of various parts of the stem, in relation to the development of flower-buds and leaf-buds on the parts above them.

The absence of annual rings of growth in wood many years old, indicates a more general vitality in the stem, or, at least, a less definite boundary between the living and dead wood; in other words, a more generally diffused activity of the stem seems necessary to the life of the plant than is usual amongst Exogens, whose inner layers of wood are virtually inactive. The very frequently woody nature of the pith-cells, which form long cylindrical rigid tubes with blunt square ends, placed above one another, would also appear to be an adaptation of that part to some modification of its usual functions; but for what special design, we have no idea.

2. The question whether the structure of Menisperms approaches that of Endogens, has been well answered by Decaisne in the negative; but as there are still two opinions on the subject, we shall view this point in another light from that excellent author. If the Endogenous stem is regarded as an imperfect development of the Exogenous, and if (as is perhaps the general opinion) an annual addition to a once-formed deposit of pleurenchyma and parenchyma, etc., be considered typical of the highest-developed Exogenous stem, then Menisperms may, inasmuch as they depart from these characteristics, be considered to tend towards Endogens; but if, on the other hand, the Endogenous stem be considered as constructed upon a totally different type from the Exogenous, and that the terms high and low are not applicable to them in any but general terms, we lose sight of any transition being indicated by Menisperms from the Exogenous to the Endogenous type; for whereas they offer all the peculiarities of the Exogen as contradistinguished from the Eudogen, they share none of the distinguishing characters of the latter. The mere resemblance of a transverse section of a Menispermeous stem, with several rather irregularly deposited zones of wood, to an Endogen, argues nothing, for the structure of the bundles thus compared is totally dissimilar, no less than their relations to one another; and whatever casual resemblance transverse sections show in these cases (and upon which so much stress is laid), a vertical section annuls.

The fundamental facts, that the vascular system of Menisperms is double, that each in many cases, and one in all, increases annually, that the wood-bundles are separated by continuous narrow medullary rays, and that on a vertical section the wood-zones are all seen traversing the stem in straight lines, and always parallel to one another, are entirely opposed to the view which would consider the Menispermeous stem as showing an approach to that typical of Endogens.

3. The Exogenous Orders to which Menisperms may be supposed to betray an affinity in the structure of their stems are mentioned above, but identity of structure is hardly to be found between *Menispermaceæ* and any of them. The greatest resemblance exists perhaps in *Myzodendron*, an erect-growing Santalaceous plant, and the horizontal rhizomes of some *Balanophoreæ*, but upon these it would be superfluous to dwell. Much stress has been laid upon the resemblance to *Aristolochia*, and Decaisne has exposed the mistaken views upon which this was founded, showing, in the first place, that this is neither constant nor of importance; and in the second, that the genus *Aristolochia* presents as many variations from a common type as *Menispermaceæ* do, and that these deviations are neither common to both Orders nor analogous in each.

In the present state of our knowledge, we cannot do better than quote Decaisne's remarks, that "no special value can be attached to characters drawn from the organs of nutrition," and that "all observations tend to prove (as Mirbel has already said) that the anatomical structure of wood offers no sure guide to affinity."

We have still a few words to add upon the individual peculiarities of Menispermeous woods. With regard to any agreement in wood structure amongst themselves, which the plants of this Order show, it is very vague; closely allied genera have often very similar woods, but so have more distantly allied ones, as *Limacia* and *Pachygone*, *Coscinium* and *Anamirta;* and closely allied genera have occasionally very different wood, as *Tinospora* and *Parabæna*. In short, the deviations from a common type presented by the various species of *Menispermaceæ* are, perhaps, greater of their kind than the deviation of the wood of the whole Order is from that of other Exogens.

Decaisne sums up these peculiarities with great neatness and precision, and with so true an appreciation of their value, that, slender as were his resources compared with ours, we have but few alterations to suggest; and these we shall accordingly append to the three heads under which he classes the peculiarities of the wood of the Order.

I. " *Menispermaceæ* differ from other Dicotyledones by the last annual deposit of " wood not being separated from that of the former year by those large vessels which, " in other Exogens, indicate the annual increase; by each wood-bundle remaining " undivided; and by the liber, once formed, not being added to."

To the above general rule we find partial exceptions in *Tinospora* and others, which have indications of annual growth in the wood, and in *Coscinium*, where there are manifest signs of increment in those of the liber. The liber of several species increases annually, as in *Pycnarrhena*, *Aspidocarya*, *Limacia*, *Tinospora*, and others. The wood-wedges become partially divided (as in *Aristolochia*) in *Tinospora*, or rather two continuous wedges become confluent.

II. "The wood-bundles of *Menispermaceæ* cannot be compared with those of Mo- " nocotyledones, because they increase annually, are disposed in regular symmetrical " circles round a defined pith, and because the liber does not form an integral part of " each bundle."

In this view we entirely concur, adding that neither do the bundles of wood follow the same course or development as in Monocotyledones. The liber does, however, appear in some species of *Limacia* to be an integral part of the wood. The great frequency of a portion of the pith being formed of woody tissue, consisting of long cells with truncate ends, and passing insensibly into ordinary medullary tissue, is a well-marked peculiarity of *Menispermaceæ*.

III. "In some species (*Cissampelos Pareira* and *Cocculus laurifolius*), after the " first-formed wood-wedges have continued to increase for several years, other wood- " wedges, altogether similar to these, only without spiral vessels and liber, are depo- " sited in a zone exterior to them, which operation being repeated, the stem finally " appears to be made up of concentric circles of wood-wedges; and further, the " liber, which is only found in the first-formed wood-zone, is placed much nearer the " centre than the circumference of the stem, and hence not in the bark."

This account is perfectly accurate, and describes a structure which is very frequent, and perhaps general, in the Order, and constitutes a remarkable deviation from the ordinary Exogenous type. Each zone is of several years' growth, and possibly the outermost is not the only one which receives additions.

The number of species of *Menispermaceæ* is probably about 150, or at most 200. They are generally widely diffused, and are with few exceptions confined to tropical and very hot subtropical countries. One inhabits Canada, and one Eastern Siberia, and a few are found in the United States, China, and Japan. In Europe they are unknown, as well as in New Zealand, Tasmania, and temperate South America. Several species extend in Australia almost to the south coast, and they occur in all parts of Africa from the Mediterranean to the Cape of Good Hope.

Like *Anonaceæ*, they are most abundant in perennially humid climates, and they occur in about equal numbers in Malabar, Ceylon, Malaya, Khasia, and Java. Less than a third of the Ceylon species are common to that island and Malaya; but this proportion being much larger than that which is found to exist in *Anonaceæ*, indicates that the species of *Menispermaceæ* are much more widely diffused. Khasia has many species common to it and Malaya; but many have their southern limit in Khasia, and are found also in Sikkim and throughout the Eastern Himalaya, and probably extend thence into the mountainous parts of West China. A few species extend west along the lower and outer Himalaya, but only one (which is also a Japanese species) is found in the middle and western parts of that chain, without extending to the eastward. In the mountains they are confined to the subtropical and lower part of the temperate region, never rising above 7000 feet. On the whole, *Menispermaceæ* are less intolerant of dryness than *Anonaceæ*, several species inhabiting the most arid parts of Hindostan, and even the Panjab and Sindh, whence they stretch across the hot belt·of Southern Asia, through Arabia and Egypt, to Senegal.

The genera and species of *Menispermaceæ* were left in a very unsatisfactory state by De Candolle, who, possessing no materials from which to study the Order in detail, and finding it impossible to reconcile with one another the chaotic descriptions of previous authors, was obliged to content himself with reproducing them as he found them, at the same time urgently recommending the study of the Order to tropical botanists. Considerable light was thrown on the structure of the fruit in a paper by Mr. Colebrooke, published in the 'Transactions of the Linnean Society' in 1822; but, his knowledge of the Order being confined to the species indigenous in Bengal, or cultivated in the Calcutta Botanic Gardens, he contented himself by establishing several new genera, all of which have been found sound.

For a long time little further progress was made in the study of the Order, though isolated observations were contributed by A. St. Hilaire, Blume, A. Richard, and others. In the 'Bijdragen,' Blume instituted the genus *Clypea*, which was afterwards discovered to be identical with *Stephania* of Loureiro. The first important step in advance was made by Wight and Arnott, who in 1832 divided the Indian species of the genus *Cocculus* into sections according to the nature of the embryo, and thus laid the foundation for the more complete study of the Order by Miers, who has devoted much time and labour to the investigation of this very difficult family, and, by making careful analyses of the flowers and fruit of all the species to which he could obtain access, has acquired a very complete knowledge of their structure, and has therefore been able to impart a degree of precision to the ordinal characters and those of the main groups, which they did not before possess.

It is much to be regretted that Mr. Miers has not made public his complete monograph of the Order, for which such ample materials are in his possession, but has confined himself to publishing a very concise sketch of his views in Taylor's 'Annals,' and in Lindley's 'Vegetable Kingdom.' We have thus been compelled to follow out for ourselves the details of structure of the Indian species, guided, of course, by the generally accurate indications contained in Mr. Miers' papers, and by the brief diagnoses there to be found. It will be seen that the result of this study has been the adoption of most of the great groups and subdivisions proposed by Mr. Miers. We have, however, arrived at different conclusions regarding the limits of genera, the number of which we think Mr. Miers has unnecessarily augmented, by placing too great reliance upon characters derived from the shape and number of the petals and stamens, and slight modifications of the putamen. Where his genera are founded upon characters derived from the seed, it will be seen that we have invariably adopted them.

Mr. Miers' views as to the limits of species can only be gathered from the notes and remarks appended to his paper in Taylor's 'Annals,' already referred to, the extreme brevity of which often makes his meaning doubtful. In several cases, however, to which we shall refer more particularly under their respective genera, we are satisfied that he regards as distinct, forms which are either certainly not so, or are so imperfectly known that their distinctness cannot be confidently asserted. In such cases we have not hesitated to dissent from his views, as we are deeply impressed with the importance of avoiding the addition of imperfectly-defined species to our lists.

Cocculus palmatus, Wall. Cat. 4953! (*Jateorhiza*, Miers), from the east coast of Africa, and *Cocculus hexagynus*, Wall. Cat. 4968! (*Cocculus ovalifolius*, DC.), from China, are not natives of British India. *Cocculus flavicans*, Wall. Cat. 4976, is a species of *Anisophyllum* (*Tetracrypta*, Gardner). Mr. Miers' genus *Antitaxis*, of which the male flower only is known, with two sepals, two petals, and four stamens, is a doubtful Menispermaceous plant; a specimen without flowers, which we have examined in the Benthamian Herbarium, having quite as much the appearance of *Euphorbiaceæ.* Several new species of *Menispermaceæ,* in addition to those described for the first time in the following pages, exist in our own collections, but in a state too imperfect to enable us to characterize them, some being without flowers, and others consisting of flowering panicles of one sex without leaves.

CONSPECTUS TRIBUUM.

A. Semina albuminosa.
   a. Cotyledones divaricatæ (*Heteroclineæ,* Miers).
      Cotyledones patentim divaricatæ . . . I. Coscinieæ.
      Cotyledones lateraliter divaricatæ . . II. Tinosporeæ.
   b. Cotyledones appositæ.
      Ovaria 3 vel plura . . . . III. Cocculeæ.
      Ovaria solitaria . . . . IV. Cissampelideæ.
B. Semina exalbuminosa . . . . V. Pachygoneæ.

CONSPECTUS GENERUM.

I. Coscinieæ . . . . 1. *Coscinium.*
II. Tinosporeæ.
   A. Stamina 6, monadelpha; antheræ circa discum peltatum horizontales.
      Putamen antice planum . . . . . 2. *Aspidocarya.*
      Putamen antice excavatum . . . . 3. *Parabæna.*
   B. Stamina 6, libera . . . . 4. *Tinospora.*
   C. Stamina numerosa monadelpha; antheræ in globum coalitæ . . . . 5. *Anamirta.*
III. Cocculeæ.
   A. Albumen ruminatum; ovaria indefinita . 6. *Tiliacora.*
   B. Albumen homogeneum; ovaria 3-6.
      a. Putaminis cavitates laterales, internæ, lamina ossea tectæ . . . . 7. *Limacia.*
      b. Putaminis cavitates laterales, externæ, nudæ.
         Stylus simplex . . . . 8. *Cocculus.*
         Stylus bipartitus . . . . 9. *Pericampylus.*
IV. Cissampelideæ.
   Fl. masc., sepala libera; fl. fœm., sepala 3 . 10. *Stephania.*
   Fl. masc., sepala libera; fl. fœm., sepalum 1 . 11. *Cissampelos.*
   Fl. masc., sepala coalita; fl. fœm., sepala 2 . 12. *Cyclea.*
V. Pachygoneæ.
   Petala 6 . . . . 13. *Pachygone.*
   Petala 0 . . . . 14. *Fibraurea.*
      Genera dubiæ tribus, fructu ignoto.
         Stamina 6, libera . . . . 15. *Tinomiscium.*
         Stamina ad medium monadelpha . . 16. *Pycnarrhena.*

Tribus I. COSCINIEÆ.

Petala sepalis majora, parum imbricata. *Albumen* irregulariter ruminatum. *Radicula* supera, a hilo remota. *Cotyledones* magnæ, patentim
divaricatæ.

1. **COSCINIUM,** Colebrooke.

Pereiria, *Lindl.*

Sepala 6, rotundata, bractea 1 conformi stipata. *Petala* 3, sepalis
majora, patentia, elliptica, æstivatione parum imbricata. MAS. *Stamina* 6, exteriora (petalis alterna) libera, interiora ad medium monadelpha. *Filamenta* cylindrica; *antheræ* adnatæ, ovales, exteriores uniloculares, interiores didymæ biloculares. FŒM. *Stamina* 6, abortiva.
Ovaria 3–6, subglobosa, stylis subulatis reflexis. *Drupæ* globosæ,
carnosæ. *Putamen* crassum, osseum, intus processum globosum et
spongiosum continens; pedicello osseo basi putaminis inserto. *Semen*
externe visum subglobosum, intus cavum et circa processum condyli-
formem convolutum. *Testa* tenuis, lævis. *Albumen* oleosum, carnosum,
hinc (quo latere hilum spectat) plicis podospermii vel membranæ exterioris seminis ruminatum. *Embryo* fere rectus; radicula parva cylindrica supera, apicem drupæ spectans. *Cotyledones* tenuissimæ, rotundatæ, margine irregulares, divaricatæ, undulatæ, secundum Gærtner
foraminibus crebris perforatæ, vel fide Miers profunde sinuato-laci-
niatæ.—Frutices *alte scandentes*, petiolis *cylindricis basi et apice incrassatis,* foliis *amplis palminerviis, junioribus saltem peltatis,* floribus *in capitula globosa dense congestis.*

The genus *Coscinium* differs so much from the rest of the Order in the comparatively large size of its petals, and in the structure of the seed, as to deserve to
be distinguished as a separate tribe. The radicle, if Gærtner's plate may be relied
on, is at the geometrical apex of the seed; and the cotyledons, which are nearly circular, expand widely, and descend one on each side of the internal process of the
putamen, which occupies the hollow in the middle of the seed.

The structure of the drupe of *Coscinium* is unfortunately as yet so imperfectly
understood, that we cannot express ourselves decidedly regarding it. The nutrient vessels pass into the seed through two canals, the external apertures of which
are conspicuous on the putamen, one on each side of the hilum. Gærtner represents and describes the woody process which rises from the hilum as forming an
integral portion of the seed, and as being gradually broken up into plates, which
penetrate into the substance of the albumen. Mr. Miers, on the other hand, thinks
that the condyloid process is quite distinct from the membrane which lines it, and
which gives off the plates by which the albumen is ruminated. The latter structure
is undoubtedly more analogous to that of the rest of the Order; but it appears to us
that the view of Gærtner is more in accordance with the specimens we have examined,
of which, however, one only was in a good state, all the others being decayed. The
putamen is very thick and hard, and is composed of columnar fibres, extending through
its whole thickness, like those of the middle coat of the seed of the nutmeg. Indeed, if the analogy of structure to other *Menispermaceæ,* especially in the tubular
canals which penetrate through the putamen, were not quite opposed to such a view,
we should be inclined to suggest the possibility of the woody coat of *Coscinium* being
an integument of the seed, and its internal process analogous to the plates (gradually
branching from the chalaza) by which the albumen of nutmegs is ruminated.

Mr. Miers is inclined to think that Gærtner is mistaken in representing the cotyledons as perforated with holes, and that they are rather lacerated at the margins.
They lie very near the surface of the albumen, and are not flat, but are irregularly
folded over undulating tubercles, produced by the ruminating plates which project
from the condyle, and are so thin as to be with difficulty detached from the albumen
without injury. This may have led Gærtner into error; but the point is still doubtful, Mr. Miers' materials, like our own, having been very scanty.

The nuts of *Coscinium* which we have seen were all deprived of the sarcocarp, so
that the position of the style and the insertion of the fruit could not be determined.

The species of *Coscinium* are entirely Indian. The wood, which has a deep yellow
colour, affords an indifferent yellow dye, and is esteemed as a drug by the natives of
Ceylon, but does not appear to be active in its qualities. A few years ago it was
imported into England in some quantity, on the supposition that it would answer
as a substitute for the Calumba root (*Jateorhiza palmata*, Miers), but the specula-
tion was unsuccessful.

The wood of *Coscinium* may be thus described:—A several years old portion of
stem is rather cellular and spongy, furrowed externally, and ⅓ inch in diameter.
Pith broad, half diameter of stem, central part of large, loose, hexagonal tissue, towards the exterior gradually becoming smaller, longer and denser, and finally passing
into a woody tissue of vertically elongated cells, with truncated apices. *Wood-wedges*
small, very numerous, 40–70, closely placed, of dotted pleurenchyma, and large hexagonal scalariform vessels, and occasionally spiral vessels towards the pith. *Liber-
bundles* very much radially elongated, annually increasing, and with obscure traces of
annual rings distant from one another. *Bark* tolerably thick, of small cellular tissues,
with a continuous very narrow zone of slender liber-tubes a short way from the circumference.

1. **C. fenestratum** (Colebrooke in Linn. Tr. xiii. 65); foliis fere
rotundatis basi cordatis vel subtruncatis subtus flavido-tomentosis, petiolis (nisi in plantis junioribus) vix peltatis, capitulis in axillis umbellatis.—*Miers in Hook. Bot. Mag. t.* 4658, *et in Pharm. Journ.* xii. 185.
—C. Wallichianum *et* C. Wightianum, *Miers in Taylor's Annals, ser.* 2.
vii. 37. Menispermum fenestratum, *Gærtn. Fr.* i. 219. *t.* 46. *f.* 5; *DC.
Syst.* i. 541, *Prod.* i. 103; *Roxb. Fl. Ind.* iii. 809. Cocculus Blumeanus, *Wall. Cat.* 4971 *partim!* Pereiria medica, *Lindl. Fl. Med. p.* 370.

HAB. In Zeylania! in Peninsula (loco non indicato), *Wight!* Penang? *Wall.!*—(*v. s.*)

Frutex alte scandens. *Ramuli* juniores dense incano-tomentosi, crassiores glabriusculi, eleganter striatuli. *Folia* ampla, basi subcordata, 7–9-nervia, coriacea, supra
glabra, subtus incana, venulis crebris reticulata, 5–7 poll. longa et fere æquilata, juniora oblongo-deltoidea, acuminata, peltata. *Petioli* 3–5-pollicares, incani, basi torti
et dilatati. *Capitula* florum pedicello pollicari suffulta, diametro ½–⅔-pollicaria, in
axillis vel ad axillas foliorum delapsorum fasciculata. *Flores* subsessiles, virides, fulvo-
tomentosi. *Petala* rotundata, acuta, intus glabra et nervosa, patentia. *Stamina*
sterilia nervosa. *Drupæ* 1–3, calyce petalisque persistentibus stipatæ, subglobosæ,
villosæ, diametro fere pollicares.

The specimen of *C. Blumeanum* from Singapur, in the Wallichian Herbarium at
the Linnean Society, contains a fragment apparently of this species, without flower,
which Mr. Miers has called *C. Wallichianum.* Mr. Miers has also distinguished
C. Wightianum as a species, without assigning any characters. Dr. Wight's specimens exhibit only unexpanded flowers, but they seem identical with the Ceylon
plant. There is evidently some confusion in Mr. Miers' remarks, as *C. Wightianum*
is not included among Dr. Wallich's 4971, not having been communicated by Dr.
Wight to Dr. Wallich, but distributed separately by him under the name of *Coscinium
fenestratum.*

2. **C. Blumeanum** (Miers in Taylor's Annals, ser. 2. vii. 37);
foliis crasse coriaceis ovalibus vel oblongis peltatis acuminatis vel obtusis basi truncatis vel subcordatis subtus niveo-tomentosis, capitulis
in axillis racemosis.—Cocculus Blumeanus, *Wall. Cat.* 4971 ! *excluso
B partim.*

HAB. Malaya: ad Penang et Singapur, *Wall.!*

Frutex alte scandens. *Caules* dense lanato-tomentosi, infra lanam fusci, striati.
Folia 7–12 poll. longa, 3–6 lata, petiolo 3–5-pollicari, supra atro-viridia, glabra,
lucida. *Racemi* fulvo-tomentosi, validi, 3–4 poll. longi, pedunculis capitulorum fere
pollicaribus. *Flores* masculi ut in *C. fenestrato.*

This species, so far as can be ascertained from the small number of specimens
which we have seen, seems very distinct from *C. fenestratum*, in the much more
rigid and more elongated leaves, which are always peltate, whereas those of *C. fenestratum* are only so in young plants. In young plants of the Ceylon species, however, the leaves are elongated like those of *C. Blumeanum.* The character derived
from the inflorescence is perhaps not constant.

Tribus II. TINOSPOREÆ.

Sepala 6. *Petala* 6, sepalis minora, rarius 0. *Ovaria* 3. *Drupæ*
styli cicatrice subterminali vel fere basilari notatæ. *Putamen* antice
planum vel excavatum vel processu interno munitum. *Semen* amphi-
tropum, rarius anatropum, albuminosum. *Embryo* axilis. *Radicula*
supera, styli cicatricem spectans. *Cotyledones* lateraliter divaricatæ,
tenues.

The genera which are associated in this tribe by means of the character of the
laterally divaricating cotyledons, form a very natural group; and, though they differ
from one another a good deal in the shape and structure of the putamen and seed,
yet in these respects also a regular gradation may be traced from one genus to
another, and they are all nearer to one another than to the other tribes of the Order.
The style is, in many of the genera, almost terminal, even in the ripe fruit, but in
Anamirta it is nearly basal. The peculiar obliquity of the cotyledons, which separate like the blades of a pair of scissors (sometimes overlapping a little at the edges
only), make the seed much broader than in the following tribes, in which it is always
nearly cylindrical. In *Aspidocarya* the seed is quite flat, but more frequently it is
curved forwards round the internal process of the putamen, when it becomes ovoid
or globose, and excavated anteriorly. The ruminated albumen of *Tinospora* is peculiar, but is not an indication of immediate affinity, as it is absent in those genera
nearest allied to *Tinospora*, and present in *Tiliacora*, which has no near relationship
with it.

There is in the Hookerian Herbarium a specimen of a Menispermaceous plant in
fruit, which probably belongs to this tribe, but which is too imperfect to admit of
proper description. It was collected in Assam by Griffith. The drupe is more than
an inch long, much compressed, with a fleshy exocarp and a thin bony putamen, very
slightly rugose externally, and with a broad, shallow, longitudinal furrow on the ventral face. On the inner surface of the same face, there is a groove extending from
the base to near the apex, from which the seed is pendulous. The seed is marked
by a distinct rhaphe, running from the hilum to the opposite extremity; it is quite
flat, but, from the decayed state of the specimen, the presence of albumen and the
structure of the embryo cannot be determined. If the cotyledons be laterally divaricated, this fruit will come near *Aspidocarya*, agreeing with it in the absence of
any internal process of the putamen, and in the anatropous seed, but differing in the
shape of the putamen.

The leaves of this interesting plant are somewhat membranous, oblong-lanceolate and acuminate, five-nerved at the base, glabrous on both sides and paler below, 6–7 inches long, and 2½–3 broad, with a long slender petiole (3–3½ inches), twisted near the base, and falsely articulate at each end. In foliage it somewhat resembles *Tinomiscium*, a genus of which the position is doubtful, the male flowers only being known: with this it agrees in the elongated petioles, but the leaves are thinner, more pointed, and five-nerved at the base; still, though not identical in species, it is certainly probable that the two are congeners.

2. ASPIDOCARYA, H.f. et T.

Sepala 6–12, ovali-oblonga, interiora sensim latiora. *Petala* 6, cuneato-obovata, sepalis breviora. MAS. *Stamina* in columnam centralem cylindricam apice antheras 6 horizontales gerentem coalita. FŒM. *Stamina* sterilia 6, clavata. *Ovaria* 3, oblonga; *stigmata* subcapitata. *Drupæ* pulposæ, oblongæ, cylindricæ, putamine compresso dorso argute carinato, ventre haud excavato. *Semen* pendulum, oblongum, antice rhaphe conspicua notatum. *Albumen* carnosum. *Radicula* brevis, hilo terminali approximata. *Cotyledones* rectæ, planæ, oblongæ, tenuissimæ, obliquæ, basi divaricatæ, dein parallelæ, marginibus oppositis tantum se invicem obtegentes.—Frutex *scandens*, petiolis *cum caule pseudo-articulatis, prope basin debilibus subtortis*, floribus *in paniculas racemiformes elongatas subcompositas axillares dispositis.*

This interesting plant comes very near *Parabæna*, but differs in many points of the structure of the female flower and fruit. The seed is attached to the top of the cell, so that the ovule must be anatropous. The putamen and seed are also quite flat anteriorly, and not excavated like those of *Parabæna*. The inflorescence, too, is very different. It therefore forms a new genus, the name of which is derived from ασπις, *a shield*, and καρυον, *a nut*.

The wood of *Aspidocarya* differs remarkably from that of other *Menispermaceæ*, in respect of the crescent-shaped bundles of tissue, altogether resembling liber, which are found at the inner end of each wood-wedge.

A piece of stem several years old, and from ⅓–¼ inch in diameter, is deeply furrowed, spongy, and much compressed. *Pith* broad, white, of hexagonal soft cellular tissue, becoming much closer, smaller, and longer towards the wood, and occupying three-fourths of the circumference of the stem. Medullary rays of dense cellular tissue. *Wedges of wood* towards circumference, about 20, broadly ovate, margined radially by a narrow crescent-shaped mass of pleurenchyma. *Wood* of dotted pleurenchyma, and numerous very large vessels, with short transverse striæ on their walls. *Liber-bundles* forming almost a horse-shoe round half the circumference of the wedge, the contiguous bundles approaching and almost cohering. The *liber* is annually added to, but not the tissue at the inner end of the wood. *Bark* of several series of cellular layers.

1. **A. uvifera** (H.f. et T.); foliis rotundato- vel ovato-cordatis subpeltatis abrupte et longe acuminatis subtus ad nervos pilosis.

HAB. In Sikkim exteriori subtropico, alt. 1–5000 ped.—(Fl. Mai.; fr. Jul.) (*v. v.*)

Frutex alte scandens. *Ramuli* cylindrici, striati, sparse strigoso-puberuli. *Folia* 4–6 poll. longa, 3–6 lata, petiolo fere æquilongo, basi leviter vel profunde cordata, lobis rotundatis vel subtruncatis rarius subsagittatis, supra ad nervos pubescentia demum glabra, subtus præsertim ad nervos pilosa, basi 5-nervia, cæterum penninervia. *Petioli* cylindrici, striati, basin versus incrassati. *Paniculæ* 4–8-pollicares, ramis alternis, inferioribus compositis, superioribus simplicibus, fulvo-pubescentes, bracteis minutis subulatis muniti. *Flores* in ordine majusculi, viridescentes. *Sepala* ciliata. *Petala* obtusa vel emarginata, concava, marginibus infra medium incrassatis involutis. *Columna staminea* petalis æquilonga. *Antheræ* profunde 4-lobæ, biloculares. *Ovaria* in spec. omnia destructa. *Drupæ* pollicares, læves, lateritiæ, edules, sapore dulci, gynophoro brevissimo insidentes. *Putamen* valde compressum, lignosum, fragile, præter carinas læve, secus margines argute lobulatum, sinubus rotundatis, apice trilobum, lobo medio laterales superante, laminam argutam tenuem transverse compressam formante. *Carina* dorsalis arguta, longitudinaliter unisulcata. *Facies ventralis* medio leviter carinata, et prope marginem serie duplici longitudinali tuberculorum munita. *Semen* plano-compressum, oblongum. *Testa* tenuis, flavida, lævis ; *rhaphe* recta ventrali ; *chalaza* in facie antica seminis subterminali. *Embryo* albus, semine ¼ brevior.

3. **PARABÆNA,** Miers.

Sepala 6, carnosula, oblonga, fere æqualia. *Petala* 6, sepalis dimidio breviora, cuneato-triloba vel obovata. MAS. *Stamina* monadelpha, columna centralis cylindrica ; *antheræ* 6 in capitulum subglobosum coalita, transverse dehiscentia. FŒM. *Stamina* sterilia 6, cylindrica. *Ovaria* 3, stylis subulatis recurvis. *Drupæ* ovales, styli cicatrice subterminali. *Putamen* superne rostratum, dorso tuberculis acicularibus fragilibus exasperatum, antice profunde excavatum. *Semen* peltatum, circa putaminis processum internum involutum. *Albumen* copiosum, carnosum, homogeneum. *Embryo* curvatus. *Radicula* supera longa, styli cicatricem spectans. *Cotyledones* ovatæ, divaricatæ, in loculis diversis albuminis sitæ.—Frutices *scandentes lactescentes*, inflorescentia *axillari dichotome cymosa*.

In *Parabæna* the ovule is amphitropous, but attached considerably above the middle of the cell, so that the micropyle is not far from the hilum. The seed is peltate, and attached to the upper part of the internal process, which is hemispherical and quite open externally when the sarcocarp is removed. In general aspect and in the male flowers it is very close to *Aspidocarya*, but the structure of the putamen clearly distinguishes it.

We have examined the wood of *Parabæna sagittata*, which we preserved in spirits in the Khasia mountains ; a specimen about ⅛ of an inch in diameter is cylindrical and spongy, consisting chiefly of a very abundant lax cellular tissue, with large areolæ. There are five principal wedges of wood, each cuneate on a transverse section, placed midway between centre and circumference, and alternating with these are 5–7 other much smaller bundles, forming an exterior zone. Each wood-bundle consists of a very little dotted pleurenchyma and a large *cambium-layer*. Exterior to the wood, and removed from it, the *liber* forms a continuous wavy zone, each arc corresponding to the position of the wood-bundles, both small and great ; this liber is added to annually. The cellular and vascular tissues are all dotted, and the bark is cellular, without any distinct cuticle.

1. **P. sagittata** (Miers in Tayl. Ann. ser. 2. vii. 39) ; foliis oblongis abrupte acuminatis rarius obtusis basi sagittatis lobis obtusis vel acutis.—P. oleracea, P. heterophylla, *et* P. ferruginea, *Miers, l. c.* Cissampelos sagittata, *Ham. ex Wall. Cat.* 4983 ! C. oleracea, *Wall. Cat.* 4984 !

HAB. In dumetis subtropicis Nipaliæ orientalis ! Sikkim ! Assam ! Khasia ! Chittagong !—(Fl. Jun.-Jul.) (*v. v.*)

Frutex scandens. *Ramuli* sulcati, glabriusculi vel molliter pubescentes. *Folia* 4–8 poll. longa, 2–4 lata, petiolo 3–4-pollicari, primordialia argute sinuato-dentata, cætera integerrima, profunde cordata vel sagittata, basi 5–7-nervia, cæterum penninervia, utrinque glabriuscula vel tenuiter pubescentia, vel subtus laxe et molliter tomentosa. *Cymæ* axillares vel paullo supra-axillares, plerumque binæ, petiolos æquantes vel breviores, pluries dichotomæ, multifloræ, bracteis ad ramificationes filiformibus. *Flores* minuti, pallidi, pubescentes. *Sepala* acutiuscula, nervosa. *Petala* obovatocuneata, superne triloba, lobo medio emarginato, lateralibus inflexis, interdum vix lobata. *Drupæ* pulposæ, viridescentes, succo viscido scatentes, ovales, læves (in sicco rostratæ).

A very variable plant. The leaves of young plants are often remarkably toothed. Mr. Miers indicates four species, but he assigns no characters. We find the form and clothing of the leaves to vary so much, even on the same specimens, that we are fully persuaded that all the forms hitherto known belong to one species.

4. TINOSPORA, Miers.

Sepala 6, biserialia, interiora majora, ovalia vel obovata, membranacea. *Petala* 6, sepalis interioribus minora, obovata vel cuneata. MAS. *Stamina* 6 ; *filamenta* cylindrica, crassa, apice subclavata ; *antheræ* biloculares, loculis oblique adnatis lateralibus. FŒM. *Stamina* sterilia 6, clavata, carnosa. *Ovaria* 3, gynophoro convexo insidentia. *Stigmata* lacera. *Drupæ* 1–3, carnosæ, dorso convexæ, ventre planæ, styli cicatrice subterminali notatæ. *Putamen* rugosum, dorso carinatum, ventre leviter excavatum. *Podospermium* in cavitatem projectum, leviter bilobum, intus cavum. *Semen* circa podospermium convolutum. *Albumen* carnosum, oleosum, antice laminis transversis ruminatum. *Embryo* subcurvatus. *Radicula* supera cylindrica ; *cotyledones* ovatæ, divaricatæ, in loculis diversis albuminis segregatæ.—Frutices *scandentes*, petiolis *basi articulatis, basin versus incrassatis,* racemis *elongatis axillaribus vel terminalibus.*

This genus and the last agree with *Aspidocarya* in the subterminal position of the style in the drupe, but differ from it in the decidedly amphitropous ovules and peltate seeds. In *Tinospora* the internal process of the putamen is much more developed than in *Parabæna,* in which it is merely a depression on the surface of the putamen, convex internally. Here (as in *Anamirta* and *Coscinium*) the condyloid process has a narrow base, and projects far into the interior of the cell, and is embraced by the overlapping edges of the seed. It is also hollow, and the interior is occupied by a gelatinous mass. The cavity of its interior communicates with the exterior of the putamen by two perforations in the latter, one on each side of the median line. These do not, as in *Anamirta* and *Coscinium,* form elongated canals in the thickened bony mass, but the structure is the same as in those genera, differing only in degree. The albumen on the ventral side of the seed is divided into irregular masses by thin transverse plates of cellular tissue, which penetrate almost to the embryo.

All the species of this genus are remarkable for their extreme vitality. When the main trunk is cut across or broken, a rootlet is speedily sent down from above, which continues to grow till it reaches the ground, and restores the connection.

Chasmanthera of Hochstetter, with the habit and inflorescence of *Tinospora,* has monadelphous stamens. The fruit is also a little different, the concavity of what we have called the podosperm forming a deep hollow on the ventral face of the putamen, conspicuous externally as soon as the sarcocarp is removed, almost as in *Calycocarpum* of Nuttall, figured in Asa Gray's genera of North American plants. The shape of the embryo and the nature of the albumen of *Chasmanthera* were not determinable in the seed examined.

In a *Tinospora* which we refer to *T. crispa*, Miers, a portion of stem, probably six to eight years old, is loose, and soft and spongy, about half an inch in diameter, and has the following structure:—*Pith* one-third the diameter of the stem, of large hexagonal utricles, full of starch. Medullary rays and bark the same. Wood-wedges small, about twenty, half-way between centre and circumference, often lobed, and with traces of annual increase, divided by broad medullary rays, broadly lanceolate on a transverse section, formed of dotted or perforated pleurenchyma, and large dotted ducts, with oblique gashes on their walls. *Liber-bundles* arcuate, rather distant from the wood, often confluent into a narrow zone. *Bark* of delicate utricular tissue, full of starch ; outer layer of many rows of parallel radially compressed cells. *Epidermis* covered with many longitudinal rimæ, each with a central furrow and prominent cellular lips.

1. **T. tomentosa** (Miers in Taylor's Annals, ser. 2. vii. 38); foliis subtrilobis subtus tomentosis.—Cocculus tomentosus, *Colebr. in Linn. Tr.* xiii. 59; *Wall. Cat.* 4956! Menispermum tomentosum, *Roxb. Fl. Ind.* iii. 813.

HAB. In dumetis Bengaliæ, *Roxb.!* Ava, *Wall.!*—(Fl. Febr. Mart.) (*v. s.*)

Frutex alte scandens, cortice cinereo, pustulis scabris tecto; partes novellæ tomentosæ. *Folia* rotundato-cordata, antice repanda, vel plus minus triloba, utrinque (subtus præsertim) tomentosa, 3–6 poll. longa et fere æquilata. *Petioli* folia fere æquantes, tomentosi. *Racemi* solitarii vel fasciculati, plerumque simplices, floribus in axillis bractearum minutarum deciduarum fasciculatis. *Filamenta* clavata. *Antheræ* bilobæ. *Drupæ* 1–3, pisi majoris magnitudine, subglobosæ, læves, aurantiacæ.

Our description is entirely taken from Roxburgh, as we have seen no specimens except those in the Wallichian Herbarium, which are very imperfect. The stem is covered with very minute granular tubercles.

2. **T. Malabarica** (Miers in Taylor's Annals, ser. 2. vii. 38); foliis cordato-ovatis subtus dense vel tenuiter pubescentibus.—Menispermum Malabaricum, *Lam. Willd.* Cocculus Malabaricus, *DC. Syst.* i. 518, *Prod.* i. 97; *Wall. Cat.* 4969!—*Rheede Mal.* vii. *t.* 19.

HAB. In Malabaria, *Rheede*; Concan, *Nimmo*; in Bengalia versus basin Himalayæ Sikkimensis, *Hamilton!* in montibus Khasia a basi ad alt. 4000 ped.! et in prov. Chittagong!—(*v. v.*)

Frutex scandens, cortice cinereo; partes novellæ pilis albicantibus obsitæ. *Petioli* teretes, basi incrassati, pilosi. *Folia* cordiformia, acuminata, subtus lanuginosa, superne pilis subaspera septemnervia, 3–6 poll. longa et fere æquilata. *Racemi* folii longitudine. *Flores* virides. *Drupæ* maturæ corallini ruboris.

There is a specimen in the Hookerian Herbarium from Ceylon, without leaves, which is probably referable to this species; but, as we cannot identify it with certainty, we do not describe it. Our Khasia and Chittagong specimens are in leaf only, and are therefore also doubtful. Careful observations are required to establish the distinctive characters of all the species of this genus.

3. **T. crispa** (Miers in Taylor's Annals, ser. 2. vii. 38); foliis cordato-ovatis vel oblongis acuminatis glabris, staminibus basi cum petalis cohærentibus, antheris tetragonis.—Menispermum crispum, *Linn. Sp.* 1468. M. verrucosum, *Roxb. Fl. Ind.* iii. 808; *Fleming in Asiat. Res.* xi. 171. Cocculus crispus, *DC. Syst.* i. 521, *Prod.* i. 97; *W. et A. Prod.* i.

12 *in adnot.; Hasskarl, Pl. Jav. Rar. p.* 166 ; *Colebr. in Linn. Tr.* xiii.
60. Cocculus verrucosus, *Wall. Cat.* 4966 *A ! B ! (non C–E)*. C. coria-
ceus, *Bl. Bijdr.* 25.

HAB. Silhet, *Colebrooke ;* Pegu, *Wall.!—*(*v. s.*)
DISTRIB. Sumatra ; Java ; ins. Molucc. et Philippin.

Frutex alte scandens, cortice lævi, distanter verruculoso; partes novellæ glabræ.
Folia ovali-oblonga, acuminata, basi leviter cordata, lobis distantibus interdum subsagittatis, integerrima vel repanda, utrinque glabra, 2–6 poll. longa, 1–4 lata, petiolis
½ brevioribus.   *Racemi* ad axillas foliorum delapsorum secus caules vetustiores, solitarii vel fasciculati, elongati, 4–8-pollicares.   *Flores* 2–3 in axilla bracteæ ovatæ
carnosæ, pedicellati, virides, campanulati, 2 lineas longi.   *Drupæ* pallide aurantiacæ
vel flavæ, olivæ magnitudine.

Colebrooke's synonym is perhaps doubtful, as he says that the cotyledons of his
plant are not divaricate, and he figures them as partially overlapping.   The specimens in the Wallichian Herbarium are very imperfect, but the glabrous bark, with
distant rough tubercles, is very conspicuous.   On the first sheet a piece of the stem
of *T. tomentosa* is fastened down along with the stems and foliage of the true plant.
We found at Chittagong and in Silhet specimens of a Menispermaceous plant without leaves or flowers, the scandent stems of which agree with the description given
of this species.   Their structure has been described above.   As we have no materials
of our own to depend upon, we have embodied in the diagnosis and description the
main points of distinction pointed out by authors between this species and the last;
but, as these are in part derived from the description of Roxburgh and De Candolle,
and partly from those of Blume and Hasskarl, all of which are not certainly specifically identical, our character is perhaps little to be relied upon.   We are, however,
inclined to believe that Roxburgh's plant is the same as that of the Javanese botanists, because he attributes to it the same medicinal (tonic) virtues as are usually
attributed to *T. crispa,* and because their descriptions agree so far as they go.   *T.
crispa* is highly esteemed by the natives of the Malayan Archipelago as a febrifuge.

4. **T. cordifolia** (Miers in Taylor's Annals, ser. 2. vii. 38); foliis
cordatis glabris, staminibus liberis, antheris ovali-oblongis.—Menispermum Malabaricum *β, Lam. Dict.* iv. 96.   M. cordifolium, *Willd.; Roxb.
Fl. Ind.* iii. 811.   Cocculus cordifolius, *DC. Syst.* i. 518, *Prod.* i. 97 ;
Colebr. in Linn. Tr. xiii. 62 ; *Wall. Cat.* 4955 ! ; *W. et A. Prod.* i. 12 ;
Wight, Ic. t. 485, 486.   C. convolvulaceus, *DC. Syst.* i. 518, *Prod.* i.
97.   C. verrucosus, *Wall. Cat.* 4966 *C ! D ! E ! (non A nec B)*.

HAB. Per Indiam tropicam in dumetis vulgaris; in Zeylania, *Thwaites !*
Carnatica ! Malabaria ! Maisor ! Dekhan, *Jacquemont !* Concan, *Graham ;* Orissa ! Bengalia ! Assam, *Jenkins !* Bahar, *Hamilton !—*(Fl. per
totum annum.) (*v. v.*)

Frutex alte scandens, cortice suberoso verruculoso; partes novellæ glabræ.   *Folia*
late cordata, acuta, vel acumine gracili terminata, 2–4 poll. longa et lata, petiolis
fere æquilongis.   *Racemi* axillares, rarius terminales, vel ex axillis foliorum delapsorum solitarii, folia sæpe longe superantes, simplices vel basi subcompositi.   *Bracteæ*
subulatæ, inferiores rarius subfoliaceæ.   *Flores* flavi ; masculi fasciculati ; fœminei
plerumque solitarii, glabri.   *Petala* cuneata, lamina triquetra vel subtriloba, demum
reflexa.   *Drupæ* cerasi parvi magnitudine, rubræ, pulpa glutinosa fœtæ.

Wight and Arnott seem disposed to attach a good deal of importance to the shape of
the petals, and to doubt the identity of the plants of Roxburgh and Wallich with that
of the Peninsula, because Roxburgh's plate differs in that respect from the specimens
before them.   We believe that this character will be found to vary much, as usual in
the Order, and that the petals embrace the filaments in the bud, and become reflexed

in the expanded flower.   *T. cordifolia* seems a very variable plant, and some forms of it approach very near to *T. Bakis*, Miers (*Cocculus Bakis*, Fl. Senegamb. t. 4), which has, however, a different habit, and often terminal flowers.   According to Ainslie and Wight, this species is equally efficacious with *T. crispa* as a tonic, and is known by the same name, *Guluncha.*

5. **ANAMIRTA,** Colebrooke.

Sepala 6, ovali-oblonga, obtusa, carnosula, bracteis 2 adpressis stipata.   *Petala* 0.   MAS.   *Filamenta* in columnam crassam centralem coalita ; *antheræ* sessiles, biloculares, transverse dehiscentes.   FŒM. *Stamina* sterilia 9, clavata, uniserialia.   *Ovaria* 3, gynophoro brevi carnoso hemisphærico insidentia.   *Stigma* depressum, reflexum, fere capitatum. *Drupæ* gynophoro apice trifido stipatæ, oblique ovales, carnosæ, dorso gibbosæ, antice styli cicatrice a hilo non longe distante notatæ.   *Putamen* lignosum, processum alte bilobum læve intus cavum in seminis cavitatem intrusum continens.   *Semen* globosum, intus cavum, funiculo inter lobos processus interni inserto.   *Testa* tenuis, membranacea. *Albumen* fere corneum, oleosum, massulis crebris albidis farinaceis plus minus rotundatis inter se discretis quasi ruminatum.   *Embryo* curvatus ; *radicula* superior, styli cicatricem spectans ; *cotyledones* anguste oblongæ, tenuissimæ, divaricatæ, in loculis diversis albuminis inclusæ.— Frutices *scandentes ;* petioli *cylindrici basi crassiores articulati ;* paniculæ *maximæ e ramis vetustioribus pendulæ, multifloræ.*

The wood of *Anamirta* appears to agree in all essential particulars with that of *Coscinium,* but the liber does not present any traces of annual growth by obscure concentric rings.

1. **A. Cocculus** (W. et A. Prod. i. 446) ; foliis cordatis glabris.— A. paniculata, *Colebr. in Linn. Tr.* xiii. 52, 66.   Menispermum Cocculus, *Linn. Sp.* 1468 ; *Gært. Fr. t.* 70. *f.* 7 ; *Wall. As. Res.* xiii. ; *Roxb. Fl. Ind.* iii. 807.   M. heteroclitum, *Roxb. Fl. Ind.* iii. 817.   Cocculus lacunosus, *DC. Syst.* i. 519, *Prod.* i. 97.   Cocculus suberosus, *DC. Syst.* i. 519, *Prod.* i. 97 ; *W. et A. Prod.* i. 11 ; *Wall. Cat.* 4954 ! ; *Colebr. in Linn. Tr.* xiii. 63.   C. populifolius, *DC. Syst.* i. 519, *Prod.* i. 97 ; *Decaisne, Tim.* 95.

HAB. In Zeylania, *Gardner! Thwaites!* Malabar, *Roxb.*, *Wight!* Concan, *Law!* Orissa, *Roxburgh ;* Khasia ! Assam, *Jenkins!*—(*v. v.*)

DISTRIB. Celebes, ins. Moluccan., Timor.

Frutex alte scandens, cortice cinereo rimoso suberoso.   *Ramuli* crassi, cylindrici, glabriusculi, striati.   *Folia* exacte cordiformia, vel ovalia basi cordata seu truncata, acuta vel acuminata rarius obtusiuscula, supra glabra, subtus pallida, et ad axillas nervorum fasciculis pilorum munita, basi trinervia cæterum penninervia, 4–8 poll. longa, et æquilata vel paullo angustiora.   *Petioli* elongati, striatuli, 2–6 poll. longi. *Paniculæ* e ramis crassioribus pendulæ, pedales vel sesquipedales, ramosæ ; ramuli 1-2-pollicares, multiflori.   *Flores* glabri, majusculi, diam. fere ¼-pollicares.   *Sepala* decidua.   *Gynophora* ½-pollicaris, lignosa.   *Drupæ* glabræ, ⅔-pollicares, nigricantes, sapore (ex Roxburgh) pessimo.

Mr. Miers mentions four species, but only names the one described above and *C. populifolius,* which we believe to be a synonym.   We see nothing in the specimens to which we have access, nor in the descriptions of authors, which implies there

2 B

being another species. In a specimen from Ceylon, not otherwise distinguishable, the leaves are acute at the base; and our Khasia specimens, which are not in flower, have very lucid, ovate, somewhat elongated, subpeltate leaves, which seem to belong to a young shoot. One of Gardner's Ceylon specimens has very similar leaves. Wight and Arnott quote also *C. flavescens*, DC. (described from Rumph. v. t. 24), and *C. orbiculatus*, DC. (Rheede, xi. t. 62). The latter synonym is very doubtful. Rheede's plate does not at all resemble the present genus, and the description in DC. Syst. i. 523, which is taken from a specimen in the Lambertian Herbarium, belongs, no doubt, to *Cissampelos Pareira*. The berries of *Anamirta Cocculus*, which are poisonous, are employed by the natives of India to kill fish. In England they are extensively used in the adulteration of beer.

Tribus III. COCCULEÆ.

Ovaria 3 vel plura. *Drupæ* obovatæ vel hippocrepiformes, styli cicatrice fere basilari, plus minus lateraliter compressæ, cavitate semini subcylindrico conformi. *Embryo* in albumine parco axilis; *cotyledones* appositæ, elongatæ.

The structure of the seed of this tribe is completely masked in the fresh drupe by the sarcocarp, but, in a dried state, the outer coat shrinks so as to display the markings and structure of the putamen. When the sarcocarp is removed, the putamen is seen to form an elongated cylinder, folded on itself, so as to bring the base and apex into contact: the concavity of the horse-shoe being filled up by a bony plate, variously perforated, along which the nutritive vessels pass to the hilum, which is situated at the apex of the sinus: in this way the radicular extremity of the seed, which is really superior, is brought down close to the base of the drupe.

The genus *Tiliacora* is placed in a distinct tribe by Mr. Miers, on account of its numerous ovaries, ruminated albumen, and valvate calyx; but as *Tinospora* among *Tinosporeæ* has ruminated albumen, which is wanting in others of the same tribe, and several species of *Limacia* have a valvate æstivation of the inner sepals, we cannot think that it is desirable to retain the tribe *Tiliacoreæ*.

6. **TILIACORA,** Colebrooke.

Sepala 6, biserialia, exteriora multo minora, interiora ovalia, æstivatione margine vix imbricata. *Petala* 6, minuta, cuneata. MAS. *Stamina* 6; *filamenta* cylindrica subcompressa; *antheræ* adnatæ, introrsæ, biloculares. FŒM. *Ovaria* 9–12, stylo brevi subulato apiculata, gynophoro brevi insidentia. *Drupæ* pedicellatæ, obovatæ, lateraliter subcompressæ, prope basin styli cicatrice notatæ. *Putamen* tenue, lignosum, obscure costatum, utrinque sulco notatum. *Semen* uncinato-incurvum. *Testa* tenuissima. *Albumen* oleosum, endospermii plicis membranaceis ruminatum. *Embryo* semen longitudine fere æquans. *Radicula* cylindrica. *Cotyledones* carnosæ, plano-convexæ.—Frutices *alte scandentes*, inflorescentia *axillari paniculata*, petiolis *gracilibus basi articulatis*.

Tiliacora is readily distinguished from all the other genera of its tribe by its ruminated albumen and numerous ovaries. One species only is known to us, which is widely diffused throughout tropical India. Mr. Miers alludes to an hermaphrodite species from Ceylon, but this we have not seen; and Mr. Thwaites's Ceylon specimens do not differ in any way from continental or Malayan ones.

In *Tiliacora* the stem, when several years old, and one-third of an inch in diameter, is cylindrical, hard, and woody, striated externally. *Pith* very dense and hard, centre softer and of hexagonal cells, becomiug cubical outwards, and then vertically elongated with thick perforated walls. *Wood-bundles* about forty, placed towards the circumference, close-set, oval, separated by narrow, dark-red medullary rays, of a little dotted pleurenchyma and some large ducts. *Liber-bundles* crescent-shaped, almost confluent, annually increasing. *Bark* a very narrow, dense, cellular zone.

1. **T. acuminata** (Miers in Taylor's Annals, ser. 2. vii. 39); foliis ovatis acuminatis glabris.—T. racemosa, *Colebr. in Linn. Tr.* xiii. 53, 67. Menispermum acuminatum *et* M. radiatum, *Lam. Dict.* iv. 101. M. polycarpum, *Roxb. Fl. Ind.* iii. 816. Cocculus acuminatus, *DC. Syst.* i. 527, *Prod.* i. 99 ; *Deless. Ic. Sel.* i. *t.* 95 ; *W. et A. Prod.* i. 12 ; *Graham, Cat. Bombay.* C. radiatus, *DC. Syst.* i. 527, *Prod.* i. 99. C. polycarpus, *Wall. Cat.* 4958 ! (*excl. K, L.*) C. Bantamensis, *Bl. Bijdr.* 26.—*Rheede Mal.* vii. *t.* 3.

HAB. Per totam Indiam tropicam et calidam, a Zeylania ! et Singapur ! ad Concan ! et Orissa ! et in planetie Gangetica ! a Bengalia ! ad Oude !—(*v. v.*)

DISTRIB. Java, *Blume.*

Frutex alte scandens, cortice cinereo striatulo glabro. *Folia* ovata, acuminata, basi interdum acuta sed sæpius truncata, rotundata vel leviter cordata, 3–6 poll. longa, 1½–3½ lata, petiolo ½–1 poll. longo, tenuia, margine undulata et subrepanda, utrinque glabra. *Paniculæ* axillares, folia vix æquantes vel longe superantes, interdum fere petales, incanæ vel demum glabrescentes ; rami pollicares, fœminei subsimplices 1-flori, masculi apice 3–7-flori. *Bracteæ* oblongæ vel subulatæ. *Flores* flavi. *Drupæ* rubicundæ, ½-pollicares.

Mr. Miers has noted that Wall. Cat. 4958 K., from Singapur, is perhaps a species of *Sabia.* It is not in flower or fruit, and is not accurately determinable.

7. LIMACIA, Lour.

Limacia *et* Hypserpa, *Miers.*

Sepala 6, biserialia, exteriora minora. *Petala* 6, sepalis interioribus multo minora, auriculata, stamina amplectentia. MAS. *Stamina* 3–9 ; filamenta cylindrica vel clavata. *Antheræ* biloculares ; loculi laterales vel subextrorsæ, adnatæ, longitudinaliter dehiscentes. FŒM. *Stamina* sterilia 6, clavata. *Ovaria* 3, gynophoro brevissimo insidentia. *Styli* breves, compressi. *Drupæ* obovatæ vel reniformes. *Putamen* vix tuberculatum, lateribus convexum, intus præter cavitatem seminiferam loculos 2 laterales vacuos continens. *Semen* elongatum, cavitati conforme. *Cotyledones* semicylindricæ, radiculam cylindricam latitudine vix superantes.—Frutices *scandentes*, petiolis *simplicibus*, floribus *paniculatis*.

Limacia, which is by Mr. Miers referred to the tribe *Pachygoneæ*, we find to be albuminous, and therefore more properly to belong to *Cocculeæ*. No character therefore remains to distinguish *Hypserpa* but the imbricate inner sepals. The sepals of *Limacia velutina* and *oblonga* are, as Asa Gray has pointed out in the Botany of Captain Wilkes' Expedition, decidedly valvate ; but, as this depends mainly on their thicker texture, we do not attach generic importance to it. We have derived our character of the fruit from two species only, *L. velutina* and *L. cuspidata* (*Hypserpa*, Miers). In both the nut presents no lateral excavations like those of *Cocculus*, but is convex on both sides ; a transverse section, however, shows two large cavities

quite distinct from that occupied by the seed, which is like that of other *Cocculeæ.*
These large cavities are separated from one another by a thin double plate, in the few
nuts we have seen perforated by a hole, so as to connect the two cavities; this is,
however, possibly artificial. The funicle or nutritive cord probably passes to the
seed between these plates. These cavities are, in the dried state, empty, and are
covered externally by a thin arch of the putamen; they communicate by very narrow
canals with its outer surface near the base of the drupe, and evidently correspond to
the deep external excavations of the putamen of *Cocculus* or *Stephania.* The bony
arch by which they are covered springs from the sides of the seed-containing cavity.
We have examined the wood of three species of this genus, and find nearly the
same structure in all.

In *L. velutina* a piece of stem, several years old, and half an inch in diameter, is
tolerably firm and woody in consistence, reddish inside, furrowed and pubescent externally. *Pith* two-thirds the diameter of the stem, central parts of soft utricular tissue,
gradually passing externally into long, narrow, woody tubes, which in a transverse
section resemble a thick zone of liber, but have square extremities, traversed by canals full of red fluid. *Medullary rays* dense. *Wedges of wood* close to circumference, about forty, broadly ovate, rounded towards the bark and pith, of very large
barred vessels and dotted pleurenchyma. *Liber-bundles* semilunar, placed at outer
extremity of each wood-bundle, and more or less entangled in it. *Medullary rays*
of dense, radially elongated mural cells. *Bark* a very thin layer of hexagonal cellular
tissue.

In *L. oblonga* the whole substance of the wood-wedges appears, in a transverse
section, to be formed of broad vessels and liber, which latter, in a vertical section,
consists of pleurenchyma, with perforated walls. The liber seems to be hardly at all
added to in these species after the first year.

In *L. cuspidata* a two or three years old portion of stem is of a dense woody consistence. *Pith* one-third the diameter of stem, of loose, hexagonal, soft, spongy
cellular tissue in the centre, passing into cubical cells towards circumference, and
then lengthening into a dense, hard woody layer of long tubes, with truncate ends.
Medullary rays large, of minute, cubical, thick-walled cells. *Wood-zones* forty narrow wedges of dotted pleurenchyma, and large transversely marked vessels. *Liber-
bundle* reniform. *Bark* a very narrow, dense zone of cellular tissue. A second
small deposit of liber is often seen outside each wood-zone.

1. **L. triandra** (Miers in Taylor's Annals, ser. 2. vii. 43); foliis
oblongo-lanceolatis acutis glabris, paniculis racemiformibus folio brevioribus, floribus triandris.—Menispermum triandrum, *Roxb. Fl. Ind.* iii.
816. Cocculus triandrus, *Colebr. in Linn. Tr.* xiii. 64; *Wall. Cat.*
4962! 4959 C! 4958 L!

HAB. Malaya ad Penang, *Roxb.!* Pegu prope Prome, *Wall.!*—(*v.s.*)

Frutex scandens; ramulis puberulis demum glabratis. *Folia* 2-4 poll. longa,
¾-1½ poll. lata, petiolis puberulis ½-pollicaribus, basi rotundata, triplinervia, apice
acuta vel acuminata cum mucrone, tenuia. *Paniculæ* ½-1½-pollicares, puberulæ;
ramuli bracteis minutis deciduis stipati, abbreviati, 3-5-flori. *Flores* flavi, minutis-
simi. *Sepala* exteriora minuta, interiora ovalia. *Petala* 6, anguste obovata, integra. *Stamina* 3, sepalis exterioribus opposita; *filamenta* carnosa, cuneato-oblonga,
erecta; *antheræ* terminales, biloculares, loculis adnatis divaricatis lateralibus.

Mr. Miers constitutes of this a distinct section, characterized by the absence of half
the number of stamens. We agree with him in considering this character not to be of
generic importance, and we further think that the species is too nearly allied in habit
and characters to the two next, both of which are hexandrous, to make it desirable to
place it in a distinct section. Mr. Miers has noted in the Wallichian collection that
the specimen of *Cocculus Wightianus,* from Prome, 4959 C, belongs to this species;
but he seems to have afterwards regarded it as distinct, as he states in Taylor's An-

uals that he is acquainted with a second triandrous species, represented by a part of Wallich's No. 4952, evidently a misprint for 4959, as the former of these numbers is not a Menispermaceous plant.

2. **L. oblonga** (Miers in Taylor's Annals, ser. 2. vii. 43); caule fulvo-pubescente, foliis oblongis vel lanceolatis utrinque glabris, paniculis elongatis petiolos pubescentes longe superantibus plerumque folio brevioribus.—Cocculus oblongus, *Wall. Cat.* 4963!

HAB. In Malaya ad Singapur, *Lobb!* Malacca, *Griffith!* et Penang, *Wall.!—(v. s.)*

Frutex alte scandens. *Folia* acuta vel longe acuminata, mucronata, basi rotundata vel acutiuscula, tenuiter coriacea, utrinque præter petiolum nervumque medium subtus pubescentes glabra, subtus pallida et nervosa, 3–8 poll. longa, 1–4½ lata, petiolo ¾–1½-pollicari, basi pseudo-articulato. *Paniculæ* elongatæ, paullo supra-axillares, fœmineæ solitariæ, masculæ plerumque 2–3 superpositæ graciliores, 3–8-pollicares, fulvo-pubescentes, ramulis 1–2 pollicaribus multifloris vel apicem versus plurifloris. *Sepala* exteriora minuta, interiora crassa, extus tomentosa, late ovalia, apiculo inflexo, æstivatione subvalvata.

The specimen of *L. scandens*, Lour., at the British Museum, has the leaves of this species, but the inflorescence is more like that of the next species, the male panicles being few-flowered, and the peduncles, from which the drupes have fallen, solitary. It may, however, prove to be an abnormal state of *L. oblonga*.

3. **L. velutina** (Miers in Taylor's Annals, ser. 2. vii. 43); caule velutino, foliis late ovalibus vel ovali-oblongis subtus vel utrinque fulvo-tomentosis, paniculis petiolos subæquantibus paucifloris, staminibus 6, drupis obovatis.—*A. Gray, Bot. Wilkes' Exp.* i. 40. Cocculus velutinus, *Wall. Cat.* 4970!

HAB. Mergui, *Griffith!* Moulmein, *Lobb!* Singapur, *Wallich!—(v.s.)* DISTRIB. Ins. Philip., *Cuming, No.* 2402!

Frutex scandens, caule dense aureo- vel fulvo-tomentoso demum glabrescente. *Folia* ovali-oblonga, oblonga vel lanceolata, forma et magnitudine valde varia, basi rotundata vel acuta, interdum obliqua, triplinervia, cæterum penninervia, acuta cum mucrone parvo, rarius obtusa, interdum obtusissima et fere rotundata, supra glabra (exceptis junioribus) sed secus costam fulvo-pubescentia, (in sicco) crebre reticulato-venosa, subtus cum petiolis (juniora dense, adulta sparse) fulvo-tomentosa, 2–6 poll. longa, ¾–4 lata, petiolo ½–1¼-pollicari. *Paniculæ* axillares vel sæpius paullo supra-axillares, solitariæ vel plures in eadem axilla, petiolo subbreviores, rarius secus ramulos axillares aphyllos dispositæ, fulvo-tomentosæ, paucifloræ. *Bracteæ* squamæformes. *Flores* fusco-villosi. *Sepala* interiora rotundata, intus glabra, æst. valvata. *Petala* obovato-spathulata, retusa vel truncata. *Pedunculi* fructiferi in specimine solitarii, pollicares. *Drupæ* obovatæ, compressæ, pollicares, glabræ. *Putamen* læve, obovatum, zona lata carinali cinctum.

4. **L. cuspidata** (H.f. et T.); foliis ovato- vel oblongo-lanceolatis acuminatis glabris, paniculis masculis petiolos parum superantibus, fœmineis subunifloris, staminibus 6–9, drupis subglobosis.—Cocculus cuspidatus, *Wall. Cat.* 4960! Hypserpa cuspidata, *Miers in Taylor's Annals, ser.* 2. vii. 40.

HAB. In Zeylania, *Walker! Gardner! Thwaites!;* in Tenasserim ad Mergui, *Griffith!;* Silhet et Khasia, *Wall.!—(v. s.)*

Frutex alte scandens. *Ramuli* eleganter striatuli, juniores pubescentes. *Folia* basi rotundata vel subcuneata, 3-nervia, tenuiter coriacea, lucida, crebre reticulata, adulta glabra, juniora subtus secus costam pubescentia, 2–3 poll. longa, 1–1¼ lata, in ramulis sterilibus interdum 5–6 poll. longa, et ultra 2 poll. lata. *Petioli* ½–1 poll. longi, superne incrassati, pubescentes, demum glabrati. *Paniculæ* pubescentes, axillares vel paullo supra-axillares, solitarii vel bini, tuberculo tomentoso insertæ; masculæ petiolis duplo longiores, parce ramosæ, vel subracemosæ, fœmineæ simplices, petiolum vix æquantes, paucifloræ, subunifloræ. *Bracteæ* minutæ, subulatæ. *Drupæ* vix ½-pollicares.

Hypserpa nitida, Miers (in Hook. Kew Journ. Bot.), does not appear to be distinct from *L. cuspidata. Cocculus cynanchoides*, Presl, is perhaps also a synonym.

8. COCCULUS, DC.

Nephroia, *Lour.;* Nephroica, Holopeira *et* Diploclisia, *Miers.*

Sepala 6, biseriatim imbricata, exteriora minora. *Petala* 6, sepalis minora, cuneata vel obovata, integra vel sæpius emarginata, v. plerumque auriculata et in masc. circa stamina involuta. MAS. *Stamina* 6; *filamenta* cylindrica vel compressa; *antheræ* terminales, subglobosæ, 4-lobæ, biloculares, loculis lateraliter dehiscentibus profunde didymis. FŒM. *Stamina* sterilia 6 vel nulla. *Ovaria* 3, gynophoro brevi insidentia. *Styli* erecti vel reflexi, ovarii longitudine, cylindrici. *Drupæ* lateraliter compressæ, obovatæ vel rotundatæ. *Putamen* fragile, hippocrepiforme, dorso carinatum et varie tuberculatum, utrinque profunde excavatum. *Semen* hippocrepiforme, cavitati putaminis conforme. *Embryo* in albumine parco carnoso homotropus, radicula brevi cylindrica, cotyledonibus linearibus planis.—Frutices *scandentes vel saltem sarmentosi, rarissime suberecti,* foliis *forma variis basi pseudo-articulatis,* petiolo *gracili cylindrico haud dilatato,* inflorescentia *axillari paniculata,* paniculis *elongatis vel sæpius paucifloris et fœmineis interdum ad florem solitarium reductis.*

We include in this genus the whole of Mr. Miers' tribe *Platygoneæ,* as we cannot attach that degree of importance to the shape of the petals (in itself indeed far from constant in each species) which Mr. Miers seems to do; nor do we think that the modifications of the structure of the putamen are either sufficiently constant or sufficiently important to be relied upon as generic distinctions.

Diploclisia of Miers, with a very different habit from the other Indian species, presents no characters by which to separate it generically, except the elongated drupe.

Cocculus is mostly an Indian genus, but several American species are no doubt correctly referred to it, and some of the most common Indian species extend across tropical Africa even to the west coast. One species (by Mr. Miers referred to *Diploclisia*)·is a native of New Holland, extending as far south as the colony of Victoria.

Cocculus ovalifolius, DC., is also a true member of the genus, and is closely allied to *C. Carolinus.* It is the *Nephroia sarmentosa* of Loureiro, of which *Menispermum hexagynum,* Roxb. (*C. hexagynus,* Wall.) is a synonym. *C. trilobus,* DC., and *Nephroia pubinervis,* Miers in Hook. Kew Journ. iii. 259, from Hongkong, are also, we believe, not distinct.

Among the *Menispermaceæ* of Dr. Hooker's East Nipal collections there is a specimen, without flower or fruit, which so closely resembles *Menispermum Dahuricum* that it is probably a congener, and perhaps not specifically distinct. The leaves are deeply three-lobed. There is also a three-lobed Menispermaceous plant among Captain Strachey's Kumaon collections, but in a very bad state. The genus *Menispermum* only differs from *Cocculus* by having 12–18 instead of 6 stamens.

Our specimens of the stem of *Cocculus Leæba* are all small; one, evidently several years old, and one-fourth of an inch in diameter, is compact, woody, and cylindrical, with only one zone of wood-wedges. These are separated by very narrow medullary rays, and extend nearly from the pith to the circumference; they are very much more numerous and closely placed than in *C. laurifolius*, but their component tissues entirely correspond with those of that plant. Older stems may present other concentric zones of wood.

The stem of *C. villosus* attains a considerable diameter, but our specimens are only small branches of the same size and apparent age as those of *C. Leæba*. The tissues of these only differ from those of the above-named plant and of *C. laurifolius* in the outer portion of the pith, which is contiguous to the wedges of wood, becoming smaller and denser, and the cells elongating vertically into woody tubes, with blunt superimposed ends. This is a very common form of pith in Menisperms, varying in proportion to the more strictly cellular pith in different species and individuals.

In *Cocculus macrocarpus* the young shoots have the same structure as those of *C. villosus*, the outer pith-cells next the wood being elongated, woody, and dense. In a shoot nearly half an inch thick there are two zones of wood-wedges analogous to those of *C. laurifolius*, as described by Decaisne, while an old trunk, more than two inches in diameter, from Chittagong, believed to belong to this species, has a succession of concentric zones of wood-wedges, irregularly arranged around an excentric axis; the number of zones on one semi-diameter being eight or ten, and on the other about twenty, owing to occasional union or suppression. The wedges of each zone are separated by thin medullary plates, which do not run in straight lines from the centre to the circumference.

1. **C. macrocarpus** (W. et A.! Prod. i. 13); foliis fere rotundatis glabris longe petiolatis, paniculis longissimis, drupis obovato-oblongis.—*Wight, Ill.* i. 22. *t.* 7. Diploclisia macrocarpa, *Miers in Taylor's Ann. ser.* 2. vii. 42.

HAB. In montibus inferioribus Zeylaniæ, *Gardner!* *Thwaites!* Malabar, *Wight!* Concan, *Graham, Law!*—(*v. s.*)

Frutex alte scandens, cortice cinereo rugoso. *Ramuli* eleganter striatuli, atrofusci vel viridescentes. *Folia* rotundata vel reniformia, basi truncata vel cordata, margine subrepanda, obtusa vel retusa cum mucrone, rarius acuta, 5-nervia, nervis lateralibus extrorse venosis, glaberrima, subtus glauca, 2–3 poll. longa et longitudine paullo latiora, petiolo gracili 2–4-pollicari. *Paniculæ* secus ramos vetustiores dispositæ vel rarius versus apices ramulorum axillares, plerumque elongatæ, haud raro pedales, ramosæ, multifloræ, ramulis 1–2-pollicaribus apice corymbosis solitariis vel fasciculatis. *Sepala* tenuissime membranacea, lineis punctisque purpureis interdum confluentibus (ut etiam petala et stamina et ovaria) notata. *Petala* late cuneata, triloba, lobo medio emarginato vel eroso-dentato, rarius acutiusculo, lateralibus circa stamina involutis. MAS. *Filamenta* planiuscula, ligulata. *Antheræ* biloculares, obliquæ, didymæ. FŒM. *Stamina* sterilia 6, carnosa, linearia, obtusa. *Ovaria* oblonga, incurva; styli fere æquilongi, recurvi. *Drupæ* pollicares, obovato-oblongæ, obtusæ, sarcocarpio parco viscido. *Putamen* tenue, lignosum, dorso leviter carinatum, utrinque sulcis transversis profundis notatum, excavatione laterali elongato, subcurvato, superne latiore, costa longitudinali per totam longitudinem notata.

A specimen, in leaf only, collected in the Khasia hills, is very like this species, but cannot in that state be identified with any certainty. A specimen collected in south China by Seemann has also very similar foliage.

2. **C. laurifolius** (DC. Syst. i. 530, Prod. i. 100); arboreus, foliis lanceolatis lucidis glabris breviter petiolatis, paniculis axillaribus folio brevioribus.—*Delessert, Ic. Sel.* i. *t.* 97; *Colebrooke in Linn. Tr.* xiii. 65; *Wall. Cat.* 4965! C. angustifolius, *Hasskarl, Hort. Bog.* 172;

Plant. Jav. Rar. 167.   Menispermum laurifolium, *Roxb. Fl. Ind.* iii. 815.

HAB. In Himalaya subtropica media et occidentali, alt. 2–5000 ped.; in Nipalia centrali, *Wall!* Kumaon! Simla! Jamu!—(Fl. vere.) (*v. v.*) DISTRIB. Japonia, *Hasskarl.*

Arbor parva, vel *frutex* trunco abbreviato. *Rami* divaricati, apice penduli, sæpe elongati et sarmentosi, angulati, striatuli, glabri; ramuli basi fasciculo pilorum circumdati. *Folia* lanceolata, coriacea, lucida, subtus pallida, acuta vel acuminata, mucronulata, trinervia, 3–6 poll. longa, 1–1¼ lata, petiolo ⅓–½-pollicari. *Paniculæ* axillares vel paullo supra-axillares, solitariæ, vel 2 superpositæ, basi fasciculo pilorum stipatæ, 1–2-pollicares, corymbosæ; masculæ plerumque majores. *Bracteæ* minutæ, cito deciduæ. *Flores* minuti. *Sepala* exteriora interioribus dimidio minora, acuta. *Petala* profunde biloba, lobis obtusis vel acutiusculis. *Stamina* sterilia in flore fœmineo 6. *Ovaria* late ovalia. *Styli* cylindrici, reflexi. *Drupæ* minimæ, rotundatæ. *Putamen* fragile, dorso obscure carinatum, rugis transversis validis notatum, utrinque profunde excavatum, imperforatum.

This plant is remarkable in the Order for its erect habit. It is, however, often decidedly sarmentose, and on the ticket attached to Wallich's Nipal specimens in the Linnean Society's collection (Cat. No. 4965 A), we find it noted that the plant is "valde similis *M. laurifolio*, Roxb., sed scandit." The axillary corymbs vary much in size. Sometimes they are very short, and eight or ten flowered, as described by De Candolle. More rarely, in very luxuriant specimens, they are expanded into compound panicles, leafy at top. A Java specimen from Mr. Lobb (we presume cultivated) is broader-leaved than the ordinary state of the Himalayan plant, but we can see no other difference.

3. **C. Leæba** (DC. Syst. i. 529, Prod. i. 99); foliis glabriusculis oblongis vel trapezoideis integris vel lobatis, floribus in axillis fasciculatis, fœmineis subsolitariis.—*A. Richard, Fl. Seneg.* i. 13; *Webb, Spic. Gorg. in Hook. Niger Flora, p.* 97. C. Cebatha, *DC. Syst.* i. 527, *Prod.* i. 99; *Miers in Hook. Niger Flora, p.* 215. C. ellipticus, *DC. Syst.* i. 527, *Prod.* i. 100. C. Epibaterium, *DC. Syst.* i. 530, *Prod.* i. 100. C. lævis, *Wall. Cat.* 4975! C. glaber, *Wight et Arn. Prod.* i. 13. Leæba *et* Cebatha, *Forsk. Fl. Ægypt. Arab.* Epibaterium, *Forster, Gen. t.* 54. Menispermum Leæba, *Del. Fl. Æg. t.* 51. *f.* 2, 3. M. edule, *Vahl, Lam., Willd.* M. ellipticum, *Poiret.*

HAB. In Carnaticæ montosis aridis prope Madura, *Wight!* et Coimbator, *Gardner!* in dumetis aridis Sindh, *Vicary!* Panjab, *Edgeworth!* usque ad Firozpur! et Lodhiana!; et in Afghanistan, *Griffith!*—(Fl. per totum annum.) (*v. v.*)

DISTRIB. Arabia! Ægyptus! ins. Cap. Viridis! Senegambia!

Frutex alte scandens, ramis glabris cinereis, ramulis elongatis vimineis gracillimis pube minuta incano-puberulis demum glabratis. *Folia* forma valde varia, oblonga aut trapezoidea angulis rotundatis, aut obscure triloba rarius obtuse 3–5-loba, interdum lineari-oblonga, plerumque obtusa cum mucrone, basi cuneata vel rarius rotundata, juniora incano-puberula vel glabrata, adulta glabra utrinque glauca, ½–1½ poll. longa, ¼–¾ lata, petiolo 2–3 lineas longo. *Flores* axillares, tuberculo piloso inserti; *masculi* in fasciculum densiflorum petiolum vix æquantem congesti, pedicellis unifloris; *fœminei* solitarii vel bini, rarius in speciminibus luxuriantibus plures, pedicellis petiolos æquantibus pubescentibus. *Drupæ* minimæ, 1–2 lin. longæ.

Of this very variable plant, which has a wide range in the hot desert regions of Africa and Asia, we have been able to compare extensive suites of specimens. *C.*

glaber, W. et A., is a very luxuriant form, with larger leaves and a more lax habit, but is certainly not distinct. To the numerous lists of synonyms already brought together by Richard and others, Mr. Miers has added the *Cebatha* of Forskål. The fruit of that species is said to be eatable, but it is so small that it can hardly be worth eating. A fermented liquor is also stated to be prepared in Arabia from the juice.

4. **C. villosus** (DC. Syst. i. 525, Prod. i. 98); foliis ovali-oblongis subdeltoideis villosis, paniculis masculis abbreviatis, floribus fœmineis in axilla 1–3.—*Wall. Cat.* 4957!; *W. et A.! Prod.* i. 13. C. sepium, *Colebr. in Linn. Tr.* xiii. 58. C. hastatus, *DC. Syst.* i. 522, *Prod.* i. 98. Menispermum villosum, *Lam.* (*non Roxb.*). M. hirsutum, *Linn.; Roxb. Fl. Ind.* iii. 814. M. myosotoides, *Linn.*

Hab. Pegu! Ava! Carnatica! Malabar! Maisor! Dekhan! Concan! Bahar, et per totam Hindustaniam et Panjab usque ad basin Himalayæ, ubique in sepibus et dumetis vulgatissimus; sed e Malaya non vidimus.—(Fl. per fere totum annum.) (*v. v.*)

Distrib. Africa occidentalis extratropica, *Curror* (*in Herb. Hook.*).

Frutex alte scandens, ramulis villosis sulcatulis. *Folia* ovalia vel ovali-oblonga, late deltoidea, angulis rotundatis, vel elongato-deltoidea, interdum subtriloba, retusa vel obtusa cum mucrone, rarius acutiuscula, basi subcordata vel truncata, juniora utrinque molliter villosa, majora ad paginam superiorem fere glabrata, ramorum 2–3 poll. longa, 1½–2 lata, pedicellis gracilibus vix semipollicaribus, ramulorum semper minora et plerumque angustiora, sæpe lineari-oblonga, sed interdum fere orbicularia, ½–1¼ poll. longa, acuta vel obtusa, setoso-mucronata, dense incano-villosa. *Paniculæ* masculæ in ramulis axillares, solitariæ vel binæ, foliis dimidio breviores, pilis patentibus laxe villosæ. *Bracteæ* lineares, minutæ. *Sepala* laxe villosa. *Petala* acuta, bidentata. *Flores fœminei* solitarii vel 2–3-fasciculati, rarius (foliis ramulorum omnino abortivis) in axillis foliorum majorum longe racemosi. *Drupæ* atro-purpureæ. *Putamen* dorso argute carinatum, tuberculatum.

A very well-marked species, which can scarcely be confounded with any other. According to Roxburgh, ink is made of the berries, and a decoction of the roots is used in Hindoo medicine as a substitute for Sarsaparilla. The first year's shoots are barren and very long, and bear large leaves, in the axils of which short flower bearing branches are produced, densely covered with small leaves. The closely-allied African species mentioned by Miers in Hooker's Niger Flora, at p. 215, we consider a state of *C. villosus.*

5. **C. mollis** (Wall. Cat. 4973!); foliis ovatis acutis vel acuminatis subtus albo-villosis, paniculis paucifloris petiolo subbrevioribus.

Hab. In Nipalia, *Wall.!* in mont. Khasia, alt. 5000 ped.!—(*v. v.*)

Frutex scandens, ramulis cylindricis striatis molliter pubescentibus demum glabratis. *Folia* ramulorum sterilium interdum obtusa, mucronata, 2–4 poll. longa, 1½–2½ lata, petiolo ¾–1-pollicari, basi cordata vel truncata, supra læte viridia, pilis adpressis sparsis pubescentia, demum fere glabra, subtus molliter tomentosa. *Pedunculi* axillares vel paullo supra-axillares, tuberculo tomentoso interposito, masculi irregulariter cymosi, 4–7-flori, bracteis paucis filiformibus, fœminei 1–3-flori. *Bracteæ* 2 calyci adpressæ. *Sepala* ovalia. *Petala* emarginata. *Fructus* plerumque secus ramulos (ob folia decidua) nudos pseudo-racemosi. *Drupæ* compressæ, pisiformes. *Putamen* dorso carinatum et lineis 4 tuberculorum notatum.

9. **PERICAMPYLUS,** Miers.

Sepala 6, biseriatim imbricata, exteriora minora. *Petala* 6. Mas.

Stamina 6; *filamenta* cylindrica; *antheræ* adnatæ, ovales, biloculares, loculis lateraliter dehiscentibus. FŒM. *Stamina* sterilia 6, subclavata. *Ovaria* 3. *Stylus* ad basin bipartitus, segmentis reflexis subulatis. *Drupæ* subglobosæ. *Putamen* hippocrepiforme, dorso cristatum, lateraliter utrinque excavatum, imperforatum. *Embryo* in axi albuminis cylindricus; cotyledonibus elongatis, radicula cylindrica vix latioribus, planis.—Frutices *scandentes*, foliis *subpeltatis*, petiolis *gracilibus basi articulatis*, cymis *dichotomis axillaribus longe pedunculatis multifloris*, *sæpe in una axilla pluribus superpositis, interdum secus ramum elongatum aphyllum paniculatis.*

Pericampylus has the fruit of *Cissampelos* or *Stephania*, with the flower of the tribe *Cocculeæ*. The bipartite style and peculiar inflorescence distinguish the genus. When there are several cymes in one axil, they are superposed, and the higher is then usually supported on a longer peduncle, and bears more numerous flowers.

The specimen of Loureiro's *Pselium* at the British Museum, though very imperfect, is clearly identical with *Pericampylus incanus;* but as Loureiro's character is very confused, the two sexes belonging evidently to different plants, the name must be rejected.

The stem of *Pericampylus incanus* is cylindrical and grooved; a portion of perhaps four or five years' growth (or more), is one-third of an inch in diameter, and presents eighteen *wood-wedges*, separated by very narrow *medullary rays;* there are evident traces of periodic deposition of wood in the wedges, marked by the number, size, and disposition of the great vessels. The *pith* is loosely cellular in the centre, passing into woody tubes near the wedges of wood, and becoming very dense and firm in the medullary rays and bark. *Liber-bundles* isolated, applied to the cambium layer, but not much increased after the first year.

1. **P. incanus** (Miers in Taylor's Annals, ser. 2. vii. 40); foliis fere orbicularibus acutis vel obtusis.—Cocculus incanus, *Colebrooke in Linn. Tr.* xiii. 57. Clypea corymbosa, *Blume, Bijdr.* 24. Cissampelos Mauritiana, *Wall. Cat.* 4980! (*non DC.*). Menispermum villosum, *Roxb. Fl. Ind.* iii. 812 (*non Lam.*).

HAB. In dumetis subtropicis Sikkim! Assam! Khasia! Silhet! Chittagong! Tenasserim! Maluya usque ad Malacca!—(*v. v.*)

DISTRIB. Java!

Frutex scandens, ramulis fulvo-tomentosis demum fere glabris. *Folia* latissima, basi leviter cordata vel subtruncata, subpeltata, mucrone subdeciduo apiculata, interdum retusa, diametro 2–4-pollicaria, petiolo 1–2-pollicari tomentoso, supra glabra, subtus albo- vel cinereo-tomentosa vel incana, rarius glabrescentia, 5-nervia. *Cymæ* bi-trichotomæ, folio breviores, pedunculis 1–2-pollicaribus. *Bracteæ* ad ramificationes subulatæ. *Sepala* dorso villosa. *Petala* trapezoidea, acuta vel obtusa, marginibus inflexis. *Drupæ* lætæ rubræ. *Putamen* dorso seriebus 3 tuberculorum obtusorum notatum.

Tribus IV. CISSAMPELIDEÆ.

Stamina in columnam centralem cylindricam coalita, apice discum planum peltatum rotundatum margine antheriferum gerentem. *Ovaria* solitaria, stylo 3–5-partito coronata. *Embryo* hippocrepiformis, cylindricus, in axi albuminis parci, cotyledonibus oppositis.

This tribe is characterized by the solitary ovary, crowned by the style, which is divided to the base into three or five divergent, almost acicular teeth. The ovule is inserted considerably below the middle of the ventral suture, and the chalazal end is rounded, while the upper end is elongated and gradually narrowed towards the apex. The inflorescence is also often different from that of the other tribes, but it is peculiar in each genus, and in *Cissampelos* the male cymes are very like those of *Pericampylus*. The leaves are generally, but not always, peltate.

10. STEPHANIA, Lour.

Clypea, *Blume ;* Stephania, Clypea, *et* Ileocarpus, *Miers.*

MAS. *Sepala* 6–10, biserialia, ovalia vel obovata. *Petala* 3–5, obovata, carnosa. FŒM. *Sepala* 3–5. *Petala* totidem, carnosa. *Drupa* solitaria ; putamen compressum, hippocrepiforme, dorso tuberculatum, ad latera utrinque excavatum et foramine circulari perforatum.—Frutices *scandentes*, foliis *plerumque peltatis*, inflorescentia *axillari umbellata*.

Aśa Gray has pointed out the inconstancy of the character derived from the number of parts in each verticil of the flower, and has accordingly reduced Mr. Miers' genus *Clypea*, which is not marked by any striking characters of vegetation or inflorescence. As they now stand, the genera of *Cissampelideæ* are all very distinct in inflorescence ; but in several species the floral characters, of the female especially, are still imperfectly known.

In *Stephania rotunda* a piece of stem, six to eight years old, is about half an inch in diameter, of a spongy consistence, with much cellular tissue. *Pith* of large, loose, elongated utricles. *Medullary rays* and *bark* the same ; all full of starch-granules. *Epidermis* smooth, covered with longitudinal rimæ of tumid cells, with projecting lips. *Wood-wedges* twelve, cuneate, with broad medullary rays, formed of punctate pleurenchyma, and large vessels whose walls are covered with very narrow, oblique, transversely elongated discs, each with a mesial dark line. *Liber* a very narrow arcuate line of pleurenchyma opposite each wood-bundle, and sometimes confluent into a narrow zone of liber ; it does not increase after the first year. *Bark* tolerably broad, cellular, with scattered masses of sclerogen cells ; circumference of many layers of radially compressed cells. It is thus almost identical in structure with *Tinospora.*

In *S. elegans* the base of a portion of stem of great length, but not many years old, and one-fourth of an inch in diameter, is moderately woody, seven-angled ; angles opposite as many wedges of wood of ordinary menispermous tissue. *Pith* narrow, of loose hexagonal cellular tissue. *Medullary rays* very large, as broad as the wood-wedges.

Among Dr. Hooker's Sikkim *Menispermaceæ* there is a specimen in young fruit which seems to constitute an undescribed species of this genus. The leaves are broad ovate, acuminate, cordate at base, not peltate, thin, pale below, palmately seven-nerved, glabrous, except the nerves, which are slightly adpressed-hairy beneath. They are 5 inches long, 4½ broad, and the slender petioles are 3 inches in length. The female inflorescence is umbellate on a long peduncle, with subulate bracts. The young fruits are subsessile, in heads, at the apex of thick fleshy rays, ⅓ inch long. There is in the Hookerian Herbarium a very similar specimen, without flower or fruit, from Garhwal, collected by Major Madden, in which the leaves are pubescent underneath.

1. **S. elegans** (H.f. et T.) ; foliis elongato-deltoideis acuminatis basi truncatis vel cordatis tenuiter coriaceis glaberrimis, umbellis longe pedunculatis, umbellulis laxifloris.

HAB. Khasia ! Assam ! Sikkim ! Kumaon, *Str. et Wint. !* a planitie ad alt. 6–7000 ped.!—(Fl. per totum æst.) (*v. v.*)

Frutex scandens, caule gracili angulato striato glabro.   *Folia* interdum apice obtusa, subtus pallida, 2½–4 poll. longa, 1–2¼ lata, petiolo gracili dimidio breviore. *Pedunculi* gracillimi, plerumque petiolis longiores, fructiferi sæpe folia superantes. *Umbellæ* multiradiatæ, interdum bis divisæ.   *Flores* purpurei vel virides, graveolentes.   *Sepala* late ovata, acuminata.   *Petala* obovata, interdum emarginata.   *Drupæ* translucentes, subglobosæ, rubræ.

This pretty little species, which is very common in the Eastern Himalaya among the lower hills, appears undescribed, unless it be the *S. longa* of Loureiro, with which, however, we cannot venture to unite it, as the description does not satisfactorily accord.   No specimen of that plant exists in the British Museum.   It may readily be known from the next by the flowers being supported on pedicels, and not collected into heads.

2. **S. hernandifolia** (Walpers, Rep. i. 96); foliis ovatis vel subdeltoideis acutis vel obtusis interdum acuminatis, umbellulis capitatis.

α; foliis subtus glabris vel secus nervos tenuiter puberulis, pedunculis glabris.—Cissampelos hernandifolia, *Willd.; DC. Syst.* i. 533, *Prod.* i. 100; *Wall. Cat.* 4977 D ! Clypea hernandifolia, *W. et A. Prod.* i. 14; *Wight, Ic. t.* 939, *Spic. Neilg. t.* 7.

β; foliorum pagina inferiore pedunculisque pubescentibus.—Cissampelos discolor, *DC. Syst.* i. 534, *Prod.* i. 101; *Blume, Bijdr.* 26. C. hexandra, *Roxb. Fl. Ind.* iii. 842.   C. hernandifolia, *Roxb. Fl. Ind.* iii. 842; *Wall. Cat.* 4977 E ! F ! G ! H ! K ! (*excl. cæt. lit.*).   Stephania discolor, *Hassk. Pl. Jav. Rar. p.* 168.

HAB. Var. α. In dumetis humidis præsertim montanis Zeylaniæ! Carnaticæ australis! Malabar! Concan! a planitie ad alt. 7000 ped. Var. β. In Malaya! Ava! Chittagong! Khasia! Bengalia! et in Himalaya subtropicâ orientali et media ab Assam! ad Nipaliam!—(Fl. per totum annum.) (*v. v.*)

DISTRIB. Abyssinia, Java, ins. Philipp., Timor, Australia tropica.

Frutex scandens, ramulis striatis glabris. *Folia* basi truncata vel leviter cordata, tenuiter coriacea, subtus pallida vel glaucescentia, glabra vel tenuiter pubescentia, 3–6 poll. longa et æquilata vel ¼ angustiora, petiolo 1½–4-pollicari. *Pedunculi* axillares, abbreviati vel petiolos superantes, apice umbellati; umbellæ radii longitudine varii, bracteis subulatis stipati, *capitula* 8–12- vel pluriflora gerentes, interdum proliferi.   *Sepala* obovata, obtusa.   *Petala* dimidio minora, 3–4.

The amount of variation to which this species is subject is so great, that we have little doubt that all the synonyms quoted above are correctly referred to it.   Indeed, we can assign no characters for the separation of *Cissampelos Forsteri*, DC. (*Stephania Forsteri*, Asa Gray, Plants Wilkes' Exp.), to which *Clypea venosa*, Bl. Bijdr., and *C. glaucescens*, Dcne. Pl. Tim., ought, in all probability, to be reduced. The Malabar and Nilghiri plant is usually more glabrous than the eastern and Himalayan form, but the latter occasionally occurs with the leaves quite glabrous below. So far as we have observed, however, the peduncles of the eastern plant are always pubescent, and those of the western plant always glabrous; but more extended observation will perhaps break down even that difference.   *Stephania Abyssinica* of Richard, Fl. Abyss. t. 4 (which is *Ileocarpus Schimperi*, Miers, or *Menispermum Schimperi*, Hochstetter), is evidently the same as the glabrous Nilghiri form.   Both varieties occur in Java, and the glabrous form extends to Timor and the Philippines, while the pubescent state occurs in tropical Australia.   Wight and Arnott quote under this species *Cocculus Roxburghianus*, Wall. Cat. 4972 A; but in the Linnean Society's collection that and all the other letters belong to *Stephania rotunda*, Lour., except a specimen from the Calcutta garden, distributed at a later period than the rest, and included in one of the supplementary lists under the letter C, Dr. Wallich having inadvertently over-looked the previous employment of that letter in the body of the work.

3. **S. rotunda** (Lour.! Fl. Coch. Chin. 747); foliis late ovatis vel fere rotundatis irregulariter sinuato-lobatis vel repandis tenuibus gla-bris longe petiolatis, umbellulis laxe cymosis.—Cocculus Roxburghia-nus, *Wall. Cat.* 4972! (*vix DC.*); *W. et A. Prod.* i. 450 *in adnot.* C. Finlaysonianus, *Wall. Cat.* 4974! *excl. spec. sinist. ad* S. hernandifo-liam *pertinens.* Cissampelos glabra, *Roxb. Fl. Ind.* iii. 840 (*et vero-similiter etiam Herb. Hamilton*). Clypea Wightii, *Arn.! in Wight, Ill.* i. 22.

HAB. In Himalaya tropica et temperata a basi ad alt. 7000 ped.; Simla! Kumaon, *Str. et Wint.!* Nipal, *Wall.!* Sikkim! Bhotan, *Grif-fith!* Assam, *Hamilton;* in montibus Khasia! et Silhet, *Wall.!;* in Pegu, *M'Clelland!;* et in montibus peninsulæ australioris ad Courta-lam, *Wight!*—(Fl. Apr., Jun.) (*v. v.*)

DISTRIB. Siam! Cochin China!

Frutex alte scandens. *Radix* tuberosa, magna, subglobosa. *Caules* vetustiores tuberculis rimosis crebris tecti, grisei vel flavicantes; juniores glaberrimi, atrofusci, striati. *Folia* obtusa vel acuta, interdum acuminata, subtus pallida, 3–7 poll. longa et æquilata. *Petioli* folia æquantes vel (præsertim in foliis majoribus) longe supe-rantes, interdum 9-pollicares, graciles, basi subarticulati. *Pedunculi* longitudine valde varii, sæpe petiolos æquantes, axillares et gracillimi, vel secus caules crassiores ad axillas foliorum delapsorum solitarii aut in ramulo abbreviato aphyllo racemosi, et tunc crassiores, fœminei sæpe carnosuli. *Umbellæ radii* abbreviati vel elongati, basi bracteolis subulatis stipati, cymosi. *Flores* majusculi, diametro interdum fere ½-pollicares, plerumque minores, flavidi vel crocei, carnosuli. *Sepala* in flore masc. 6–10, biserialia, anguste cuneata, obtusa, dorso furfuracea vel puberula. *Pe-tala* 3–5, late cuneata, sepalis ½ breviora. *Drupæ* glabræ.

Loureiro's specimen in the British Museum, though very imperfect, evidently be-longs to the species now described. We refrain from quoting De Candolle's *C. Rox-burghianus*, because he describes the peduncles as "adpresse velutini," and the leaves as quite entire; his description is also otherwise unintelligible. *C. Wightii*, Arn., is stated to have the male flowers in a simple capitulum, but the specimen before us (which bears ripe fruit) agrees so exactly with *S. rotunda* that we cannot doubt the identity of the two; probably, therefore, the male umbels are very young, in which state those of *S. rotunda* appear to form a simple head. Roxburgh describes the female flower with one sepal, and two petals longer than the sepal, and of a deep orange-yellow colour. This is evidently the structure of the genus *Cyclea*, but Loureiro describes the perianth of the female flower as consisting of six leaves. We do not possess the means of determining this point. Possibly Roxburgh may have had sent to him specimens of *Cyclea populifolia*, as his description of the female in-florescence does not agree with our specimens, in which it is the same as in the male. According to Roxburgh the tuberous roots of this species are very acrid, and are used in medicine. Loureiro says they are very bitter, and have similar qualities to those of *Aristolochia rotunda.*

11. CISSAMPELOS, Linn.

MAS. *Sepala* 4. *Petala* 4, in corollam cupuliformem margine fere indivisam coalita. FŒM. *Sepala* 2 in squamam carnosulam sæpius binervem emarginatam vel indivisam bractea antica suffultam coalita.

Drupæ subglobosæ.   *Putamen* compressum, dorso tuberculatum, lateribus utrinque excavatum.—Frutices *scandentes vel suberecti,* inflorescentia mascula *cymosa,* fœminea *racemosa floribus ad axillas bractearum
rotundarum fasciculatis.*

The female flowers of this plant have usually been described very differently, as
possessing a *lateral* calyx of one sepal, with one petal in its axil, and no one seems
to have adverted to the anomalous nature of such an arrangement. The composition of the so-called petal is generally indicated by the existence of two nerves. It
is often emarginate, and we have several times seen it bipartite to the very base.
The female flower is thus evidently analogous in structure to that of *Cyclea.* With
regard to the lateral position usually ascribed to the bract and sepal, they are certainly opposite the ovary, and are, therefore, more probably anterior.

The only Indian *Cissampelos* is very widely diffused throughout the tropics.
Except one West African plant, the remainder of the genus is American. Several
species are erect or suberect, and quite distinct; but many of the scandent ones
have very slender claims to be considered species, and will probably be reduced on a
careful revision of the genus.

Cissampelos acuminata, DC., is not determinable with certainty without an authentic specimen. From the description, it is evidently no *Cissampelos.* It may be
Limacia triandra or *cuspidata,* but the very short petioles (two to three lines long)
seem rather to point to *Cocculus laurifolius.*

Our Indian *Cissampelos* has rather soft and spongy wood ; a section of a stem one-
fourth of an inch in diameter, and several years old, presents twelve to fifteen large
irregularly formed *wood-wedges,* separated by narrow *medullary rays;* these wedges
reach about the same distance from the bark, but do not all advance inwards to the
pith. *Bark* a dense compact mass of homogeneous tissue, applied close to the liber,
and transversed by longitudinal canals full of a red secretion. *Liber-bundles* isolated,
opposite each wedge of wood, and in contact with the cambium layer, separated from
one another by broad wedge-shaped terminations of the medullary rays. *Pith* spongy
in the centre, and traversed with canals like the bark, becoming denser towards the
wood, and very dense and opaque in the medullary rays. We have no large specimens of this species to compare with Decaisne's description and figure of *C. Pareira,*
but the structure of the stem differs in no respect from that of the first wood-zone of
that plant.

1. **C. Pareira** (Linn. Spec. Pl. 1473) ; scandens, foliis reniformibus vel rotundatis vel late cordatis plus minus pubescentibus, cymis
masculis longe pedunculatis multifloris pilosis, racemis fœmineis bracteas rotundatas amplas gerentibus, drupis subglobosis hirsutis.—*Lam.
Ill. t.* 830 ; *DC. Syst.* i. 533, *Prod.* i. 100; *Macfadyen, Fl. Jamaic.* i.
16 ; *Blanco, Fl. Filip.* 815. C. Caapeba, *Linn. Sp.* 1473 ; *DC. Syst.*
i. 536, *Prod.* i. 101 ; *Roxb. Fl. Ind.* iii. 842 ; *Rich. Cub.* 58.   C. Cocculus, *Poir. Dict. v.* 9 (*excl. syn. paucis*). C. convolvulacea, *Willd.;
DC. Syst.* i. 536, *Prod.* i. 101 ; *Wall. Cat.* 4979 ; *W. et A. Prod.* i. 14;
Roxb. Fl. Ind. iii. 842 ; *Hassk. Pl. Jav. Rar.* 107. C. Mauritiana,
Thouars, Journ. Bot. 1809. ii. *t.* 3–4; *DC. Syst.* i. 535, *Prod.* i. 100,
(*non Wall. Cat.*). C. pareiroides, *DC. Ess. Méd.* C. orbiculata,
DC. Syst. i. 537, *Prod.* i. 101. C. hirsuta, *DC. Syst.* i. 535, *Prod.* i.
101. C. tomentosa, *DC. Syst.* i. 535, *Prod.* i. 101. C. microcarpa,
DC. Syst. i. 534, *Prod.* i. 101; *Macf. Fl. Jam.* i. 17. C. hernandifolia,
Wall. Cat. 4977 *A! B partim! C! I!* C. obtecta, *Wall. Cat.* 4981!
C. gracilis, *St. Hil. Fl. Bras. Mer.* i. 54. C. mucronata, *A. Rich. Fl.*

Seneg. i. 11.   C. acuminata, *Benth.! Pl. Hartweg.*   C. nephrophylla, *Bojer, Ann. Sc. Nat. ser.* 2. xx. 55.   C. comata, *Miers! in Hook. Niger Fl.* 215.   C. Vogelii, *Miers, l. c.* 214 (*quoad plant. masc.!*).   C. discolor, *A. Gray, Bot. Wilkes Exped.* i. 38 (*non DC.*).   Menispermum orbiculatum, *Linn. Sp.* 1468.   Cocculus orbiculatus, *DC. Syst.* i. 523, *Prod.* i. 98.   C. villosus, *Wall. Cat.* 4957 *G partim! (spec. masc.*).   C. membranaceus, *Wall. Cat.* 4967!

Quoad foliorum formam variat—

a. foliis peltatis rotundato-deltoideis basi truncatis.

β. foliis plus minus peltatis rotundatis vel late ovatis basi cordatis.

γ. foliis haud peltatis rotundatis vel late ovatis basi cordatis.

δ. foliis haud aut vix peltatis reniformibus obtusissimis sæpe emarginatis.

Quoad indumentum variat—

a. foliis firmis conspicue nervosis laxe hirsutis vel tomentosis.

b. foliis tenuibus supra puberulis subtus adpresse sericeis.

c. foliis supra glabriusculis subtus pubescentibus.

d. foliis etiam junioribus utrinque glabriusculis.

HAB. Per totam Indiæ planitiem et in montibus utriusque peninsulæ et Himalayæ inferioris usque ad Jelam flumen vulgatissima, exceptis provinciis aridissimis Afghanistan, Beluchistan, Sindh, Marwar, Panjab.—(Fl. per totum annum.) (*v. v.*)

DISTRIB. In zona tropica et temperata calidiore utriusque orbis.

Frutex alte scandens, ramulis striatulis tomentosis vel pubescentibus, rarius subglabris. *Folia* forma et magnitudine valde varia, plerumque obtusa cum mucrone, rarius acuta, rarissime acuminata, basi truncata vel cordata, sinu aperto vel profundo, diametro 1–4-pollicaria. *Petioli* folia plerumque vix æquantes, interdum duplo superantes. *Cymæ masculæ* axillares vel paullo extra-axillares, interdum in ramulo axillari corymbosæ, ½–1½ poll. longæ, plerumque 2–3 superpositæ in eadem axilla, decompositæ, multifloræ, bracteolis minutis subulatis rarissime una alterave subfoliosa. *Pedunculi* graciles, pubescentes vel tomentosi, vel pilis patentibus hirsuti, plerumque petiolo dimidio breviores, interdum eum superantes.   Loco cymæ inferioris vel omnium, in plantis luxuriantibus, ramus microphyllus elongatus, cymas parvulas in axillis gerens, haud raro evolvitur. *Racemi fœminei* solitarii vel bini, axillares, florigeri folia vix æquantes, fructiferi plerumque elongati. *Bracteæ* subsessiles, dense vel laxe imbricatæ, rotundatæ vel reniformes, breviter vel longe mucronatæ, cinereo-incanæ, plerumque ciliato-villosæ, interdum glabræ, in speciminibus luxuriantibus petiolatæ, vel in folia parva evolutæ, in fructu persistentes, vel rarius post florescentiam deciduæ, in fructu immutatæ vel submembranaceæ. *Pedicelli* florum fœmineorum brevissimi, pubescentes vel laxe villosæ. *Ovaria* tomentosa, rarius glabriuscula. *Drupæ* compressæ, subrotundatæ, coccineæ, circa 2 lineas latæ, hirsutæ vel demum subglabræ.

The long list of synonyms which we have given above will show the view we take of this species.   Wight and Arnott having reduced all the Indian forms to *Cissampelos convolvulacea*, it became necessary to find some character, if possible, to distinguish that species from its American and African congeners.   The two characters indicated by Wight and Arnott, namely, the length of the staminal column and the shape of the male sepals, proved very unsatisfactory, and an examination of the floral organs, male and female, has indicated a degree of variability for which we were not prepared.   The size of the flowers, both male and female, the shape of the male sepals and of the cup of the corolla, the shape of the bract in the female inflorescence, and the number, size, and degree of hairiness of the flowers, are all extremely variable. The shape of the sepal of the female flower, we thought, might afford characters of importance, but it may be seen to vary to a great extent even in the different flowers in the axil of the same bract, not only in size, but in shape, being sometimes spathulate obovate, sometimes broader than long, and quite entire, at other times emarginate, or even bipartite to the base; it is either very fleshy or almost membranous, nerveless, or one- two- or three-nerved, with or without red oblong dots (intercellular spaces full of a coloured juice), hairy, or almost glabrous; the bract is equally variable in absolute and relative size.

An examination of many hundred flowers having shown that no reliance can be placed on characters derived from them, and the shape of the leaves being manifestly of no value as a character, we have been compelled to conclude that the American and Indian plants are not distinct. Poiret had already anticipated us in this conclusion; and though some of the synonyms which he quotes are undoubtedly erroneous, it is evident that the Indian plant which he declared so positively to be the same with the *Pareira* and *Caapeba* of America, was the *C. convolvulacea* of Willd. and Wight and Arnott, and that his mistake lay, not in this conclusion as to the specimens before him, but in supposing that these produced the Cocculus berries of commerce.

We have carefully compared a very extensive series of specimens of this species from India, Africa, and America, and can with confidence declare that many American and African specimens are identical with others from India. There are, no doubt, one or two forms from America for which we have not been able to find exact representatives among our Indian specimens; but their differences are so trifling as to be quite within the limits of variation in this very variable Order.

It will be seen that we have not quoted many synonyms of American authors. We believe the number might have been very much extended, but the want of authentic specimens has prevented us from enlarging the list. The characters dwelt upon for the discrimination of the numerous species described are notoriously variable in all the species of this genus, and we think it very doubtful if more than one scandent species exists in America at all liable to be confounded with *C. Pareira*.

Cissampelos Vogelii, Miers, from tropical West Africa, is perhaps distinct, the shape of the drupe being much more elongated. The male specimen, which Mr. Miers considered to belong to that species, we have above referred to *C. Pareira*. *C. Vogelii* is also an American plant; at least specimens of *C. fasciculata*, Bentham, which is perhaps the same as *C. denudata*, Miers, and is only a tomentose state of *C. andromorpha*, DC., are undistinguishable from it.

Cocculus membranaceus, Wallich, is a curious diseased state of *C. Pareira*, in which the branches are covered with multitudes of little pale-coloured leaves. It is not uncommon in shady jungles in the damper parts of the Himalaya.

12. **CYCLEA,** Arnott.

Cyclea *et* Rhaptomeris, *Miers*.

MAS. *Sepala* 4–8 in calycem campanulatum vel inflato-subglobosum coalita. *Petala* totidem, plus minus coalita. *Antheræ* horizontales, sepalis numero æquales, uniloculares, transverse dehiscentes. FœM. *Sepala* 2, lateralia, bractea antica suffulta. *Ovarium* solitarium, anticum. *Stigma* in segmenta 3–5 subulata radiatim divergentia fissum. *Drupæ* subglobosæ. *Putamen* hippocrepiforme, dorso varie tuberculatum, lateraliter convexum, haud excavatum, sed intus loculos 2 vacuos iis *Limaciæ* similes continens.—Frutices *scandentes*, inflorescentia *paniculata axillari*.

This genus, which was originally proposed and characterized by Arnott, in Wight's 'Illustrations,' has been adopted by Mr. Miers, and appears very natural. It is at once distinguished from *Stephania* by the paniculate inflorescence. The characters of the female flowers are not yet perfectly ascertained, as we have only been able to examine those of *C. Burmanni* and *C. populifolia* in a very advanced state, in which they appear to us to present the characters given above. The putamen differs from that of *Cissampelos* and *Stephania* precisely as do those of *Limacia* and *Cocculus* from one another. *Rhaptomeris* appears to differ chiefly in the number of parts, and is therefore not tenable as a genus. The degree of combination of the petals varies so much even in the same species, that we cannot venture to employ it as a generic character.

In *Cyclea populifolia* a portion of stem, of five or six years' growth, and half an inch in diameter, is soft, with a spongy bark, rather small pith, and sixteen rather narrow wood-wedges. *Pith* of elongated delicate cells, superimposed by their square apices, and full of starch. *Medullary rays* of mural cells. *Wood-wedges* (on a transverse section) linear-clavate, of much dotted pleurenchyma, and large vessels with transverse bars and gashes. *Liber-bundles* free from the wood, rather small, and apparently not increasing after the first year. *Bark* of soft hexagonal cellular tissue, surrounded by several dense layers of radially compressed cells. The wood of the other species is very similar.

1. **C. Burmanni** (Miers, in Taylor's Annals, l. c., non Arnott); foliis peltatis elongato-deltoideis acuminatis basi cordatis sagittato-lobatis, lobis rotundatis, margine subrepandis, calyce inflato subgloboso 6–8-lobo, corolla dimidio minore urceolata vix lobata.—Cocculus Burmanni, *DC. Syst.* i. 517, *Prod.* i. 96. Clypea Burmanni, *W. et A. Prod. ex parte.*—*Burm. Zeyl. t.* 101.

HAB. In Zeylania, *Walker!* (prope Peradenia, alt. 2600 ped., *Gardner!*) Concan, *Gibson! Stocks!*—(*v. s.*)

Frutex scandens. *Caules* sulcati, pilosiusculi vel glabrescentes. *Folia* tenuiter coriacea, iis *C. peltatæ* longiora et angustiora, supra nitida, subtus dense pubescentia vel rarius subglabra, 2–4 poll. longa, basi ¾–2 poll. lata, petiolo ½–1¼-pollicari. *Paniculæ* folia æquantes vel longe superantes, laxe ramosæ, multifloræ, pubescentes. *Flores* masculi iis *C. peltatæ* duplo majores, subglobosi, hispiduli. This species is readily distinguished from the next by the shape and size of the male flowers, but much confusion exists in the synonymy of the two. Wight and Arnott distributed specimens of both, but their description (of the flower at least) seems to have been taken entirely from the next. Burmann's plate, however, corresponds in foliage with the present species, and though the flowers are represented much smaller, they are described as 6-fid. We have therefore followed Mr. Miers in referring Burmann's figure to the plant described above. A fragment of *C. Burmanni* exists in the Wallichian Herbarium at the Linnean Society, under No. 4982, collected by Heyne, and communicated by Wight, whose own specimens belong to the next species, under which, therefore, we have quoted Wallich's synonym. As the characters derived from the shape of the leaves cannot be considered certain till confirmed by careful study of the living plant, we do not describe in detail the female plant, though we have a specimen before us from Ceylon which agrees in foliage with the male, and is therefore probably referable here. Judging from this specimen, the female inflorescence does not differ from that of *C. peltata.*

2. **C. peltata** (H.f. et T.); foliis peltatis deltoideis basi subcordatis, petalis calyce campanulato 4-lobo dimidio brevioribus in cyathum irregulariter 4-lobum coalitis.—Menispermum peltatum, *Lam.* Cocculus peltatus, *DC. Syst.* i. 516, *Prod.* i. 96. Cissampelos discolor, *Wall. Cat.* 4982! (*ex parte*), non *DC. nec Blume.* C. barbata, *Wall.*

2 D

Cat. 4978 ! Clypea Burmanni, *W. et A. Prod.* i. 14 (*ex parte*), *non DC.*
Cyclea Burmanni, *Arnott in Wight Ill.* i. 22.—*Rheed. Mal.* vii. *t.* 49.
HAB. In Zeylania, *Walker!* etc.; Malabaria et Carnatica australiori,
Wight! Concan, *Law!* Assam, *Jenkins!* Khasia ! Chittagong ! Ava et
Pegu, *Wall.!* Malacca, *Griff.!* Singapur, *Wall.!*—(*v. v.*)
DISTRIB. Java, *Spanoghe!* (*in Herb. Hook.*)

Frutex scandens. *Caules* sulcati, pilis laxis stramineis subreflexis sparsis hispiduli,
rarius glabrescentes. *Folia* late deltoidea, margine subrepanda, acuta vel obtusiuscula, mucronata, tenuiter coriacea, 3–6 poll. longa, 2–4 lata, petiolo 1–2½-pollicari,
supra glabra vel pilis paucissimis sparsis tecta, et margine ciliata, subtus pubescentia,
tenuiter tomentosa vel rarius glabra. *Paniculæ* paullo supra-axillares, folia æquantes
vel superantes, puberulæ; *masc.* plerumque longiores, graciles, interdum pedales, ramulis elongatis vel subcontractis multifloris; *fœm.* strictiores, folia vix æquantes,
3–6-pollicares, ramis rigidis 1–2-pollicaribus. *Bracteæ* oblongæ vel subulatæ. *Flores*
masculi hispiduli vel glabrescentes. *Drupæ* reniformes, lateraliter compressæ, pilosæ.

We are not quite satisfied that all the synonyms above referred to belong to one
species, because our specimens from the Madras peninsula bear only female flowers,
while those from the eastward are all male. We have not, however, been able to
find any characters upon which to found a diagnosis between the two. We know
from our own Khasia specimens that the degree of branching of the male panicle
varies very much, as well as the amount of hairiness of the stem, and the pubescence
of the under-surfaces of the leaves. The degree of division of the lobes of the corolla
is also a very variable character.

We possess panicles of fruit of another species of *Cyclea*, collected in the Khasia
hills, without leaves, and preserved in spirits. These drupes are quite glabrous, arranged in panicles 3–4 inches long, of which the branches are very short. They
resemble very closely the female panicles of Mr. Miers' *Cyclea deltoidea* from Hongkong.

3. **C. populifolia** (H.f. et T.); foliis cordatis acuminatis haud
peltatis.—Menispermea, *Griff. Itin. Notes, pp.* 114, 165.

HAB. In dumetis subtropicis Sikkim ! Bhotan, *Griffith!* Khasia !—
(*v. v.*)

Frutex alte scandens, cortice ramorum lactea, ramulorum cinereo-pubescente.
Folia late cordata, 7–9-nervia, subtus crebre reticulata, coriacea, firma, supra glabra,
subtus pilis rigidiusculis pubescentia, 4–6 poll. longa, 3–6 lata, petiolo 2–4-pollicari
pubescente cylindrico, basi et apice incrassato. *Paniculæ* axillares, tomentosæ, ex
eadem axilla plures, fœmineæ plerumque e caulibus crassioribus; *masc.* in speciminibus vix evolutæ, decompositæ, multifloræ, calyce campanulato 4-lobo, anthera peltata disciformi 4-loculari; *fœm.* 2–6-pollicares, ramosæ; *flores* in spec. fere omnibus
delapsi, *sepalis* 2, lateralibus, carnosulis, glabris, subcucullatis. *Drupæ* glabræ.

Tribus III. PACHYGONEÆ, *Miers.*

Ovaria 3 vel plura. *Drupæ* ovales vel hippocrepiformes, styli cicatrice fere basilari. *Putamen* haud tuberculatum. *Semen* uncinatum,
exalbuminosum. *Embryo* semini conformis, cotyledonibus semicylindricis carnosis amygdalinis.

13. **PACHYGONE,** Miers.

Sepala 6, biserialia, exteriora minora. *Petala* 6, auriculata, sepalis ½
breviora, stamina amplectentia. MAS. *Stamina* 6; *filamenta* cylindrica,

apice incurva; *antheræ* subglobosæ, didymæ, biloculares. *Ovaria* rudimentaria 3, minutissima. Fœm. *Stamina* 6, ananthera, breviter clavata. *Ovaria* 3. *Styli* crassi, horizontales. *Drupæ* reniformes, lateribus leviter excavatæ. *Semen* hippocrepiforme, radicula brevissima, stylo subbasilari approximata, cotyledonibus semicylindricis vel subclavatis fere corneis.—Frutices *scandentes*, floribus *axillaribus racemosis*.

A piece of the stem of *Pachygone*, one-fourth of an inch in diameter, is firm and woody, faintly sulcate externally. *Pith* occupying the greater part of the stem, wholly composed of vertically elongated, parallel woody cells, with square superimposed ends, the inner shortest. *Medullary rays* of mural cells. *Wood-wedges* about thirty, placed near the circumference, of dotted pleurenchyma and few large vessels. *Liber-bundles* closely applied to wood, semilunar, rather broad. *Bark* a narrow cellular zone of long cells, with very thick transparent walls.

1. **P. ovata** (Miers, mss.); foliis ovato-oblongis subtrapezoideis, racemis masculis folia superantibus.—Cissampelos ovata, *Poir.; DC. Syst.* i. 537, *Prod.* i. 102. Cocculus leptostachyus *et* brachystachyus, *DC. Syst.* i. 528, *Prod.* i. 99; *Decaisne, Timor* 96. C. Plukenetii, *DC. Syst.* i. 520, *Prod.* i. 97; *W. et A.! Prod.* i. 14; *Wight, Ic. t.* 824, 825. Cocculus Wightianus, *Wall. Cat.* 4959 A!

HAB. In Zeylania, *Thwaites!* et in Carnaticæ arenosis præsertim prope mare, *Wight!*—(*v. s.*)

DISTRIB. Timor.

Frutex alte scandens, ramosissimus. *Ramuli* eleganter striatuli, tomento flavescente incani, demum glabrescentes. *Folia* basi cuneata vel rotundata, apice obtusa vel retusa cum mucrone, 3–5-nervia, crasse coriacea, utrinque glabra, 1½–2 poll. longa, ¾–1¼ lata, petiolo ½–¾-pollicari incano-pubescente, apice incrassato et subarticulato; petioli et racemi basi fasciculo pilorum circumdati. *Racemi* graciles, pubescentes vel tomentosi, masculi folia superantes, fœminei folia vix æquantes vel breviores. *Pedicelli* flores vix superantes. *Bracteæ* subulatæ. *Flores* minuti, masculi in axillis bractearum fasciculati, fœminei solitarii. *Sepala* interiora ovalia vel obovata, membranacea. *Petala* acute vel obtuse bidentata. *Drupæ* subcompressæ, pisi magnitudine, læviusculæ. *Putamen* rugulosum.

Decaisne has pointed out the probable identity of *C. brachystachyus* and *C. leptostachyus* with one another, and with *C. Plukenetii*; and as his description, as well as a specimen of *C. brachystachyus* from Timor in Mr. Bentham's Herbarium, agrees precisely with the Ceylon plant, which varies with three or five nerves, we have not hesitated to unite them all. It is curious to remark the recurrence of this plant, which is a native of the drier parts of the Carnatic and Ceylon, in Timor, which has a drier climate than Sumatra and Java.

14. **FIBRAUREA,** Lour.

Sepala 6, biserialiter imbricata, ovalia vel obovata. *Petala* 0. MAS. *Stamina* 6; *filamenta* crassa, carnosa, angulata, infra antheram linea elevata obliqua cincta, et antice et postice subcristata; *antheræ* late ovales, biloculares, lateraliter dehiscentes. Fœm. *Ovaria* 3–6. *Drupæ* totidem ovoideæ, læves.—Frutices *scandentes*, foliis *crasse coriaceis*, petiolis *basi et apice articulatis et incrassatis*, paniculis *axillaribus ramosissimis*.

In Mr. Miers' paper this genus is placed among *Heteroclineæ*, but he had not seen perfect seeds. We are induced to transfer it to the exalbuminous tribe, from its resemblance to a plant found by ourselves in the Khasia hills, in fruit only. This we have described below, as a second species.

Mr. Miers considers that the peculiar structure of the filaments of *Fibraurea* is due to their being enveloped by, and consolidated, as it were, into one mass with the petals. We have sought in vain, however, for any evidence of the accuracy of this view, in the structure of the filament or the mode of insertion of the anther, and are therefore disposed to regard the flowers as truly apetalous.

According to Loureiro, the wood of this genus yields a yellow dye, which is permanent, but not bright; it is also used as a dye in Borneo, according to Mottley.

In *Fibraurea tinctoria* the wood is firm, shrinks very little, and the stems remain cylindrical when dry. *Bark* formed of several layers of papery epidermis, beneath which is a dense zone of closely packed, transparent, woody cubical cells, with very small cavities. The *wood-zone* is formed of about twenty narrow wedges, separated by narrow medullary rays : the latter consist of a very dense cellular tissue of thick woody cells, that form a complete indurated mass surrounding the pith and wood, and appear to be confluent with the liber, but are not so. The *liber-bundles* are isolated, small, and placed in contact with the cambium-layers. The *pith* is loose and spongy in the centre, becoming firmer, denser, and woody towards the medullary rays.

1. **F. tinctoria** (Lour. Fl. Coch. Chin. ed. Willd. 769); foliis ovalibus ovatis vel oblongis obtuse acuminatis coriaceis, floribus paniculatis.—*Cocculus Fibraurea, DC. Syst.* i. 525, *Prod.* i. 99.

HAB. In Malaya ad Penang, *Phillips!* et Malacca, *Griffith!*—(*v. s.*)

DISTRIB. Cochin China, *Loureiro;* Borneo, *Mottley!*

Frutex alte scandens, glaberrimus, cortice cinereo vel albido laxo rugoso, ramulorum nitido subvelutino. *Folia* 4–7 poll. longa, 2–4 lata, petiolo 1½–3-pollicari striato subangulato, utrinque glaberrima, supra lucida, subtus pallida, crasse coriacea, basi triplinervia, marginibus in sicco subreflexis. *Paniculæ* axillares, a basi ramosæ, ramosissimæ, multifloræ, foliis breviores vel longiores. *Alabastri* globosi. *Flores* (fide Loureiro albi) bracteolis 2 minutis calyci adpressis stipati. *Sepala* glabra, dorso puberula, carnosula, margine tenuia. *Panicula.* fructifera elongata, in spec. fere pedalis, pedunculis strictis lignosis ½–1-pollicaribus. *Drupæ* pollicares, ex Loureiro flavæ.

In the Benthamian Herbarium a few detached drupes accompany a specimen from Malacca, but they are empty. The specimens before us are considered by Mr. Miers to belong to three species, all different from that of Loureiro. We have, on the contrary, no doubt that all belong to one, and we think that Loureiro's description agrees well enough with the specimens to make it probable that they are the same. The specimen in the British Museum from that author is very imperfect, and scarcely determinable; it has neither flower nor fruit.

2. **F.? Hæmatocarpa** (H.f. et T.) ; foliis oblongis obtuse acuminatis crasse coriaceis margine recurvis, pedunculis fructus axillaribus abbreviatis, drupis obovato-oblongis atro-purpureis stipitatis.

HAB. In montibus Khasia !—(*v. v.*)

Frutex scandens, cortice cinereo vel pallido striatulo. *Folia* (in specimine solitario) 3 poll. longa, 1⅔ lata, petiolo cylindrico gracili ¾-pollicari, pallide viridia, subtus albida, basi triplinervia, cæterum penninervia, nervis basilaribus ad apicem usque extensis et cum lateralibus arcus formantibus. *Racemi* fructus 1–3-pollicares, vestigia 1–5 florum gerentes. *Torus fructus* globosus, cicatricibus magnis 4–6 notatus. *Drupæ* totidem, pedicello crasso ⅓-pollicari stipitatæ, obovato-oblongæ, utrinque rotundatæ, læves, parum obliquæ, styli cicatrice basilari notatæ, 1½–2 poll. longæ. *Sarcocarpium* e stratis 2 compositum, exterius dense carnosum, sanguineum, interius fibrosum pallidius, fibrillis e putamine ortis. *Putamen* tenue, crustaceum, bicrure, leviter trisulcum, sulcis longitudinalibus, vasa nutrientia continentibus, intus tricostatum, inter crura tenuissimum. *Semen* putamini conforme. *Testa* tenuissima, membranacea, fusca, facie interna crassiuscula. *Embryo* semini conformis, amygdalinus, suberoso-carnosus, leviter sulcatus. *Radicula* styli cicatricem spectans, brevissima. *Cotyledones* elongatæ, plano-convexæ, semicylindricæ, uncinatæ, apice obtusæ, longitudinaliter leviter sulcatæ.

We have placed this very remarkable but unfortunately little-known plant provisionally in the genus *Fibraurea*, on account of the resemblance of the leaves and general aspect. We obtained only one fruiting branch, which was brought to us soon after our arrival in the Khasia, from an elevation of about 3000 feet, and every effort to procure more was unsuccessful. The fruit of *Fibraurea* is still almost unknown, but immature imperfect specimens in Mr. Bentham's Herbarium resemble what the young fruit of this plant may be assumed to be.

F. Hæmatocarpa is undoubtedly one of the most interesting plants of this family which have yet been found. The very large size of the fruit, and its peculiar structure, are alike unique in the Order. It is nevertheless, though exalbuminous, an undoubted Menispermaceous plant. The two arms of the putamen are not united by a bony plate, as in all the other elongated-seeded plants of the Order, but the nutrient vessels pass from the base of the drupe to the bottom of the sinus of the curved seed, just as in *Cocculus* or *Pachygone*.

A piece of stem several years old, and ¼ inch in diameter, is firm and woody, not shrinking in drying. *Bark* smooth, polished, searcely furrowed. *Pith* one-fifth the diameter of stem, very firm and woody, wholly formed of long tubular cylindrical thick-walled cells, with square extremities placed end to end. *Medullary rays* about forty, of very much radially elongated compressed mural cells. *Bark* a very thin cellular layer. *Wedges of wood* long, narrow, gradually broader outwards, of numerous dotted pleurenchyma tubes and large vessels, whose walls are covered with innumerable transverse bars; there are also a few spiral vessels towards the axis. *Liber-bundle* semilunar, placed in contact with the wood.

GENERA DUBIÆ TRIBUS, FRUCTU IGNOTO.

15. **TINOMISCIUM,** Miers.

MAS. *Sepala* 9; 3 exteriora parva, ovata, acuta, bracteis 1–2 minimis conformibus stipata; 6 interiora conformia, exterioribus paullo latiora. *Petala* 6, sepalis interioribus parum breviora, oblonga, membranacea, marginibus inflexis. *Stamina* 6; *filamenta* planiuscula; *antheræ* oblongæ, adnatæ, extrorse biloculares.—Frutex *scandens lactescens,* petiolis *elongatis basi incrassatis et flexuosis, pseudo-subarticulatis,* foliis *basi trinerviis cæterum penninervis,* floribus *racemosis.*

There is nothing in the male flower of this plant to guide us as to its immediate affinity, for, though the technical character agrees with *Tinospora*, the appearance of the flowers and the whole habit are very different. Mr. Miers has conjectured that it belongs to his tribe *Heteroclineæ*, and we have, at p. 179, described a fruit which we think probably belongs to a nearly allied species.

The wood of *Tinomiscium* is hard, and does not contract much in drying. A section half an inch in diameter presents a broad pith, and twenty-five to thirty wood-wedges, divided by moderately broad medullary rays. The general arrangement is as in *Pericampylus*, but the liber-bundles evidently increase annually, and there are no traces of periodic deposits of wood.

1. **T. petiolare** (Miers in Taylor's Annals, ser. 2. vii. 44); foliis ovali-oblongis acuminatis glabris, racemis elongatis fusco-tomentosis.—Cocculus petiolaris, *Wall. Cat.* 4964!

Hab. In Malaya ad Penang, *Wall.! Phillips!* (in Herb. Hook.).—
(*v. s.*)

Frutex alte scandens, cortice cinereo longitudinaliter rimoso ;ʼ partes novellæ fusco-
tomentosæ. *Folia* basi fere truncata, obtusa vel plerumque abrupte acuminata, coriacea, utrinque inter nervos (in sicco) eleganter striatula, subtus pallida, basi 3–5-
nervia, 4–6 poll. longa, 2½–4 lata, petiolo 3–4-pollicari glabro striatulo. *Racemi* in
axillis super tuberculum fasciculati, aut secus ramulum abbreviatum foliosum vel
aphyllum alterni, 4–8 poll. longi. *Flores* minuti, subremoti, fasciculati vel solitarii,
bractea subulata suffulti. *Pedicelli* abbreviati. *Sepala* extus puberula. *Petala*
apice emarginata.

16. **PYCNARRHENA,** Miers.

Mas. *Sepala* 6 ; interiora majora, rotundata. *Petala* 6, parva, rotundata, varie lobata. *Stamina* 9, monadelpha ; *filamenta* apice libera ;
antheræ adnatæ, biloculares, loculis lateralibus sutura continua transverse dehiscentibus. — Frutex *forsan scandens,* petiolis *brevibus basi
pseudo-articulatis superne incrassatis,* foliis *coriaceis,* floribus *in axillis
fasciculatis.*

The female flower of this plant being unknown, its claim to a place among *Menispermaceæ* is still doubtful. The structure of the petiole differs somewhat from that
of the typical *Menispermaceæ,* and approaches to that of *Codiæum,* and of many
species of *Croton* among *Euphorbiaceæ,* and the inflorescence has no parallel in the
Order. The wood is identical in structure with that of *Tiliacora.* In the meantime, as it has been referred hither by Wallich and Miers, we place it provisionally
at the end of the Order.

1. **P. planiflora** (Miers in Taylor's Annals, ser. 2. vii. 44) ; foliis
oblongo-lanceolatis coriaceis, floribus in axillis congestis, pedicellis
brevibus.—Cocculus planiflorus, *Wall. Cat.* 4961 !

Hab. In prov. Silhet, *Wall.!*—(*v. s.*)

Frutex scandens ? vel saltem sarmentosus. *Ramuli* tenuiter puberuli, eleganter
striatuli, pallide straminei vel fusci. *Folia* oblonga vel oblongo-lanceolata, obtuse
acuminata, basin versus angustata, tenuiter coriacea, utrinque glabra, sed subtus secus
costam tenuiter puberula, 5–7 poll. longa, 1¾–2½ lata ; nervi subtus conspicui, longe
intra marginem arcuati, venulæ (in sicco) eleganter reticulatæ. *Petioli* ¾-pollicares,
puberuli, basin versus cylindrici, striatuli, versus apicem incrassati et antice profunde
sulcati, apice cum folio pseudo-articulati, et paullo intra marginem laminæ (igitur-
que subpeltatim) inserti. *Flores* in axillis dense congesti ; *pedicelli* graciles, 2–3
lineas longi, pubescentes, basi bracteati et medio bracteolam minutam squamæformem
gerentes.

IX. SABIACEÆ.

Flores hermaphroditi, rarius polygami. *Sepala* 5 (rarissime 4), parva,
basi coalita, subpersistentia, æstivatione imbricata, 2 exteriora, basi
bractea minuta antica adpressa suffulta. *Petala* 5 (vel 4), sepalis opposita, hypogyna, lineis coloratis pellucido-punctata, decidua vel marcescenti-persistentia, æstivatione imbricata. *Stamina* petala numero
æquantia iisque opposita, disci dentibus alterna ; *filamenta* compressa,
carnosula vel plana, ligulata vel subulata, apice angustata ; *antheræ*
didymæ, connectivo lateraliter adnatæ vel subliberæ, ovoideæ, bilocu-

lares, extrorsæ vel introrsæ, longitudinaliter dehiscentes, valvis a connectivo solutis, quapropter antheræ post dehiscentiam uniloculares fiunt.
Discus hypogynus ,columnæ brevi insidens, 5-lobus, lobis carnosis cum
petalis sepalisque alternantibus. *Ovaria* 2, rarissime 3, in axi subcohærentia, biovulata; *ovula* suturæ ventrali inserta, superposita, inferius
descendens, campylotropum, superius fere horizontale, suborthotropum.
Styli 2, erecti, terminales, cylindrici, secus faciem ventralem subco-
hærentes, sed facile separabiles. *Stigmata* simplicia, obtusiuscula.
Carpella 2 vel abortu solitaria, drupacea, dorso gibbosa, intus stylo
subpersistente fere basilari rostrata. *Endocarpium* lignosum, irregulariter rugosum. *Semen* solitarium, reniforme, prope basin insertum,
campylotropum. *Testa* coriacea, punctis coloratis notata. *Endopleura*
crassiuscula, alba. *Embryo* exalbuminosus, radicula infera horizontali
cylindrica, cotyledonibus ovalibus incurvis planiusculis carnosis.—
Frutices scandentes foliosi, ramulis *basi squamis gemmæ persistentibus
stipatis,* foliis *alternis integerrimis exstipulatis, cum petiolo haud articulatis,* floribus *axillaribus, solitariis cymosis vel paniculatis, mediocribus
vel parvis, viridibus flavis vel purpureis, plerumque cum foliis nascentibus
evolutis.*

The genus *Sabia* was first described by Colebrooke[*] in the year 1820, with a
somewhat erroneous generic character, and a plate which accurately represents the
habit and general appearance of the plant, but is accompanied by a very imperfect
figure of the flower. In 1824, Wallich[†] published excellent descriptions of two additional species, giving at the same time a corrected generic character, and referring
the genus to *Terebinthaceæ*. In 1825, Blume,[‡] unaware of what had previously
been done, added another species under the generic name *Meniscosta*, which he
placed at the end of *Menispermaceæ*. Endlicher and Meisner, adopting Wallich's
suggestion, placed *Sabia* at the end of *Anacardiaceæ*. In 1842, Falconer[§] published
an excellent account of the genus, under the name of *Enantia*, which he indicated as
the type of a distinct Order, pointing out the resemblance of the fruit to *Menisper-
maceæ*, but not pronouncing definitely on its affinities. In 1851, Blume,[||] who had
discovered the identity of his genus *Meniscosta* with *Sabia*, constituted the Natural
Order *Sabiaceæ*, the place of which he fixed in the immediate vicinity of *Menispermaceæ;* and in 1853 Miers[¶] adopted that Natural Order, taking the same view of
its affinities. He has, however, fallen into an error in describing the ovules as solitary, and has overlooked the remarkable character of the opposition of the petals and
sepals.

The structure of the genus *Sabia* is so remarkable, that its claims to form a distinct Order are unquestionable; but, as in the case of many Orders of limited extent,
the characters point in so many different directions that it is not easy to determine
the position which it ought to occupy in our systems. If the ovary of *Sabiaceæ* be
considered syncarpous, the presence of a well-marked hypogynous disc, and many
other characters, would seem to indicate the Rhamnal alliance as that to which they
are most nearly allied. Among its Orders, *Chailletiaceæ*, which have a two-celled
ovary, containing two collateral pendulous ovules in each cell, a simple style, and
exalbuminous seeds, appear to exhibit the greatest amount of resemblance to *Sabiaceæ*.
There are, however, many obvious differences, such as the structure of the petals, the
drupaceous fruit, and the curved embryo with inferior radicle, and this affinity is
probably a distant one.

---

* Linn. Tr. xii. 355.   § In Hook. Journ. Bot. iv. 75.
† Roxb. Fl. Ind. ed. Wall. ii. 308. || Mus. Lugd. Bat. i. 368. t. 44.
‡ Bijdr. p. 28.   ¶ Lindley's Veg. Kingd. 3rd ed. p. 467.

The cohesion of the-carpels in *Sabiaceæ* is so very slight, even in the ovary, and disappears so rapidly as the fruit advances towards maturity, that the connection is probably chiefly with apocarpous orders. Blume and Miers, as we have seen, place the Order in the immediate neighbourhood of *Menispermaceæ*, indicating at the same time an affinity with *Lardizabalaceæ*. To us it appears intermediate between *Schizandraceæ* and *Menispermaceæ*, agreeing with the former in the subscandent habit, in the persistence of the bud-scales at the base of the branches, in the synchronous evolution of flowers and leaves from the same buds, the dotted flowers, two-celled ovaries, and the amphitropous or campylotropous ovules, and with the latter in the oblique development of the ovary, by which the style becomes basilar, and the drupaceous fruit, and differing from the ordinary structure of both in the pentamerous flowers, in the opposition of the sepals and petals, the presence of a disc, the partial cohesion of the ovaries and styles, the inferior radicle, and the exalbuminous seeds. The last character, however, is present in some *Menispermaceæ*.

The quinary arrangement of the flowers at first sight appears a great obstacle to the association of *Sabiaceæ* with *Menispermaceæ* or *Schizandraceæ;* but this difficulty loses much of its force in consequence of the occurrence of pentamerous flowers in *Odontocarya* in the one Order and in *Schizandra* in the other. *Odontocarya*, from Mr. Miers' description and drawing, appears to have many points in common with *Sabia*, and deviates considerably from the normal structure of the Order to which he has referred it. *Schizandra* has been well illustrated by Dr. Asa Gray, who has shown that, though the number of stamens is always five, the petals and sepals vary from five to six.

The frequent transition from trimerous to pentamerous flowers in certain genera of *Ranunculaceæ*, and the close affinity of *Ranunculaceæ*, which are usually pentamerous, to *Berberideæ*, which are always trimerous or tetramerous, tend still further to weaken the force of this objection. It may be observed that the transition is usually from trimerous flowers arranged in three or more rows, to pentamerous flowers in two rows only. This is also the case with the similar transition in *Polygonaceæ*, in which Order some genera have pentamerous flowers in a single series, while others have trimerous flowers in a double verticil. An exception, however, occurs in *Helleborus* and some other pentamerous *Ranunculaceæ*, in which the petals are about twice (or three times) as numerous as the sepals.

The most remarkable character of *Sabiaceæ* is undoubtedly the opposition of the sepals and petals, because the alternation of succeeding verticils both of leaves and flowers is so universal, that any exception has come to be regarded as next to impossible. To this rule, indeed, we believe it will be found that *Sabia* offers rather an apparent than a real exception ; for though the opposition of each member of the two verticils is very evident, we believe the explanation to be that a portion only of the outer verticil belongs to the calyx, the two outer segments being lateral bractlets.

In all the species of *Sabia* which we have examined, a single anterior bract is found usually in close contact with the calyx. The two lateral sepals (as they are usually termed) are exterior in æstivation, and are in most of the species a little longer and broader than the three inner sepals. The æstivation of the petals is the same as that of the sepals, the two lateral being exterior, one anterior, and two posterior, interior and overlapping each other by one margin.

A very similar structure exists in *Helianthemum*, but the small size of the two bract-like lateral sepals, or more properly bracteoles, and the great breadth of the three inner sepals, prevent the opposition of the two verticils from being so decided as in *Sabia*. In *Cistus*, where the lateral bractlets are wanting, no evident relation can be traced between the position of the sepals and petals. In *Amaranthaceæ*, where the calyx is usually very much imbricated, the structure is possibly analogous, as is indicated by the reduction of the number of sepals to three in several species of *Amaranthus*.

Sabia is entirely an Indian genus. Blume indicates three species, all seemingly quite distinct from those described below, so that the number known amounts to ten. The dehiscence of the anthers requires to be observed carefully in the living plant. In

S. paniculata and *S. lanceolata* it is certainly introrse, but by Blume it is described as extrorse in the genus, and it has appeared so to us in several species. The filament is generally hooked at the apex; and as the anther looks downwards and forwards, and dehisces close to the connective, a very slight increase of obliquity in the position of the anther will produce the change from introrse to extrorse dehiscence.

1. SABIA, Colebrooke.

Meniscosta, *Blume.* Enantia, *Falconer.*

Character ordinis.

1. **S. campanulata** (Wall.! in Roxb. Fl. Ind. ed. prior. ii. 311, Cat. 1002!); foliis oblongis acuminatis basi acutis puberulis, pedunculis unifloris, petalis ovalibus nervosis, filamentis subulatis petalis multo brevioribus.

Hab. In Himalaya temperata : Sikkim, alt. 9–10,000 ped.! Nipal, *Wall.!* Kumaon, *Str. et Wint.!* Garhwal, *Edgeworth!* Simla! Jamu, alt. 5000 ped.!—(Fl. Apr. Mai.) (*v. v.*)

Ramuli striatuli, glabri, stram.inei, juniores puberuli. *Folia* tenuiter membranacea, supra pallide viridia, subtus pallida, utrinque petiolisque puberula, margine subciliata, 2–3 rarius 4 poll. longa, ¾–1½ poll. lata; nervi in sicco conspicue reticulati. *Pedunculi* axillares, solitarii, subclavati, puberuli, demum glabrescentes, 1–2 poll. longi. *Flores* virides, subglobosi vel campanulati, magni, inodori. *Sepala* 5, rotundata. *Petala* sepalis duplo majora, ⅓–½ pollicaria, ovalia vel obovata, glabra, interdum post anthesin aucta, ⅔-pollicaria, marcescenti-persistentia, nervosa. *Filamenta* erecta, stigmata fere æquantia. *Antheræ* ovales, extrorsæ. *Styli* ovariis duplo longiores. *Drupæ* pallide cærulescentes, succo concolore, lateraliter compressæ, rotundatæ, ½-pollicares.

This is the largest-flowered species of the genus. The petals enlarge after the sepals fall away, and are sometimes persistent round the ripe fruit ; but this is by no means a constant character, as they are often deciduous very soon after the fall of the stamens.

2. **S. leptandra** (H.f. et T.); foliis ovalibus vel oblongis acuminatis basi rotundatis vel acutis glaberrimis, pedunculis unifloris, petalis ovali-oblongis obtusiusculis, filamentis elongatis petala demum superantibus.

Hab. In Himalaya orientali temperata : Sikkim, alt. 5–7000 ped.! —(Fl. Apr.) (*v. v.*)

Ramuli striatuli, pallidi, glabri. *Folia* tenuiter coriacea, subtus pallida, magnitudine valde varia, plerumque 3–4 poll. longa, 1–1¾ lata, sed interdum sexpollicaria et fere 3 poll. lata. *Nervi* pauci, obliqui ; venulæ in sicco eleganter reticulatæ. *Pedunculi* graciles, apice subclavati, 1–2 poll. longi, glaberrimi. *Flores* e viridi purpurascentes, campanulati. *Sepala* 5, rotundata, glabra, basi subcoalita. *Petala* glabra, glanduloso-punctata, ¼-pollicaria. *Filamenta* anguste ligulata, superne vix attenuata. *Antheræ* extrorsæ, late ovales. *Styli* elongati, graciles. *Drupæ* fere *S. campanulatæ.*

3. **S. purpurea** (H.f. et T.); foliis oblongis longe attenuatis basi plerumque rotundatis, junioribus subpuberulis, pedunculis axillaribus 3–5-floris, petalis acutis, filamentis abbreviatis late subulatis.—S. parviflora, *Wall. Cat.* 1001 *ex parte.*

2 E

HAB. In montibus Khasia, alt. 4–6000 ped.—(Fl. Mart. Apr.) (*v. v.*)

Ramuli striatuli. *Folia* tenuiter coriacea, glabra, 2–3 poll. longa, $\frac{3}{4}$–1 poll. lata, in ramulis sterilibus plerumque majora, interdum sex pollices longa, 2$\frac{1}{4}$ lata. *Nervi* obliqui, longe intra marginem arcubus connexi. *Cymæ* longe pedunculatæ, foliis dimidio breviores, irregulariter ramosi, purpurascentes, glabri. *Flores* parvi, purpurei. *Sepala* ovata, acutiuscula. *Petala* ovato-lanceolata, 5-nervia, nervis subsimplicibus. *Stamina* stylos æquantia. *Drupæ* priorum.

Under the No. 1001 of Wallich's Catalogue, specimens of *S. purpurea, S. parviflora,* and *S. campanulata* are mixed.

4. S. parviflora (Wall. in Roxb. Fl. Ind. ed. prior. ii. 310, Cat. 1001! ex parte); foliis ovatis vel oblongis acuminatis tenuiter coriaceis margine undulatis, pedunculis axillaribus dichotome cymosis 7–11-floris, staminibus inæquilongis stylo brevioribus, carpellis ovoideis compressis.

HAB. In Himalaya temperata et subtropica, alt. 3–6000 ped.: Kumaon, *Str. et Wint.! Nipal, Wall.!* Sikkim!—(Fl. Mart. Apr.) (*v. v.*)

Ramuli gracillimi, striatuli, glabri; partes novellæ puberulæ. *Folia* utrinque glabra, subtus pallida, 2–4 poll. longa, $\frac{3}{4}$–1$\frac{1}{2}$ poll. lata. *Nervi* fere transversi, intra marginem folii arcubus connexi. *Pedunculi* folia fere æquantes, gracillimi. *Bracteæ* minutæ ciliatæ ad ramificationes cymæ. *Flores* minimi. *Sepala* ovata, ciliata. *Petala* anguste oblonga, obtusiuscula, 5-nervia. *Stamina* inæquilonga, breviora petalis dimidio breviora, longiora stylum fere æquantia. *Filamenta* anguste ligulata, apice subulata. *Drupæ* ovales, obovatæ vel rotundatæ, $\frac{1}{3}$–$\frac{1}{2}$ poll. longæ, putamine compresso.

This is a very delicate small-flowered species, which is more nearly allied in the shape of the carpels to *S. lanceolata* than to the species already described.

5. S. lanceolata (Colebrooke in Linn. Tr. xii. 355. t. 14); foliis oblongo-lanceolatis basi acutis vel obtusis glaberrimis tenuiter coriaceis, cymis folio brevioribus longe pedunculatis corymbosis multifloris, petalis ovato-lanceolatis, staminibus inclusis, drupis ovoideis.—*Wall. in Roxb. Fl. Ind. ed. prior.* ii. 309, *Cat.* 999 ; *Blume, Mus. Lugd. p.* 368.

HAB. In sylvis montanis Assam, *Griffith!* Khasia, a planitie ad alt. 4000 ped.! Silhet, *Colebrooke!*—(Fl. Oct. Nov.) (*v. v.*)

Frutex alte scandens. *Ramuli* glabri, striatuli, cortice fusco vel flavido. *Folia* 4–7 poll. longa, 1$\frac{1}{4}$–2$\frac{1}{2}$ lata, petiolo $\frac{1}{4}$–$\frac{1}{2}$-pollicari, glaberrima, supra lucida, subtus pallida, glaucescentia. *Nervi* fere transversi, arcubus conspicuis intramarginalibus connexi. *Pedunculi* graciles, axillares vel paullo supra-axillares, 1$\frac{1}{2}$–2 poll. longi, ramis alternis aut plerumque verticillatis, ramulis irregulariter divisis. *Bracteæ* bracteolæque minutæ, deciduæ. *Flores* viridescentes, suaveolentes. *Sepala* ovata, acuta. *Petala* acutiuscula, 2 lineas longa. *Stamina* petalis dimidio breviora, carnosula, subcompressa, subulata. *Antheræ* introrsæ. *Styli* stamina æquantes. *Drupæ* compressiusculæ, pulposæ, cærulescentes, $\frac{2}{3}$ poll. longæ. *Putamen* rugosum, compressum. *Testa* maculis elongatis rubris notata.

Colebrooke describes the cotyledons as "folded one within the other, plaited once longitudinally and thrice transversely."

6. S. Limoniacea (Wall. Cat. 1000!); foliis lanceolatis vel oblongis acutis vel acuminatis crasse coriaceis glaberrimis, paniculis elongatis folia æquantibus vel superantibus aphyllis vel foliosis breviter ramosis, petalis late ovalibus obtusis, staminibus haud exsertis, drupis compressis rotundatis.—Celastrinea, *Wall. Cat.* 9015 !

HAB. Sikkim, ad basin Himalayæ! Assam, *Griffith !* Khasia, a basi ad alt. 3000 ped.! Silhet, *Wall.!* Chittagong!—(Fl. Sept. Oct.) (*v. v.*)

Frutex alte scandens, cortice lævigato vel vix striatulo fusco. *Folia* 3–7 poll. longa, 1¼–2½ lata, basi rotundata vel acuta, subtus in sicco crebre reticulata. *Nervi* obliqui, incurvi. *Paniculæ* rubescentes, glaberrimæ, interdum foliosæ et elongatæ, sæpius aphyllæ et folia æquantes vel iis paullo breviores, axillares vel paullo supra-axillares, basi squamulis gemmaceis persistentibus stipatæ, alterne ramosæ; ramuli vix ½-pollicares, rarius in paniculis foliosis fere pollicares, 2–5-flori. *Flores* minuti, flavidi. *Sepala* subciliata, rotundata. *Petala* obovata, brevissime unguiculata, $\frac{1}{10}$ poll. longa, 5-nervia, carnosula, basi intus sulco longitudinali exarata. *Stamina* demum petala æquantia; *filamenta* carnosa, subcompressa, apice incurva. *Ovaria* parum compressa. *Styli* abbreviati.

This species comes nearer to Blume's *S. Meniscosta* than any other of our continental species.

In some collections, but not in the Linnean Society's Herbarium, a species of *Photinia* occurs as Wall. Cat. 1000 B.

7. **S. paniculata** (Edgeworth! mss. in Herb. Benth.); foliis elliptico- vel oblongo-lanceolatis acutis basi rotundatis vel acutis coriaceis, paniculis elongatis folia fere æquantibus vel superantibus laxe pilosulis a basi ramosis ramis irregulariter cymosis, antheris introrsis.

HAB. In Himalaya subtropica, infra 3000 ped. alt.; Garhwal, *Edgeworth !* Kumaon, *Madden! Str. et Wint.!*—(*v. s.*)

Ramuli glabri, fusci, striatuli. *Folia* 5–8 poll. longa, 1¼–3 lata, petiolo ⅔–1-pollicari, glabra, juniora vix puberula. *Nervi* obliqui, arcuati, venulis in sicco crebre reticulatis. *Paniculæ* plerumque aphyllæ sed interdum foliosæ, et tunc folia floralia cæteris conformia aut parva ovato-lanceolata 1½–2-pollicaria. *Ramuli paniculæ* 1–2-pollicares, apicem versus cymose multiflori. *Sepala* ovalia, 1-nervia, dorso dense pilosa. *Petala* oblonga, vix acuta, 3–5-nervia, 1½ lin. longa. *Stamina* petalis dimidio breviora. *Filamenta* lignlata, apice subcontracta.

X. LARDIZABALEÆ.

Flores abortu unisexuales vel polygami. *Sepala* 6, serie duplice disposita, rarius 3, hypogyna, caduca, æstivatione valvata vel parum imbricata. *Petala* 6, rarius nulla, sepalis opposita et sæpe multo minora, squamæformia. *Stamina* 6, in floribus masculis sepalis petalisque opposita; *filamenta* libera vel in tubum coalita; *antheræ* liberæ, adnatæ, extrorsæ, connectivo apiculatæ : in hermaphrodito-fœmineis parva, semper libera, sæpissime polline vacua (in *Decaisnea* pollinifera). *Ovaria* 3, sepalis exterioribus opposita, rarissime 6–9, lineari-oblonga, unilocularia ; ovula numerosa, secus suturam ventralem biserialiter disposita, anatropa vel supra totam superficiem ovarii sparsa, orthotropa, serius anatropa vel campylotropa. *Carpella* magna, tot quot ovaria, pulposa, indehiscentia vel intus longitudinaliter dehiscentia, follicularia, pulpa repleta, polysperma, sæpissime endospermii processubus ad axin fere productis pseudo-multilocularia. *Semina* anatropa vel campylotropa, testa nitida crustacea vel cartilaginea. *Albumen* copiosum, oleosum ; embryonis minuti radicula ad hilum versa.—Frutices *ut plurimum volubiles, rarius erecti, glabri.* Ramuli *basi gemmæ squamis sub-*

persistentibus vestiti. Folia *alterna, digitata vel pinnata, exstipulata, foliolis articulatis.* Inflorescentia *racemosa, racemis axillaribus vel terminalibus interdum corymbosis.* Flores *albi viridescentes vel purpurei.* Fructus *pulposus, edulis.*

This small but curious group was originally indicated as a distinct Order by Brown, and has been admirably illustrated by Decaisne in a paper published in the 'Archives du Musée,' in 1837, since which time no addition has been made to our knowledge of the Order. *Lardizabaleæ* are quite intermediate between *Menispermaceæ* and *Berberideæ,* but possess in common a number of striking characters, which entitle them to be regarded as a very distinct family. In the number and arrangement of the parts of the perianth the flowers agree with both Orders; but their form, and especially the shape of the stamens, which are often monadelphous, and have elongated anthers, readily distinguish them from both. The polyspermous fruit is also a peculiar character, shared only by *Podophyllum* amongst *Berberideæ.* The abnormal arrangement of the ovules over the whole surface of the ovary was formerly considered a universal distinguishing mark, but in *Decaisnea* the ordinary type reappears.

In the unisexual flowers and scandent habit of the majority of the Order, *Lardizabaleæ* agree with *Menispermaceæ,* but the indefinite ovules and the whole structure of the androecium at once distinguish them, and compound leaves do not occur in *Menispermaceæ,* except in the imperfectly known genus *Burasaia,* which, as we have already mentioned, is in that respect quite intermediate, but seems to have the embryo of *Menispermaceæ.* To *Berberideæ* they approach through *Lardizabala,* which has flowers and leaves more like those of a Berberry than those of the Asiatic genera of the Order, and especially through *Decaisnea,* which has the simply pinnated leaves, and leaflets articulating with the petiole, of the section *Mahonia,* and through *Podophyllum,* which has a fleshy pericarp, broad placenta, and the seeds imbedded in pulp. The solitary carpels of *Berberideæ,* however, at once distinguish them.

The number of species known is very small, and, except two, which are natives of western South America, beyond the tropic, the group is entirely confined to the Himalayo-Chinese region, the species occurring throughout the Himalaya and in the Khasia, and in the hilly regions of China and of Japan. None are known in Ava, in the Malayan Peninsula, or in the Indian Archipelago.

1. **DECAISNEA,** H.f. et T.

Sepala 6, lineari-subulata, æst. subimbricantia. *Petala* 0. *Stamina* in fl. masc. monadelpha, tubo cylindrico, antheris oblongis, connectivo in processum longum attenuatum producto; in hermaphroditis parva, antheris masculorum similibus sed minoribus, filamento brevissime libero suffultis. *Ovaria* 3, lineari-oblonga, stylo disciformi oblique obovato-oblongo intus sulcato. *Ovula* placentis 2 filiformibus parallelis suturæ ventrali approximatis sed ab ea discretis inserta, indefinita, numerosissima, anatropa. *Folliculi* pulpa repleti; *semina* indefinita, prope suturam ventralem biserialia, horizontalia, compressa, obovata, testa crustacea atro-fusca nitida lævi.—Frutex *erectus subsimplex,* foliis *pinnatis,* inflorescentia *racemosa terminali,* floribus *viridescentibus.*

This remarkable genus makes a very unexpected and valuable addition to our knowledge of the Natural Order to which it belongs, and will therefore most appropriately have the name of M. Decaisne*, in whose admirable monograph we have a model of

---

* Two Orchideous genera have already been dedicated to M. Decaisne, one by Brongniart, the other by Lindley; but, by an unfortunate mischance, in both cases a previous name supersedes that of *Decaisnea.*

botanical investigation. The floral characters, and even the fruit of *Decaisnea*, establish in the clearest manner its close affinity to *Stauntonia* and *Lardizabala*, while the more normal arrangement of its ovules and seeds constitutes a remarkable transition from their abnormal insertion in these genera to the ordinary mode of placentation.

The ripe fruit is entirely filled with a cellular pulp, which is developed from the growing walls of the whole surface of the pericarp, and forms a complete homogeneous mass, leaving no cavity anywhere. This is firmly attached to the seeds all round, but we cannot find that the adhesion is organic, except at the hilum, where there is a broad organic attachment between the testa and pulp. Vessels originating from all parts of the surface of the pericarp ramify through the pulp, but do not meet in the axis of the fruit. This structure is very different from that of *Holl-böllia*, in which the ovules are imbedded in cavities of the walls of the ovary, and the seeds are consequently included in separate loculi of the walls of the pericarp, and in which the pulpy septa do not meet in the axis, nor contract any adhesion with the surface of the testa. Torrey describes the arillus of *Podophyllum*, a genus allied to *Lardizabaleæ* in several important characters, as a pulpy expansion of the very broad placenta, filling the cavity of the fruit, and enveloping the seeds, but not contracting any further adhesion with the walls of the pericarp ; this is a third modification of the development of pulp which is only partially comparable with the two described.

The genus *Decaisnea* is even more interesting on account of its peculiar habit than its placentation. It is erect and nearly simple, resembling at first sight one of the shrubby *Araliaceæ* which are so characteristic of the humid forests of the eastern Himalaya. The soft stem, with large pith, and the very large pinnated leaves, which disarticulate between each pair of leaflets, increase this resemblance, which is another curious instance of the analogy in general aspect between *Araliaceæ* and *Umbelliferæ*, on the one hand, and the group of *Apocarpous Thalamifloræ* on the other, long ago indicated by Lindley.

1. **D. insignis** (H.f. et T. in Proc. Linn. Soc. ii. Dec. 1854).— Slackia insignis, *Griffith*, *Itin. Notes*, 187, *No.* 977 (*non ejusdem in Palm. Bot. Ind.* 161).

HAB. In Himalaya orientali interiori temperata, alt. 6–10,000 ped. ; Sikkim ! Bhotan, *Griffith !*—(Fl. Mai.; fr. Oct.) (*v. v.*)

Frutex erectus, robustus, subsimplex, medulla crassissima, apicem versus carnosulus, herbaceus, foliosus, glaber. *Folia* alterna, patentia, imparipinnata, 2–3-pedalia, petiolo cylindrico subangulato striato, superne non sulcato, basi articulato. *Foliola* opposita, 6–8-juga, ovata vel ovato-lanceolata, plerumque longe acuminata, 3–5 poll. longa, 1½–3 lata, basi acuta, petiolulo ¼–½-poll., submembranacea, subtus glauca, secus costam nervosque sparse puberula, demum glabrata. *Racemi* plures, terminales vel laterales, elongati, fere pedales, multiflori, erecto-patentes. *Bracteæ* minutæ, subulatæ, cito deciduæ. *Pedunculi* graciles, pollicares, flores longitudine æquantes. *Sepala* lineari-lanceolata, longissima, angustata, tenuiter membranacea (in vivo subcarnosa), multinervosa, tenuiter puberula. *Folliculi* 3 poll. longi, diam. ¾-poll., cylindrici, divaricati, recurvi, utrinque obtusi, irregulariter rugosi, sutura ventrali dehiscentes, crasse coriacei, pulpa solida dulci repleti. *Semina* circa 40, placentis binis paullo intra folliculi margines sitis a sutura ¼ vel ½ poll. distantibus inserta, obovato-ovalia, compressa, pulpo nidulantia. *Testa* fragilis, basi suboblique hili cicatrice lineari-oblonga notata, intus rhaphe marginali pericarpio aversa percursa; *chalaza* apicalis ; *endospermium* tenue ; *albumen* flavum, carnosum, oleosum; *embryo* albus ; *radicula* hilo versa.

The fruit of this species, which is eaten by the Lepchas of Sikkim, is very palatable, and might probably be improved by cultivation.

2. PARVATIA, Decaisne.

Sepala 6, biserialia, ext. æst. valvata. *Petala* 6, lanceolata, sepalis multo minora. *Stamina* in masculis monadelpha, connectivo ultra antheras oblongas apiculato, in fœmineis minima libera abortiva. *Ovaria* 3, ovoidea, stylo oblongo acuto apiculata. *Ovula* parieti affixa, sparsa, pilis immersa.—Frutex *scandens*, foliis *trifoliolatis,* inflorescentia *axillari racemosa,* floribus *parvulis ex albo viridescentibus.*

This little genus, of which only one species is known, is closely allied to *Stauntonia,* from which, however, it is at once distinguished by the presence of petals and by the trifoliolate (not digitate) leaves.

1. **P. Brunoniana** (Decaisne, Arch. Mus. i. 190. t. 12 A); foliolis ternis lanceolato-ovatis acuminatis supra nitidis subtus glaucescentibus, floribus racemosis laxis, pedunculis subfasciculatis.—Stauntonia Brunoniana, *Wall. Cat.* 4592!

HAB. In montibus Khasia, alt. 3–4000 ped.!—(Fl. Oct. Nov.) (*v. v.*)

Frutex alte scandens, ramis teretibus cortice rugoso suberoso pallido. *Ramuli* purpurei, striati, subangulati. *Folia* longe petiolata, petiolis basi incrassatis; foliola ovata vel ovato-lanceolata, obtuse vel argute acuminata, rarius obtusa, basi rotundata, glabra, supra lucida, subtus glauca, 3–5 poll. longa, 1–2½ lata, petiolulis angulatis medio 1–1½ poll. longo, lateralibus dimidio brevioribus. *Pedunculi* axillares, fasciculati, tuberculo squamigero inserti, 2–4 poll. longi, rigidiusculi, flexuosi, graciles. *Pedicelli* patentes, bracteola lineari-membranacea suffulti, longi. *Flores* fœminei masculis fere duplo majores ⅔-pollicares. *Sepala* tenuia, tenuiter nervosa. *Carpella* (1 tantum visum) ovoidea, utrinque obtusa, granulosa, 1½ poll. longa. *Semina* in pulpa nidulantia, undique affixa.

We have before us two specimens from the valley of Assam, one collected by Griffith and one by Mr. Simons, which are probably referable to this species, as they only differ by the leaves being obtuse and thinner in texture, both very variable characters in this Order. The flowers, which are male, are identical with those of *P. Brunoniana.*

3. HOLLBÖLLIA, Wall. (non Hook.)

Sepala 6, biserialia, ext. æst. valvata, int. subimbricata. *Petala* 6, minuta, squamæformia, rotundata. *Stamina* 6, libera; *filamenta* (in fœmineis minima effœta) crassiuscula, cylindrica; *antheræ* lineares, extrorse biloculares, connectivo apiculatæ. *Ovaria* (in masculis rudimentaria) lineari-oblonga, pulpa repleta, stigmate oblongo terminata. *Ovula* numerosa, parietibus undique affixa, pilis immersa, orthotropa, demum anatropa. *Carpella* indehiscentia, baccata, polysperma, septis pulposis a pariete ortis medium fere attingentibus pseudo-multilocularia. *Semina* in loculis solitaria, anatropa vel semi-anatropa, testa fusca cartilaginea.—Frutices *alte scandentes,* foliis *digitatis* 3–9*-foliolatis,* racemis *axillaribus corymbiformibus,* floribus *purpureis vel viridescentibus.*

This genus was originally founded by Wallich in his Tentamen Fl. Nep., but afterwards abandoned by him on the supposition that it was not distinct from *Stauntonia,* DC. Decaisne has, however, clearly shown that unless all the digitate plants of the Order are to be reduced to one genus, a course which does not seem to us advisable, these two genera must remain separate, the distinct stamina of *Hollböllia* being abundantly sufficient to characterize it. It has a very wide range in the Himalaya, extending from the Satlej to Assam. In the extreme west the species are rare, occurring only in very humid woods, but to the eastward they are very abundant, forming immense climbers, whose branches ascend lofty trees, and hang down in dense masses.

The leaves are at first very thin and membranous, but become finally very thick and coriaceous; and the flowers do not accompany one form of leaf only, but occur with every state, from those of the recently expanded shoot to the most rigid and leathery. The pulpy fruit of both species are eatable.

1. **H. latifolia** (Wall. Tent. Nep. 24. t. 16); foliolis 3–5 ovatis vel oblongis, seminibus rectis obovatis.—*Decaisne, Arch. Mus.* i. 194. *t.* 12. *f. B.* H. acuminata, *Lindl. Journ. Hort. Soc.* ii. 313. Stauntonia latifolia, *Wall. Cat.* 4950 !

HAB. In Himalaya temperata, alt. 5–9000 ped., a Simla! ad Bhotan! et in montibus Khasia supra alt. 4000 ped.!—(Fl. Apr. Mai.) (*v. v.*)

Frutex alte scandens, glaberrimus, cortice cinereo vel flavicante. *Folia* 3–5-foliolata; *petioli* foliola æquantes, angulati, striatuli. *Foliola* basi trinervia, coriacea, rigida, magnitudine valde varia, minora 2 poll. longa, ¾ lata, majora 6 poll. longa, fere 2 lata, petiolis partialibus utrinque articulatis ½–1½-pollicaribus, intermedio longiore, lateralibus (dum quinque) gradatim brevioribus. *Racemi* versus basin ramulorum fasciculati, elongati (folia fere æquantes), vel abbreviati, pauciflori. *Flores* ½–⅔ poll. longi, suaveolentes, albi vel viridescentes, purpurascentesve.

This is a very variable plant, but we are unable to distinguish more than one species. The shape of the leaves is very variable, and the colour of the flowers seems unimportant. The fruit may perhaps afford characters of importance, though we have failed to detect any.

2. **H. angustifolia** (Wall. Tent. Nep. 25. t. 17); foliolis 7–9 anguste- vel lineari-lanceolatis.—*Decaisne, Arch. Mus.* i. 194. Stauntonia angustifolia, *Wall. Cat.* 4951 !

HAB. In Himalaya temperata: Nipal, *Wallich!* Kumaon, *Strachey et Winterbottom!*—(*v. v.*)

Habitus prioris sed gracilior. *Folia* longius petiolata. *Foliola* tenuiora, lanceolata, 3–6 poll. longa, ½–1 lata, 2 exteriora brevissime petiolata. *Semina* ovato-reniformia, minora quam in specie præcedente.

We have not ourselves found this species in good state, and can therefore add nothing to the characters given by Wallich. The shape of the seed is perhaps the only important distinction between this and the last species, but we must leave the decision of the validity of the species to those who have an opportunity of studying this and the last together in a living state. Many specimens, which we cannot otherwise distinguish from *H. latifolia*, have the leaves very narrow, oblong, or almost linear, and therefore differ from *H. angustifolia* only in the number of leaflets. Those of *H. angustifolia* are, however, much thinner in texture. The shape of the fruit seems the same in both.

XI. BERBERIDEÆ.

Sepala et *petala* 2-3-4-mera, triplici vel multiplici serie alternatim imbricata. *Stamina* definita, petalis opposita, rarius indefinita; *antheræ* loculis plerumque valvulis sursum revolutis dehiscentes. *Ovarium* solitarium, monocarpellare; *ovula* pauca v. plurima; *stylus* brevissimus. *Fructus* baccatus, rarius capsularis v. transverse dehiscens.   *Semina* erecta v. horizontalia, umbilico prope basin sublaterali.   *Albumen* carnosum v. corneum.   *Embryo* axilis, orthotropus.   *Cotyledones* appositæ, germinatione foliaceæ.—Frutices, *rarius* herbæ, *pleræque glaberrimæ*, foliis *alternis simplicibus compositisve stipulatis v. exstipulatis*, floribus *axillaribus solitariis v. fasciculatis racemosis v. subcorymbosis*, pedicellis *basi bracteatis.*

   Berberideæ are pretty uniformly scattered over the north temperate zone, excepting in Europe, where the species are very few.   They abound in the Himalaya, and in the mountains of America from the latitude of Canada to Cape Horn, and are also found in the Malayan Archipelago.   Within the Arctic zone they are unknown, as also in Australasia, Polynesia, and Africa, except in the Mediterranean region. *Berberis* itself is the only widely spread genus of the Order, and is most fully developed in the Himalaya and South American Andes.   *Podophyllum* has one North American and one Himalayan species.   *Epimedium* is confined to the north temperate zone, and its maximum occurs in Japan.   *Leontice* and *Bongardia* are oriental genera.
   The affinities of *Berberideæ* are very evident, and the limits of the Order are pretty well marked.   They are immediately allied to *Lardizabaleæ* through *Decaisnea*, which has simply pinnated leaves and articulated petioles, and to *Menispermeæ;* also to *Ranunculaceæ* through *Berberis*, which has nectarial glands on the petals, also through an American genus, *Jeffersonia*, which has 4–5-merous flowers, and through *Podophyllum*, whose anthers open by longitudinal slits, and in one species of which the stamens are numerous.   Other points of affinity may be pointed out with *Anonaceæ, Magnoliaceæ,* and *Fumariaceæ*, but these are what are more or less common to the whole group of Orders to which it belongs.   In its cotyledons being closely applied to one another, it differs from many of these Orders, and in its anthers opening by valves from all except *Atherospermeæ.*
   Berberideæ we consider to have no striking affinity with any Orders but *Apocarpous Thalamifloræ*, except *Fumariaceæ* and their allies, though the valvate anthers have been considered to ally them to *Lauraceæ*, and both Auguste St. Hilaire, and latterly Lindley, have endeavoured to show that they are most closely allied to Vines. In the ' Vegetable Kingdom,' indeed, they are classed in the same alliance with Vines, *Droseraceæ, Fumariaceæ, Pittosporaceæ, Olacaceæ,* and *Cyrillaceæ*, with none of which, except *Fumariaceæ*, do we regard them as holding any direct affinity.   It is there said that Vines and *Berberideæ* "so nearly agree in fructification, that if a Berbery had two consolidated carpels, and anthers opening longitudinally, it would almost be a Vine."   But, though not inclined to lay much stress on the anthers, we cannot overlook the importance of the characters of the floral organs, nor the habit of Vines, the number of parts of their flower, their disc, and the valvate æstivation of their perianth, points which, if disregarded, leave few upon which to systematize amongst Dicotyledons; added to which, the affinities of Vines are so manifestly with other Orders, *Meliaceæ* (and perhaps *Araliaceæ*), of *Pittosporeæ* with *Violaceæ* and *Tremandreæ*, of *Olacaceæ* with *Santalaceæ*, and of *Cyrillaceæ* with still further removed Orders, that it appears to us impossible to bring these families together without in each case substituting analogical resemblances for affinities.

1. **BERBERIS,** L.

Mahonia, *Nutt., DC.*

   Sepala 6, extus 2–3-bracteolata.   *Petala* 6, concava, intus plus minusve biglandulosa.   *Stamina* 6.   *Stigma* peltatum.   *Bacca* oligosperma, seminibus erectis. *Embryo* majusculus.—Frutices *ligno flavo*, foliis *pinnatis v. suppressione pinnarum lateralium simplicibus*, foliolis *stipulisque sæpe in spinas abeuntibus*, floribus *flavis*.

Berberis, including *Mahonia*, is a perfectly natural and well-defined genus, whose species, however, are so singularly sportive in habit and all characters, that it is impossible to form any accurate estimate of its extent. One hundred have been enumerated, which number may no doubt be reduced by one-half. Both botanical authors and horticulturists have long been aware of the extreme difficulty of limiting the species of this genus. Of its sportive character the European *B. vulgaris* is a good example, upon which we are the more anxious to dwell, both because this plant occurs in its normal English form, and in many abnormal states in the Himalaya, and because it is of the utmost advantage to us, who press upon the attention of our fellow-botanists an amount of variation in mountain and tropical plants which they are slow to believe, to have such an example of variation in Europe to quote. With the *B. vulgaris*, in its ordinary north of Europe form, most botanists are familiar; but this is so unlike the Mediterranean forms, that two were described as different, one by Linnæus and Sibthorp, under the name of *B. Cretica*, and another by Rœmer and Schultes as *B. Ætnensis*, species that are now considered, by some of even the most critical European botanists (Boissier and Cosson and Gussone), as forms of *B. vulgaris;* and it is this prominent fact to which we desire to draw attention at the outset, that none of the Himalayan forms we here reduce to *B. vulgaris* differ more from the typical state of that plant than do *B. Ætnensis* and *B. Cretica*. The *B. cratægina*, DC., of Asia Minor, and *B. emarginata*, Willd., of Siberia, appear to us to have still less claims to specific distinction than *Ætnensis* and *Cretica*, and indeed they have been reduced by some authors already; and if to these be added the *B. Canadensis* of North America, the geographical range of the species will then be from Siberia westward to the lakes of Canada.

In the Himalaya Dr. Wallich distinguished nine species, all differing widely in general appearance from one another, and from *B. vulgaris;* many of them also in specific characters. To these (three of which are founded on error) others have been added, which, being found further west than Dr. Wallich's species, approached nearer to the European types, without, however, so resembling the common state of *B. vulgaris* as to suggest a comparison with any of the varieties of that plant which inhabit a similar climate; these were consequently described as new.

The first impression conveyed by reviewing the whole Himalayan genus, by laying out our very large suites of specimens collected with a view to show variations, was the strong resemblance between the West Himalayan deciduous-leaved forms and the European *B. vulgaris*, amounting, in Kashmir and Kishtwar specimens, to absolute identity; and that, proceeding eastwards and southwards, the more coriaceous-leaved species prevailed, and soon replaced the others, in the form of *B. aristata* and its varieties; that in Tibet and in the drier regions of the lofty Himalayan valleys, we everywhere found small, stunted, excessively spinous species, with small, extremely coriaceous leaves, and racemes often reduced to umbels, and even to axillary single-flowered pedicels; and that, descending lower in the same valleys and to the foot of the hills along the whole length of the Himalaya, many of these appeared to pass by insensible gradations into the large-leaved bushy form of *aristata*, with coriaceous foliage. It is very true, that both in the dry lofty regions and in the lower humid valleys, we could distinguish several well marked forms and species, often growing side by side; but the specimens from intermediate elevations, of intermediate temperature and humidity, appeared to combine all these into an inextricable plexus of species or forms that admitted of no absolute characters; and the more complete and extensive our materials, the more did the species blend.

If from our collections we turn to the labours of others, we find that they have terminated in an equally unsatisfactory manner. So long as botanists had few specimens, these were easily divided into species; but the characters attributed to

2 F

them broke down under every successive author's hands, so that each, thinking his own species new, because not agreeing with the descriptions of his predecessors, described them as such accordingly. Lastly, we have compared our notes and observations with the results arrived at by Madden, Strachey and Winterbottom, Wallich, Edgeworth, Royle, and others, and find that none of these botanists agree with one another nor with us in their views of the limits of the forms.

Under these circumstances we have felt it incumbent upon us to devote a great deal of time to studying the variations of each organ, and the result has been to reduce the species to a few well-marked forms; under these we have ranged the spurious species as varieties, retaining, however, the specific names they bore, so that they may be applied as such by those who take a different view of the value of specific characters to ourselves. We have also pointed out, under each variety, its relations to the other varieties of the same species, and to those of other species.

The following remarks on the variations of organs, etc., may be useful.

As regards habit, the species, without exception, vary extremely, many of them from tall bushes with twiggy branches to prostrate stunted shrubs, according to cold, and the degree of exposure to winds and drought; a reduction of leaves and stipules to spines, of racemes to fascicles of flowers, a shortening of the peduncles and pedicels, a reduction in the size of the flowers and of the leaves, with additional coriaceousness, and sometimes the development of glandular pubescence and glaucous bloom, are all characters more or less directly attributable to elevation, exposure, cold, or drought; it is however to be remarked, that an increased size and fleshiness of berry often accompanies these changes. The spines are more usually 5-fid in the dry country forms than in those from humid localities.

There is no natural or constant distinction into evergreen and deciduous-leaved species; for, though some species or forms, as *B. Lycioides, Asiatica,* and *Nepalensis* are always persistent-leaved, and the common form of *B. vulgaris* is always deciduous-leaved, the forms *Cretica* and *Ætnensis* of the latter have often very persistent foliage, and the duration of the leaves of *B. aristata* entirely depends on the depth of forest, and the amount of light, heat, and moisture to which it is consequently exposed. The many forms of this plant which have been raised in Kew Gardens, from seeds sent home by ourselves and others, we find to present every variety in amount of persistence; and after three years' observation we conclude, that in certain seasons some are wholly deciduous which in others are quite persistent, and that the period at which the cold arrives has a different effect on different varieties. We also observe that much depends on the age of the plant, and that different parts of the shrub are very differently affected.

The size, toothing, and cutting of the leaves, and of the opposite sides of each leaf, vary extremely in all the species, as does the number of leaves in each fascicle, in all parts of the individuals. The rapidity with which they colour is equally variable; those alpine species which are in the upper temperate Himalayan regions exposed to sudden frosts, redden rapidly, converting green mountain-slopes into bright-red in two nights. The racemes of flowers are often more or less cymose, the pedicels being more or less fasciculate; these and the peduncles vary extremely in robustness, and are sometimes almost fleshy and very glaucous. We have been unable to connect the various forms of inflorescence with habit, further than that, as stated above, there is a reduction of all parts in alpine forms. Though the extreme states of *B. aristata*, with racemose and cymose inflorescence, are extremely unlike, we have gathered specimens on which these occur on one and the same branch; we have also found stunted specimens of the same plant with solitary axillary pedicels, wholly resembling *B. angulosa* in this respect, which is typically one-flowered.

We have devoted especial attention to the variations of the flowers and fruit, because, in all polypetalous genera, in which there is a gradual transition from bracts to petals, the floral envelopes all vary extremely in relative size and form. The petals themselves are notched, entire, or bifid sometimes in the same species, specimen, and even flower, and vary from being larger than, to smaller than the sepals. The size, position, and prominence of the glands at the base of the petals is a most fallacious character: these glands originate in the thickened bases of the nerves of the petals, and in the bud almost surrounding the bases of the filaments.

The varieties of *B. vulgaris* show many forms, and every colour of fruit,—black, white, violet, and red,—as indeed was long ago pointed out by De Candolle; the size and number of seeds and colour of the testa also vary much, as does the length of the style and breadth of the stigma, though to a less extent.

Amongst the peculiarities of *Berberis* the leaf is the most remarkable. It was originally explained by Linnæus (Proleps. Plant. Amœn. Acad. v. p. 330) that the spines originate in reduced leaves, and represent three nerves. At first the spines are simple, and have a small tooth on each side (or two in some alpine forms) towards the base, which teeth elongate and produce the triple spine. In a seedling *Berberis* the petiole of the leaf will always be found to be long, slender, articulate at the base, and there furnished with two minute stipules, and bearing one articulate leaflet; the latter is often contracted above the joint into a partial petiole. As the plant grows older the petiole shortens, and finally becomes obliterated, but in all cases the leaf will be found to be articulate with the stem. The minute stipules at the base of the slender petiole of most species is replaced by an expanded auricled sheath in the pinnate-leaved species.

The uses of the species of *Berberis* are few and unimportant; the yellow wood can be used as a dye, and the fruit of some is acid and eatable; *B. Lycium* is considered by Royle to be the *Lycium* of Dioscorides, and its extract is found useful in India in inflammation of the eyes, under the name of *Rasot*.

Sect. 1. MAHONIA.—*Folia* imparipinnata.

1. **B. Nepalensis** (Spr. Syst. ii. 120); foliis pinnatis, petiolo articulato basi dilatato vaginante utrinque stipula subulata, foliolis 2–12-jugis spinuloso-dentatis, floribus in racemos erectos simplices v. basi divisos dispositis.—*Wall. Cat.* 1480! B. Miccia, *Ham. mss. ex Don, Prod.* 205. B. acanthifolia, *Wall.! mss. Don, Syst. Gard.* i. 118. B. Leschenaultii, *Wall. Cat.* 1479!; *Wight et Arn. Prod.* i. 16; *Wight, Icones, t.* 940, *Spicil. Neilgh.* i. 7. *t.* 8. B. pinnata, *Roxb. mss.* Mahonia Nepalensis, *DC. Syst. Veg.* ii. 21, *Prod.* i. 109; *Deless. Ic. Sel.* ii. *t.* 4. Ilex Japonica, *Thunb. Jap.* 79. *ejusd. Ic. t.* 32 (*fid. Don*).

HAB. In sylvis Himalayæ exterioris temperatæ, alt. 6–8000 ped.: a Bhotan! usque ad Garhwal! vulgatiss.; in montibus Khasia, 4–5000 ped.!; in montibus Nilghiri et Travancor, alt. 5–8000 ped.!—(Fl. Oct. –Mart.) (*v. v.*)

DISTRIB. Japan?

Frutex 3–6-pedalis (arbor parva in montibus peninsulæ, fide Wight). *Caulis* erectus, superne parce ramosus, ramis strictis erectis apice foliosis. *Folia* patentia, 6 unc. ad 1½-pedalia; *foliola* 1–6 unc. longa, ovata, lanceolata, v. rotundata, recta v. falcata, interdum basi cordata, inferiora minora et rotundata, valde coriacea, nervis basi flabellatim dispositis; *petiolus* strictus, rigidus, ad insertionem foliolorum articulatus, basi in vaginam semiamplexicaulem v. amplexicaulem dilatatus; *vagina* utrinque stipula subulata aucta; vaginæ superiores lamina et petiolo orbatæ in bracteas seu squamas gemmarum transeunt. *Bracteæ* 1–2 unc. longæ, apice dentatæ, interiores lineares membranaceæ. *Racemi* plurimi, erecti, multiflori, 1 unc. ad pedales, glauci v. rubicundi, interdum subglanduloso-puberuli. *Bracteolæ* coriaceæ, persistentes, oblongæ v. late ovatæ, in pedunculum decurrentes, obtusæ v. acuminatæ. *Pedicelli* erecti v. ascendentes, bracteis æquilongi v. longiores, ½ unc. longi. *Flores* flavi, ¼–⅓ unc. longi. *Sepala* exteriora parva. *Petala* oblonga, bifida, nervo centrali apice furcato (in exemplaribus Sikkimensibus). *Bacca* oblonga v. globosa, violacea, glauca, carnosa, acerba, ¼–½ unc. longa, in exempl. Nipalens. elliptica, in exempl.

Sikkimens. et Kumaonens. latiora, denique Khasianis et mont. peninsulæ fere globosis. *Semina* 2–4.

We have no hesitation in uniting the Peninsular and Khasia with the Himalayan species, notwithstanding the difference in the shape of the berries and leaflets between the extreme states of each. Dr. Wight informs us that he has cultivated the Himalayan one side by side in his garden with that of the Nilghiri, and finds them to be undistinguishable. Specimens of the Sikkim plant, cultivated for a good many years at Dorjiling, acquired longer racemes, larger flowers, and more slender pedicels than the wild specimens in the adjacent woods. The bracts are very variable organs.

Sect. 2. *Folia* simplicia (nempe unifoliolata).

§ 1. *Flores racemosi v. subcorymbosi* (*interdum in* B. Asiatica *fasciculati*).

2. **B. vulgaris** (L.); foliis plus minusve deciduis, racemis elongatis v. abbreviatis non umbellatis, petalis subintegris, baccis stigmate sessili discoideo coronatis.

α. *normalis;* ramulis teretibus, foliis gracile petiolatis amplis membranaceis argute serratis oblongis lanceolatis obovatisve acutis v. apice rotundatis, racemis foliis longioribus pendulis simplicibus non glaucis, floribus majusculis, baccis ovato-oblongis rubris compressis, stigmate subsessili, seminibus 2–5.—B. vulgaris, *Linn. Sp.* 472; *DC. Prodr.* i. 105; *Led. Fl. Ross.* 79; *Thunb. Fl. Jap.* i. 146; *Reich. Ic. Fl. Germ. t.* 18. B. Altaica *et* B. Dahurica, *Hort.* (*fid. Herb. Lindl.*).

β. *cratægina;* foliis magis coriaceis rigidis et persistentibus 1–2½-uncialibus integerrimis v. spinuloso-serratis, racemis elongatis, baccis oblongis subsphæricisve.—B. cratægina, *DC. Syst.* ii. 9, *Prodr.* i. 106. B. emarginata, *Willd.* i. 395; *DC. Prodr.* i. 105. B. Canadensis, *Mill. Dict. n.* 2; *DC. Prodr.* i. 106; *Torrey et Gray, Fl. Bor. Am.* i. 50. B. sphærocarpa, *Kar. et Kir.! En. Pl. Fl. Alt. n.* 46; *Led. Fl. Ross.* i. 742. B. heteropoda, *Schrenk.! En. Pl. Nov. Soong.* 102; *Led. Fl. Ross.* i. 742. B. Turcomanica, *Karel. mss.; Led. Fl. Ross.* i. 79.

γ. *Ætnensis;* rigidior, robustior, humilis, ramis validis crassis, foliis 1–1½-uncialibus obovatis obtusis mucronatisve rarius lanceolatis grosse v. crebre spinuloso-serratis rarius integerrimis subcoriaceis nervis prominulis opacis nitidisve, racemis suberectis v. nutantibus foliis paullo longioribus.—B. vulgaris, *var.* macroacantha, *Gussone, Fl. Prod. Flor. Sic.* i. 426. B. Ætnensis, *Presl, Flor. Sic.* i. 28; *R. et S.* vii. 2; *Moris, Fl. Sardoa,* i. *t.* 5. B. Kunawarensis, *Royle ? Ill.* 64.

δ. *brachybotrys;* ramulis robustis sæpius glaucis, foliis vix coriaceis ½–1½-uncialibus obovatis lanceolatisve aristatis spinuloso-serrulatis integerrimisve, racemis abbreviatis multifloris subcorymbosis.—B. brachybotrys, *Edgeworth! in Linn. Soc. Trans.* xx. 29.

ε. *Cretica;* fruticulus humilis v. prostratus, robustus, dense ramosus, foliis parvulis ½–1-uncialibus rigide coriaceis angulatis spinuloso-serratis v. lobatis obovatis cuneato-lanceolatisque margine incrassatis nervis conspicuis, racemis elongatis abbreviatisve.—B. Cretica, *Linn. Sp. Pl.* 472; *Thunb. Fl. Jap.* i. 146; *Sibth. Fl. Græc. t.* 342; *DC. l. c.* B. vulgaris, *var.* australis, *Boissier in Ann. Sc. Nat. ser.* 2. xvi. 371. B. Thunbergii, *DC. Syst.* ii. 9, *Prodr.* i. 106.

HAB. In Himalaya præcipue occidentali temperata et subalpina, rarius in orientali ; in montibus Beluchistan. — *a. normalis.* Kashmir, Kishtwar, alt. 5–10,000 ped.!—*β. cratægina.* Balti et Kashmir, alt. 8–10,000 ped.!; Beluchistan ad Kelat, *Stocks!*—*γ. Ætnensis.* In Himalaya temperata et subalpina, a Simla usque ad Balti vulgatissima, alt. 6–12,000 ped.!—*δ. brachybotrys.* In Himalaya temperata et subalpina, a Simla ad Kashmir frequens, alt. 6–12,000 ped.! Sikkim, vallibus interioribus, alt. 9–11,500 ped.!—*ε. Cretica.* Garhwal! Kunawar! Kashmir, alt. 9–11,000 ped.! Balti, 10,000 ped.!—(Fl. vere.) (*v. v.*)

DISTRIB. *a.* In Europa boreali! et media! Podolia! Persia boreali! Asia minore!—*β.* In Europa centrali! et orientali! Rumelia! Turcomania! in Asia occidentali et centrali, Soongaria! necnon in montibus Americæ borealis!—*γ.* In mont. Hispaniæ australis! et Siciliæ. — *ε.* In montibus Hispaniæ australis! et insularum maris Mediterranei! Asiæ Minoris! et in Japonia (*Thunb.*).

Our Kashmir specimens are in no way distinguishable from the common English form of *B. vulgaris;* they have obovate membranous leaves, narrowed into rather long petioles, long pendent racemes, with subfasciculate pedicels, and obovate-oblong, compressed, scarlet berries, with two to five seeds, and sessile stigmata. The bark is attacked by a minute fungus, giving it a dotted appearance, as in England, and which led Torrey and Gray (Fl. N. Am. p. 50) to give the dotted bark as a distinguishing character between *B. vulgaris* and *B. Canadensis.* Proceeding eastward from Kashmir the form gradually changes. The glands or rather thickened nerves on the petals are very variable ; sometimes there are two diverging thickened lines or tubercles, and at others these divide, and in some cases the two lateral nerves unite with the central into a fleshy opaque mass.

β. cratægina. Although there are some differences in the fruits of the specimens brought together under this variety or form, we do not find that they are constant or accompanied with any other characters whatever. It is hardly distinguishable, except by the want of a style, from vars. *normalis* and *floribunda* of *B. aristata.* Stocks considered his Kelat specimens as undoubted *B. vulgaris.*

B. emarginata and *B. Turcomanica* have finely toothed leaves. *B. Canadensis* has the toothing of *Turcomanica,* with berries like those of *emarginata* and *a, normalis;* some specimens of it in the Hookerian and Smithian Herbaria, from Asa Gray, Boott, Hort. Paris, etc., are absolutely undistinguishable from *B. vulgaris.* With regard to some individuals of this variety, they are more nearly allied in habit to *B. aristata* amongst the Himalayan Berberies, than to *vulgaris,* for they exhibit the large size of leaf and coriaceous texture of that plant. But a moment's reflection will show that this is what should be expected, the hot summers of Western Asia, Siberia, and the Canadas being more favourable to the foliage becoming coriaceous, than the damper climate of Western Europe is ; and the same thing happens in North-west India, where the forms of Berbery belonging to this group have more membranous leaves in humid localities than in dry. Torrey and Gray indeed say that *B. Canadensis* is "very distinct from *B. vulgaris,* with which it has in some degree been confounded" (*Fl. Bor. Am.* i. *p.* 50) ; but these authors give no characters that are not common to both European and Asiatic specimens of *B. vulgaris,* and authentic specimens from Dr. Gray show them to be specifically identical. In a letter Dr. Gray informs us that, as seen growing in America, they appear very distinct, but that no definite characters are observable ; and the same may be said of many forms of *B. vulgaris* in this country, as any good nursery-garden proves.

γ. Ætnensis. Our extensive suites of specimens accord perfectly with Moris' figure and description, and vary a great deal in the amount of toothing of the leaves and in the length of the racemes (in which there is less tendency to become abbreviated and

umbellate than in the following). The branches are not glaucous. We have seen
no Ætna specimens of this plant, which Moris describes as intermediate between *B.
Cretica* and *vulgaris,* and adds that it retained its characters of habit for five years
when grown side by side with *B. vulgaris* in the Turin Botanical Garden. Philippi
also, in his account of the vegetation of Ætna (Comp. Bot. Mag. i. 92), states that
it is the same as *B. vulgaris.* There are also in Herb. Hook. specimens of this
species amongst Bourgeau's South Spanish plants, labelled *B. Ætnensis,* R.S., with
the synonym of *B. vulgaris,* var. *australis,* Boiss., appended by M. Cosson. These
specimens have the leaves less serrated than in the Himalayan form, but they are
very variable in this respect. Royle's description of *B. Kunawarensis* is erroneous
in describing the panicle as leafy and pedicels as 3–5-flowered.

δ. *brachybotrys.* In this, which is hardly distinguishable in many cases from var.
γ, *Ætnensis,* the old leaves are very coriaceous. The flowers are abundantly produced.
The fruit in Kashmir specimens is large or small, reddish-black or covered with blue
bloom, on stiff and horizontal or pendulous pedicels of variable length. The Sikkim
specimens are extremely coriaceous-leaved, and some of them, not being in fruit, are
perhaps referable to *B. umbellata* or var. *floribunda* of *B. aristata,* which has brown
and polished (not glaucous) branchlets.

ε. *Cretica.* The European state of this plant, from which we cannot distinguish our
Indian ones, has been described by Boissier as a southern variety of *B. vulgaris* in
the body of his 'Voyage Botanique dans le Midi de l'Espagne,' but in the appendix
he suspends his opinion in deference to Grisebach, who (Fl. Rumel.) says that it is
perfectly distinct. It appears under favourable conditions to grow into *B. cratægina,*
which again is not to be distinguished from var. *floribunda* of *B. aristata* except by
the fruit; indeed, Griffith's Bhotan specimens of *B. aristata* accord in habit and
foliage entirely with Aucher-Eloy's *B. Cretica* (391) from Libanus and the Grecian
Archipelago, having very small, nearly entire, lanceolate leaves, but differ in fruit
and the long pendulous raceme. Our Kunawar specimens accord perfectly with the
plate in Sibthorp's 'Flora Græca.'

3. **B. aristata** (DC. Syst. Veg. ii. 8); foliis valde coriaceis plerumque persistentibus obovatis oblongis lanceolatisve venosis varie
grosse spinuloso-serratis integerrimisve acutis obtusis aristatisve sessilibus v. in petiolum angustatis, floribus racemosis subpaniculatis v.
subcymosis pendulis suberectisve, baccis stylo brevi stigmateque parvo
terminatis.

a. *normalis;* foliis amplis obovatis oblongis ellipticisve acutis
aristatisve (1–3-pollicaribus) apicem versus hic illic spinuloso-dentatis
utrinque viridibus v. subtus glaucis, racemis compositis multifloris v.
subcorymbosis, floribus magnis, pedicellis rubris glaucisve.—*DC. Prod.*
i. 106; *Hook. Exot. Flor. t.* 98; *Royle, Ill.* 64; *Wall. Cat.* 1474 *et*
1475 *ex parte!* B. tinctoria, *Lesch. in Mem. Mus.* ix. 306; *Wight et
Arn. Prod.* i. 16; *Deless. Ic. Sel.* ii. *t.* 2; *Wight, Ill.* i. *t.* 8; *Vanhoutte,
Flore des Serres,* vi. *t.* 75; *Lindley et Paxton, Fl. Garden,* i. 18. *f.* 5;
Wall. Cat. 1476! B. Chitria, *Ham. mss.; Ker in Bot. Reg. t.* 729;
Don, Prodr. 204. B. angustifolia, *Roxb. Hort. Beng.* 87.

β. *floribunda;* foliis obovatis oblongis lanceolatisve integerrimis v.
varie spinuloso-serratis subtus glaucis v. concoloribus, floribus racemosis
umbellatisve, pedunculis sæpius elongatis simplicibus, pedicellis breviusculis elongatisve interdum valde glaucis.—B. floribunda *et* B. petiolaris, *Wall. mss. sub* 1474, *et Don, Syst. Gard.* i. 115; *Lindley in Penny
Cyclop.* iv. 261. B. aristata, *Wall. Herb.* 1474, *ex parte!* B. affinis
et B. ceratophylla, *Don, Syst. Gard.* B. coriaria, *Royle, mss.; Lindley*

in Bot. Reg. N.S. xiv. *t.* 46. B. umbellata, *Lindl. in Bot. Reg.* 1844. *t.* 44, *non Wall. mss.*

γ. *micrantha;* foliosa, foliis valde coriaceis obovato-lanceolatis lanceolatisve (1–3-pollicaribus) grosse spinuloso-dentatis, racemis elongatis nutantibus, floribus parvis.—*Wall. Cat. sub* 1474!

HAB. Per totam Himalayam temperatam, a Bhotan usque ad Kunawar, alt. 6–10,000 ped. ; et in montibus Nilghiri et Zeylaniæ, alt. 6–7000 ped.—*a. normalis.* Vulgatissima a Nipalia! ad Sirmur! sed non in Sikkim visa.—β. *floribunda.* Kumaon et Garhwal, alt. 7700–9500 ped., *Str. et Wint.!* Simla, alt. 9000 ped.! Kunawar, *Munro!*—γ. *micrantha.* Nipal, *Wall.!* Garhwal! Sikkim, alt. 9000 ped.! Bhotan, *Griffith!*—(Fl. vere.) (*v. v.*)

This plant we regard as only less variable than *B. vulgaris*, from which its generally much more coriaceous leaves, more fascicled flowers of the raceme, and the long style and small stigma, best distinguish it. Several forms are known in our gardens, of which *B. Chitria* and *B. aristata* are the most marked, but these are certainly not specifically distinct. In the Himalaya we find far too many intermediate states to admit of our separating them even as varieties, and we believe that they are chiefly due to humidity for their characters. The *B. tinctoria* of the Nilghiri mountains and Ceylon is another form which sometimes appears distinct, but we have many specimens from those countries wholly undistinguishable from the Himalayan ones.

a. normalis. The leaves vary much in size, and the small-leaved specimens from Simla, having often smaller flowers too (and which might as well have been included under var. *micrantha*), are identical with both Nilghiri and Ceylon individuals. The handsomest state of this variety is the Kumaon one known in gardens as *B. Chitria*, Ham., with broad elliptical, almost entire, green, veined leaves, often 3½ inches long, and racemose panicles 4 inches long, bearing fascicles of flowers ½ inch in diameter ; it has dark berries ⅓–¾ inch long, often thickly covered with bloom. The state figured by Ker in the 'Botanical Magazine,' with lanceolate spinulose leaves and numerous pendulous racemes, is a very slight deviation from this. This form (*B. Chitria*) inhabits Nepal, Kumaon, Garhwal, and Sirmore; we have it not from Sikkim, nor from the peninsula. It is scarcely an evergreen, though the leaves remain for a long time. In the peninsular and Ceylon plant (*B. tinctoria*) the leaves are 1½–2 inches long, veined, vary from orbicular to obovate and lanceolate, are all aristate and more or less spinulose, and often very glaucous below. It was originally referred to *B. aristata* by Lindley in the 'Penny Cyclopædia;' its style is sometimes a line long. It is frequently an evergreen. A host of subvarieties of var. *normalis*, often grafted on *vulgaris*, but which keep their habit for a certain length of time in gardens, are referable to states of the Nipal and large-leaved form called *Chitria*, of the very glaucous evergreen peninsular plant called *tinctoria*, and of the small lanceolate-leaved Simla one, the *angustifolia* of Roxburgh.

β. *floribunda.* Many specimens of this appeared to be so distinct from var. *a, normalis,* that we at first hesitated about uniting them ; we find, however, not only that they are connected by every intermediate grade, but that several Himalayan botanists well acquainted with their forms have preceded us in uniting them. The very regularly racemose disposition of the flowers is its best character, but on some of Strachey and Winterbottom's and Wallich's specimens both fasciculate and corymbose and racemose flowers occur, and sometimes on the same specimen. The pedicels of the flowers also vary extremely, from ¼–¾ inch long, are either slender, or stouter and almost fleshy, and are green or very glaucous. The flowers are usually pale, the petals bifid; berry shortly oblong, very glaucous, its style distinct. The *B. coriaria* of Royle appears to us undoubtedly this plant, differing only in the lanceolate leaves and red fruit without bloom, characters of no importance. The name *Chitria* was intended by Hamilton (fid. Penny Cyclopædia) to have been applied to this plant. Wallich's *B. petiolaris*, mss., has membranaceous leaves, and exactly resembles *B. umbellata*, except in having a long style.

γ. *micrantha*. This retains its leaves in the moist forests of Sikkim throughout a great part of the year, and is probably perfectly evergreen in many places. In Sikkim specimens the leaves are concolorous below, but they are glaucous in some of Wallich's from Nipal, and in some of Griffith's Bhotan ones. One of Wallich's specimens entirely resembles the 'Botanical Register' plate of *B. Chitria*, except in the smaller flowers, which are less corymbose. In some Sikkim specimens the leaves are not an inch long, and are nearly entire; in Nepal ones three inches long, and grossly spinulose. Small states of this are not distinguishable except by the fruit from var. *Cretica* of *B. vulgaris*, and others in all respects resemble forms of *B. Lycium*, Royle.

4. **B. umbellata** (Wall. Cat. sub 1475 !) ; ramulis gracilibus virgatis, foliis plerumque deciduis obovatis submembranaceis varie spinuloso-serratis in petiolum angustatis subtus glaucis concoloribusve, pedunculo elongato, floribus paucis longe pedicellatis pendulis subumbellatis, baccis oblongis, stigmate subsessili discoideo.—*Don, Syst. Gard.* i. 116. B. aristata, *Bot. Mag. t.* 2549; *Wall. Cat.* 1474 ! *ex parte.*

HAB. In Himalaya temperata et subalpina, alt. 9–11,000 ped. : Bhotan, *Griffith !* Sikkim, in vallibus interioribus ! Nepal, *Wall./* Kumaon et Garhwal, *Str. et Wint.!*—(Fl. vere.) (*v. v.*)

Frutex virgatus, 8–10-pedalis, ramis gracilibus sparse foliosis. *Folia* ½–2 unc. longa, vix coriacea. *Flores* et fructus *B. vulgaris*, sed inflorescentia diversa. *Baccæ* fusco-rubræ.

Intermediate in many respects between *B. vulgaris* and *B. aristata*, and possibly only a variety of *B. vulgaris*. It is a slender-branched plant, 8–10 feet high, with scattered obovate leaves, hardly glaucous and sparingly toothed. It is not uncommon in the interior valleys of Sikkim, where it perfectly resembles the *B. vulgaris* in habit.

We have found it quite impossible to give any satisfactory references to Wallich's Herbarium in the case of the species of *Berberis*. Specimens of this occur under *B. aristata* and *B. angulosa*, and these names, together with those of *B. umbellata* and *B. petiolaris*, have been used almost indiscriminately for the different forms of the species we retain as *B. aristata, umbellata,* and *angulosa*, and have been distributed with them to the Linnean Society's and other Herbaria. The 'Botanical Magazine' plate of *B. aristata* (2549) well represents Wallich's and our *umbellata*.

5. **B. Asiatica** (Roxb. in DC. Syst. ii. 13) ; cortice pallido, spinis mediocribus parvisve foliis multoties brevioribus, foliis duris lacunoso-reticulatis orbiculatis obovatis obovato-lanceolativse grosse sinuato-spinosis integerrimisve subtus glaucis, pedicellis dense confertis v. in racemum dispositis, ovario lagenæformi, stylo subelongato, baccis ovoideis stylo distincto.—*DC. Prodr.* i. 107 ; *Roxb. Flor. Ind.* ii. 182 ; *Deless. Ic. Sel.* ii. *t.* 1 ; *Wall. Cat.* 1477 ! (*excl. syn.* B. tinctoriæ). B. hypoleuca, *Lindl. Hort. Soc. Journ.* ii. 246 ! *cum ic. xyl.*

HAB. In Himalayæ vallibus exterioribus siccis : Bhotan, *Griffith !* Nipal, *Wall !* Kumaon et Garhwal, alt. 3–7500 ped.! Afghanistan, *Griff'./* monte Parasnath prov. Bahar, alt. 3500 ped., *Edgeworth !*— (Fl. Feb. Mar.) (*v. v.*)

Frutex robustus, 3–6-pedalis, e basi ramosus, ramis rigidis crassis sæpius tortuosis. *Spinæ* pro genere parvæ, 3–5-crures. *Folia* breve petiolata, subconferta, crasse coriacea, $\frac{3}{4}$–3 unc. longa, aristata v. apice inermi, varie grosse spinuloso-dentata v. integerrima, subtus glauça, alba, sicco utrinque pallida. *Flores* parvuli, $\frac{1}{4}$–$\frac{1}{8}$ unc. diametro, in eodem ramulo fasciculati et corymboso-racemosi, fasciculis racemisve foliis brevioribus, pedicellis rubris glaucisve rigidis $\frac{1}{8}$–1-pollicaribus. *Stamina* ut in *B. vulgari*. *Baccæ* rubræ v. nigræ, glaucæ v. nitidæ, magnitudine variæ, stylo distincto stigmateque discoideo terminatæ.

Though difficult to define by words, this species may be distinguished in all states from *B. aristata* by its pale bark, smaller, often 5-fid spines, extremely hard, coriaceous, strongly nerved and reticulated leaves, that are lacunose on the surface, pale and very glaucous below, and by the much shorter racemes or fascicles of more numerous flowers.

B. Asiatica affects dry rocky places, seldom attaining a great elevation, and is found neither in Sikkim, the Khasia, nor the peninsula, whereas it abounds on the summit of Parasnath in Bahar, and occurs in the dry Himalayan valleys of Bhotan and Nepal, and thence westward to Afghanistan, though we have seen no specimens from the country between the Indus and Satlej. The berries are often large and eatable.

6. **B. Lycium** (Royle! Ill. 64); spinis mediocribus trifidis, foliis anguste v. obovato-lanceolatis integerrimis v. spinoso-dentatis pungentibus pallidis subtus glaucis, floribus corymboso-racemosis, pedicellis elongatis, baccis ovoideis stylo distincto.—*Royle, in Linn. Soc. Trans.* xvii. 94.

HAB. In apricis Himalayæ subtropicæ et temperatæ vulgaris : Garhwal, 3500 ped., *Royle! Str. et Wint.!* Simla, 3–9000 ped.! Jamu, 3–4000 ped.! Kishtwar, 2500–9000 ped.! Kashmir, 5000 ped.; Marri, *Fleming!*—(Fl. Apr. Mai.; fr. Jun. Jul.) (*v. v.*)

Fruticulus rigidus, ramulis virgatis, cortice pallido. *Folia* 6–8 fasciculata, 1$\frac{1}{2}$–2$\frac{1}{2}$-pollicaria, $\frac{1}{8}$ vix $\frac{1}{2}$ poll. lata, pallida, laxe venosa, subtus glauca, plerumque integerrima, pungentia, rarius varie spinuloso-dentata. *Racemi* sæpius folio longiores, multiflori, longe pedunculati, erecti v. nutantes, demum penduli, pedicelli elongati, solitarii v. fasciculati. *Fructus* violaceus, glaucus, 2–4-spermus.

This is a very distinct-looking form, of which we have a profusion of specimens from all the localities indicated. It frequents sunny places at elevations between 2500 and 9000 feet, whence the specimens at the lowest elevations are often fruiting whilst those at the upper are in flower. The narrow, entire, not lacunose leaves, pale colour, and copious small flowers, well distinguish it from the ordinary state of *B. Asiatica ;* but there are states with broader, more coriaceous, and more reticulated leaves, that are difficult of discrimination. Other states resemble forms of *B. aristata,* var. *micrantha,* and still others the *B. vulgaris,* var. *Cretica,* from which, however, the style and stigma always distinguish it. The broader, paler leaves chiefly (in the absence of fruit) distinguish it from the *B. Chinensis. B. coriaçea,* Royle, mss., which we have included under *B. aristata,* var. β, *floribunda,* may be referable to this, but we are inclined to think not.

§ 2. *Pedicelli fasciculati, uniflori* (*vide* B. Asiaticam *in* § 1).

7. **B. Wallichiana** (DC. Prodr. i. 107); sempervirens, spinis gracilibus 3–5-fidis, foliis fasciculatis late orbiculari- v. oblongo-ellipticis lanceolatisve utrinque acuminatis varie grosse spinuloso-serratis utrinque lucidis, pedicellis plurimis aggregatis brevibus, bacca stigmate subsessili terminata.

a. atroviridis; ramulis angulatis, foliis 1$\frac{1}{2}$–4-pollicaribus lanceolatis v. anguste obovatis spinuloso-serratis.—B. atroviridis, *Wall. mss.*

B. Wallichiana, *Wall. Pl. As. Rar.* iii. 23. *t.* 243 ; *Lindl. et Paxt. Fl. Gard.* i. 79. *f.* 58 ; *Don, Prod.* 204 ; *Wall. Cat.* 1478!

β. *microcarpa ;* ramis angulatis v. profunde sulcatis, foliis ut in var. *a* sed interdum integerrimis, baccis parvis elliptico-oblongis vix pulposis stylo brevi stigmateque parvo terminatis.

γ. *latifolia ;* foliis late obovatis elliptico-oblongisve 1–2-pollicaribus, floribus ut in var. *a.*

δ. *pallida ;* foliis anguste lanceolatis 2–3-pollicaribus spinulosodentatis subtus pallidis glaucisve, fasciculis paucifloris.

Hab. In sylvis Himalayæ temperatæ mediæ et orientalis et mont. Khasia.—*a.* Nipal, *Wall.!* Sikkim, alt. 8–10,000 ped.! Bhotan, *Griffith!*—β. Khasia, alt. 5–6000 ped.!—γ. Sikkim, alt. 10,000 ped.!—δ. Bhotan, *Griffith !*—(Fl. vere.) (*v. v.*)

a. The common Sikkim and Nepal form of this species is a small evergreen bush, with shining glossy foliage, never glaucous below, and fascicles of 3–20 flowers, variable in size, as are their pedicels in length and stoutness. *Berries* very variable in size and colour ; those of our Sikkim specimens are fleshy and very fair eating, of a black purple colour, without bloom. It is found in the inner valleys only of Sikkim.

Var. β has altogether the habit and appearance of *a,* but the berries are remarkably different, being much shorter, smaller (¼ inch long), scarcely fleshy, with a short style, small stigma, and one or two seeds. It is found in the Khasia alone, and there inhabits a much lower elevation than the other varieties do in the Himalaya.

Var. γ is probably only a state of *a,* with very broad leaves. It was found in exposed skirts of woods, at a great elevation, and 1000 feet above the level at which the common state of the plant grew. In form of leaves it resembles some states of *B. aristata,* but the serratures point upwards, and the habit is different.

δ. Of this variety we have two forms from Griffith, of which one differs conspicuously from the ordinary form of *B. Wallichiana* in the distinctly glaucous under surface of the leaves, approaching *B. Asiatica* in this respect, from which it differs in the long slender spines and lanceolate leaves, which are not lacunose. It is very probable that its glaucous hue is due to the bushes having grown in dry places. The other specimens have not the glaucous under-surface, but agree in every other respect ; and, indeed, considering how variable the glaucous character is, it is quite possible that these two forms grew on the same bush.

A very fine Javanese *Berberis,* collected at 9000 feet elevation, by Mr. Lobb, has been alluded to as *B. Wallichiana* by Moore (in Gard. Mag. i. 168), who says that it bears the name of *B. macrophylla* in gardens. The flowers and fruit are unknown, but the foliage differs a good deal from any known state of *B. Wallichiana.* It is possibly *B. Xanthoxylon,* Hasskarl, Hort. Bogor.

8. **B. insignis** (H.f. et T.); sempervirens, ramulis sæpissime inermibus subteretibus, foliis solitariis binisve amplis breve petiolatis elliptico- v. lineari-lanceolatis utrinque lucidis spinuloso-dentatis spinulis divergentibus, pedicellis confertis crassis brevibus, baccis ovoideis stigmate sessili.

Hab. In vallibus humidis Himalayæ temperatæ : Bhotan, *Griffith !* Sikkim! et Nipal orient.! alt. 7–10,000 ped.—(Fl. vere.) (*v. v.*)

Frutex 4–6-pedalis, ramosus, virgatus, ramulis elongatis cortice rufo-fusco. *Folia* alterna, subremota, rarius bina, rarissime fasciculata et spina imperfecta suffulta, lætissime viridia, nitida, valde coriacea, sinuato-dentata, dentibus spinosis patentibus, folia *Ilicis Aquifolii* referentia, 3–7-pollicaria, petiolo brevi cum ramulo articulata. *Flores* 3–20 fasciculati, pedicellis brevibus validis, ½–1-pollicaribus, curvis. *Peri-*

anthium coriaceo-carnosum. *Petala* bifida. *Stamina* breviuscula. *Baccæ* nigræ, ⅕ unc. longæ, pulposæ, stylo brevissimo, stigmate parvo, 2–4-spermæ, carne aureo.

A native of the lofty damp forests of Sikkim and East Nepal, where it forms a most beautiful evergreen bush, with leaves closely resembling those of holly, and clusters of pale golden blossoms. The rarity of spines, terete branches, solitary and very large leaves, are all remarkable characters, but no doubt susceptible of great modifications by climate, of which we have indications in the occasional development of 3–7-fid spines, and in the leaves becoming smaller, with a tendency to be fasciculate, in the drier-more northern valleys. The leaf-spines, too, which are generally divergent from the margin, sometimes point upwards towards the apices of the leaves which then strongly resemble those of *B. Wallichiana*, var. γ.

There are leaves of a Javanese species in Herb. Hook. much resembling this, but they are broader, more membranous, finely toothed, rather glaucous beneath, and reticulated above.

9. **B. ulicina** (H.f. et T.); fruticulus glaucus robustus horridus, spinis validis 3-partitis basi latis, foliis fasciculatis spinis æquilongis lineari-lanceolatis cuneatis obovatisve pungentibus marginibus incrassatis, floribus parvis brevissime pedicellatis dense congestis, ovariis subglandulosis, ovulis 4, baccis parvis.

Hab. In Tibetia occidentali; Nubra, in petrosis aridis, alt. 14–16,000 ped.!—(Fl. Jul. Aug.; fr. Sept.) (*v. v.*)

Fruticulus 1–2-pedalis, *Ulicem* referens, conferte foliosus et creberrime spinosus, glaucescens, ramis strictis crassis, cortice rufo-brunneo. *Spinæ* rigidæ, validæ, patentes, basi elongata crassa subdilatata. *Folia* ½-pollicaria, valde coriacea, rigida, conferta, omnia conformia, utrinque uni-bispinuloso-dentata v. rarius lobata. *Flores* inter folia densissime fasciculati, parvi, vix ¼-pollicares, aurantiaco-flavi. *Petala* bifida. *Stamina* brevia. *Ovarium* obscure glandulosum, 4-ovulatum. *Baccæ* breviter pedicellatæ, nigræ, glaucæ; *stigmate* sessili; *seminibus* 1–4.

A very remarkable-looking little species, and by far the most alpine of any; it is also the smallest leaved and flowered, most rigid, woody, and densely armed of any Indian species. All our specimens are very uniform in appearance. The branches are clothed throughout their whole length with spines, flowers, and leaves.

§ 3. *Pedicelli solitarii, rarissime bini, uniflori, rarissime biflori; sepala exteriora majuscula, interdum interioribus majora.*

10. **B. angulosa** (Wall. Cat. 1475! in parte); ramis sulcatis novellis puberulis, spinis 3–5-fidis, foliis parvis obovatis obovato-lanceolatisve aristatis integerrimis v. sinuato-dentatis marginibus incrassatis minute puberulis, pedicellis validis curvis foliis longioribus, floribus majusculis nutantibus, sepalis exterioribus interiora æquantibus, baccis 5–7-spermis, stigmate sessili v. stylo brevi.

β. pedicellis fasciculatis interdum 2-floris.

Hab. In Himalaya temperata; Nipal, *Wallich!* Sikkim, 11–13,000 ped.!—β. Sikkim, alt. 10,000 ped.!—(Fl. Jun. Jul.; fr. Sept.) (*v. v.*)

Frutex 4-pedalis, e basi ramosus; ramis strictis elongatis rigidis patulis foliosis, ramulis crassis sæpe pubescentibus novellis subtomentosis; spinis 3–5-fidis, interdum dorso puberulis, gracilibus, foliis longioribus brevioribusve. *Folia* 1–1½-pollicaria, fasciculata, sessilia, obovata, coriacea sed non crassa, plerumque integerrima, nunc spinuloso-sinuata v. déntata, apice rotundata, apiculata, superne opaca, papillis minimis puberula, subtus subnitida, costa prominula, nervis subparallelis. *Pedicelli* solitarii (in var. β fasciculati et interdum divisi), validi, glanduloso-puberuli. *Flores* majusculi, flavi. *Sepala* exteriora ampla. *Petala* sepalis paullo minora, integra.

Stamina brevia. *Baccæ* globosæ v. late oblongæ, $\frac{1}{4}-\frac{1}{2}$ unc. longæ, pedicello incrassato curvo pendulæ, stylo brevi terminatæ, rubræ, edules.

This species, of which we have many specimens from different localities, may be best known by its puberulous branches, and especially by the subglandular, stout, curved pedicels, minute, short, transparent, microscopic hairs on the foliage, and large sepals. The seldom-toothed, narrow, obovate, small leaves, and pendulous broad fruits, are all good characters.

11. **B. macrosepala** (H.f. et T.); humilis, glaberrimus, ramulis sulcatis, spinis 3-fidis gracilibus, foliis obovato-oblongis grosse spinuloso-dentatis coriaceis margine incrassatis, pedicellis gracilibus, floribus majusculis, sepalis exterioribus interiora æquantibus, baccis magnis ovoideis polyspermis stigmate sessili.

HAB. In Himalaya temperata interiori : Sikkim, alt. 12–13,000 ped.!
—(Fl. Jun.; fr. Nov.) (*v. v.*)

Fruticulus 2–4-pedalis, ramis patentibus. *Spinæ* graciles, foliis æquilongæ v. breviores. *Folia* fasciculata, $\frac{1}{2}$–1 unc. longa, crasse marginata, subtus plerumque glauca. *Pedicelli* graciles, glaberrimi, curvi. *Flores* majusculi. *Baccæ* $\frac{1}{2}$–$\frac{3}{4}$ unc. longæ, rubræ. *Semina* sæpe 6–10, latiuscula, compressa.

The flowering and fruiting specimens of this species were gathered at different places, but we have no doubt of their specific identity ; in the flowering specimen the branches are more slender and divergiug, the leaves smaller, less toothed, and more glaucous, all signs of being in a younger state. In this, as in the last species, the pedicels are sometimes fascicled and sometimes two-flowered.

This species approaches more nearly to the *B. Sibirica* than any other Himalayan one ; the Siberian plant, however, differs remarkably in its broad, almost palmate, 5–7-fid spines, shorter pedicels, and smaller flowers.

12. **B. concinna** (H.f. Bot. Mag. t. 4744); fruticulus ramosissimus, ramulis gracilibus, spinis gracilibus 3-fidis, foliis obovatis spinuloso-dentatis margine incrassatis subtus albo-glaucis, pedicellis gracilibus, sepalis exterioribus interioribus dimidio minoribus, baccis magnis oblongis polyspermis stigmate sessili.

? β. *cæspitosa;* fruticulus 6–8-uncialis cæspitosus, foliis irregulariter subangulato-lobatis spinuloso-dentatisve.

HAB. In Himalaya alpina in vallibus interioribus : Sikkim, alt. 12–13,000 ped ! (Fl. Jun.; fr. Nov.) (*v. v.*) Var. β. Kumaon, *Str. et Wint.*, 12,500 ped.! Garhwal, 9–10,000 ped., *Madden!*

Fruticulus 1–3-pedalis, plerumque terræ appressus, ramis rubris erectis patentibus prostratis v. demissis sulcatis gracilibus. *Spinæ* foliis æquilongæ v. breviores. *Folia* $\frac{1}{2}$–$\frac{3}{4}$ poll. longa, apice rotundata v. subtruncata, margine incrassato, subtus valde glauca albida, interdum quasi albo picta. *Pedicelli* graciles, folio longiores. *Flores* mediocres. *Baccæ* pendulæ, $\frac{1}{2}$–$\frac{3}{4}$ unc. longæ, compressæ, oblongæ, polyspermæ, stylo nullo ; *seminibus* parvis.

The most beautiful of all the species of its size, from the abundance of dark-green leaves with snow-white undersides, and the profusion of pale-yellow flowers aud red berries. In Sikkim it forms a small low bush, generally pressed on the ground, but in Kew Gardens it has altered its habit entirely, and grows more diffusely. It often accompanies the *B. angulosa*, which forms a bush over it.

The plant which we have ventured to include under this with a mark of doubt, differs iu its smaller angular leaves, with fewer larger teeth, and much longer spines. Our specimens are unfortunately insufficient to determine its identity, or the contrary, satisfactorily.

There is also in our Sikkim collections (from alt. 9000 ped.) a *Berberis* belonging apparently to this section, but which, from want of fruit, we have not been able to reduce to any of the above, it being in flower and young leaf only. The flowers are small, otherwise like those of *B. concinna* and *macrosepala*, but they are fasciculate or subumbellate on a slender peduncle. The leaves are obovate lanceolate, entire, aristate, and in the young state membranous.

2. LEONTICE, L.

Sepala 6, colorata. *Petala* 6, sepalis opposita, breviora, unguiculata; ungue squamula aucto. *Stamina* 6, petalis opposita; *antheris* extrorsis valvulis a basi sursum revolutis dehiscentibus. *Ovarium* 1-loculare; *ovulis* basilaribus. *Stylus* brevis rectus; *stigmate* simplici. *Capsula* vesicaria, membranacea, irregulariter rupta. *Semina* subglobosa, basi excavata, umbilicata. *Embryo* in albuminis dense carnosi basi endopleuræ duplicatura vaginatus, minimus; cotyledonibus brevissimis subdivaricatis; radicula infera.—Herbæ *glaberrimæ*, rhizomate *tuberoso perennante*, caulibus *annuis*, foliis *radicalibus sectis*.

The nearest ally of this genus is the North American *Caulophyllum thalictroides*, Mich., which agrees with it in most characters, but differs in habit and inflorescence, in the bracts external to the sepals, in the fleshy sarcocarp of its fruit, and in the latter becoming ruptured long before the ripening of the seeds. Several species of *Leontice* are enumerated besides the *L. Leontopetalum*, some of which may occur in Tibet, or the provinces west of India proper; but of these the *L. Altaica*, which ranges from Odessa to Tarbagatai in Soongaria (near the confines of Western Tibet), is the only one of which we have an accurate knowledge. Of the *L. Vesicaria*, Pal., and *L. Eversmannii*, Bunge, we have seen only imperfect specimens, which we cannot distinguish from small states of *L. Leontopodium*.

The induplication of the inner coat of the seed, which forms a sheath to the radicle of the embryo, is a very remarkable and hitherto unexplained fact, which requires a careful study of the ovule in all stages of growth.

1. **L. Leontopetalum** (Linn. Sp. Pl. 448); foliis biternatim sertis, foliolis petiolatis obovatis obtusis coriaceis, bracteis oblongis subfoliaceis pedicellis gracilibus multoties brevioribus.—*Lam. Ill. t.* 254. *f.* 1; *DC. Syst.* ii. 25, *Prod.* i. 109; *Led. Fl. Ross.* i. 81; *Griff. It. Notes in Affghan. Journ. No.* 235.

HAB. In montibus Afghanistan, *Griffith!* Beluchistan, *Stocks!*—(Fl. vere.) (*v. s.*)

DISTRIB. Etruria, Apulia, Creta (*DC.*), Grecia! Asia media (*Ledebour*) et minore! Syria! Mesopotamia! Persia!

Herba robusta, ½–1½-pedalis, glaucescens. *Radix* tuberosa. *Caulis* crassus, medullosus. *Folia* radicalia 1–2, caulina parva, longe petiolata, petiolo basi vaginante amplexicauli, late deltoidea, biternata, 3–7 poll. lata; foliolis ½–1½-pollicaribus integerrimis reticulatim venosis, supremis lobatis partitisve. *Racemus* strictus, erectus, crassus, simplex v. basi ramosus, pedicellis inferioribus folio ternato bracteatis, bracteis superioribus 2 lin. ad ½ poll. longis, orbiculatis oblongisve, obtusis. *Pedicelli* graciles, patentes, 1–2-pollicares. *Flores* plurimi, aurei, ½ poll. diametro. *Sepala* 6, obovata. *Petala* parva, carnosa, pedicellata, late orbiculata, subtriloba, filamentum crassiusculum amplectentia. *Ovarium* oblique ovatum, in stylum crassum truncatum attenuatum; *stigmate* terminali; *ovulis* 2–3. *Capsula* inflata, diametro pollicari, membranacea, reticulatim venosa, oblique apiculata, demum obconica, apice irregulariter rupta. *Semina* 3, basilaria, globosa, bruunea v. glauca, diametro pisi minoris.

Albumen corneum.   *Embryo* axilis ; *radiculæ* vagina spongiosa ; *cotyledones* plano-
concavæ, hiantes.

3. BONGARDIA, C. A. Meyer.

Sepala 3–6.   *Petala* 6, sepalis opposita, breviora, vix unguiculata,
basi exappendiculata, poro nectarifero instructa.   *Stamina* 6, petalis
opposita ; antheris extrorsis longitudinaliter valvulis a basi sursum revolutis dehiscentibus.   *Ovarium* 1-loculare ; *ovulis* basilaribus ; *stylo*
brevi, disco foliaceo plicato margine stigmatoso terminato.   *Capsula*
vesicaria, membranacea, indehiscens.   *Semina* 1–4 ut in *Leontice.*—
Herbæ *glaberrimæ*, rhizomate *perennante*, caulibus *annuis*, foliis *pinnatisectis.*

1. **B. Rauwolfii** (C. A. Meyer, Veg. d. Pflz. Am. Caucas. 174).—
Led. Fl. Ross. i. 80 ; *Floral Cabinet,* iii. 33. *t.* 98 ; *Henslow in Botanist,*
i. *t.* 50.   Leontice Chrysogonum, *Linn. Sp. Pl.* 447 ; *DC. Syst.* i. 24,
Prod. i. 109 ; *Griff. It. Notes, p.* 237, *No.* 286.

HAB. Montibus Afghanistan prope Quettah, alt. 5500 ped., *Griff.!*
Beluchistan, *Stocks!*—(Fl. vere.)

DISTRIB. Græcia, *DC.;* insula Rhoda ! Georgia ! Syria, *DC.;* Persia !

Herba 1-2-pedalis, laxe ramosa.   *Folia* radicalia longe petiolata, petiolo 2–6-
pollicari, flexuoso, ad pinnulas subarticulato (basi, fide DC., stipula scariosa aucto) ;
pinnulæ numero variæ, 2–10-jugæ, solitariæ v. binæ, ½–1½ poll. longæ, sæpius
glaucæ, late v. anguste oblongæ v. lineares, lobatæ v. dentatæ, ex schedis Griffithii
brunneo-fasciatæ.   *Scapus* (seu caulis pars superior) aphyllus, teres, glaucus, paniculatim ramosus ; ramis elongatis, bracteis appressis membranaceis suffultis, laxe
subdichotome divisis, ramulis apice floriferis, pedicellis elongatis ebracteatis, fructiferis strictis rigidis.   *Flores* ½ unc. diametro, ebracteati, perianthio 9–12-phyllo.
Sepala plerumque 6 ; 3 exteriora inæqualia, rotundata, late concava, membranacea,
venosa : interiora minora, oblonga.   *Petala* late obovata, sepalis interioribus latiora,
membranacea, basi saccata, apice truncata, erosa v. sinuata.   *Stamina* filamentis
brevibus, antheris elongatis, per totam longitudinem utrinque introrsum dehiscentibus, demum e basi subvalvatim ruptis, connectivo apiculatis.   *Ovarium* oblique
lagenæforme, membranaceum, plicatum, in stylum brevem attenuatum, stigmate sub-
3-lobo anfractuoso, lobis plicatis.   *Ovula* 4–8, funiculis rigidis erectis inæquilongis.
Capsula membranacea, plicata, elliptica, ½–¾ poll. longa, demum apice irregulariter
rupta.   *Semina* 1–3, globosa, glauca ; *testa* brunnea, coriacea ; *endopleura* subspon-
giosa, rufa, albumini adhærente ; *albumen* corneum.   *Embryo* cavitate basilari albuminis rectus axilis ; *radicula* hilo proxima, endopleuræ duplicatura vaginata.   *Cotyledones* breves.

The structure of the seed is remarkable ; it consists of a firm testa, within which
is a delicate endopleura adhering to the albumen.   The embryo lies in a cylindrical
cavity of the albumen, with its radicle exposed, but sheathed in a thin fold of the
endopleura.   Ledebour (Fl. Ross. l.c.) describes the petals as unguiculate, which appears hardly to be the case.   The anthers are truly introrse and dehisce longitudinally, but the fissure, which extends the whole length of each cell, is towards its
margin, and after dehiscence a rupture takes place along the connective also, from
the filament upwards, indicating an approach to the valvular dehiscence of *Berberis* and *Leontice.*   The *stigma* resembles that of *Podophyllum* to a considerable
degree.

B. Olivieri, considered another species by Meyer, is described as having the segments of the leaves (leaflets) solitary and opposite, which is the case with Stocks'
specimens of *B. Rauwolfii,* and with the upper leaves only of others from Georgia.

In Griffith's specimens, again, there are twin linear leaflets on opposite sides of the petiole below, and solitary ones above, so that no importance can be attached to this character. The appearance of twin (or binate) leaflets arises from the splitting of one leaflet. Stocks' specimens show all degrees of division, from the leaflet being oblique, toothed on one side, lobed, bifid and bipartite to the base.

4. EPIMEDIUM, L.

Sepala 4, bibracteolata. *Petala* 8, sepalis biseriatim opposita, exteriora plana, interiora cucullata v. calcarata. *Stamina* 4, petalis opposita; *antheris* introrsis, valvulis a basi sursum revolutis deciduis dehiscentibus. *Ovarium* oblongum, ovulis plurimis juxta placentam unilateralem adscendentibus 2–3-seriatis. *Stylus* lateralis; stigmate subcapitato. *Capsula* siliquæformis, bivalvis, valvula altera sterili, altera medio seminifera. *Semina* pauca; testa subcrustacea, umbilico supra basin laterali, rhaphe incrassato-inflata arillæformi. *Embryo* in basi albuminis dense carnosi incurvus; cotyledonibus brevissimis obtusis; radicula umbilico parallele contigua, infera.—Herbæ *habitu* Thalictri, rhizomate *elongato perennante,* foliis *ternatis biternatisve,* foliolis *dentatis ciliatis,* floribus *oppositifoliis racemosis v. paniculatis.*

1. **E. elatum** (Decaisne, Ann. Sc. Nat. ser. 2. ii. 356); elatum, ramosum, foliis 2–3-ternatis, foliolis oblique ovatis integerrimis dentatis ciliatisque, sepalis ovato-lanceolatis acutis, filamentis ovario æquilongis, ovulis 2–3.—*Decaisne in Jacq. Voy. Bot. 9. t. 8.*

HAB. Himalaya occidentali temperata; Kashmir, alt. 6–7000 ped., *Jacquemont!* Banahal! Kishtwar, alt. 6–8000 ped.!—(Fl. Jun.) (*v. v.*)

Herba 2–3-pedalis, gracilis, paniculatim ramosa. *Caulis* teres, glaucescens. *Folia* spithamæa et ultra, foliolis gracile petiolulatis, 1½–2½-pollicaribus, membranaceis, acutis, obtusis retusisve. *Panicula* ampla, ramis paucis gracillimis, pilis longis apice glanduloso-incrassatis conspersa. *Flores* pallide flavi, ¼ poll. diametro. *Sepala* biserialia, ovata, concava, puberula, interiora majora. *Petala* tenuissime membranacea, interiora cucullata. *Antheræ* lineares. *Ovarium* lineare, stylo elongato; *Folliculus* membranaceus, ⅓ poll. longus, stylo recto æquilongo terminatus, 2–3-spermus. *Semina* (immatura) elongato-reniformia, arillo carnoso majusculo bilabiato inclusa, ventre basi lata inserto.

We regret not having ripe seeds of this fine species, the arillus or expansion of the rhaphe of which is as fleshy as that of *E. alpinum,* and affords a proof of the affinity of the *Berberideæ* with the Papaveraceous Alliance on the one hand, and perhaps with the Dilleniaceous on the other. Decaisne points out the length of the filaments as a good distinguishing character, to which we may add the length of the ovary and the few ovules.

5. PODOPHYLLUM, L.

Sepala 6, caducissima. *Petala* 6–9. *Stamina* petalis numero æqualia v. dupla; *antheris* longitudinaliter dehiscentibus. *Ovarium* ovatum, ovulis plurimis juxta placentam latam parietalem pluriseriatis, stigmate peltato subsessili margine crispato. *Bacca* ovata v. oblonga, carnosa. *Semina* plurima, ascendentia; testa membranacea, umbilico basilari. *Embryo* basi albuminis dense carnosi brevissimus, cotyledonibus semicylindricis, radicula crassa infera.—Herbæ *rhizomate horizontali perennante,* caule erecto tereti, foliis *ad apicem cuulis* 2 *longe petiolatis peltatis, lobatis partitisve;* floribus *solitariis axillaribus v. supra-axillaribus albis.*

1. **P. Emodi** (Wall. Cat. 814); pedunculis supra-axillaribus, floribus hexandris.—P. hexandrum, *Royle, Ill.* 64; *Decaisne in Jacq. Voy. Bot.* 11. *t.* 9.

HAB. In Himalaya interiore temperata et subalpina : Sikkim 10–14,000 ped.! Nipal, *Wall.! Kumaon,* etc., 9–14,000 ped.! in Kashmir ad alt. 6000 ped. descendens !—(Fl. Apr. Mai.) (*v. v.*)

Herba scapigera. *Radix* e fibris crassis. *Caulis* solitarius, longe nudus, basi vaginatus, herbaceus, teres, glaber. *Folia* 2, alterna, petiolata, late orbiculari-reniformia, palmatim 3–5-loba, 6–10 unc. lata, viridia, sæpius purpureo-maculata, segmentis vernatione deflexis, cuneatis, supra medium lobatis et argute serratis, junioribus subtus tomentosis. *Pedunculus* validus. *Flos* erectus, primo vere evolutus, erectus, demum nutans, albus v. roseus, cyathiformis, 1–1½-pollicaris. *Sepala* 3, late oblonga. *Petala* 6, obovato-oblonga. *Stamina* 6, ovario æquilonga ; antheris elongatis. *Ovarium* ampullaceum ; stylo brevissimo ; stigmate cristato ; ovulis in placenta laterali multiseriatis. *Bacca* oblonga v. elliptica, 1–2-pollicaris, rubra, carnosa, edulis, seminibus dense farcta. " *Semina* subellipsoidea, brunnea, 2 lin. longa. *Integumentum* duplex, exterius membranaceum ; interius pellucidum. *Albumen* album, carnosum. *Embryo* parvulus, hilo proximus, radicula crassa, obtusa, hilo spectante ; cotyledones parvulæ, semicylindricæ,"—*Decaisne, l.c.*

A very remarkable plant, one of the earliest spring flowers in the Himalaya. The leaflets, or segments of the leaf, are plicate, and folded downwards on to the petiole in bud, and the whole plant has much the habit of *Eranthis hyemalis,* though its being a true member of the *Berberideæ* is, we think, indisputably proved by the structure of the fruit. The broad placenta, with many rows of ovules, is an approach to the structure of *Nymphæaceæ.* The pulpy covering of the seeds in *P. peltatum* of North America, is described by Torrey, Flora of the State of New York, i. 35, as an arillus developed from the whole surface of the placenta ; a modification of this we have shown to take place in some *Lardizabaleæ.* The supra-axillary peduncle is a singular feature, which is, however, not shared by its American congener. We find it repeated in many *Menispermeæ, Anonaceæ,* and amongst the allies of *Berberideæ,* and in *Capparideæ* and *Solaneæ,* and other Orders having little direct affinity with these. The pulpy tasteless fruit is eaten, as is that of the North American *P. peltatum,* L., whose leaves are poisonous and the root a drastic cathartic.

XII. NYMPHÆACEÆ.

Cabombeæ, *Rich.*

Sepala 3–6, libera v. basi inter se et cum toro connata, interdum cum ovariis cohærentia. *Torus* nullus, v. carnosus, cum sepalis petalisque adnatus, v. cum sepalis in tubum apice stamina et petala gerentem coalitus. *Petala* 3–6 v. plerumque indefinita, multiseriata, seriebus alternantibus oppositisve, interiora sæpissime in stamina transeuntia, rarissime in corollam gamopetalam coalita. *Stamina* definita v. indefinita, sæpissime perplurima, multiseriata, petalis opposita v. opposita et alterna. *Antheræ* innatæ, longitudinaliter dehiscentes. *Carpella* 3 v. sæpius indefinita, libera v. sæpissime verticillata et mediante toro in fructum multilocularem coalita ; *stigmatibus* sessilibus, linearibus, radiantibus, appendiculatis v. inappendiculatis. *Ovula* pauca v. plurima, anatropa, per totam cavitatem sparsa, rarius 2–3 sutura dorsali inserta. *Carpella* pauca, libera, v. plurima in baccam multilocularem polyspermam putredine dehiscentem mediante toro coalita, carpellis rarius dorso obscure dehiscentibus. *Semina* libera v. in pericarpii pulpa immersa, arillata v. exarillata ; *testa* coriacea crustacea v. subossea, scabra v. lævi ; *tegmine* membranaceo ; *albumen* farinaceum v. subcarnosum, axi plerumque canale percursum. *Embryo* orthotropus, sacculo nuclei inclusus, albuminis cavitate prope hilum semi-immersus ; *cotyledonibus* crassis, plerumque intus cavis, plumulam foventibus ; *radicula* brevi.— Herbæ *aquaticæ,* rhizomate *crasso prostrato folia et scapos rarius ramos foliiferos et floriferos gerente,* foliis· *natantibus peltatis hastatis cordativse rarius demersis sectisque,* petiolo *stipulato* v. *exstipulato,* pedunculis *extra-axillaribus,* floribus *natantibus nuptiis peractis plerumque demersis.*

The true position of this Order we believe to be between *Berberideæ* and *Papaveraceæ,* as far as this can be shown in a linear series. Before proceeding to discuss its affinities, it is necessary to enter into the conflicting statements and opinions of some able botanists who have studied its organization and relationship.

Brown long ago announced it as his opinion (' Flinders' Voyage,' ii. 598, and latterly, Plant. Jav. Rar. 108), that the *Cabombeæ* are only a section of *Nymphæaceæ,* a conclusion in which he has been followed by none, though Asa Gray (Gen. Plants United States, i. 91) has, under the former Order, recorded his adhesion to this opinion, and we know it to be Bentham's also ; and, after a very careful examination of the structure of all the genera, we have no hesitation in adopting it too.

The Orders *Nymphæaceæ, Cabombeæ,* and *Nelumbiaceæ* have long been considered as forming one group or alliance ; which has been called *Nymphææ** by Salisbury (Ann. Bot. ii. 70), *Hydropeltideæ* by Bartling, *Vitelligeræ* by Martius, *Nymphæineæ* by Brongniart, *Nymphales* by Lindley, *Chlamydoblastæ* by Adrien de Jussieu, *Nelumbia* by Endlicher, and *Nymphæoideæ* by Meisner (including in the last two cases the *Sarraceniaceæ*).

It is useful to quote these terms, for they show how uniformly all systematic botanists have regarded the alliance as natural. Much difference of opinion has, however, existed, as to whether its members should be referred to Monocotyledons or to Dicotyledons, and very recently an eminent botanist and accomplished anatomist has endeavoured to prove that it should be divided, *Nelumbiaceæ* being retained in Dicotyledons, and *Nymphæaceæ* perhaps referred to Endogens.

It is not necessary to do more than allude to the opinions of some of the earlier botanists, of whom Cæsalpinius, Magnolius, and Bernard de Jussieu referred *Nymphæa* to *Papaveraceæ;* or of their followers, who, being ignorant of the structure and development of the embryo and young plant, were led astray by analogies, and classed *Nymphæa* with *Hydrocharideæ* and other Monocotyledons ; such were Gærtner, A. L. Jussieu, Claude Richard, and J. St. Hilaire : their views have been discussed at length by De Candolle and others. Of the modern systematic authors who have studied the subject we believe that the following consider the place of *Nymphæa* to be where we retain it—Arnott, Brown, Brongniart, Bartling, Bentham, De Candolle, Endlicher, Asa Gray, A. de Jussieu, Meisner, Salisbury, Spach, Wight ; those who incline to consider it Monocotyledonous are Lindley, and perhaps Planchon ; Trécul, who discusses the question in an anatomical and physiological point of view only,

---

* For the dates and relative merits of these names see Planchon's excellent ' Études sur les Nymphéacées' (Ann. Sc. Nat. ser. 3. xix. 17), which contains by very far the best systematic account of the Order that has hitherto appeared.

2 H

considers the seeds as truly Dicotyledonous, but the rhizome as Endogenous; lastly, Henfrey, who confines his attention solely to the rhizome, and of *Victoria* only, considers this to be more Endogenous than Exogenous.

For our own parts, we consider that these Orders are truly Dicotyledonous, and that the rhizome, though not strictly speaking Exogenous, is by no means Endogenous, that there are no Monocotyledonous Orders to which they have any affinity, and that the arguments hitherto adduced to the contrary are based upon what appear to us to be very feeble analogies.

In stating our reasons for these opinions, we need hardly say that we do so with the utmost deference to the great authorities from whom we differ, especially our friend Dr. Lindley (to whose profound knowledge of structure and affinities we are in the habit of resorting in cases of difficulty), and M. Trécul, whose admirable essays on the anatomy of *Nuphar, Victoria,* and *Nelumbium* (Annales des Sciences Naturelles, ser. 3. iv. 286; ser. 4. i. 145, 291) are no less elaborate than lucid and exhaustive of the subject. Wherever it has been possible, we have followed the observations of the last-named author on the living plants; but whilst bearing willing testimony to his accuracy and skill as a phytotomist, we must also record our dissent from the conclusions he draws from the facts observed. In removing *Nymphæaceæ* to a distance from *Nelumbiaceæ,* he has overlooked structural and morphological considerations, and attached undue importance to anatomical and physiological details; and whilst we admit that in an abstract point of view the value of such details cannot be over-estimated, in a systematic one we believe that they will be found capable of a very different interpretation. In illustration of our meaning, we have only to refer to what has been demonstrated under *Menispermaceæ,* where closely allied genera and species have wood of so totally different an anatomical structure, that in a physiological point of view they could never be supposed to be allied. Similar instances, indeed, abound in the vegetable kingdom: witness the structure of the embryo, the germination and anatomy of *Cuscuta,* a genus which totally differs in all these respects from other *Convolvulaceæ,* but which is an undoubted member of that Order; the wide departure from the normal structure and mode of growth of *Scrophularineæ* displayed by *Orobanche, Lathræa,* and *Melampyrum;* the structural, anatomical, and functional differences between terrestrial and epiphytical *Orchideæ;* between *Ambrosinia* and other *Aroideæ* (see Griffith in Linn. Soc. Trans. xx. 263); and lastly, between the species of *Corydalis* belonging to the sections *Capnites* and *Bulbocapnos,* the germination of one of which is apparently Monocotyledonous, and of the other Dicotyledonous. In these and all similar cases we cannot but conclude that the value of the physiological differences implied by the extreme diversity of anatomical details is to be explained by morphological and structural laws, and is not real but apparent. If such remarkable differences occur in closely allied genera and species, it follows that we may expect as great resemblances to occur in plants belonging to the most widely different natural families; and we believe the similarity of the rhizome of *Nymphæaceæ* to that of Endogens, and the partial resemblance of the habit and foliage of this Order to that of *Hydrocharideæ,* are instances; and of such as these every large Natural Order presents us with examples.

We shall now examine—1, embryo; 2, germination; and 3, rhizome of *Nymphæaceæ.*

1. *Embryo.* The peculiarities of this organ are detailed in the ordinal character. Its truly Dicotyledonous structure was first shown by De Candolle, and shortly afterwards by Mirbel and Salisbury, and their conclusions have been assented to by almost every subsequent observer, except Lindley, who expresses himself doubtfully; and perhaps Planchon. The latest views of the latter author we only gather from Trécul's paper on *Victoria,* which states (l. c. p. 145) that Planchon has announced the embryo of that plant to be Monocotyledonous, adding, however, that Planchon's plate represents a Dicotyledonous embryo, "le mieux conformé que l'on peut imaginer." And we may add that in M. Planchon's 'Études des Nymphéacées' (Ann. Soc. Nat. ser. 3, xix. 3, 31), he describes the embryos of both *Nymphæa* and *Victoria* as truly Dicotyledonous. Lindley (Veg. Kingd. 409) discusses the subject fully

in all its bearings: he considers—1. That the two cotyledons may be regarded as one split cotyledon; against which we would urge, that the plumule ascends directly from between them, that the first pair of leaves are at right angles to them, and that the relation of the plumule to these lobes differs in no way from what is seen in other Dicotyledons, and is not like that of any Monocotyledon known to us.   2. He suggests a comparison of the embryo with those of *Aponogeton, Cymodocea,* and *Posidonia.* This we have made; they are exalbuminous seeds, with strictly monocotyledonous coleorhizal embryos, not contained in the sac of the amnios.   Of these, *Aponogeton,* the germination of which we have studied (see also Edgeworth in Hook. Journ. Bot. 1844, p. 405. t. xvii. and xviii.), has a linear plumule parallel to the cotyledon, and lying in a narrow slit or fold of the latter. In *Posidonia* also the cotyledon is longitudinally cleft on one side, and the plumule, which is lodged in the slit, is inflexed. In *Cymodocea* the plumule is enclosed in the acute cotyledon. 3. Another supposed anomaly is founded on the cotyledons not being contracted at their bases, and the plumule having an oblique position relatively to them; the latter observation, however, is not confirmed, and very many dicotyledonous embryos are continuous with the radicle in diameter, or even taper upward from it.

2. *Germination.* This we have studied in three species of *Nymphæa,* in *Euryale* and *Victoria,* all of which present the same appearance, with little modification. The radicle and bases of the cotyledons protrude through an orifice at the micropylar end of the seed, caused (as explained by Trécul) by the falling away of a little operculum opposite the radicle. The radicle turns downwards, and becomes a filiform rootlet, or is sometimes altogether arrested. The body of the cotyledons remains within the seed, and the plumule ascends from between their exserted bases, attains a considerable length, and gives off two strictly opposite leaves at right angles to the cotyledons; of these leaves one has a vaginate petiole, with adventitious rootlets developed at its base, and a lanceolate lamina with reticulate venation; the other is reduced to a mere filiform subulate petiole, and has no rootlets.   Within these first pair of leaves two others are developed at right angles to them, the sheathing base of the petiole of the lower embracing that of the upper, which is much the smallest; the first pair of leaves we hence consider to be opposite, and the following alternate. Trécul, on the other hand, by calling the leaf reduced to a petiole the *first,* and the larger one the *second,* would seem to imply that the first two leaves are alternate, or developed at different epochs; but they are so strictly opposite (at the apex of the terete tigellus) in the numerous specimens we have examined, and in all three genera, that we are inclined to consider their dissimilarity in size to be due to unequal development. In *Nuphar lutea,* however (which we have not examined), Trécul describes the first leaf as springing at once from between the cotyledons, and the second from the axil of the first.

Near the cotyledonary end of the radicle of *Nymphæa* is a swollen ring, which Lindley suggests may be analogous to a coleorhiza; but this never forms a sheath to the radicle, is not developed till the radicle germinates, and, as Trécul has shown, it performs the office of adventitious rootlets, and hence its function commences when that of a coleorhiza ceases. In *Euryale* and *Victoria* it sends forth horizontal processes, in all respects like rootlets, which perform the office of the radicle, which most frequently in these genera does not elongate. The radicle itself invariably decays soon after the leaves are formed, with the tigellus and remains of the seed, and the plant is nourished by the adventitious rootlets at the bases of the petioles. These rootlets emerge enclosed in a cellular sheath, which elongates considerably, and at last tears away, leaving a tubular sheath at the base, and calyptra at the apex of the rootlet. The formation of this and of the vascular bundles in the rhizome, rootlet, cotyledons, etc., are beautifully demonstrated in Trécul's papers, to which we refer for their minute anatomy. Trécul considers that the cotyledons being retained within the seed, and the radicle not becoming the root of the future plant, are both indications of an approach to Monocotyledons. This is a point which we are not prepared to discuss. We cannot, however, withhold an impression that neither of these phenomena are confined to Monocotyledons; but the point has not, so far as we are aware, been worked out in a comprehensive manner,—that is, with reference to the germination of all Natural Orders. Lindley, on the other hand, cites the fact of the bases of the cotyledons elongating and emerging in germination, as "perhaps one of the strongest arguments in favour of the lobes of the embryo being really cotyledons."

3. *Rhizome.* The true anatomy and structure of this organ is one of the most difficult possible to demonstrate, nor do we profess to understand it thoroughly. We have attempted to trace the courses of the vascular bundles in *N. pygmæa, Lotus,* and *stellata,* both before reading Trécul's paper and since, but without being able to give the necessary time, of which some idea may be formed from Trécul's having devoted more than a year to the study of *Nuphar lutea* alone, the result of which, so far as the rhizome was concerned, brought him no further towards a definite conclusion than that "the structure of the stem, and of some other parts of the plant, is what prevails in the greater number of plants that have one cotyledon." More recently, however, after the study of *Victoria,* he expresses himself more positively, and is "confirmed in his opinion of the *analogy* of structure that exists between *Nymphæaceæ* and Monocotyledons."

Commencing with our own analysis, we found that the rhizomes presented a central medullary mass, surrounded by a tolerably well-defined zone of vascular bundles. They differ from Exogens in wanting liber, wood-wedges, and medullary rays, and in the confused arrangements of the vascular tissue; and from Endogens in the vascular zone surrounding a column of pith, in the arrangement of the vascular fascicles, and in their composition. Our conclusion was, that this structure was quite reducible to a very low and deranged type of Exogenous stem, such as might be expected to occur in an axis of which all the internodes are crowded into the smallest possible compass, and in a plant the habit and general arrangement of whose organs of support and nutrition differ so widely from that of ordinary Exogens. In this opinion we were strengthened by some peculiarities in the structure of the abbreviated rhizomes of other Exogens, by the fact that vascular bundles often do form a confused plexus at the nodes, and that their arrangement in these is hence not reducible to the Exogenous type which prevails in other parts of the same stem. The great deviations from the normal type in *Menispermaceæ,* and very many other plants of less peculiar habit than *Nymphæaceæ,* further confirmed us in this opinion, no less than the fact that there are no Endogenous rhizomes known to us with which those of *Nymphæaceæ* can at all be compared. We may also repeat here what we have alluded to under *Menispermaceæ,* that in our opinion a mere reduction of the Exogenous stem, by the successive obliteration of its medullary rays and liber, and the confused arrangement of its vascular bundles, by no means implies a transition to the Endogenous class. We consider that there are other and far more important anatomical differences between these two great classes, and that, to establish an Endogenous affinity for the rhizomes of such very anomalous plants as *Nymphæaceæ,* it is necessary to prove the existence of some, at any rate, of the absolute characters of Endogens, as the courses of the vascular bundles and their composition.

Turning to Trécul's beautiful analysis of the rhizome of *Nuphar lutea,* we do not find our opinion altered; these show the courses of the vascular bundles, and their relations to the petioles, peduncles, and axis, with a precision that we failed to attain, and we have full confidence in their accuracy; but there is nothing in these that appears to us to establish an Endogenous affinity, and much that is seen in other Exogens.

Henfrey's careful observations on the rhizome of *Victoria* differ from Trécul's on *Nuphar,* and he treats the subject rather differently. The rhizome of *Victoria* presents an almost solid axis of vascular bundles, not a zone of them. Its points of affinity with Endogens Henfrey states to be:—1. The apparently continuous development of a terminal bud. To this we would object that the real nature of the growing point is not likely to be easily demonstrable in an abbreviated axis of so many internodes, and that other manifestly Exogenous rhizomes present a similar appearance. 2. That the roots are all adventitious. This is perhaps the strongest point of any, but its value in relation to the laws of germination in general cannot be said to be established; and we have seen somewhat analogous instances in the growth of *Fici* and *Loranthaceæ* and *Rhizophoreæ*, the plants of which are nourished by adventitious roots, having no connection with that originally developed, which has died away. *Cuscuta* offers another analogous case, as do those parasites which are supposed to be developed first on other plants, but which afterwards are nourished by terrestrial roots. 3. The absence of a cambium-layer, of bark, pith, and of a circular arrangement of vascular structures. Of these points, the absence of pith and of the vascular bundles forming a zone is exceptional in *Victoria*. The absence of a cambium-layer is not a strong point, for there are many Exogens in which we have failed to trace it in a normal condition, and there is as much a bark in *Nymphæaceæ* as there is in a great many other Exogens. 4. The isolated condition of the vascular bundles. This perhaps requires confirmation, as it appeared to us that the bundles often united, and, at any rate, there are various Exogens with isolated vascular bundles both in the pith and bark. 5. There being no analogue to wood and liber. This appears to militate equally against their Endogenous affinity, for the vascular bundles of Endogens are composed of wood and liber, while those of *Nymphæaceæ* are not; added to which, we have seen that in *Menispermeæ* and *Aristolochieæ*, and other Orders, the liber is constantly absent, and in very many Orders of Exogens the wood is wholly replaced by vascular tissue.

Our great objection, however, to all the above arguments, is their not bearing strongly upon the question; all appear to argue an anomalous condition of Exogenous stem, none at all approach to positive indications of the Endogenous, and we need hardly say, that in a case of this kind the tendency is always to magnify the importance of small deviations from a normal type, and to seek to attach an absolute value to them. Henfrey, however, states several objections to the Endogenous affinity of *Victoria*, which, in an abstract point of view, seem as unanswerable as the arguments in favour of the same affinity, but to which we do not attach any importance, simply because their value as physiological and structural facts is as much unknown as that of the others. These are:—1. The vascular cord of each root-bundle has not a central woody cylinder. 2. There is no fibrous layer between the cortical and central substances. 3. The composition of the vascular bundles is formed exclusively of ducts and unrollable spiral fibres. 4. The frequent anastomosis of the vascular bundles, which is not commonly the case in Monocotyledones.—To these we may add, as of far greater weight, the arrangements of the vascular bundles on a longitudinal section, and that many of these run completely round the stem.

Before dismissing this difficult subject, there are two theoretical considerations which, we think, should not be overlooked :—1. That assuming the rhizome of *Nymphæa* to be that of a Dicotyledon, a consideration of its habit, development, and mode of growth would lead us to expect that its structure would deviate widely from the type upon which it is formed; but that, assuming it to be a Monocotyledon, the considerations in question would not lead us to expect in its rhizome so total a departure from the type of that class. 2. That in a case of this kind, where the class to which a group belongs is indicated clearly by the general structure and development of its embryo, leaves, flowers, fruit, and germination, and by direct affinity with individual members of that class, it is much more philosophical to regard an apparent exception in one organ as reducible to an anomaly of the class with which the group has a direct affinity, rather than an indication of affinity to that with which it has otherwise none. We hence urge, as a fatal objection to the Endogenous affinity of *Nymphæaceæ*, that there is no Order amongst Monocotyledons to which Trécul or Henfrey has allied them, whilst there are many amongst Dicotyledons, with which they accord in the structure of their foliage, perianth, fruit, and seed.

We sum up our reasons for considering *Nymphæaceæ* to be true Dicotyledons as follows :—

1. The structure of the embryo is truly Dicotyledonous, and resembles nothing amongst Monocotyledons.

2. The germination is strictly Dicotyledonous and Exorhizal. The primary leaves are an opposite pair, alternating with the cotyledons.

3. The structure of the rhizome does not deviate more from the Exogenous type than that of many other Dicotyledons. It does not belong to the Endogenous type, and no Monocotyledon is known to have a similar rhizome.

4. The venation of the leaves is reticulated, and their vernation is involute.

5. The floral organs are generally arranged upon a quaternary or quinary plan.

6. *Nymphæaceæ* present many direct affinities with both apocarpous and syncarpous *Thalamifloræ*, as *Ranunculaceæ, Berberideæ, Magnoliaceæ,* and *Papaveraceæ,* and they present no affinity whatever with any Monocotyledonous Orders.

7. Systematic botanists are almost unanimously inclined to the above view of their immediate affinities.

There are very many interesting and curious points in the structure of *Nymphæaceæ* quite apart from those we have dwelt upon, for which we must again refer to Trécul and Planchon, confining our attention to such only as have a systematic value. The floral envelopes usually form an uninterrupted spiral from the sepals to the inner stamens, the transition being gradual from one class of organs to the other, as in *Magnoliaceæ.* In *Nymphæeæ* the prevalent numbers are four sepals, succeeded by several whorls of eight petals, four opposite to and four alternate with the sepals, and the stamens are similarly disposed; but in some American species the eight leaves of each whorl of stamens and petals are all opposite one another : this arrangement of parts is eminently characteristic of the allied Orders *Menispermaceæ, Berberideæ, Sabiaceæ,* and *Lardizabaleæ.*

The disc or torus of *Nymphæaceæ* is a most remarkable modification of the bases of the perianthial leaves and apex of the peduncle. We cannot agree with Trécul in denying the presence of a disc, though it is difficult to assign its limits and origin. The fact dwelt upon by that author, that in the earliest state of development of the flower, when the stamens and carpels appear as mere points, there is no space between the latter, appears to us to have no weight in this case, for the carpels are congenitally imbedded in it, and it appears adherent to the walls of the ovary as these are developed ; it is not a free organ, like the perigynous ring of *Alsineæ,* and does not arrive at its full development till the floral organs are fully formed. Its structure was first clearly explained to us by Bentham, who has shown us that in the fully formed fruit of all *Nymphææ* the carpels are imbedded in the disc, which rises in the centre of the compound ovary in the form of a cone or mamilla. The ovaries are hence gynobasic. The stamens are inserted into the disc at the base of the ovaries, or all round the whorl of carpels; or in *Victoria* the disc is carried up above the carpels, forming a ring upon which the stamens and petals are inserted. In *Cabombeæ* there is no disc, the carpels are free, and the stamens hypogynous. In *Barclaya* the four sepals are inserted at the base of the flower, and the petals and stamens carried up upon the disc, which is adherent with the carpels to their summit, whence the calyx is inferior and the corolla superior, as in some species of the curious Himalayan genus *Codonopsis* of *Campanulaceæ.* Lastly, in *Euryale* and *Victoria* the whole perianth is superior, which may perhaps be explained by supposing the flower to be sunk in the expanded apex of the peduncle, as in *Rosa* and perhaps the *Pomaceæ,* and to which there is a tendency in *Eschscholtzia* amongst *Papaveraceæ.*

Between the stamens and carpels there are in *Nymphæa* organs that have been regarded as incomplete stamina, as appendices to the stigmata, or as prolongations of the stigmata themselves. These are always opposite to the stigmatic lines, and are continuous with the disc below, so that their real nature is not apparent in some cases; they appear in some to be rudimentary stamina, as in the American *Nymphææ* of the *blanda* group, though the stigmatic surface is prolonged on to their bases. In *N. Lotus* they are very large, and are generally regarded as stigmatic appendages ; in *N. cærulea* and its allies they form short horns to the stigmatic rays, and can only be theoretically, if at all, attributed to the presence of rudimentary stamina ; they may be analogous to the stigmatic appendages of *Eschscholtzia* and *Fumariaceæ,* or to the appendages to the carpels of some other *Papaveraceæ.* This point wants a systematic study.

The fact of the placenta being spread over the whole surface of the cavity of the carpels is a well known one, to which we only call attention as indicating an affinity with *Berberideæ* through *Podophyllum*, with *Lardizabaleæ* through *Hollböllia* and all the typical genera of that Order, and with *Papaveraceæ* through *Papaver* itself, which has broad placentæ, and especially through the Mexican genus *Romneya*, the ovules of which are distributed over the whole cavity of the ovary. In *Cabombeæ* the ovules are few, and confined to the dorsal suture of the carpels; and these are free, indicating an affinity to *Nelumbiaceæ* on the one hand and *Platystemon* on the other, a genus of *Papaveraceæ* with two free carpels.

The seeds of *Nymphæaceæ* are sometimes arillate, when the arillus forms an elongated fleshy cup, arising from towards the base of the funiculus and completely enveloping the seed. In most species the seeds are completely imbedded in a cellular pulp derived from the walls of the carpels and placental surfaces, affording a strong analogy to the pulp of *Lardizabaleæ* and *Podophyllum*. The fact of the embryo being enclosed in the amniotic sac is well known to be common to this Order, and to some very far removed from it, as *Piperaceæ* and *Saurureæ;* but we have indicated a very analogous structure in *Monimiaceæ*, and we would further call attention to the strong resemblance between the canal in the axis of the farinaceous albumen of *Nymphæaceæ* and the cellular mass occupying the axis of the fleshy albumen of *Hortonia* and *Boldoa*. The relation of these to the amniotic sac is not made out, but we may remark that they are certainly part of the nucleary sac of the ovule, and that in *Hortonia* but little albumen is developed in that part, which remains cellular in the ripe seed, whilst in *Nymphæa*, owing to the cellular tissue itself being absorbed, an open canal remains. The fact of the embryo lying in a cavity at the apex of the albumen, and not immersed in it, is repeated in *Leontice* and *Bongardia*, genera of *Berberideæ*, where we have further indicated the sheath of the radicle as an important modification of embryo-coverings, and requiring explanation.

Other peculiarities of *Nymphæaceæ*, indicating their affinity, are that *Cabombeæ* differ little from the ternary-sepaled *Ranunculi*, except in the insertion, etc., of their ovules and their amniotic sac, and that they closely imitate in habit the *Ranunculi* of the *Batrachium* section. The great disc of *Nymphæa* is represented by that of *Pæonia*, as indicated by De Candolle. In form the stigmata strongly resemble those of *Papaver*, as do the seeds to a great extent. The whorl of carpels of *Nymphæa* further resembles in some degree that of *Illicium* and *Dillenia*, to which may be added that Trécul describes the carpels of *Nuphar* as exhibiting a tendency to a dorsal dehiscence.

We have thus a multitude of most important structural and physiological characters connecting *Nymphæaceæ* with the Orders amongst which we place them, besides many minor ones which are individually of little importance, but which together establish an accumulation of affinities all pointing in the same direction; to this we may add, that we doubt if they agree with any other Natural Orders but the immediate allies of these, in any characters of systematic importance.

Suborder I. NYMPHÆEÆ.

Stamina plurima. *Carpella* in ovarium pluriloculare concreta. *Ovula* plurima, parietibus ovarii undique affixa.

1. **NYMPHÆA**, L.

Sepala 4, imo toro inserta. *Petala* 12–20, 2–4-seriata. *Stamina* 40–60, multiseriata. *Ovarium* 6–8-loculare; stigmatibus sessilibus linearibus radiatis. *Bacca* spongiosa, irregulariter rupta. *Semina* in pulpa nidulantia, arillo sacciformi apice aperto induta; *testa* coriacea.

To any one who has studied a numerous suite of specimens of the Indian species of this beautiful genus, and the published descriptions of them, it will not be a matter of surprise that we find it necessary to unite a considerable number.

Several authors have asserted that *Nymphæaceæ* form an exception to the rule that water-plants are widely diffused, a statement we cannot confirm, for a detailed study of the Asiatic varieties assures us that they afford a remarkable confirmation of that rule. The species, however, are exceedingly variable, exhibiting that tendency to sport which so many thalamiflorous polypetalous plants do; and this circumstance, together with that of there being few badly preserved specimens in Herbaria, sufficiently accounts for the prevalent but most erroneous impression that the genus contains many species, and that these are confined to narrow areas. Of the amount of variation to which they are subject, few botanists appear to have any idea; but we have been accustomed in India to see the same species assume several varieties in one tank, differing in leaf and flower, size, colour, number of petals, stamens, and stigmata, and we much doubt if there be more than four decidedly distinct species within the limits of our Flora.

Upwards of sixty species have been recorded by Lehmann, in his recent enumeration ('Ueber die Gattung Nymphæa'), of which eleven are said to be Indian, the latter estimate being quite at variance with our experience. Planchon again curtails the genus to thirty-eight species, including eight or ten doubtful ones, and nine Indian, of which four are doubtful. This also exceeds our estimate, and evidently Planchon's too, for that author indicates with great judgment a considerable number of the described forms as being possibly varieties, but these he is not able to reduce for want of materials. We are perfectly aware that, in reducing almost all the Indian species, except *N. alba* and *N. pygmæa*, to the well-known *N. Lotus* and *stellata*, we are exposing ourselves to a most severe criticism on the part of both botanical authors and horticulturists; we must, however, in accordance with our principles, do so, admitting, at the same time, that we shall be only too glad to revise our opinion when botanists with equal means of judging shall point out some structural peculiarities that may afford tangible characters whereby to discriminate them. We cannot, in the meantime, withhold the result of our very long and detailed study of the species in a wild, cultivated, and dried state, nor hesitate to impress upon botanists the obvious bearings of the facts,—that all authors who have written on this genus are at variance with one another,—that it is impossible to distinguish their species in a dried state,—that the characters hitherto published as specific are those of individuals, and not of species,—that all water-plants are variable, and have wide ranges,—that all polypetalous flowers with a gradual transition from sepals to stamens are notoriously variable,—and that, though no single author has grouped all those species under two which we now have, there is not one of the species we have so reduced that has not been referred to *Lotus* or *stellata* by some author of note, excepting the most recent species of Lehmann and Edgeworth, and these we have ourselves fortunately examined in the living state. Lastly, we are glad to be able to give the authority of J. Smith, whose botanical knowledge and experience in the Royal Gardens at Kew entitle his opinion to the greatest respect, for saying that all the species we have referred to *N. Lotus* and *N. stellata* present no specific characters whatever under cultivation, the differences amongst them being all of degree and inconstant throughout. Except, indeed, considerable allowance be made for variation in the species of this genus, there are no limits to them, for twelve have been made out of the European *N. alba* alone, excluding the Indian *N. Cachemiriana*, which is the same plant, as is probably the *N. odorata** of North America also.

1. **N. alba** (L. Sp. Pl. 729); foliis cordatis integerrimis, floribus albis, sepalis obtusis tenuiter nervosis, antheris muticis, stigmatis radiis

---

* Professor Henslow, who has both plants in cultivation in the same pond, fails to find any characters whereby to distinguish them. De Candolle says it is often confounded with *N. alba*, but certainly distinct: he gives no distinctive characters, however.

sub-16, appendiculis brevibus cylindraceis, seminibus minutis.—*DC. Syst.* i. 56; *Led. Fl. Ross.* i. 83.

β. *Kashmiriana;* ovario pubescente v. villoso.—N. Cachemiriana, *Cambess. in Jacq. Voy. Bot.* 11. *t.* 10.   N. Kosteletzkyi, *Palliardi mss. in Lehm. Hamb. Garten und Blumenz.* viii. 369.   N. alba, var. Kosteletzkyi, *Planchon, Etudes sur les Nymph. Ann. Sc. Nat. ser.* 3. xix. 53.

HAB. Kashmir, alt. 5300 ped., *Winterbottom!*—Var. β. Kashmir, *Jacquemont.*—(Fl. Apr. Mai.) (*v. v.*)

DISTRIB. Europ. tota! Sibiria! Am. Bor.!

Folia suborbiculata, coriacea, vix aut non peltata, subtus tenuiter venosa, integerrima, lobis parallelis v. subdivergentibus. *Sepala* lineari- v. ovato-oblonga, reticulatim nervosa. *Petala* sub-10, exteriora sepalis æquilonga, lineari-oblonga. *Stamina* perplurima, filamentis subdilatatis. *Stigmatis* appendices suberecti. *Pollen* echinulatum. *Semina* puuctis minutis leviter striata.

Our specimens are certainly referable to the common white Water-lily of Europe. Planchon remarks, under *N. Cachemiriana,* which he had not seen, that it is too nearly allied to *N. alba,* of which we do not doubt that it is a form, though the plate in Jacquemont's Voyage is in many points unlike that plant, the petals being too narrow and acute, and the fruit different-looking. It is, however, impossible to figure *Nymphææ* from dried specimens. Cambessèdes describes the fruit as lanate; and we find, from a memorandum by Lehmann in the Hookerian Herbarium, that the ovary of *N. Kosteletzkyi* is villous. Planchon has referred *N. Kosteletzkyi* to *N. alba.*

2. **N. Lotus** (L. Sp. Pl. 729); foliis argute sinuato-dentatis, sepalis oblongis obtusis 5–7-costatis, petalis lineari- v. ovato-oblongis, filamentis basi late dilatatis, antheris inappendiculatis, stigmatis appendicibus cylindraceo-clavatis.

a. Lotus; foliis subtus dense pubescentibus orbiculatis reniformibusve lobis divergentibus approximatisve, floribus amplis rubris roseis albidisve.—N. Lotus, *Delile, Fl. Ægypt. t.* lx. *f.* 1; *DC. Syst.* ii. 53. N. rubra, *Roxb. Fl. Ind.* ii. 576; *DC. Syst.* ii. 52; *Wight et Arn. Prod.* i. 17; *Wight. Ill.* i. *t.* 10; *Andr. Bot. Rep. t.* 503; *Bot. Mag. t.* 1280 *et* 1364; *Wall. Cat.* 7255! N. Devoniensis, *Hook. Bot. Mag. t.* 4665. N. edulis, *DC. Syst.* ii. 52; *Roxb. Fl. Ind.* ii. 578; *Wall. Cat.* 7254! Castalia magnifica, *Sal. Par. Lond. t.* 14. C. mystica, *Ann. Bot.* ii. 73. N. semisterilis, *Lehmann, Ueber die Gattung Nymphæa,* 23.

β. *cordifolia;* foliis subtus dense pubescentibus cordato-ovatis lobis divergentibus, floribus mediocribus albis v. carneis.

γ. *pubescens;* foliis subtus puberulis pubescentibusve, floribus minoribus albis roseis rubrisve.—N. pubescens, *Willd. Sp. Pl.* ii. 1154; *DC. Syst.* ii. 52, *Prod.* i. 115; *Blume, Bijdr.* i. 48; *Wight et Arn. Prod.* i. 17 *et* 447; *Wall. Cat.* 7256!; *Planchon, l.c.* 35. N. sagittata, *Edgew. in Linn. Soc. Trans.* xx. 29.

HAB. *a* et γ. Per totam Indiam calidam vulgaris.—β. Chittagong! —(Fl. per totum annum.) (*v. v.*)

DISTRIB. Africa borealis! et tropica! Hungaria! Java! ins. Philip.!

Folia 6-unc. ad pedalia, juniora subsagittata. *Flores* 2-10 unc. lati. *Pollen* læve. *Semina* elliptico-rotundata, papillosa v. subscaberula.

Among the Indian varieties of this plant, we believe that we have seen specimens

similar to all the figures quoted above. It is quite impossible to reconcile the descriptions of authors with all the plants we have brought under *N. Lotus*, whether in
a state of nature, cultivation, or in the Herbarium. De Candolle describes *N. Lotus,
pubescens*, and *rubra*, as distinct species, but gives no diagnostic character, except the
spots of the leaves of *N. pubescens*, which we do not find to be constant even on individuals. Andrews (Bot. Rep.) says of *N. rubra* that it is allied to *N. Lotus*, but
is certainly specifically distinct in the colour of the flowers. Sims, in the 'Botanical Magazine,' figures *N. rubra*, var. *rosea*, with spotted leaves; and De Candolle
quotes the plate under his *N. rubra*, whose diagnostic character is "foliis immacu-
latis." Lehmann (Ueber die Gattung Nymphæa) enumerates *N. Lotus* of Roxburgh's
'Flora Indica' as the plant of Linnæus, and retains also *N. rubra*, Roxb., and *pubescens*, Willd., as distinct; whereas Planchon, who publishes, in the same year with
Lehmann, his 'Etudes sur les Nymphéacées,' quotes *N. Lotus*, Roxb., under *L. pubescens*, Willd., and keeps *N. Lotus*, L., and *N. rubra*, Roxb., distinct; he also
quotes the var. *rosea* under *rubra*, but remarks its spotted leaves. Wight and Arnott distinguish *N. pubescens*, Willd., from *N. rubra*, Roxb., by its spotted leaves
and white flowers. Planchon lays some stress upon the colour of the stamens; these,
however, vary from white to red, with often an orange-yellow shade, and when much
pollen is scattered about, they appear still more yellow, whence probably the yellow
stamens of Wight's figure. Roxburgh says of *N. Lotus*, that it differs from *N. rubra*
in the colour of the flowers only, which are white or pink, and yet he describes a
variety of *rubra* as having rose-coloured flowers. These contradictory statements
are of themselves suggestive of all belonging to one species; and that such is the
case we are perfectly satisfied, after an attentive study of all the states, living and
dried.

With regard to Edgeworth's *N. sagittata*, it is founded on a young leaf of *N.
rubra*: we have from Assam a perfectly similar leaf attached to the same rhizome
with an older leaf of the ordinary form. In Royle's Herbarium we find one specimen labelled "*N. Lotus, rosea*, and *pubescens*," indicating that these are considered
one species by him; and another specimen, called "*N. Lotus flore rubro*," is Roxburgh's
N. rubra. With regard to the *N. Devoniensis* of the 'Botanical Magazine,' it is a
common Bengal state of *N. rubra*, as described by Roxburgh, and not, as some suppose, a hybrid. We have most carefully compared the Indian plant with many
African specimens of *N. Lotus*, from the Nile, Senegal, and Sierra Leone, and confidently pronounce them the same, as indeed Roxburgh supposed. Planchon characterizes the Egyptian variety of *N. Lotus* as having all the anthers shorter than the
filaments, but this is certainly not the case in Damietta specimens. Under *N. pubescens*, Willd., he says that, except by the locality, it is difficult to distinguish it
from *N. Lotus*, but that, whereas the dense pubescence is constant in *N. pubescens*,
it is accidental in *N. Lotus;* this appears to us to be saying, in other words, that one
of these is an accidental variety of the other, for if it varies in pubescence in Egypt,
and is always pubescent in India, we cannot avoid the conclusion that the pubescent
state is the typical.

Lehmann's *N. semisterilis* is the common form of the *N. Lotus* of Linnæus and
Roxburgh, as we ascertained on collecting it; nor can we doubt that Waldstein
and Kitaibel were right in referring the Hungarian plant to *N. Lotus*, from which it
does not appear to be distinguished by any character of importance. To ourselves,
indeed, it appears very remarkable that it should not differ as a strongly marked variety at least, considering that Hungary is far north of its usual habitat, and that it
is dependent on the thermal springs for its existence. We have very carefully compared dried specimens and the plate with our Indian and Egyptian plant. We have
not seen other authentic specimens of *N. edulis*, DC., than those in Wallich's Herbarium.

Planchon says of the section *Lotus*, "anthesi nocturna." This is a subject requiring investigation. In India we have found *N. Lotus* expanded during the day,
but cannot say whether the weather had any influence. Sims (Bot. Mag.) states that,
though the Marquis of Blandford's specimens and those in Kew Gardens blossomed

at night, and closed at 10 A.M., his own, from Hungary, did not.　Pliny (as quoted by Salisbury) says that the flowers retire under water at night.

3.　**N. stellata** (Willd. Sp. Pl. ii. 1153); foliis orbiculatis v. elliptico-orbiculatis obtuse sinuato-dentatis integerrimisve, sepalis nervosis (sed noñ costatis), petalis lineari-oblongis lanceolatisve acutis v. apice angustatis, antheris longe appendiculatis, stigmatis radiis in cornua brevia productis inappendiculatis, seminibus substriatis.

a. cyanea; floribus mediocribus cyaneis non aut vix odoris.—N. cyanea, *Roxb. Fl. Ind.* iii. 577; *Wight et Arn. Prod.* i. 17; *Wall. Cat.* 7253 *A! et D!*　N. stellata, *β, Bot. Mag. t.* 2058; *Planchon, Etudes, l. c.* 40.

β. parviflora; floribus plerumque minoribus cæruleis.—N. stellata, *Willd. Sp. Pl.* ii. 1153; *Andr. Bot. Rep. t.* 330; *DC. Syst.* ii. 51; *Prodr.* i. 115; *Wight et Arn. Prod.* i. 17; *Wall. Cat.* 7253 *C! et E!* N. stellata, *β?* major, *Planchon, Etudes, l. c.*

γ. versicolor; floribus majoribus albis cæruleis carneis pallide purpureisve, staminibus perplurimis.—N. versicolor, *Roxb. Hort. Beng.* 41; *Fl. Ind.* ii. 577; *Sims, Bot. Mag. t.* 1189; *Planchon, Etudes, l. c.* 39; *Wall. Cat.* 7257!; N. punctata, *Edgew. in Linn. Soc. Trans.* xx. 29.　N. Hookeriana, *Lehmann, Ueber die Gattung Nymphœa,* 21; N. Edgeworthii, *Lehm. l. c.* 7.

HAB.　Per totam Indiam calidam vulgatissima.—(Fl. per totum annum.) (*v. v.*)

DISTRIB.　Var. *a.* Africa borealis! tropica! et australis?; ins. Philip.!

Folia submersa (dum adsunt) membranacea, natantia coriacea, omnia integerrima v. sinuato-dentata, plerumque per totam superficiem grosse v. minute impresso-punctata, subtus obscure maculata, rarius omnino lævia v. disco punctato; lobis acutis v. obtusis divaricatis parallelis v. incumbentibus.　*Flores* 1–10 unc. diametro, cærulei, albidi, rosei, v. purpurei, in stirpibus Ægyptiacis odori, in Indicis vix odori. *Sepala* lineari-ovata v. oblonga, petalis æquilonga v. longiora, viridia, lineolis purpureis sæpius notata, multinervia sed non costata.　*Petala* 10–30, versus apices plerumque sensim acutata, interiora exemplaribus grandifloris sæpe in stamina transeuntia.　*Stamina* 10–50, 2–4-seriata, in stirpibus minoribus pauciora, longe acute v. obtuse appendiculata, appendice albida v. cærulea.　*Pollen* læve.　*Stigmatis* radii 10–30, apicibus obtusis v. in cornua longitudine varia erecta incurva producti, inappendiculati.

The *N. stellata,* var. *β,* of the 'Botanical Magazine,' is referred by De Candolle (Systema) to *N. cærulea;* and this is the only allusion we find to an opinion we have long entertained, that the Blue Water-lily of the Nile and India are (like their white congener *Lotus*) specifically the same.　The most prominent difference we find between them is the sweet scent of the African plant, whether wild or cultivated, and its usually more numerous petals and stamina, and, according to De Candolle, the smallness of the parts of *N. stellata,* the leaves not being purple below, its lobes being divaricated, and the petals and stigmata being only eight to twelve.　We have had abundant proof in India, that, except the odour, not one of these characters is of the smallest value.　Whether the South African *N. scutifolia* (which has many petals) and one of the two Madagascar species (also found in the Mauritius) be the same, we do not venture to say, never having compared living specimens; but we find them both marked *N. cærulea* by Planchon (in Herb. Hook.), and except in the greater number of petals and stamens they do not appear to differ from that plant, to which *N. scutifolia* was referred by Dryander, Andrews, and Sims.

With regard to the three varieties we have included under the Indian *N. stellata,* we have been quite unable to distinguish them in India, or in our stoves, the differences between them being of degree only, except the colour of *versicolor.* The carpels vary in number from eight to twenty and even thirty, and the length to which the apices of the stigmatic rays are extended is also extremely variable: they are sometimes merely blunt points, and in other cases produced into long incurved points: the latter are the appendiculate stigmata of Roxburgh's *versicolor*, and, as Planchon rightly supposes, are very different organs from the true stigmatic appendices of *N. Lotus.* *N. Hookeriana* of Lehmann we collected at Chittagong, and again at the mouth of the Megna; its flowers varied from rose-coloured to pale purple and light blue, and it entirely accords with Roxburgh's *N. versicolor.*

Edgeworth's *N. punctata* is founded on the erroneous idea that the leaf of *N. stellata* is not punctate, which it almost invariably is in all its varieties, though described as impunctate by De Candolle. One of Edgeworth's three flowers (in Herb. Hook.) is of the variety *versicolor*, the two others of *N. stellata*,—a fair proof in itself of these being but one species. Planchon, whose views of the affinities of the species are always correct, has already suggested its being *N. versicolor.* In all the varieties the leaves vary from being quite entire to toothed along their whole circumference; all the varieties agree in the arrangement of the air-canals in the peduncles and petioles.

4. **N. pygmæa** (Ait. Hort. Kew. ed. alt. iii. 293); minima, foliis oblongo-orbiculatis integerrimis lobis acutis, staminibus inappendiculatis, stigmatibus 4–8 late ovatis cochleariformibus.—*Bot. Mag.* 1525; *DC. Syst.* ii. 58; *Prod.* i. 116; *Led. Fl. Ross.* i. 84.

Hab. Assam, *Jenkins!* montibus Khasia, ad Nonkrem in paludibus, alt. 5600 ped. !—(Fl. Aug.) (*v. v.*)

Distrib. Sibiria! China borealis!

Rhizoma subperpendiculare, diametr. pollicis, pilis atris mollibus lanatum. *Petioli* graciles. *Folia* 1½–2 poll. longa, elliptico- v. obovato-orbiculata, lobis divergentibus acutis, nervis filiformibus. *Flores* albi, inodori (valde odori, fid. DC.), 1½–2 poll. diametro. *Calyx* basi quadratus; sepalis lineari-oblongis obtusis. *Petala* sub-10, sepalis paullo longiora v. iis æquilonga, lineari-oblonga, obtusa. *Stamina* 3–4-seriata, brevia, antheris connectivo æquilatis, filamentis late dilatatis intimis ad apicem ovarii insertis; polline subgranuloso. *Stigmatis radii* breves, obtusi.

This curious and well-marked little species is one of the many proofs of the intimate relation between the Khasian and Chinese Floras, to which we have alluded at p. 105 of our Introductory Essay; we are unable to find any character by which to distinguish this plant from the Siberian and Chinese, except the inodorous flowers, which tends to weaken that analogous mark of difference between the *N. cærulea* of Egypt and *N. stellata* of India, and the *N. alba* of Europe and *N. odorata* of North America.

2. **EURYALE,** Salisb.

Sepala 4, margini tori ultra ovarium producti inserta, erecta. *Petala* indefinita, sepalis breviora, 3–5-seriata. *Stamina* indefinita, multiseriata, seriebus 8-meris, filamentis linearibus; pollen sphæricum, 3-nucleatum. *Ovarium* 8-loculare, toro apice dilatato immersum; stigmate discoideo obscure globoso depresse concavo, tubo tori accreto. *Ovula* pauca, parietibus affixa. *Bacca* spongiosa, irregulariter rupta, sepalis persistentibus coronata. *Semina* 8–20, arillo pulposo involuta; testa atra crassa. —Herba *aculeis horrida*, rhizomate *crasso fibras crassas emittente*, foliis *orbicularibus primum corrugatis demum bullatis marginibus planis*, floribus *purpureo-violaceis suaveolentibus*, seminibus *edulibus.*

A very remarkable plant, closely allied to the *Victoria* of the South American rivers. We have, in the observations under the Natural Order, indicated the morphological differences between the structure of the flower of *Nymphæa* and *Euryale.* A detailed description of its mode of germination will be found in Roxburgh's 'Flora Indica,' according to which, and to Planchon's and our own observations at Kew, the process is exactly that of *Victoria regia,* and differs from *Nymphæa* in the radicle being even less developed perpendicularly, but sending out short, horizontal, often branched arms, that perform the office of rootlets to the radicle. The elongating plumule bears two strictly opposite primary leaves, one of which remains as a subulate petiole and the other bears a very long linear lamina, with a hastate base, and gives off adventitious rootlets from its petiole: within the first pair a third is developed sheathed in an opposite stipule, which much resembles the vaginate petiole of one of the second pair of leaflets of *Nymphæa.*

The only known species is also a native of China, where it has been cultivated for its edible seeds, from time immemorial. Planchon has made a second species of this, founded on a description of *E. ferox,* the fruit of which Salisbury describes as being 80–100-seeded, which is no doubt a misprint for 8–10, the number I find in the original specimens from which his description was drawn up. The seeds vary exceedingly in size, from a small pea to a nut, and the starch grains of the albumen are so minute as to exhibit the " Brownian motion " under a sufficiently high power. The testa is always hard and almost bony, and smooth or wrinkled.

The large fruits of this plant are sold in the markets of Eastern Bengal, stripped of their spiny pericarp; and the seeds are roasted and eaten as food and medicine. These seeds have been found by Dr. Falconer in tertiary beds of peat near Calcutta, a district the plant does not now inhabit.

1. **E. ferox** (Salisb.Ann.Bot. ii. 73).—*DC. Syst.* ii. 40, *Prod.* i.114; *Roxb. Plant. Cor.* iii. *t.* 244; *Bot. Mag. t.* 1447; *Planchon, Etudes, l.c.* 29. E. Indica, *Planchon, l. c.* Anneslea spinosa, *Roxb. Fl. Ind.* ii. 573; *Andrews, Bot. Rep. t.* 618.

Hab. In paludibus Chittagong, *Roxburgh !* Bengaliæ orientalis !; in provincia Oude planitiei Gangeticæ superioris, *Royle ;* Kashmir !—(Fl. hieme et vere.) (*v. v.*)

Distrib. China !

Rhizoma breve. *Folia* ovalia v. orbicularia, 1–4 ped. diametro, supra viridia, subtus puberula, læte purpurea v. rubra. *Flores* 1–2 poll. longi. *Sepala* et *ovarium* aculeis horrida. *Bacca* 2–4 unc. diametr. *Semina* magnitudine pisi parvi vel cerasi; testa crassa, lævi v. subrugosa.

Royle mentions that the *Euryale* is found, but no doubt in a cultivated state, in the plains near Saharanpur.

3. **BARCLAYA,** Wall.

Sepala 5, basi ovarii inserta. *Petala* membranacea, apici tori ovario accreti cum staminibus inserta, supera. *Stamina* alternatim multiseriata, annulo tori intus inserta, e filamentis brevibus incurvis pendula, superiora sterilia. *Ovarium* e carpellis sub-10 arcte concretis, apice conicum; stigmatibus totidem conniventibus in conum apice fissum coadunatis, intus stigmatiferis. *Ovula* plurima, parietibus ovarii undique inserta. *Bacca* globosa, annulo tori et corollæ coronata. *Semina* sphærica, echinata; testa subcoriacea. *Albumen* et *embryo* ut in *Nymphæa.*—Herba *aquatica* Potamogetonis *facie,* rhizomate *brevi erecto villoso,* pedunculis *elongatis,* foliis *anguste lineari-oblongis obtusis basi hastato-bilobis membranaceis penninerviis glaberrimis* v. *subtus pu-*

berulis, pedunculis *extra-alaribus*, floribus *extus luride viridibus intus rubris v. purpureis inodoris*, bacca *magnitudine cerasi pulposa putredine dehiscente.*

1. **B. longifolia** (Wall. Linn. Soc. Trans. xv. 442. *t.* 18.)—*Hook. Ic. Pl. t.* 809, 810, *et in Ann. Sc. Nat. Ser.* 3. xvii. 301. *t.* 21; *Griffith, Not. Pl. Asiat.* i. 218. *t.* 57. *f.*; *Planchon, Etudes des Nymph. Ann. Sc. Nat. Ser.* 3. xix. 56.

HAB. In Pegu ad Rangoon, *Wallich!* Tenasserim ad Martaban, *Lobb!* et Mergui, *Griffith!*—(Fl. hieme.) (*v. s.*)

Rhizoma (ex sicco) breve, perpendiculare, ½–⅔ unc. longitudine, fibras plurimas crassas demittente, pilis erectis mollibus dense intertextis villosum. *Petioli* spithamæi, graciles. *Folia* petiolis æquilonga, 1–1½ unc. lata. *Flores* ½–1½ unc. longi. *Sepala* lineari-oblonga, costa crassa exserta percursa. *Petala* sub-3-seriata, brevia, obtusa, inæqualia. *Stamina* superiora ad filamenta brevia hamata reducta.

Suborder II. CABOMBEÆ.

Sepala et *petala* definita, libera. *Stamina* toro inserta, hypogyna. *Ovaria* 3–18, disco v. toro explanato inserta. *Ovula* 2–3, suturæ dorsali inserta.

4. **BRASENIA,** Schreb.

Sepala 3. *Petala* 3, sessilia, linearia, sepalis alterna. *Stamina* 12–18; *antheris* linearibus, rimis lateralibus dehiscentibus. *Ovaria* 6–18, cylindracea, apice breviter angustata, dein in stigmata intus longitudinaliter villosa subdilatata.—Herba *aquatica*, rhizomate *repente*, caule ramoso, pedunculis *petiolisque mucilagine indutis*, foliis *alternis peltatis elliptico-oblongis penninerviis*, pedunculis *axillaribus apice subincrassatis*, floribus *rubris.*

The remarkable little water-plant upon which this genus is founded is a native of the United States of North America and Canada, and was found early in the present century by Mr. Brown in Australia, and latterly by Griffith in the Khasia Mountains and Bhotan. Being inconspicuous, it is probably not so rare as is supposed to be the case. We are quite unable to detect any difference between our specimens which we have preserved in spirits, and the excellent analysis in Gray's ' Genera of United States Plants,' except that the filaments and dorsum of the sepals are puberulous.

Asa Gray observes that the curious mucilaginous covering of the peduncles and petioles is formed by the rapid formation and rupturing of successive epithelial cells, as mucilage is formed on the surfaces of animal mucous membranes ; we may observe that the gelatinous coat of the seeds of various *Compositæ* and *Cruciferæ* is quite analogous. Gray further states that the rhizoma contains oblong transversely annulated starch-grains of unusual size, the larger being ₇⅟₅ inch long.

1. **B. peltata** (Pursh, Fl. Bor. Am. ii. 389).—*Torrey et Gray, Fl. N. Am.* i. 55. Hydropeltis purpurea, *Richard, in Mich. Fl. Bor. Am.* i. 324. *t.* 29, *et in Ann. Mus.* xvii. 230. *t.* 5. *f.* 22 ; *Bot. Mag. t.* 1147 ; *DC. Syst.* ii. 37, *Prodr.* i. 112.—*Griff. Itin. Notes, p.* 160.

HAB. Khasia prope Nonkrem, *Griffith*, et ad Joowye, alt. 4500 ped.! Bhotan, ad Santagoung prope Panaka, alt. 6000 ped., *Griffith.*—(Fl. æstate.) (*v. v.*)

Distrib. America borealis, a Canada ad flum. Mississippi! Australia orientalis.

Pedunculus pubescens, apice infra florem incrassatus. *Flos* ½ unc. longus. *Sepala* 3, lineari-oblonga v. lineari-obovata, obtusa, dorso setulis carnosulis conicis pubescentia, concava, basi crassa. *Petala* 4, sepalis ⅓ longiora, perigyna, rosea, basi distantia, anguste lineari-oblonga, apicibus incurvis obtusis, dorso basi puberula. *Stamina* sub-12, obscure biseriata, hypogyna, carpellis opposita et alterna, filamentis cylindricis demum elongatis puberulis; *antheris* linearibus glabris, rimis lateralibus. *Pollen* (in alcohol conservatum) irregulariter globosum, opacum, obscure granulosum, disco pellucido angustissimo circumdatum. *Carpella* 10, disco plano inserta, 2-seriata, sessilia, linearia, cylindracea, puberula. *Ovula* 2, pendula, anatropa, raphe ad suturam versa. *Carpella* matura 3 v. plura, turgida, coriacea, indehiscentia, stigmate persistente cuspidata, submonosperma. *Semen* magnum, ovoideum; testa crustacea, lævi. *Albumen* farinaceum, et *embryo* ut in *Nymphæa*.

XIII. NELUMBIACEÆ.

Sepala 4–5, imo toro inserta, decidua. *Petala* plurima, multiseriata, libera, decidua. *Stamina* plurima, cum petalis imo toro multiplici serie inserta; *filamentis* supra antheram in appendicem productis; *antheris* introrsis, loculis adnatis. *Torus* carnosus, obconicus, apice lato truncato. *Ovaria* plurima, foveolis apicis plani tori singillatim basifixis, unilocularia; stigmate discoideo subsessili. *Ovulum* solitarium v. 2 collateralia, suspensum, funiculo filiformi parieti ovarii affixo; *raphe* dorsali. *Nuces* subglobosæ, stylo superatæ, coriaceo-corneæ, e tori foveolis semi-emersæ, longitudinaliter obscure dehiscentes. *Semen* inversum, testa spongiosa; *embryo* exalbuminosus, orthotropus; *cotyledones* crasse carnosæ, plumulam diphyllam valde evolutam foventes, petiolis inflexis vagina stipulari inclusis; *radicula* brevissima.—Herbæ, rhizomate *elongato horizontali*, foliis *longe crasse petiolatis*, lamina *peltata integerrima nervis radiantibus marginibus vernatione involutis*, floribus *amplis*.

We have, under the Order *Nymphæaceæ*, considered *Nelumbium* as a member of the group *Nymphales*, and stated some of our objections to M. Trécul's opinion, that these two Orders have nothing in common, but their numerous petals and stamens, and the medium they inhabit. The most prominent differences between them reside in the form and structure of the rhizome, the development of the leaves, the deciduous perianth and stamina, and the remarkable development of the torus, the sessile small carpels, with one (rarely two collateral) pendulous ovulum, and the exalbuminous seeds, with a very highly-developed plumule. Though these distinctions appear so great, they are much diminished in value by a study of *Brasenia*, which, in its rhizomes and mode of growth, is as different from *Nymphæa* as *Nelumbium* is, and whose ovaria are of exactly intermediate structure. The great torus of *Nelumbium* is a peculiar development of that of *Nymphæa;* and, as Asa Gray has demonstrated, the embryo of *Nymphæeæ* and *Cabombeæ* is just that of *Nelumbium* on a smaller scale. If the germinating seed of *Nymphæa* be compared with the embryo of *Nelumbium*, the affinity is very obvious; the principal modifications being the inflexed petioles of the plumule of the latter plant, and the stipulary sheath enclosing it, which last is perhaps analogous to the sheath enclosing the first leaf within the primary pair of leaves of *Nymphæa*. Trécul has admirably illustrated the anatomy and development of *Nelumbium codophyllum* (Ann. Sc. Nat. Ser. i. 291), and made some most important and interesting observations on the mode of growth of the leaves and petioles; of which the latter have interrupted vascular bundles. These peculiarities he quotes as grave reasons against associating *Nelumbium* and *Nymphæa* together; they are of great interest in a physiological point of view, but of no weight in a systematic one, especially as they accompany a very different habit.

Amongst the many minor points of affinity between *Nelumbiaceæ* and the contiguous Orders, which have not been alluded to under *Nymphæaceæ*, are the milky juice, which they have in common with *Nymphæaceæ* and *Papaveraceæ*, the resemblance of the imbricated, deciduous sepals and petals to those of *Berberideæ*, the adnate anthers, and the appendage to the filaments. In addition it may be remarked that *Nelumbiaceæ* are not allied, even distantly, to any other natural family whatever. Trécul remarks a tendency to sutural dehiscence in the carpels.

Several species of this genus have been described, but it is doubtful whether there are more than two, an American yellow-flowered one, and the pink or white Indian one (the Lotus, or Sacred Bean of India), which is said by Herodotus and Theophrastus to have been a native of Egypt, where it is not now found. The seeds and rhizomes are eatable.

1. NELUMBIUM, Juss.

Character ordinis.

1. **N. speciosum** (Willd. Sp. Pl. ii. 1258); floribus albis roseisve. —*DC. Syst.* ii. 44, *Prod.* i. 311; *Bot. Mag. t.* 903; *Led. Fl. Ross.* i. 83; *Wight et Arn. Prod.* i. 16; *Roxb. Fl. Ind.* 647; *Wight, Ill. t.* 9. N. Asiaticum, *Rich. Ann. Mus.* xvii. 249. *t. 9, semen, etc.* Nelumbo nucifera, *Gærtn. Fruct.* i. 73. *t.* 19. *f.* 2. N. Indica, *Poir. Dict.* iv. 453. Cyamus Nelumbo, *Smith, Exot. Bot.* i. 59. *t.* 31–32. C. mysticus, *Salisb. Ann. Bot.* ii. 75. Nymphæa Nelumbo, *Linn. Sp. Pl.* 730.

HAB. Per totam Indiam calidam divulgatum, sed sæpe (an semper?) introductum: in Kashmir in lacu prope urbem, alt. 5300 ped., vulgare! —(Fl. tempore pluvioso.) (*v. v.*)

DISTRIB. Mare Caspicum! et Aral; Persia!; ins. Malayanis et Philippinis! China! Japonia; Australia tropica!

Petioli et *pedunculi* supra aquam exserti, tuberculis retrorsis scaberuli v. læves, vasis spiralibus repleti, succo lacteo scatentes. *Folia* 1–2 ped. diam., exacte peltata, glabra, margine subundulata, subtus pallidiora, nervis prominulis. *Flores* ampli, 4–6 unc. diam. *Antheræ* connectivo in appendicem subclavatam producto. *Torus* fructus 2–4 unc. diam. *Nuces* magnitudine pisi vel cerasi parvi.

XIV. PAPAVERACEÆ.

Sepala 2, rarius 3, decidua. *Petala* 4, rarius 6, hypogyna, æstivatione plerumque plicata. *Stamina* libera indefinita, rarius definita, hypogyna; *antheris* liberis 2-locularibus longitudinaliter dehiscentibus. *Ovarium* liberum, e carpellis 2 v. pluribus compositum (rarissime carpellis discretis); ovula plurima (rarissime solitaria), placentis latiusculis inserta, anatropa v. amphitropa. *Stylus* terminalis v. nullus. *Stigmata* radiantia, sæpe bicrura et ob crura connata quasi placentis opposita. *Fructus* siccus, capsularis, rarius baccatus, 1-locularis v. septis incompletis multilocularis, indehiscens v. valvis brevibus dehiscens. *Semina* plurima, exarillata, funiculo brevi. *Albumen* copio-

sum, oleosum. *Embryo* parvus, hilum versus albumine inclusus ; cotyledonibus 1–4, plerumque 2, radicula ab hilo remota centrifuga.

We commence with *Papaveraceæ* the series of polypetalous *Thalamifloræ* with
consolidated carpels, parietal placentation, and anthers not adnate with the filament to
that degree that they are in all the previously described families. Its affinities are
not doubtful: they have been alluded to under *Nymphæaceæ* and *Berberideæ*, but
are so much more nearly related to the following Orders, *Fumariaceæ*, *Cruciferæ*,
and *Capparideæ*, that they are by some authors included with them into one great
alliance, the *Rhœades* of Endlicher and Meisner. Endlicher unites *Fumariaceæ* and
Papaveraceæ into one Order, and Brongniart classes them together as *Papaverineæ*.
Hypecoum, indeed, amongst *Fumariaceæ*, being quite intermediate in structure, is the
connecting link between these Orders, and *Platystemon*, a Papaveraceous genus with
free ovaries, is the passage between the two groups of apocarpous and syncarpous
families, more especially showing the affinity of *Papaveraceæ* with *Nymphæaceæ* on
the one hand, and with *Ranunculaceæ* on the other. With *Cruciferæ* this Order is
allied not only by the structure of the fruit of many species, but by the quaternary
arrangement of the sepals and petals.

Papaveraceæ are almost entirely natives of the northern hemisphere and of extratropical regions. They are numerous in Northern India, but attain their maximum
in Western North America. Their properties are narcotic, and their seeds usually
yield a bland oil.

1. PAPAVER, L.

Sepala 2, rarius 3, concava. *Petala* 4, rarius 6. *Stamina* indefinita.
Ovarium e carpellis 4 v. pluribus, stigmatibus radiantibus coronatum.
Capsula placentis parietalibus in cavitatem projectis polysperma, poris
v. valvis brevibus infra stigmata dehiscens.—Herbæ *succo lacteo, sæpe
hispidæ*, radicibus *fibrosis*, foliis *plerumque lobatis dentatisque*, pedunculis
axillaribus solitariis unifloris nudis.

About twelve species of *Papaver* are known, of which all but *P. nudicaule* are
confined to the Old World, and almost entirely to the north temperate zone, one
only being found in Australia, and another in South Africa.

1. **P. nudicaule** (Linn. Sp. Pl. 725); scapo unifloro, flore croceo.
—*Elkan, Monog. Pap.* 17 ; *Sims, Bot. Mag. t.* 1633 ; *DC. Syst.* ii. 71,
Prod. i. 117. P. alpinum, *Linn. Sp. Pl.* 725 ; *Led. Fl. Ross.* i. 87 ; *DC.
l. c.* P. Pyrenaicum, *DC. l. c., et* P. microcarpum, *DC. l. c.* P. aurantiacum, *Lois.; DC. Fl. Fr. Suppl.* 585. P. croceum, *Led. Fl. Alt.* ii. 271.

HAB. Tibetia occidentalis alpina : in summis montibus Ladak et
Nubra, alt. 16–17,000 ped.! Afghanistan, 15,000 ped., *Griff.!*—(Fl.
Aug.) (*v. v.*)

DISTRIB. Per totam zonam arcticam ad lat. bor. 78°! in alpibus
Norvegiæ! Helvetiæ! Pyrenæis! Dahuriæ! et Altai! in montibus scopulosis Americæ borealis !

Spithamæum v. pedale. *Folia* radicalia petiolata, 2–4-pollicaria, lineari-obovata
v. oblonga, pinnatifida, lobis paucis oblongis acutis utrinque pilosis. *Scapi* 3–5, graciles, patentim hispido-pilosi. *Flores* 1–3 poll. diam. *Sepala* hirsuta. *Filamenta*
capillaria. *Capsula* late obovata, strigoso-hispida, stigmate profunde inciso.

We have followed Elkan in uniting the *P. alpinum, nudicaule, Pyrenaicum, croceum* and *aurantiacum*, amongst which we can find no specific characters. Our
Tibetan specimens perfectly accord with Arctic American and Siberian ones.

2. **P. dubium** (Linn. Sp. Pl. 726); caule folioso multifloro setoso hispido v. glabro, foliis pinnatipartitis v. bipinnatifidis, capsula oblongoclavata.—*DC. Syst.* ii. 75.

β. *lævigatum* (Elkan, Monog. Pap. 25); caule foliisque glabris.— P. dubium, *var.* subglabrum, *Led. Fl. Ross.* i. 89. P. lævigatum, *Bieb. Fl. Taur. Cauc.* iii. 364; *DC. Syst.* ii. 78, *Prod.* i. 119. P. glabrum, *Royle, Ill.* 67. P. Decaisnei, *Hochst. et Steud. mss.; Dcne. in Ann. Sc. Nat. ser.* iii. 269; *Webb, Fragm. Flor. Æthiop. Ægypt.* 2.

HAB. Var. *lævigatum.* In arvis Himalayæ occidentalis temperatæ, alt. 5–7000 ped.!: a Kumaon! ad Kashmir! Afghanistan, *Griffith!* Beluchistan, *Stocks!*—(Fl. vere.) (*v. v.*)

DISTRIB. Var. *lævigatum.* Tauria! Rumelia! Asia Minor! Ægyptus! Persia! Caucasus!

Herba 1–3-pedalis, simplex v. ramosa. *Folia* interdum glauca, utrinque sparse pilosa, pilis subappressis v. glabrata, laciniis integris v. inciso-dentatis. *Scapi* et sepala sparse pilosa. *Flores* magnitudine valde varii. *Filamenta* subulata. *Capsula* ½–1 unc. longa. *Stigma* 5–8-radiatum, ambitu crenatum.

This appears a very variable plant in India, though perhaps not more so than its allies, or indeed than most annuals. Some of our specimens are hardly distinguishable from *P. dubium* itself in amount of hairiness, but it is usually nearly glabrous in India. The perfectly glabrous and glaucous specimens have been named *P. Decaisnei* by Hochstetter and Steudel, and form the var. ζ. of Elkan.

3. **P. somniferum** (Linn. Sp. Pl. 726); caule simplici v. diviso, foliis oblongis amplexicaulibus grosse lobato-dentatis serratisve lobis dentatis, capsula globosa glaberrima, stigmate 5–12-radiato.—*DC. Syst.* ii. 81, *Prod.* i. 119; *Roxb. Fl. Ind.* ii. 571; *Wight et Arn. Prodr.* 17; *Wall. Cat.* 8118!; *Engl. Bot. t.* 2145. P. amœnum, *Lindl. in Bot. Reg. N. S.* xii. 56. *No.* 80.

HAB. Per totam Indiam præcipue borealem cultum, et in ruderatis quasi spontaneum!—(Fl. hieme.) (*v. v.*)

DISTRIB. Europa temperata! Africa borealis! Asia subcalida!

Caulis 2–4-pedalis, simplex (rarius divisus), uniflorus v. pedunculos 3–4 ·gerens. *Folia* in stirpibus Indicis glaberrima, 4–8 unc. longa, late ovata, oblonga v. lineari-oblonga, basi cordata, sæpe duplicato-dentata. *Flores* ampli, albi, pallide purpurei v. coccinei. *Sepala* glaberrima. *Filamenta* superne paullo dilatata. *Capsula* 1 unc. diam., fere sphærica, pedicellata. *Semina* plerumque nigra.

This, the common Opium Poppy, is not known in a wild state in India, but is found occasionally in roadsides and in waste places.

4. **P. cornigerum** (Stocks, in Lond. Journ. Bot. iv. 142); sparse hispido-pilosum, caule basi ramoso, foliis pinnati- v. bipinnatisectis, sepalis pilosis dorso infra apicem cornigeris, capsula globosa subangulata ad angulos setis rigidis sparsis arcuatis hispida, stigmate 4–5-radiato.

HAB. Panjab ad Peshawar, *Vicary!* Afghanistan, *Griffith!* Beluchistan, *Stocks!*—(Fl. vere.) (*v. s.*)

Spithamæum v. pedale. *Folia* pleraque radicalia 2–4-pollicaria, petiolata, in segmenta linearia secta. *Caules* v. scapi 3–6, ascendentes, parce foliosi, simplices v. divisi, cum pedunculis appresse setoso-pilosi. *Flores* coccinei, 1½–2 poll. diam., petalis basi nigris. *Sepala* late ovato-oblonga, sparse setosa. *Filamenta* subulata. *Capsula* ¼ poll. longa, breve stipitata, obscure v. manifeste 4–5-gona, præcipue ad angulos hispida, stigmatis parvi radiis crassis.

A very remarkable little species, well characterized by the short horn or spur towards the apex of the sepals, and by the fruit.

P. orientale, L. (Wall. Cat. 8119 !), *P. Rhœas,* L. (Wall. Cat. 8120 ?), and *P. Argemone,* L., are all common garden plants in India, and hence occur in various collections.

2. ARGEMONE, L.

Sepala 2–3. *Petala* 4–8. *Stamina* perplurima. *Stigmata* 4–7, subsessilia v. breve stipitata, radiata, libera. *Capsula* obovata, apice valvulis inter placentas parietales dehiscens. *Semina* scrobiculata, raphe nuda.—Herbæ *ramosæ, aculeatæ, glaucescentes, omnes Americanæ, succo flavo,* foliis *inciso-pinnatifidis dentibus spinulosis,* alabastris *erectis.*

An American genus, of which one species is naturalized all over India, and in many other parts of the world, abounding on roadsides and in waste places, but never seen far from habitations. The seeds partake of the acrid properties of the plant, and are employed in America, as a substitute for Ipecacuanha, and as a purgative. The flowers, which are always yellow in India, are sometimes white in other countries.

1. **A. Mexicana** (L. Sp. Pl. 727); foliis sessilibus semiamplexicaulibus sinuato-pinnatifidis albo variegatis, capsula setosa.—*DC. Syst.* ii. 85, *Prod.* i. 120; *Torr. et Gray, Fl. N. Am.* i. 61; *Wight et Arn. Prod.* i. 18; *Roxb. Fl. Ind.* ii. 571; *Wight, Ill. t.* 11; *Wall. Cat.* 8126!

HAB. Per totam Indiam calidam in ruderatis vulgatissima, sed certe introducta.—(Fl. Feb. Mar.) (*v. v.*)

Herba suffruticosa, 2–4-pedalis. *Caules* divaricatim ramosi, teretes, fistulosi v. intus spongiosi. *Folia* 3–7 unc. longa, secus nervos primarios albo-variegata. *Sepala* apice cornuta. *Flores* 1–3 unc. diametro, aurei. *Capsula* oblonga v. elliptico-oblonga, ¾–1½ poll. longa, teres, setosa v. rarius inermis. *Semina* brunnea, turgida, multicostata, profunde cancellata.

3. MECONOPSIS, Vig.

Sepala 2. *Petala* 4 (rarius plura). *Stamina* perplurima. *Stylus* distinctus, sæpius tortus, *stigmatibus* 4–8 radiantibus cum placentis alternantibus coronata. *Capsula* obovata v. elliptica, interdum linearis, cylindracea, apice valvis brevibus dehiscens; *placentis* plus minus versus axin capsulæ productis. *Semina* raphe tumida cristata.—Herbæ *perennantes, simplices v. rarius ramosæ, interdum acaules et scapigeræ, succo flavo,* foliis *radicalibus v. radicalibus et caulinis integris lobatis pinnatifidisve,* alabastris *nutantibus,* floribus *amplis,* capsulis *erectis.*

A small genus, the Himalayan species of which are all confined to the upper temperate zone, some ascending almost to the limits of phænogamic vegetation. The sepals, which are described as valvate by Endlicher, are decidedly imbricated in the Himalayan species. *Stylophorum* hardly appears to be different generically, except in the valves of the capsule being dehiscent to the base; for the style is present in all the species of *Meconopsis,* and varies extremely in length, and in amount of torsion. The only other character attributed to *Stylophorum* is the crested seeds, but this also is a variable character, the raphe in all being accompanied with a more or less thickened testa, which sometimes expands into a crest. In both genera the placentæ are described as filiform and not projecting far into the cavity of the ovary; but in *M. Nepalensis* and others these almost meet in the axis, forming spurious dissepiments. The hairs or setæ of the stem are simple, or branched and scabrid. The stamens are described as extrorse by Endlicher in *Stylophorum*, and lateral in *Meconopsis*. Gray says that those of the former genus are slightly extrorse; in all our species they are as nearly lateral as possible. The capsule is three-valved to the base in the American *S. diphyllum*, Nutt., according to Gray, incompletely so according to Endlicher, whereas in all the Himalayan species the valves are free only at the upper part of the capsule (as in *Papaver*), and are quite confluent below.

The roots of some of the Himalayan species are said to be virulent poisons.

§ 1. *Scapigeri.*

1. **M. simplicifolia** (H.f. et T.); patentim hispido-pilosa, setis scapi decurvis, foliis omnibus radicalibus lanceolatis, scapo subsolitario 1-floro, capsula lineari-clavata.—Papaver simplicifolium, *Don, Prodr.* 196; *Wall. Cat.* 8125!

HAB. In Himalaya alpina centrali et orientali : Nepal ad Gossainthan, *Wallich!* Sikkim, alt. 12–14,000 ped.!—(Fl. Mai. Jun.) (*v. v.*)

Radix crassa, fusiformis, collo pilis fulvis dense barbato. *Folia* 2–6 unc. longa, in petiolum laminæ æquilongum angustata, acuta v. obtusa, integerrima v. rarissime utrinque 1–2-dentata, pilosa v. glabrata. *Scapi* 1–3, robusti, pedales, fructiferi 2-pedales, pilis patentibus decurvisque hispidi, juniores hispido-tomentosi. *Flos* cernuus, 2–3 unc. diam., pulcherrimus, purpureo-cæruleus. *Sepala* hispidissima, pilis patulis. *Petala* late obovato-rotundata v. cuneata. *Stamina* filamentis lineari-subulatis; antheris lineari-oblongis. *Ovarium* cylindricum; stylo crasso; stigmate subgloboso, 5–7-lobo; lineis stigmaticis crassis, papilloso-tomentosis. *Capsula* 1–2-pollicaris, lineari-clavata, patentim hispido-pilosa v. glabrata, stylo ¼-pollicari; placentis 5–8. *Semina* testa cellulosa, profunde cancellata.

One of the most beautiful and conspicuous plants in the alpine regions of Sikkim. Don describes the anthers as spirally twisted, and the capsules as oblong, which is hardly the case in our specimens.

2. **M. horridula** (H.f. et T.); foliis lanceolatis scapisque setis validis elongatis aculeatis, scapis plurimis unifloris, capsulis obovato-clavatis.

HAB. In Himalaya orientali alpina : Sikkim, locis petrosis, alt. 14,000–17,000 ped.—(Fl. Jun. Jul.) (*v. v.*)

Spithamæa, ubique setis rigidis patentibus pungentibus ¼–¾ unc. longis horrida. *Folia* 3–5-pollicaria, lanceolata, obtusa v. acuta, integerrima v. sinuato-dentata. *Scapi* 8–12, rigidi, interdum basi coaliti, 4–8 unc. longi, virides v. glauci. *Flores* purpurei v. cæruleo-purpurascentes, 1½ poll. lati. *Sepala* setis aculeata. *Petala* 4, late obovata, floribus monstrosis plurima linearia. *Stamina* perplurima, antheris subtortis. *Stylus* crassus. *Capsula* ½–1 poll. longa, setis patentibus ascendentibusve aculeata, lineari-obovata v. elliptico-oblonga, rarius late ovata, stylo ¼ unc. longo, stigmateque conico terminata. *Semina* ⅓ minora quam in præcedente, curva; testa submembranacea, reticulata, cancellata.

A very remarkable and distinct-looking little species, the smallest and most alpine of the genus, so aculeate that it cannot be conveniently gathered with the naked hand. It differs in size, the great aculei on all its parts, the number of scapes, form of petals and capsules, and seeds, from *M. simplicifolia*. On the other hand, though so very dissimilar from *M. aculeata* in size and habit, we should not be surprised at its proving a variety of that plant.

§ 2. *Caules foliosi ; flores racemosi paniculative.*

3. **M. aculeata** (Royle, Ill. 67. t. 15); sparse hispido-aculeata, foliis radicalibus lineari-oblongis lanceolatisve remote irregulariter pinnatifido-lobatis lobis varie lobulatis, floribus racemosis purpureo-cæruleis, capsulis brevibus setoso-echinatis.—*Wall. Cat.* 8122!

HAB. In Himalaya occidentali subalpina et alpina: Kumaon, *Wallich!* alt. 11,000 ped., *Str. et Wint.!* Sirmur, *Royle!* Kunawar, *Munro!* Zanskar et Kishtwar, 10–14,000 ped.! Kashmir, *Winterbottom!*—(Fl. Jun.) (*v. v.*)

Herba pedalis et ultra, subglauca, caule folioso. *Folia* radicalia 4–8 unc. longa, 1–1½ lata, varie pinnatifida v. lobata, lobis latis obtusis acutisve, utrinque una cum caulibus et pedunculis aculeis rigidis sparsis horrida, rarius glabrata ; caulina angustiora, decurrentia. *Flores* gracile pedicellati, 1½–8 unc. lati, pulchre cærulei v. purpurei (non rubri ut in ic. Roylei). *Sepala* glaberrima, aculeata. *Petala* late obovata v. obcuneato-rotundata. *Antheræ* breviter oblongæ. *Capsula* late obconica, oblonga v. obovata, rarius clavata, cum stylo crasso 1 unc. longa, 5–7-valvis ; *stigmate* breviter conico.

We have described the flowers of this plant as blue-purple, on the testimony of various collectors, who have never seen the colour to be as represented in the plate quoted.

4. **M. robusta** (H.f. et T.); elata, glaucescens, paniculatim ramosissima, setis paucis sparsis scaberulis flexuosis mollibus longe patentibus sparse crinita, rarius glabrata, foliis pinnatifido-lobatis, pedunculis apice pubescentibus, sepalis setosis, capsula lineari-oblonga 7–8-valvi crassa patentim setosa demum glabrata stylo crasso conico terminata. —*Wall. Cat.* 8124!

HAB. In Himalaya centrali temperata: Kumaon, *Wall.!* alt. 8000 ped., *Str. et Wint.!*—(Fl. æstate.) (*v. s.*)

Herba 4–6-pedalis, pilis laxis longe patentibus crinita ; *caule* crassitie pollicis, setis flexuosis ¼–⅜ unc. longis. *Folia* caulina lineari-oblonga, 4–6 unc. longa, sinuatov. pinnatifido-lobata. *Rami* fructiferi glabrati, 6 unc. ad pedales, floriferi ad apices pubescentes v. setosi. *Sepala* ¾ unc. longa, setosa. *Flores* 2 unc. diam. *Capsula* immatura setis plurimis elongatis patentissimis laxe vestita, stylo tenui æquilongo terminata ; matura glabrata, elliptico-oblonga, una cum stylo ½ unc. longo valido basi conico 1½–1¾ unc. longa, ⅓–⅔ unc. lata, 7–8-valvis ; *stigmate* capitato, costis placentiferis crassis. *Semina* testa brunnea, celluloso-cancellata.

All the specimens of this plant in Wallich's and Strachey and Winterbottom's Herbaria are indifferent. It appears to be a very large species, allied to *M. Nepalensis* and *M. Wallichii*. From *M. aculeata* it differs in size, in the branched stem, the scattered, very long, soft, deciduous bristles, and in the much larger capsules ; from *M. Nepalensis* in being more glabrous, and the hairs being very much longer, spreading, and flexuose. Wallich's 8124 is in fruit and glabrous ; 8126 is very crinite in some parts. Strachey and Winterbottom's specimens are quite intermediate in character ; the colour of the flowers is unknown. De Candolle's description of *M. Nepalensis* does not materially differ from this plant.

5. **M. Nipalensis** (DC. Prod. i. 121); elata, robusta, tota setis patentibus crinita pubeque stellata sicco aurea obtecta, foliis caulinis sessilibus linearibus lineari-oblanceolatisve sinuato-lobatis, floribus aureis racemosis, pedicellis elongatis patentibus, capsula 8–10-valvi setis appressis pubeque stellata dense obsita.—Papaver paniculatum, *Don,* Prod. 197; *Wall. Cat.* 8123 *A.*!

HAB. In sylvis Himalayæ centralis et orientalis temperatæ: Nipal ad Gossainthan, *Wallich!* Sikkim, alt. 10–11,000 ped.!—(Fl. Mai. Jun.) (*v. v.*)

Caulis simplex v. parce ramosus, 3–5-pedalis, basi fere 2 unc. diam. *Folia* radicalia petiolata, ½–1½-pedalia, lineari-lanceolata v. oblonga, sinuato-pinnatifida. *Racemi* laxiflori, erecti, 1–2-pedales, conspicui; *pedicelli* distantes, inferiores interdum biflori. *Flos* 2½–3½ unc. diametro, aureus. *Sepala* pollicaria, pube stellata et setis brevibus omnino obtecta. *Ovarium* late oblongum, setis erectis flavis densissime obtectum, stylo 2 lin. longo terminatum. *Capsula* pedicello elongato erecto suffulta, una cum stylo 1¾–2 unc. longa, setis patentibus, obovato-oblonga v. subclavata, inclinata, interdum elliptico-oblonga, 8–10-valvis, stylo ½ unc. longo. *Stigma* globoso-capitatum, 8–10-lobum. *Semina* testa cancellata, cellulosa.

This is one of the handsomest plants in Sikkim, resembling a young Hollyhock in its size and general appearance. Of Wallich's specimens under this number, the 8123 B is much branched, and appears to us to belong to *M. Wallichii:* 8123 A is more robust, exactly resembling the Sikkim individuals, which are always simple, with racemose flowers. All these specimens are however very bad, and we are rather at a loss to know which was intended by Don as his *P. paniculatum;* the colour of the flower and shape of the fruit in Don's character applying only to *M. Nepalensis,* whilst his name of *paniculatum* would refer either to *M. robusta* or to *M. Wallichii.*

2. **M. Wallichii** (Hook. Bot. Mag. t. 4668); tota setis mollibus scaberulis pubeque substellatim ramosa vestita, caule gracili erecto paniculatim ramoso, foliis oblongo- v. obovato-lanceolatis pinnatifido-lobatis subtus glaucis, floribus breve pedicellatis paniculatis purpureis, capsulis dense setosis 5-valvibus.—*Wall. Cat.* 8123 *B.*

HAB. In sylvis Himalayæ temperatæ centralis et orientalis ad Nipal, *Wallich!;* Sikkim, alt. 9–10,000 ped.!—(Fl. June.) (*v. v.*)

Herba 6-pedalis, valde ramosa; *caulis* crassitie pollicis. *Folia* plerumque profunde pinnatifida, lobis brevibus v. elongatis, integris lobativse, obtusis. *Flores* secus ramos paniculæ perplurimi, in pedunculos graciles breve pedicellati, nutantes, 1½ unc. diam. *Sepala* dense pubescentia, non setosa. *Petala* late obovata. *Antheræ* oblongæ. *Capsula* elliptico-oblonga, subcylindrica, cum stylo semipollicari gracili 1½ poll. longa. *Semina* ut in *M. Nepalensi.*

A very beautiful plant, conspicuous for its height, much branched stem, and very numerous, pendulous, beautiful pale blue-purple blossoms. It is closely allied to the *M. Nipalensis,* but differs in the want of setæ on the sepals, in the smaller blue-purple flowers, in the more cylindrical capsule with only five valves, slender style, branched stem, and many-flowered peduncles, producing a paniculate inflorescence. The root is reputed to be very poisonous by the natives of Sikkim. Wallich's 8123 B appears to be referable to this species.

4. **CATHCARTIA**, H.f.

Sepala 2, imbricata. *Petala* 4. *Stamina* indefinita. *Ovarium* sessile, cylindraceum, 4–6-sulcatum; placentis crassiusculis. *Stigma* hemisphæricum, amplum, sessile, 4–6-lobum, radiis lamellæformibus placentis oppositis. *Capsula* erecta, stricta, teres, ab apice ad basin complete 5–6-valvis, valvis linearibus. *Semina* scrobiculata, strophiolata, cristata.—Herba *pilis mollibus patentibus fulvis villosa, succo flavo, caule terete simplici v. diviso, foliis radicalibus longe petiolatis cordatis rotundatis lobatis, caulinis superioribus sessilibus, pedunculis terminalibus axillaribusque, floribus amplis cernuis aureis.*

1. **C. villosa** (H.f. in Bot. Mag. t. 4596).
HAB. In Himalaya orientali temperata, alt. 10–12,000 ped.!—(Fl. Jul.) (*v. v.*)

Herba spithamæa vel pedalis, perennis. *Folia* radicalia plurima, rotundata, 3–5-loba, lobis crenato-lobulatis, basi profunde cordata, 1–3 poll. lata, petiolo 3–5-pollicari; *caulina* media brevius pedicellata, suprema oblongo-pinnatifido-lobulata. *Flores* in racemum laxum pauciflorum caulem terminantem disposita, 2–3 poll. lata, pedicellis curvis, alabastris cernuis. *Petala* rhombeo- v. obovato-rotundata. *Stamina* aurantiaca. *Capsula* 2–3 unc. longa, cylindrica, gracilis, erecta, valvis membranaceis.

This beautiful plant was named in honour of the late J. F. W. Cathcart, judge in the Bengal Civil Service, who devoted several years to forming, by means of native artists, a most important collection of illustrations of Sikkim plants, which are now deposited in the Museum of the Royal Gardens of Kew.

5. DICRANOSTIGMA, H.f. et T.

Sepala 2, imbricata. *Petala* 4. *Stamina* indefinita. *Ovarium* stipitatum, lagenæforme; *stylo* brevi; *stigmate* furcato, cruribus erectis placentis 2 oppositis.—Herba *perennans, glauca, sparse subglanduloso-pilosa*, foliis *radicalibus perplurimis sinuato- v. lobato-pinnatifidis*, scapis *caulibusve gracilibus supra medium paucifoliatis v. bracteolatis 2–3-floris, floribus* aureis.

This remarkably distinct genus has only been found by Strachey and Winterbottom. It differs from *Chelidonium* (its nearest ally) in the shape of the ovary and form of stigma, which presents two erect arms, alternating with the placentæ, each being simple, and consisting of the confluent arms of contiguous stigmata. The habit is very peculiar, and much resembles a lactucoid plant, and the *Stylophorum diphyllum* of North America, which has similar small weak soft hairs.

1. **D. lactucoides** (H.f. et T.)—Meconopsis, *Herb. Str. et Wint.* 3!

HAB. In Himalaya temperata ad Rogila in Garhwal, alt. 11,000 ped., *Str. et Wint.!*— (Fl. æstate.) (*v. s.*)

Herba spithamæa et ultra, tota plus minus pilis laxis compressis sub lente articulatis subpaleaceis conspersa. *Folia* omnia radicalia, cum petiolo dilatato 4 unc. longa, lineari-oblonga, ¾–1 unc. lata, lobis pinnisve sub-5-jugis late ovatis, grosse et irregulariter dentatis, dentibus acutis, subtus glaucis, superne albo-variegatis. *Scapi* 3–4, foliis duplo longiores, graciles, ascendentes, longe nudi, supra medium foliis 1–2-uncialibus sessilibus pinnatifidis oppositis alternisve bracteati. *Alabastri* ovoidei, abrupte acuminati, erecti? *Flores* gracile pedicellati. *Sepala* patentim laxe pilosa, ½ unc. longa, marginibus imbricatis late membranaceis, apice in acumen v. cornu producta. *Petala* 1-pollicaria, late obovata. *Antheræ* lineari-oblongæ. *Ovarium* breviter stipitatum, ¼ unc. longum, pilis mollibus hispidulo-pubescens; *stylo* 1 lin. longo; *stigmata* incrassata, cruribus late subulatis intus marginibusque puberulis.

We have fewer specimens of this plant than are desirable for drawing a complete specific character; the genus is, however, a most distinct one, and cannot be confounded with any other.

6. GLAUCIUM, Tourn.

Sepala 2. *Petala* 4. *Stamina* indefinita. *Ovarium* lineare; *stig-*

mate bilobo v. bilamellato sessili. *Capsula* elongata, complete 2-valvis, 2-locularis, placentis nempe dissepimento spongioso conjunctis, stigmate sæpe stipitato. *Semina* foveolis dissepimenti subimmersa, scrobiculata. —Herbæ *biennes v. perennes, pleræque austro-Europææ et orientales, glaucæ, succo croceo,* foliis *radicalibus petiolatis,* caulinis *amplexicaulibus incisis lobatisve,* pedunculis *axillaribus terminalibusque solitariis unifloris,* floribus *flavis v. phœniceis.*

No species of this genus has hitherto been found in India proper; and of the two Afghanistan ones here described, one is Persian, and the other appears to be identical with the British species, which has a very wide range in western and southern Europe, and in western Asia.

1. **G. elegans** (Fisch. et Meyer, Ind. Sem. Hort. Petrop. 1835, 29); divaricatim ramosa, foliis latissime amplexicaulibus cordatis ovato-oblongis sinuato-lobatis, capsulis gracilibus torulosis strictis curvis v. subcircinatis.—*Led. Fl. Ross.* i. 93. G. contortuplicatum, *Boiss. Ann. Sc. Nat.* xvi. 376.

HAB. Afghanistan, *Griffith!*—(Fl. æstate.) (*v. s.*)
DISTRIB. Persia borealis !

Caulis gracilis, erectus v. decumbens, pluries dichotome ramosus, ramis divaricatis glabris setulosisve. *Folia* caulina coriacea, 1–2 unc. longa. *Siliquæ* perplurimæ, 2–3 unc. longæ, 1 lin. latæ, stigmate late trigono terminatæ, teretes, patentim setulosæ, curvatæ, tortuosæ v. strictæ et erectæ. *Semina* parva, clathratim cancellata.

Our plant perfectly agrees in all essential characters, and the often tortuose pod, with Aucher-Eloy's specimens of *G. elegans* (4042), which latter, however, present a denser ramification, longer and more arcuate pedicels, more glabrous pods, and more turgid seeds than our plant does,—characters which do not appear to us to be of much value in other species of the genus. The *G. pumilum,* Boiss., of Persia (Kotschy), appears to us, judging from our small specimen of it, to be a state of the same plant.

2. **G. corniculatum** (Linn. Sp. Pl. 724) ; caule gracili ramoso setoso v. glabrato, foliis caulinis latissime ovato-oblongis cordatis amplexicaulibus, capsulis rectis curvisve setosis.—*DC. Syst.* ii. 96 ; *Led. Fl. Ross.* i. 92.

HAB. Afghanistan, *Griffith !*—(Fl. æstate.) (*v. s.*)
DISTRIB. Europ. bor. occ.! regio Mediterranea ! ins. Canariens.! Persia !

Exemplar mancum. *Folia* parva, coriacea, latè cordata. *Pedicelli* florum graciles, fructiferi validi elongati. *Siliquæ* 7 unc. longæ, valvæ setis aculeisve sparsis erectis ornatæ. *Semina* oblonga, profunde cancellata.

Our specimens are very variable, and do not appear to be specifically distinct from the common European one to which we have referred it, and which in our opinion includes a good many modern species of the genus.

7. **RŒMERIA,** DC.

Sepala 2. *Petala* 4. *Stamina* indefinita. *Ovarium* lineare; stigmate 2–4-lobo sessili. *Capsula* elongata, complete 2–4-valvis, 1-locularis, valvis ab apice deorsum dehiscentibus, placentis foveolatis liberis. *Semina* reniformia, scrobiculata.—Herbæ *annuæ, succo flavo,* foliis *petiolatis pinnatipartitis, lobis multifidis,* pedunculis *solitariis oppositifoliis,* floribus *violaceis.*

The distribution of this genus very nearly coincides with that of *Glaucium,* and the only species included within our flora is not found east of the Indus.

1. **R. hybrida** (DC. Syst. ii. 92).—*Led. Fl. Ross.* i. 92. R. refracta, *DC. Syst.* ii. 93. R. pinnatifida, *Boivin. in Belang. Voy. Ic. ined.* R. rhœadiflora, *Boiss. Diagn.* vi. 7. R. orientalis, *Boiss. Ann. Sc. Nat. ser.* ii. xvi. 374.

α. sepalis hirsutis setosisve, siliquis pedunculis longioribus hispidosetosis, valvis obtusis.

β. sepalis glaberrimis, siliquis brevibus pollicaribus, pedunculis brevioribus glaberrimis, valvis acutis v. in aculeum productis.

γ. sepalis glaberrimis, siliquis ¾-pollicaribus, pedunculis brevioribus hispido-setosis, valvis acuminatis.

HAB. Afghanistan, alt. 10–12,000 ped., *Griffith!* et Beluchistan, *Stocks!*—(Fl. vere.) (*v. s.*)

DISTRIB. Europa australis! regio Mediterranea! Ægyptus! Asia occidentalis!

Herba spithamæa et ultra, ramosa, foliosa, glabra pilosa v. hispido-pilosa, gracilis v. robusta, statura et habitu valde polymorpha. *Folia* 2–4 unc. longa, bipinnatisecta, lobis segmentisve linearibus late oblongisve integerrimis v. dentatis obtusis acutis subaristatisve. *Pedicelli* breves v. valde elongati, robusti v. graciles. *Sepala* obtusa, glabra, pubescentia v. setosa. *Flos* diam. variabilis, phœniceus violaceus v. ruber, alabastra oblonga v. fere globosa. *Siliqua* longe v. brevius pedicellata, erecta v. (pedunculo curvo) refracta, 1½–3 unc. longa, 3–4-valvis; valvæ obtusæ, acutæ v. in cornua ultra stigmata productæ, angustatæ, glaberrimæ v. setosæ.

After a very careful comparison of our Indian specimens with Boissier's *R. orientalis* and *rhœadiflora,* and De Candolle's *refracta,* we are unable to find any character by which these species are to be distinguished even as constantly marked varieties. Stocks, in his notes on the Beluchistan species, points out the invalidity of the characters of *R. refracta,* which are taken from the curved peduncle and breadth of the lobes of the leaves. The number and size of the valves of the pods, their sharp or blunt apices, and more or less hairy or setose valves, are characters that vary with every European and oriental specimen that has fallen under our observation.

XV. FUMARIACEÆ.

Sepala 2, lateralia, decidua. *Petala* 4, cruciata, libera v. varie connata, irregularia, postico sæpissime calcarato. *Stamina* sæpissime 6, diadelpha, rarius 4 et libera, phalanges petalis antico et postico oppositæ, antheræ 6, laterales cujusve phalangis uniloculares. *Ovarium* uniloculare, ovulis 1 v. pluribus horizontalibus amphitropis. *Stylus* filiformis. *Stigma* simplex v. lobatum. *Fructus* indehiscens v. capsularis, 1-polyspermus. *Semina* testa sæpissime nitida; arillo parvo lacero v. lobato, rarius 0. *Albumen* carnosum. *Embryo* minutus, plerumque excentricus, rectus v. curvus.—Herbæ *erectæ decumbentes v. scandentes, succo aqueo scatentes,* foliis *alternis rarius oppositis pinnatisectis.*

We have few remarks to offer upon this well known family, which, though it at-

tains its maximum in point of development of species in the Himalaya, is far from
rich in generic forms in that country. Its affinities we conceive to be undoubtedly
with *Papaveraceæ*, under which it is included by Endlicher as a suborder. From
that Order, however, the majority differ remarkably in their irregular perianth, definite diadelphous stamens, two of which in each bundle have one-celled anthers, and
in their arillate seeds. The curious genus *Hypecoum* combines both Orders, having
unequal petals, but together forming a nearly regular corolla, and free stamens.
Some remarks on the affinities of the Order, as indicated by the structure of *Hypecoum*, will be found under that genus.

1. FUMARIA, L.

Sepala 2. *Petala* 4, ringentia, anticum carinatum, posticum obtuse
calcaratum, cum 2 lateralibus interioribus inferne coalitum. *Stamina*
6, diadelpha. *Ovulum* 1, parietale. *Stylus* deciduus. *Stigma* biparti-
tum. *Fructus* carnosus, demum siccus, subglobosus. *Semen* reniforme,
opacum, umbilico nudo.—Herbæ, foliis *multifidis*, floribus *racemosis*.

We agree with Bentham in considering that most of the numerous European forms
of *Fumaria*, including *F. parviflora*, Lam., may be reduced to one variable plant,
F. officinalis, L., which, with larger or smaller flowers, variously cut leaves, an erect
or decumbent habit, large or small, more or less cut sepals, and very many forms of
fruit, frequents waste places throughout Europe and a great part of temperate Asia.
The only Indian state of the plant abounds in waste places, corn-fields, etc., and
differs in no respect from the form that bears the same name in Europe.

1. F. parviflora (Lam. Dict. ii. 567).

Var. *Vaillantii ;* foliorum laciniis linearibus planis, bracteis pedicellum fructiferum fere æquantibus, sepalis parvis petalis multoties
angustioribus, fructu globoso lævi.—F. Vaillantii, *Loisel. Not.* 102 ;
DC. Syst. ii. 137. F. parviflora, *Wight et Arn. Prod.* 18 ; *Wight, Ill.
Gen. t.* 11 ; *Roxb. Fl. Ind.* iii. 217 ; *Wall. Cat.* 1436 ! ; *Led. Fl. Ross.*
i. 105.

HAB. In India extratropica in planitie et montibus subtropicis vulgaris (in Sikkim non occurrit). In peninsulæ montibus temperatis :
Nilghiri, *Wight !* et in montibus Afghanistan, *Griffith !*—(Fl. hieme et
vere.) (*v. v.*)

DISTRIB. Europa et Asia temperata et calidior.

Caulis diffuse ramosus, spithamæus v. 2-pedalis. *Folia* multifida.

2. CORYDALIS, DC.

Sepala 2, decidua, plerumque squamulæformia. *Petala* 4, anticum
planum v. concavum, posticum basi gibbum v. calcaratum, 2 lateralia
interiora antico subconformia. *Stamina* 6, diadelpha, synemate postico extus basi processu calcariformi aucto. *Ovula* juxta placentas intervalvulares plurima. *Stigma* bilobum. *Capsula* siliquosa, bivalvis,
valvis a replo persistente placentifero solutis. *Semina* lenticularia,
rostellata, arillo carnosulo v. 0. *Embryo* linearis, brevis.—Herbæ *erectæ*,
foliis *caulinis interdum oppositis*, floribus *racemosis*.

Corydalis is one of the few genera containing many species which we have hitherto described, in which the majority of the species are upon the whole remarkably well marked and distinct from one another; amongst the Himalayan ones, at any rate, there is none of that interlacement of forms that has rendered the disentanglement of the species of *Ranunculaceæ* and *Berberideæ* so laborious and unsatisfactory.

We have not adopted the sectional groups proposed by De Candolle, as they do not seem to be altogether natural, and some of the best characters by which they are limited (those of the root, for instance) are practically unavailable. Many of the species have tuberous roots, but in a considerable number these are so deeply buried in the earth or lodged in crevices of rocks, that it is impossible to prove their existence in the living plant. A knowledge of the roots of the species is a great desideratum, which we often in vain attempted to supply, and the more to be regretted because the characters they afford are eminently natural. With regard to the character taken from the length of the spur of the posticous bundle of filaments, that seems to depend mainly upon the length of the spur of the posticous petal itself; and where it does not, a strict adhesion to its proportional length would sunder very closely allied species. The persistence of the style is a very inconstant character, and that drawn from the lobing of the stigmata is not available in dried specimens, and of doubtful value. The arillus varies extremely in form and relative size during different stages of the growth of the seed, and is not quite constant in each species. A much more important character is drawn from the development of the young plant; the seed in the section *Bulbocapnos* being described as germinating by a single cotyledon, whose radicle forms a perennial tuber, which sends up a primordial leaf in the following year, and a flowering stem in subsequent ones: the other sections, again, have opposite cotyledonary leaves. It is evident, however, that it must be many years before observations on this point can be verified on even a few species of the genus, and until done for the majority, the value of the characters they afford must be quite problematical. Lastly, the sections *Capnoides* and *Capnites* are hardly distinguishable by any character, and we find species placed in each that should certainly stand very close together. Under these circumstances we have not hesitated to take definite characters drawn from the pod for the primary divisions, and others from the perianth, etc., for those of secondary value. These, however, are in a great measure arbitrary, and are proposed as provisional only.

The maximum of the genus *Corydalis* is certainly to be sought in the Himalaya, where the species of the western mountains differ so much from those of the eastern, that there are no doubt others to be discovered, especially in Bhotan, Abor, and Mishmi. In the mountains of western China also they probably abound, and there are a considerable number of known but undescribed species even in the eastern and drier parts of that empire. With the exception of one species, and that a common Himalayan and Siberian one, found in the Khasia, the genus finds its southern limit in the Himalaya.

Of the 24 species we have described, 9 are new, a much larger proportion than in any other genus hitherto described in this work. In this respect *Corydalis* is rivalled by very few, except *Rhododendron, Impatiens,* and *Astragalus.* We have also added 2 Siberian and 1 European species not hitherto supposed to be Himalayan. Of the Himalayan species 12, or one-half, are found to the eastward of the valley of Nipal, and 7, of which 6 are new, are confined to the eastern Himalaya. On the other hand, 16 are found to the westward of the valley of Nipal, of which 10 are confined to the western ranges, and only 3 are new. If, however, we exclude the more strictly Tibetan species of the western regions, some or most of which probably occur in eastern Tibet also, we have 10 western forms, of which only 4 are not found east of the valley of Nipal. Hence we may infer that the damp regions of the eastern Himalaya are the most favourable to the development of species of this beautiful genus.

Sect. 1.—*Siliqua* longe lineari-elongata. *Semina* 1-seriata.—Herbæ elatæ ramosæ foliosæ, radice *fibrosa.*

1. **C. ophiocarpa** (H.f. et T.); gracilis, ramosa, foliis bipinnati-

sectis subtus glaucis, racemis oppositifoliis fructiferis elongatis, petalo postico apice spathulato-obcordato planiusculo sub-bifido calcar latum obtusum æquante v. superante, antico lineari concavo, lateralibus liberis, siliquis tortuosis.

HAB. In vallibus humidis Himalayæ temperatæ : Sikkim, alt. 9000 ped.!—(Fl. Jun.) (*v. v.*)

Caulis 2-3-pedalis, debilis, laxe ramosus. *Folia* 4 unc. vel spithamæa, lineari-oblonga; *pinnis* laxis, alternis, superioribus pinnatifidis; *pinnulis* obovato-oblongis apiculatis grosse obtuse lobatis, pinnatifidisve, *petiolo* basi vix dilatato. *Racemi* secundi, foliis æquilongi, fructiferi graciles, pedales, multiflori; *bracteæ* subulatæ. *Sepala* minima, squamulæformia, orbicularia, fimbriato-lacera. *Petalum* posticum curvum, apice dilatatum, emarginato-bifidum, lateribus recurvis; anticum angustum, concavum, apice rotundatum emarginatum; petala lateralia apice crassa, oblique rotundata, infra apicem coadunata; synematis appendice libera, curva, calcaris $\frac{1}{4}$ æquante. *Ovarium* lineare, stylo gracili æquilongo. *Ovula* plurima, 2-seriata. *Siliquæ* admodum singulares, gracile pedicellatæ, $\frac{3}{4}$-1 poll. longæ, 1 lin. latæ, flexuosæ et tortæ, stylo gracili elongato terminatæ; *valvæ* membranaceæ, torulosæ. *Semina* parva, atra, nitida, reniformi-globosa, subcompressa; testa punctata crassiuscula; funiculo magno elongato crasso apice acuminato.

A most remarkable species, easily recognized by the tortuous pod and the curious funicle of the seed. The broad posticous petal allies it to *C. diphylla* and others, and the habit to *C. chærophylla.* The spur varies a good deal in length, and the appendix is quite free. The lateral petals are united at a point some way below their apices, producing a gibbosity which no other Himalayan species known to us possesses.

2. **C. flaccida** (H.f. et T.); elata, ramosa, foliis subtriternatim pinnatisectis, pinnulis oblongis ovatis v. rotundatis, racemis brevibus, sepalis late ovatis acutis erosis, petalo postico apice spathulato emarginato-bilobo calcar curvum superante, antico apice rotundato.

HAB. In sylvis humidis Himalayæ temperatæ orientalis : Sikkim, alt. 11-12,000 ped.!—(Fl. Jun. Jul.) (*v. v.*)

Herba robusta, 2-3-pedalis, foliosa, ramosa; *caule* crassiusculo. *Folia* 6-poll. vel pedalia, circumscriptione ovato-oblonga v. subdeltoidea, valde membranacea, vix glauca; *foliola* $\frac{1}{4}$-$\frac{3}{4}$ unc. lata, petiolulata, ultima sessilia, interdum cordata, varie inciso-lobata v. subpinnatifida; segmentis obtusis, apiculatis; *caulinorum petioli* basi vix dilatati, radicalium subvaginantes. *Racemi* terminales et axillares, 4-6-pollicares, basi foliolis sessilibus decompositis bracteati. *Flores* subconferti, breve et gracile pedicellati, $\frac{1}{2}$-$\frac{3}{4}$ poll. longi, pallide fusco-purpurei; *bracteæ* lineares, apice dilatati, inferiores laciniati v. foliacei. *Sepala* pro genere ampla. *Petala* exteriora fere recta, planiuscula, consimilia, dorso apices versus anguste carinata, appendice crassa libera calcar $\frac{2}{3}$ æquantes, interiora infra medium superiori adnata. *Ovarium* elongatum, stylum gracilem superans; ovulis plurimis. *Siliqua* immatura anguste linearis, recta, stylum elongatum superans, matura (e replis persistentibus tantum visis) $1\frac{1}{2}$ poll. longa.

This very handsome species agrees in the structure of the flower and habit with *C. ophiocarpa*, but differs in the more compound not glaucous leaves, smaller, broader, more laciniate pinnules, axillary or terminal short racemes, which have not secund flowers, larger, very different bracts, purplish flowers, large sepals, keeled upper and lower petals, the lateral petals being adnate to the upper, and in the form of the ovary and pods. Our specimens are in flower, but we have a few with very old pods, the valves and seeds of which have fallen away: from these remains, however, we confidently place it in the present section.

3. **C. leptocarpa** (H.f. et T.); caule breviusculo debili vage ramoso,

foliis radicalibus caulinisque gracile petiolatis laxe biternatim pinnatisectis, pinnis longe petiolulatis, pinnulis late obovatis oblongisve lobatis,
segmentis latis, racemis oppositifoliis paucifloris bracteatis, sepalis parvis, petalo postico longe calcarato, antico apice subspathulato acuminato, siliquis elongato-linearibus rectis torulosis.

HAB. In sylvis Himalayæ temperatæ orientalis interioris: Bhotan,
Griffith! Sikkim, alt. 8000 ped.!—(Fl. Jul. Aug.) (*v. v.*)

Herba diffusa, rhizomate sæpius elongato, prostrato, fibras crassas emittente. *Rami*
prostrati, debiles, spithamæi. *Folia* alterna v. subverticillata, longe et gracile petiolata, 3–6 unc. longa, petiolo basi dilatato, pinnis subalternis longe et gracile petiolulatis,
pinnulis ¼–1 unc. longis membranaceis obovatis varie sectis rarius integerrimis.
Racemi semper oppositifolii, apice 2–5-flori, pedunculo foliis æquilongo. *Bracteæ*
pedicello longiores, obovato- v. anguste elongato-cuneatæ, apice laceræ. *Flores*
pallide sordide purpurei, angusti, 1 poll. longi. *Sepala* parva, oblonga, scariosa,
lacera. *Petalum* posticum concavum, acuminatum, dorso alatum, in calcar flore
longius elongatum curvum angustum productum; appendice gracili, ½ calcaris
æquante. *Ovarium* stylum gracilem superans. *Siliqua* 1–1½-poll., recta v. paullo
curva, valvis torulosis. *Semina* uniseriata, atra, compressa, rotundato-reniformia,
nitida, impunctata, arillo bilobo latiusculo instructa.

In general appearance this species resembles branching specimens of the *C. longiflora,* Bunge, of the Altai; it has, however, a more branched prostrate stem, and has
not a bulbous root, nor the sheathing scales at the base of the petioles; it has also
much larger sepals, winged upper petals, and more slender pods. Griffith's specimens are in an exceedingly unsatisfactory state, and may possibly be made up of
more than one species. Of these his n. 1752 is neither in flower nor fruit, and
n. 1152 has very young flowers and imperfect pods: the flowers are smaller than
in the Sikkim specimens, their spurs shorter, and the wing on the lateral petals
broader.

Sect. 2.—*Siliqua* late elliptico-ovata v. globosa, inflata. *Semina* biseriata. *Radix* fibrosa.

4. **C. crassifolia** (Royle, Ill. 69); crasse coriacea, glauca, caule
simplici, foliis late oblongis reniformibusve 3-sectis pinnatisectisve, segmentis cuneatis, racemo multifloro bracteato, petalis exterioribus calcar
obtusum æquantibus.

a. crassissima; foliis caulinis sessilibus petiolatisve trilobis.—C.
crassissima, *Cambess. in Jacq. Voy. Bot.* 12. *t.* 11; *Thomson in Hook.
Journ. Bot.* 1853, v. *p.* 17 & iv. *t.* 9.

β. *physocarpa;* foliis petiolatis pinnatisectis. — C. physocarpa,
Cambess. in Jacq. Voy. l. c. t. 12.

HAB. In Himalaya occidentali interiori et Tibetica, alt. 14–16,000
ped.: Piti, *Jacquemont!* Kunawar, *Munro!* Kishtwar! Ladak! Zanskar!—(Fl. Jul.) (*v. v.*)

Rhizoma crassum, elongatum, spithamæum et ultra, fibras simplices crassas emittens. *Caulis* simplex, spithamæus vel pedalis, basi nudus, superne foliosus, apice
floriferus. *Folia* radicalia pauca v. nulla; caulina 1–3, interdum solitaria, sessilia,
latissime reniformia, 2–5 poll. lata, varie grosse crenata et lobata, v. 2–3-secta; in
var. β. petiolata, oblonga, pinnatisecta; pinnis 1–3-jugis, oblongis obcuneatisve, varie
obtuse lobatis incisisve. *Racemus* 1–2-pollicaris, bracteatus, simplex v. basi ramosus.
Bracteæ coriaceæ, flores superantes v. iis breviores, lanceolatæ, obovatæ, integerrimæ

v. varie lobatæ. *Pedicelli* floriferi breves; fructiferi elongati. *Flores* albi v. flavidi, purpureo variegati, $\frac{3}{4}$–$1\frac{1}{2}$ poll. longi. *Sepala* squamæformia. *Petala* exteriora apice rotundata, lata, marginibus recurvis; calcare petalum æquante, curvo, apice subhamato, appendice calcar $\frac{2}{3}$ æquante, apice clavata, libera; synematum margines superiores petali postici marginibus utrinque adnati. *Ovarium* latum, stylo brevius; ovulis plurimis. *Capsula* vesicularis, $\frac{3}{4}$ poll. diametr., ovata v. globosa, stylo persistente terminata. *Semina* orbiculari-reniformia, compressa, testa atra.

We have no hesitation in uniting the two species of Cambessèdes with that of Royle, having compared authentic specimens of all, and finding the same varieties amongst our own, together with every intermediate form.

Sect. 3.—*Siliqua* elliptico-ovata obovata v. lanceolata. *Semina* 2-seriata.

a. CAPNITES.—*Radix bulbosa. Caulis v. scapus simplex, basi aphyllus v. foliis paucis radicalibus, supra medium foliosus, foliis oppositis alternis verticillatisve.*

5. **C. rutæfolia** (Sibth. Fl. Græc. t. 667); foliis oppositis verticillatisve 2–3-ternatim sectis, racemo 6–8-floro, bracteis integris, petalis exterioribus apice dilatatis (postico rarius parvo acuto) ecarinatis calcar obtusum æquantibus v. brevioribus.—C. rutæfolia *et* oppositifolia, *DC. Syst.* ii. 114, *Prod.* i. 126. C. diphylla, *Wall. Cat.* 1430! *Tent. Flor. Nep.* 54. C. pauciflora, *Edgeworth, in Linn. Soc. Trans.* xx. 30. C. Ledebouriana, *Kar. et Kiril. En. Plant. Fl. Alt.* 54; *Led. Fl. Ross.* i. 745. C. longipes, *Don, Prod.* 198 (*non DC.*). C. Hamiltoniana, *Don, Syst. Gard.* i. 142.

β. petalo superiore minore concavo acuto.—C. verticillaris, *DC. Syst.* ii. 114, *Prod.* i. 126.

HAB. In Himalaya occidentali temperata, alt. 6–10,000 ped.: a Kumaon! ad Kashmir! et in montibus Afghanistan, *Griffith!*—β. Kashmir, 9500 ped., *Winterbottom!* Marri, 9700 ped., *Fleming!*—(Fl. Apr. Jun.) (*v. v.*)

DISTRIB. Montibus Cretæ! Græciæ! Tauriæ! Asiæ minoris! Syriæ! Persiæ australis! et borealis! et Soongariæ!

Rhizoma elongatum, crassitie pennæ anserinæ. *Caulis* simplex, erectus, 3 poll. vel spithamæus. *Folia* opposita, terna v. rarius verticillata, insigniter varia, membranacea v. subcoriacea, sessilia v. longe petiolata, ternati-biternatisecta, foliolis breve v. longe petiolulatis, oblongis obovatis linearibusve, simplicibus lobatis tripartitisve. *Racemus* solitarius (rarius 2), erectus, interdum 10-florus. *Bracteæ* parvæ v. magnæ, obovato-oblongæ v. lanceolatæ, v. rotundatæ, integerrimæ v. dentatæ. *Flores* læte purpurei, $\frac{1}{2}$–1 poll. longi. *Sepala* parva, squamæformia. *Petala* majora plerumque ampla, apice rotundata v. retusa cum mucrone, rarius bifida, marginibus recurvis, postico in var. β parvo acuto v. subacuto non explanato; calcare curvo apice obtuso deflexo. *Siliqua* longe pedicellata, late elliptico-ovata.

An extremely variable plant, common from the Levant to Kumaon, but not found further east. The appearance of whorled leaves is perhaps due to their being sessile, and what appears as separate leaves being the primary divisions only of these. The flowers are extremely variable in form, size, depth of colour, and the breadth of the dilated apices of the outer petals. Griffith's and Wallich's specimens unite the characters of *verticillaris* and *rutæfolia*. Edgeworth's *C. pauciflora* (altered to *oligantha* in MSS.) is a very luxuriant state, with flaccid leaves, sometimes two racemes, and few flowers; it is certainly, however, the same species; we have it also from Strachey and Winterbottom, and from Kashmir, where it frequents damp woods. Hohenacker's North Persian specimens of *C. verticillaris* have minute glaucous leaflets, and very long spurs to the flowers, whilst those from South Persia have shorter spurs.

The flower of var. β looks very different from that of the ordinary states of *C. ru-tæfolia*, having a narrow upper petal, which is very concave and acute; we find, however, various intermediate states, and the foliage, fruit, and all other parts of the plant being identical, we are unable to make a distinct species of it. A similarly narrow short upper petal occurs in Kotschy's n. 15 from Taurus, in *C. Ledebouriana* (Karel. and Kir. 66) from Tarbagatai, also conspicuously in Cretan specimens (Sieber's *C. uniflora*), and others from Boissier, named *C. rutæfolia*, var., and in Syrian ones from Aucher-Eloy (402); also in Kotschy's *C. verticillaris*, DC., from north and south Persia (107 and 471), which in foliage and appearance approach very near Griffith's Afghanistan specimens, both having exceedingly small flowers. The foliage of β is quite as variable as that of *rutæfolia*, and the variations are entirely similar in each; Winterbottom's specimens having delicate, flaccid, broad, green, obtuse leaflets, and large flowers; Fleming's having very patent, much divided, linear, glaucous, more coriaceous ones.

6. **C. Kashmiriana** (Royle, Ill. 69. t. 16. f. 1); foliis radicalibus ternatis, foliolis trilobis lobis sectis, caule filiformi simplici supra medium 1-3-foliato, foliis 3-multisectis lobis oblongo- v. anguste linearibus, bracteis inferioribus 3-sectis, floribus subumbellatis, petalo postico apice concavo acuminato calcari æquilongo, inferiore dilatato trullæformi rhomboideo v. obscure trilobo.

Hab. In Himalaya temperata et subalpina : Sikkim, alt. 12–14,000 ped.! Kumaon, 10,000 ped., *Str. et Wint.!* Kishtwar, 12,000 ped.! Kashmir, *Royle!* 8500 ped., *Winterbottom!*—(Fl. Jun. Jul.) (*v. v.*)

Species parvula, pulcherrima. *Caulis* gracilis, capillaceus, basi bulbilliferus, supra terram 2-8-pollicaris, flexuosus, apicem versus 2-3-foliatus. *Folia* radicalia 0 v. pauca, longe petiolata, palmatim 3-5-partita, pinnis longe petiolulatis; caulina alterna, rarius opposita, sessilia, rarius petiolata, in segmenta 3-6 linearia profunde secta, ½-1 poll. longa, lobis intermediis elongatis trifidis v. pinnatisectis, omnibus obtusis v. mucronulatis. *Racemus* 3-8-florus. *Bracteæ* inferiores laciniatæ, superiores integerrimæ. *Pedicelli* elongati, graciles. *Sepala* 0, v. squamæformia. *Flores* cælestini, ½-1 poll. longi. *Petalum* posticum fornicatum, acutum v. acuminatum, calcari curvo gracili v. latiusculo æquilongum v. brevius, anticum apice dilatatum, obtusum; appendice versus apicem calcaris producta. *Ovarium* stylo æquilongum, multiovulatum. *Siliquæ* immaturæ pendulæ v. deflexæ, lineari-oblongæ.

A beautiful little plant, the smallest of its genus, easily recognized by the colour of the flowers, which appear in May and June in the north-west Himalaya, but not till July and August in Sikkim. The cauline leaves are variable in number, shape, and length of the petiole, though generally sessile. The flowers also vary in the length of the spur, shortness of the upper petal, which is sometimes obscurely keeled above, and in the breadth of the lower petal, which is generally very broad, membranous, and entire or three-lobed.

7. **C. polygalina** (H.f. et T.); caule gracili simplici superne 1-3-foliato, foliis pinnatisectis segmentis linearibus subcoriaceis acutis nervis parallelis, racemo basi ramoso v. racemis 1-3 5-10-floris, pedicellis brevibus, petalo postico apice fornicato acuto dorso breviter alato, inferiore apice cucullato dorso alato, calcare flore longiore rectiusculo obtuso.

Hab. In Himalaya orientali alpina, alt. 14–16,000 ped.: Sikkim !— (Fl. Aug.) (*v. v.*)

Herba facie et habitu *C. rutæfoliæ*, a qua differt statura majore, caule subrigido, foliis coriaceis pinnatisectis, racemis 2 v. pluribus, floribus brevius pedicellatis, calcare fere recto, petalo postico dorso alato, antico apice angusto cucullato, et siliquis ut videtur latioribus.—*Radix* ignota. *Caulis* spithamæus, longe nudus, flexuosus, rigidus. *Folia* alterna v. subopposita, plerumque sessilia, 1–1½ poll. longa, æquilata, laciniis (pinnulisve) 3–6-jugis, inter se consimilibus, subremotis, ¾ poll. longis, ⅓ poll. latis, acutis, integerrimis v. paucidentatis. *Racemi* divisi v. in caule apice bis terve diviso axillares. *Bracteæ* integræ v. sectæ. *Flores* ½ poll. longi, flavescentes, purpureo maculati. *Sepala* squamæformia.

We have but few specimens of this very distinct-looking species. It has many characters in common with *C. rutæfolia*, but differs much in size and habit, the narrow apices of the outer petals, winged posticous petal, and in the short pedicels of the flowers. The petioled leaves, shape of petals, the flowers not being umbellate, and their colour, distinguish it at once from *C. Kashmiriana*.

8. **C. juncea** (Wall. Tent. Fl. Nep. 54. t. 42); aphylla, scapo gracili 1–2-bracteato v. nudo, racemo multifloro, bracteis linearibus, pedicellis gracilibus, calcare ascendente flore æquilongo v. breviore, petalis exterioribus cucullatis dorso alatis.—*Wall. Cat.* 1429!

HAB. In Himalaya centrali et orientali alpina, alt. 12–14,000 ped.: Nipal, *Wall.!* Sikkim!—(Fl. Jul. Sept.) (*v. v.*)

Species distinctissima.—*Radix* ignota. *Caulis* v. *scapus* pedalis et ultra, gracilis, erectus, subflexuosus, omnino nudus v. bracteis paucis ornatus. *Racemus* 1½–3-pollicaris, multiflorus, subcylindraceus. *Bracteolæ* pedicellis gracilibus ⅓–⅔ poll. longis bis terve breviores. *Flores* breves, latiusculi, ⅓–½ poll. longi, flavi, macula purpurea ad apices petalorum superiorum utrinque notati. *Sepala* squamæformia.

A leafless slender species, which cannot be confounded with any other. The raceme is generally many-flowered, but in weak specimens only a few flowers are produced.

b. *Radix fusiformis.   Calcar flore æquilongum v. longius.*

* *Caulibus scapisve simplicibus rarius divisis parce foliatis.*

9. **C. crithmifolia** (Royle, Ill. 68); foliis omnibus radicalibus bitripinnatisectis, segmentis linearibus acutis integris v. varie sectis, racemo multifloro, bracteis elongatis linearibus pedicellos superantibus, calcare florem superante.—C. epithymifolia (errore typographico), *Walpers, Rep.* i. 120.

HAB. In Himalaya occidentali temperata, Garhwal, *Munro!* Kunawar, *Jacquemont! Royle!*—(Fl. Apr. Mai.) (*v. s.*)

Radix fusiformis, crassa. *Caulis* scapusve 3-poll. ad pedalem, crassiusculus, omnino aphyllus et ebracteatus. *Folia* radicalia scapo æquilonga, petiolus basi vaginans, lamina 1–4 unc. longa, circumscriptione late ovato-rotundata, subirregulariter bitripinnatisecta, segmentis paucis crassiusculis, maguitudine variis, exemplaribus in Garhwal lectis cæteris multoties latioribus; pinnis primariis longe petiolulatis. *Racemus* subdensiflorus. *Bracteæ* interdum 1½-pollicares, racemum totum superantes, rarius pedicellis breviores. *Flores* pollicares, sulphurei. *Petala* apice purpurea; exteriora apice cucullata, acuta, dorso incrassata v. alata.

Apparently a rare species, easily recognized by its long-petioled radical leaves, leafless scape, and very long, linear, entire, green bracts. Munro's specimens have wings on the back of the upper and lower petals, which are not apparent in the Kunawar ones, and which, with the much greater breadth of the leaflets, probably are the effect of the damper climate of Garhwal.

10. **C. elegans** (Wall. Cat. 1435 !); caule debili, foliis radicalibus longe petiolatis irregulariter ternati- v. pinnatisectis, pinnulis paucis petiolulatis amplis lobatis, scapo aphyllo v. 1-foliato, racemo 6–8-floro, bracteis late obovato-lanceolatis acuminatis, floribus (in genere maximis) calcare obtuso, petalo superiore cucullato dorso late alato, ala secus calcar producta.

HAB. In Himalaya occidentali alpina : Kumaon, *Blinkworth !* alt. 13,500 ped., *Str. et Wint.!*—(Fl. æstate.) (*v. s.*)

Radix elongata, simplex v. divisa. *Folia* radicalia petiolo basi vaginante gracili, limbo 2–3-pollicari, 1½–2 poll. lato, segmentis pinnulisve ½ poll. latis late obovatis cuneatisve, varie grosse lobatis, lobulis obtusis mucronatisve. *Scapus* foliis æqui-longus, plerumque nudus. *Racemus* 1–2-pollicaris. *Bracteæ* magnæ, pedicellis lon-giores brevioresve. *Flores* pollicares, ob alam latam petalorum latiores quam in con-generibus. *Sepala* late ovata. *Petala* lata, subacuta, dorso apice late alata. *Siliqua* immatura stylo brevior.

The large broad leaflets of this plant resemble those of *C. Marschalliana* and *C. pæoniæfolia.* The flowers are the largest of any species known to us. Our speci-mens are not very good, and the species may possibly be more properly referable to the section including *C. rutæfolia.*

11. **C. Govaniana** (Wall. Tent. Fl. Nep. 55) ; radice crassa bi-tricipite, foliis radicalibus plurimis longe petiolatis decomposite pinna-tisectis glaucis segmentis cuneato-lanceolatis acutis, scapo nudo v. pau-cifoliato, racemo dense multifloro, floribus bracteis foliaceis obcuneatis laceris occlusis.—*Wall. Cat.* 1431 ! an *Royle, Ill.* 69. *t.* 15. *f.* 2 (*mala*)?

HAB. In Himalaya occidentali temperata, alt. 8–12,000 ped.: a Ku-maon, *Govan!* ad Kashmir!—(Fl. Jun.) (*v. v.*)

Herba robusta, palmaris v. bipedalis, erectus, carnosus, crassus, glaucus. *Radix* fusiformis, crassitie pollicis, apice vaginis subrigidis nitidis foliorum vetustorum or-nata. *Folia* radicalia scapo æquilonga v. breviora, petiolo elongato, lamina pinna-tim decomposita, in pinnulas plurimas lobatas latitudine varias secta. *Scapus* cras-sus, aphyllus, v. basi seu medio 1–2-foliatus, foliis interdum oppositis. *Racemus* 2–6-pollicaris. *Bracteæ* glaucæ, foliaceæ, late cuneatæ v. cuneato-lanceolatæ, apice la-ceræ v. varie sectæ. *Pedicelli* bracteis breviores. *Flores* pollicares, lutei. *Calcar* curvum, flore longius. *Petala* exteriora cucullata. *Sepala* squamæformia. *Siliqua* immatura obovato- v. lineari-lanceolata v. elliptica, stylo longior, ⅓–⅔-pollicaris. *Se-mina* splendentia, arillo parvo lobato.

A handsome species, very variable in stature and in the size of all its parts, espe-cially of the bracts ; it has been compared with the Siberian *C. nobilis* and *C. brac-teata,* from the former of which it differs in the flowering stem being more of a scape, with no cauline leaves, or few and small ones, and in the winged outer petals and narrower spur. *C. bracteata,* again, belongs to the same section as *C. rutæ-folia,* and has no near affinity with this.

12. **C. Tibetica** (H.f. et T.) ; humilis, glauca, carnosula, foliis plurimis decomposite pinnatisectis, petiolo subscarioso vaginante, pinnis petiolulatis varie sectis, pinnulis acutis obtusisve, scapo rarissime diviso 1-phyllo, racemo terminali 3–5-floro, bracteis integris lobatis pinnati-fidisve, calcare subrecto flore æquilongo; petalis exterioribus apice cu-cullatis acutis carinatis alatisve.

HAB. In Himalaya Tibetica occidentali alpina, alt. 14–17,000 ped.:

2 M

Guge, *Str. et Wint.!* Kunawar, *Jacquemont!* Ladak! Zanskar!—(Fl. Jul.) (*v. v.*)

Herba parvula, sicca flaccida, spithamæa. *Radix* crassa, elongata, bitriceps, superne vaginis angustis foliorum delapsorum longe vaginata. *Folia* lineari-oblonga, lamina 1½-3-pollicari, petiolo elongato, vaginis elongatis sulcatis nitidis pallidis. *Pinnæ* v. *pinnulæ* breviter petiolatæ, magnitudine variæ, ⅙-¾ poll. longæ, varie lobatæ v. sectæ. *Racemus* brevis; pedicellis breviusculis. *Sepala* squamæformia, lacera. *Flores* ¾-pollicares, sulphurei, petalis dorso fusco-brunneo viridique variegatis. *Calcar* apice paullo dilatatum v. subacutum. *Petala* exteriora carinata v. ala lata membranacea cristata. *Ovarium* immaturum lineari-ellipticum, stylo æquilongum, pendulum, glaucum, ⅓ poll. longum. *Semina* pauca, testa lævi, arillo parvo.

** *Caule ramoso folioso (rarius in* C. Moorcroftiana *et* Gortschakovii *simplici*); *racemo sæpius diviso.*

13. **C. Moorcroftiana** (Wall. Cat. 1432!); erecta, robusta, glauca, superne glanduloso-puberula v. pulverea, foliis radicalibus plurimis bitripinnatisectis, pinnis pinnatifidis, pinnulis varie incisis, racemo interdum basi diviso densifloro, bracteis lanceolatis integerrimis lobatisve, petalis exterioribus obtusis apice late alatis.—C. Griffithsii, *Boiss. Diag. ser.* ii. 14?

HAB. In Himalaya et Tibetia occidentali, alt. 10–17,000 ped.: Guge, *Str. et Wint.!* Kunawar, *Munro!* Piti! Ladak, *Moorcroft!* Nubra! Pangong, *H. Strachey!* Afghanistan, *Griffith!*—(Fl. Aug.) (*v. v.*)

Herba plerumque elata, variabilis, plus minusve minute glanduloso-puberula; varietates parvæ scaposæ, scapo aphyllo, ad sectionem posteriorem pertinent. *Radix* crassa, fusiformis, bi-multiceps, crassitie pollicis et ultra, vaginis nitidis sulcatis submembranaceis petiolorum vetustorum sæpe coronata. *Folia* radicalia perplurima, carnosa, spithamæa v. pedalia, pinnatisecta, pinnis primariis distantibus approximatisve, petiolulatis, late ovatis, profunde pinnatifidis bipinnatisectisve, folia caulina minora sparsa. *Racemus* simplex v. plerumque basi divisus, densus, multiflorus. *Bracteæ* (superiores saltem) lineares, integerrimæ, acutæ, glandulosæ, inferiores v. omnes pinnatifido-lobatæ. *Flores* speciosi, sulphurei, ¾ poll. longi. *Sepala* minima, lacera. *Petala* exteriora apice late alata, ala antice sæpius producta. *Siliquæ* lineari-ellipticæ, ¼-pollicares. *Semina* reniformia, aterrima, testa nitida crustacea, minutissime punctulata, arillo parvo bilobo.

A very handsome species, peculiar to the dry climates of the Western Himalaya, Tibet, and Afghanistan. The whole plant is covered with a minute glandular pubescence; but this is only visible with a lens in Strachey and Winterbottom's Guge specimens, which have more glaucous racemes. Small states have simple, scape-like, leafless stems, a span high, whilst typical ones have tall branching stems. It is very nearly allied to *C. Gortschakovii*, of which it may be a form, but in that plant the bracts are pinnatisect, and it is never glandular.

Boissier's *C. Griffithsii* we have referred to this, though it is not the n. 1419 of Griffith in Leman's Herbarium; the name would claim priority (that of *Moorcroftiana* not being published) were we certain of the identity, but no allusion is made by Boissier to the glandular pubescence, which is evident in Griffith's specimens, and the name should further have been *Griffithii*.

14. **C. Gortschakovii** (Schrenk. En. Plant. Nov. 100); erecta, robusta, ramosa (rarius simplex), foliosa, glauca, radice crassa, foliis radicalibus amplis pinnatisectis bi-tri-pinnatisectisve, pinnis primariis petiolulatis, segmentis ovatis varie sectis lobulis acutis, racemis densifloris,

bracteis pinnatisectis lobis linearibus, petalis exterioribus apice obtusis
dorso late v. anguste alatis.—*Karel. et Kiril. En. Plant. Alt.* 59, *Herb.*
1188!; *Led. Fl. Alt.* i. 746.

HAB. In Himalaya Tibetica alpina, alt. 10–15,000 ped.; Guge, *Str.
et Wint.!* Kunawar et Kashmir, *Jacquemont!*—(Fl. Jul. Aug.) (*v. v.*)
DISTRIB. Soongaria !

Statura variabilis, caule simplici v. ramoso; habitus *C. Moorcroftianæ,* sed differt racemis non glanduloso-puberulis, et bracteis magis pinnatisectis. *Folia* radicalia simpliciter pinnatisecta v. decomposito-bi-tripinnatisecta. *Petala* exteriora apice
alata, ala angusta v. lata, interdum ultra apicem petali producta. *Siliqua et semina
C. Moorcroftianæ.*—Exemplar authenticum a Karel. et Kiril. missum caulem simpli-
cem 1-2-foliatum exhibet, foliaque simpliciter pinnatisecta.

We have but one authentic specimen of the Soongarian *C. Gortschakovii,* which
agrees perfectly with a small unbranched state of our Tibetan plant. Amongst a few
fragments of plants brought to us (when in Sikkim) from Nepalese Tibet, are racemes
of a *Corydalis* apparently intermediate between *C. Moorcroftiana* and *C. Gortschakovii,* having the glandular pubescence of the former plant, and the bracts of the
latter.

15. **C. ramosa** (Wall. Cat. 1434 !); humilis v. elata, glauca, carnosula v. membranacea, caule gracili flexuoso ramoso, foliis petiolatis in
segmenta linearia acuta flabellatim v. ternatim bi-tri-pinnatisectis, racemis ramos divaricatos terminantibus, calcare latiusculo obtuso flore
æquilongo, petalis exterioribus apice cucullatis subacutis dorso alatis
ala angusta v. lata integra v. lacera, siliqua late elliptica, seminibus
splendentibus.

a. ramosa; caule elongato debili ramoso, foliorum lobis plerumque
angustis, bracteis pinnatifidis, ala petali dorsalis lata subintegra, pedicellis elongatis.

β. vaginans; caule elongato debili ramoso, foliorum lobis latioribus
membranaceis, bracteis pinnatifidis, ala petali dorsalis lata laciniata,
pedicellis elongatis.—C. vaginans, *Royle, Ill.* 68.

γ. nana; pumila, glauca, caule brevi simplici v. ramoso, foliis flabellatim pinnatisectis palmatisectisve, petalo dorsali vix alato.—C. nana,
Royle, Ill. 68.

HAB. *a.* et *β.* Per totam Himalayam temperatam, alt. 6–12,000 ped.;
a Sikkim ! ad Kashmir ! *a.* locis editioribus ; *β.* depressioribus.—*γ.* In
Himalaya Tibetica alpina, alt. 12–15,000 ped.; Kumaon et Guge, *Str.
et Wint.!* Kunawar, *Jacquemont!*—(Fl. æstate.) (*v. v.*)

Species plerumque humifusa, caulibus elongatis v. locis alpinis abbreviatis, 2-polli-
caris v. bipedalis. *Radix* fusiformis, elongata. *Caulis* sæpissime debilis, laxe vage divaricatim ramosus, glaucus, sicco viridis v. atro-fuscus. *Folia* radicalia in var. *nana*
plurima, in var. *vaginante* pauca, omnia petiolata. *Racemi* 1-5 poll. longi. *Flores*
½ poll. longi, subsecundi, flavi. *Pedicelli* ¼-½-pollicares, fructiferi decurvi. *Sepala*
squamæformia, crassiuscula, opaca, fimbriata. *Petalum* exterius apice cucullatum,
plerumque dorso ala lata lacera cristatum, sed interdum vix carinatum. *Siliqua* pedicello brevi v. elongato curvo subpendula, ¼-⅔ unc. longa, stylo breviore terminata,
obovato-elliptica, planiuscula. *Semina* pauca, minutissime punctulata, splendentia.

All our dried specimens, and these are exceedingly numerous, from almost every
province between Sikkim and Kashmir, are of a peculiar grey glaucous hue; they
are excessively variable in stature and habit, and the size, depth, and lobing of the

keel or wing of the dorsal petal. *C. nana* is rather a dwarf alpine state than a
marked variety; its stems are sometimes excessively branched from the base. The
common state closely resembles the Siberian *C. Gebleri,* differing in the much broader,
shorter pod. It is also nearly allied to *C. Sibirica* in habit, but the pod and spur are
very different, the latter being neither so broad nor turned up; also to *C. cornuta,*
Royle, which has opaque seeds.

16. **C. Sibirica** (Pers. Syn. ii. 70); caulibus gracilibus vage de-
cumbentibus elongatis ramosis foliosis, foliis longe petiolatis membranaceis bi-tri-pinnatisectis segmentis latiusculis 3–5-fidis, bracteis inferioribus lobatis sectisve, calcare lato flore æquilongo ascendente, petalis exterioribus cucullatis acutis, siliquis parvis linearibus lineari-obovatisve,
seminibus splendentibus.—*DC. Syst.* ii. 124, *Prod.* i. 128. C. Sibirica
et C. impatiens, *Fisch. in DC. Prod. l. c.;* Led. *Fl. Ross.* i. 103. C.
longipes, *DC. Prod. l. c.; Wall. Cat.* 1433!, *Tent. Flor. Nep. t.* 42
mala; non Don, Prod. 198. C. filiformis, *Royle, Ill.* 65.

HAB. In Himalaya temperata et subalpina, alt. 7–14,000 ped.: Sikkim! Nipal, *Wallich!* Garhwal, *Royle!* et in mont. Khasia, alt. 6000
ped., *Griffith!*—(Fl. Jun. Jul.) (*v. v.*)

DISTRIB. Sibiria Baikalensis! et trans-Baikalensis; Dahuria; Kamtchatka.

Herba diffusa, gracilis, ramosa, statura variabilis. *Caulis* 6 unc. v. bipedalis, divaricatim ramosus. *Folia* varie secta, segmentis late linearibus cuneato-obovatisve
3–5-fidis, lobis obtusis apiculatisve integerrimis v. 2–3-crenatis. *Bracteæ* inferiores
lobatæ v. sectæ, superiores integræ v. lobatæ. *Pedicelli* $\frac{1}{3}$–$\frac{2}{3}$ poll. longi. *Sepala*
squamæformia, lacera, membranacea. *Calcar* lente v. abrupte ascendens, appendice
brevi v. elongata. *Siliquæ* $\frac{1}{4}$–$\frac{1}{2}$-pollicares, anguste lineari-obovatæ v. lineares, $\frac{1}{10}$–$\frac{1}{8}$
poll. latæ. *Semina* splendentia.—A *C. ramosa* differt, caulibus gracilioribus, foliis
minus sectis, segmentis latioribus, sed præcipue calcare ascendente breviore et latiore,
et siliquis angustioribus stylo brevi terminatis.

This is a very distinct but variable plant. We have examined a multitude of specimens, especially from the Khasia (where it is the only species known, and inhabits
a much lower level than in the Himalaya) and from Sikkim, where it is extremely
common, and may be followed up any of the valleys continuously from 10,000 nearly
to 15,000 feet elevation, gradually changing its habit and appearance a good deal, but
retaining the marked character of the spur, and all the general features of the species
in a greater or less degree. We have also examined very carefully all Royle's and
Wallich's specimens, and compared these together and with the Siberian ones. Wallich's specimens have pods exactly intermediate in character between those of *C. impatiens* and *C. Sibirica.* Royle's *C. filiformis* was probably inadvertently proposed as
new, for it is identical with Wallich's plant. The Khasia individuals have larger
flowers and broader wings to the outer sepals than the Sikkim, but not than Ledebour's Siberian specimens. Wallich's figure of *C. longipes* (Tent. Fl. Nep.) represents
a very much larger plant than his specimens, with the spurs not at all ascending,
which they manifestly are in his Herbarium; his quotations of *Fumaria bulbosa,*
Thunb. Jap. 277, and *C. decumbens,* Pers. Ench. 269, both with a mark of doubt,
we cannot confirm, never having seen authentic specimens, and the descriptions being
insufficient.

Ledebour, in the 'Flora Rossica,' states of *C. impatiens,* that if at all different
from *C. Sibirica,* its characters depend on the diffuse stem, narrow pod, and short
pedicel, all which we find so variable in every locality, that we cannot even propose
to make a variety of it.

17. **C. cornuta** (Royle, Ill. 69); caule debili ramoso folioso, foliis

glaucis bi-tri-pinnatisectis, bracteis inferioribus lobatis pinnatisectisve, racemo elongato, pedicellis brevissimis, calcare flore sublongiore recto apice recurvo v. decurvo, siliquis brevissime pedicellatis, seminibus opacis granulatis.—C. debilis, *Edgew.! in Linn. Soc. Trans.* xx. 30.

HAB. In Himalaya occidentali temperata, alt. 8–10,000 ped. : Kumaon, *Str. et Wint.! Garhwal, Edgeworth! Sirmur, Royle !*—(Fl. Jul. Aug.) (*v. v.*)

Planta glauca, vage diffuse ramosa, habitu omnino *C. Sibiricæ* et *C. ramosæ.* *Folia* longe petiolata, membranacea ; segmentis late cuneato-obovatis oblongisve, 3- 5-fidis, lobis obtusis apiculatisve, integerrimis 2–3-crenatisve. *Racemi* stricti, pollicares. *Bracteæ* varie lobatæ v. sectæ, rarius integerrimæ. *Sepala* minima, squamæformia. *Petalum* posticum apice concavum, acutum, dorso alatum, ala nunquam lobata, v. nudum. *Siliqua* lineari-obovata v. late obovato-oblonga.—A *C. Sibirica* differt, racemis elongatis, calcare longiore non ascendente, pedicellis brevioribus, siliquis majoribus et latioribus, et præcipue ab omnibus seminibus opacis granulatis.— A *C. Bungeana* (e China orta) differt statura, segmentis foliorum amplis, racemo elongato, bracteis multoties minoribus et minus sectis, floribus majoribus flavis (non purpureis), forma petalorum, et siliquis minoribus in stylum non attenuatis.

Royle describes the spur as erect, but it is not so in his specimens, which are not, however, in seed, and therefore cannot be satisfactorily identified.

18. **C. chærophylla** (DC. Prod. i. 128) ; erecta, robusta, foliosa, ramosa, foliis amplis bi-triternatis pinnatisectisve subtus glaucis, lobis decurrenti-coadunatis ultimis divaricatis varie obtuse incisis, racemis ramosis multifloris, floribus secundis, bracteis parvis lobatis, pedicellis brevissimis, calcare gracili, siliquis parvis lineari-obovatis, seminibus splendentibus.—*Wall. Cat.* 1428 !; *Tent. Fl. Nep.* 52. *t.* 40; *Don, Prod.* 198.

HAB. In Himalaya temperata, alt. 6–10,000 ped. : Sikkim ! Nipal, *Wallich!* Kumaon, *Madden!*—(Fl. Jun. Jul.) (*v. v.*)

Radix valida, lignosa, fusiformis. *Caules* elati, foliosi, carnosuli, glaucescentes, 2–4-pedales. *Folia* longe petiolata, pedalia, lamina spithamæa, trisecta, lobis primariis longe petiolulatis, demum bi-triternatis pinnatifidisque, laciniis varie incisis, ultimis linearibus obtusis divaricatis. *Racemi* compositi, e basi divisi, ramis strictis, basi foliosis. *Flores* plurimi, aurei, graciles, ¾ poll. longi. *Sepala* minima. *Calcar* elongatum, rectum v. ascendens, flore longius, appendice filiformi. *Petala* exteriora apice concava, acuminata, dorso breviter cristata. *Siliquæ* parvæ, patentes, ⅛ poll. longæ, breve pedicellatæ, lineari-obovatæ. *Semina* pauca.

A remarkably handsome, most distinct species, well figured in Wallich's 'Tentamen,' where, however, the specimen represented is young, and the spur is very much shorter than in any of his own or our individuals.

19. **C. geraniifolia** (H.f. et T.) ; suberecta, ramosa, foliosa, foliis deltoideis ternatim sectis, segmentis bipinnatifidis v. basi pinnatisectis, laciniis lineari-oblongis decurrenti-coadunatis inciso-lobatis, racemo ramoso, bracteis foliaceis incisis, calcare gracili flore longiore, petalis exterioribus concavis acutis dorso vix alatis, siliquis secundis patulis breve pedicellatis lineari- v. elliptico-obovatis, seminibus splendentibus.

HAB. In Himalayæ orientalis exterioris temperatæ sylvis : Sikkim, alt. 8–9000 ped. !—(Fl. Sept.) (*v. v.*)

A *C. chærophylla,* quæ affinis, differt, lobulis foliorum acuminatis, racemis breviori-

bus, floribus laxius racemosis, longius pedicellatis, bracteisque multoties majoribus.—
Folia longe petiolata, subtus pallida.   *Flores* aurei.

c. *Radix fusiformis.*   *Calcar breve, saccatum.*

20. **C. latiflora** (H.f. et T.) ; caulibus v. scapis foliisque e radice
v. caule brevi plurimis, foliis longe petiolata glaucis bi-tri-pinnatisectis, segmentis petiolulatis ultimis lineari-oblongis, scapis medio 2-folia-
tis rarius subumbellatim ramosis, floribus subumbellatis longe pedicellatis, bracteis magnis linearibus, calcare brevi obtuso, petalis exterioribus latis dorso late alatis.

HAB. In Himalaya orientali alpina, alt. 12–15,000 ped.: Sikkim, ad
Tankra !—(Fl. Aug.) (*v. v.*)

Herba pusilla, 2–4-uncialis, glaucescens, cæspitosa, carnosula, sicco nigra, flaccida.   *Rhizoma* simplex, gracile, 6-poll. v. pedale, apice foliosum.   *Folia* radicalia
plurima, late ovata v. ovato-deltoidea, 1–2-pollicaria, apicibus segmentorum acutis,
petiolo gracili 1–2-pollicari, basi late membranaceo-vaginante.   *Scapi* basi nudi,
interdum ad axillas foliorum umbellatim ramosi, ramo intermedio florifero, lateralibus 2-phyllis, supra medium 2-foliatis, foliis oppositis petiolatis, radicalibus consimilibus interdum ad petiolum vaginantem squamæformem reductis.   *Flores* 3–6, pro
planta magni, pedicello breviores, pallide cærulei, apice flavidi.   *Bracteæ* elongatæ,
simplices, lineares.   *Sepala* squamæformia, lacera.   *Petala* exteriora brevia, lata,
dorso late alata, ala interdum secus calcar producta.   *Ovarium* lineari-ellipticum,
multiovulatum ; *stylo* brevi ; *stigmate* reniformi.

A very remarkable and distinct little species, readily distinguished by its long rhizome, many scapes and radical leaves, pairs of opposite leaves on the scape, which
is simple or branched, long linear erect bracts, large, very broad, pale blue-grey
flower, and very short broad spur.

21. **C. Astragalina** (H.f. et T.); erecta, robusta, subrigida,
glauca, caule sulcato subsimplici, foliis carnosulis bipinnatisectis, segmentis petiolulatis parvis pinnatifidis lobatis aristato-acuminatis, racemo densifloro, bracteis subulatis membranaceis albis, calcare brevissimo obtuso, petalis exterioribus planiusculis, siliquis magnis pendulis
linearibus.

HAB. In Tibetia occidentali, alt. 14–16,000 ped.: Nari, prope Bekar,
Jacquemont! Ladak !—(Fl. Jul.) (*v. v.*)

Herba robusta, glauca, 1–2-pedalis.   *Caulis* basi reliquiis suberosis vaginarum cir-
cumdatus, collo diametro pollicari, superne crassitie pennæ anserinæ, sulcatus, simplex
v. divisus.   *Folia* omnia conformia, ovato-oblonga, radicalia plurima, caulina alterna
petiolata, 3–6 poll. longa, laciniis crassis coriaceisque, caulinis petiolo basi obscure
dilatato, segmentis ⅙–⅓ poll. longis, latis angustisve.   *Racemus* brevis, densiflorus,
floribus imbricatis subpendulis.   *Bracteæ* parvæ, deciduæ, ¼ poll. longæ.   *Pedicelli*
breves, crassi, curvi.   *Flores* ½–¾ poll. longi, flavi, anguste elongati.   *Sepala* lanceolata, subulata, basi rotundata, fimbriato-lacera.   *Petala* subcarnosa, exteriora apice
abrupte acuminata, apicibus recurvis, marginibus membranaceis, lateralia libera ; *calcare* brevi obtuso incurvo.   *Ovarium* lineari-lanceolatum, stylum gracilem æquans ;
stigmate parvo ; ovulis plurimis.   *Siliquæ* magnæ, strictæ, pendulæ v. cernuæ, 1–1½
poll. longæ, ⅙ poll. latæ, in stylum pugioniformem angustatæ ; valvæ tumidæ.   *Semina* splendentia.

A most distinct species, conspicuous for its size, robust habit, glaucous hue, brittle
texture when old, and its curious subulate bracts, lanceolate-subulate sepals, recurved
tips to the outer petals, very large siliquæ with rigid persistent pungent styles, and
large seeds.   It varies extremely in the size and cutting of the leaf-lobes.   Jacquemont's specimens were accidentally mixed with *C. Gortschakovii.*   It is nearly

allied to *C. stricta*, Led., Fl. Alt., but the leaves differ, and in the absence of the fruit of the latter plant we are unable to unite them.

22. C. meifolia (Wall. Tent. Fl. Nep. 52. t. 41); robusta, erecta, ramosa, foliosa, foliis supradecomposite pinnatisectis, segmentis lineari-oblongis capillaceisve, racemis brevibus densifloris, bracteis pectinatis superioribusve integris, calcare obtuso flore ½ breviore.—*DC. Prod.* i. 128; *Wall. Cat.* 1427!

HAB. In Himalaya alpina et subalpina : Sikkim, alt. 12–15,000 ped.! Nipal, *Wall.!* Kumaon, *Blinkworth!* Kunawar, *Jacquemont!*—(Fl. Jul. Aug.) (*v. v.*)

Herba crassa, glauca, 3 unc. v. 3-pedalis. *Radix* valida, fusiformis. *Folia* radicalia perplurima, suberecta v. patentia, segmentis confertis, forma variis, semper angustis, acuminatis. *Racemi* ad apices ramulorum 1–2 poll. longi, breves, densiflori, ob bracteas foliaceas quasi foliosi. *Bracteæ* magnæ v. parvæ. *Pedicelli* validi, superiores axillares elongati, 1–2-pollicares, curvi. *Flores* sordide aurei apicibus purpureis, ½ unc. longi, apice ob alam petalorum latam dilatati. *Sepala* squamæformia, subcoriacea v. carnosa. *Ovarium* breve, ellipticum. *Siliqua* ½ poll. longa, latiuscula; semiuibus 4–6, biseriatis (atris, lucidis, *Wall.*).

23. C. flabellata (Edgew. in Linn. Soc. Trans. xx. 30); erecta, rigida, robusta, glauca, valde ramosa, caule striato, foliis longe petiolatis pinnàtisectis, pinnis petiolulatis obcuneato-flabellatis, bracteis parvis setaceis, pedicellis brevissimis, floribus curvis, calcare flore ½ breviore, siliquis linearibus utrinque acuminatis.

HAB. In Himalaya et Tibetia occidentali alpina : Kumaon, 11,500 ped., *Strachey et Wint.!* Garhwal, 9–10,000 ped., *Edgeworth!* Ladak et Zanskar glareosis, alt. 10–12,000 ped.! Gilgit, 5000 ped., *Winterbottom!*—(Fl. Jul. Aug.) (*v. v.*)

Herba glauca, 3-pedalis, caule crassitie pennæ olorinæ. *Folia* forma varia, linearia v. lineari- v. ovato-oblonga, spithamæa et ultra; pinnulis remotis 1½–2 poll. latis, simplicibus lobatisve, margine exteriore crenulato v. integerrimo, interdum 2–3-partitis. *Racemi* ad apices ramulorum subpaniculatim ramosi, stricti, densiflori. *Bracteæ* pusillæ. *Pedicelli* breves. *Flores* horizontales, ½–¾ poll. longi, flavi, curvi, superne concavi, calcare et apicibus petalorum ascendentibus. *Sepala* scariosa, dentata, subulata. *Petala* angusta, exteriora apice cucullata apiculata, dorso nuda v. carinata; calcare inflato, flore ½ breviore, apice dilatato decurvo. *Siliquæ* ½–1 poll. longæ, juniores elliptico-ovatæ, maturæ lineares, stylo recto terminatæ; valvæ subconcavæ. *Semina* 8–10, biseriata, punctulata, nitida.

Edgeworth describes the siliquæ as ovate, which is not the case in his or our specimens.

24. C. adiantifolia (H.f. et T.); suberecta, e basi ramosa, ramis crassiusculis glaucis, foliis remote pinnatisectis, segmentis petiolulatis cuneato- v. reniformi-flabellatis crasse coriaceis, racemis basi ramosis apicibus bracteis elongatis setaceis capillaceo-acuminatis crinitis, calcare brevi inflato, sepalis basi fimbriato-laceris, siliquis linearibus utrinque acuminatis.

HAB. In glareosis Himalayæ occidentalis alpinæ : Kishtwar, alt. 12–14,000 ped.!—(Fl. vere.) (*v. v.*)

Radix fusiformis, crassa, denique vaginis suberosis persistentibus foliorum delapsorum coronata. *Caules* 6 poll. v. 2-pedales, basi sæpius decumbentes, glauci.

Folia circumscriptione lineari-oblonga. *Pinnulæ* 2–5-jugæ, sparsæ, ½–¾ poll. latæ, varie crenato-lobatæ, rarius integerrimæi, basi sæpius cordatæ. *Racemi* breves, densiflori. *Pedicelli* brevissimi. *Bracteæ* floribus ½ breviores, patentes, membranaceæ, anguste subulatæ. *Flores* fere recti v. calcare et apicibus petalorum sensim ascendentibus, ¾ poll. longi, rosei. *Sepala* elongato-subulata, membranacea. *Petala* exteriora apice concava, abrupte acuminata, interioribus apice emarginatis; calcare brevi inflato apice obtuso incurvo. *Siliquæ* immaturæ ¾ poll. longæ, stylo recto valido terminatæ.— *C. flabellatæ* affinis, differt præcipue statura humiliore, foliorum pinnis paucijugis, floribus majoribus roseis minus curvis, et bracteis elongatis subulatis.

3. DICENTRA, Bork.

Diclytra, *DC.*, Dactylicapnos, *Wall.*, Macrocapnos, *Royle.*

Sepala 2, decidua. *Petala* 4, libera, anticum et posticum basi saccata v. calcarata. *Stamina* 6; filamenta libera v. basi distincta, superne coalita, intermedio cujusvis synematis basi processu calcariformi aucto. *Ovula* juxta placentas intervalvulares plurima. *Stigma* bilobum. *Capsula* siliquosa v. baccata, valvis linearibus membranaceis v. ovatis carnosisque; placentis seminibusque ut in *Corydali.*—Herbæ *glaberrimæ, Indicæ, plerumque scandentes,* radicibus *perennantibus,* caule *tereti ramoso gracili carnosulo,* foliis *oppositis decompositis,* petiolis *cirrhosis,* pinnulis *3-nerviis ovatis membranaceis,* racemis *oppositifoliis nutantibus.*

All the Indian species of this pretty genus are scandent, whereas most of the American and Siberian ones have bulbous roots, radical leaves, and the flowers on erect scapes. There appears to us to be no grounds for dividing the genus, either on this account or on the form and structure of the pods, as the same differences of habit occur in *Corydalis,* and there is a transition in the structure of the pod from the membranous linear valves of the first section, to the more or less fleshy, broad, and almost indehiscent ones of *D. thalictrifolia.*

Sect. 1.—*Siliqua* linearis; *valvis* membranaceis.

1. **D. torulosa** (H.f. et T.); corymbis 6–8-floris, siliquis anguste linearibus torulosis, seminibus uniseriatis opacis granulatis basi strophiolo cinctis.

HAB. In montibus Khasia, graminosis alt. 5–6000 ped., *Griffith!*— (Fl. Aug.) (*v. v.*)

Herba tenella, glaberrima, glauca, 8–10-pedalis. *Caules* scandentes, debiles, angulati. *Folia* 2–4 poll. longa, e basi bi-tri-pinnatisecta, pinnis primariis longissime petiolulatis, divaricatis; pinnulis paucis, remotis, ½–¾ poll. longis, ellipticis, acutis acuminatisve, membranaceis. *Petiolus* basi simplex, apice sæpius in ramos capillares cirrhosos dichotome divisus. *Pedunculi* oppositifolii, ½–2 poll. longi, graciles, apice subumbellatim corymbosi. *Bracteæ* membranaceæ, lineares, pedicellos subæquantes, marginibus laceris. *Pedicelli* 6–10, ⅓–⅔ poll. longi, graciles, stricti, superne subincrassati. *Flores* penduli, ½ poll. longi, aurei. *Sepala* lineari-subulata, flore ¼–⅓ breviora, membranacea, margine plus minusve lacera, basi dilatata fimbriata. *Petala* exteriora apice breviter cucullata, abrupte acuminata, basi saccata, interiora infra apicem in rostrum breve bicuspidatum ultra petala exteriora breviter porrectum coalita, lamina cujusvis brevi panduriformi, ungue capillari elongato libero. *Synemata* basi calcare brevi incurvo instructa. *Ovarium* lineare, ovulis plurimis; stylo gracili; stigmate subquadrato, bilobo, basi utrinque in cornu producto. *Siliqua* 1½–2 poll. longa, curva, torulosa, in stylum validum angustata. *Semina* 15–20, alterna, majuscula, basi strophiolo bilamellato inclusa.

2. **D. Roylei** (H.f. et T.); corymbis 2-3-floris, siliquis late linearibus, seminibus biseriatis atris nitidis, strophiolo mediocri.

HAB. In Himalaya temperata, alt. 5–6000 ped.: Bhotan, *Griffith!*
Sikkim! Garhwal, *Royle!* Simla, *D^{na} Dalhousie!* et in Mont. Khasia,
graminosis alt. 5–6000 ped.!—(Fl. Mai.–Jul.) (*v. v.*)

Herba 3-pedalis, decumbens v. subscandens, habitu foliis floribusque priori simillima, differt præcipue corymbis pauci(2-3)floris non subumbellatis, sepalis brevioribus, siliqua breviore et latiore, seminibus biseriatis nitidis, strophioloque parvo.—*Sepala* ovato-subulata, flore multoties breviora. *Petala* exteriora calcare brevi lata,
apice brevissime acuminata, interiorum lamina late spathulato-orbiculata, basi bifida,
ungue capillari. *Stigma* lunatum. *Siliqua* 1¼–1¾ poll. longa, ½ poll. lata, valvis
planiusculis non torulosis.

Sect. 2. DACTYLICAPNOS.—*Siliqua* ovata, elliptica v. oblonga; seminibus utrinque biseriatis perplurimis; valvis carnosis membra-
naceisve.—Dactylicapnos, *Wall. Tent. Fl. Nep.* 51.

3. **D. scandens** (Walp. Rep. i. 118); racemis sub-10-floris longe
pedunculatis, sepalis subulatis v. triangulari-ovatis, siliquis ellipticis,
valvis membranaceis, seminibus basi lævibus ambitu granulatis.—Diclytra scandens, *Don, Prod.* 198. Macrocapnos, *Royle in Lindl. Introd.
Nat. Ord. ed.* ii. 439.

HAB. In Himalaya temperata, alt. 5–6000 ped.: Nipal, *Wallich!*
Kumaon, *Str. et Wint.!* Garhwal, *Edgeworth.*—(Fl. æstate.) (*v. s.*)

Herba tenella, alte scândens, caule tenui flexuoso angulato. *Folia* alterna, 3–6
poll. longa, ab ima basi tripinnatisecta, pinnis primariis longe gracile petiolulatis,
forma et magnitudine variis, plerumque pollicaribus, late ovatis, obtusis apiculatis
acutisve, subtus glaucis. *Pedunculi* interdum ramosi et foliosi, plerumque simplices,
graciles, 3-pollicares, apice racemosi floriferi. *Bracteæ* parvæ, subulatæ. *Pedicelli*
inferiores ½-pollicares, filiformes. *Flores* immaturi tantum dissecti. *Siliqua* pollicaris, ½ poll. lata, anguste elliptica v. elliptico-ovata, in stylum validum ⅓-pollicarem
angustata. *Valvæ* subconcavæ, membranaceæ. *Semina* atra, subnitida.

This is the plant mentioned by Royle (Ill. 68), which he says so closely resembles
D. thalictrifolia in all respects but the pod as to be otherwise undistinguishable, adding that Wallich's specimens of the two are confounded in the Linnæan Herbarium.
We have adopted Don's name for it, rather than propose a new one, though Don's
character is wholly insufficient, and applies equally to both. Royle's character of
the five-winged stem we are unable to verify. Our doubts as to the validity of the
character drawn from the pods are expressed under the following species.

4. **D. thalictrifolia** (H.f. et T.); siliqua ovato-oblonga v. late
elliptica, valvis carnosis, seminibus basi granulatis ambitu tuberculatis v. asperis.—Dactylicapnos thalictrifolia, *Wall. Tent.* 51. *t. 39, Cat.*
1426!; *Sweet, Brit. Fl. Gard. ser.* ii. *t.* 127.

Variat conspicue forma et magnitudine pinnularum et sepalorum, valvis
siliquæ carnosis v. submembranaceis, seminibus granulatis asperisve.

HAB. In Himalaya temperata centrali et orientali, alt. 4–8000 ped.:
Bhotan, *Griffith!* Sikkim, umbrosis! Nipal, *Wallich!* et in mont. Khasia, alt. 5000 ped.!—(Fl. Sept.) (*v. v.*)

Priori simillima, et verosimiliter non distincta. Forma major locis depressis occurrit, pinnulis amplis 2-pollicaribus basi cordatis, floribus pollicaribus et siliqua valde
carnosa; locis editioribus humilis et omnibus partibus minor evadit. *Petala* exte-

riora basi in saccos orbiculatos producta, apicibus concavis obtusis v. brevissime acuminatis; interiora lamina obovato-rotundata, basi contracta, oblique biloba v. cordata,
apice rostrata; appendice dilatata subinflata. *Stigma* lunatum. *Siliquæ* ¾–1¼-polli-
cares, plerumque exacte ellipticæ v. elliptico-ovatæ, cylindraceæ v. compressæ, rarius
basi truncatæ v. cordatæ (cf. Ic. Tent. Fl. Nep.), in stylum validum ⅓–½-poll. angustatæ. *Valvæ* nunquam omnino indehiscentes, sæpissime facile solutæ, rubræ, vallibus humidis succulentæ, collibus siccioribus submembranaceæ. *Semina* oblique
obovata, subgibba, basi utrinque areola minus granulata notata, ambitu plerumque
subhispido-granulata.

This is a very abundant Sikkim plant, whose extreme forms we have in vain attempted to separate by any constant characters; whilst yet in that country, however,
we convinced ourselves that they all belong to one highly variable plant, and our
subsequent examinations, with the aid of Wallich's and our Khasia specimens, have
confirmed that conclusion. In the latter country we found it at the Kala Pani Bun-
galow only, where it is abundant. We further much doubt whether *D. scandens*
be distinct from this; in the absence of perfect flowers we cannot pronounce positively, but the membranous valve of the pod is of itself not a sufficient character,
and the markings of the surface of the seed vary so much in the Sikkim plant, that
we cannot lay much stress on them.

4. HYPECOUM, Tourn.

Chiazospermum, *Bernh.*

Sepala 2, decidua. *Petala* 4, exteriora anticum et posticum, obtusa,
triloba, subunguiculata, interiora trifida, lobo medio cochleariformi.
Stamina 4, petalis opposita, basi nuda v. utrinque glandula stipata;
antheræ biloculares. *Ovarium* 1-loculare, ovulis in placentis interval-
vularibus pluribus, isthmis transversis sejunctis. *Capsula* siliquæformis,
intus articulata, articulis monospermis indehiscens v. dehiscens. *Semina* compressa, umbilico ventrali lineari.—Herbæ *Mediterraneæ orientales et Sibiricæ, paucæ Indicæ, annuæ,* succo *aqueo,* radice *fusiformi,*
foliis *glaucis pinnatisectis,* scapis caulibusve *pluribus simplicibus vel divisis,* floribus *terminalibus.*

This curious genus is intermediate in many respects between *Papaveraceæ* and
Fumariaceæ, having the flower much more regular than in most *Fumariaceæ,* but
not so regular as in *Papaveraceæ;* in the characters of its petals it resembles *Epimedium* and *Bongardia* amongst *Berberideæ,* as also in its definite stamina being
opposite the petals. The glands described by Endlicher at the bases of the filaments
are hardly visible in the species we have examined; when developed they probably
represent the appendix within the spur of *Corydalis,* and are possibly also analogous
to the glands of *Cruciferæ,* and remotely to the glandular bases of the petals of *Berberis.* The middle lobe of the inner petals resembles a deformed anther, and is said
by some authors to be occasionally polliniferous, an observation we cannot confirm.

The opposition of the four stamens to the petals in this genus would seem to confirm
Lindley's suggestion, that the corresponding lateral one-celled anthers of each bundle
in *Fumaria, Corydalis,* etc., are the half-anthers of one stamen, for this would reduce
the staminal series of those genera to the same numerical formula as occurs in *Hypecoum, Epimedium,* and *Aceranthus:* and the two central perfect stamina of *Corydalis,
Fumaria,* etc., being opposite the outer petals, the abnormal fission of the two lateral
stamina may theoretically be supposed to result from the tendency to cohesion of all
the filaments in that Order being partially overcome by the great irregularity of their
corolla. Supposing that the disposition of the stamens and petals of *Hypecoum* had
been the prevalent one in *Fumariaceæ,* and that of *Fumaria, Corydalis,* etc., exceptional, the correctness of the above explanation would probably never have been

questioned. The same argument we consider opposed to Lindley's view of the sepals of *Fumariaceæ* being bracts, the outer petals, sepals, etc., the inner alone true petals, chiefly because the relation of the stamens to the petals is, as above stated, exactly what occurs in the Berberideous genera mentioned above, and because *Meconia* in *Papaveraceæ* has also only four petals and as many stamens.

De Candolle (Syst. ii. 101) says, that according to Schultes, Obs. Bot. 26, *H. procumbens* has sometimes four sepals and six stamens. This increase of the number of sepals is further opposed to the view of the sepals being bracts, and the presence of six stamens reduces the genus to the hexandrous type common to all other *Fumariaceæ*. The position of the two additional sepals and stamens is not given by De Candolle; but if the six latter consist of two pairs, opposite to the smaller petals, and consequently to the outer sepals, it affords an additional point of affinity between *Fumariaceæ* and *Cruciferæ*, and favours M. Gay's views of the identity of the staminal formula in these two Orders (Ann. Sc. Nat. ser. 2. xviii. 216).

Grisebach (Fl. Rumel.) describes the anthers as eight, combined into four phalanges, as follows :—" Antheræ 8 in phalangibus 4 distinctis petalo oppositis binatim conjunctæ;" and again, "Stamina tetradelpha per paria petalo opposita" (Grundriss der Syst. Bot. 70), but in our specimens (and all others we have examined) the stamens are as described by other authors, namely, four opposite the petals, with sometimes a real or perhaps only apparent cohesion of the anthers. The anthers are completely enclosed within the middle lobes of the lateral petals during impregnation, when the lobes retain the pollen in contact with the stigma, exactly as the lateral petals do in other *Fumariaceæ*. Indeed, the bud of *Hypecoum* scarcely differs from that of a spurless *Corydalis* in appearance, the outer petals being sharply keeled at their connivent apices, and the inner enclosing the anthers in their tips.

Bernhardi (Linnæa, viii.) regards the sepals as bracts, the two outer petals as sepals, and the two inner as stamens and petals combined: but this appears to us to be a purely hypothetical view, and not supported by any anatomical or morphological facts in the structure of these organs or their relations.

We have in vain endeavoured to find any character whereby to separate *Chiazospermum* generically from *Hypecoum*; the articulation of the pod varies greatly in amount, and is shared by *Hypecoum* to a considerable extent. The forms of this genus are, like all cut-leaved herbaceous annuals, excessively variable in habit and foliage, and there are probably extremely few good species.

1. **H. procumbens** (Linn. Sp. Pl. 181); siliquis arcuatis subcompressis, sepalis ovatis, petalis exterioribus late obcuneato-trilobis, interioribus alte trilobis, lobo intermedio spathulato margine subciliato. —*Sibthorp, Flor. Græc.* ii. 46. *t.* 155 ; *Schkuhr, Handbuch,* i. 90. *t.* 27; *DC. Syst.* ii. 102, *Prodr.* i. 123.

HAB. Panjab ad Peshawar, *Vicary!* Multan, *Edgeworth!* "Salt range," *Fleming!* Beluchistan, *Stocks!* Afghanistan, *Griffith!*—(Fl. vere.) (*v. s.*)

DISTRIB. Regio Mediterranea! et Caspica! Asia Minor! Mesopotamia! Persia!

Herba procumbens, annua, glauca, spithamæa. *Folia* radicalia in lacinias angustas bitripinnatisecta. *Caules* v. scapi 2–4, decumbentes v. ascendentes, teretes, læves. *Bracteæ* verticillatæ, $\frac{1}{2}$–1-pollicares, laceræ. *Flores* pauci, pedicellati; alabastri nutantes, aurei, $\frac{1}{2}$ poll. lati. *Sepala* fugacia, late ovata v. ovato-lanceolata, acuminata, petalis $\frac{1}{2}$ breviora. *Petala* exteriora late obovato- v. cuneato-rhombea, plus minusve sinuato-triloba, lobis lateralibus lineari oblongis obtusis, intermedio spathulato, forma varia, oblonga, retusa v. bifida, marginibus fimbriato- v. sinuato-dentatis. *Antheræ* connectivo plus minusve ultra loculos producto; filamenta dilatata. *Stigma* bicrure. *Siliqua* linearis, $1\frac{1}{2}$–$2\frac{1}{4}$ poll. longa, 2 lin. lata, utrinque attenuata, compressa, spongiosa. *Semina* oblique obovata, arillo parvo instructa, testa lævi brunnea.

Variable in foliage and size of flower, as are all the species of the genus; also in the form of the sepals and petals; the inner petals have the middle lobe larger or smaller than the lateral, and more or less fimbriated.

2. **H. leptocarpum** (H.f. et T.); floribus pallide purpureis, sepalis ovato-lanceolatis, petalis extcrioribus late obovatis interioribus trifidis lobo intermedio spathulato, siliquis gracillimis.

HAB. In Tibetia occidentali, frequens arvis alt. 9–12,000 ped.! in Himalayæ Tibeticæ arenosis : Sikkim, alt. 12–14,000 ped. !—(Fl. Jul.) (*v. v.*)

Herba gracilis, annua, diffusa, statura valde variabilis, 3-poll. v. bipedalis. *Folia* radicalia plurima, patula, lineari-oblonga, pinnatisecta, 2–4 poll. longa, pinnis ½ unc. longis, remotis, late ovatis, pinnatisectis, lobis dentatis acuminatis. *Caules* plurimi, basi decumbentes, sæpius elongati, simplices v. pluries dichotome ramosi, bracteis sectis. *Pedicelli* filiformes, bracteolis setaceis involucrati. *Flores* pallide purpurei v. lilacini, ⅓–½ poll. lati. *Sepala* petalis ½ breviora. *Petala* exteriora apice subcoriacea, viridia, interiora minora, vix ad medium fissa, lobis lateralibus obtusis, intermedio late oblongo, sessili, cucullato, marginibus incurvis integerrimis. *Stigmata* 2, recurva. *Siliquæ* pollicares, vix ½ lin. latæ, gracillimæ, compressæ, 8–10-spermæ, articulis indehiscentibus, facile solutis. *Semina* oblique oblonga, loculos implentia; testa subcoriacea, brunnea. *Albumen* carnosum.

Very closely allied to the Siberian *Chiazospermum erectum*, Bernh., in habit, colour of flowers, form of sepals, and slender siliqua, but the latter are very different in structure, showing no trace of valvular dehiscence, but breaking across at the joints even before the ripening of the seeds, which adhere firmly to the cavity of the pericarp. The *C. lactiflorum*, Karel. et Kiril., of Soongaria, seems the same as *C. erectum*, the inner petals in our (authentically named) specimens of it differing in no way from those of other species of *Hypecoum ;* and the character of their middle lobe being antheriferous, is either inconstant or founded in error, and possibly arises from the pollen being sometimes adherent to its cucullate face.

Our Sikkim specimens are very much smaller than most of our western Tibetan ones, but agree in all essential characters.

INDEX.

Synonyms in italics. An asterisk (*) signifies that the species is not Indian, but alluded to at the page.

END OF VOLUME I.

JOHN EDWARD TAYLOR, PRINTER,
LITTLE QUEEN STREET, LINCOLN'S INN FIELDS.